What Is Life? A Guide to BIOLOGY

What Is Life? A Guide to BIOLOGY

Jay Phelan

University of California, Los Angeles

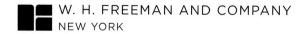

W. H. FREEMAN AND COMPANY
NEW YORK

Publisher: Peter Marshall

Publisher: Sara Tenney

Senior Development Editors: Elizabeth Howe, Jane Tufts

Associate Director of Marketing: Debbie Clare

Managing Editor for First Editions: Elaine Palucki, PhD

Senior Media Editor: Patrick Shriner

Associate Editor for First Edition Supplements: Beth McHenry

Editorial Assistant: Andy Lawler

Art Development Editor: Tommy Moorman

Illustrations: Tommy Moorman, Erin Daniel

Managing Editor: Philip McCaffrey

Senior Project Editor: Mary Louise Byrd

Art Director: Diana Blume

Text Design: Lissi Sigillo

Photo Editor: Christine Buese

Photo Researchers: Elyse Rider, Deborah Anderson

Production Manager: Julia DeRosa

Composition and Layout: Sheridan Sellers

Printing and Binding: Quebecor Dubuque

Library of Congress Control Number: 2009920535

Student Edition: ISBN 10: 1-4292-2318-9
                   ISBN 13: 978-1-4292-2318-8
Loose Leaf Edition: ISBN 10: 1-4292-3858-5
                      ISBN 13: 978-1-4292-3858-8

Printed in the United States of America

First Printing

W. H. Freeman and Company

41 Madison Avenue, New York, NY 10010

www.whfreeman.com

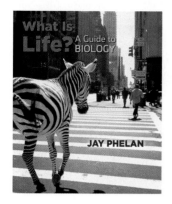

What Is Life? A Guide to BIOLOGY

JAY PHELAN

At a Glance

Intriguing Q Questions Motivate Readers

Jay Phelan connects with his students by posing questions that intrigue or challenge them. He brings this technique to *What Is Life?* through a system of **Q** questions that **apply the key biological concepts to a novel situation** and require the student to think. Enough information is provided in the text for the student to answer the question.

> "Every single **Q Question** made me want to flip to the section where the answer would be and figure it out. You can easily tell that the author **wants non-biology-minded people to figure things out** about the ways of life."
>
> —SARAH MOFFAT,
> UNIVERSITY OF NORTH CAROLINA-CHARLOTTE,
> BUSINESS MAJOR

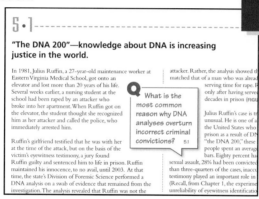

5•1

"The DNA 200"—knowledge about DNA is increasing justice in the world.

In 1981, Julius Ruffin, a 27-year-old maintenance worker at Eastern Virginia Medical School, got onto an elevator and lost more than 20 years of his life. Several weeks earlier, a nursing student at the school had been raped by an attacker who broke into her apartment. When Ruffin got on the elevator, the student thought she recognized him as her attacker and called the police, who immediately arrested him.

Ruffin's girlfriend testified that he was with her at the time of the attack, but on the basis of the victim's eyewitness testimony, a jury found Ruffin guilty and sentenced him to life in prison. Ruffin maintained his innocence, to no avail, until 2003. At that time, the state's Division of Forensic Science performed a DNA analysis on a swab of evidence that remained from the investigation. The analysis revealed that Ruffin was not the

Q What is the most common reason why DNA analyses overturn incorrect criminal convictions? 5.1

attacker. Rather, the analysis showed th matched that of a man who was alread serving time for rape. F only after having serve decades in prison (FIGU

Julius Ruffin's case is tr unusual. He is one of a the United States who prison as a result of DN "the DNA 200," these people spent an average bars. Eighty percent ha sexual assault, 28% had been convicted than three-quarters of the cases, inaccu testimony played an important role in (Recall, from Chapter 1, the experime unreliability of eyewitness identificatio

Collaborative Art Program Works Seamlessly with the Text

Jay Phelan and Tommy Moorman, a scientific illustrator with degrees in biology and illustration, worked collaboratively to conceptualize and prepare each illustration. Reviewers describe the art program as **fresh, simply drawn, consistent** across the book, seamlessly tied to the text, and easy to understand. Students who class-tested the art rated it very highly.

> "The **visuals** are amazing. They show what is mentioned in the chapter instead of something completely irrelevant."
>
> —MELISSA COLE,
> UNIVERSITY OF NORTH CAROLINA, CHARLOTTE,
> PSYCHOLOGY MAJOR

THE PHOTOSYSTEM (WATER-SPLITTING)

1 Light energy excites electrons in a pigment molecule (such as a chlorophyll molecule). The energy from the excited electrons is transferred to nearby pigment molecules.

2 When transferred energy excites electrons in the chlorophyll a molecule, the primary electron acceptor grabs them and sends them to the electron transport chain.

3 To replace electrons sent to the electron transport chain, water molecules are split and oxygen and hydrogen are released as by-products.

Sun
Primary electron acceptor
To electron transport chain
Chlorophyll a molecule
Oxygen released into the atmosphere
Pigment molecule
Water
Oxygen needed for all animal life is produced here!
Chloroplast Thylakoid
AREA OF DETAIL

Examples Engage and Show Relevance

What Is Life? threads **fascinating, relevant, contemporary examples** throughout every chapter. For example, the DNA chapter in *What Is Life?* (Chapter 5) opens with a discussion of the exoneration of Julius Ruffin with forensic evidence, using the Ruffin case to introduce the discussion of biotechnology.

Vivid Photos Show the Story of Biology

Jay Phelan has filled *What Is Life?* with hand-picked favorites from his extensive lecture photo database. These striking images appear as unit openers and are combined with illustrations of biological processes, concepts, and experimental techniques to form a **notably original art program.**

Street Bio: Knowledge You Can Use Tackles Practical Biological Topics

Each chapter concludes with **StreetBio: Knowledge You Can Use**. Presented in a question and answer format, each StreetBio centers on an intriguing, high-interest application rooted in the chapter concepts, and takes learning a step further, by helping students "own" the information.

> "The Street Bio was something I've never seen before. Nice!"
>
> —SANDI SUNDSTROM, SOUTHWESTERN UNIVERSITY, EXERCISE SCIENCE MAJOR

Take-Home Messages Provide a Quick Summary

All chapter sections end with a simple **Take-Home Message**, plainly restating the section's main ideas. These messages provide a concise, memorable summing up of the section, and are also useful when students review a chapter in preparation for quizzes or exams.

> "The Take-Home Messages are very informative and relevant to everyday life."
>
> —GUILLERMO R. FERNANDEZ, SOUTHWESTERN UNIVERSITY, COMMUNICATIONS MAJOR

TAKE-HOME MESSAGE 8·1

The characteristics of individuals in a population can change over time. We can observe such change in nature and can even cause such change to occur.

BRIEF CONTENTS

CONTENTS

3 • Cells

THE SMALLEST PART OF YOU 76

4 • Energy
FROM THE SUN TO YOU IN JUST TWO STEPS

5 • DNA, Gene Expression, and Biotechnology

WHAT IS THE CODE AND HOW IS IT HARNESSED?

6 • Chromosomes and Cell Division

CONTINUITY AND VARIETY

7 • Mendelian Inheritance

FAMILY RESEMBLANCE: HOW TRAITS ARE INHERITED

8 • Evolution and Natural Selection

DARWIN'S DANGEROUS IDEA

9 • Evolution and Behavior

COMMUNICATION, COOPERATION, AND CONFLICT IN THE ANIMAL WORLD 328

10 • The Origin and Diversification of Life on Earth

11 • Animal Diversification

12 • Plant and Fungi Diversification

13 • Evolution and Diversity Among the Microbes

PART 4 Ecology and the Environment

14 • Population Ecology

PLANET AT CAPACITY: PATTERNS OF POPULATION GROWTH 514

15 • Ecosystems and Communities

ORGANISMS AND THEIR ENVIRONMENTS...548

16 • Conservation and Biodiversity

HUMAN INFLUENCES ON THE ENVIRONMENT 586

To Julia

Dear Readers,

Biology is about you and it touches on your life every day, in dozens of ways. It's creative. It's fun. And understanding it is increasingly essential to your life. I wrote this book because I want to help you become biologically literate: to speak confidently and knowledgeably about science with your friends and to understand and appreciate both the beauty and the utility of this subject.

You will find here an overview of the key themes in biology as well as the specifics about topics meaningful to your life. You'll find answers to the questions that you're curious about, whether you are a future lawyer, teacher, entrepreneur, parent, consumer, citizen, or all of the above:

Why are humans one of the only species to have friendships?
Is DNA fingerprinting foolproof?
How can it be that almost everyone in the United States consumes genetically modified foods regularly without knowing it?
Why do dieters lose large amounts of "water weight" during the first few days of a diet?

But as much as this book is about answers, it is also about questions. Long after your biology course is complete and you can no longer remember the exact phases of mitosis, it is my hope that you will approach the issues in your life from a scientific perspective, and that in doing so, you will be a more active participant in the ongoing, ever-changing dialogue about life.

In this book, as in life, there are interesting questions everywhere. As you read, red **Q**s point you toward specific passages in the text that provide interesting answers to questions with real relevance for you. If you look at those bits of the text and don't think the answer is there, look again and think some more. You may be surprised at what you know.

There's much more to this book than just words. Flip through its pages and look at the photographs. I believe that great photography is one of the most effective ways to learn biology. For twenty years, I have searched for the best photos to illustrate biological concepts. I hand-picked every photo here with a special purpose in mind—each photo should provoke, engage, and inspire you, while helping you to make connections between complex ideas.

Amid all the beautiful pictures and varied subjects, I don't want you to lose sight of "the big picture." So in organizing each chapter, I have broken down the topics into discrete sections. At the end of each topic, I highlight a Take-Home Message in clear and concise terms. Although there may be a lot of material, in these manageable chunks you can master it.

The impact of science on your life is increasing every day. I hope you embrace this exciting world of knowledge and gain as much satisfaction and enjoyment reading this book as I have received in putting it together for you.

Sincerely,

Jay Phelan

About the Author

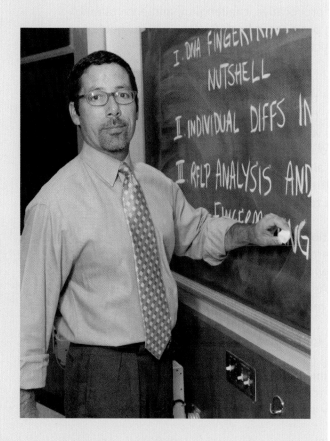

Jay Phelan teaches biology at UCLA, where he has taught introductory biology to more than 8,000 majors and non-majors students over the past twelve years. He received his Ph.D. in evolutionary biology from Harvard in 1995, and his master's and bachelor's degrees from Yale and UCLA. His primary area of research is evolutionary genetics, and his original research has been published in *Evolution*, *Experimental Gerontology*, and the *Journal of Integrative and Comparative Biology*, among other journals. His research has been featured on *Nightline*, CNN, the BBC, and National Public Radio; in *Science Times* and *Elle*; and in more than a hundred newspapers. He is the recipient of more than a dozen teaching awards.

With economist Terry Burnham, Jay is the co-author of the international best-seller *Mean Genes: From Sex to Money to Food—Taming Our Primal Instincts*. Written for the general reader, *Mean Genes* explains in simple terms how knowledge of the genetic basis of human nature can empower individuals to lead more satisfying lives.

Acknowledgments

As a new graduate student at Harvard, I heard from experienced teaching fellows that if you were interested in learning how to be an effective teacher, it was essential to seek out extraordinary mentors. Based on word of mouth, I became involved with E. O. Wilson's course in Evolutionary Biology and Irven DeVore's course in Human Behavioral Biology. Both were known to be unusually provocative, challenging, and entertaining classes for non-science majors. I aggressively pursued teaching positions in both classes—which I held onto tightly for 12 semesters. Working under these legendary instructors, I was set on a course that inspired and prepared me to write this book.

The two courses were quite different from each other, but at their core both were built on two beliefs that are central to this book and to my thinking about education: (1) Biology is creative, interesting, and fun. (2) Biology is relevant to the daily life of every person. There was a palpable sense that, in teaching non-science majors especially, we had a responsibility to provide our students with the tools to thrive in a society increasingly permeated by scientific ideas and issues, and that one of our most effective strategies would be to convey the excitement we felt for biology and the enormous practical value it has to help us understand the world. I thank Professors Wilson and DeVore for all that they have shared with me.

My development as a scientist and, particularly, my appreciation for rigorous and methodical thinking have been shaped by the kind support and wise guidance of Richard Lewontin. I have also been fortunate to have as a long-time mentor and collaborator Michael Rose, who has instilled in me a healthy skepticism about any observation in life that is not fivefold replicated. And for almost daily insightful input on matters relating to scientific content, teaching, writing, and more, I thank Terry Burnham.

There are many other friends and colleagues I wish to thank for helping me with *What Is Life?*

In researching and writing the book and in developing the numerous courses I teach, I have benefited from more than a decade of perceptive and valuable contributions too numerous to list from Glenn Adelson, Alon Ziv, Michael Cooperson, and Alicia Moretti. I am tremendously appreciative of all they have done for me.

For a project covering so many topics and years, it is essential to have a close group of trusted, tolerant, and knowledgeable colleagues, I am grateful to Harold Owens, Greg Graffin, Jeff Egger, Andy Tobias, Elisabeth Tobias, Joshua Malina, Melissa Merwin-Malina, Bill U'ren, Chris Bruno, Michelle Richmond, and Meredith Dutton, who have offered advice, guidance, and support, far beyond the call of duty.

Numerous colleagues at UCLA provided assistance and support, including Steve Strand, Cliff Brunk, Fred Eiserling, Emil Reisler, Deb Pires, Lianna Johnson, Gaston Pfluegl, Bob Simons, Frank Laski, Jeff Thomas, and Tracy Newman. As a result of their commitment to excellence in the UCLA Life Science Core Curriculum, I have been able to acquire a wealth of experiences that have helped me continue improving as a teacher.

I owe a tremendous debt to editor and publisher Sara Tenney, without whose vision and knowledge this project could never have been begun or completed. Thank you, Sara.

W. H. Freeman is an extremely author-centric publisher. Throughout the process of creating this book, from the first inception of the idea through the production of all the supplementary materials, Liz Widdicombe,

Brian Napack, and John Sargent have been tremendously supportive. I am grateful for them welcoming me into their publishing family. Publisher Peter Marshall has been a tenacious, versatile, and skillful manager of the entire team. I am very fortunate to have such a wise leader overseeing all aspects of this project.

The team of editors that worked with me on this book—and two people in particular—improved it immeasurably. I cannot adequately convey my gratitude to development editor Jane Tufts, whose meticulous attention to detail, commitment to accuracy, and almost obsessive drive to create a thorough and readable book are apparent on every page. And development editor Beth Howe, who oversaw every aspect of the writing and production of the book, attending to issues of content and design while making insightful contributions throughout, expertly managed the thousand details necessary to put everything together.

It is impossible to teach biology without illustrations. My deepest gratitude goes to Tommy Moorman for creating such innovative and effective figures for the book. Tommy's vision for an elegant and beautiful art program completely integrated with the text is apparent on every page. Working with him to develop each illustration in this book has been one of my most enjoyable and satisfying professional collaborations. Thanks also go to Erin Daniel for assisting with the creation of the illustrations. For their input into the design of the book, I thank Lissi Sigillo and Anna Palchik. And for excellent assistance with photo research, thanks to Deborah Anderson, Elyse Rieder, and Christine Buese.

I wish to thank Harvey Pough for providing excellent drafts of Chapters 11–13 and contributing to Chapter 16.

For creating the innovative media and print materials that accompany the book, I am thankful for the vision of my excellent media editor, Patrick Shriner, and for the extensive input of supplements editor Beth McHenry. I thank all of the contributors and advisors who helped create the student and instructor resources; your efforts have been invaluable. I thank Jennifer Warner and Meredith Norris for their work on the *Student Success Guide,* and for Jennifer's willingness to share her classroom experiences with the sales force. I also appreciate the contributions of Renee Altier, Reid Sherline, and Larry Jankovic to Prep-U. Elaine Palucki, with her broad knowledge of biology, coordinated the class-testing with exceptional skill and efficiency. Sheri Snavely provided significant guidance in developing pedagogical strategies throughout the book; I also appreciate her thoughtful and smart advice at nearly every step in the publishing process.

Copyeditor Linda Strange helped to ensure consistency and readability throughout the text. I thank Chris Hunt for compiling the thorough index. The rest of the life sciences editorial team at W. H. Freeman, too, have been knowledgeable and supportive, particularly Kate Ahr, Marc Mazzoni, Jerry Correa, Susan Winslow, Matt Tontonoz, Susan Moran, Lisa Samols, Karen Horton, and Andy Lawler.

For their efficiency and commitment to producing a beautiful book, I am most grateful to the W. H. Freeman production team: Mary Louise Byrd, Sheridan Sellers, Diana Blume, Julia DeRosa, Philip McCaffrey, and Ellen Cash.

The people on the marketing team at W. H. Freeman have contributed enormously in helping with the challenging task of introducing a new book to students and instructors across the country. Debbie Clare, Steve Rigolosi, and John Britch have been enthusiastic and dedicated in creating materials and strategies to assist instructors in evaluating the ways that *What Is Life?* can aid them as they develop their own courses and strategies for success.

Finally, I thank my family—Kevin Phelan, Patrick Phelan, Erin Enderlin, and my parents—for their unwavering support and interest as I wrote this book. Reading draft after draft and following each revision, they made valuable contributions at every stage. I thank Jack, Charlie, and Sam, too. Most of all, for her generous and passionate support of this project from day one, her substantive contributions to both the content and presentation of ideas, and so much more, I thank Julia.

Contact the author with your feedback.

The content of this book has been greatly improved through the counsel of reviewers and students. Your comments, suggestions, and criticism are also welcome; they are essential in guiding its ongoing evolution. Please contact the author at **jay@jayphelan.com**.

We thank the many reviewers who aided in the development of this text.

Text Development

Stephanie J. Aamodt, University of Louisiana–Shreveport
M. Stephen Ailstock, Anne Arundel Community College
Sylvester Allred, Northern Arizona University
Jessica K. Baack, Montgomery Community College–Rockville
Sarah Barlow, Middle Tennessee State University
Rebecca A. Bartow, Western Kentucky University
Brian J. Baumgartner, Trinity Valley Community College
José Bava, El Camino Community College
Lisa L. Boggs, Southwestern Oklahoma State University
Charlotte Borgeson, University of Nevada–Reno
Peggy Brickman, University of Georgia
Jason Brown, Young Harris College
Carole L. Browne, Wake Forest University
Neil J. Buckley, State University of New York–Plattsburgh
Anne E. Bunnell, East Carolina University
Joseph Burdo, Bridgewater State University
David Byres, Florida Community College–Jacksonville
William Caire, University of Central Oklahoma
Michael S. Carr, Oakton Community College
Kelly Carrillo Burke, College of the Canyons
Thomas Chen, Santa Monica College
Thomas F. Chubb, Villanova University
Nira Clark, Southwestern College
Erica Cline, University of Washington
George Clokey, University of Wisconsin–Whitewater

Anthony D. Cornett, Valencia Community College–Osceola
Karen Curto, University of Pittsburgh
Don C. Dailey, Austin Peay State University
Douglas W. Darnowski, Indiana University–Southeast
Garry Davies, University of Alaska
Renee Dawson, University of Utah
Lewis E. Deaton, University of Louisiana–Lafayette
Tom Deaton, Auburn University
Elizabeth Anne Desy, Southwestern Minnesota State University
Doreen R. Dewell, Whatcom Community College
Mary E. Dominiecki, Slippery Rock University
Charles Dunn, Austin Community College
Frank DuRoy, Essex County College
Kari Eamma, Tarrant County Community College–Northeast
Heidi Engelhardt, Brandon University
Marirose T. Ethington, Genesee Community College
Rebecca A. Fetherson, University of Dayton
Robert C. Frankis, University of New Mexico
Monica C. Frazier, Columbus State University
Anne Galbraith, University of Wisconsin–La Crosse
Sandra Gibbons, Moraine Valley College
Philip G. Gibson, Gwinnett Technical College
Andrew Goliszek, North Carolina A & T State University
Stephen M. Gómez, Central New Mexico Community College
Becky C. Graham, University of West Alabama

Cara Gubbins, Butte College
Charles (Billy) Gunnels IV, Florida Gulf Coast University
Janelle M. Hare, Morehead State University
Carla Hass, Pennsylvania State University
Colleen Hatfield, California State University–Chico
J. L. Henriksen, Bellevue Community College
John A. Hnida, Peru State College
Thomas Horvath, State University of New York–Oneonta
Adam W. Hrincevich, Louisiana State University
Dianne Jennings, Virginia Commonwealth University
Mitrick A. Johns, Northern Illinois University
Robert D. Johnson Jr., Pierce College
Martin A. Kapper, Central Connecticut State University
Arnold J. Karpoff, University of Louisville
Todd Kelson, Brigham Young University–Idaho
Stephen T. Kenny, Yakima Valley Community College
Jennifer M. Kilbourne, Community College of Baltimore County
Joanne Kilpatrick, Auburn University–Montgomery
Brenda Knotts, Eastern Illinois University
Michael Koban, Morgan State University
Catherine Koo, Caldwell College
Jerome A. Krueger, South Dakota State University
Jim Krupa, University of Kentucky
Ellen Shepherd Lamb, University of North Carolina–Greensboro
Thomas Landefeld, California State University–Dominquez Hills
Leah LaPerle Larkin, University of New Mexico
Lynn Larsen, Portland Community College
Kathleen H. Lavoie, State University of New York–Plattsburgh
Gregory P. Lewis, Furman University
Harvey Liftin, Broward Community College
Tammy J. Liles, Lexington Community College
Cynthia W. Littlejohn, University of Southern Mississippi
Jonathan Lochamy, Georgia Perimeter College
Melanie Loo, California State University–Sacramento
Juan M. Lopez-Bautista, University of Alabama
Blasé Maffia, University of Miami
Jeffrey M. Marcus, Western Kentucky University
Betsy Maxim, Austin Community College
Karen McFarland, Austin Peay State University
Mark A. McGinley, Texas Technical University
John Mersfelder, Sinclair Community College
Shahroukh Mistry, Westminster College
Jeanne Mitchell, Truman State University
Beth A. Montelone, Kansas State University
Cynthia L. Morin, Lincoln University of Missouri
Royden Nakamura, California Polytechnic–San Luis Obispo
Rocky Nation, Southern Wesleyan University
Allan D. Nelson, Tarleton State University
Tania Nezrick, Kishwaukee College
David Niebuhr, College of William and Mary
Meredith Sommerville Norris, University of North Carolina–Charlotte
Alexander E. Olvido, Virginia State University

Randi Papke, Southwestern Illinois College
Mary Paulson, Central Washington University
John M. Pleasants, Iowa State University
Gregory J. Podgorski, Utah State University
Melissa Presch, California State University–Fullerton
Eric Rabitoy, Citrus College
Karen Raines, Colorado State University
Erin Rempala, San Diego Mesa College
Dawn Ranish, Nova Southeastern University
Melody Ricci, Victor Valley College
Kathleen Richardson, Portland Community College
Dave Rintoul, Kansas State University
Steven Rissing, The Ohio State University
Robert R. Robbins, Utah Valley State College
Bill Rogers, Ball State University
Troy T. Rohn, Boise State University
Susan T. Rouse, Southern Wesleyan University
Lynette Rushton, South Puget Sound Community College
Michael L. Rutledge, Middle Tennessee State University
Shamili Ajgaonkar Sandiford, College of DuPage
Fayla Schwartz, Everett Community College
Erik P. Scully, Towson University
Juanita C. Sharpe, Chicago State University
Marilyn Shopper, Johnson County Community College
Michele Shuster, New Mexico State University
Neil E. Simister, Brandeis University
Anu Singh-Cundy, Western Washington University
Jennifer Skillen, College of Southern Nevada
Kerri M. Skinner, University of Nebraska–Kearny
Marc Allen Smith, Sinclair Community College
Nancy G. Solomon, Miami University of Ohio
Roberta Lynn Soltz, Central Washington University
Bryan G. Spohn, Florida Community College at Jacksonville–Kent Campus
Amanda Starnes, Emory University
Leo Sternberg, University of Miami
Joan C. Stover, South Seattle Community College
Sukanya V. Subramanian, Collin County Community College
Franklyn Tan Te, Miami Dade Community College
William J. Thieman, Ventura College
Pamela Thinesen, Century College
Michael W. Thompson, Middle Tennessee State University
Ronald Toth, Northern Illinois University
Martin A. Vaughan, Indiana University Purdue University at Indianapolis
Jack Waber, West Chester University of Pennsylvania
Jennifer M. Warner, University of North Carolina–Charlotte
John E. Whitlock, Hillsborough Community College
Elizabeth Willott, University of Arizona
Clifford Wilson, Chicago State University
Mark Woelfle, Vanderbilt University
Calvin Young, Fullerton College
Brenda Zink, Northeastern Junior College

Art Development

Ann Aguanno, Marymount Manhattan College
Kelly Carrillo Burke, College of the Canyons
William Caire, University of Central Oklahoma
Debra Chapman, Wilkes University
Genevieve C. Chung, Broward Community College
Jeffrey Scott Coker, Elon University
Michael S. Dann, Pennsylvania State University
Renee Dawson, University of Utah
Francisco Delgado, Pima Community College
Doreen R. Dewell, Whatcom Community College
Hartmut G. Doebel, George Washington University
Diane Lynn Doidge, Grand View College
James Doyle, College of the Holy Cross
Dave Eakin, Eastern Kentucky University
Paul Farnsworth, University of New Mexico
Brandon L. Foster, Wake Technical and Community College
Patrick Galliart, North Iowa Community College
David Gordon, Pittsburg State University
Charles (Billy) Gunnels IV, Florida Gulf Coast University
Bernard Hauser, University of Florida
Jane J. Henry, Baton Rouge Community College
Mark A. Holland, Salisbury University
Virginia E. Irintcheva, Black Hawk College–Quad City Campus
Philip Jardim, City College of San Francisco
Denim M. Jochimsen, University of Idaho
Judy Kaufman, Monroe Community College
Ariel R. Krakowski, Laney College
Dan E. Krane, Wright State University
Maria Kretzman, Glendale College
Kathleen H. Lavoie, State University of New York–Plattsburgh
Mary LeFever, Columbus State Community College
Stephen Matheson, Calvin College
James McCaughern-Carucci, Quinnipiac University
Lori A. Pitkofsky, Rampano College
Michael P. Robinson, Miami University
Wendy F. Rothwell, University of California–Santa Cruz
Jason F. Schreer, State University of New York–Potsdam
Jennifer J. Scoby, Illinois Central College
Dawn Sherry, Macon State University
Cara Shillington, Eastern Michigan University
Patricia L. Smith, Valencia Community College
Jeffrey Thomas, California State University–Northridge
Bruce Tomlinson, State University of New York–Fredonia
Joy B. Trauth, Arkansas State University
Alan R. Wasmoen, Metropolitan Community College
Michael Windelspecht, Appalachian State University
David E. Wolfe, American River College
Kerry Yurewicz, Plymouth State University

Accuracy Reviewers

Sylvester Allred, Northern Arizona University
Christine Bezotte, Elmira College
Kelly Carrillo Burke, College of the Canyons

Michael Carr, Oakton Community College
Erica Champion
William Crampton, University of Central Florida
Michael S. Dann, Pennsylvania State University
Renee Dawson, University of Utah
Janelle M. Hare, Morehead State University
Bernard Hauser, University of Florida
John A. Hnida, Peru State College
Kelly Hogan, University of North Carolina–Chapel Hill
Anne Houtman, California State University–Fullerton
Ariel R. Krakowski, Laney College
Michael Silva, El Paso Community College
Pamela Thinesen, Century College
Jeffrey Thomas, California State University–Northridge
Martin Vaughn, Indiana University Purdue University–Indianapolis
Michael Wenzel, California State University–Sacramento
John E. Whitlock, Hillsborough Community College

Focus Group Attendees

Text and Content Development
(2004, Los Angeles)

Garen Baghdasarian, Santa Monica College
Thomas Chen, Santa Monica College
Elizabeth Ciletti, Pasadena City College
Mary Colavito, Santa Monica College
Ana Ester Escandon, Los Angeles Harbor College
Phyllis Hirsh, East Los Angeles College
Melody Ricci, Victor Valley College

(2004, Memphis)

T. Wayne Barger, Tennessee Technical University
Don Baud, University of Memphis
Jay Blundon, Rhodes College
Martha P. Brown, University of Memphis
Dave Eakin, Eastern Kentucky University
Stan Eisen, Christian Brothers University
Jack Grubaugh, University of Memphis
Min-Ken Lao, Furman University

(2004, Chicago)

Sandra Bobick, Community College of Alleghany
Christopher Dobson, Grand Valley State University
Sondra Dubowsky, Allen County Community College
Eileen Gregory, Rollins College
Tracy Harris, Florida Keys Community College
Robert Krasner, Providence College
Bernard A. Marcus, Genesee Community College
Laurel B. Roberts, University of Pittsburgh
Lyndell P. Robinson, Lincoln Land Community College
Brian Schmaefsky, Kingwood College
Janet Vigna, Grand Valley State University
Daniel W. Ward, Waubonsee Community College

(2006, Florida)

Peggy Brickman, University of Georgia
Anne E. Bunnell, East Carolina University

Cara Gubbins, Butte College
Janelle M. Hare, Morehead State University
John A. Hnida, Peru State College
Mitrick A. Johns, Northern Illinois University
Kathleen H. Lavoie, State University of New York–Plattsburgh
Blasé Maffia, University of Miami
Gregory J. Podgorski, Utah State University
Susan T. Rouse, Southern Wesleyan University
Jennifer M. Warner, University of North Carolina–CharlottE

Art Development
(2007, Atlanta)

Renee Dawson, University of Utah
Charles (Billy) Gunnels IV, Florida Gulf Coast University
Bernard Hauser, University of Florida
Arnold J. Karpoff, University of Louisville
Stephen R. Kelso, University of Illinois–Chicago
Meredith Sommerville Norris, University of North Carolina–Charlotte
Therese M. Poole, Georgia State University
Mary Celeste Reese, Mississippi State University
Bill Rogers, Ball State University
Kim Cleary Sadler, Middle Tennessee State University
Amy S. Wernette, Hazard Community College

Media and Supplements
(2007, Santa Monica)

Michael Bucher, College of San Mateo
Kelly Carrillo Burke, College of the Canyons
John A. Hnida, Peru State College
Anne Houtman, California State University–Fullerton
Hinrich Kaiser, Victor Valley College
Melanie Loo, California State University–Sacramento
Gregory J. Podgorski, Utah State University
Melissa Presch, California State University–Fullerton
Calvin Young, Fullerton College

(2008, Miami)

Barbara Blonder, Flagler College
Charlotte Borgeson, University of Nevada–Reno
Kris Curran, University of Wisconsin–Whitewater
Kari Eamma, Tarrant Community College
Evelyn Frazier, Florida Atlantic University
John Janovy, University of Nebraska–Lincoln
Jennifer Schramm, Chemeketa Community College
Greg Sievert, Emporia University
Heather Vance Chalcraft, East Carolina University
Martin Vaughn, Indiana University Purdue University–Indianapolis

Comparative Reviewers

Zulfiqar Ahmad, East Tennessee State University
Julie H. Aires, Florida Community College at Jacksonville
William Anyonge, Xavier University
Tami Asplin, North Dakota State University
Randy Brewton, University of Tennessee–Knoxville
Dustin Brisson, University of Pennsylvania

Jeannie Chapman, University of South Carolina–Upstate
Thomas Chen, Santa Monica College
Craig Clifford, Northeastern State University
William Gareth Richard Crampton, University of Central Florida
Chris Davison, Long Beach City College
Tiffany Doan, Central Connecticut State University
Ernest F. DuBrul, University of Toledo
Eugene Fenster, Metropolitan Community College–Longview
Rebecca Fetherson, University of Dayton
Teresa G. Fischer, Indian River State College
Anita Pardue Flick, North Carolina State University
Paul Florence, Jefferson Community and Technical College
Evelyn Frazier, Florida Atlantic University
Diane Wilkening Fritz, Northern Kentucky University
Dennis W. Fulbright, Michigan State University
Douglas C. Gayou, University of Missouri–Columbia
Betsy Gerbec, University of Wisconsin–River Falls
Oliver Ghobrial, Santa Fe College
Julie Gibbs, College of DuPage
Florence K. Gleason, University of Minnesota
Bruce Griffis, Kentucky State University
Tim Grogan, Valencia Community College
Charles J. Grossman, Xavier University
Luis S. Guerra, South Texas College
Pieter de Haan, Berkeley City College
Mary F. Haskins, Rockhurst University
Keith R. Hench, Kirkwood Community College
Sherry Hickman, Hillsborough Community College
Mark Hollier, Georgia Perimeter College
Robert A. Holmes, Hutchinson Community College
Laurie Host, Harford Community College
Robert Iwan, Inver Hills Community College
Philip Jardim, City College of San Francisco
Dianne Jennings, Virginia Commonwealth University
David Knowles, East Carolina University
Kim Lackey, University of Alabama
Ellen Lamb, University of North Carolina at Greensboro
David Loring, Johnson County Community College
Bill Mackay, Edinboro University of Pennsylvania
Barbara Mania-Farnell, Purdue University–Calumet
Tom J. McConnell, Ball State University
Wallace M. Meyer, College of the Redwoods
Thelma Miller-Anderson, Hillsborough Community College
Michael R. Millward, University of Cincinnati
Brenda Moore, Truman State University
Michael Muller, University of Illinois–Chicago
Zia Nisani, Antelope Valley College
Tanya Noel, York University
James Nolan, Georgia Gwinnett College
Laura Palmer, Pennsylvania State University–Altoona
Louis Pech, University of Wisconsin Centers–Marathon County
Joel B. Piperberg, Millersville University
Karen Plucinski, Missouri Southern State University
Aggie Posthumus, Olivet Nazarene University
Bonnie Ripley, Grossmont College

Troy Rohn, Boise State University
Matthew Rowe, Sam Houston State University
Yelena Rudayeva, Palm Beach Community College
Arthur Sandquist, Washburn University
Donald Slish, Plattsburgh State University
Christine Stracey, University of Florida
Anthony J. Stancampiano, Oklahoma City Community College
Tim Strakosh, State University of New York–Fredonia
Richard P. Stringer, Harrisburg Area Community College
Mark Sturtevant, Oakland University
Sukanya Subramanian, Collin County Community College
Rob Swatski, Harrisburg Area Community College
Kimberly Taugher, Diablo Valley College
Jeffrey Thornsberry, Northwest Missouri State

Nina Thumser, Indiana University of Pennsylvania
Jonathan Titus, State University of New York–Fredonia
Mike Tveten, Pima Community College Northwest
Katherine M. Van de Wal, Community College of Baltimore County–Essex Campus
Leslie VanderMolen, Humboldt State University
Fred Vogt, Elgin Community College
William A. Wehbi, University of Pennsylvania
John E. Whitlock, Hillsborough Community College
Jennifer Wiatrowski, Pasco-Hernando Community College
Christina Willis, Rockhurst University
Donald S. Wood, Odessa College
Aimee Wurst, Lincoln University of Missouri
Lan Xu, South Dakota State University

We thank the hundreds of students and faculty at the schools listed below for class-testing elements of the text.

Q Questions

Borough of Manhattan Community College
Broward Community College
Central New Mexico Community College
Colorado Mountain College
Florida Community College–Jacksonville
Heartland Community College
Johnson County Community College
Kutztown University of Pennsylvania
Lincoln Land Community College
Metropolitan Community College
Mt. San Jacinto College
Portland State University
Rowan University
Seattle Central Community College
Suffolk County Community College–Ammerman
Tarrant County Community College–Northeast
University of Northern Colorado
Wake Technical Community College

Chapters
(2007)
College of San Mateo
Georgia Perimeter College
Georgia State University
Hazard Community College
Southwestern College
University of Arizona
University of Louisville
University of North Carolina–Charlotte
University of Utah
Utah State University
Virginia Commonwealth University

(2008)
Bluegrass Community College
California State University–Northridge
Chaffey College
Clark Atlanta University
Columbus State University
Daytona Community College
Edinboro University
Genesee Community College
George Washington University
Glendale College
Lake Land College
Long Beach Community College
Los Angeles Harbor College
Macon State University
McPherson College
Millersville University of Pennsylvania
Milwaukee Area Technical College
Owensboro Community College
Oxnard College
Philadelphia University
Prince George's Community College
Quinnipiac University
Rock Valley College
Sacramento City College
Salem State College
San Jose State University
Sinclair Community College
Southwestern College
St. Clair Community College
University of New Mexico
University of Rhode Island
Valencia Community College

The media and supplements package was created with the input of hundreds of non-major biology instructors and through the hard work of our experienced and committed media and supplements contributor team.

Contributors and Advisors

Instructor's Manual—Classroom Catalysts
Verona A. Barr, Heartland Community College
Stephen M. Gómez, Central New Mexico Community College
Paul H. Marshall, Northern Essex Community College

eBook Study Tools
Robert Iwan, Inver Hills Community College
Eric Stavney, Pierce College

Figure Conversion Engine
Ann Aguanno, Marymount Manhattan College

Image Bank Advisor
Jane J. Henry, Baton Rouge Community College

Keynote Lecture Presentation
Michael C. Bucher, College of San Mateo

PowerPoint Lecture Outlines
Kristen L. Curran, University of Wisconsin–Whitewater
Mark S. Manteuffel, St. Louis Community College

Prep-U Instructor's Test Bank and Prep-U for Students
Glenn Adelson, Lake Forest College
Jay Phelan, University of California, Los Angeles
Alon Ziv, Prep-U

Q Animations
Anne Bunnell, East Carolina University
Julie V. Gibbs, College of DuPage

Student Success Guide for Phelan's *What Is Life?*
Meredith S. Norris, University of North Carolina–Charlotte
Jennifer M. Warner, University of North Carolina–Charlotte

Student Worksheets
Michael C. Bucher, College of San Mateo

Take-Home Message Podcasts
Robert Iwan, Inver Hills Community College

Reviewers
Ann Aguanno, Marymount Manhattan College
Charlotte E. Borgeson, University of Nevada–Reno
Kelly Carillo Burke, College of the Canyons
Michelle Cawthorn, Georgia Southern University
Michael S. Dann, Pennsylvania State University
Debra B. Folsom, Pasadena City College
Julie V. Gibbs, College of DuPage
Stephen M. Gómez, Central New Mexico Community College
Jane J. Henry, Baton Rouge Community College
Jane Horlings, Saddleback College
Carina Endres Howell, Lock Haven University
John Janovy, Jr., University of Nebraska–Lincoln
Hinrich Kaiser, Victor Valley College
Judy Kaufman, Monroe Community College
Charlease Kelly-Jackson, Claflin University
Patrick J. Lewis, Sam Houston State University
Cindy S. Malone, California State University–Northridge
Meredith S. Norris, University of North Carolina–Charlotte
Greg Podgorski, Utah State University
Lawrence F. Roberge, Florida Community College–Jacksonville
William D. Rogers, Ball State University
Georgianna Saunders, Missouri State University
Jason Schreer, SUNY–Potsdam
Marilyn Shopper, Johnson County Community College
Carol St. Angelo, Hofstra University
Eric Stavney, Pierce College
Alicia Steinhardt, Hartnell College
Jamey Thompson, Hudson Valley Community College
Helen Walter, Diablo Valley College
Jennifer M. Warner, University of North Carolina–Charlotte
Michael Windelspecht, Appalachian State University

What Is Life? Media and Supplements

Prep-U Online Test Bank and Student Quizzing

www.prep-u.com
A first for non-majors biology—a test bank/student quizzing system developed by the textbook author.

Prep-U Instructor's Test Bank and Student Quizzing System offers more than 4,000 class-tested, instructor-vetted multiple-choice questions. It is the only comprehensive online assessment system for a non-majors biology textbook developed by the book's author.

Prep-U Instructor's Test Bank
For each chapter, *What Is Life?* adopters have access to:

- **200 Practice Questions** for homework or quizzing
- **75 Private Reserve Questions,** hidden from student access, to be used in testing
- **5–10 Short-Answer Questions** with answers to help with grading
- **4 Essay Questions,** also with suggested answers

Prep-U contains questions from all levels of the Bloom's Taxonomy scale, requiring students to know, understand, apply, analyze, synthesize, or evaluate. The system allows instructors to select, sequence, and modify existing questions or add their own.

Prep-U's Data System enables instructors to evaluate how well each student is doing, how students compare with a national average, and which questions are difficult for students. **Prep-U** categorizes questions by a misconception index, a scale used to reveal common misunderstandings. Instructors can revisit these topics in class and even create clicker questions to address them.

Prep-U for Students: *Free with the purchase of each new book*
Prep-U Personalized Student Quizzing measures student scores against nationwide results, then generates follow-up practice quizzes geared to the student's specific needs. Efficacy studies show that students who use **Prep-U** quizzing score higher on tests than those who do not.

Faculty Lounge for Non-Majors Biology

www.whfreeman.com/facultylounge/nonmajors
The first publisher-provided Web resource linking the non-majors biology community

At the **Faculty Lounge for Non-Majors Biology,** instructors can download materials contributed by their counterparts nationwide—as well as contribute their own. Most important, non-majors biology instructors can communicate and share freely with colleagues nationwide who face the same challenges in the classroom.

Materials are categorized by topic and rated using a 5-star system. **The Faculty Lounge** is continually updated and vetted by non-majors biology instructors to ensure that there is always something new to see. The site includes:

 Videos and Animations suitable for enriching lectures

 Supplemental Lecture Images providing the best visual support

 News Stories highlighting the relevance of biology in students' lives

 Lecture Activities to get students talking and thinking about biology

The Faculty Lounge is also a great source for classroom activities and teaching tips.

The complete set of **Phelan Instructor Resources and Student Resources** is also available for download to *What Is Life?* adopters through the **Faculty Lounge.**

Instructor Resources
A robust set of instructor resources is available to adopters via download through the **Phelan Faculty Lounge for Non-Majors Biology** or on the **Instructor's Resource DVD**:

- Image Bank
- PowerPoint Lecture Outlines
- Overhead Transparency Set
- Phelan **Q** Animations in PowerPoint
- Instructor's Manual including chapter outlines, learning objectives, Classroom Catalyst activities, and student worksheets

Textbook Conversion Tools
Correlation Grids and a **Figure Conversion Engine** ease the transition to *What Is Life?*

What Is Life? eBook
The eBook includes the complete text, all student media, and dynamic, interactive study tools. To view a demo chapter of the eBook, go to **http://ebooks.bfwpub.com.**

BioPortal and Course Management System Cartridges
BioPortal is a course management system that brings together the eBook and all the online resources for *What Is Life?* All resources are also offered in Blackboard and WebCT. Other course management system cartridges are available on request.

Student Resources

Online resources
On the Book Companion Site (www.whfreeman.com/phelan)
and the **eBook (ebooks.bfwpub.com/phelan)**:

- **Q** Animations
- Take-Home Message Audio and Video Podcasts
- Lecture Art Notebook
- Chapter Quizzes

Printed resources
Student Success Guide for Phelan's *What Is Life?*
Meredith Somerville Norris, University of North Carolina–Charlotte
Jennifer Warner, University of North Carolina–Charlotte
Based on a study tool used successfully with over 10,000 students.

Questions About Life Reader
Heather Vance-Chalcraft, East Carolina University
Biological questions and answers from the *Scientific American* feature "Ask the Experts."

Lab Solutions

Freeman LabPartner Biology

www.whfreeman.com/labpartner

Create a custom manual from a wide selection of experiments published by W. H. Freeman and Hayden-McNeil Publishing. Choose from a variety of traditional and inquiry-based labs and add content on lab techniques and safety. Reorder the labs to fit your syllabus and include your original experiments.
Minimum order: 50 copies.

Freeman Custom Publishing

 W.H. FREEMAN CUSTOM PUBLISHING
custom with care

If you prefer to publish only original labs for your course, we can publish your work in style, bound in a full-color cover customized to your college or university. All Freeman Custom Publishing authors retain their own copyright. Add graph paper or a pocket folder. Choose from paperback, spiral, or loose-leaf formats.
Minimum annual enrollment: 100 copies.

"We chose W. H. Freeman as our custom publisher because they were able to deliver specialized custom content, which was relevant to our course and our students, in a highly professional package at a competitive price." —H. M. Pylypiw and T. Luersen, Quinnipiac University

Hayden-McNeil

 Hayden-McNeil, Freeman's sister company, sets the gold standard in service and quality for qualifying customers, creating dynamic partnerships with educators and producing exceptional course materials at a fair price to students.

All Hayden-McNeil authors retain their own copyrights. H-M's exemplary service offerings include:

- A dedicated editor devoted to shepherding the production of your lab manual from beginning to end
- Professional typesetting and design
- Customized scientific illustrations
- Access to a searchable online image database
- Full-color cover

Minimum annual enrollment: 1,000 copies
visit www.whfreeman.com/phelanpreview

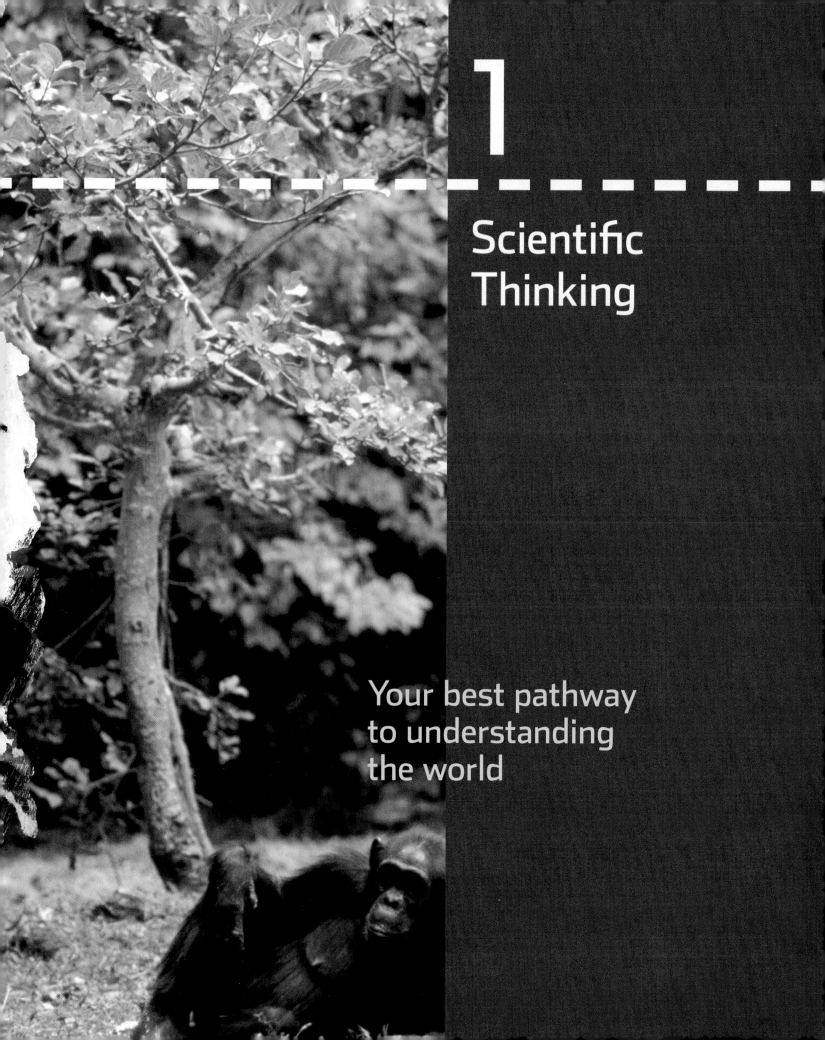

1

Scientific
Thinking

Your best pathway
to understanding
the world

① Science is a collection of facts and a process for understanding the world.

Already a scientist? It starts with curiosity.

1·1

What is science? What is biology?

You are already a scientist. You may not have realized this yet, but it's true. Because humans are curious, you have no doubt asked yourself or others questions about how the world works and wondered how you might find the answers.

• Does the radiation released by cell phones cause brain tumors?

• Are antibacterial hand soaps better than regular soap?

• Do large doses of vitamin C reduce the likelihood of getting a cold?

These are all important and serious questions. But you've probably also pondered some less weighty issues, too.

• Why is morning breath so stinky? And can you do anything to prevent it?

• Why is it always windy on streets with tall buildings?

• Does taking aspirin before drinking alcohol lead to faster intoxication?

And if you really put your mind to the task, you will start to find questions all around you whose answers you might like to know (and, like those above, whose answers you will learn as you read this book).

• Why is it that seaside towns are not as hot in the summer or as cold in the winter as inland communities?

• Why do so few women get into barroom brawls?

• What is "blood doping" and does it really improve athletic performance?

• Why is it so much easier for an infant to learn a complex language than it is for a college student to learn biology?

Still not convinced you're a scientist? Then here's some good news: science doesn't require advanced degrees or secret knowledge dispensed over years of technical training. It does, however, require an important feature of our species—a big brain—as well as curiosity and a desire to learn about the world around us and inside us. But curiosity, casual observations, and desire can only take you so far. In today's world there are important and pressing issues that require some understanding of science, what it can and can't do, and what it can and can't explain. Nutritional claims on foods and dietary supplements, human behavior, health and disease, how drugs can cure or poison you, the interactions of plants, animals, and their environments, global warming, the continuity and diversity of life in all its forms—by learning more about science, you can think about, learn about, and understand issues like these and things about yourself, your life, and the world around you.

To satisfy your curiosity and to explain seemingly unexplainable, mysterious, or magical things, curiosity and the desire to learn are not enough. Explaining how something

What Is Science?   Scientific Method   Experimental Design   Decision Making

works or why something happens requires methodical, objective, and rational observation and analysis that are not clouded with emotions or preconceptions about what is being studied and observed.

Science is not simply a body of knowledge or a list of facts to be remembered. It is an intellectual activity, encompassing observation, description, experimentation, and explanation of natural phenomena. Put another way, science is a pathway by which we can come to discover and better understand the world around us. Later in this chapter, we explore specific ways in which we can most effectively use scientific thinking in our lives. But first let's look at how our understanding of the world can be enhanced by asking the single question that underlies scientific thinking:

Q Can we trust the packaging claims that companies make?

How do you know that is true?

Once you begin asking this question—of others and of yourself—you are on the road to a better understanding of the world.

The following story about a popular and successful over-the-counter medicine shows the importance of questioning the truth of many "scientific" claims you see on merchandise packages or read in newspaper or on the internet. For more than 10 years, a product called "Airborne" was marketed and sold to millions of customers. On the packaging and in advertisements, the makers asserted that Airborne tablets could ward off colds and boost your immune system (**FIGURE 1-1**). Not surprisingly, Airborne was a great success and generated

more than $100 million in revenue, and became one of the fastest selling health products ever. Then some consumers posed a reasonable question to the makers of Airborne: how do you know that it wards off colds?

To prove their claims, the makers of Airborne pointed to the results of a "double-blind, placebo-controlled study" conducted by a company specializing in clinical drug trials. We'll discuss exactly what those terms mean later in the chapter; for now we just need to note that as a result of a class-action lawsuit, it became clear that no such study had been conducted and that there was *no* evidence to back up Airborne's claims. The Airborne company removed the claims from the packaging and agreed to refund the purchase price to anyone who had bought Airborne. It also removed any reference to its "clinical trials," with the company's CEO saying that people "are really not scientifically minded enough to be able to understand a clinical study."

Are you insulted by the CEO's assumption about your intelligence? You should be. Did you or your parents (along with millions of other people) fall for Airborne's false claims? Possibly. But here's some more good news: you don't have to be at the mercy of slick packaging, false advertising, cranks, or charlatans that claim a product can ward off colds, make you lose 10 pounds in 7 days, make you healthier by cleansing your intestines, or increase or decrease the size of various body parts. You can learn to be skeptical and suspicious (in a good way) of such claims. You can learn exactly what it means to have scientific proof or evidence that something is absolutely true. And you can learn this by learning what it means to think scientifically.

Scientific thinking is an important and productive element in the study of a wide variety of topics: it can help you understand economics, psychology, history, and many other subjects. Our focus in this book is **biology,** the study of living things. Taking a scientific approach, we will investigate the facts and ideas in biology that are already known and will study the process by which we come to learn new things. As we move through the four parts of the book, we explore the most important questions in biology.

- What is the chemical and physical basis for life and its maintenance?
- How do organisms use genetic information to build themselves and to reproduce?
- What are the diverse forms that life on earth takes and how has that diversity arisen?
- How do organisms interact with each other and with their environment?

FIGURE 1-1 **Some products claim to improve our health, but how do we know if they work?**

In this chapter, we explore how to think scientifically and how to use the knowledge we gain to make wise decisions. Although we generally restrict our focus to biology, scientific thinking can be applied to nearly every endeavor, and so in this chapter we use a wide range of examples—including some examples from beyond biology—as we learn how to think scientifically. Although the examples will vary greatly, they all convey a message that is key to scientific thinking: it's okay to be skeptical and you don't need to take things on faith.

Fortunately, learning to think scientifically is not difficult—and it can be fun, particularly because it is so empowering. **Scientific literacy,** a general, fact-based understanding of the basics of biology and other sciences, is increasingly important in our lives, and literacy in matters of biology is especially essential. As we see in the next section, issues hinging on biological ideas and processes are turning up in nearly every facet of our lives, from medicine to nutrition to personal relationships to politics, and even to crime and justice.

TAKE-HOME MESSAGE 1·1

Through its emphasis on objective observation, description, and experimentation, science is a pathway by which we can discover and better understand the world around us.

1·2

Biological literacy is essential in the modern world.

In the past, it was possible to get through life with little or no knowledge of science, but now scientific literacy has become a necessity. A brief glance at any newspaper will reveal just how many important health, social, medical, political, economic, and legal issues pivot on complex scientific data and theories.

- Why are unsaturated fats healthier for you than saturated fats?

- What are allergies? Why do they strike children from clean homes more than children from dirty homes?

- Why is friendship so common among humans yet extremely rare elsewhere in the animal kingdom?

- Why do new agricultural pests appear faster than new pesticides?

- What does scientific testing reveal about the accuracy of eyewitness testimony in courtrooms?

Biological literacy is the ability to (1) use the process of scientific inquiry to think creatively about real-world issues that have a biological component, (2) communicate these thoughts to others, and (3) integrate these ideas into your

January 24, 2007

With DNA From Exhumed Body, Man Finally Wins Freedom

By FERNANDA SANTOS

AUBURN, N.Y., Jan. 23 — Roy Brown, who spent 15 years in prison on a murder conviction and uncovered evidence while there that linked another man to the crime, was released from prison on Tuesday after DNA tests on the other man's exhumed body matched saliva on a nightshirt at the crime scene.

FIGURE 1-2 In the news. Every day, news sources report on social, political, medical, and legal issues related to science.

4 CHAPTER 1 • SCIENTIFIC THINKING

What Is Science?   Scientific Method   Experimental Design   Decision Making

decision making. Biological literacy doesn't involve just the big issues facing society or just abstract ideas (**FIGURE 1-2**). It also matters to you personally. Should you take aspirin when you have a fever? Are you using the wrong approach if you try to lose weight and, after some initial success, you find your rate of weight loss diminishing? Is it a good idea to consume moderate amounts of alcohol?

> " Scientific issues permeate the law. I believe [that] in this age of science we must build legal foundations that are sound in science as well as in law. The result, in my view, will further not only the interests of truth but also those of justice. "
>
> — U.S. Supreme Court Justice Stephen Breyer, at the annual meeting of the American Association for the Advancement of Science, February 1998

Biological issues—including global warming, fossil fuel use, stem cell research, and the proliferation of genetically modified foods—have also become important in political campaigns. And all around you, you will encounter products bearing too-good-to-be-true claims and technical-sounding language designed to lure you into purchasing them. Lack of biological literacy will put you at the mercy of "experts" who may try to confuse you or convince you of things in the interest of (their) personal gain. Scientific thinking will help you make wise decisions for yourself and for society.

TAKE-HOME MESSAGE 1·2

Biological issues permeate all aspects of our lives. To make wise decisions, it is essential for individuals and societies to attain biological literacy.

1·3

The scientific method is a powerful approach to understanding the world.

> " If science proves some belief of Buddhism wrong, then Buddhism will have to change. "
>
> — The 14th Dalai Lama, *New York Times*, December 2005

It's a brand new age, and science, particularly biology, is everywhere. We are called upon more and more frequently to make decisions that hinge on our abilities to grasp biological information and to think scientifically. To illustrate the value of scientific thinking in understanding the world, let's look at what happens in its absence, by considering some unusual behaviors in the common laboratory rat.

Rats can be trained without much difficulty to push a lever to receive a food pellet from a feeding mechanism (**FIGURE 1-3**). When the mechanism is altered so that there is a 10-second delay between the lever being pushed and the food pellet being dispensed, however, strange things start to happen. In one cage, the rat will push the lever and then, very methodically, run and push its nose into one corner of the cage. Then it moves to another corner and again pushes its nose against the cage. It repeats this behavior at the third and fourth corners of the cage, after which the rat stands in front

of the feeder and the pellet is dispensed. Each time the rat pushes the lever it repeats the nose-in-the-corner sequence before moving to the food tray.

In another cage, with the same 10-second delay before the food pellet is dispensed, a rat pushes the lever and then

FIGURE 1-3 "In the absence of the scientific method . . ." Rats develop strange superstition-like behaviors if there is a 10-second delay between when they push a lever and when food is delivered.

proceeds to do three quick back-flips in succession. It then moves to the food tray for the food pellet when the 10 seconds have elapsed. Like the nose-in-the-corner rat, the back-flip rat will repeat this exact behavior each time it pushes the lever.

In cage after cage of rats with these 10-second-delay food levers, each rat eventually develops its own peculiar series of behaviors before moving to the food dish to receive the pellet. Why do they do this? Because it seems to work! They have discovered a method by which they can get a food pellet. To some extent, the rats' behaviors are reasonable. They associate two events—pushing the lever and

> **Q** Why and when do people develop superstitions? Can animals be superstitious?

engaging in some sequence of behaviors—with another: receiving food. In a sense, they have taken a step toward understanding their world, even though the events are not actually related to each other.

Humans can also mistakenly associate actions with outcomes in an attempt to understand and control their world. The irrational belief that actions that are not logically related to a course of events can influence its outcome is called **superstition.** In the absence of scientific thinking, individuals can develop incorrect ideas, such as superstitions about how the world works (**FIGURE 1-4**). For example, Nomar Garciaparra,

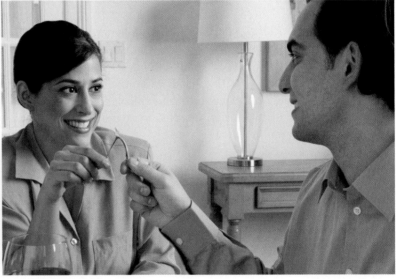

FIGURE 1-4 Superstitions abound. As comforting as myths and superstitions may be, they are no substitute for really understanding how the world works.

What Is Science?   Scientific Method   Experimental Design   Decision Making

shortstop for the Los Angeles Dodgers baseball team, engages in a precise series of toe taps and adjustments to his batting gloves before he bats.

Beyond superstitions, humans have developed a variety of ways of understanding their world. Ancient Greeks, for example, understood that the evils of the world flew out of Pandora's box when, despite Zeus's strict orders not to open the box, her curiosity got the best of her and she lifted the lid. In the Kalahari desert, the African Bushmen tell the tale that their creator, Kaang, ordered them not to build fires. Because their ancestors forgot these orders and built a fire, Kaang punished them by destroying their friendship with animals. Although Bushmen and animals could communicate with each other in ancient times, now the animals simply fear the Bushmen.

Thousands of different narratives from all over the world exist to help people understand the world around them. These stories explain everything from birth and death to disease and healing. On the creation of the earth and universe alone, more than 4,000 narratives have been catalogued.

As helpful and comforting as stories and superstitions may be (or as helpful as people *think* they are!), they are no substitute for really understanding how the world works: for really understanding, for example, that you are sick not because the gods are displeased with you but because the water you are drinking is contaminated, and that, if the water can be purified, then you won't get sick. This kind of understanding does not come all at once by some magical power; instead, it begins when someone wonders about why something is the way it is and then decides to try to find out the answer. This process of examination and discovery is called the **scientific method.**

The scientific method usually begins with someone observing a phenomenon and proposing an explanation for it. Next, the proposed explanation is tested through a series of experiments. If the experiments reveal that the explanation is accurate, and if the experiments can be done by others with the same result,

then the explanation is considered to be valid. If the experiments do not support the proposed explanation, then the explanation must be revised or alternative explanations that more closely reflect experimental results must be proposed and tested. This process continues as better, more accurate explanations are found.

While the scientific method reveals much about the world around us, it doesn't explain everything. There are many other methods through which we can gain an understanding of the world. For example, much of our knowledge about plants and animals does not come from the use of the scientific method, but rather comes from systematic, orderly observation, without the testing of any explicit hypotheses. Other disciplines also involve understandings of the world based on non-scientific processes. Knowledge about history, for example, comes from the systematic examination of past events as they relate to humans, while the "truths" in other fields, such as religion, ethics, and even politics, often are based on personal faith, traditions, and mythology.

Scientific thinking can be distinguished from these alternative ways of acquiring knowledge about the world in that it is **empirical.** Empirical knowledge is based on experience and observations that are rational, testable, and repeatable. The empirical nature of the scientific approach makes it self-correcting: in the process of analyzing a topic, event, or phenomenon with the scientific method, incorrect ideas are discarded in favor of more accurate explanations. In the next sections, we look at how to put the scientific method into practice.

TAKE-HOME MESSAGE 1•3

There are numerous ways of gaining an understanding of the world. Because it is empirical, rational, testable, repeatable, and self-correcting, the scientific method is a particularly effective approach.

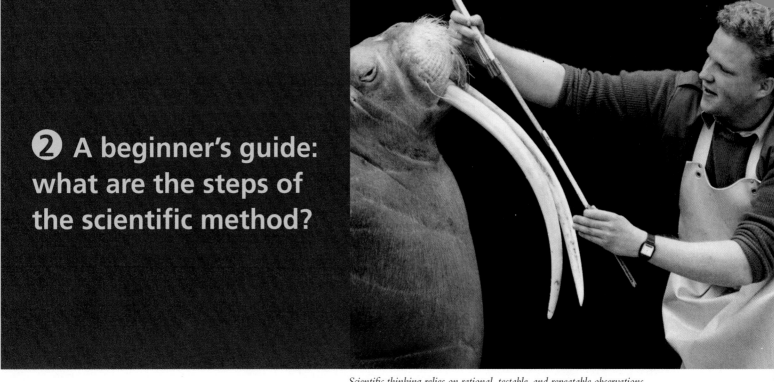

❷ A beginner's guide: what are the steps of the scientific method?

Scientific thinking relies on rational, testable, and repeatable observations.

1•4 ---

Thinking like a scientist: how do you use the scientific method?

"Scientific method"—this term sounds like a rigid process to follow, much like following a recipe. In practice, however, the scientific method is not a single pathway that is always rigidly followed from start to finish. Rather, it is an adaptable process that includes many different methods. This flexibility makes the scientific method a more powerful process that can be used to explore a wide variety of thoughts, events, or phenomena, not only in science, but in other areas as well.

The basic steps in the scientific method are:

Step 1. Make observations.

Step 2. Formulate a hypothesis.

Step 3. Devise a testable prediction.

Step 4. Conduct a critical experiment.

Step 5. Draw conclusions and make revisions.

Once begun, though, the process doesn't necessarily continue linearly through the five steps until it is concluded (**FIGURE 1-5**). Sometimes, observations made in the first step can lead to more than one hypothesis and several testable predictions and

experiments. And the conclusions drawn from experiments often suggest new observations, refinements to hypotheses, and, ultimately, increasingly precise conclusions.

THE SCIENTIFIC METHOD

STEP 1	STEP 2	STEP 3	STEP 4	STEP 5
Make observations.	Formulate a hypothesis.	Devise a testable prediction.	Conduct a critical experiment.	Draw conclusions and make revisions.

 The scientific method rarely proceeds in a straight line. Conclusions, for example, often lead to new observations and refined hypotheses.

FIGURE 1-5 The scientific method: five basic steps and one flexible process.

What Is Science?   Scientific Method   Experimental Design   Decision Making

What should you do when something you believe in turns out to be wrong?

An especially important feature of the scientific method is that its steps are self-correcting. As we continue to make new observations, a hypothesis about how the world works might change. If our observations do not support our current hypothesis, that hypothesis must be given up in favor of one that is not contradicted by any observations. This may be the most important feature of the scientific method: *it tells us when we should change our minds.*

Because the scientific method is a general strategy for learning, it needn't be used solely to learn about nature or scientific things. In fact, we will investigate how it works by looking at a specific question that has recently begun to benefit from analysis using the scientific method:

•   How reliable is eyewitness testimony in criminal courts?

For more than 200 years, courts in the United States have viewed eyewitness testimony as unassailable. Few things are seen as more convincing to a jury than an individual testifying that she can identify the person she saw commit a crime (**FIGURE 1-6**). Common sense has told us that events we see with our own two eyes are as they appear. But is eyewitness identification always right? Can the scientific method tell us whether this

FIGURE 1-6 **"With your own two eyes . . ."** How reliable is eyewitness testimony in criminal courts?

belief—or some other belief—is true? As we describe how to use the scientific method to answer questions about the world, it will become clear that the answer is a resounding *yes*. To show how the scientific method can be used to examine a wide variety of issues, we will also look at how the scientific method can be used to answer three additional questions:

•   Does echinacea reduce the intensity or duration of the common cold?

•   Does chemical runoff give rise to hermaphrodite fish?

•   Does shaving hair from your face, legs, or anywhere else cause it to grow back coarser or darker?

TAKE-HOME MESSAGE 1•4

The scientific method (observation, hypothesis, prediction, test, and conclusion) is a flexible, adaptable, and efficient pathway to understanding the world, because it tells us when we must change our beliefs.

Hold the fries. We apply an understanding of science when we choose foods from the menu that have fewer calories and less saturated fat.

Step 1: Make observations.

Scientific study always begins with observations. At the first stage in the scientific method, we simply look for interesting patterns or cause-and-effect relationships. This is where a great deal of the creativity of science comes from. In the case of eyewitness testimony, we know now that new technologies have made it possible to assess whether tissue such as hair or blood from a crime scene came from a particular suspect. Armed with these tools, the Justice Department recently reviewed 28 criminal convictions that had been overturned by DNA evidence. They found that in most of the cases, the strongest evidence against the defendant during the trial had been eyewitness identification. The observation here is that many defendants who are later found to be innocent were initially convicted based on eyewitness testimony.

Opportunities for other interesting observations are unlimited. Using the scientific method, we can (and will) answer our three other questions.

Many people have claimed that consuming extracts of the herb echinacea can reduce the intensity or duration of symptoms of the common cold. For this reason, echinacea is a widely used herbal treatment (**FIGURE 1-7**). Returning to our fundamental question underlying the scientific method, we can ask: how do you know that is true?

- Does taking echinacea reduce the intensity or duration of the common cold?

Some people have noted that chemicals in sewage runoff—particularly those related to the hormone estrogen—seem to cause male fish to turn into hermaphrodites, organisms that have the reproductive organs of both sexes. Is this true?

- Does chemical runoff give rise to hermaphrodite fish?

And finally, some people have suggested that shaving hair from your face, legs, or anywhere else causes the hair to grow back coarser and darker. Is this true?

STEP 1   STEP 2   STEP 3   STEP 4   STEP 5

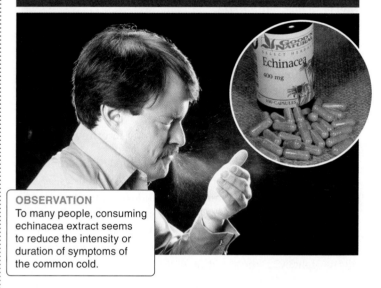

STEP 1: MAKE OBSERVATIONS

OBSERVATION
To many people, consuming echinacea extract seems to reduce the intensity or duration of symptoms of the common cold.

FIGURE 1-7 The first step of science: making observations about the world.

- Does hair that is shaved grow back coarser or darker?

Using the scientific method, we can answer all of these questions.

TAKE-HOME MESSAGE 1•5

The scientific method begins by making observations about the world, noting apparent patterns or cause-and-effect relationships.

Step 2: Formulate a hypothesis.

Based on observations, we can develop a **hypothesis** (*pl.* **hypotheses**), a proposed explanation for observed phenomena. What hypotheses could we make about the eyewitness testimony observations described in the previous section? We could start with the hypothesis: "Eyewitness testimony is always accurate." We may need to modify our hypothesis later, but this is a good start. At this point, we can't draw any conclusions. All we have done is summarize some preliminary observations into a possible explanation for what we have observed.

To be most useful, a hypothesis must accomplish two things.

1. It must clearly establish mutually exclusive alternative explanations for a phenomenon. That is, it must be clear that if the proposed explanation is not supported by evidence or further observations, a different hypothesis is a more likely explanation.

2. It must generate testable predictions (**FIGURE 1-8**). This characteristic is important because we can only evaluate the validity of a hypothesis by putting it to the test. A hypothesis that can never be shown, by some form of observation, to be untrue is not a useful hypothesis. Consider, for example, the hypothesis that your dog loves you. If, for any observation someone makes, you can explain how that observation is consistent with your hypothesis, then the hypothesis is not useful (even if your dog really does love you).

For example we could disprove the "Eyewitness testimony is always accurate" hypothesis by demonstrating that, in certain circumstances, individuals who have witnessed a crime might misidentify the criminal when asked to select the suspect from a lineup.

Often researchers will pose a hypothesis as a negative statement proposing that there is no relationship between two factors, such as "Echinacea has no effect on the duration and severity of cold symptoms." Or "There is no difference in the coarseness or darkness of hair that has been shaved." A hypothesis that states a *lack* of relationship between two factors is called a **null hypothesis.** These hypotheses are equally valid but are easier to disprove. This is because a single piece of evidence or a single new observation that contradicts a null hypothesis is sufficient to reject it and conclude that an alternative hypothesis is true, or that it is highly probable that an alternative is true. So, once you have one piece of solid evidence that your null hypothesis is not true, you gain little by collecting further data. Conversely, it is impossible to prove a hypothesis is absolutely and permanently true: all evidence or further observations that support a hypothesis are valuable, but they do not rule out the possibility

STEP 1   STEP 2   STEP 3   STEP 4   STEP 5

STEP 2: FORMULATE A HYPOTHESIS

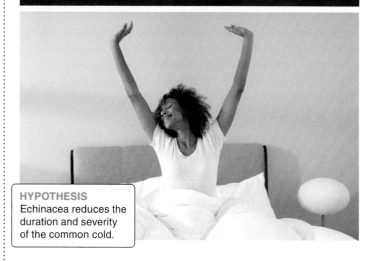

HYPOTHESIS
Echinacea reduces the duration and severity of the common cold.

FIGURE 1-8 Hypothesis: the proposed explanation for a phenomenon.

that some future evidence or observation might show that the hypothesis is not true.

For our other observations, we could state our hypotheses in two different ways:

Hypothesis: Echinacea reduces the duration and severity of the symptoms of the common cold.

Null hypothesis: Echinacea has no effect on the duration or severity of the symptoms of the common cold.

Hypothesis: Estrogens in sewage runoff turn fish into hermaphrodites.

Null hypothesis: Estrogens in sewage runoff have no effect in turning fish into hermaphrodites.

Hypothesis: Hair that is shaved grows back coarser and darker.

Null hypothesis: There is no difference in the coarseness or color of hair that is shaved relative to hair that is not shaved.

TAKE-HOME MESSAGE 1·6

A hypothesis is a proposed explanation for an observed phenomenon.

Step 3: Devise a testable prediction.

This step of the scientific method really is part of the previous step. Formulating a hypothesis is important, but not all hypotheses are created equal. Some, in fact (such as the hypothesis about your dog loving you), are of no use at all when it comes to helping us better understand the world. For a hypothesis to be useful, it must generate a prediction. That is, it must suggest that under certain conditions we will be able to observe certain outcomes. Put another way, a good hypothesis helps us make predictions about novel situations. This is a powerful feature of a good hypothesis: it guides us to knowledge about new situations.

All of this is rather abstract. Let's get more concrete with the four hypotheses we are considering. Keep in mind that when you do not understand some aspect of the world, any one of several possible explanations could be true. In devising a testable prediction from a hypothesis, the goal is to propose a situation that will give a particular outcome if your hypothesis is true, but will give a different outcome if your hypothesis is not true.

Hypothesis: Eyewitness testimony is always accurate.

Prediction: Individuals who have witnessed a crime will correctly identify the criminal regardless of whether multiple suspects are presented one at a time or all at the same time in a lineup.

This is a good, testable prediction because if our hypothesis is true, then our prediction will always be true. On the other hand, if one method of presenting suspects consistently causes incorrect identification of the criminal, our hypothesis cannot be true and must be revised or discarded.

Hypothesis: Echinacea reduces the duration and severity of the symptoms of the common cold (**FIGURE 1-9**).

Prediction: If echinacea reduces the duration and severity of the symptoms of the common cold, then individuals taking echinacea should get sick less frequently than those not taking it, and when they do get sick, their illness should not last as long.

Hypothesis: Estrogens in sewage runoff turn fish into hermaphrodites.

Prediction: If estrogens in sewage runoff turn fish into hermaphrodites, then exposing fish to estrogens should cause more fish to turn into hermaphrodites than when fish are not exposed to estrogens.

STEP 1   STEP 2   **STEP 3**   STEP 4   STEP 5

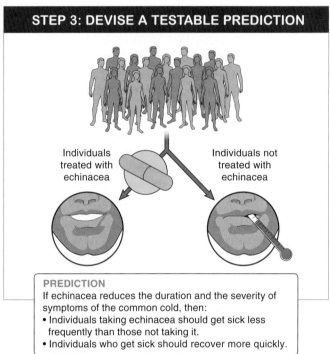

STEP 3: DEVISE A TESTABLE PREDICTION

Individuals treated with echinacea

Individuals not treated with echinacea

PREDICTION
If echinacea reduces the duration and the severity of symptoms of the common cold, then:
• Individuals taking echinacea should get sick less frequently than those not taking it.
• Individuals who get sick should recover more quickly.

FIGURE 1-9 Devising a testable prediction. For a hypothesis to be useful, it must generate a testable prediction.

Hypothesis: Hair that is shaved grows back coarser and darker.

Prediction: If shaving leads to coarser, darker hair growing back, then if individuals shaved one leg only, the hair that re-grows on that leg should become darker and coarser than the hair growing on the other leg.

As you begin to think scientifically, you will find yourself making a lot of "if . . . then" types of statements: "*If* that is true," referring to some hypothesis or assertion someone makes, or perhaps to a claim made by the manufacturer of a new health product, "*then* I would expect that . . ." proposing your own prediction about a related situation. Once you've made a testable prediction, the next step is to go ahead and test it.

TAKE-HOME MESSAGE 1·7

For a hypothesis to be useful, it must generate a testable prediction.

12   CHAPTER 1 • SCIENTIFIC THINKING

What Is Science?   Scientific Method   Experimental Design   Decision Making

1·8

Step 4: Conduct a critical experiment.

Once we have formulated a hypothesis that generates a testable prediction, we conduct a **critical experiment,** an experiment that makes it possible to decisively determine whether a particular hypothesis is correct. There are many crucial elements in designing a critical experiment, and Section 1-11 covers the details of this process. For now, it is important just to understand that, with a critical experiment, if the hypothesis being tested is not true, we will make observations that compel us to reject that hypothesis.

In this step of the scientific method, a bit of cleverness can come in handy. Suppose we were to devise a critical experiment to test our hypothesis "Eyewitness testimony is always accurate." First, we stage a mock crime such as a purse snatching in front of a group of observers who do not know the crime is staged. Next, we ask observers to identify the criminal. To one group of observers we might present six "suspects" all at once in a lineup. To another group of observers, we might present the six suspects one at a time. The beauty of this experiment is that we actually know who the "criminal" is. With this knowledge, we can evaluate with certainty whether an eyewitness's identification is correct or not.

This exact experiment was done with an additional, devious little twist: the researcher did not include the actual "criminal" in any of the lineups of six "suspects." If eyewitness testimony is always accurate, however, this slight variation should not matter. We would predict that the observers in both groups (the lineup group and the one-at-a-time group) would indicate that the criminal was not present. In the next section, we will see what happened. Right now, let's devise critical experiments for our other hypotheses.

Hypothesis: Echinacea reduces the duration and severity of the symptoms of the common cold. The critical experiment for the echinacea hypothesis has been performed, and it is about as close to a perfect experiment as possible. Researchers began with 437 people who volunteered to be exposed to viruses that cause the common cold. Exposure to the cold-causing viruses was a bit unpleasant: all of the volunteers had cold viruses (in a watery solution) dripped into their noses. Research subjects were then secluded in hotel rooms for five days, and doctors examined them for the presence of the cold virus in their nasal cavity and for any cold symptoms (**FIGURE 1-10**).

The subjects were randomly divided into four groups. In two of the groups, each individual began taking a pill each day for

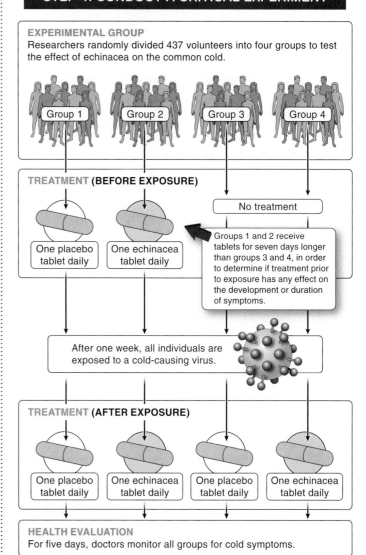

FIGURE 1-10 "When you need to know . . ." A critical experiment makes it possible to conclusively determine whether a hypothesis is correct.

a week prior to exposure to the cold virus. Those in one group received echinacea tablets while those in the other took a **placebo,** a pill that looked identical to the echinacea pill but contained no echinacea or other active ingredient. Neither

FIGURE 1-11 **Does shaving hair cause it to grow back coarser or darker?**

hypothesis, researchers placed cages containing male trout downstream from sewage treatment plants on several rivers in England. They then examined the fish for the presence of eggs in their testes and tested their blood for the presence of an egg-production hormone that is not normally found in males.

Hypothesis: Hair that is shaved grows back coarser and darker. A critical experiment does not have to be complex or high-tech. All that matters is that it can decisively determine whether or not a hypothesis is correct. This point is illustrated by an experiment published in the scholarly journal *Archives of Facial Plastic Surgery.* Researchers decided to test the hypothesis that shaving hair causes it to grow back coarser or darker. A group of volunteers had one of their eyebrows selected at random and completely shaved off. The subjects were then evaluated for eyebrow re-growth over the course of the next six months, with observers also analyzing photographs of the individuals (**FIGURE 1-11**).

Q Does shaving or cutting hair make it grow back more thickly?

In the next section we'll see how our hypotheses survive being confronted with the results of the critical experiments described above and how we can move toward drawing conclusions.

TAKE-HOME MESSAGE 1·8

A critical experiment is one that makes it possible to decisively determine whether a particular hypothesis is correct.

the subject nor the doctor administering the pills (and later checking for cold symptoms) knew what the pills contained. In the other two groups, the individuals did not begin taking the pills until the day on which they were exposed to the cold virus. Again, one group got the echinacea pill and the other group got the placebo.

Hypothesis: Estrogens in sewage runoff turn fish into hermaphrodites. To try to determine the validity of this

1·9 -

Step 5: Draw conclusions, make revisions.

Once the results of the critical experiment are in, they are pulled apart, examined, and analyzed. Researchers look for patterns and relationships in the evidence they've gathered from their experiments; they draw conclusions and see whether their findings and conclusions support their hypotheses. If an experimental result is not what you expected, that does not make it a "wrong answer." Science includes a great deal of trial and error, and if the conclusions do not support the hypothesis, then you must

Q Is eyewitness testimony in courts always right?

revise your hypothesis, which often spurs you to conduct more experiments. This step is a cornerstone of the scientific method because it demands that you must be open-minded and ready to change what you think.

The results of the purse-snatching experiment were surprising. When the suspects (which, as you'll recall, did not include the actual "criminal") were viewed together in a lineup, the observers/witnesses erroneously identified someone

14 CHAPTER 1 • SCIENTIFIC THINKING

What Is Science?   Scientific Method   Experimental Design   Decision Making

STEP 1 STEP 2 STEP 3 STEP 4 **STEP 5**

STEP 5: DRAW CONCLUSIONS AND MAKE REVISIONS

FIGURE 1-12 Drawing conclusions and making revisions.

as the purse snatcher about a third of the time. When the suspects were viewed one at a time, the observers made a mistaken identification less than 10% of the time.

In this or any other experiment, it does not matter whether we can imagine a reason for the discrepancy between our hypothesis and our results. What is important is that we have demonstrated that our initial hypothesis—"Eyewitness testimony is always accurate"—is not supported by the data. Our observations suggest that, at the very least, the accuracy of an eyewitness's testimony depends on the method used to present the suspects. Based on this result, we might then adjust our hypothesis to: "Eyewitness testimony is more accurate when suspects are presented to witnesses one at a time."

We can then devise new and more specific testable predictions to try to further refine a hypothesis. In the case of eyewitness testimony, further investigation suggests that when suspects or pictures of suspects are placed side by side, witnesses compare them and tend to choose the suspect that *most resembles* the person they remember committing the crime. When viewed one at a time, suspects can't be compared in this way and witnesses are less likely to make misidentifications.

Q Does echinacea help prevent the common cold?

Hypothesis: Echinacea reduces the duration and severity of the symptoms of the common cold (**FIGURE 1-12**). In the echinacea study, the results were definitive. Those who took the echinacea were just as likely to catch a cold, and, once they caught the cold, the symptoms lasted for the same amount of time. In short, echinacea had no effect at all. Several similar studies have been conducted, all of which have shown that echinacea does not have any beneficial effect. As one of the researchers commented afterward, "We've got to stop attributing any efficacy to echinacea."

Although it seems clear that our initial hypothesis that echinacea prevents people from catching colds and reduces the severity and duration of cold symptoms is not correct, further experimentation might involve altering the amount of echinacea given to the research subjects or the length of time they take echinacea before exposure to the cold-causing viruses.

Hypothesis: Estrogens in sewage runoff turn fish into hermaphrodites. The results of the sewage experiment showed that, in every single river tested, male fish downstream from the runoff developed eggs in their testes and had significant levels of the egg-production hormone in their blood. Further research stimulated by these results has investigated a variety of common industrial chemicals to see whether they can stimulate production of the egg-development hormone in male fish under laboratory conditions, in an attempt to identify which particular chemical component in the sewage causes the male fish to change.

Hypothesis: Hair that is shaved grows back coarser and darker. In the hair-shaving experiment, the observers discovered that it was impossible to distinguish which eyebrow had been shaved. This was not a surprise to dermatologists evaluating the study because all of the living parts involved in hair growth are below the surface of the skin. (Plucking hairs can damage the root of the hair, potentially affecting future hair growth, but that is another story, awaiting further study.)

The outcomes in all of these studies show that after the results of a critical experiment have been gathered and interpreted, it is important not just to evaluate the initial hypothesis but also to consider any necessary revisions or refinements to it. This revision is an important step; by revising a hypothesis, based on the results of experimental tests, we can explain the observable world with greater and greater accuracy.

TAKE-HOME MESSAGE 1·9

Based on the results of experimental tests, we can revise a hypothesis and explain the observable world with increasing accuracy. A great strength of scientific thinking, therefore, is that it helps us understand when we should change our minds.

1·10

When do hypotheses become theories, and what are theories?

It's an unfortunate source of confusion that, among the general public, the word "theory" is often used to refer to a hunch or a guess or speculation—that is, something we are not certain about—while to scientists, the word means nearly the opposite: a hypothesis of which they are most certain. To reduce misunderstandings, we examine two distinct levels of understanding that scientists use in describing our knowledge about natural phenomena.

Hypothesis As we have seen, hypotheses are at the very heart of scientific thinking. A hypothesis is a proposed explanation for a phenomenon. A good hypothesis leads to testable predictions. Commonly, when non-scientists use the word "theory"—as in, "I've got a theory about why there's less traffic on Friday mornings than on Thursday mornings"—they actually mean that they have a hypothesis.

Theory A **theory** is an explanatory hypothesis for natural phenomena that is exceptionally well supported by the empirical data. A theory can be thought of as a hypothesis that has withstood the test of time and is unlikely to be altered by any new evidence. Like a hypothesis, a theory is testable; but because it has already been repeatedly tested and no

observations or experimental results have contradicted it, a theory is viewed by the scientific community with nearly the same confidence as a fact. For this reason, it is inappropriate to describe something as "just a theory" as a way of asserting that it is not likely to be true.

Theories in science also tend to be broader in scope than hypotheses. In biology, two of the most important theories (which we explore in more detail in Chapters 3 and 8) are *cell theory,* that all organisms are composed of cells and all cells come from pre-existing cells, and the *theory of evolution by natural selection,* that species can change over time and all species are related to each other through common ancestry.

TAKE-HOME MESSAGE 1·10

Scientific theories do not represent speculation or guesses about the natural world. Rather, they are hypotheses—proposed explanations for natural phenomena—that have been so strongly and persuasively supported by empirical observation that the scientific community views them as very unlikely to be altered by new evidence.

16 CHAPTER 1 • SCIENTIFIC THINKING

What Is Science?   Scientific Method   Experimental Design   Decision Making

❸ Well-designed experiments are essential to testing hypotheses.

Controlled experiments increase the power of our observations. Here, flasks containing Panax (Asian ginseng) are prepared.

1·11 ---

Controlling variables makes experiments more powerful.

From our earlier discussion of critical experiments, you have a sense of how important it is to have a well-planned, well-designed experiment. Some experiments are just better than others when it comes to figuring out how the world works. At their most basic level, experiments help us figure out the cause-and-effect relationship between two things. In the previous sections, for example, we hypothesized about the relationship between witnessing a crime and being able to identify the person who committed it, between taking echinacea and catching a cold and having it for a period of time, between water pollution and hermaphrodite fish, and between shaving and the subsequent coarseness and color of hair. In performing experiments, our goal is to figure out whether one thing influences another thing; if an experiment enables us to draw a correct conclusion about that cause-and-effect relationship, it is a good experiment.

In our initial discussion of experiments, we just described what the experiment was without examining why the researchers chose to perform the experiment the way they did. In this section, we explore some of the ways to maximize an experiment's power, and we'll find that, with careful planning, it is possible to increase an experiment's ability to discern causes and effects.

First, let's consider some elements common to most experiments. These include:

1. Treatment. This is any experimental condition applied to the research subjects. It might be the shaving of one of an individual's legs, or the pattern used to show "suspects" (all at once or one at a time) to the witness of a staged crime, or a dosage of echinacea given to an individual.

2. Experimental group. This is a group of subjects who are exposed to a particular treatment—for example, the individuals given echinacea rather than placebo in the experiment described above. It is sometimes referred to as the "treatment group."

3. Control group. This is a group of subjects who are treated identically to the experimental group, with one exception: they are not exposed to the treatment. An example would be the individuals given placebo rather than echinacea.

17

4. Variables. These are the characteristics of an experimental system that are subject to change. They might be, for example, the amount of echinacea a person is given, or a measure of the coarseness of an individual's hair. When we speak of "controlling" variables—the most important feature of a good experiment—we are describing the attempt to minimize any differences (which are also called "variables") between a control group and an experimental group other than the treatment itself. That way, any differences in the outcomes we observe between the groups are most likely due to the treatment.

Let's look at an example that illustrates the importance of considering all these elements when designing an experiment.

Stomach ulcers are erosions of the stomach lining that, due to the highly acidic condition of that part of the digestive tract, can be very painful. In the late 1950s, a doctor reported in the *Journal of the American Medical Association* that stomach ulcers could be effectively treated by having a patient swallow a balloon connected to some tubes that circulated a refrigerated fluid. He argued that by super-cooling the stomach, acid production was reduced and the ulcer symptoms relieved. He had convincing data to back up his claim: in all 24 of his patients who received this "gastric freezing" treatment, their condition improved. As a result, the treatment became widespread for many years (**FIGURE 1-13**).

Although there was a clear hypothesis ("Gastric cooling reduces the severity of ulcers") and some compelling observations (all 24 patients experienced relief), this experiment falls short of qualifying as a good example of the scientific method. Why? Because it was not designed well. In particular, there was no clear group with whom to compare the patients who received the treatment. In other words, who is to say that just going to the doctor or having a balloon put into your stomach doesn't improve ulcers? The results of this doctor's experiment do not rule out these interpretations.

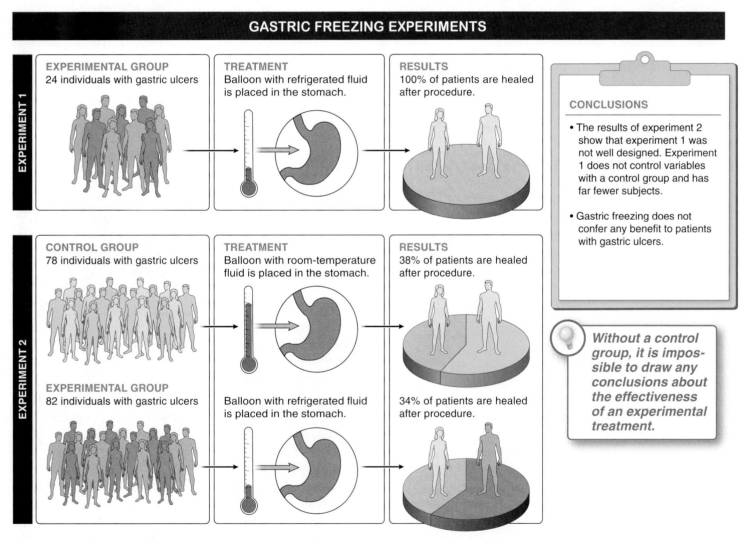

GASTRIC FREEZING EXPERIMENTS

EXPERIMENT 1

EXPERIMENTAL GROUP
24 individuals with gastric ulcers

TREATMENT
Balloon with refrigerated fluid is placed in the stomach.

RESULTS
100% of patients are healed after procedure.

EXPERIMENT 2

CONTROL GROUP
78 individuals with gastric ulcers

TREATMENT
Balloon with room-temperature fluid is placed in the stomach.

RESULTS
38% of patients are healed after procedure.

EXPERIMENTAL GROUP
82 individuals with gastric ulcers

Balloon with refrigerated fluid is placed in the stomach.

34% of patients are healed after procedure.

CONCLUSIONS

• The results of experiment 2 show that experiment 1 was not well designed. Experiment 1 does not control variables with a control group and has far fewer subjects.

• Gastric freezing does not confer any benefit to patients with gastric ulcers.

Without a control group, it is impossible to draw any conclusions about the effectiveness of an experimental treatment.

FIGURE 1-13 No controls. A poorly designed experiment: gastric freezing and stomach ulcers.

18 CHAPTER 1 • SCIENTIFIC THINKING

What Is Science?   Scientific Method   Experimental Design   Decision Making

A few years later, another researcher decided to do a more carefully controlled study. He recruited 160 ulcer patients and gave 82 of them the gastric freezing treatment. The other 78 received a similar treatment in which they swallowed the balloon but had room-temperature water pumped in. The latter was an appropriate control group because the subjects were treated exactly like the experimental group, with the exception of only a single difference between the groups—whether they experienced gastric freezing or not. The new experiment could test for an effect of the gastric freezing, while controlling for the effects of other, lurking variables that might affect the outcome.

Surprisingly, although the researcher found that 34% of those in the gastric freezing group improved, he also found that 38% of those in the control group improved. These results indicated that gastric freezing didn't actually confer any benefit when compared with a treatment that did not involve gastric freezing. Not surprisingly, the practice was abandoned. (We will discuss below why the control group might have had such a high rate of improvement.)

Q Is arthroscopic surgery for arthritis beneficial for the 300,000 people in the United States who have it each year? How do we know?

More recently, a well-controlled experiment demonstrated that arthroscopic knee surgery, a treatment for osteoarthritis performed on more than 300,000 people in the United States each year, produced moderate benefits to the patient. But the study demonstrated even greater benefits to patients in a control group who were subjected not to the arthroscopic procedure but rather to "sham" surgery, in which the surgeon made three superficial incisions in the knee, manipulated the knee a bit, and clanged the instruments around in the operating room. As a consequence of this study, many doctors have begun abandoning the use of arthroscopic surgery for the treatment of osteoarthritis—which once accounted for more than a billion dollars in surgical procedures each year. (This type of surgery is still useful for the treatment of some other knee problems, including the repair of ligament and cartilage damage.)

A surprising result from both the gastric freezing studies and the arthroscopic surgery study was that they demonstrated the **placebo effect,** the frequently observed, poorly understood, phenomenon in which people respond favorably to *any* treatment, regardless of whether it is as simple as swallowing a sugar pill or as unpleasant as having to swallow a balloon and have it inflated in their stomach. The placebo effect highlights the need for comparison of treatment effects with an appropriate control group. We want to know whether the treatment is actually responsible for any effect seen; if the control group receiving the placebo or sham treatment has an outcome like that of the experimental group, we can conclude that the treatment itself does not have an effect.

Another pitfall to be aware of in designing an experiment is to ensure that the person(s) conducting the experiment don't influence the experiment's outcome. An experimenter can often unwittingly influence the results of an experiment. For example, eyewitnesses to crimes more frequently identify someone that the police suspect rather than other individuals in the lineup when the police officer conducting the lineup knows which individual is the suspect.

This phenomenon is seen in the story of a horse named Clever Hans. Hans was considered clever because his owner claimed that Hans could perform remarkable intellectual feats, including multiplication and division. When given a numerical problem, the horse would tap out the answer number with his foot. Controlled experiments, however, demonstrated that Hans was only able to solve problems when he could see the person asking the question and when that person knew the answer (**FIGURE 1-14**). It turned out that the questioners revealed, unintentionally and through very subtle body language, the answers.

FIGURE 1-14 **Math whiz or ordinary horse?** The horse Clever Hans was said to be capable of mathematical calculations, until a controlled experiment demonstrated otherwise.

The Clever Hans phenomenon highlights the benefits of instituting even greater controls when designing an experiment. In particular, it highlights the value of **blind experimental design,** in which the experimental subjects do not know which treatment (if any) they are receiving, and **double-blind experimental design,** in which neither the experimental subjects nor the experimenter knows which treatment a subject is receiving.

Another hallmark of an extremely well designed experiment is that it combines the blind/double-blind strategies we've just described in a **randomized,** controlled, double-blind study. In this context, "randomized" refers to the fact that, as in the echinacea study described above, the subjects are randomly assigned into experimental and control groups. In this way, researchers and subjects have no influence on the composition of the control and treatment groups.

The use of randomized, controlled, double-blind experimental design can be thought of as an attempt to imagine all the possible ways that someone might criticize an experiment and to design the experiment so that the results cannot be explained by anything but the effect of the treatment. In this way, the experimenter's results either support the hypothesis or invalidate it—in which case, the hypothesis must be rejected. If multiple explanations can be offered for the observations and evidence from an experiment, then it has not succeeded as a critical experiment.

Suppose you want to know whether a new drug is effective in fighting the human immunodeficiency virus (HIV), the virus that leads to AIDS. Which experiment would be better, one in which the drug is added to HIV-infected cells in a test tube under carefully controlled laboratory conditions, or one in which the drug is given to a large number of HIV-infected individuals? There is no definitive answer. In laboratory studies, it is possible to control nearly every environmental variable. In their simplicity, however, lab studies may introduce difficulties. Complex and naturally varying factors in human subjects, such as nutrition and stress, may have important interactions with the experimental drug that influence its effectiveness. These interactions will not be present and taken into account in the controlled lab study. Good experimental design is more complex than simply following a single recipe. The only way to determine the quality of an experiment is to assess how well the variables that were not of interest were controlled, and how well the experimental treatment tested the relationship of interest.

TAKE-HOME MESSAGE 1·11

To draw clear conclusions from experiments, it is essential to hold constant all those variables we are not interested in. Control and experimental groups should vary only with respect to the treatment of interest. Differences in outcomes between the groups can then be attributed to the treatment.

1·12

Repeatable experiments increase our confidence.

Q *Can science be misleading? How can we know?*

In 2005, a study showed that some patients being treated for HIV infection who were also taking an epilepsy drug called valproic acid had significantly reduced numbers of HIV particles in their blood. Newspaper headlines announced this finding—"AIDS Cure Possible, Study Suggests"—and raised many people's hopes. Two years later, however, a study of people who were already taking valproic acid and anti-HIV drugs concluded that those on valproic acid had not benefited at all from the drug (**FIGURE 1-15**). It's not certain why the later study didn't produce the same results as the first study, but the second study did address two serious shortcomings of the first study—namely, that the first study

was much smaller, involving only four patients, and that no control group was used. This pair of studies reveals the importance of repeatability in science.

A powerful way to demonstrate that observed differences between a treatment group and a control group truly reflect the effect of the treatment is for the researchers to conduct the experiment over and over again. Even better is to have other researcher groups repeat the experiment and get the same results. Researchers describe this desired characteristic of experiments by saying that an experiment must be "reproducible" and "repeatable."

Q *Do megadoses of vitamin C reduce cancer risk?*

What Is Science?   Scientific Method   Experimental Design   Decision Making

HOME | NEWS | WEATHER | HEALTHWISE | SPORTS | COURTS & CRIME

✉ Email   🖶 Print   A A A Text Size

8/7/08
Texas Tech Researchers May Have Found AIDS Cure
Posted: Aug 7, 2008 09:59 PM

Texas Tech University in partnership with
Harvard University is on the research forefront
of a possible vaccine for HIV, the virus that
causes AIDS.

To give you some idea of how big this story
is, on Thursday it was the number one story on
Newsweek's website and it will be the lead in
their next magazine.

NewsChannel 11 met with the president of the
Texas Tech Health Sciences Center, Doctor John
Baldwin and Texas Tech Chancellor Kent
Hance. Both men are optimistic about this
research and its possible impact on the world
and Texas Tech.

Repeatability is essential! Scientific conclusions are more reliable when experiments have been repeated (and modified, if necessary).

Valproic Acid Does Not Reduce Latent HIV Reservoir in Resting CD4 Cells

By Liz Highleyman

Some prior studies have suggested that valproic acid
(Depakote), a drug used to treat epilepsy and bipolar
disorder, might help eradicate HIV from the body. Two
studies published in the June 19, 2008 issue of *AIDS*,
however, indicate that it has little long-term effect.

Valproic Acid
(Depakote)

Study 1

David Margolis and colleagues from the University of North Carolina
previously reported that valproic acid plus intensified antiretroviral therapy
could "flush" HIV out of long-lived latently infected CD4 T-cells and potentially
reduce the resting CD4 cell reservoir. They later conducted a study to test the
ability of valproic acid to deplete resting CD4 cell infection in patients receiving
standard HAART.

Infection of resting CD4 cells was measured in 11 patients with stable
undetectable HIV viral load using a standard HIV RNA assay twice rior to, and
twice after, adding 1000 mg valproic acid to their standard antiretroviral
regimen. Resting CD4 T-cell infection frequency was measured by outgrowth
assay, and low-level viremia was quantified in 9 patients using a highly
sensitive single-copy plasma HIV RNA assay.

FIGURE 1-15 Once is not enough. Experiments and their outcomes must
be repeatable for their conclusions to be valid and widely accepted.

An experiment that can be done over and over again by a
variety of researchers with the same results is an effective
defense against biases (which we discuss in the next section)
and reflects a well-designed experiment. Experiments whose
results cannot be confirmed by repeated experiments or by
experiments performed by other researchers are the downfall
of many dramatic claims. Even though chemist Linus Pauling
had not one, but two Nobel prizes, he was never able to
convince the scientific community that megadoses of vitamin
C are an effective treatment for cancer. Every time other
individuals tried to repeat Pauling's studies, properly matching
the control and treatment groups, they found no difference.
Vitamin C just isn't effective against cancer—a conclusion that
the vast majority of the medical community now believes.
The scientific method is profoundly egalitarian: more
important than a scientist's credentials are sound and
reproducible results.

When a study is repeated (also referred to as "replicating" a
study), sometimes a tiny variation in the experimental design
can lead to a different outcome; this can help us isolate the
variable that is primarily responsible for the outcome of the
experiment. Alternatively, when experiments are repeated and
the same results are obtained, our confidence in them is
increased.

TAKE-HOME MESSAGE 1·12

Experiments and their outcomes must be repeatable for their
conclusions to be considered valid and widely accepted.

We've got to watch out for our biases.

Q Can scientists be sexist? How would we know?

In 2001, the journal *Behavioral Ecology* changed its policy for reviewing manuscripts that were submitted for publication. Its new policy instituted a double-blind process, whereby neither the reviewers' nor the authors' identities were revealed. Previously, the policy had been a single-blind process in which reviewers' identities were kept secret but the authors' identities were known to the reviewers. In an analysis of papers published between 1997 and 2005, it turned out that after 2001, when the double-blind policy took effect, there was a significant increase in the number of published papers in which the first author was female (**FIGURE 1-16**). Analysis of papers published in a similar journal that maintained the single-blind process over that period revealed no such increase.

This study reveals that people, including scientists, may have biases—sometimes subconscious—that influence their behavior. It also serves as a reminder of the importance of proper controls in experiments. If knowing the sex of the author influences a reviewer's decision on whether a paper should be published, it is possible that researchers' biases can creep in and influence their collection of data and analysis of results. (Even the decision of what—and what not—to study can be influenced by our biases.)

It can be hard to avoid biases. Consider a study that required precise measuring of the fingers of the left and right hands—comparing the extent to which people in different groups were physically symmetrical. The researcher noted that when she measured individuals from the group she predicted would be more symmetrical, she felt a need to re-measure if the reading on her digital ruler indicated a big asymmetry. Because she was thinking that the person's fingers *should* be

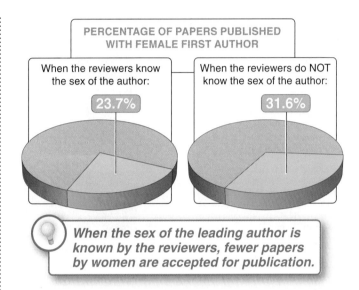

PERCENTAGE OF PAPERS PUBLISHED WITH FEMALE FIRST AUTHOR

When the reviewers know the sex of the author: **23.7%**

When the reviewers do NOT know the sex of the author: **31.6%**

When the sex of the leading author is known by the reviewers, fewer papers by women are accepted for publication.

FIGURE 1-16 Bias against female scientists? The journal *Behavioral Ecology* accepted more papers from female authors when the reviewers were not aware of the author's sex.

symmetrical, she assumed that she had made an error. She felt no such need to re-measure for subjects in the other group, when asymmetries confirmed her hypothesis. To control for this bias, the researcher connected her digital ruler to a computer and, without ever having the number displayed, transmitted the measurement directly to the computer when she pushed a button. In this way, she was able to make each measurement without introducing any regular bias.

TAKE-HOME MESSAGE 1·13

Biases can influence our behavior, including our collection and interpretation of data. With careful controls, it is possible to minimize such biases.

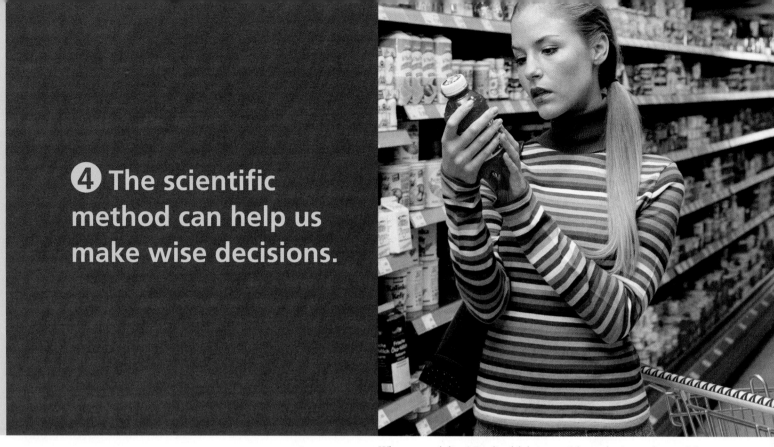

❹ The scientific method can help us make wise decisions.

What can you believe? Reading labels is essential to evaluating products and the claims about them.

1·14

Statistics can help us in making decisions.

In Section 1-12 we saw that researchers repeatedly found, through experimentation, that megadoses of vitamin C do not reduce cancer risk. If you put yourself in the researchers' shoes, you might wonder how you'd figure out that the vitamin C did not reduce cancer risk. Perhaps there were 100 individuals in the group receiving megadoses of vitamin C: some of them developed cancer and some of them did not. And among the 100 subjects in the group not receiving the megadoses, some of the individuals developed cancer and some did not. How do you decide whether the vitamin C actually had an effect? This knowledge comes from a branch of mathematics called **statistics,** a set of analytical and mathematical tools designed to help researchers gain understanding from the data they gather. To understand statistics, let's start with a simple situation.

Suppose you measure the height of two people. One is a woman who is 5 feet 10 inches tall. The other is a man who is 5 feet 6 inches tall. If these were your only two observations of human height, you might conclude that female humans are taller than males. But suppose you measure the height of 100 women and 100 men chosen randomly from a population. Then you can say, "of the 100 men, the average man is 5 feet 9.5 inches, and of the 100 women, the average woman is 5 feet 4 inches." Better still, the data can illuminate for you not only the average, but also some measure of how much variation there is from one individual to another. Statistical analysis can tell you not only that the average man in this study is 5 feet 9.5 inches tall, but that two-thirds of the men are between 5 feet 6.5 inches tall (3 inches less than the average) and 6 feet 0.5 inches tall (3 inches more than the average). You will often see this type of range stated as "5 feet 9.5 inches ± 3 inches" ("plus or minus 3 inches"). Similarly, the data might show that the female subjects in the study are 5 feet 6 inches ± 3 inches," indicating that two-thirds of the women are between 5 feet 3 inches and 5 feet 9 inches tall. As we discussed in Section 1-11, on experimental design, and as this example shows, larger numbers of participants are

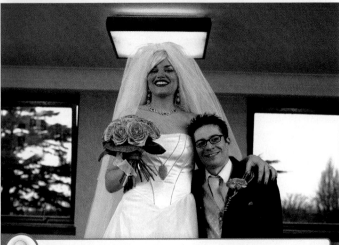

HUMAN HEIGHT AND STATISTICS

Don't get fooled. When drawing conclusions about a natural phenomenon, such as height, it is best to choose subjects at random and to include as many as possible.

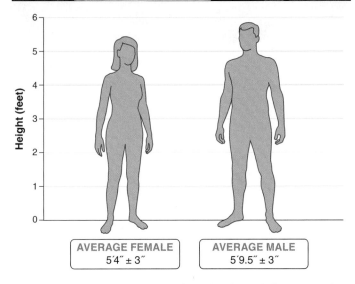

AVERAGE FEMALE
5′4″ ± 3″

AVERAGE MALE
5′9.5″ ± 3″

FIGURE 1-17 Drawing conclusions based on limited observations is risky. Measuring a greater number of people will generally help us draw more accurate conclusions about human height.

better than smaller numbers if you want to draw general conclusions about natural phenomena such as the height of men and women (**FIGURE 1-17**).

Using data to describe the characteristics of individuals participating in a study is useful, but often we want to know whether data support (or do not support) a hypothesis. If the scientific method is to be effective in helping us understand the world, it must help us make wise decisions about concrete things. For example, suppose we want to know whether having access to a textbook helps a student perform

better in a biology class. Statistics can help us answer this question. After conducting a study, let's say we find that students who had access to a textbook scored an average of 81% ± 8% on their exams, while those who did not scored an average of 76% ± 7%.

In this example, it is difficult to distinguish between the two possible conclusions.

Possibility 1: Students having access to a textbook DO perform better in biology classes. In other words, our sampling of this class revealed a true relationship between the two variables, textbook access and class performance.

Possibility 2: Students having access to a textbook DO NOT perform better in biology classes. The variation in the scores in the two groups may be too large to allow us to notice any effect of having access to a textbook. Instead, the difference in average scores for the two groups may mean that more of the high-performing students just happened, through random chance, to be in the group given access to a textbook.

But what if the students with access to a textbook scored, on average, 95% ± 5%, while those without access scored only 60% ± 5%. In this case, we would be much more confident that there is a significant effect of having access to a textbook (**FIGURE 1-18**), because even with the large variation in the scores seen in each group, the averages are still very different from each other.

Statistical methods help us to decide between these two possibilities and, importantly, to state how confident we are that one or the other is true. The greater the difference between two groups (95% versus 60% is a greater difference than 81% versus 76%), and the smaller the variation in each group (± 5% in the 95% and 60% groups, versus ± 8% and ± 7% in the 81% and 76% groups), the more confident we are of the conclusion that there is a significant effect of the treatment (having access to a textbook, in this case). In other words, in the case where the groups of students scored 95% or 60%, depending on whether or not they had access to a textbook, it is possible that having a textbook did not actually improve performance and that this observed difference was just the result of chance. But this conclusion is very, very unlikely.

Statistics can also help us identify relationships (or the lack of relationships) between variables. For example, we might note that when there are more firefighters at a fire, the fire is larger and causes more damage. This is a **positive correlation,** meaning that when one variable (the number of firefighters) increases, so does the other (the severity of the fire). Should we conclude that firefighters make fires worse? No. While correlations can reveal relationships between variables, they

What Is Science?　　　Scientific Method　　　Experimental Design　　　Decision Making

INTERPRETING STATISTICS: A HYPOTHETICAL STUDY

Students without textbooks score an average of 60% ± 5%

Students with textbooks score an average of 95% ± 5%

Proportion of students in population

Average exam scores

Statistics can quantify and summarize large amounts of data, making it possible to draw more accurate conclusions.

don't tell us *how* the variables are related or whether change in one variable causes change in another. (You may have heard or read the phrase "correlation is not causation," which refers to this sort of situation.) Before drawing any conclusions about more firefighters causing larger fires, we need to know about the type of fire and its size when the firefighters arrived, because those factors will significantly influence the ultimate amount of damage. To estimate the effect of the number of firefighters on the amount of damage, we would need to compare the amount of damage from fires of similar size that are fought by different numbers of firefighters.

Ultimately, statistical analyses can help us organize and summarize the observations that we make and the evidence we gather in an experiment. These analyses can then help us decide whether any differences we measure between experimental and control groups are likely to be the result of the treatment, and how confident we can be in that conclusion.

TAKE-HOME MESSAGE 1·14

Because much variation exists in the world, statistics can help us evaluate whether any differences between a treatment group and control group can be attributed to the treatment rather than random chance.

1·15

Pseudoscience and misleading anecdotal evidence can obscure the truth.

One of the major benefits of being familiar with the scientific way of thinking is that it can prevent you from being taken in or fooled by false claims. There are two types of "scientific evidence" that frequently are cited in the popular media and are responsible for people erroneously believing that links between two things exist, when in fact they do not. These are:

1. **Pseudoscience,** in which individuals make scientific-sounding claims that are not supported by trustworthy, methodical scientific studies.

2. **Anecdotal observations,** in which, based on only one or a few observations, people conclude that there is or is not a link between two things.

Pseudoscience is all around us, particularly in the claims made on the packaging of consumer products and food (**FIGURE 1-19**). Beginning in the 1960s, for example, consumers encountered the assertion by the makers of a sugarless gum that "4 out of 5 dentists surveyed recommend sugarless gum for their patients who chew gum." If you ask yourself the question

25

FIGURE 1-19 Pseudoscientific claims are often found on food products.

"How do they know what they know?" and can't answer it, you may be looking at pseudoscience. Maybe the statement is factually true, but the general relationship it implies may not be. How many dentists were surveyed? If the gum makers surveyed only five dentists, then the statement may not represent the proportion of *all* dentists who would make such a recommendation. And how were the dentists sampled? Were they at a shareholders meeting for the sugarless gum company? What alternatives were given—perhaps gargling with a tooth-destroying acid? You just don't know. That's what makes it pseudoscience.

Pseudoscience capitalizes on a belief shared by most people: that scientific thinking is a powerful method for learning about the world. The problem with pseudoscience is that the scientific bases for a scientific-sounding claim are not clear. The claims generally sound reasonable, and they are persuasive in convincing people to purchase one product over another. But one of the beauties of real science is that you never have to just take someone's word about something. Rather, you are free to evaluate people's research methods and results and decide for yourself whether their conclusions and claims are appropriate.

We are all familiar with anecdotal evidence. Striking stories that we read or hear, or our own experiences, can shape our views of cause and effect. We may find compelling parallels between suggestions made in horoscopes and events in our

lives, or we may think that we have a lucky shirt, or we may be moved by a child whose cancer went into remission following treatment involving eating apricot seeds. Yet, despite lacking a human face, data are more reliable than anecdotes, primarily because they can illustrate a broader range of observations, capturing the big picture.

Anecdotal observations can seem harmless and can be emotionally powerful. But because they do not include a sufficiently large and representative set of observations of the world, they can lead people to draw erroneous conclusions, often with disastrous consequences. One important case of anecdotal evidence being used to draw general conclusions about a cause-and-effect relationship between two things involves autism, a developmental disorder that impairs social interaction and communication, and the vaccination for measles, mumps, and rubella (commonly called the MMR vaccine) that is given to most children.

In 1998, the prestigious medical journal *The Lancet* published a report by a group of researchers that described a set of symptoms (diarrhea, abdominal pain, bloating) related to bowel inflammation in 12 children who exhibited the symptoms of autism. The parents or physicians of 8 of the children in the study said that the behavioral symptoms of autism appeared shortly after the children received MMR vaccination. For this reason, the authors of the report recommended further study of a possible link between the MMR vaccine, the bowel problems, and autism.

Q Does the measles, mumps, and rubella vaccine cause autism?

In a press conference, one of the paper's authors suggested a link between autism and the MMR vaccine, recommending single vaccines rather than the MMR triple vaccine until it could be proved that the MMR vaccine did not trigger autism. This statement caused a major health scare that only recently has begun to subside. Noting that there had been a significant increase in the incidence of autism in the 1990s and early 2000s, the press and many people took the claims made by the researcher at the press conference as evidence that the MMR vaccine causes autism. Over the course of the next few years, the number of children getting the MMR vaccine dropped significantly, as parents sought to reduce the risk of autism in their children (**FIGURE 1-20**).

Unfortunately, this is a notable case of poor implementation of the scientific method, with numerous flaws leading to an incorrect conclusion. Most important among these is that

What Is Science?   Scientific Method   Experimental Design   Decision Making

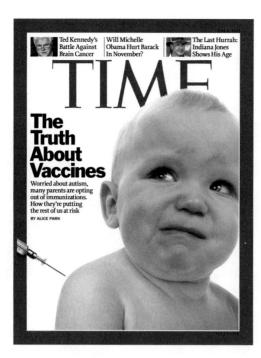

FIGURE 1-20 Headline news. A misguided fear of "catching" autism has caused some parents to decline immunizations for their children.

the study was small (only 12 children), the sample was carefully selected rather than randomized (that is, the researchers selected the study participants based on the symptoms they showed), and no control group was included for comparison (for example, children who had been vaccinated but did not exhibit autism symptoms). On top of this, the study, even flawed as it was, did not actually find or even report a link between autism and the MMR vaccine. That purported link only came from the unsupported, suggestive statements made by one of the study's authors at the press conference.

On their own, the design flaws in the study would be sufficient to invalidate the one author's claims about the link between autism and the MMR vaccine, but there are more weaknesses in the research. It later was discovered that before the paper's publication, the study's lead author (the one who spoke at the press conference) had received large sums of money from lawyers seeking evidence to use in lawsuits against the MMR vaccine manufacturers. It also came to light that he had applied for a patent for a vaccine that was a rival to the most commonly used MMR vaccine. In the light of these undisclosed biases, 10 of the paper's 12 authors published a retraction of their original interpretation of their results.

In the years since this paper was published, many well-controlled large-scale studies have conducted critical experiments on this subject, and all have been definitive in their conclusion that there is no link between the MMR vaccine and autism. Here are a few of these studies:

1. A study of all children born in Denmark between 1991 and 1998 found no difference in the incidence of autism among the 440,655 children who were vaccinated with the MMR vaccine and the 96,648 children who were not vaccinated.

2. A 2005 study in Japan showed that after use of the MMR vaccine was stopped in 1993, the incidence of autism continued to increase.

3. A study of 1.8 million children in Finland who were followed up for 14 years after getting the MMR vaccine found no link at all between the occurrence of autism and the vaccine.

At this point, the consensus of the international scientific community is that there is no scientific evidence for a link between the MMR vaccine and autism.

 Science is a way to call the bluff of those who only pretend to knowledge. It is a bulwark against mysticism, against superstition, against religion misapplied to where it has no business being. If we're true to its values, it can tell us when we're being lied to.

— Carl Sagan, in *The Demon-Haunted World: Science as a Candle in the Dark,* 1997

So what explains the observation that there are more autism cases now than in the past? It seems that the increased number of autism cases is a function of better identification of the condition by doctors and of changes in the process by which autism is diagnosed. Another reason for the perceived link between autism and the MMR vaccine was simply the coincidence that most children receive the vaccine at around 18–19 months of age, which happens to be the age at which the first symptoms of autism are usually noticed. In the end, what we learn from this is that we must be wary that we do not generalize from anecdotal observations or let poorly designed studies obscure the truth.

TAKE-HOME MESSAGE 1•15

Pseudoscience and anecdotal observations often lead people to believe that links between two phenomena exist, when in fact there are no such links.

There are limits to what science can do.

The scientific method is a framework that helps us make sense of what we see, hear, and read in our lives. Science and scientific thinking guide us in recognizing facts and help us in the interpretations of data, analyses of hypotheses, and drawing of conclusions. In doing so, scientific thinking reveals and illuminates explanations about how to think about various events and phenomena, and it can help us make decisions in diverse areas of our lives, not just "scientific" areas. There are, however, limits to what science can do.

The scientific method will never prove or disprove the existence of God. Nor is it likely to help us understand the mathematical elegance of Fermat's last theorem or the beauty of Shakespeare's sonnets. As one of several approaches to the acquisition of knowledge, the scientific method is, above all, empirical. It differs from non-scientific approaches such as mathematics and logic, history, music, and the study of artistic expression in that it relies on *measuring* phenomena in some way. The generation of value judgments and other types of non-quantifiable, subjective information—such as religious assertions of faith—falls outside the realm of science. Despite all of the intellectual analyses the scientific method gives rise to and the objective conclusions it makes possible, it does not, for example, generate moral statements and it cannot give us insight into ethical problems. What "is" (i.e., what we observe in the natural world) is not necessarily what "ought" to be (i.e., what is morally right). It may or may not be.

Further, much of what is commonly considered to be science, such as the construction of new engineering marvels or the heroic surgical separation of conjoined twins, is not scientific at all. Rather, these are technical innovations and developments. While they frequently rely on sophisticated scientific research, they represent the application of research

The application of science to solve a problem can result in sophisticated technical innovations, like this thought-controlled prosthetic arm.

findings to varied fields such as manufacturing and medicine to solve problems.

As we begin approaching the world from a more scientific perspective, we can gain important insights into the facts of life, yet must remain mindful of the limits to science.

TAKE-HOME MESSAGE 1·16

Although the scientific method may be the most effective path toward understanding the observable world, it cannot give us insights into the generation of value judgments and other types of non-quantifiable, subjective information.

28 CHAPTER 1 • SCIENTIFIC THINKING

What Is Science?   Scientific Method   Experimental Design   Decision Making

❺ On the road to biological literacy: what are the major themes in biology?

Finding unity in all the diversity: the Hall of Biodiversity, American Museum of Natural History, New York.

1·17

A few important themes tie together the diverse topics in biology.

Driven forward by the power of the scientific method, modern biology is in the midst of an explosion of new and exciting developments. Now more than ever, biology is influencing and even transforming our lives. Its increasing relevance outside the world of scientists is felt in a multitude of areas, including breakthroughs in agriculture, innovative developments in criminology, and increased understanding and awareness of environmental issues and the need for widespread changes in behavior.

In this guide to biology, as we explore the many facets of biology and its relevance to life in the modern world, you will see two unifying themes recurring throughout.

- **Hierarchical organization.** Life is organized on many levels within individual organisms including atoms, cells, tissues, and organs. And in the larger world, organisms themselves are organized into many levels: populations, communities, and ecosystems within the biosphere.

- **The power of evolution.** Evolution, the change in genetic characteristics of individuals within populations over time, accounts for the diversity of organisms, but also explains the unity among them.

These two major themes connect the diverse topics, as we make our way through four chief areas of focus:

1. The chemical, cellular, and energetic foundations of life

2. The genetics, evolution, and behavior of individuals

3. The staggering diversity of life and the unity underlying it

4. Ecology, the environment, and the subtle and important links between organisms and the world they inhabit

Armed with an appreciation for the scientific method and the broader themes you'll encounter in your study of biology, let's continue our exploration of life!

TAKE-HOME MESSAGE 1·17

Although the diversity of life on earth is tremendous, the study of life is unified by the themes of hierarchical organization and the power of evolution.

Knowledge You Can Use

When should you change what you think? Risky behaviors while driving.

Q: Should drunk driving be illegal? Why? Almost everyone believes that people should not drive when they've been drinking. Even after just one or two drinks, a driver is more than twice as likely to cause a fatal traffic accident as a sober driver. With three to five drinks, the risk is almost 10 times greater than when sober. People see the data and understand; the conclusion is not difficult: you put yourself and others at risk when driving drunk.

Q: What about people who claim that they can drive safely even when they are drunk? People believe that their own risks are much lower than they really are. Based on an analysis of U.S. National Highway Traffic Safety data, traffic risks among 16- to 21-year-olds after consuming alcohol are 2–10 times greater than their perceptions of the risks. For instance, people often feel that their reaction times are just as quick whether they are sober or drunk.

Q: If observations and analyses revealed an equally risky and preventable behavior on the road, would it be advisable to prohibit that behavior, too? They have! Such observations and analyses have been made and have resulted in many states banning the use of handheld cell phones and text-messaging while driving. Let's consider two studies.

In a study using a driving simulator, individuals were evaluated when talking on a cell phone or after they had consumed alcohol to the point of intoxication (they used vodka and orange juice in this University of Utah study). The results? The driving of individuals using cell phones was as much impaired as the driving of those who were drunk. One very interesting and perhaps surprising result: handheld and hands-free phones impaired driving equally; it seems that it is inattention, rather than reduced dexterity, that is responsible for the impairment.

Another study, analyzing the cell phone records of individuals who were in real traffic collisions, came to the same conclusion, finding that cell phone users were 4–5 times more likely to get into an accident than non-distracted drivers. When you use a cell phone while driving, you put yourself and others at the same level of risk as if you were driving drunk.

Q: What observations could be made that would cause *you* not to use your cell phone while driving?

What can you conclude?
Scientific observations can reveal unexpected truths about the world and, in doing so, can cause us to change what we think and how we behave. If we consider one risk (such as driving drunk) unacceptable, should we consider an equal risk (such as driving while using a cell phone) also unacceptable? Science can't answer that, but it can help you to understand the world better and to think more consciously about decisions you may be making without even realizing it.

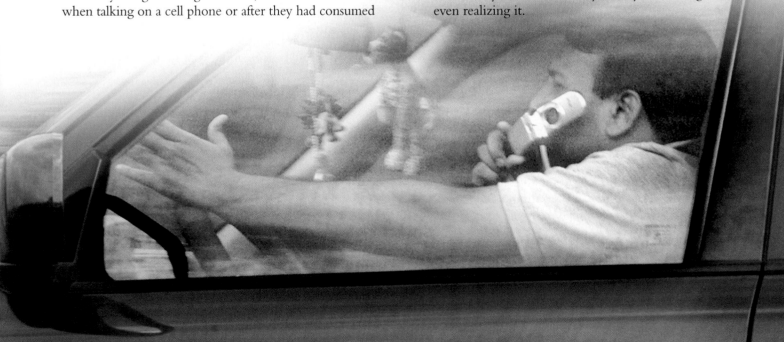

1 Science is a collection of facts and a process for understanding the world.

Through its emphasis on objective observation, description, and experimentation, science is a pathway by which we can discover and better understand the world around us. Biological issues permeate all aspects of our lives. To make wise decisions, it is essential for individuals and societies to attain biological literacy. There are numerous ways of gaining an understanding of the world. Because it is empirical, rational, testable, repeatable, and self-correcting, the scientific method is a particularly effective approach.

2 A beginner's guide: what are the steps of the scientific method?

The scientific method (observation, hypothesis, prediction, test, and conclusion) is a flexible, adaptable, and efficient pathway to understanding the world because it tells us when we must change our beliefs. The scientific method begins by making observations about the world, noting apparent patterns or cause-and-effect relationships. A hypothesis is a proposed explanation for a phenomenon. For a hypothesis to be useful, it must generate a testable prediction. A critical experiment is one that makes it possible to decisively determine whether a particular hypothesis is correct. Based on the results of experimental tests, we can revise our hypothesis and explain the observable world with increasing accuracy. A great strength of scientific thinking, therefore, is that it helps us understand when we should change our minds. Scientific theories do not represent speculation or guesses about the natural world. Rather, they are hypotheses—proposed explanations for natural phenomena—that have been so strongly and persuasively supported by empirical observation that the scientific community views them as very unlikely to be altered by new evidence.

3 Well-designed experiments are essential to testing hypotheses.

To draw clear conclusions from experiments, it is essential to hold constant all those variables we are not interested in. Control and experimental groups should vary only with respect to the treatment of interest. Differences in outcomes between the groups can then be attributed to the treatment. Experiments and their outcomes must be repeatable for their conclusions to be valid and widely accepted. Biases can influence our behavior, including our collection and interpretation of data. With careful controls, we can minimize such biases.

4 The scientific method can help us make wise decisions.

Because much variation exists in the world, statistics can help us evaluate whether differences between a treatment group and control group can be attributed to the treatment rather than random chance. Pseudoscience and anecdotal observations often lead people to believe that links between two phenomena exist, when in fact there are no such links. Although the scientific method may be the most effective path toward understanding the observable world, it cannot give us insights into the generation of value judgments and other types of non-quantifiable, subjective information.

5 On the road to biological literacy: what are the major themes in biology?

Although the diversity of life on earth is tremendous, the study of life is unified by the themes of hierarchical organization and the power of evolution.

KEY TERMS

anecdotal observation, p. 25
biological literacy, p. 4
biology, p. 3
blind experimental design, p. 20
control group, p. 17
critical experiment, p. 13

double-blind experimental design, p. 20
empirical, p. 7
experimental group, p. 17
hypothesis (*pl.* hypotheses), p. 11
null hypothesis, p. 11

placebo, p. 13
placebo effect, p. 18
positive correlation, p. 24
pseudoscience, p. 25
randomized, p. 30
science, p. 3
scientific literacy, p. 4

scientific method, p. 7
statistics, p. 23
superstition, p. 6
theory, p. 16
treatment, p. 17
variable, p. 18

1. Science is:
   a) a field of study that requires certain "laws of nature" to be taken on faith.
   b) both a body of knowledge and an intellectual activity encompassing observation, description, experimentation, and explanation of natural phenomena.
   c) a process that can be applied only within the scientific disciplines, such as biology, chemistry, and physics.
   d) the only way to understand the natural world.
   e) None of the above.

2. All of the following are elements of biological literacy except:
   a) the ability to use the process of scientific inquiry to think creatively about real-world issues having a biological component.
   b) reading the most important books in biology.
   c) the ability to integrate into your decision making a consideration of issues having a biological component.
   d) the ability to communicate with others about issues having a biological component.
   e) All of the above are elements of biological literacy.

3. Superstitions are:
   a) held by many humans, but not by any non-human species.
   b) just one of many possible forms of scientific thinking.
   c) true beliefs that have yet to be fully understood.
   d) irrational beliefs that actions not logically related to a course of events influence its outcome.
   e) proof that the scientific method is not perfect.

4. In a recent study, patients treated with a genetically engineered heart drug were able to walk on a treadmill 26 seconds longer than those not receiving the drug, and they showed no side effects from taking the drug. Can we conclude that this drug is an effective treatment for heart disease?
   a) No. Genetically engineered drugs cannot be tested by the scientific method; they require comparative observations.
   b) No. It is not clear how many subjects were in the study.
   c) No. It is not clear that the drug is not a placebo.
   d) Yes.
   e) No. It is not clear that the proper controls were included.

5. Empirical results:
   a) rely on intuition.
   b) are generated by theories.
   c) are based on observation.
   d) cannot be replicated.
   e) must support a tested hypothesis.

6. In a well-designed experiment:
   a) the prediction will be highly probable if the experiment shows the explanation is correct.
   b) the prediction will be highly improbable if the experiment shows the explanation is incorrect.
   c) the null hypothesis will not be tested.
   d) the prediction will most likely be correct.
   e) Both a) and b) are correct.

7. Which of the following statements is correct?
   a) A hypothesis that does not generate a testable prediction is not useful.
   b) Common sense is usually a good substitute for the scientific method when trying to understand the world.
   c) The scientific method can be used only to understand scientific phenomena.
   d) It is not necessary to make observations as part of the scientific method.
   e) All of the above are correct.

8. The placebo effect:
   a) is the frequently observed, poorly understood phenomenon that people tend to respond favorably to any treatment.
   b) reveals that sugar pills are generally as effective as actual medications in fighting illness.
   c) reveals that experimental treatments cannot be proven effective.
   d) demonstrates that most scientific studies cannot be replicated.
   e) is an urban legend.

9. Before experimental drugs can be brought to market, they must undergo many rigorous trials to ensure they deliver their medical benefits effectively and safely. One method commonly used in this process is to compare the effect of a drug with that of a placebo in double-blind tests. Which of the following correctly describes a double-blind test?
   a) The researchers apply two-layered blindfolds to the study participants so they cannot see whether they are receiving the drug or a placebo.
   b) Neither the researchers nor the study participants know who is receiving the drug and who is receiving the placebo.
   c) The researchers know who is receiving the drug and who is receiving the placebo, but do not know what the supposed effects of the drug should be.
   d) The researchers do not know who receives the drug or the placebo, but the participants know and tell them later.
   e) None of the above.

10. In controlled experiments:
   a) one variable is manipulated while others are held constant.
   b) all variables are dependent on each other.
   c) all variables are held constant.
   d) all variables are independent of each other.
   e) all critical variables are manipulated.

11. If a researcher collects data by using the same experimental setup as in another study, but using different research subjects, the process is considered:
   a) an uncontrolled experiment.
   b) intuitive reasoning.
   c) extrapolation.
   d) replication.
   e) exploration.

12. You hear the males of a particular bird species calling in their natural environment and question why they call. You design an experiment to try to answer your question. You cage one male bird and record his calling rate in response to four treatments, varying the number of conspecific (of the same species) males and the number of conspecific females that the male subject can see. The treatments are: no conspecific birds; 10 conspecific males; 10 conspecific females; 5 conspecific males and 5 conspecific females. Your results are: the rate at which the male subject calls is the same across all four treatments. Which of the following is a null hypothesis that your experimental design could reject?
    a) Males of the species do not call.
    b) Males of the species call more often than do females.
    c) Males of the species call at a rate that is independent of the sex composition of the audience.
    d) Males of the species call less often than do females.
    e) Males of the species call when called to.

13. Statistical methods make it possible to:
    a) prove any hypothesis is true.
    b) determine how likely it is that certain results have occurred by chance.
    c) unambiguously learn the truth.
    d) reject any hypothesis.
    e) test non-falsifiable hypotheses.

14. Anecdotal evidence:
    a) is a more efficient method for understanding the world than the scientific method.
    b) tends to be more reliable than data based on observations of large numbers of diverse individuals.
    c) is a necessary part of the scientific method.
    d) is often the only way to prove important causal links between two phenomena.
    e) can seem to reveal links between two phenomena, but the links do not actually exist.

15. A relationship between phenomena that has been established based on large amounts of observational and experimental data is referred to as:
    a) a theory.
    b) a fact.
    c) an assumption.
    d) a conjecture.
    e) a hypothesis.

16. Which of the following issues would be least helped by application of the scientific method?
    a) developing more effective high school curricula
    b) evaluating the relationship between violence in videogames and criminal behavior in teens
    c) determining the most effective safety products for automobiles
    d) formulating public policy on euthanasia
    e) comparing the effectiveness of two potential antibiotics

17. What is the meaning of the statement "correlation does not imply causation"?
    a) Just because two variables vary in a similar pattern does not mean that changing one variable causes a change in the other.
    b) It is not possible to demonstrate a correlation between two variables.
    c) When a change in one variable causes a change in another variable, the two variables are not necessarily related to each other in any way.
    d) It is not possible to prove the cause of any naturally occurring phenomenon.
    e) Just because two variables vary in a similar pattern does not mean that they have any relationship to each other.

Short-Answer Questions

1. A pharmaceutical company plans to test a potential anti-cancer drug on human subjects. The drug will be administered in pill form. How should this study be designed so that appropriate controls are in place?

2. Lately there have been many claims concerning the health benefits of green tea. Suppose you read a claim that alleges drinking green tea causes weight loss. You are provided with the following information about the studies that led to this claim.

    People were weighed at the beginning of the study.

    People were asked to drink two cups of green tea every day for 6 weeks.

    People were weighed at the end of the study.

    People who drank green tea for 6 weeks lost some weight by the end of the study.

    It was concluded that green tea is helpful for weight loss.

    This study obviously had some holes in its design. Assuming no information other than that provided above, indicate at least four things that could be done to improve the experimental design.

3. Use of the word "theory" can imply different things to different people. Statements referring to evolution as "just a theory" imply that a scientific theory is much like a hunch, something that can easily be disproved. How is a scientific theory different from any other type of "theory"? What requirements must be met before a hypothesis is considered to be a scientific theory?

See Appendix for answers. For additional study questions, go to www.prep-u.com.

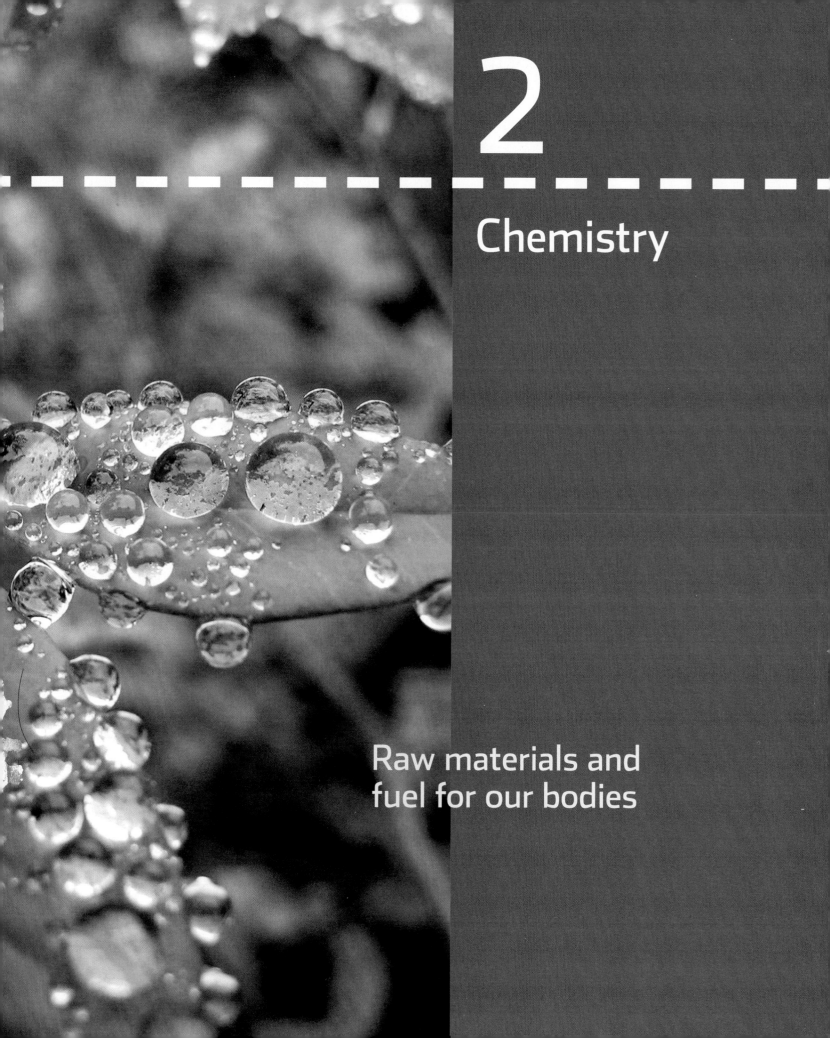

2
Chemistry

Raw materials and
fuel for our bodies

❶ Atoms form molecules through bonding.

A collection of salt crystals, made from cesium chloride.

2•1

Everything is made of atoms.

A little bit of chemistry goes a long way in the study of biology and in understanding a great deal about your everyday life. Will eating those beans in your soup keep you up all night in gastrointestinal distress? Just knowing whether the beans are lentils or lima beans—each contains slightly different types of sugar molecules—will give you an answer. Will the butter you spread on your toast sabotage your efforts to lose weight? Understanding a little about the carbon-hydrogen connections in the fat molecules can help you decide.

The chemistry that is most important in biology revolves around a few important elements. An **element** is a substance that cannot be broken down chemically into any other substances. Gold and carbon and aluminum are elements you might be familiar with. Whatever the element, if you keep cutting it into ever smaller pieces, each of the pieces behaves exactly the same as any other piece. The smallest piece of pure gold will still have the softness, reflectivity, and malleability characteristic of that element (**FIGURE 2-1**).

If you could continue cutting, you would eventually separate the gold into tiny pieces that could no longer be divided

Gold

Carbon

Copper

Although we associate pennies with the element copper, all pennies minted after 1982 are made with a zinc core and are plated with copper.

FIGURE 2-1 Familiar elements. Elements are substances that cannot be broken down chemically into any other substances.

Atoms   Water   Carbohydrates   Lipids

without losing their gold-like properties. These are the individual component pieces of an element, called **atoms.** An atom is a bit of matter that cannot be subdivided any further without losing its essential properties. The word "atom" is from the Greek for "indivisible."

Everything around us, living or not, can be reduced to atoms. All atoms—whether of the element gold or some other element such as oxygen or aluminum or calcium—have the same basic structure. At the center of an atom is a **nucleus,** which is usually made up of two types of particles, called protons and neutrons. **Protons** are particles that have a positive electrical charge and **neutrons** are particles that have no electrical charge. The amount of matter in a proton or neutron, its **mass,** is about the same (**FIGURE 2-2**).

Whirling in a cloud around the nucleus of every atom are negatively charged particles called **electrons.** An electron weighs almost nothing—less than one-twentieth of one percent of the weight of a proton. (The mass of an atom—its **atomic mass**—is made up of the combined mass of all of its protons and neutrons; for our purposes here, electrons are so light that their mass can be ignored.) Particles that have the same charge repel each other; those with opposite charges are attracted to each other. Because all electrons have the same charge, the electrons in an atom repel each other. But because they are negatively charged, they are attracted to the positively charged protons in the nucleus. This attraction holds electrons close enough to the nucleus to keep them from flying away,

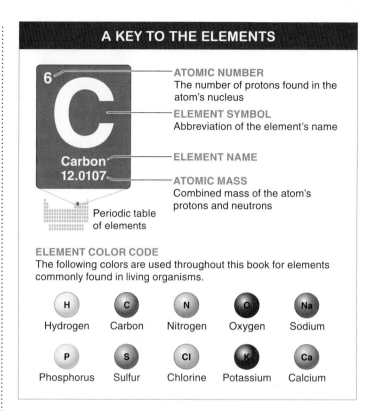

A KEY TO THE ELEMENTS

ATOMIC NUMBER
The number of protons found in the atom's nucleus

ELEMENT SYMBOL
Abbreviation of the element's name

ELEMENT NAME

ATOMIC MASS
Combined mass of the atom's protons and neutrons

Periodic table of elements

ELEMENT COLOR CODE
The following colors are used throughout this book for elements commonly found in living organisms.

Hydrogen   Carbon   Nitrogen   Oxygen   Sodium

Phosphorus   Sulfur   Chlorine   Potassium   Calcium

FIGURE 2-3 The vital statistics of atoms. A guide to reading the periodic table.

while the energy of their fast movement keeps them from collapsing into the nucleus.

Atoms are tiny. Enlarge an atom by a billion times and it would only be the size of a grapefruit. Paradoxically, most of the space taken up by an atom is empty. That is, because the nucleus is very small and compact, the electrons zip about relatively far from the nucleus. If the nucleus were the size of a golf ball, the electrons would be anywhere from half a mile to six miles away.

What distinguishes one element, such as chlorine, from another, such as neon or oxygen? Atoms of different elements have a different number of protons in the nucleus. A chlorine atom has 17 protons, a neon atom has 10 protons, and an oxygen atom has 8 protons. Each element is given a name (and an abbreviation, such as O for oxygen and C for carbon) and an **atomic number** that corresponds to how many protons it has (**FIGURE 2-3**). All the known elements can be arranged in a scheme, in the order of their atomic number, called the "periodic table." So far, about 90 elements have been discovered that are present in nature and about 25 others can be made in the laboratory. Everything you see around you is made up of some combination of those naturally occurring elements.

THE ATOM: BASIC STRUCTURE

Nucleus

Hydrogen atom
1 Proton
0 Neutrons
1 Electron

Carbon atom
6 Protons
6 Neutrons
6 Electrons

Forces of attraction between positive and negative charges hold the fast-moving electrons (negative) close to the nucleus (positive).

Nucleus:

Proton   Neutron   Electron

FIGURE 2-2 The atom. At the center of an atom is a nucleus containing protons and (in all elements except hydrogen) neutrons. The nucleus is surrounded by electrons whirling about in a cloud.

Proteins   Nucleic Acids

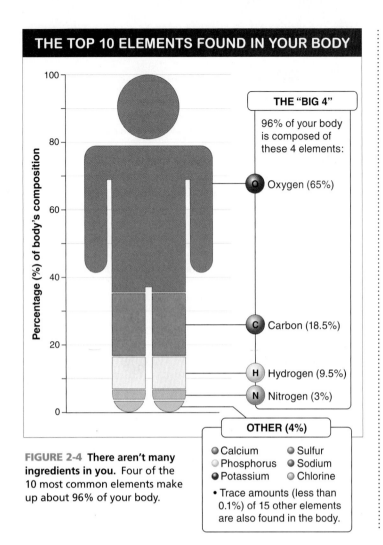

THE TOP 10 ELEMENTS FOUND IN YOUR BODY

Percentage (%) of body's composition

THE "BIG 4"

96% of your body is composed of these 4 elements:

O Oxygen (65%)

C Carbon (18.5%)

H Hydrogen (9.5%)

N Nitrogen (3%)

OTHER (4%)

- Calcium
- Phosphorus
- Potassium
- Sulfur
- Sodium
- Chlorine

• Trace amounts (less than 0.1%) of 15 other elements are also found in the body.

FIGURE 2-4 There aren't many ingredients in you. Four of the 10 most common elements make up about 96% of your body.

Of all the elements found on earth, only 25 are found in your body. The "Top 10" most common of these make up 99.9% of your body (**FIGURE 2-4**). And just four elements make up more than 96% of your body. These "Big 4" are oxygen, carbon, hydrogen, and nitrogen. With knowledge about the Big 4, you can understand a huge amount about nutrition and physiology (how your body works), so we'll focus on the properties of these four elements later in this chapter.

The number of protons in an atom's nucleus determines what element it is. The number of neutrons in the nucleus is usually equal to the number of protons. Since protons and neutrons have approximately the same mass, the mass of the atom is about double its atomic number. The element oxygen, for example, has the atomic number 8 because it has 8 protons, so we expect it to have an atomic mass of 16, simply the mass of the 8 protons and 8 neutrons added together—which it generally does. (As noted above, we can ignore the mass of the electrons.)

TAKE-HOME MESSAGE 2·1

Everything around us, living or not, is made up of atoms, the smallest unit into which material can be divided without losing its essential properties. All atoms have the same general structure. They are made up of protons and neutrons in the nucleus, and electrons, which circle far around the nucleus.

2·2

An atom's electrons determine how (and whether) the atom will bond with other atoms.

While the number of protons identifies an element, it is an atom's electrons that determine how (and whether) it bonds with other atoms. Electrons move so quickly that it is impossible to determine, at any given moment, exactly where an electron is. Electrons are not just moving about haphazardly, though. Speeding around the nucleus, they tend to stay within a prescribed area. This area is called an "electron shell," and an atom may have several shells, each shell occupied by its own set of electrons. Within a shell, the electrons stay far apart because their negative charges repel one another.

The first electron shell is closest to the nucleus and can hold two electrons (**FIGURE 2-5**). If an atom has more than two

electrons, as most atoms do, the other electrons are arranged in other shells. The second shell is a bit farther away from the nucleus and can hold as many as eight electrons. There can be as many as seven shells in total, holding varying numbers of electrons.

Atoms become stable when their outermost shell is filled to capacity. Those with this configuration behave like loners, neither reacting nor combining with other atoms. On the other hand, when atoms have outer shells with vacancies, they are likely to interact with other atoms, giving, taking, or sharing electrons to achieve that desirable state: a full outer shell of electrons. In fact, based on the number of electron

Atoms   Water   Carbohydrates   Lipids

ELECTRON SHELLS

Electrons move around the nucleus in designated areas called electron shells. An atom can have as many as seven electron shells in total.

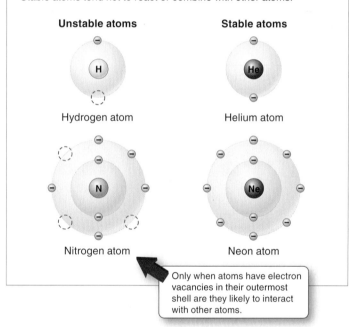

First electron shell (capacity: 2 electrons)

Second electron shell (capacity: 8 electrons)

Vacancy

Oxygen atom

The chemical characteristics of an atom depend upon the number of electrons in its outermost shell.

ATOM STABILITY

Atoms become stable when their outermost shell is filled to capacity. Stable atoms tend not to react or combine with other atoms.

Unstable atoms

Hydrogen atom

Nitrogen atom

Stable atoms

Helium atom

Neon atom

Only when atoms have electron vacancies in their outermost shell are they likely to interact with other atoms.

FIGURE 2-5 Electrons and the shells they inhabit.

vacancies in the outermost shell of an atom, it's possible to predict how amenable to bonding that atom will be, and even which other atoms its likely bonding partners will be.

An atom that loses one or more electrons becomes positively charged, while an atom that acquires electrons becomes negatively charged. This transfer of electrons is driven by the fact that atoms with full outer electron shells are more stable.

Donated electron

Sodium ion
⊕ 11 Protons
11 Neutrons
⊖ 10 Electrons

Chloride ion
⊕ 17 Protons
17 Neutrons
⊖ 18 Electrons

NET CHARGE ➕ Positive ➖ Negative

FIGURE 2-6 Ions are electrically charged atoms.

Normally, an atom has the same number of electrons as protons. Sometimes an atom may have one or more extra electrons or may lack one or more electrons relative to the number of protons. An atom with extra electrons becomes negatively charged, and an atom lacking one or more electrons is positively charged. Such a charged atom is called an **ion** (**FIGURE 2-6**). Due to their electrical charge, ions behave very differently from the atoms that give rise to them. As we'll see later in the chapter, ions are more likely to interact with other, oppositely charged ions.

TAKE-HOME MESSAGE 2·2

The chemical characteristics of an atom depend on the number of electrons in its outermost shell. Atoms are most stable and least likely to bond with other atoms when their outermost electron shell is filled to capacity.

Proteins   Nucleic Acids

Atoms can bond together to form molecules or compounds.

When you eat a meal, it is like filling your car's tank with gasoline; you have a source of energy that can be used to fuel activities like running, thinking, building muscle, and maintaining the machinery of life. That energy initially comes from the sun and is captured and stored by plants. When we eat plants or eat other animals that eat plants, we ingest the energy stored in the plant material. But how exactly is energy stored in plants? It is stored in bonds that join atoms together. Eventually the bonds are broken, the energy is released, and it can be used to fuel the body's activities.

Groups of atoms held together by bonds are called **molecules.** It usually requires energy to build bonds and "glue" atoms together. Conversely, when molecules are broken down—such as when animals (including humans) consume molecules in their diet—bonds are broken, releasing the energy that was used to create them. In a sense, molecules are created as a short-term storage of energy that can be harnessed later.

There are three principal types of bonds that hold multiple atoms together. The type of bonding that any atom is likely to take part in depends almost entirely on the number of electrons in its outermost shell.

Covalent Bonds **Covalent bonds** are strong bonds formed when two atoms share electrons. The simplest example of a covalent bond is the bonding of two hydrogen atoms to form H_2 (**FIGURE 2-7**), the simplest of all molecules. A hydrogen atom has an atomic number of 1: it has a single proton in its nucleus and a single electron circling around the nucleus in the first shell. Because the atom is most stable when the first shell has two electrons, two hydrogen atoms can each achieve a complete outermost shell by sharing electrons. The nuclei come close together (but not too close, since they are positively charged) and the two electrons circle around both of the nuclei, almost in a figure 8. The new H_2 molecule is very stable because, now that both atoms have two electrons in their outermost shell, they are no longer likely to bond with other atoms.

Oxygen, with an atomic number of 8, has two electrons in its innermost shell and six in its outermost shell. Consequently it needs to form two covalent bonds to fill its outermost shell. Sometimes two oxygen atoms join together to form O_2. Each oxygen atom shares two electrons with the other, filling the outermost shell of both. The sharing of two electrons between two atoms is called a **double bond** (see Figure 2-7). O_2 is the most common form in which we find oxygen in the world.

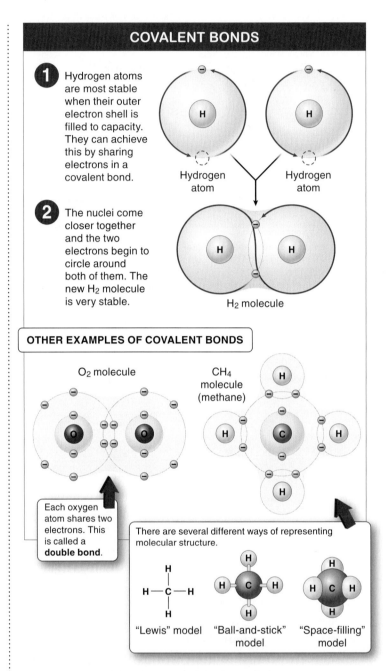

COVALENT BONDS

1 Hydrogen atoms are most stable when their outer electron shell is filled to capacity. They can achieve this by sharing electrons in a covalent bond.

Hydrogen atom   Hydrogen atom

2 The nuclei come closer together and the two electrons begin to circle around both of them. The new H_2 molecule is very stable.

H_2 molecule

OTHER EXAMPLES OF COVALENT BONDS

O_2 molecule

CH_4 molecule (methane)

Each oxygen atom shares two electrons. This is called a **double bond**.

There are several different ways of representing molecular structure.

"Lewis" model   "Ball-and-stick" model   "Space-filling" model

FIGURE 2-7 Covalent bonds: strength through electron sharing. The molecular structures formed by covalent bonding can be represented on paper in several different ways.

Carbon is a particularly extroverted molecule. It has an atomic number of 6, meaning that two electrons fill its first shell and four remain to occupy the second shell. Because four electron vacancies are left, carbon can, and frequently does, form four covalent bonds, joining up with other atoms in a wide variety

Atoms   Water   Carbohydrates   Lipids

of molecules. Methane, the chief component of natural gas, is formed when one atom of carbon covalently bonds with four atoms of hydrogen (see Figure 2-7). It has the chemical formula CH_4.

Ionic Bonds
Atoms can also bond together without sharing electrons. In **ionic bonds,** one atom transfers one or more of its electrons completely to another. As a result of this transfer, each atom becomes an ion, since each has an unequal number of protons and electrons. The atom gaining electrons becomes negatively charged, while the atom losing electrons becomes positively charged. Rather than forming molecules, ionic bonds form **compounds,** in which the two oppositely charged ions attract each other. Unlike the much stronger covalent bonds, in ionic bonds each electron circles around a single nucleus. Ions

of equal and opposite charges are attracted to each other and the compound is neutral—that is, it has no charge (**FIGURE 2-8**).

Hydrogen Bonds
Ionic and covalent bonds link two or more atoms together. **Hydrogen bonds,** on the other hand, are important in bonding multi-atom molecules together. A hydrogen bond is formed between a hydrogen atom in one molecule and another atom, often an oxygen or nitrogen atom, in another molecule (or, in a very large molecule, in another part of the same molecule). This bond is based on the attraction between positive and negative charges (**FIGURE 2-9**).

The atoms taking part in hydrogen bonds are not ions, so where do the electrical charges come from? The hydrogen atom is already covalently bonded to another atom in the

IONIC BONDS

1 One atom transfers one or more electrons completely to another. The result is two oppositely charged ions.

Donated electron

Sodium ion ⊕   ⊖ Chloride ion

OPPOSITE CHARGES ATTRACT

2 The two oppositely charged ions attract each other, forming a compound.

NaCl compound

NaCl crystals (table salt)

NaCl crystal structure

When multiple Na⁺ ions and Cl⁻ ions are attracted to each other, a NaCl crystal structure forms.

FIGURE 2-8 Ionic bonds: transfer of electrons from one atom to another. The resulting charged atoms (ions) attract each other.

HYDROGEN BONDS

In water, the hydrogen nucleus, with only a single positively charged proton, doesn't attract the negatively charged electrons as much as does the larger nucleus of the oxygen atom (which has eight positively charged protons). As a consequence, all of the electrons spend more time near the oxygen. This makes that side of the molecule slightly positive and the hydrogen side slightly negative.

Polarity

Covalent bonds

Water molecule

Hydrogen bonds are formed between the slightly positively charged hydrogen atoms of one water molecule and the slightly negatively charged atoms of another.

Hydrogen bond

FIGURE 2-9 Hydrogen bonds: attraction between a polar atom or molecule and a hydrogen atom. As we'll see later in the chapter, the hydrogen bonding between H_2O molecules gives water some of its characteristic properties.

Proteins   Nucleic Acids

SUMMARY: THREE TYPES OF BONDS

1 COVALENT BOND
A strong bond formed when atoms share electrons in order to become more stable, forming a molecule.

H₂ molecule

2 IONIC BOND
An attraction between two oppositely charged ions, forming a compound.

NaCl compound

3 HYDROGEN BOND
An attraction between the slightly positively charged hydrogen atom of one molecule and the slightly negatively charged atom of another.

H₂O molecule   H₂O molecule

Strongest

Bond Strength

Weakest

FIGURE 2-10 Three ways in which atoms and molecules are bound.

same molecule and shares its electron. That electron circles both the hydrogen nucleus and the nucleus of the other atom, but the electron is not shared equally. Given that the other atom always has more than the one proton found in the hydrogen nucleus, the electron contributed by hydrogen spends more of its time near the other, more positively charged nucleus than near the hydrogen nucleus. Having an extra electron nearby causes the larger atom to be slightly negatively charged, while the hydrogen atom becomes slightly positively charged.

In a sense, the covalently bonded molecules become like a magnet, with distinct positive and negative sides (see Figure 2-9). Magnet-like molecules with distinct positive and negative regions like this are said to be "polar." They are attracted to other polar molecules, lining up in particular orientations such that the positive regions of one molecule are near the negative regions of another. These attractions between polar molecules are called hydrogen bonds. Although they are only about one-thirtieth as strong as covalent bonds, hydrogen bonds are responsible for many of the unique characteristics of water that make it one of the most important molecules for life on earth. See **FIGURE 2-10** for a review of the three types of bonds discussed in this section.

TAKE-HOME MESSAGE 2·3

Atoms can be bound together in three different ways. Covalent bonds, in which atoms share electrons, are the strongest. In ionic bonds, the next strongest, one atom transfers its electrons to another and the two oppositely charged ions are attracted to each other, forming a compound. Hydrogen bonds, the weakest of the three types, involve the attraction between a hydrogen atom and another polar atom or molecule.

2·4

A molecule's shape gives it unique characteristics.

When many atoms bond together, the resulting molecule or compound acquires a "personality" that results from the accumulated properties of the atoms and their interactions with each other. The arrangement of atoms within a molecule also determines the molecule's shape: the atoms may be lined up in a row or may stack in a pyramid or may pack in a more globular formation, like a piece of popcorn. Whatever form

the atoms take, a molecule's shape will determine many of its physical properties. Let's consider two important consequences of shape: how a molecule tastes and how it smells.

When you put a piece of chocolate into your mouth, your brain quickly registers a sweet taste. What exactly does that mean? Why does chocolate taste sweet, while a piece of fish

Atoms   Water   Carbohydrates   Lipids

may taste salty? The human tongue has various types of taste receptors; some detect bitter tastes, some sweet, some sour, some salty, and others savory. A molecule of the sugar glucose in that piece of chocolate triggers a sensation of sweetness when it binds to one of our sweet-taste receptors.

How sweet something tastes depends on the shape of its molecules—how well the molecules fit into the sweet-sensing receptors. Glucose molecules fit very well—slightly more snugly than the sucrose molecules in table sugar (**FIGURE 2-11**). Molecules of fructose, a sugar found in fruits and some vegetables, fit even better.

Food chemists have been able to create artificial sweeteners that have a better and better fit with receptors. Aspartame and saccharin, for example, are 150 times and 450 times sweeter than table sugar. Splenda is 600 times sweeter. This means that with very tiny amounts of artificial sweetener, we can sense a sweetness equivalent to a much greater amount of table sugar.

Insect mating behavior is also heavily influenced by molecular interactions that depend on molecular shape. Male gypsy moths, for example, will fly as far as two miles just because a molecule they encounter fits into one of their receptors and tells them that a sexually receptive female is nearby (**FIGURE 2-12**). At the heart of this behavior is something chemically similar to how taste receptors sense

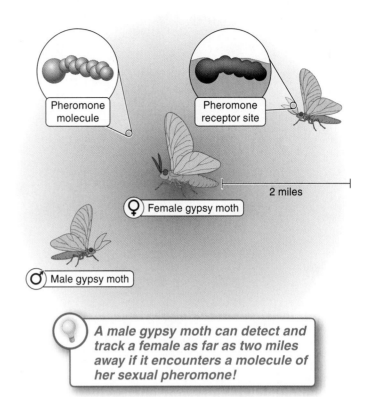

A male gypsy moth can detect and track a female as far as two miles away if it encounters a molecule of her sexual pheromone!

FIGURE 2-12 Molecular shape and function. A molecule's shape can give it a particular smell—and even be essential to its function as a sexual attractant.

? **Why does your brain register chocolate as tasting sweet and potato chips as salty?**

FIGURE 2-11 The shape of a molecule matches the shape of its taste receptors.

the sweetness of a piece of chocolate. Rather than being in a piece of food, though, the molecules the insects "taste" are airborne.

Taste is just one special case illustrating the chemical importance of a molecule's shape. More generally, every interaction between molecules is influenced by the fit between the molecules. Next we will look at the shape and bonding characteristics of one of the most important biological molecules, water, and how they give that molecule superhero-like properties.

TAKE-HOME MESSAGE 2·4

The chemical and physical properties of a molecule—such as how the molecule tastes or smells, or its likelihood of bonding with other molecules—depend on the shape in which the atoms are linked together and on the electrons exposed in their outermost shells.

Proteins   Nucleic Acids

❷ Water has features that enable it to support all life.

Water flowing down a Japanese waterfall.

2·5

Hydrogen bonds make water cohesive.

Predators chasing after green basilisk lizards in Central America sometimes get the shock of a lifetime (and lose their lunch in the process). When chased, the lizards run quickly across rocks and dirt—nothing unusual for a lizard. Then, when they come to the edge of a stream or river, these fleeing lizards show their uniqueness and earn their nickname, the "Jesus lizard." Without stopping, a Jesus lizard runs right across the surface of the water, leaving stunned predators behind at the water's edge. How do these animals do it?

The Jesus lizard, like numerous insects such as the water strider, makes use of the fact that water molecules have tremendous cohesion. That is, they stick together with unusual strength. This molecular cohesiveness is due to hydrogen bonds between the water molecules.

Each water molecule is V-shaped (**FIGURE 2-13**; see also Figure 2-9). The hydrogen atoms are at the ends of the two arms and the oxygen is at the bottom end of the V, between the two

hydrogen atoms. Oxygen's strongly positively charged nucleus pulls the circling electrons toward itself and holds on to them for more than its fair share of the time. Consequently, the oxygen at the bottom of the V has a slight negative charge and the end of the water molecule containing the hydrogen atoms has a slight positive charge.

Because of their unequally shared electrons, water molecules are polar; like a magnet, they have a positively charged and a negatively charged side (see Figure 2-9). Consequently, large numbers of water molecules orient themselves in an orderly arrangement that positions the negative side of one molecule near the positive side of another. Hydrogen bonds form between the relatively positively charged hydrogen atoms and the relatively negatively charged oxygen atoms of *adjacent* water molecules.

Hydrogen bonds are much weaker than covalent bonds and they don't last very long. Nonetheless, to the Jesus lizard, the

Atoms   Water   Carbohydrates   Lipids

Pressure applied to water surface

Hydrogen bond

V-shaped water molecules are held together by hydrogen bonds. The bonds are just strong enough to give water a surface tension with net-like properties.

? How are these animals able to walk on water?

FIGURE 2-13 Walking on water! Hydrogen bonds make this possible for some animals.

cumulative effect of all the hydrogen bonds in water is to link together all the water molecules in the stream just enough to give the water a surface tension with some net-like properties. If the lizard were too fat or too slow, its weight would overwhelm the hydrogen bonds and the lizard would push the molecules apart and sink like a rock.

TAKE-HOME MESSAGE 2•5

Water molecules easily form hydrogen bonds, giving water great cohesiveness.

2•6

Water has unusual properties that make it critical to life.

All life on earth depends on water; organisms are made up mostly of water and require it more than any other molecule. Hydrogen bonding among water molecules gives water several important properties that contribute to its crucial role in the biology of all organisms.

1. Cohesion. We saw in the previous section how the connection of water molecules through hydrogen bonds makes water cohesive, resulting in, for example, high surface tension. The cohesiveness of water molecules also makes it possible for tall trees to exist (**FIGURE 2-14**). Leaves need water. Molecules of water in the leaves are continually lost to the atmosphere through evaporation or are used up in the process of photosynthesis. In order to get more water, plants must pull it up from the soil. The problem for many plants,

such as the giant sequoia trees, is that the soil may be 300 feet below the leaf. Hydrogen bonds, however, allow water molecules to pull up adjacent water molecules to which they have hydrogen-bonded. The chain of linked molecules extends all the way down to the soil, where another water molecule is pulled in via the roots each time a water molecule evaporates from a leaf far above.

2. Large heat capacity. Walking across a sandy beach on a hot day, you can feel how easily sand heats up. By comparison, stepping from the beach into the cooler ocean reveals that water resists warming. It takes a lot of energy to change the temperature of water even a small amount. Why? Again, we must look to hydrogen bonding for our answer.

Proteins   Nucleic Acids

WATER: STRONG COHESIVENESS

Because of the cohesive properties of water, trees such as the giant sequoia are able to transport water molecules from the soil to their leaves 300 ft. above.

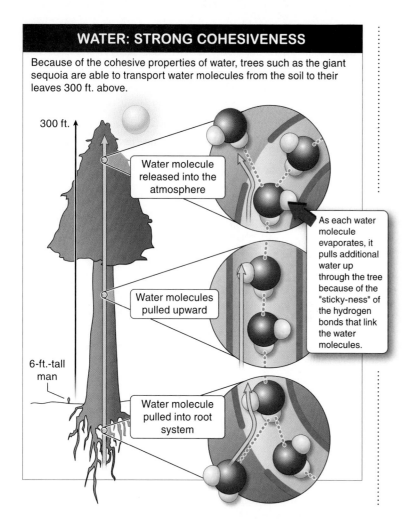

300 ft.

Water molecule released into the atmosphere

Water molecules pulled upward

6-ft.-tall man

Water molecule pulled into root system

As each water molecule evaporates, it pulls additional water up through the tree because of the "sticky-ness" of the hydrogen bonds that link the water molecules.

FIGURE 2-14 **Like a giant straw.** Hydrogen bonds cause water molecules to "stick" together, so that they can be pulled up through the giant sequoia.

The temperature of a substance is a measure of how quickly all of the molecules are moving. The molecules move more quickly when energy is added in the form of heat. When we heat water, the added energy doesn't immediately increase the movement of the individual water molecules. Rather, it disrupts some of the hydrogen bonds between the molecules (**FIGURE 2-15**). As quickly as they can be disrupted, though, hydrogen bonds form again somewhere else. And since the water molecules themselves don't increase their movement, the temperature doesn't increase. The net effect is that even if you release a lot of energy into water, the temperature doesn't change much. For this reason, because so much of your body is water, you are able to maintain a relatively constant body temperature.

Large bodies of water, especially oceans, can absorb huge amounts of heat from the sun during warm times of the year, reducing temperature increases on the land. Similarly, during cold times of the year the ocean slowly cools, giving off heat that reduces the temperature drop on shore.

Q Why do coastal areas have milder, less variable climates than inland areas?

WATER: HIGH HEAT CAPACITY

1 Heat (energy) from the sun disrupts some of the hydrogen bonds between water molecules.

2 New hydrogen bonds are formed almost as quickly as they are disrupted.

3 Because heat energy from the sun is used up breaking and re-forming hydrogen bonds, the water temperature doesn't increase by much.

4 Molecules that make up sand, on the other hand, don't have so many hydrogen bonds, so the heat of the sun just increases the temperature.

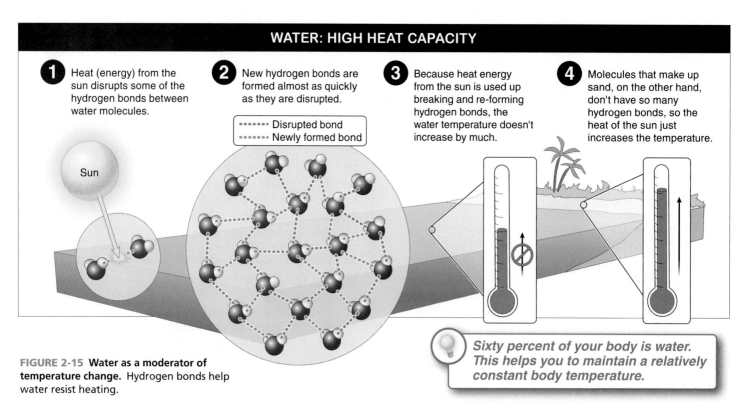

Sun

········· Disrupted bond
········· Newly formed bond

FIGURE 2-15 **Water as a moderator of temperature change.** Hydrogen bonds help water resist heating.

Sixty percent of your body is water. This helps you to maintain a relatively constant body temperature.

Atoms   Water   Carbohydrates   Lipids

3. Low density as a solid. Ice floats. This is unusual, since most substances *increase* in density when frozen; as the molecules slow down, they pack together more and more efficiently—and densely. Consequently, the solid sinks. Water, however, becomes less dense and, as you might expect by now, this is due to hydrogen bonding. As the temperature drops and water molecules slow down, rather than becoming more and more tightly packed, they become less so. Each V-shaped water molecule bonds with four partners, via hydrogen bonds, forming a crystalline lattice in which the molecules are held slightly farther apart than in the liquid, causing ice to be less dense than water (**FIGURE 2-16**).

4. Good solvent. If you put a pinch of table salt into a glass of water, it will quickly dissolve. This means that all the charged sodium (Na^+) and chloride (Cl^-) ions that were ionically bonded together become separated from one another. The sodium and chloride ions were initially attracted to each other because they are polar molecules (ions), each carrying a slight charge. Water is able to pry them apart because, as a polar molecule, it, too, carries charges. The positively charged sodium ions are attracted to the negatively charged side of the water molecule, and the negatively charged chloride ions are

NaCl crystal (table salt)

When placed in water, ionic compounds such as NaCl are broken down into separate Na^+ and Cl^- ions. Positively charged Na^+ ions are attracted to the negatively charged side of the H_2O molecule, while negatively charged Cl^- ions are attracted to the positively charged side.

Dissolved Na^+ and Cl^- ions

Na^+

Cl^-

FIGURE 2-17 Solutions. Water pries apart ionic bonds, dissolving ionic compounds.

attracted to the positively charged side (**FIGURE 2-17**). Many substances are, like water, polar. That is why, like salt, they easily dissolve in water.

Non-polar molecules (such as oil) have neither positively charged regions nor negatively charged regions. Consequently, the polar water molecules are not attracted to them. Instead, when oil is poured into a container of water, the water molecules distance themselves from the oil, leaving the oil molecules in isolated aggregations that never dissolve.

Because so much salt is dissolved in the oceans, many of the water molecules have their positively charged sides facing Cl^- ions. Simultaneously, many molecules of water are turned the other way, with their negatively charged sides facing Na^+ ions. Consequently, the orderly lattices of hydrogen bonds found in ice cannot form in salt water, and it does not freeze well.

Q Why don't oceans freeze as easily as freshwater lakes?

FROZEN WATER
Hydrogen bonding arranges water molecules into a crystalline lattice, keeping them slighty farther apart and, therefore, less dense.

LIQUID WATER
Water molecules move about freely, allowing them to be closer to one another.

RELATIVE AREA OCCUPIED BY THE SAME NUMBER OF H_2O MOLECULES: Frozen water   Liquid water

FIGURE 2-16 Ice floats. When frozen, water becomes less dense.

TAKE-HOME MESSAGE 2·6

The hydrogen bonds between water molecules give water several of its most important characteristics, including cohesiveness, a reduced density as a solid, the ability to resist temperature changes, and broad effectiveness as a solvent.

Proteins   Nucleic Acids

Living systems are highly sensitive to acidic and basic conditions.

There's a lot more going on in water than meets the eye. Most of the molecules are present as H_2O, but at any instant some of them break into two parts: H^+ and OH^-. In pure water, the amount of H^+ and OH^- must be exactly the same, since every time a molecule splits, one of each type of ion is produced. But in some fluids containing other dissolved materials, this balance is lost: the fluid can have more H^+ or more OH^-.

The amount of H^+ or OH^- in a fluid gives it some important properties. In particular, the amount of H^+ in a solution is a measure of its acidity and is called **pH.** The greater the number of free hydrogen ions floating around, the more acidic the solution is.

Pure water is in the middle of the pH scale, with a pH of 7.0. Any fluid with a pH below 7.0 has more H^+ ions (and fewer OH^- ions) and is considered an **acid.** Any fluid with a pH above 7.0 has fewer H^+ ions (and more OH^- ions) and is considered a **base.** The pH scale, like the Richter scale for earthquakes, is logarithmic, although in the case of pH, the *lower* the number the *greater* the acidity: a decrease of 1 on the scale represents a 10-fold increase in the hydrogen ion concentration (**FIGURE 2-18**). A decrease of 2 represents a 100-fold increase. This means that a cola, with a pH of about 3.0, is 10,000 times (!) more acidic than a glass of water, with a pH of 7.0.

H^+ ions are essentially free-floating protons. Acids can donate their H^+ ions to other chemicals. In fact, H^+ is a very reactive little ion. Its presence gives acids some unique properties. For instance, the hydrogen ions in acids can bind with atoms in metals, causing them to corrode. That's why you can dissolve nails by dumping them in a bucket of acid (or cola).

Your stomach produces large amounts of hydrochloric acid (HCl) and has a pH between 1 and 3. (HCl is acidic because most of the hydrogen ions split off from the chlorine, raising the H^+ concentration of the fluid.) The acid in your stomach helps to kill most bacteria that you ingest. It also greatly enhances the breakdown of the chemicals in the food you eat and the efficiency of digestion and absorption. You may have

THE pH SCALE

H^+ ion   Water OH^- ion

ACIDS

Acids are fluids that have a greater proportion of H^+ ions to OH^- ions.

- H^+ ions are very reactive.
- Strong acids are corrosive to metals.
- Acids break down food in your digestive tract.
- Acids are generally sour in taste.

BASES

Bases are fluids that have a greater proportion of OH^- ions to H^+ ions.

- OH^- ions bind with H^+ ions, neutralizing acids.
- Strong bases are caustic to your skin.
- Bases can be found in many household cleaners.
- Bases are generally bitter in taste and soapy.

0 1 2 3 4 5 6 7 8 9 10 11 12 13 14

Battery acid | Soda | Beer | Coffee | Water | Blood | Baking soda | Ammonia | Bleach

Soda, with a pH of about 3.0, is 10,000 times more acidic than a glass of water, with a pH of 7.0!

FIGURE 2-18 pH is a measure of acidity.

Atoms   Water   Carbohydrates   Lipids

BUFFERS IN BLOOD STABILIZE pH

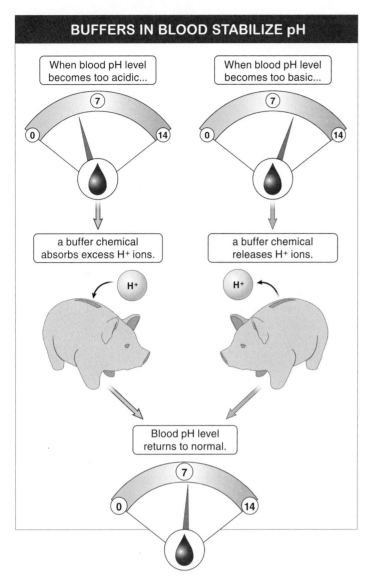

When blood pH level becomes too acidic...

7
0   14

a buffer chemical absorbs excess H⁺ ions.

H⁺

When blood pH level becomes too basic...

7
0   14

a buffer chemical releases H⁺ ions.

H⁺

Blood pH level returns to normal.

7
0   14

FIGURE 2-19 Maintaining a constant internal environment. Buffers in blood prevent potentially damaging pH swings.

Buffers keep the pH of blood close to 7.4.

learned firsthand of the high acidity of your stomach fluids if you have experienced heartburn or the sour taste of vomit.

Bases have very low concentrations of H^+ and relatively high concentrations of OH^-. Baking soda is a common basic substance. Some bases are called "antacids" because the OH^- ions in bases can bind with excess H^+ ions in acidic solutions, neutralizing the acid. Base-containing products such as Alka-Seltzer and Milk of Magnesia reduce the unpleasant feeling of

heartburn and acid indigestion that sometimes arise from the overproduction of acids by the stomach. Bases are commonly used in household cleaning products and generally have a bitter taste and a soapy, slippery feel.

The pH of blood is usually 7.4. Given that most cellular reactions produce or consume H^+ ions, there ought to be great swings in the pH of our blood. Unfortunately, our bodies can't tolerate such swings. Most of the chemicals that aid in the chemical reactions in our blood or cells stop functioning properly if the pH swings up or down by less than half a point. Fortunately, our bodies contain some chemicals that act like bank accounts for H^+ ions (**FIGURE 2-19**). Called **buffers,** these chemicals can quickly absorb excess H^+ ions to keep a solution from becoming too acidic, and they can quickly release H^+ ions to counteract any increases in OH^- concentration. Therefore, buffers are chemicals that act to resist changes in pH.

TAKE-HOME MESSAGE 2·7

The pH of a fluid is a measure of how acidic or basic the solution is and depends on the concentration of dissolved H^+ ions present. Acids, such as vinegar, can donate protons to other chemicals; bases, including baking soda, bind with free protons.

Proteins   Nucleic Acids

❸ Carbohydrates are fuel for living machines.

Worldwide, more corn is produced (by weight) than any other grain.

2•8

Carbohydrates include macromolecules that function as fuel.

Hardly a day goes by without an item in a magazine or newspaper or on TV talking about whether carbohydrates are good or bad for us. Sports drinks such as Gatorade are filled with carbs, and in the days before a big game or race, athletes often "carbo-load" by eating bowls of pasta. Yet other dietary supplements and "power" bars tout that they are low-carb and effective in weight loss by causing the body to resort to using fat for energy. Meanwhile, nutritionists, doctors, and many diet programs exhort people to increase the amount of fiber—another type of carbohydrate—in their diet. What exactly are they all talking about?

There are four types of **macromolecules**—large molecules made up from smaller building blocks or subunits—in living organisms: carbohydrates, lipids, proteins, and nucleic acids. **Carbohydrates** are molecules that contain mostly carbon, hydrogen, and oxygen: they are the primary fuel for running all of the cellular machinery and also form much of

FIGURE 2-20 All carbohydrates have a similar structure and function.

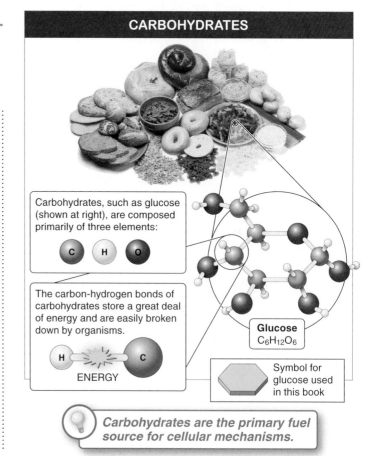

CARBOHYDRATES

Carbohydrates, such as glucose (shown at right), are composed primarily of three elements:

C H O

The carbon-hydrogen bonds of carbohydrates store a great deal of energy and are easily broken down by organisms.

H ✦ C

ENERGY

Glucose
$C_6H_{12}O_6$

Symbol for glucose used in this book

Carbohydrates are the primary fuel source for cellular mechanisms.

Atoms   Water   Carbohydrates   Lipids

the structure of cells in all life forms. Sometimes they contain atoms of other elements, but they must have carbon, hydrogen, and oxygen to be considered a carbohydrate (**FIGURE 2-20**). Further, a carbohydrate generally has approximately the same number of carbon atoms as it does H_2O units. For instance, the best-known carbohydrate, glucose, has the composition $C_6H_{12}O_6$ (6 carbons and, as a little math will show us, 6 H_2O units; notice that $6 \times H_2 = H_{12}$ and $6 \times O = O_6$). A carbohydrate called maltose has the composition $C_{12}H_{22}O_{11}$.

Carbohydrates function well as fuels because their many carbon-hydrogen bonds store a great deal of energy. These

C–H bonds are easily broken, and organisms can capture the energy released and put it to use.

TAKE-HOME MESSAGE 2·8

Carbohydrates are the primary fuel for running all cellular machinery and also form much of the structure of cells in all life forms. Carbohydrates contain carbon, hydrogen, and oxygen, and generally have the same number of carbon atoms as they do H_2O units. The C–H bonds of carbohydrates store a great deal of energy and are easily broken by organisms.

2·9

Simple sugars are an effective, accessible source of energy.

Carbohydrates are classified into several categories, based on their size and their composition. The simplest carbohydrates are the **monosaccharides** or **simple sugars.** These simple sugars contain anywhere from three to seven carbon atoms and, when they are broken down, the products usually are not carbohydrates. Two common monosaccharides are glucose, found in the sap and fruit of many plants, and fructose, found primarily in fruits and vegetables, as well as in honey. Fructose is the sweetest of all naturally occurring sugars. The suffix -*ose* tells us that a substance is a carbohydrate.

SOME COMMON MONOSACCHARIDES

Glucose
$C_6H_{12}O_6$

Fructose
$C_6H_{12}O_6$

Galactose
$C_6H_{12}O_6$

The carbohydrate of most importance to living organisms is glucose. This simple sugar is found naturally in most fruits, but most of the carbohydrates that you eat, including table sugar (called sucrose) and the starchy carbohydrates found in bread and potatoes, are converted into glucose in your digestive system. The glucose then circulates in your blood at a concentration of about 0.1%. Circulating glucose, also called "blood sugar," has one of three fates (**FIGURE 2-21**):

1. **Fuel for cellular activity.** Once it arrives at and enters a cell, glucose can be used as an energy source. Through

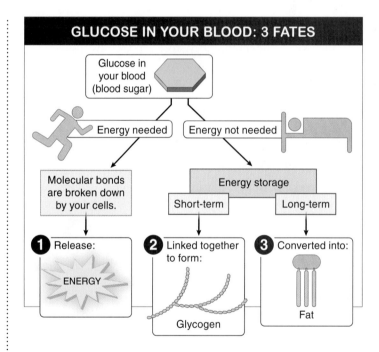

FIGURE 2-21 What happens to sugar in your blood?

a series of chemical reactions, the cell breaks the bonds between the atoms of the glucose molecule (a process explained in detail in Chapter 4) and then uses the released energy to fuel cellular activity, including the muscle contractions that enable you to move and the nerve activities that enable you to think.

2. **Stored temporarily as glycogen.** If there is more glucose circulating in your bloodstream than is necessary

Proteins   Nucleic Acids

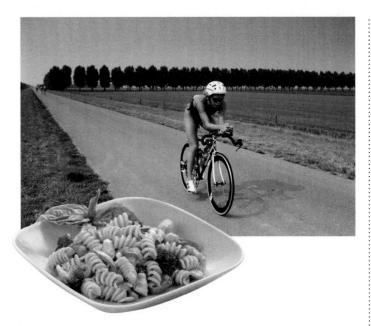

FIGURE 2-22 **Some competitive athletes carbo-load to maintain high energy levels.**

? **Why do dieters lose large amounts of "water weight" during the first few days of a diet?**

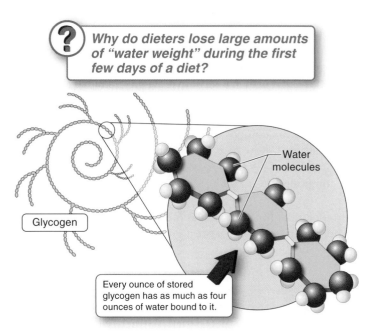

Water molecules

Glycogen

Every ounce of stored glycogen has as much as four ounces of water bound to it.

FIGURE 2-23 **Water weight.** Water molecules bound to glycogen account for much of the weight lost early in a diet.

to meet your body's current energy needs, the excess glucose can be temporarily stored in various tissues, primarily your muscles and liver. The stored glucose molecules are linked together to form a large web of molecules called **glycogen.** When you need energy later, the glycogen can easily be broken down to release glucose molecules back into your bloodstream. Glycogen is the primary form of short-term energy storage in animals.

3. **Converted to fat.** Finally, additional glucose circulating in your bloodstream can be converted into fat, a form of long-term energy storage.

"Carbo-loading" is a method by which athletes can, for a short time, double or triple the usual amount of glycogen stored in their muscles and liver, increasing the store of fuel available for extended exertion and delaying the onset of fatigue during an endurance event (**FIGURE 2-22**).

Q What is "carbo-loading"?

Carbo-loading is usually done in two phases, a depletion phase and a loading phase. The depletion phase begins six or seven days before a competition. In this phase, the combination of a super-low carbohydrate intake and exhaustive exercise depletes glycogen in the muscles. The loading phase takes place during the two days before the competition. During this phase, a super-high carbohydrate diet is combined with reduced exercise to achieve a higher

blood glucose than is necessary, so that much of the excess glucose is stored as glycogen.

Glycogen also plays a role in the initial rapid weight loss people experience when dieting. If you reduce your caloric intake—perhaps as part of a low-carbohydrate diet—such that your body is burning more calories than you are consuming, your body must use stored forms of energy. The first, most accessible molecules that can be broken down for energy in the absence of sufficient sugar in your bloodstream are glycogen molecules in your muscles and liver. Large amounts of water are bound to glycogen. In fact, every ounce of stored glycogen has as much as 4 ounces of water bound to it. As that glycogen is removed from your tissue, so, too, is the water. This accounts for the initial dramatic weight loss that occurs before your body resorts to using stored fat, at which point the rate of weight loss slows considerably (**FIGURE 2-23**).

TAKE-HOME MESSAGE 2·9

The simplest carbohydrates, including glucose—the most important carbohydrate to living organisms—are monosaccharides or simple sugars. They contain from three to seven carbon atoms.

Atoms   Water   Carbohydrates   Lipids

2·10

Complex carbohydrates are time-released packets of energy.

In contrast to the simple sugars, **polysaccharides** contain more than one sugar unit or building block. For example, two simple sugars can be joined together into a **disaccharide,** such as sucrose (table sugar) and lactose (the sugar found in milk). When more than two simple sugars are joined together—sometimes as many as 10,000—the resulting polysaccharide is called a **complex carbohydrate** (**FIGURE 2-24**). Depending on how the simple sugars are bonded together, polysaccharides may function as "time-release" stores of energy or as structural materials that may be completely indigestible by most animals. An example of such a structural material is the polysaccharide cellulose—the primary component of plant cell walls.

Like simple sugars, many disaccharides and polysaccharides are important sources of fuel. Unlike simple sugars, however, disaccharides and polysaccharides must undergo some preliminary processing before the energy can be released from their bonds. Let's look at what happens when we eat some sucrose, common table sugar. Sucrose is the primary carbohydrate in plant sap. It is a disaccharide composed of two simple sugars, glucose and fructose, linked together. Because humans can't directly use sucrose, the body must first break the bond linking the glucose and fructose. Only then can the individual monosaccharides be broken down into their component atoms and the energy from the broken bonds be harvested and used. Similarly, lactose is a disaccharide made up of a molecule of glucose and a molecule of galactose bound together. As with sucrose, we must break the bond before we can extract any usable energy from a molecule of lactose.

Energy can also be stored in a complex carbohydrate called **starch,** which consists of a hundred or more glucose molecules joined together in a line. In plants, starch is the primary form of energy storage, found in roots and other tissues (see Figure 2-24). Commonly cultivated grains such as barley, wheat, and rye are high in starch content, and corn and rice are more than 70% starch. Although it is composed exclusively of glucose molecules linked together, starch does not taste sweet. Because of its molecular shape it does not stimulate the sweetness receptors on the tongue. The glycogen that stores energy in your muscles and liver is also a complex carbohydrate, so it is sometimes referred to as "animal starch" (although it has a more branched structure than starch and carries more glucose units linked together).

Complex carbohydrates are like "time-release" fuel pellets. The glucose molecules become available slowly, one by one, as the bonds between glucose units are broken. As an interesting demonstration of the gradual release of glucose from complex carbohydrates, put a piece of potato or bread into your mouth and let it rest on your tongue. Initially, it does not have a sweet taste. After a few minutes, though, you will begin to taste some sweetness as the chemicals in your saliva break the starch down into glucose.

The relative amounts of complex carbohydrates and simple sugars in foods cause them to have very different effects when you eat them. Oatmeal (along with rice and pasta), for example, is rich in complex carbohydrates. Fresh fruits, on the other hand, are rich in simple sugars, such as fructose.

FIGURE 2-24 Chains of sugars. Polysaccharides are made from simple sugars bound together.

Proteins   Nucleic Acids

Q Before heading to the library for a long study session, students would be wise to consume oatmeal rather than fresh fruits. Why?

Consequently, although the fruit will give a quick burst of energy as the sugars are almost immediately available, the fuel will soon be gone from the bloodstream. The simple sugars in the oatmeal will become available only gradually, as the complex carbohydrates of the oats are slowly broken down into their simple sugar components (**FIGURE 2-25**).

TAKE-HOME MESSAGE 2·10

Multiple simple carbohydrates are sometimes linked together into more complex carbohydrates. Types of complex carbohydrates include starch, which is the primary form of energy storage in plants, and glycogen, which is a primary form of energy storage in animals.

Depending on their structure, dietary carbohydrates can lead to quick-but-brief or slow-but-persistent increases in blood sugar.

Fructose

Complex carbohydrates

FIGURE 2-25 Short-term versus long-term energy? Complex carbohydrates and simple sugars differ in the way they make energy available to you.

2·11

Not all carbohydrates are digestible.

Despite their general importance as a fuel source for humans, not all carbohydrates can be broken down by our digestive system. Two different complex carbohydrates—both indigestible by humans—serve as structural materials for invertebrate animals and plants: **chitin** (pronounced KITE-in) and **cellulose** (**FIGURE 2-26**). Chitin forms the rigid outer skeleton of most insects and crustaceans (such as lobsters and crabs). Cellulose forms a huge variety of plant structures that are visible all around us. We find cellulose in trees and the wooden structures we build from them, in cotton and the clothes we make from it, in leaves and in grasses. In fact, it is the single most prevalent compound on earth.

Surprisingly, cellulose is almost identical in composition to starch. Nonetheless, because of one small difference in the chemical bond between the simple-sugar units, cellulose has a slightly different three-dimensional structure. And, as we noted earlier, even tiny differences in the shape of a molecule can have a huge effect on its behavior. In this case, the

FIGURE 2-26 Carbohydrates can serve as structural materials.

Though indigestible in humans, fiber still aids in digestion and has numerous health benefits.

FIGURE 2-27 Fiber. It's not digestible but it's still important for our diet.

difference in shape makes it impossible for humans to digest cellulose as they can starch. Consequently, the cellulose we eat passes right through our digestive system unused.

Although it is not digestible, cellulose is still important to human diets. The cellulose in our diet is known as "fiber" (**FIGURE 2-27**). It is also appropriately called "roughage" because, as the cellulose of celery stalks and lettuce leaves passes through our digestive system, it scrapes the wall of the digestive tract. Its bulk and the scraping stimulate the more rapid passage of food and the unwanted, possibly harmful products of digestion through our intestines. That is why fiber reduces the risk of colon cancer and other diseases (but it is also why too much fiber can lead to diarrhea.)

Unlike humans, termites have some microorganisms living in their gut that are able to break down cellulose. That's why they can chew on wood and, with the help of the cellulose-digesting boarders in their gut, can break down the cellulose and extract usable energy from the freed glucose molecules.

TAKE-HOME MESSAGE 2•11

Some complex carbohydrates, including chitin and cellulose, cannot be digested by most animals. Such indigestible carbohydrates in the diet, called fiber, aid in digestion and have many health benefits.

Proteins   Nucleic Acids

4 Lipids store energy for a rainy day.

A well-insulated harbor seal, in Alaska.

2·12

Lipids are macromolecules with several functions, including energy storage.

Lipids are a second group of macromolecules important to all living organisms. Lipids, just like carbohydrates, are made primarily from atoms of carbon, hydrogen, and oxygen, but the atoms are in different proportions. In particular, while carbohydrates usually have two hydrogen atoms for every oxygen, lipids have more hydrogen atoms for each carbon. As a result, lipids have significantly more C–H bonds and contain significantly more stored energy.

Q Why does a salad dressing made with vinegar and oil separate into two layers shortly after you shake it?

What exactly is a lipid? That's not as easy to answer as you might expect. Lipids come in a wide variety of structures. They don't have any unique subunits (such as the simple sugars that make up disaccharides and polysaccharides) or particular ratios of atoms that serve as defining features. Consequently, lipids are defined based on their physical characteristics. Most notably, lipids do not dissolve in water

and are greasy to the touch—think of salad dressings.

Lipids are insoluble in water because, in sharp contrast to water, they consist mostly of hydrocarbons, which are non-polar. Non-polar molecules (or parts of molecules) tend to minimize their contact with water and are considered **hydrophobic** ("water-fearing"). Lipids cluster together when mixed with water, never fully dissolving. Molecules that readily form hydrogen bonds with water, on the other hand, are considered **hydrophilic** ("water-loving").

One familiar type of lipid is *fat,* the type most important in long-term energy storage and insulation (**FIGURE 2-28**). (Penguins and walruses can maintain relatively high body temperatures, despite living in very cold habitats, due to

| Atoms | Water | Carbohydrates | Lipids |

FIGURE 2-28 **The many purposes of lipids.**

TYPICAL FEATURES OF LIPIDS
• Non-polar molecules that do not dissolve in water
• Greasy to the touch
• Significant source of energy storage

THREE TYPES OF LIPIDS

FATS	STEROLS	PHOSPHOLIPIDS
FUNCTION Long-term energy storage and insulation	**FUNCTION** Regulate growth and development	**FUNCTION** Form the membranes that enclose cells

their thick layer of insulating fat.) Lipids also include *sterols,* which include *cholesterol* and many of the sex hormones that play regulatory roles in animals, and *phospholipids,* which form the membranes that enclose cells. We examine each type of lipid below.

TAKE-HOME MESSAGE 2·12

Lipids are insoluble in water and greasy to the touch. They are valuable to organisms for long-term energy storage and insulation, in membrane formation, and as hormones.

2·13

Fats are tasty molecules too plentiful in our diets.

All fats have two distinct components: they have a "head" region and two or three long "tails" (**FIGURE 2-29**). The head region is a small molecule called **glycerol.** It is linked to "tail" molecules known as **fatty acids.** A fatty acid is simply a long hydrocarbon—that is, a chain of carbon molecules, often a dozen or more, linked together with one or two hydrogen atoms attached to each carbon.

The fats in most foods we eat are **triglycerides,** which are fats having three fatty acids linked to the glycerol molecule. For this reason, the terms "fats" and "triglycerides" are often used interchangeably. Triglycerides that are solid at room temperature are generally called "fats," while those that are liquid at room temperature are called "oils."

Fat molecules contain much more stored energy than carbohydrate molecules. That is, the chemical breakdown of fat

FIGURE 2-29 **Triglycerides have glycerol heads and fatty acid tails.**

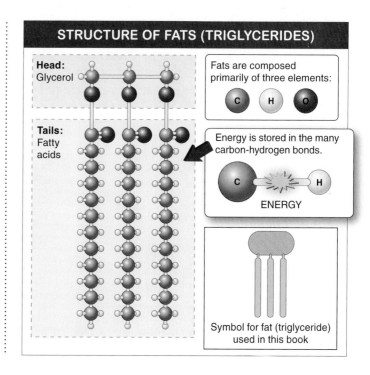

STRUCTURE OF FATS (TRIGLYCERIDES)

Head: Glycerol

Tails: Fatty acids

Fats are composed primarily of three elements:
C H O

Energy is stored in the many carbon-hydrogen bonds.

C · · · H

ENERGY

Symbol for fat (triglyceride) used in this book

Proteins   Nucleic Acids

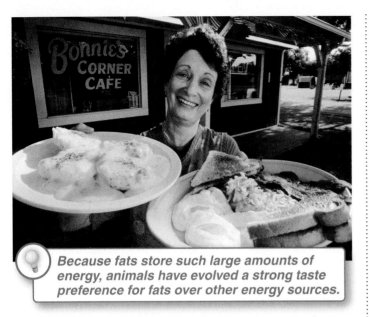

Because fats store such large amounts of energy, animals have evolved a strong taste preference for fats over other energy sources.

FIGURE 2-30 **Animals (including humans!) prefer the taste of fats.**

molecules releases significantly more energy. A single gram of carbohydrate stores about 4 calories of energy, while the same amount of fat stores about 9 calories—not unlike the difference between a $5 bill and a $10 bill. Because fats store such a large amount of energy, animals have evolved a strong taste preference for fats over other energy sources (**FIGURE 2-30**). Organisms evolving in an environment of uncertain food supply will build the largest surplus by consuming molecules that hold the most amount of energy in the smallest mass. This feature helped the earliest humans to survive, millions of years ago, but today puts us in danger from the health risks of obesity now that fats are all too readily available.

An important distinction is made between "saturated" and "unsaturated" fats (**FIGURE 2-31**). These terms refer to the hydrocarbon chain in the fatty acids. If each carbon atom in the hydrocarbon chain of a fatty acid is bonded to two hydrogen atoms, the fat molecule carries the maximum

SATURATED FATS

In saturated fats, each carbon in the hydrocarbon chain is bound to two hydrogen atoms.

Straight fatty acids can be packed together tightly. As a result, saturated fats are solid at room temperature.

UNSATURATED FATS

Double bond

In unsaturated fats, at least one carbon in the hydrocarbon chain is bound to just one hydrogen, causing the fatty acid to have a crooked shape.

Crooked fatty acids cannot be packed together tightly. As a result, unsaturated fats are liquid at room temperature.

FIGURE 2-31 **Degrees of saturation.** Fatty acids (and thus the fats that contain them) can be unsaturated or saturated; unsaturated fats can be saturated artificially, making them tastier but less healthful.

Atoms   Water   Carbohydrates   Lipids

Q Chocolate chip cookie recipes call for some lipids. How will the "chewy-ness" of the cookies differ depending on whether you use butter or vegetable oil as the lipid? Which cookies will be better for your health?

number of hydrogen atoms and is said to be a **saturated fat.** Most animal fats, including those found in meat and eggs, are saturated. They are not essential to your health and, because they accumulate in your bloodstream and can narrow the vessel walls, they can cause heart disease and strokes.

An **unsaturated fat** is one in which some of the carbon atoms are bound to only a single hydrogen. Most plant fats are unsaturated. Unsaturated fats may be *mono-unsaturated* (if a fatty acid hydrocarbon chain has only one pair of neighboring carbon atoms in an unsaturated state—that is, has only one double bond) or *polyunsaturated* (if more than one pair of carbons is unsaturated—there's more than one double bond). Unsaturated fats are still high in calories, but because they can lower cholesterol, they are generally preferable to saturated fats. Foods high in unsaturated fats include avocados, peanuts, and olive oil. Relative to other animals, fish tend to have less saturated fat.

The shapes of unsaturated fat molecules and saturated fat molecules are different. When saturated, the hydrocarbon tails of the fatty acids all line up very straight and the fat molecules can be packed together tightly. The tight packing causes the fats, such as butter, to be solid at room temperature. When unsaturated, the fatty acids have kinks in the hydrocarbon tails and the fat molecules cannot be packed together as tightly (see Figure 2-31). Consequently, unsaturated fats, such as canola oil and vegetable oil, do not solidify as easily and are liquid at room temperature.

The ingredient list for many snack foods includes "partially hydrogenated" vegetable oils. The hydrogenation of an oil means that a liquid, unsaturated fat has had hydrogen atoms added to it so that it becomes more saturated. This can be useful in creating a food with a more desirable texture, since increasing a fat's degree of saturation changes its consistency and makes it more solid at room temperatures. By attaining just the right degree of saturation, it is possible to create foods, such as chocolate, that are near the border of solid and liquid and "melt in your mouth" (**FIGURE 2-32**). Unfortunately, hydrogenation also makes the food less

Q Many snack foods contain "partially hydrogenated" vegetable oils. Why might it be desirable to add hydrogen atoms to a vegetable oil?

HYDROGENATION

Hydrogenation is the artificial addition of hydrogen atoms to an unsaturated fat. This can improve a food's taste, texture, and shelf-life.

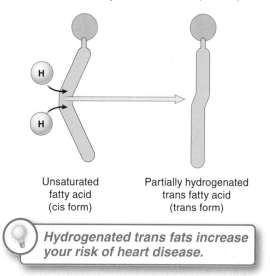

Unsaturated fatty acid (cis form)

Partially hydrogenated trans fatty acid (trans form)

Hydrogenated trans fats increase your risk of heart disease.

FIGURE 2-32 Hydrogenation improves a food's taste, texture, and shelf-life (but at a cost).

healthful because saturated fats are less reactive—your body is less likely to break them down—and so are more likely to accumulate in your blood vessels, increasing the risk of heart disease.

Hydrogenation of unsaturated fats is problematic from a health perspective because it creates **trans fats,** the "trans" referring to the unusual orientation of the added hydrogen atoms, which differs from that in other dietary fats—which have their hydrogens in an orientation called "cis." Trans fats cause your body to produce more cholesterol, further raising the risk of heart disease, and they also reduce your body's production of a type of cholesterol that protects against heart disease.

Proteins   Nucleic Acids

Q *Olestra is a recently developed "fake fat" chemical that gives foods the taste of fat, without adding the calories of fats. What chemical structure might make this possible?*

Because of the well-documented links between dietary fats and heart disease, many people are trying to reduce their fat intake. "Fake fats" make this possible. They are designed to be similar to fats in taste and texture, but have one big difference: they cannot be digested by humans. One such "fake fat" is olestra. Olestra, instead of being a triglyceride fat with three fatty acid tails linked to a glycerol molecule, has eight separate hydrocarbon fatty acids attached to a molecule of sucrose. This octopus–like molecule allows the fatty acids to stimulate their usual taste buds on your tongue, telling your brain that you are eating a fat. The complex shape of the molecule, however, prevents your body's digestive chemicals from grabbing onto it and breaking it down. As a consequence, it passes through your digestive system without being digested. It's not a perfect solution, however. Olestra reduces absorption of some vitamins, and in some people causes abdominal cramping.

TAKE-HOME MESSAGE 2·13

Fats, including the triglycerides common in the food we eat, are one type of lipid. Characterized by long hydrocarbon tails, fats effectively store energy in the many carbon-hydrogen and carbon-carbon bonds. Their caloric density is responsible for humans' preferring fats to other macromolecules in the diet, and is responsible for their association with obesity and illness in the modern world.

2·14

Cholesterol and phospholipids are used to build sex hormones and membranes.

Not all lipids are fats, nor do lipids necessarily function in energy storage. A second group of lipids, called the **sterols,** plays an important role in regulating growth and development (**FIGURE 2-33**). This group includes some very familiar lipids:

STEROLS

CHOLESTEROL
- Important component of cell membranes in animals
- Can attach to vessel walls and cause them to thicken, which may lead to high blood pressure, stroke, and heart attack

Sterols are all based on a structure featuring four carbon rings.

STEROID HORMONES
- Regulate sexual development, maturation, and sex cell production
- Estrogen influences memory and mood
- Testosterone stimulates muscle growth

Estrogen

Testosterone

FIGURE 2-33 Not all lipids are for energy storage. Cholesterol, estrogen, and testosterone are all lipids.

cholesterol and the steroid hormones such as testosterone and estrogen. These molecules are all modifications on one basic structure formed from four interlinked rings of carbon atoms.

Cholesterol is an important component of most cell membranes. For this reason, it is an essential molecule for living organisms. It has a bad reputation in most Western cultures, though, that is mostly well deserved. When we ingest too much cholesterol (present in animal-based foods such as egg yolks, red meat, and cream) and high levels of cholesterol circulate in our bloodstream, the cholesterol can attach to blood vessel walls and cause them to thicken. In turn, this thickening can lead to high blood pressure, a major contributor to strokes and heart attacks. For these reasons, nutritionists advise limiting the consumption of foods high in cholesterol.

Surprisingly, however, most of the cholesterol in our blood doesn't come directly from the food we consume. Instead, cells in our liver produce almost 90% of the circulating cholesterol. How do they do it? By transforming the saturated fats in our diet. The best way to reduce cholesterol levels, then, may be to consume fewer saturated fats.

The steroid hormones estrogen and testosterone are built through slight chemical modifications to cholesterol. These hormones are among the primary molecules that direct and regulate sexual development, maturation, and sperm and egg production. In both males and females, estrogen influences memory and mood, among other traits. Testosterone has numerous effects, one of which is to stimulate muscle growth. As a consequence, athletes often take synthetic variants of testosterone to increase their muscularity. But the use of these supplements is often accompanied by dangerous side effects, including extreme aggressiveness (" 'roid rage"), high cholesterol, and, following long-term use, cancer. As a consequence, nearly all athletic organizations have banned their use (**FIGURE 2-34**).

Phospholipids and waxes are also lipids. **Phospholipids** (**FIGURE 2-35**) are the major component of the membrane that surrounds the contents of a cell and controls the flow of chemicals into and out of the cell. They have a structure similar to fats, but with two differences: they contain a phosphorus atom (hence *phospho*lipids) and they have two fatty acid chains rather than three. We will explore the

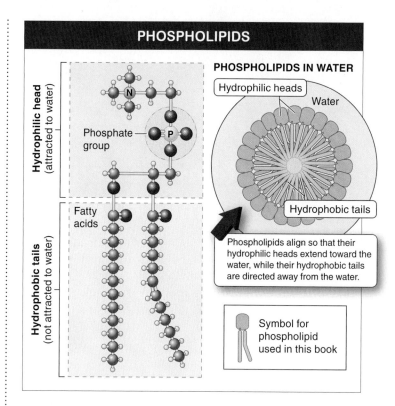

FIGURE 2-35 **Lipid versatility.** Phospholipids have important roles in many organisms.

significant role of phospholipids in cell membranes in the next chapter.

Waxes resemble fats but have only one long-chain fatty acid linked to the glycerol head of the molecule. Because the fatty acid chain is highly non-polar, waxes are strongly hydrophobic; that is, these molecules do not mix with water but repel it. Their water resistance accounts for their presence as a natural coating on the surface of many plants and in the outer coverings of many insects. In both cases, the waxes prevent the plants and animals from losing the water essential to their life processes. Many birds, too, have a waxy coating on their wings, keeping them from becoming water-logged when they get wet.

TAKE-HOME MESSAGE 2·14

Cholesterol and phospholipids are lipids that are not fats. Both are important components in cell membranes. Cholesterol also serves as a precursor to steroid hormones, important regulators of growth and development.

FIGURE 2-34 **Dangerous bulk.** Steroids can increase muscularity, but with serious health consequences.

Proteins   Nucleic Acids

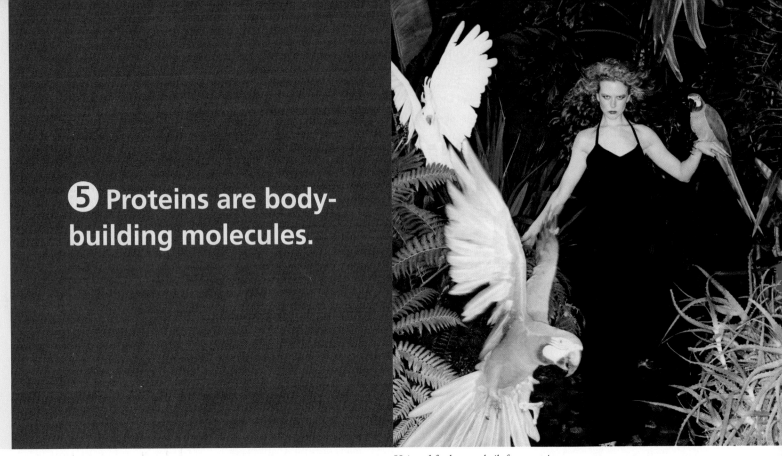

⑤ Proteins are body-building molecules.

Hair and feathers are built from proteins.

2·15

Proteins are versatile macromolecules that serve as building blocks.

You can't look at a living organism and not see proteins (FIGURE 2-36). Inside and out, **proteins** are the chief building blocks of all life. They make up skin and feathers and horns. They make up bones and muscles. In your bloodstream, proteins fight invading microorganisms and stop you from bleeding to death from a shaving cut. Proteins control the levels of sugar and other chemicals in your bloodstream and carry oxygen from one place in your body to another. And in just about every cell in every living organism, proteins called **enzymes** initiate and assist every chemical reaction that occurs.

Although proteins perform several very different types of functions, all are built in the same way and from the same raw materials in all organisms. In the English language, every sentence is made up of words and every word is formed from one or more of the 26 letters of the alphabet. With 26 letters we can write anything, from sonnets to cookbooks to biology textbooks. Proteins, too, are constructed from a

sort of alphabet. Instead of 26 letters there are 20 molecules, known as **amino acids.** Unique combinations of these 20 amino acids are strung together, like beads on a string, and the resulting protein has a unique structure and chemical behavior.

Let's look more closely at the structure of the amino acids in the protein alphabet. They all have the same basic two-part structure: one part is the same in all 20 amino acids, and the other part is unique, differing in each of the 20 amino acids.

Proteins contain the same familiar atoms as carbohydrates and lipids—carbon, hydrogen, and oxygen—but differ in an important way: they also contain nitrogen. At the center of every amino acid is a carbon atom, with its four covalent bonds (FIGURE 2-37). One bond attaches the carbon to something called a **carboxyl group,** which is a carbon bonded to two oxygen atoms. The second bond attaches the central carbon to

Atoms   Water   Carbohydrates   Lipids

Proteins perform a variety of different functions. They all, however, are built the same way and from the same raw materials in organisms.

STRUCTURAL
Hair, fingernails, **feathers**, horns, cartilage, tendons

PROTECTIVE
Help fight invading microorganisms, coagulate **blood**

REGULATORY
Control cell activity, constitute some **hormones**

CONTRACTILE
Allow **muscles** to contract, heart to pump, sperm to swim

TRANSPORT
Carry molecules such as **oxygen** around your body

FIGURE 2-36 Proteins everywhere! Proteins are the chief building blocks of all organisms.

a single hydrogen atom. The third bond attaches the central carbon to an **amino group,** which is a nitrogen atom bonded to three hydrogen atoms. These components—the central carbon with its attached hydrogen atom, carboxyl group, and

FIGURE 2-37 Amino acid structure. Amino acids are made up of an amino group, carboxyl group, and a side chain.

amino group—are the foundation that identifies a molecule as an amino acid and, as multiple amino acids are joined together, forms the "backbone" of the protein.

The fourth bond of the central carbon attaches to a functional group or side chain. This side chain is the unique part of each of the 20 amino acids. In the simplest amino acid, glycine, for example, the side chain is simply a hydrogen atom. In other amino acids, the side chain is a single CH_3 group or three or four such groups. Most of the side chains include both hydrogen and carbon, and a few include nitrogen or sulfur atoms. The side chain determines an amino acid's chemical properties, such as whether the amino acid molecule is polar or non-polar.

TAKE-HOME MESSAGE 2•15

Unique combinations of 20 amino acids give rise to proteins, the chief building blocks of the physical structures that make up all organisms. Proteins perform myriad functions, from assisting chemical reactions to causing blood clotting to building bones to fighting microorganisms.

Proteins are an essential dietary component.

The atoms present in the plant and animal proteins we eat—especially the nitrogen atoms—are essential to the constant growth, repair, and replacement that take place in our bodies. As we eat protein and break it down into its amino acids through digestion, our bodies are collecting the amino acids needed for various building projects. Proteins also store energy in their bonds and, like carbohydrates and lipids, they can be used to fuel living processes.

The amount of protein we need depends on the extent of the building projects underway at any given time. Most individuals need 40–80 grams of protein per day. Bodybuilders, however, may need 150 grams a day or more to achieve the extensive muscle growth stimulated by their training; similarly, the protein needs of pregnant or nursing women are very high.

Q Food labels indicate an item's protein content. Why is this insufficient for determining whether you are protein deficient, even if your protein intake exceeds your recommended daily amount?

Contrary to the impression you might get from the labels you see on food packaging, all proteins are not created equal (**FIGURE 2-38**). Every different protein has a different composition of amino acids. And while our bodies can manufacture certain amino acids as they are needed, many other amino acids must come from our diet. Those that we must get from our diet—about half of the 20 amino acids—are called "essential amino acids." For this reason, we shouldn't just speak of needing "x grams of protein per day." We need to consume all of the essential amino acids every day.

Many foods, containing "complete proteins," have all of the essential amino acids. Animal products such as milk, eggs, fish, chicken, and beef tend to provide complete proteins. Most vegetables, fruits, and grains, on the other hand, more often contain "incomplete proteins," which do not have all

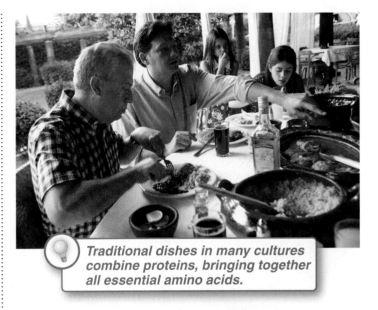

Traditional dishes in many cultures combine proteins, bringing together all essential amino acids.

FIGURE 2-38 All proteins are not created equal. Some foods have "complete proteins" with all the essential amino acids. Other foods have "incomplete proteins" and we must consume proteins from multiple sources to get all the essential amino acids.

the essential amino acids. If you consume only one type of incomplete protein in your diet, you may be deficient in one or more of the essential amino acids. But two incomplete proteins that are "complementary proteins," when eaten together, can provide all the essential amino acids. Traditional dishes in many cultures often include such pairings (see Figure 2-39). Examples are corn and beans in Mexico, rice and lentils in India, and rice and black-eyed peas in the southern United States.

TAKE-HOME MESSAGE 2 • 16

Twenty amino acids make up all the proteins necessary for growth, repair, and replacement of tissue in living organisms. Of these amino acids, eight are essential for humans: they cannot be synthesized by the body so must be consumed in the diet. Complete proteins contain all eight essential amino acids, while incomplete proteins do not.

Atoms   Water   Carbohydrates   Lipids

Proteins' functions are influenced by their three-dimensional shape.

Proteins are formed by linking individual amino acids together with a **peptide bond,** in which the amino group of one amino acid is bonded to the carboxyl group of another. Two amino acids joined together form a *dipeptide* and several amino acids joined together form a *polypeptide.* The sequence of amino acids in the polypeptide chain is called the **primary structure** of the protein and can be compared to the sequence of letters that spells a specific word (**FIGURE 2-39**).

Amino acids in a polypeptide chain don't remain in a simple straight line like beads on a string. The chain begins to fold as side chains come together and hydrogen bonds form between various atoms in the chain. The two most common patterns of hydrogen bonding between amino acids cause the chain to either twist in a corkscrew-like shape or form a zigzag folding pattern. This hydrogen bonding between amino acids gives a protein its **secondary structure.**

The protein eventually folds and bends upon itself, and additional bonds continue to form between atoms in the side chains of amino acids that are near each other. Eventually, the protein folds into a unique and complex three-dimensional shape called its **tertiary structure.** The exact form comes about as the secondary structure folds and bends, bringing together amino acids that then form bonds such as hydrogen bonds or covalent sulfur–sulfur bonds (see Figure 2-39).

Some protein molecules have a **quaternary structure** in which two or more polypeptide chains are held together by bonds between amino acids in the different chains. Hemoglobin, the protein molecule that carries oxygen from the lungs to the cells where it is needed, is made from four polypeptide chains, two "alpha" chains and two "beta" chains.

Some proteins are attached to other types of macromolecules. *Lipoproteins,* for example, circulate in the bloodstream carrying fats. They are formed when cholesterol and a triglyceride (both lipids) combine with a protein. *Glycoproteins* are combinations of carbohydrates and proteins. These are found on the surfaces of nearly all animal cells and play a role in helping the immune system to distinguish between your own cells and foreign cells. (We learn more about glycoproteins in the next chapter, which discusses cells.)

STRUCTURE OF PROTEINS

PRIMARY STRUCTURE
The sequence of amino acids in a polypeptide chain, similar to the sequence of letters that spell out a specific word

SECONDARY STRUCTURE
The corkscrew-like twists or pleated folds formed by hydrogen bonds between amino acids in the polypeptide chain

TERTIARY STRUCTURE
The complex three-dimensional shape formed by multiple twists and bends in the polypeptide chain based on interactions between the side chains

QUATERNARY STRUCTURE
Two or more polypeptide chains bonded together

FIGURE 2-39 **Protein structure.** The functions of proteins are influenced by their three-dimensional shape.

Proteins   Nucleic Acids

Q Egg whites contain a lot of protein. Why does beating them change their texture, making them stiff?

The overall shape of a protein molecule determines its function—how it behaves and the other molecules it interacts with. For proteins to function properly, they must retain their three-dimensional shape. When their shapes are deformed, they usually lose their ability to function. We can see proteins deformed when we fry an egg. The heat breaks the hydrogen bonds that give the proteins their shape. The proteins in the clear egg white unfold, losing their secondary and tertiary structure. This disruption of protein folding is called **denaturation** (**FIGURE 2-40**).

Almost any extreme environment will denature a protein. Take a raw egg, for instance, and crack it into a dish containing baking soda or rubbing alcohol. Both chemicals are sufficiently extreme to turn the clear protein opaque white, as in fried egg whites.

Q Why is wet hair easier to style than dry hair?

Hair is a protein whose shape most of us have modified at one time or another. Styling hair—whether curling or straightening it—involves altering some of the hydrogen bonds between the amino acids that make up the hair protein, changing its tertiary structure. When your hair gets wet, the water is able to disrupt some of the hydrogen bonds, causing some amino acids in the protein to form hydrogen bonds with the water molecules instead. This enables you to change your hair's shape—making it straighter or, if you manipulate it around curlers, making it curlier—if you style it while it's wet. The hair can then hold this shape when it dries as the hydrogen bonds to water are replaced by other hydrogen bonds between amino acids of

FIGURE 2-41 **Curly or straight?** Proteins determine it!

the hair protein as the water evaporates. Once your hair gets wet again, however, unless it is combed, brushed, or wrapped in a different style, it will return to its natural shape.

Whether your hair is straight or curly or somewhere in between also depends on your hair protein's amino acid sequence and the three-dimensional shape it confers (**FIGURE 2-41**). This

Q Why do some people have curly hair and others have straight hair?

amino acid sequence is something you're born with (that is, it's genetically determined). The chains are more or less coiled, depending on the extent of covalent and hydrogen bonding between different parts of the coil. Many hair salons make use of the ability to alter covalent bonds to change hair texture semi-permanently. They are able to do this in three simple steps. First, the bonds are broken chemically. Second, the hair is wrapped around curlers to hold the polypeptide chains in a different position. And third, chemicals are put on the hair to create new covalent bonds between parts of the polypeptide chains. The hair thus becomes locked in a new position. (New hair will continue to grow with its genetically determined texture, of course, requiring the procedure to be repeated regularly.)

TAKE-HOME MESSAGE 2·17

The particular amino acid sequence of a protein determines how it folds into a particular three-dimensional shape. This shape determines many of the protein's features, such as which molecules it will interact with. When a protein's shape is deformed, the protein usually loses its ability to function.

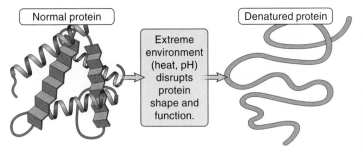

FIGURE 2-40 **Denaturation.** When proteins are unfolded, they lose their function.

Atoms   Water   Carbohydrates   Lipids

Enzymes are proteins that initiate and speed up chemical reactions.

Protein shape is particularly critical in **enzymes,** molecules that help initiate and accelerate the chemical reactions in our bodies. Enzymes emerge unchanged—in their original form—when the reaction is complete and thus can be used again and again. Here's how they work.

Think of an enzyme as a big piece of popcorn. Its tertiary or quaternary structure gives it a complex shape with lots of nooks and crannies. Within one of those nooks is a small area called the "active site" (**FIGURE 2-42**). Based on the chemical properties of the atoms lining this pocket, the active site provides a place for the participants in a chemical reaction, the reactants or **substrate** molecules, to nestle briefly.

Enzymes are very choosy: they bind only with their appropriate substrate molecules, much like a lock that can be opened with only one key (see Figure 2-42). The exposed atoms in the active site have electrical charges that attract rather than repel the substrate molecules, and only the substrate molecules can fit into the active-site groove.

Once the substrate is bound to the active site, a reaction can take place—and usually does so very quickly. An enzyme can help to bring about the reaction in a variety of ways. These include:

1. Stressing, bending, or stretching critical chemical bonds, increasing the likelihood of their breaking.

2. Directly participating in the reaction, perhaps temporarily sharing one or more electrons with the substrate molecule, thereby giving it chemical features that increase its ability to make or break other bonds.

3. Creating a "micro-habitat" that favors the reaction. For instance, the active site might be a water-free, non-polar environment, or it might have a slightly higher or lower pH than the surrounding fluid. Both of these slight alterations might increase the likelihood that a particular reaction occurs.

4. Simply orienting or holding substrate molecules in place so that they can be modified.

Sometimes a protein "word" is misspelled—that is, the sequence of amino acids is incorrect. If an enzyme is altered even slightly, the active site may change and the enzyme no longer functions. Slightly modified, non-functioning enzymes

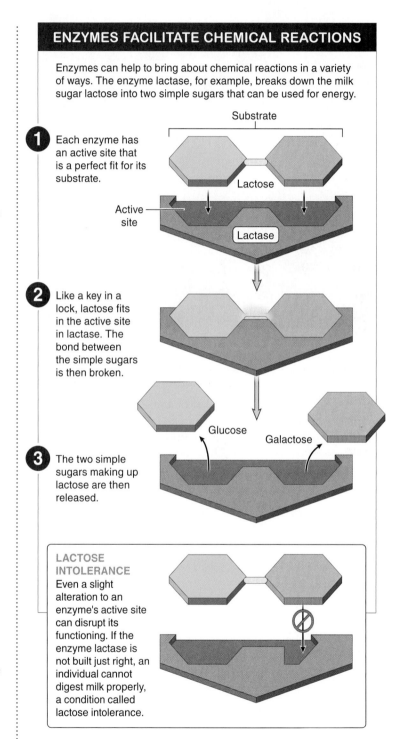

ENZYMES FACILITATE CHEMICAL REACTIONS

Enzymes can help to bring about chemical reactions in a variety of ways. The enzyme lactase, for example, breaks down the milk sugar lactose into two simple sugars that can be used for energy.

1. Each enzyme has an active site that is a perfect fit for its substrate.

Substrate
Lactose
Active site
Lactase

2. Like a key in a lock, lactose fits in the active site in lactase. The bond between the simple sugars is then broken.

3. The two simple sugars making up lactose are then released.

Glucose
Galactose

LACTOSE INTOLERANCE
Even a slight alteration to an enzyme's active site can disrupt its functioning. If the enzyme lactase is not built just right, an individual cannot digest milk properly, a condition called lactose intolerance.

FIGURE 2.42 Lock and key. Enzymes are *very* specific about which molecules and reactions they will catalyze.

Proteins   Nucleic Acids

Without a functioning version of the enzyme lactase, some people are unable to break down the disaccharide lactose in milk.

are responsible for a large number of diseases and physiological problems; an example is the inability to break down the amino acid phenylalanine (in a condition known as phenylketonuria).

Q *Why do some adults get sick when they drink milk?*

One protein "misspelling" is responsible for the condition called lactose intolerance. Normally, during digestion, the lactose in milk is broken down into its component parts, glucose and galactose (see Figure 2-42). These simple sugars are then used for energy. But some people are unable to break the bond linking the two simple sugars because they lack a functioning version of the enzyme lactase that assists in this process. Consequently, the lactose passes through their stomach and small intestine undigested. Then, when it reaches the large intestine, bacteria living there consume the lactose. The problem is that, as they break down the lactose, they produce some carbon dioxide and other gases. These gases are trapped in the intestine and lead to severe discomfort.

These unpleasant symptoms can be avoided by not consuming milk, cheese, yogurt, ice cream, or any other dairy products, but they can also be avoided by taking a pill containing the enzyme lactase. It doesn't matter how the enzyme gets into your digestive system; as long as it's there the lactose in the milk can be broken down.

TAKE-HOME MESSAGE 2·18

Enzymes are proteins that help initiate and speed up chemical reactions. They aren't permanently altered in the process, but rather can be used again and again.

Atoms   Water   Carbohydrates   Lipids

6 Nucleic acids store the information on how to build and run a body.

Three turns of the DNA double helix (from a scanning tunneling micrograph).

2·19

Nucleic acids are macromolecules that store information.

We have examined three of life's macromolecules: carbohydrates, lipids, and proteins. We turn our attention now to the fourth: **nucleic acids,** macromolecules that store information and are made up of individual units called **nucleotides.** All nucleotides have three components: a molecule of sugar, a phosphate group (containing a phosphorus atom bound to four oxygen atoms), and a nitrogen–containing molecule (**FIGURE 2-43**).

There are two types of nucleic acids: **deoxyribonucleic acid (DNA)** and **ribonucleic acid (RNA).** Both play central roles in directing the production of proteins in living organisms, and by doing so play a central role in determining all of the inherited characteristics of an individual. In both types of nucleic acids, the molecule has the same type of backbone: a sugar molecule attached to a phosphate group attached to another sugar, then another phosphate, and so on. Attached to each sugar is one of the nitrogen–containing molecules called DNA **bases** (so named because of their chemical structure).

FIGURE 2-43 **The molecules that carry genetic information.** The structure of nucleic acids.

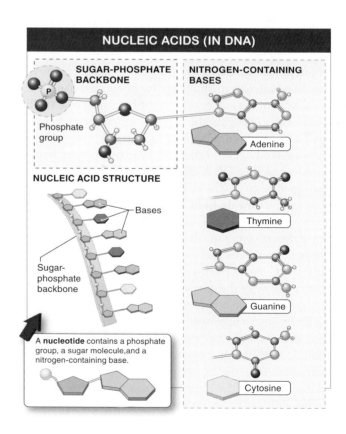

NUCLEIC ACIDS (IN DNA)

SUGAR-PHOSPHATE BACKBONE

Phosphate group

NUCLEIC ACID STRUCTURE

Bases

Sugar-phosphate backbone

A **nucleotide** contains a phosphate group, a sugar molecule, and a nitrogen-containing base.

NITROGEN-CONTAINING BASES

Adenine

Thymine

Guanine

Cytosine

Proteins   ▼   Nucleic Acids

A 10-unit nucleic acid therefore would have 10 bases, one attached to each sugar within the sugar-phosphate-sugar-phosphate backbone. But the base attached to each sugar is not always the same. It can be one of several different bases. For this reason, a nucleic acid is often described by the sequence of bases attached to the sugar-phosphate-sugar-phosphate backbone.

Nucleic acids are able to store information by varying which base is attached at each position in the molecule. At each position in a molecule of DNA, for example, the base can be any one of four possible bases: adenine (A), thymine (T), guanine (G), or cytosine (C). Just as the meaning of a sentence is determined by which letters are strung together, the information in a molecule of DNA is determined by its sequence of bases. One molecule may have the sequence adenine, adenine, adenine, guanine, cytosine, thymine, guanine—abbreviated as AAAGCTG. Another molecule may have the sequence CGATTACCCGAT. Because the information differs in each case, so, too, does the protein for which the sequence codes, as we'll see.

TAKE-HOME MESSAGE 2·19

The nucleic acids DNA and RNA are macromolecules that store information in their unique sequences of nucleotides, their building-block molecules. Both nucleic acids play central roles in directing protein production in organisms.

2·20

DNA holds the genetic information to build an organism.

A molecule of DNA has two strands, each a sugar-phosphate-sugar-phosphate backbone with a base sticking out from each sugar molecule. The two strands wrap around each other, each turning in a spiral. Although each strand has its own sugar-phosphate-sugar-phosphate backbone and sequence of bases, the two strands are connected by the bases protruding from them.

You can picture a molecule of DNA as a ladder. The two sugar-phosphate-sugar-phosphate backbones are like the long vertical sides of the ladder that give it height. A base sticking out represents a rung on the ladder. Or, more accurately, half a rung. The bases protruding from each strand meet in the center and bind to each other (via hydrogen bonds), holding the ladder together. DNA differs from a ladder slightly, in that it has a gradual twist. The two spiraling strands together are said to form a **double helix** (FIGURE 2-44).

The two intertwining spirals fit together because only two combinations of bases pair up together. The base A always pairs with T, and C always pairs with G. Consequently, if the base sequence of one of the spirals is CCCCTTAGGAACC, the base sequence of the other must be GGGGAATCCTTGG. That is why researchers working on the Human Genome Project describe only one sequence of nucleotides when presenting a DNA sequence—even though that DNA is double-stranded in our bodies. With that one sequence, we can infer the identity of the bases in the complementary sequence and thus we know the exact structure of the nucleic acid.

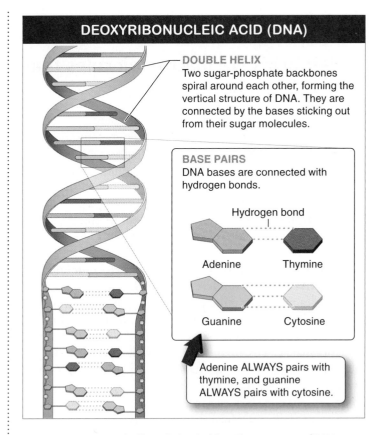

DEOXYRIBONUCLEIC ACID (DNA)

DOUBLE HELIX
Two sugar-phosphate backbones spiral around each other, forming the vertical structure of DNA. They are connected by the bases sticking out from their sugar molecules.

BASE PAIRS
DNA bases are connected with hydrogen bonds.

Hydrogen bond

Adenine   Thymine

Guanine   Cytosine

Adenine ALWAYS pairs with thymine, and guanine ALWAYS pairs with cytosine.

FIGURE 2-44 **A gradually twisting ladder.** The structure of DNA.

Atoms   Water   Carbohydrates   Lipids

The sequences of bases containing the information about how to produce a particular protein have anywhere from a hundred to several thousand bases. In a human, all of the DNA in a cell, containing all of the instructions for every protein that a human must produce, contains about 3 billion base pairs. This DNA is generally in the nucleus of a cell.

2·21

RNA is a universal translator, reading DNA and directing protein production.

The process of building a protein from a DNA sequence is not a direct one. Rather, it incorporates a middleman, RNA, that is also a nucleic acid (FIGURE 2-45). Segments of the DNA are read off, directing the production of short strips of RNA that contain the information, taken from the DNA, about the amino acid sequence in a protein. The RNA moves to another part of the cell and then directs the piecing together of amino acids into a three-dimensional protein. We explore this in greater detail in Chapter 5.

RNA differs from DNA in three important ways. First, the sugar molecule of the sugar-phosphate-sugar-phosphate backbone differs slightly, containing an extra atom of oxygen. Second, RNA is single stranded. The sugar-phosphate-sugar-phosphate backbone is still there, as are the bases that protrude from each sugar. The bases, however, do not bind with anything else—do not form base pairs with another RNA strand. And third, while RNA has the bases A, G, and C, it replaces the thymine with a similar base called uracil (U).

Whether we're looking at the nucleotides that make up RNA and DNA or the lipids used to build sex hormones and cell membranes, we see a recurring theme in the construction of biological macromolecules: from relatively simple sets of building blocks linked together, infinitely complex molecules can be formed. Complex webs of one simple sugar, bonded together as glycogen, for instance, provide fuel for organisms. Similarly, sequences of amino acids of 20 different types, joined together, specify the structure of all of the proteins found in every species on earth.

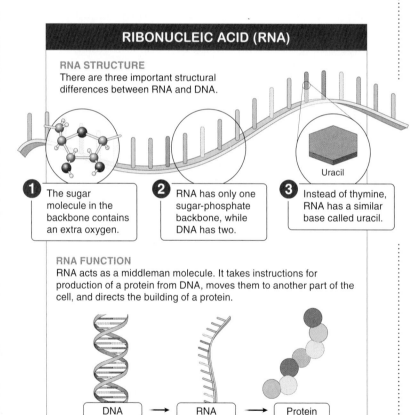

RIBONUCLEIC ACID (RNA)

RNA STRUCTURE
There are three important structural differences between RNA and DNA.

Uracil

1 The sugar molecule in the backbone contains an extra oxygen.

2 RNA has only one sugar-phosphate backbone, while DNA has two.

3 Instead of thymine, RNA has a similar base called uracil.

RNA FUNCTION
RNA acts as a middleman molecule. It takes instructions for production of a protein from DNA, moves them to another part of the cell, and directs the building of a protein.

DNA → RNA → Protein

FIGURE 2-45 The middleman between DNA and protein. The structure of RNA.

Knowledge You Can Use

Did you know? Melt-in-your-mouth chocolate may not be such a sweet idea.

Food chemists have figured out how to make chocolate that melts in your mouth. Is that a good thing?

Q: Why are some fats "liquidy," like oil, and others solid? The less saturated a fat is, the more "liquidy" it is at room temperature. Most animal fats are saturated and are solid at room temperature. Best example: butter. Most plant fats are polyunsaturated fats and are liquid at room temperature. Best example: vegetable oils.

Q: Can oils be made more solid? It's possible to increase the saturation of plant fats. Just heat them up and pass hydrogen bubbles through the liquid. In creating partially hydrogenated plant oils, this process reduces the number of carbon–carbon double bonds and **makes the oil more solid.** (It's easy, it's cheaper than just using butter, and it increases foods' shelf-life.)

Q: Does that improve their taste? By precisely controlling the level of saturation in plant fats, **it is possible to create foods** that are solid but have such a low melting point **that they quickly melt on contact with the warmth of your mouth.** This seems great, but . . .

Q: Is there a downside? The saturation of vegetable fats creates trans fats, due to the position taken by the newly added hydrogen atoms in the molecule. Trans fats increase levels of LDL ("bad") cholesterol and decrease HDL ("good") cholesterol, narrowing blood vessel walls and increasing the risk of heart disease and strokes.

Conclusion: Partially hydrogenated vegetable oils can give food a perfect texture and a pleasing feel in your mouth. But the **creaminess comes with a high cost** when it comes to your health.

① Atoms form molecules through bonding.

 An atom is the smallest unit into which material can be divided without losing its essential properties. Atoms are made up of protons and neutrons in the nucleus and electrons that circle far around that nucleus. Molecules are atoms linked together.

② Water has features that enable it to support all life.

 Water molecules easily form hydrogen bonds, giving water great cohesiveness and the ability to resist temperature changes, and making it a versatile solvent. The pH of a fluid is a measure of how acidic or basic a solution is.

③ Carbohydrates are fuel for living machines

 Carbohydrates, made up of carbon, oxygen, and hydrogen, are the primary fuel for all cellular machinery and also form much of the structure of cells. The simplest carbohydrates, including glucose, are called monosaccharides or simple sugars. Multiple simple carbohydrates sometimes link together into more complex carbohydrates, including starch, which is the primary form of energy storage in plants, and glycogen, which is a primary form of energy storage in animals. Some carbohydrates are not digestible by humans.

④ Lipids store energy for a rainy day.

 Lipids are macromolecules—made up primarily of carbon, hydrogen, and oxygen—that are insoluble in water. Lipids are important for energy storage, hormones, and membrane structure. Their breakdown of dietary fats releases more energy per gram than other macromolecules.

⑤ Proteins are body-building molecules.

 Cells and tissues are primarily built from proteins, sequences of amino acids that fold into complex, three-dimensional shapes. The atoms, especially nitrogen, present in the plant and animal proteins that an organism eats are essential to the organism's growth and repair. The amino acid sequence of a protein determines how it folds into a three-dimensional shape. This shape determines many of the protein's features, such as which molecules it will interact with. Enzymes are proteins that help initiate and speed up chemical reactions. They aren't permanently altered in the process.

⑥ Nucleic acids store the information on how to build and run a body.

 The nucleic acids DNA and RNA are macromolecules that store information by having unique sequences of molecules. Both play central roles in directing protein production in organisms. The sequence of nucleotide bases contains the information about how to produce a particular protein. RNA acts like a universal translator of the genetic code into proteins. It reads DNA sequences and directs the production of a sequence of amino acids.

BIG IDEAS IN CHEMISTRY

KEY TERMS

acid, p. 48
amino acid, p. 62
amino group, p. 63
atom, p. 37
atomic mass, p. 37
atomic number, p. 37
base, p. 48
base (of DNA), p. 69
buffer, p. 49
carbohydrate, p. 50
carboxyl group, p. 62
cellulose, p. 54
chitin, p. 54
cholesterol, p. 60
complex carbohydrate, p. 53
compound, p. 41

covalent bond, p. 40
denaturation, p. 66
deoxyribonucleic acid (DNA), p. 69
disaccharide, p. 53
double bond, p. 40
double helix, p. 70
electron, p. 37
element, p. 36
enzyme, pp. 62, 67
fatty acid, p. 57
glycerol, p. 57
glycogen, p. 52
hydrogen bond, p. 41
hydrophilic, p. 56
hydrophobic, p. 56

ion, p. 39
ionic bond, p. 41
lipid, p. 56
macromolecule, p. 50
mass, p. 37
molecule, p. 40
monosaccharide, p. 51
neutron, p. 37
nucleic acid, p. 69
nucleotide, p. 69
nucleus, p. 37
peptide bond, p. 65
pH, p. 48
phospholipid, p. 61
polysaccharide, p. 53
primary structure, p. 65

protein, p. 62
proton, p. 37
quaternary structure, p. 65
ribonucleic acid (RNA), p. 69
saturated fat, p. 59
secondary structure, p. 65
simple sugar, p. 51
starch, p. 53
sterol, p. 60
substrate, p. 67
tertiary structure, p. 65
trans fat, p. 59
triglyceride, p. 57
unsaturated fat, p. 59
wax, p. 61

73

1. The atomic number of carbon is 6. Its nucleus must contain:
   a) 6 neutrons and 6 protons.
   b) 3 protons and 3 neutrons.
   c) 6 neutrons and no electrons.
   d) 6 protons and no electrons.
   e) 6 protons and 6 electrons.

2. The second orbital shell of an atom can hold ___ electrons.
   a) 2   d) 6
   b) 3   e) 8
   c) 4

3. A covalent bond is formed when:
   a) two non-polar molecules associate with each other in a polar environment.
   b) a positively charged particle is attracted to a negatively charged particle.
   c) one atom gives up electrons to another atom.
   d) two atoms share electrons.
   e) two polar molecules associate with each other in a non-polar environment.

4. When you put a piece of chocolate on your tongue, your brain registers a sensation of sweetness. What aspect of molecules is responsible for their having a particular taste?
   a) the total number of protons in the molecule
   b) the number of hydrogen bonds in the molecule
   c) the total number of electron shells in the atoms of the molecule
   d) the ratio of covalent bonds to ionic bonds joining the atoms of the molecule
   e) the molecule's shape

5. Which of the following phenomena is most likely due to the high cohesiveness of water?
   a) Lakes and rivers freeze from the top down, not the bottom up.
   b) The "Jesus lizard" can run across the surface of liquid water for short distances.
   c) Adding salt to snow makes it melt.
   d) The temperature of Santa Monica Bay, off the coast of Los Angeles, fluctuates less than the air temperature throughout the year.
   e) All of the above are due to the cohesiveness of water.

6. Water can absorb and store a large amount of heat while increasing only a few degrees in temperature. Why?
   a) The heat must first be used to break the hydrogen bonds rather than raise the temperature.
   b) The heat must first be used to break the ionic bonds rather than raise the temperature.
   c) The heat must first be used to break the covalent bonds rather than raise the temperature.
   d) An increase in temperature causes an increase in adhesion of the water.
   e) An increase in temperature causes an increase in cohesion of the water.

7. A chemical compound that releases H^+ into a solution is called:
   a) a proton.   d) a hydroxide ion.
   b) a base.   e) a hydrogen ion.
   c) an acid.

8. Which of the following foods is not a significant source of complex carbohydrates?
   a) fresh fruit
   b) rice
   c) pasta
   d) oatmeal
   e) All of the above are significant sources of complex carbohydrates.

9. Sucrose (table sugar) and lactose (the sugar found in milk) are examples of:
   a) naturally occurring enzymes.
   b) simple sugars.
   c) monosaccharides.
   d) disaccharides.
   e) polyunsaccharide sugars.

10. Which of the following statements about starch is incorrect?
    a) Starch is the primary form of energy storage in plants.
    b) Starch consists of a hundred or more glucose molecules joined together in a line.
    c) Starch tastes sweet because it is made from glucose.
    d) Starch is a polysaccharide.
    e) All of the above statements about starch are correct.

11. Which of the following statements about fiber is incorrect?
    a) Dietary fiber reduces the risk of colon cancer.
    b) Fiber in the diet slows the passage of food through the intestines.
    c) Humans are unable to extract any caloric value from fiber.
    d) The cellulose of celery stalks and lettuce leaves is fiber.
    e) Fiber passing through the digestive system scrapes the wall of the digestive tract, stimulating mucus secretion and aiding in the digestion of other molecules.

12. In an unusually detailed dissection of your dinner, you isolate an unknown fatty acid. It is a liquid at room temperature (that is, has a low melting point) and contains carbon-carbon double bonds. What are you most likely to be eating?
    a) a plant   d) a chicken
    b) a cow   e) a lamb
    c) a pig

13. An unsaturated fatty acid is one in which:
    a) carbon-carbon double bonds are present in the hydrocarbon chain.
    b) the hydrocarbon chain has an odd number of carbons.
    c) the hydrocarbon chain has an even number of carbons.
    d) no carbon-carbon double bonds are present in the hydrocarbon chain.
    e) not all of the carbons in the hydrocarbon chain are bonded to hydrogen atoms.

14. Which statement about phospholipids is incorrect?
    a) Because their phosphate groups repel each other, they are used as organisms' chief form of short-term energy.
    b) They are hydrophobic at one end.
    c) They are hydrophilic at one end.
    d) They are a major constituent of cell membranes.
    e) They contain glycerol linked to fatty acids.

15. Proteins are an essential component of a healthy diet for humans (and other animals). Their most common purpose is to serve as:
    a) raw material for growth.
    b) fuel for running the body.
    c) organic precursors for enzyme construction.
    d) long-term energy storage.
    e) inorganic precursors for enzyme construction.

16. Dietary proteins:
    a) are considered "complete" only if they contain the eight essential amino acids required by humans.
    b) consist of all 20 amino acids required in the human body.
    c) are considered "complete" only if they contain the 12 non-essential amino acids required by humans.
    d) are nutritionally identical, since all are broken down into their constituent amino acids in the digestive system.
    e) can be obtained from animal sources but not plant sources.

17. The primary structure of proteins is often described as amino acids connected like beads on a string. In this same vein, which of the following images best describes a protein's quaternary structure?
    a) threads in a cloth
    b) needles in a haystack
    c) rungs on a ladder
    d) links on a chain
    e) coils in a spring

18. Which of the following statements about enzymes is incorrect?
    a) Enzymes can initiate chemical reactions.
    b) Enzymes speed up chemical reactions.
    c) Enzymes often induce conformational changes in the substrates that they bind.
    d) Enzymes contain an active site for binding of particular substrates.
    e) Enzymes undergo a permanent change during the reactions they promote.

19. Which of the following nucleotide bases are present in equal amounts in DNA?
    a) adenine and cytosine
    b) thymine and guanine
    c) adenine and guanine
    d) thymine and cytosine
    e) adenine and thymine

20. Which type of macromolecule contains an organism's genetic information?
    a) polysaccharide
    b) monosaccharide
    c) fatty acid
    d) DNA
    e) phospholipid

21. All of the following are features of RNA except:
    a) a sugar-phosphate-sugar-phosphate backbone.
    b) a coiled double-stranded structure.
    c) a different type of sugar than is found in DNA.
    d) an ability to direct protein production.
    e) a different type of base than is found in DNA.

SHORT-ANSWER QUESTIONS

1. An atom of sodium has an atomic number of 11 and an atom of chlorine has an atomic number of 17.
   a) Indicate the number of electrons that are present in the outermost electron shell of each atom.
   b) Predict what type of bond would form between these atoms, and explain why.

2. Explain how hydrogen bonding plays a part in each of the unique properties of water and, in essence, allows for life on earth.

3. There are many cleaning products on the market that advertise their ability to get rid of stubborn, oily dirt. Why is water alone not sufficient for removing oily dirt?

4. Some who are watching what they eat may opt for a salad when they are dining out. A dinner salad may include lettuce, celery, strawberries, walnuts, chicken, and a salad dressing of oil and vinegar. Explain which category of major macromolecules each of these individual salad components falls into. Then explain the general purpose or function of each major macromolecule.

See Appendix for answers. For additional study questions, go to www.prep-u.com.

3

Cells

The smallest part of you

① What is a cell?

Human cell, packed with organelles.

3·1

All organisms are made of cells.

Where do we begin if we want to understand how organisms work? Given their complexity, this task can be daunting. Fortunately, as with most complex things, a strategy of divide and conquer comes in handy. In the case of organisms, whether we are studying a creature as small as a flea or as large as an elephant or giant sequoia, they can be broken down into smaller units that are more easily studied and understood (**FIGURE 3-1**). The most basic unit of any organism is the **cell,** the smallest unit of life that can function independently and perform all the necessary functions of life, including reproducing itself. Understanding cell structure and function is the basis for our understanding of how complex organisms are organized.

The term "cell" was first used in the mid-1600s by Robert Hooke, a British scientist also known for his contributions to philosophy, physics, and architecture. When he was made Curator of Experiments for the Royal Society of London, Hooke suddenly had access to many of the first microscopes available, and he began to examine everything he could get his hands on. Because Hooke thought the close-up views of a very thin piece of cork resembled a mass of small, empty rooms, he named these compartments *cellulae,* Latin for "small rooms."

Today, we know that a cell is a three-dimensional structure, like a fluid-filled balloon, in which many of the essential chemical reactions of life take place (such as the breakdown of carbohydrates for energy, the modification of cholesterol to create testosterone and estrogen, and the translation of the genetic code into protein production). Generally, these reactions involve transporting raw materials and fuel into the cell and exporting finished materials and waste products out of the cell. In addition, nearly all cells contain DNA (deoxyribonucleic acid), a molecule that contains the information that directs the chemical reactions in a cell, the formation of various cellular products within the cell, and the cell's ability to reproduce itself. (A few types of cells, including mammalian red blood cells and cells called "sieve tube elements" that form part of the plant circulatory system, lose their nuclei after they are created and are unable to divide.) We explore all of these features of cell functioning in this and the next three chapters.

To see a cell, you don't have to work in a lab or use a microscope. Just open your refrigerator. Chances are you've got a dozen or so visible cells in there: eggs. Although most cells are too small to see with the naked eye, there are a few exceptions, including hens' eggs from the supermarket. As long as they are unfertilized, which most store-bought hens' eggs are, each egg tends to contain just one cell. The ostrich egg, weighing about three pounds, contains the largest of all animal cells. (It should be noted, however, that by the time the ostrich lays a fertilized egg, the embryo inside has already gone through multiple divisions.) In addition to being among the largest cells around, eggs are also the most valuable. Almas caviar, eggs from the beluga sturgeon, sells for nearly $700 per ounce. This value is exceeded only by that

FIGURE 3-1 **What do these diverse organisms have in common? Cells.**

of human eggs, which currently fetch as much as $25,000 for a dozen or so eggs on the open market (**FIGURE 3-2**). (Human sperm cells command only about a penny per 20,000 cells!)

Most cells are much smaller than hens' eggs and ostrich eggs. Consider that, at this very moment, there are probably more than seven billion bacteria in your mouth—even if you just brushed your teeth! This is more than the number of people on earth. It is possible to squeeze so many bacteria in there because most cells (not just bacteria) are really, really tiny—so tiny that 2,000 red blood cells, lined up end to end, would just extend across a dime.

After sufficient improvements were made to early microscopes in the 19th century, the central role of the cell in biology could be understood. As scientists began putting everything they studied under their microscopes, the importance and universality of the cell finally dawned on them. By the 1830s, scientists realized that all plants and animals were made entirely from cells. Subsequent studies revealed that every cell seemed to arise from the division of another cell. You, for example, are made up of at least 60 trillion cells, all of which came from just one cell: the single fertilized egg produced when an egg cell from your mother was fertilized by a sperm cell from your father.

The facts that (1) all living organisms are made up of one or more cells and (2) all cells arise from other, pre-existing cells are the foundations of **cell theory,** one of the unifying theories in biology, and one that is universally accepted by all biologists. As we will see in Chapter 10, the origin of life on earth was a one-time deviation from cell theory: the first cells on earth probably originated from free-floating molecules in the oceans early in the earth's history (about 3.5 billion years ago). Since that time, however, all cells and thus all life have been produced as a continuous line of cells, originating from these initial cells.

In this chapter, we investigate the two different kinds of cells that make up all of the organisms on earth, the processes by which cells control how materials move into and out of the cell, and how cells communicate with each other. We also explore some of the important landmarks found in many cells. We look, too, at the specialized roles these structures play in a variety of cellular functions and learn about some of the health consequences that occur when they malfunction.

TAKE-HOME MESSAGE 3·1

The most basic unit of any organism is the cell, the smallest unit of life that can function independently and perform all of the necessary functions of life, including reproducing itself. All living organisms are made up of one or more cells, and all cells arise from other, pre-existing cells.

Hummingbird egg: 0.02 ounce
Elephant bird (extinct) egg: two gallons

Beluga sturgeon eggs: $700 per ounce

Human eggs: Thousands of dollars per egg

FIGURE 3-2 **Not all cells are tiny.** And some cells are extremely valuable!

Eggs are the largest and most expensive cells in the world!

79

Prokaryotic cells are structurally simple, but there are many types of them.

Although there are millions of diverse species on earth, billions of unique organisms alive at any time, and trillions of different cells in many of those organisms, every cell on earth falls into one of two basic categories.

A **eukaryotic cell** (from the Greek for "good" and "kernel") has a central control structure called a nucleus, which contains the cell's DNA. Organisms composed of eukaryotic cells are called **eukaryotes.**

A **prokaryotic cell** (from the Greek for "before" and "kernel") does not have a nucleus; its DNA simply resides in the middle of the cell. An organism consisting of a prokaryotic cell is called a **prokaryote.**

We begin our study of the cell by exploring the prokaryotes. Prokaryotes were the first cells on earth, making their appearance about 3.5 billion years ago, and, for a long time (1.5 billion years), they had the planet to themselves.

All prokaryotes are one-celled organisms and are thus invisible to the naked eye. Prokaryotes have four basic structural features (**FIGURE 3-3**).

1. A **plasma membrane** encompasses the cell (and is sometimes simply called the "cell membrane"). For this reason, anything inside the plasma membrane is referred to as "intracellular," and everything outside the plasma membrane is "extracellular."

2. The **cytoplasm** is the jelly-like fluid that fills the inside of the cell.

3. **Ribosomes** are little granular bodies where proteins are made; thousands of them are scattered throughout the cytoplasm.

4. Each prokaryote has one or more circular loops or linear strands of DNA.

In addition to the characteristics common to all prokaryotes, some prokaryotes have additional, unique features. Many have a rigid **cell wall,** for example, that protects and gives shape to the cell. Some have a slimy, sugary capsule as their outermost layer. This sticky outer coat provides protection and enhances the prokaryotes' ability to anchor themselves in place when necessary.

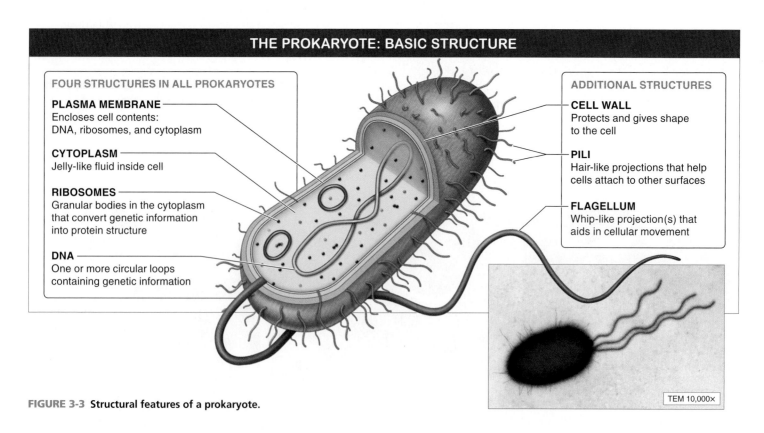

THE PROKARYOTE: BASIC STRUCTURE

FOUR STRUCTURES IN ALL PROKARYOTES

PLASMA MEMBRANE
Encloses cell contents:
DNA, ribosomes, and cytoplasm

CYTOPLASM
Jelly-like fluid inside cell

RIBOSOMES
Granular bodies in the cytoplasm
that convert genetic information
into protein structure

DNA
One or more circular loops
containing genetic information

ADDITIONAL STRUCTURES

CELL WALL
Protects and gives shape
to the cell

PILI
Hair-like projections that help
cells attach to other surfaces

FLAGELLUM
Whip-like projection(s) that
aids in cellular movement

TEM 10,000×

FIGURE 3-3 Structural features of a prokaryote.

The Cell   Cell Membranes   Crossing the Membrane   Cell Connections

Many prokaryotes have a **flagellum** (*pl.* **flagella**), a long, thin, whip-like projection that rotates like a propeller and moves the cell through the medium in which it lives. Other appendages include **pili** (*sing.* **pilus**), much thinner, hair-like projections that help prokaryotes attach to surfaces.

From our human perspective, it is easy to underestimate the prokaryotes. Although they are smaller, evolutionarily older, and structurally more simple than eukaryotes, prokaryotes are fantastically diverse metabolically (i.e., in the way they break down and build up molecules). Among many other innovations, bacteria can fuel their activities in the presence or absence of oxygen, using almost any energy source on earth, depending on the type of bacteria—from the sulfur in deep-sea hydrothermal vents to hydrogen to the sun.

You may think you've never encountered a prokaryote, but you are already familiar with the largest group of prokaryotes: the bacteria. All bacteria are prokaryotes, from those such as *Escherichia coli* (*E. coli*), which live in your intestine and help your body make some essential vitamins, to those responsible for illness, such as *Streptococcus pyogenes,* which causes strep throat. Another recently discovered group of organisms, the archaea, are also prokaryotes.

TAKE-HOME MESSAGE 3·2

Every cell on earth is either a eukaryotic or a prokaryotic cell. Prokaryotes, which have no nucleus, were the first cells on earth. They are all single-celled organisms. Prokaryotes include the bacteria and archaea and, as a group, are characterized by tremendous metabolic diversity.

3·3

Eukaryotic cells have compartments with specialized functions.

Eukaryotes showed up about 1.5 billion years after prokaryotes, and in the 2 billion years that they have been on earth, they have evolved into some of the most dramatic and interesting creatures, such as platypuses, dolphins, giant

sequoias, and the Venus fly trap. Because all prokaryotes are single-celled and thus invisible to the naked eye, every organism that we see around us is a eukaryotic organism. (**FIGURE 3-4**). All fungi, plants, and animals are eukaryotes, for

FIGURE 3-4 Diversity of the eukaryotes. Every organism that we can see without magnification is a eukaryotic organism.

Nine Cell Landmarks

TYPICAL EUKARYOTIC CELL FEATURES
• DNA contained in nucleus
• Internal structures organized into compartments
• Larger than prokaryotes—usually 10 times bigger
• Cytoplasm contains specialized structures called organelles

TYPICAL PROKARYOTIC CELL FEATURES
• No nucleus—DNA is in the cytoplasm
• Internal structures not organized into compartments
• Much smaller than eukaryotes

Compartments   Nucleus   Organelles

TEM 6,000×

TEM 10,000×

FIGURE 3-5 **Comparison of eukaryotic and prokaryotic cells.**

instance. Not all eukaryotes are multicellular, however. There is a huge group of eukaryotes, called the Protista (or protists), nearly all of which are single-celled organisms visible only with a microscope.

The chief distinguishing feature of eukaryotic cells is the presence of a **nucleus,** a membrane-enclosed structure that contains linear strands of DNA. In addition to a nucleus, eukaryotic cells usually contain in their cytoplasm several other specialized structures, called **organelles,** many of which are enclosed separately within their own lipid membranes. Eukaryotic cells are also about 10 times larger than prokaryotes. All of these physical differences make it easy to distinguish eukaryotes from prokaryotes under a microscope (**FIGURE 3-5**).

Organelles enable eukaryotic cells to do many things that a prokaryotic cell cannot do. Most important, the creation of physically separate compartments means that there can be distinct areas within the same cell in which different chemical reactions can occur simultaneously. In the non-compartmentalized interior of a prokaryotic cell, random molecular movements quickly blend the chemicals throughout the cell, reducing the ease with which different reactions can occur simultaneously.

FIGURE 3-6 illustrates a generalized animal cell and a generalized plant cell. Because they share a common,

eukaryotic ancestor, they have much in common. Both can have a plasma membrane, nucleus, cytoskeleton, and ribosomes, and a host of organelles, including rough and smooth endoplasmic membranes, Golgi apparatus, and mitochondria. Animal cells have centrioles, which are not present in most plant cells. Plant cells have a rigid cell wall (as do fungi and many protists) and chloroplasts (also found in some protists). Plants also have a vacuole, a large central chamber (only occasionally found in animal cells). We explore each of these plant and animal organelles in detail later in this chapter.

When you compare a complex eukaryotic cell with the structurally simple prokaryotic cell, it's hard not to wonder how eukaryotic cells came about. How did they get filled up with so many organelles? We can't go back 2 billion years to watch the initial evolution of eukaryotic cells, but we can speculate about how it might have occurred. One very appealing idea, called the **endosymbiosis theory,** has been developed to explain the presence of two organelles in eukaryotes: chloroplasts in plants and algae, and mitochondria in plants and animals. Chloroplasts help plants and algae convert sunlight into a more usable form of energy. Mitochondria help plants and animals harness the energy stored in food molecules. (Chapter 4, on energy, explains the details of both of these processes.)

According to the theory of endosymbiosis, two different types of prokaryotes may have set up close partnerships with each

The Cell   Cell Membranes   Crossing the Membrane   Cell Connections

THE ANIMAL CELL: BASIC STRUCTURE

THE PLANT CELL: BASIC STRUCTURE

STRUCTURES FOUND IN BOTH CELLS

Nucleus
Plasma membrane
Ribosomes
Mitochondria
Rough endoplasmic reticulum
Smooth endoplasmic reticulum
Cytoplasm
Cytoskeleton
Golgi apparatus
Lysosome

STRUCTURE NOT FOUND IN PLANT CELLS

Centriole

STRUCTURES NOT FOUND IN ANIMAL CELLS

Chloroplast
Cell wall
Vacuole (occasionally found in animal cells)

TEM 3,500×

TEM 1,500×

FIGURE 3-6 Structures found in animal and plant cells.

other. For example, some small prokaryotes capable of performing photosynthesis (the process by which plant cells capture light energy from the sun and transform it into the chemical energy stored in food molecules) may have come to live inside a larger "host" prokaryote. The photosynthetic "boarder" may have made some of the energy from its photosynthesis available to the host.

After a long while, the two cells may have become more and more dependent on each other until neither cell could live without the other (they became "symbiotic"), and they became a single, more complex organism. Eventually, the photosynthetic prokaryote evolved into a **chloroplast,** the organelle in plant cells in which photosynthesis occurs. A similar scenario might explain how a prokaryote unusually efficient at converting food and oxygen into easily usable energy took up residence inside another, host prokaryote and evolved into a **mitochondrion,** the organelle in plant and

Q Humans, deep down, may be part bacteria. How can that be?

animal cells that converts the energy stored in food into a form usable by the cell (**FIGURE 3-7**).

The idea of endosymbiosis is supported by several observations.

1. Chloroplasts and mitochondria are similar in size to prokaryotic cells.

2. Chloroplasts and mitochondria have small amounts of circular DNA, similar to the circular DNA in prokaryotes and in contrast to the linear DNA strands found in a eukaryote's nucleus.

3. Chloroplasts and mitochondria divide by splitting (fission), just like prokaryotes.

4. Chloroplasts and mitochondria have internal structures called ribosomes that are similar to those found in bacteria.

In addition to the theory of endosymbiosis as an explanation for the existence of chloroplasts and mitochondria, another

ENDOSYMBIOSIS

ANCESTRAL EUKARYOTE

ANCESTRAL PROKARYOTE proficient at converting food and oxygen into energy

1 Ancestral eukaryote engulfs prokaryote.

2 Ancestral eukaryote and prokaryote merge.

Mitochondrion

3 Over time, the engulfed prokaryote evolves into an organelle, such as a mitochondrion or a chloroplast.

INVAGINATION

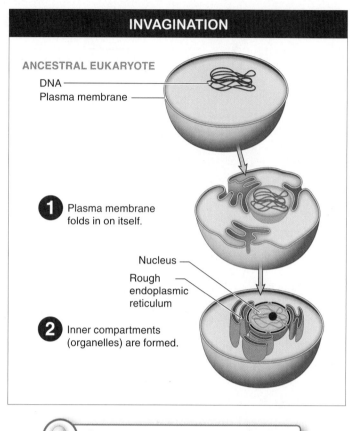

ANCESTRAL EUKARYOTE
DNA
Plasma membrane

1 Plasma membrane folds in on itself.

Nucleus
Rough endoplasmic reticulum

2 Inner compartments (organelles) are formed.

Organelles may have developed by endosymbiosis or invagination or a combination of the two.

FIGURE 3-7 How did eukaryotic cells become so structurally complex? Two theories.

theory about the origin of organelles in eukaryotes is the idea that the plasma membrane around the cell may have folded in on itself (a process called **invagination**) to create the inner compartments, which subsequently became modified and specialized (see Figure 3-7). It may turn out that organelles arose from both processes. Perhaps mitochondria and chloroplasts originated from endosymbiosis, and all of the other organelles arose from plasma membrane invagination. We just don't know.

TAKE-HOME MESSAGE 3·3

Eukaryotes are single-celled or multicellular organisms consisting of cells with a nucleus that contains linear strands of genetic material. The cells also commonly have organelles throughout their cytoplasm; these organelles may have originated evolutionarily through endosymbiosis or invagination, or both.

The Cell   Cell Membranes   Crossing the Membrane   Cell Connections

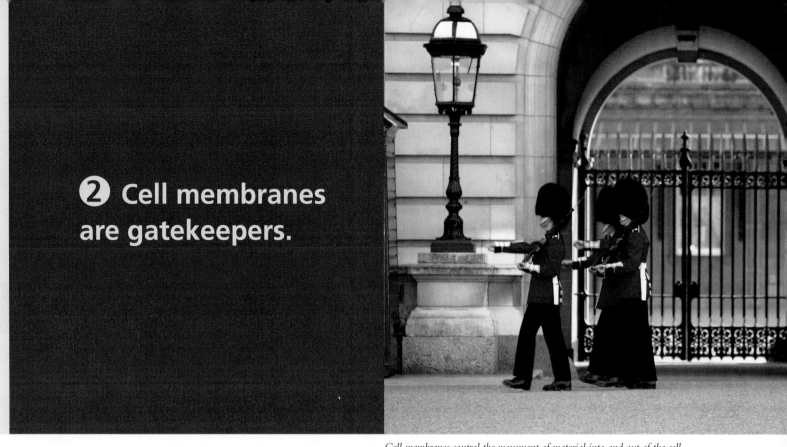

② Cell membranes are gatekeepers.

Cell membranes control the movement of material into and out of the cell.

3·4

Every cell is bordered by a plasma membrane.

Just as skin covers our bodies, every cell of every living thing on earth, whether a prokaryote or a eukaryote, is enclosed by a plasma membrane, a two-layered membrane that holds the contents of a cell in place and regulates what enters and leaves the cell. Plasma membranes are thin (a stack of a thousand would be only as thick as a single hair) and flexible, and in photos or diagrams the membranes often resemble simple plastic bags, holding the cell contents in place. This image is a gross oversimplification, however. Membranes are indeed thin and flexible, but they are far from simple: a close look at a plasma membrane will reveal that its surface is filled with pores, outcroppings, channels, and complex molecules floating around within the two layers of the membrane itself (**FIGURE 3-8**).

Why are plasma membranes such complex structures? It's because they perform several critical functions beyond simply acting as the boundary for a cell's interior contents. Cells, after all, are perpetually interacting with their external

environment. They take in food and nutrients and dispose of waste products. They take in water. They build and export molecules needed elsewhere in the body. Sometimes they absorb heat from the outside environment, whereas at other times they dissipate excess heat generated by cell activities. Like the booths or checkpoints at a country's borders that control the flow of people into and out of the country, plasma membranes serve as gatekeepers that control the flow of molecules into and out of the cell.

The foundation of all plasma membranes is a layer of lipid molecules all packed together. These are a special type of lipid, called **phospholipids,** which, as you may recall from Chapter 2, have what appear to be a head and two long tails. The head consists of a molecule of **glycerol** linked to a molecule containing phosphorus (**FIGURE 3-9**). This head region is said to be **polar,** because it has an electrical charge. As you may also recall from Chapter 2, water is also a polar molecule and, for this reason, other polar molecules mix easily with water.

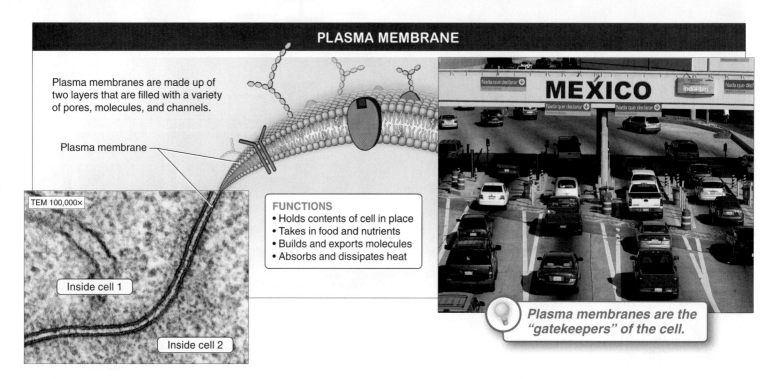

Plasma membranes are made up of two layers that are filled with a variety of pores, molecules, and channels.

Plasma membrane

TEM 100,000×

Inside cell 1

Inside cell 2

FUNCTIONS
• Holds contents of cell in place
• Takes in food and nutrients
• Builds and exports molecules
• Absorbs and dissipates heat

MEXICO

Plasma membranes are the "gatekeepers" of the cell.

FIGURE 3-8 **More than just an outer layer.** The plasma membrane performs several critical functions beyond simply enclosing a cell's interior contents.

Molecules that can mix with water are described as **hydrophilic** ("water-loving") molecules. The two tails of the phospholipid are long chains of carbon and hydrogen atoms. Because they have no electrical charge, the carbon-hydrogen chains are **non-polar.** And because they are non-polar, these tails do not mix with water and are said to be **hydrophobic** ("water-fearing"). The chemical structure of phospholipids gives them a sort of split personality: their hydrophilic head

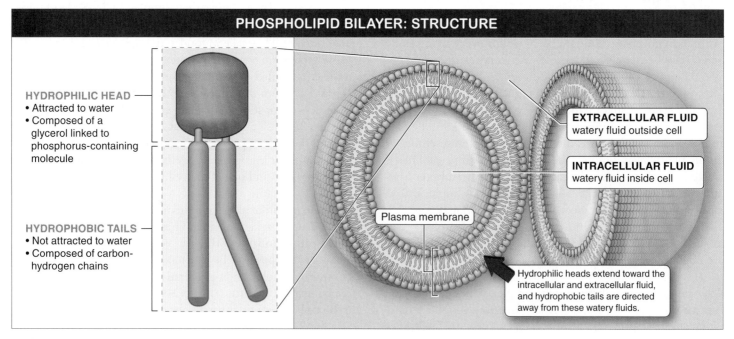

PHOSPHOLIPID BILAYER: STRUCTURE

HYDROPHILIC HEAD
• Attracted to water
• Composed of a glycerol linked to phosphorus-containing molecule

HYDROPHOBIC TAILS
• Not attracted to water
• Composed of carbon-hydrogen chains

EXTRACELLULAR FLUID
watery fluid outside cell

INTRACELLULAR FLUID
watery fluid inside cell

Plasma membrane

Hydrophilic heads extend toward the intracellular and extracellular fluid, and hydrophobic tails are directed away from these watery fluids.

FIGURE 3-9 **Good membrane material.** The phospholipid bilayer of the plasma membrane prevents fluid from leaking out of the cell.

The Cell   Cell Membranes   Crossing the Membrane   Cell Connections

region mixes easily with water, while their hydrophobic tail region does not mix with water.

The split personality of phospholipids makes them good membrane material. Once a large number of phospholipids are packed together with all of their heads facing one way and their tails the other, we have a sheet with one side that is hydrophilic and one that is hydrophobic. In a cell's plasma membrane, two of these sheets of phospholipids are arranged so that the hydrophobic tails are all in contact with one another and the hydrophilic heads are in contact with the watery solution outside and inside the cell (see Figure 3-9). This arrangement gives us another way to describe the structure of the plasma membrane: as a **phospholipid bilayer.**

The phospholipids are not locked in place in the plasma membrane; they just float around in their side of the bilayer. They cannot pop out of the membrane, or flop from one side to the other, because their hydrophobic tails always line up away from any watery solution. Just as similarly charged sides of two magnets push away from each other, so do the hydrophobic tails in the center of the membrane push away from and avoid coming into contact with water molecules. Because the center part of the bilayer membrane is made up of hydrophobic lipids, the solution on one side of the membrane cannot leak across into the solution on the other side. In this way, the plasma membrane forms a boundary around the cell's contents.

TAKE-HOME MESSAGE 3•4

Every cell of every living organism is enclosed by a plasma membrane, a two-layered membrane that holds the contents of a cell in place and regulates what enters and leaves the cell.

3•5

Molecules embedded in the plasma membrane help it perform its functions.

While the plasma membrane's phospholipid bilayer construction forms the cell's basic boundary with its environment, there is much more to a plasma membrane. Embedded within or attached to the phospholipid bilayer are different types of protein, carbohydrate, and lipid molecules (**FIGURE 3-10**).

The proteins found in the plasma membrane enable it to carry out most of its gatekeeping functions. For every 50 to 100 phospholipids in the membrane, there is one protein molecule. Some of these proteins, called **transmembrane proteins,** penetrate right through the lipid bilayer, from one side to the other. Others, called **surface proteins,** reside primarily on the inner or outer surface of the membrane.

What determines whether a protein resides on the surface or extends through the bilayer? Its tertiary structure. Remember from Chapter 2 that all the amino acids that make up each protein have side chains that differ from one another chemically. Some of these side chains are hydrophobic, others are hydrophilic, and as a protein is assembled into its final shape, these side chains can cause parts of the protein to be attracted to hydrophobic or hydrophilic regions. Because a

MOLECULES WITHIN THE PLASMA MEMBRANE

Extracellular fluid

Hydrophilic region
Hydrophobic region

Carbohydrates

Plasma membrane

Transmembrane protein

Lipid

Surface proteins

Intracellular fluid

Hydrophobic and hydrophilic forces determine the orientation of proteins in the plasma membrane.

FIGURE 3-10 Protein and carbohydrate molecules are embedded in the plasma membrane.

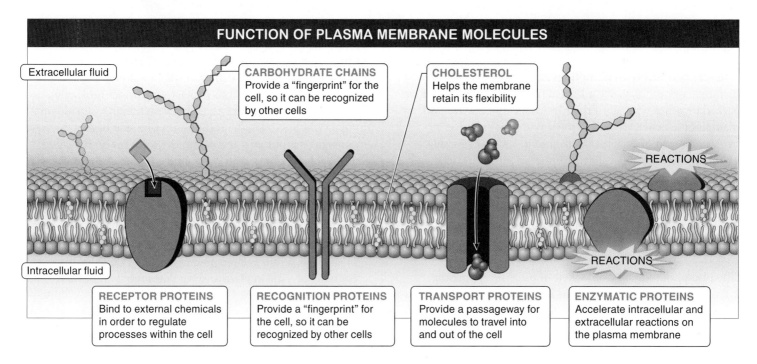

FUNCTION OF PLASMA MEMBRANE MOLECULES

Extracellular fluid

CARBOHYDRATE CHAINS
Provide a "fingerprint" for the cell, so it can be recognized by other cells

CHOLESTEROL
Helps the membrane retain its flexibility

REACTIONS

Intracellular fluid

REACTIONS

RECEPTOR PROTEINS
Bind to external chemicals in order to regulate processes within the cell

RECOGNITION PROTEINS
Provide a "fingerprint" for the cell, so it can be recognized by other cells

TRANSPORT PROTEINS
Provide a passageway for molecules to travel into and out of the cell

ENZYMATIC PROTEINS
Accelerate intracellular and extracellular reactions on the plasma membrane

FIGURE 3-11 Plasma membrane molecules serve diverse roles.

transmembrane protein has both hydrophobic and hydrophilic regions, part of the protein can be positioned in the hydrophobic region in the center of the membrane while the other part resides in the hydrophilic regions. Peripheral membrane proteins, on the other hand, reside on the membrane surface and have an entirely hydrophilic structure and so can bind only to the head regions of the phospholipids. As a consequence, they can be positioned on either the outer or the inner side of the membrane.

Once membrane proteins are in place, the hydrophobic and hydrophilic forces keep them properly oriented. Because all of the components of the plasma membrane are held in the membrane in this manner, they can float around without ever popping out.

There are four primary types of membrane proteins, each of which performs a different function (**FIGURE 3-11**).

1. Receptor proteins bind to chemicals in the cell's external environment and, by doing so, regulate certain processes within the cell. Cells in the heart, for example, have receptor proteins that bind to adrenaline, a chemical released into the bloodstream in times of extreme stress or fright. When adrenaline binds to these heart cells, the cells increase the heart's rate of contraction to pump blood through the body more quickly. You have experienced this reaction if you've ever been startled and felt your heart start to pound.

2. Recognition proteins give each cell a "fingerprint" that makes it possible for the body's immune system (which

fights off infections) to distinguish the cells that belong inside your body from those that are invaders and need to be attacked. (Note that carbohydrates also play a role in recognition.) Recognition proteins also can help cells bind to or adhere to other cells or molecules.

3. Transport proteins are transmembrane proteins that help large and/or strongly charged molecules pass through the plasma membrane. Transport proteins come in a variety of shapes and sizes, making it possible for a wide variety of molecules to be transported.

4. Enzymatic proteins (enzymes) accelerate chemical reactions on the plasma membrane's surface (a variety of enzymatic proteins exist, with some accelerating reactions on the inside of the plasma membrane and others accelerating reactions on the outside of the plasma membrane).

In addition to the various kinds of membrane proteins, two other types of molecules can be incorporated in a cell's plasma membrane.

1. Short, branched carbohydrate chains that are attached to proteins or to phospholipid heads on the outside of the cell membrane serve as part of a membrane's fingerprint, along with recognition proteins. This fingerprint allows the cell to be recognized by other cells, such as those of the immune system.

2. The plasma membrane also can contain **cholesterol,** a lipid that helps the membrane maintain its flexibility. It prevents the membrane from becoming too fluid or

The Cell   Cell Membranes   Crossing the Membrane   Cell Connections

floppy at moderate temperatures and acts as a sort of antifreeze, preventing the membrane from becoming too rigid at freezing temperatures. The membranes of some cells are about 25% cholesterol; other plasma membranes, such as those of most bacteria and plants, have no cholesterol at all.

As we've seen, the plasma membrane is made up from several different types of molecules, like a mosaic, and many of those molecules float around, held in a proper orientation by hydrophobic and hydrophilic forces, but not always anchored in place. For these reasons, the plasma membrane is often described as a **fluid mosaic.**

3•6

Faulty membranes can cause disease.

With all the complexity of plasma membranes, it's not surprising that there are many ways in which they can malfunction. One disease that results from an improperly functioning membrane is cystic fibrosis, the single most common fatal inherited disease in the United States. At any given time, about 30,000 people in the United States have cystic fibrosis.

Cystic fibrosis occurs when an individual inherits from both parents incorrect genetic instructions for producing one type of transmembrane protein. This protein occurs primarily in the membranes of cells in the lungs and digestive tract. When functioning normally, the protein serves as a passageway that allows one type of molecule—chloride ions—to get into and out of cells.

There are more than a thousand different ways in which these genetic instructions can be defective, but in every case the result is the same: the lack of properly working chloride passageways in a cell's membrane. This defect leads to gradual accumulation of chloride ions within cells. Although it is not clear exactly how the chloride ion accumulation leads to the symptoms of cystic fibrosis, in nearly all cases two primary effects occur: an improper salt balance in the cells and a buildup of thick, sticky mucus—particularly in the lungs. Normal mucus helps to protect the lungs by trapping dust and bacteria. This mucus is then moved out of the lungs (helped along by coughing). The mucus produced by someone with cystic fibrosis, however, is too thick and sticky to be moved out of the lungs, so it collects there, where it impairs lung function and increases the risk of bacterial infection. Because of the improper cellular salt balance, one way to test for cystic fibrosis is to measure the concentration of salt in the sweat—

abnormally high concentrations indicate that the person has the disease.

Although many high-tech treatments have been promised for the sufferers of cystic fibrosis and a great deal of research is being done on this disease, one of the most common treatments is decidedly low-tech. Parents help their children with cystic fibrosis clear the mucus out of their lungs by holding them on a steep slant, almost upside down, and vigorously patting or thumping their chest and back to shake loose the mucus in their lungs and move it to a place where they can cough it up. With careful treatment, the life expectancy of someone with cystic fibrosis can be 35—40 years or longer (**FIGURE 3-12**).

FIGURE 3-12 Moving mucus manually, or through an inhalation vest. The thick and sticky mucus produced by someone with cystic fibrosis collects in the lungs, impairing lung function and increasing the risk of bacterial infection. Thumping on the chest and back can loosen the mucus. The vest, by inflating and deflating rapidly, can have a similar effect in the course of a 20-minute session.

Nine Cell Landmarks

1 In stressful situations, the adrenal glands pump out adrenaline.

2 Adrenaline binds with beta-receptors on cells, causing a faster heartbeat and increased blood pressure.

3 Beta-blocker chemicals bind to receptors and prevent adrenaline from binding to the cell.

FIGURE 3-13 Reducing anxiety through beta-blockers.

By binding to adrenaline receptors, beta-blockers reduce anxiety symptoms.

While faulty membranes can cause a disorder as serious as cystic fibrosis, pharmaceutical tinkering with cell membranes can alter cellular function in ways that have beneficial effects. For example, a group of drugs called "beta-blockers" is extremely effective at reducing anxiety. This effect was discovered almost accidentally—the drugs were actually developed as a treatment for high blood pressure.

Q Why do "beta-blockers" reduce anxiety?

Here's how beta-blockers work. Many cells in your body, particularly the cells of the heart, have receptor proteins on their plasma membranes that can bind to *adrenaline,* a chemical manufactured in the adrenal glands (located atop the kidneys) that helps your body cope with stressful situations. These receptor proteins are called "beta-receptors." In stressful situations, your adrenal glands pump out adrenaline (**FIGURE 3-13**). On reaching cells in your heart (among other locations in your body) and binding to beta-receptors, the adrenaline promptly causes your heart to beat faster and more forcefully, increasing your blood pressure in the process. This reaction is fine in a short-term fight-or-flight situation, but it is not healthy over the long run because the increased pressure can damage blood vessels. Depending on its severity, this reaction can also be problematic if you are giving a presentation or taking a test, or if you're in any other anxiety-producing situation.

When you take a beta-blocker pill, the pill dissolves and the chemicals travel throughout your body until they encounter the beta-receptors. They bind to the receptors, hold on, and block the adrenaline from doing its job. This outcome slows your heart rate, causes a reduction in blood pressure, and can bring great relief to those suffering from the sweating and trembling associated with anxiety.

TAKE-HOME MESSAGE 3·6

Normal cell functioning can be disrupted when cell membranes—particularly the proteins embedded in them—do not function properly. Such malfunctions can cause health problems, such as cystic fibrosis. But disruption of normal cell membrane function can also have beneficial, therapeutic effects, such as in the treatment of high blood pressure and anxiety.

The Cell   Cell Membranes   Crossing the Membrane   Cell Connections

Membrane surfaces have a "fingerprint" that identifies the cell.

As mentioned earlier, every cell in your body has a "fingerprint" made from a variety of molecules on the outside-facing surface of the cell membrane. Some of these membrane molecules vary, depending on the specific function of the cell. Others are common to all of your cells and convey to your immune system: "I belong here." Cells with an improper fingerprint are recognized as foreign and are attacked by your body's defenses.

Throughout our evolutionary history, this system has been tremendously valuable in helping our bodies fight infection. In some cases, however, this vigilance is a problem, like a car alarm that goes off even when you don't want it to. Suppose you receive a liver (or any other organ) transplant. Even if the donor is a close relative, the molecular fingerprint on the cells of the donated liver is not identical to your own. Consequently, your body sees the new organ as a foreign object and puts up a fight against it (FIGURE 3-14). Because your body will naturally try to reject the new organ, doctors must administer drugs that suppress your immune system. Immune suppression helps you tolerate the new liver, but, as you can imagine, it leaves you without some of the defenses to fight off other foreign invaders, such as bacteria that may cause infection. The presence or absence of certain molecular markers on plasma membranes is also responsible for a person's blood type and can lead to problems with simple blood transfusions.

Q Why is it extremely unlikely that a person will catch HIV from casual contact—such as shaking hands—with an infected individual?

The AIDS-causing virus, HIV, uses the molecular markers on plasma membranes to infect an individual's cells. These same molecular markers are also the reason that it is extremely unlikely that you can catch an HIV infection from casual contact with an infected individual, such as shaking his or her hand. The specific molecular markers involved in infection by HIV belong to a group of identifying markers called "clusters of differentiation." Abbreviated as "CD markers" and having names such as CD1, CD2, and CD3, these marker molecules are proteins embedded in the plasma membrane that enable a cell to bind to outside molecules and, sometimes, transport them into the cell.

One CD marker, called the CD4 marker, is found only on cells deep within the body and in the bloodstream, such as immune system cells and some nerve cells. It is the CD4 marker, in conjunction with another receptor, that is targeted by HIV. If the virus can find a cell with a CD4 marker, it can infect you, and because the CD4 markers *never* occur on the surface of your skin cells, casual contact such as touching is very unlikely

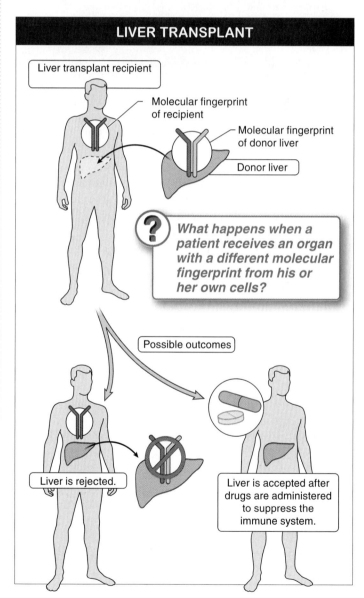

LIVER TRANSPLANT

Liver transplant recipient

Molecular fingerprint of recipient

Molecular fingerprint of donor liver

Donor liver

? *What happens when a patient receives an organ with a different molecular fingerprint from his or her own cells?*

Possible outcomes

Liver is rejected.

Liver is accepted after drugs are administered to suppress the immune system.

FIGURE 3-14 Mismatched molecular fingerprints can cause difficulty in organ transplantation.

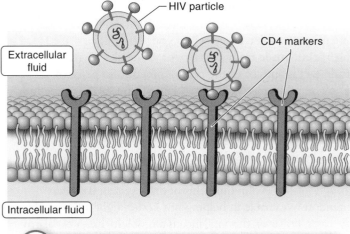

Cells within the human body, and not those on the surface, have CD4 markers. HIV infects the body by binding to the CD4 markers on these cells.

HIV is not spread through casual contact; the virus cannot bind to skin cells because they do not have CD4 markers.

FIGURE 3-15 HIV is not transmitted through casual contact.

to transmit the virus (**FIGURE 3-15**). Even if millions of HIV particles are present on a person's hands, they just can't gain access into any of the other person's surface cells.

How does HIV get transmitted? Far and away, the most common methods of transmission involve the transfer of blood, semen, vaginal fluid, or breast milk from one individual to another. Particles of the virus are present in these fluids, as are cells already infected by the virus. As a consequence, the chief routes of transmission are from an infected mother to her child in breast milk, from an infected mother to her baby at birth, the use of contaminated needles, and unprotected sexual intercourse. Because an open sore or

cut might expose some of the cells in your bloodstream to the outside world, however, it is not impossible for casual transmission of HIV to occur.

TAKE-HOME MESSAGE 3•7

Every cell in your body has a "fingerprint" made from a variety of molecules on the outside-facing surface of the cell membrane. This molecular fingerprint is key to the function of your immune system.

The Cell   Cell Membranes   Crossing the Membrane   Cell Connections

❸ Molecules move across membranes in several ways.

Passage of proteins between cells can be tracked (junctions between these human cancer cells are stained red and green).

3·8

Passive transport is the spontaneous diffusion of molecules across a membrane.

To function properly, cells must take in food and/or other necessary materials and must move out metabolic waste and molecules produced for use elsewhere in the body. In some cases, this movement of molecules requires energy and is called active transport. (We cover active transport later in this chapter.) In other cases, the molecular movement occurs spontaneously, without the input of energy, and is called **passive transport.** There are two types of passive transport: diffusion and osmosis. (Osmosis is the topic of the next section.)

Diffusion is passive transport in which a particle, called a **solute,** is dissolved in a gas or liquid (a **solvent**) and moves from an area of high solute concentration to an area of lower concentration (**FIGURE 3-16**). (When we are talking about cells, "diffusion" usually refers to the situation in which the solute moves across a membrane.) In the absence of other forces, molecules of a substance will always tend to move from

FIGURE 3-16 **Diffusion: a form of passive transport that results in an even distribution of molecules.**

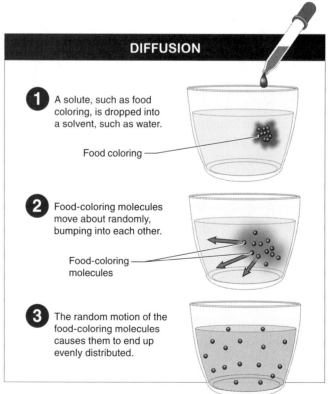

DIFFUSION

1 A solute, such as food coloring, is dropped into a solvent, such as water.

Food coloring

2 Food-coloring molecules move about randomly, bumping into each other.

Food-coloring molecules

3 The random motion of the food-coloring molecules causes them to end up evenly distributed.

93

where they are more concentrated to where they are less concentrated. This movement occurs because molecules move randomly and are equally likely to move in any direction. And when certain molecules are highly concentrated, they keep bumping into each other and eventually end up evenly distributed. For this reason, we say that molecules tend to *move down their concentration gradient*. For a simple illustration, drop a tiny bit of food coloring into a bowl of water and wait for a few minutes. The molecules of dye are initially clustered together in a very high concentration. Gradually, they disperse down their concentration gradient until the color is equally spread throughout the bowl.

Molecules such as oxygen (O_2) and carbon dioxide (CO_2) that are small and carry no charge can pass directly through the phospholipid bilayer of the membrane without the assistance of any other molecules, in a process called **simple diffusion.** Each time you take a breath, for example, there is a high concentration of O_2 molecules in the air you pull into your lungs. That oxygen diffuses across the plasma membranes of the lung cells and into your bloodstream, where red blood cells pick it up and deliver it to parts of your body where it is needed. Similarly, because CO_2 in your bloodstream is at a higher concentration than in the air in your lungs, it diffuses from your blood into your lungs and is released to the atmosphere when you exhale (**FIGURE 3-17**).

Most molecules, however, can't get through plasma membranes on their own. They may be electrically charged (polar) and, hence, repelled by the hydrophobic middle region of the phospholipid bilayer. Or the molecules may be too big to squeeze through the membrane. Nonetheless, when there is a concentration gradient from one side of the membrane to the other, these molecules may still be able to diffuse across the membrane, down their concentration gradient, with the help of a carrier molecule—one of the transport proteins, which we discussed in Section 3-5. Often, this transport protein spans the membrane and functions like a revolving door. When spontaneous diffusion across a plasma membrane requires a transport protein, it is called **facilitated diffusion** (see Figure 3-17).

Defects in transport proteins can reduce facilitated diffusion or even bring it to a complete stop, with serious health consequences. Many genetic diseases are the result of inheriting incorrect genetic instructions for building transport proteins. In the disease cystinuria, for example, incorrect genetic instructions result in a malformed transport protein in the plasma membrane. When structured and functioning properly, this transport protein facilitates the diffusion of some amino acids (including cystine, from which the disease gets its name) out of the kidneys. When the proteins are malformed, they

PASSIVE TRANSPORT

Passive transport occurs when molecules move across a membrane without energy input. Molecules move down their concentration gradients.

SIMPLE DIFFUSION
Molecules pass directly through the plasma membrane without the assistance of another molecule.

FACILITATED DIFFUSION
Molecules move across the plasma membrane with the help of a carrier molecule.

FIGURE 3-17 Simple and facilitated diffusion.

cannot facilitate this diffusion and these amino acids build up in the kidneys, forming painful and dangerous kidney stones.

Diffusion across membranes doesn't just occur in animals, we see it in all organisms on earth. As we'll see in Chapter 4, one of the most important biological processes on earth is the diffusion of CO_2 from the atmosphere (an area of relatively high concentration) into the leaf cells of plants (areas of relatively low concentration), where it can be attached to other molecules, forming sugars through photosynthesis. At the same time, O_2 diffuses out of the leaves and into the atmosphere.

TAKE-HOME MESSAGE 3•8

For proper functioning, cells must acquire food molecules and/or other necessary materials from outside the cell. Similarly, metabolic waste molecules and molecules produced for use elsewhere in the body must move out of the cell. In passive transport—which includes simple and facilitated diffusion and osmosis—the molecular movement occurs spontaneously, without the input of energy. This generally occurs as molecules move down their concentration gradient.

The Cell   Cell Membranes   Crossing the Membrane   Cell Connections

Osmosis is the passive diffusion of water across a membrane.

Just as solute molecules will passively diffuse down their concentration gradients, water molecules will also move from areas of high concentration to areas of low concentration to equalize the concentration of water inside and outside the cell. The diffusion of water across a membrane is a special type of passive transport called **osmosis** (**FIGURE 3-18**). As solute molecules diffuse across a plasma membrane, molecules of water also move across the membrane, equalizing the water concentration inside and outside the cell.

Osmosis can have some dramatic effects on cells. It turns out that many molecules just can't move across a cell membrane. They may be too big or too charged, or there may not be any carrier molecules in the membrane that can move them. Let's imagine, for instance, that there is a high concentration of a large molecule inside a cell and that it can't pass through the cell membrane. Although the molecule can't move out of the cell and down its concentration gradient, water can move into the cell down *its* concentration gradient. As water diffuses into the cell, the cell will get larger.

When a cell is in a solution, we describe the concentration of solutes outside the cell relative to that inside the cell as the **tonicity** of the solution (see Figure 3-18). A **hypertonic** solution has a greater concentration of solutes than the solution inside the cell. As a consequence, if the cell membrane is not permeable to the solutes, water will move out of the cell into the surrounding solution, equalizing the solute concentration inside and out, and the cell shrivels. A **hypotonic** solution has a lower solute concentration than the solution inside the cell. If the cell membrane is not permeable to the solutes, water will move into the cell and it will swell. In an **isotonic** solution, the concentration of solutes is the same inside and outside the cell.

You can see osmosis in action in your own kitchen (**FIGURE 3-19**). Take a stalk of celery and leave it on the counter for a couple of hours. As water evaporates from the cells of the celery stalk, it will shrink and become limp. You can make it crisp again by placing it in a solution of distilled water. Why distilled water? Because it contains very few dissolved molecules—it is a hypotonic solution. The cells in the celery stalk, on the other hand, contain many dissolved molecules, such as salt. Because those dissolved molecules can't easily pass across the plasma membrane of the celery stalk cells, the water molecules move into the cells (moving from the pool of distilled water where they are at super-high concentration to the interior of the celery cells where water is scarcer). Would the celery regain its crispness if you placed it in concentrated salt water? No, because in the hypertonic salt solution there are more dissolved molecules (and fewer water molecules) outside the celery cells, so what little water remains in the celery cells actually moves, via osmosis, out of the celery, causing it to shrivel even more.

OSMOSIS

Osmosis is a type of passive transport by which water diffuses across a membrane, in order to equalize the concentration of water inside and outside the cell. The direction of osmosis is determined by the total amount of solutes on either side of the membrane. Water will always move toward the side that has a greater concentration of solutes.

Extracellular fluid —   **PLANT CELL**   **ANIMAL CELL (RED BLOOD CELL)**

HYPERTONIC SOLUTION
- Solute concentrations are higher in the extracellular fluid.
- Water diffuses out of cells.

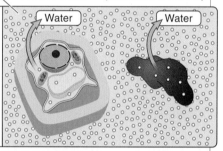

HYPOTONIC SOLUTION
- Solute concentrations are lower in the extracellular fluid.
- Water diffuses into cells.

ISOTONIC SOLUTION
- Solute concentrations are balanced.
- Water movement is balanced.

 Unlike animal cells, plant cells generally do not explode in hypotonic solutions, because their rigid cell walls limit cellular expansion.

FIGURE 3-18 Osmosis overview.

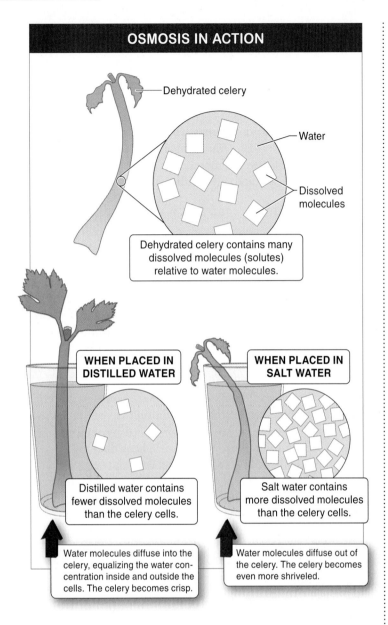

OSMOSIS IN ACTION

Dehydrated celery

Water

Dissolved molecules

Dehydrated celery contains many dissolved molecules (solutes) relative to water molecules.

WHEN PLACED IN DISTILLED WATER

WHEN PLACED IN SALT WATER

Distilled water contains fewer dissolved molecules than the celery cells.

Salt water contains more dissolved molecules than the celery cells.

Water molecules diffuse into the celery, equalizing the water concentration inside and outside the cells. The celery becomes crisp.

Water molecules diffuse out of the celery. The celery becomes even more shriveled.

FIGURE 3-19 Osmosis in your kitchen.

Q How do laxatives relieve constipation?

A more practical use of osmosis is seen with the laxative Milk of Magnesia. This product contains magnesium salts, which are poorly absorbed from the digestive tract. As a consequence, after a dose of the laxative, water moves via osmosis from the surrounding cells into the intestines. The water softens the feces in the intestines and increases the fecal volume, thereby relieving constipation.

It's important to note that the direction of osmosis is determined only by a difference in the *total concentration* of all the molecules dissolved in the water: it does not matter what the solutes are, only how many molecules of solutes there are. To determine which way the water molecules will move, you need to determine the total amount of "dissolved stuff" on either side of the membrane. The water will move toward the side with more stuff dissolved in it. For this reason, even small increases in the salinity of lakes can have disastrous consequences for the organisms living there, from fish to bacteria. Conversely, putting animal cells—such as red blood cells—in distilled water will cause them to explode, because water will diffuse into the cell (which contains more solutes) and the cell will swell and burst. This will not generally happen to plant cells because (as we will see later in this chapter) the plant cell plasma membrane is surrounded by a rigid cell wall that limits the amount of cell expansion possible when water moves in by osmosis.

TAKE-HOME MESSAGE 3·9

The diffusion of water across a membrane is a special type of passive transport called osmosis. Water molecules move across the membrane until the concentration of water inside and outside the cell is equalized.

3·10 ---

In active transport, cells use energy to move small molecules into and out of the cell.

Molecules can't always move spontaneously and effortlessly in and out of cells. Sometimes their transport needs energy, in which case the process is called **active transport.** Such energy expenditures may be necessary if the molecules to be

moved are very large or if they are being moved *against* their concentration gradient. In all active transport, proteins embedded in the membrane act like motorized revolving doors, pushing molecules into the cell regardless of the

Active transport occurs when the movement of molecules into and out of a cell requires the input of energy. For example, in response to eating, the cells lining your stomach use ATP to pump large numbers of H^+ ions into the stomach.

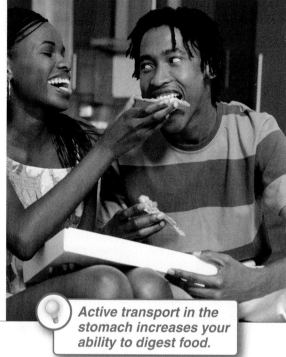

Active transport in the stomach increases your ability to digest food.

FIGURE 3-20 **Active transport.** Some energy expenditure is required to move molecules into and out of cells when the molecules are very large or are being moved against their concentration gradient.

concentration of those molecules on either side of the membrane. There are two distinct types of active transport, primary and secondary, which differ only in the source of the fuel that keeps the revolving doors spinning.

Every time you eat a meal, you start the motors that spin the revolving-door transport proteins in your stomach (**FIGURE 3-20**). To help break down the food into more digestible bits, the cells lining your stomach create an acidic environment by pumping large numbers of H^+ ions (also called protons) into the stomach contents, against their concentration gradient. (That is, there are more H^+ ions in the stomach contents than there are inside the cells lining the stomach.) All of this H^+ pumping increases your ability to digest the food but comes at a great energetic cost—in the form of usage of the high-energy molecule ATP—because the protons would not normally flow into a region against their concentration gradient. When active transport uses energy directly from ATP to fuel the revolving door, the process is called **primary active transport.** (We explore ATP in more detail in Chapter 4.)

Many transporter proteins use an indirect method of fueling their activities rather than using energy released directly from ATP. In the process of **secondary active transport,** the

transport protein simultaneously moves one molecule against its concentration gradient while letting another flow down its concentration gradient. Although no ATP is used directly in this process, at some other time and in some other location, energy from ATP was used to pump one of the types of molecule involved in the secondary transport process against its concentration gradient. The process is a bit like using energy to pump water up to the top of a high water tower. Later, the water can be allowed to run out of the tower over a water wheel, which can, in turn, power a process such as grinding wheat into flour. Our bodies frequently use the energy from one reaction that occurs spontaneously to fuel another reaction that requires energy.

TAKE-HOME MESSAGE 3·10

In active transport, moving molecules across a membrane requires energy. Active transport is necessary if the molecules to be moved are very large or if they are being moved against their concentration gradient. Proteins embedded in the plasma membrane act like motorized revolving doors to actively transport (pump) the molecules.

97

Endocytosis and exocytosis are used for bulk transport of particles.

Many molecules needed by cells are just too big to get into a cell by passive or active transport. After all, a protein embedded in a thin plasma membrane can only be so big, and some cells in your immune system, for example, must even ingest (and destroy) entire bacterial cells that are invading your body. To absorb such large particles, cells engulf them with their plasma membrane in a process called **endocytosis.** Cells also have another method besides passive and active transport for moving molecules out of the cell. Cells that manufacture molecules (such as digestive enzymes) for use elsewhere in the body must get those molecules out

of the cell, and they often use the process of **exocytosis** to do this.

There are three types of endocytosis: phagocytosis, pinocytosis, and receptor-mediated endocytosis. All three involve the basic process of the plasma membrane oozing around an object that is outside the cell, surrounding it, forming a little pocket called a **vesicle,** and then pinching off the vesicle so that it is inside the cell but separated from the rest of the cell contents.

Phagocytosis and Pinocytosis **Phagocytosis** is the process by which relatively large particles are engulfed by cells (**FIGURE 3-21**). Amoebas and other unicellular protists, as well as white blood cells, use phagocytosis to consume entire organisms either as food or as their way of defending against pathogens (disease-causing organisms or substances). Whereas "phagocytosis" comes from the Greek for "eat" and "container," the term **pinocytosis** comes from the Greek for "drink" and "container" and describes the process of cells taking in dissolved particles and liquids. The two processes are largely the same, except that the vesicles formed during pinocytosis are generally much smaller than those formed during phagocytosis.

Receptor-Mediated Endocytosis The third type of endocytosis, **receptor-mediated endocytosis,** is much more specific than either phagocytosis or pinocytosis. In this process, receptor molecules on the surface of a cell sit waiting until the one type of molecule they recognize bumps into them. For one receptor it might be insulin, for another it might be cholesterol. Many receptors of the same type are often clustered together in a cell's plasma membrane. When the appropriate molecule binds to each of the receptor proteins, the membrane begins to fold inward, first forming a little pit and then completely engulfing the molecules, which are still attached to their receptors.

One of the most important examples of receptor-mediated endocytosis involves cholesterol (**FIGURE 3-22**). Most cholesterol that circulates in the bloodstream is in the form of particles called low-density lipoproteins, or LDL. Each molecule of LDL is a cholesterol globule coated by phospholipids. Proteins embedded within the LDL's phospholipid coat are recognized by receptor proteins built into the plasma membranes of liver cells. Once bound to the receptors, the LDL molecule is consumed by the cell via endocytosis. Inside the cell, the

PHAGOCYTOSIS

TEM 4,700×

Phagocytosis is a type of endocytosis by which cells engulf large particles.

Plasma membrane

Large particle

Extracellular fluid

Intracellular fluid

1 The plasma membrane forms a pocket-like vesicle around a large particle.

2 The particle is transported into the cell in a vesicle.

Vesicle

FIGURE 3-21 Through phagocytosis, amoebas and other unicellular protists, as well as white blood cells, consume other organisms for food or for defense.

The Cell   Cell Membranes   Crossing the Membrane   Cell Connections

RECEPTOR-MEDIATED ENDOCYTOSIS

Receptor-mediated endocytosis is a type of endocytosis by which cells engulf specific particles.

Liver cells use receptor-mediated endocytosis to ingest low-density lipoprotein, or LDL.

LDL PARTICLE **LIVER CELL**

Cholesterol

Protein

Plasma membrane

Extracellular fluid

Intracellular fluid

1 LDL binds to receptor proteins that reside in the plasma membrane of liver cells.

LDL receptor protein

Vesicle

2 The plasma membrane forms a vesicle and engulfs the LDL.

3 LDL is broken down into useful molecules such as estrogen and testosterone.

Estrogen

Testosterone

? *Why are individuals with fewer LDL receptors more likely to develop cardiovascular disease?*

FIGURE 3-22 Receptor proteins aid in endocytosis. This process is important in the ability of your liver to remove cholesterol from your bloodstream.

cholesterol is broken down and used to make a variety of other useful molecules, such as the hormones estrogen and testosterone.

Circulating cholesterol often builds up on the walls of arteries, reducing the area (that is, the diameter of the blood vessel) available for blood flow and causing the artery to harden. Because too much circulating cholesterol in LDL molecules can lead to cardiovascular disease and death (**FIGURE 3-23**), it's good to remove as much cholesterol from your bloodstream as quickly as possible. Individuals lucky enough to have large numbers of LDL receptors on the plasma membranes of their liver cells have a significantly lower risk of cardiovascular disease.

Conversely, individuals who consume food laden with too much cholesterol (such as egg yolks, cheese, and sausages), or those who have the misfortune of inheriting genes that code for faulty liver cell membranes that have only a few LDL receptors, can have a heavy load of circulating blood cholesterol. This outcome can result in the early onset of cardiovascular disease. In the extreme case where an individual is born with no LDL receptors at all (a disorder called familial hypercholesterolemia), circulating cholesterol accumulates in the arteries so rapidly that cardiovascular disease begins to develop even before puberty, and death from a heart attack can occur before the age of 30.

Q *Faulty cell membranes are a primary cause of cardiovascular disease. What modification to these membranes might be an effective treatment?*

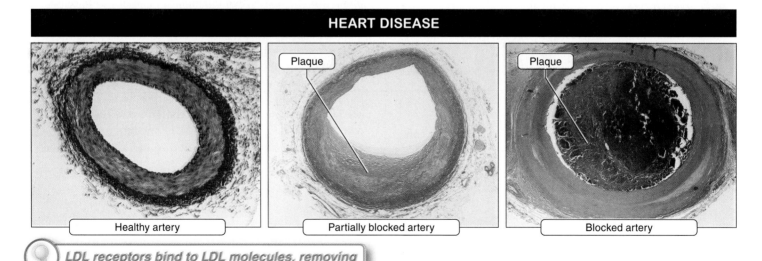

HEART DISEASE

Healthy artery

Plaque

Partially blocked artery

Plaque

Blocked artery

💡 *LDL receptors bind to LDL molecules, removing cholesterol from the bloodstream and reducing the risk of cardiovascular disease.*

FIGURE 3-23 Comparison of a healthy artery and arteries choked by the buildup of cholesterol.

Exocytosis Cells in the pancreas produce a chemical called insulin that moves throughout the circulatory system, informing body cells that there is glucose in the bloodstream that ought to be taken in and utilized for energy. The insulin molecule is much too large to pass out through the plasma membranes of the cells in which it is manufactured. As a result, after molecules of insulin are produced, they are coated with a phospholipid membrane to form a vesicle. The insulin-carrying vesicle then moves through the cytoplasm to the inner surface of the cell's plasma membrane. Once there, the phospholipid membrane surrounding the insulin and the phospholipid membrane of the cell fuse together, dumping the contents of the vesicle out of the cell and into the bloodstream (**FIGURE 3-24**).

Exocytosis, the movement of molecules out of a cell, occurs throughout the body and is not restricted to large molecules. In the brain and other parts of the nervous system, for

After drinking a glass of orange juice, your blood concentrations of glucose increase, causing the cells of your pancreas to secrete insulin through exocytosis.

example, communication between cells occurs as one cell releases large numbers of very small molecules, called neurotransmitters, via exocytosis.

TAKE-HOME MESSAGE 3•11

When materials cannot get into a cell via diffusion or a pump (for example, when the molecules are too big), cells can engulf the molecules or particles with their plasma membrane in a process called endocytosis. Similarly, molecules can be moved out of a cell via exocytosis. In both processes, the plasma membrane moves to surround the molecules or particles and forms a little vesicle that is pinched off inside the cell (endocytosis) or fuses with the plasma membrane and dumps its contents outside the cell (exocytosis).

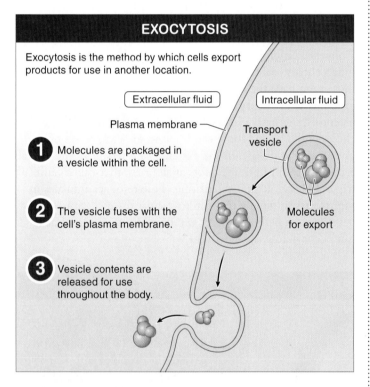

EXOCYTOSIS

Exocytosis is the method by which cells export products for use in another location.

Extracellular fluid   Intracellular fluid

Plasma membrane   Transport vesicle

1 Molecules are packaged in a vesicle within the cell.

2 The vesicle fuses with the cell's plasma membrane.

Molecules for export

3 Vesicle contents are released for use throughout the body.

FIGURE 3-24 Exocytosis moves molecules out of the cell. This is how insulin is exported from the cells where it is synthesized.

The Cell   Cell Membranes   Crossing the Membrane   Cell Connections

❹ Cells are connected and communicate with each other.

Cells reach out and connect; no telephone lines required.

3 • 12

Connections between cells hold them in place and enable them to communicate with each other.

So far in this chapter we have examined the cell as a free-living and independent entity. The majority of cells in any multicellular organism, however, are connected to other cells. Let's consider how they are connected and how they communicate with each other. There is a variety of methods by which cells adhere to other cells, involving numerous types of protein and glycoprotein adhesion molecules. We examine three primary types of connections between animal cells: (1) tight junctions, (2) desmosomes, and (3) gap junctions (**FIGURE 3-25**); and one type of connection between plant cells: plasmodesmata.

Tight junctions form continuous, water-tight seals around cells and also anchor cells in place. Much like the caulking around a tub or sink that keeps water from leaking into the surrounding walls, tight junctions prevent fluid flow between cells. Tight junctions are particularly important in the small intestine, where digestion occurs. Cells lining the small intestine absorb nutrients from the watery fluid moving through your gut. If the fluid—and the resident bacteria—inside the intestine were to leak between the cells and into your body cavity, you would not be able to extract sufficient energy and nutrients

from your food, and the bacteria would make you sick. The tight junctions instead force fluid to pass *into* the cells that line the intestine, where the nutrients can be utilized.

Desmosomes are like spot welds or rivets that fasten cells together into strong sheets. They occur at irregular intervals and function like fastened Velcro: they hold cells together but are not water-tight, allowing fluid to pass around them. Desmosomes are found in much of the tissue that lines the cavities of animal bodies. They also are found in muscle tissue, holding fibers together. Genetic disorders that reduce cells' ability to form desmosome proteins or lead to destruction of desmosomes by the immune system result in the formation of blisters, where layers of skin separate from each other.

Finally, **gap junctions** are pores surrounded by special proteins that form open channels between two cells (see Figure 3-25). Functioning like secret passageways, these junctions are large enough for salts, sugars, amino acids, and the chemicals that carry electrical signals to pass through, but are too small for the passage of organelles or very large molecules such as proteins and nucleic acids. Gap junctions

1 TIGHT JUNCTIONS
Form a water-tight seal between cells, like caulking around a tub

2 DESMOSOMES
Act like Velcro and fasten cells together

3 GAP JUNCTIONS
Act like secret passageways and allow materials to pass between cells

Cell 1

Cell 2

TEM 38,000×

TEM 70,000×

TEM 36,000×

FIGURE 3-25 Cell connections: tight junctions, desmosomes, and gap junctions.

are an important mechanism for cell-to-cell communication. In the heart, for example, the electrical signal telling muscle cells to contract is passed from cell to cell through gap junctions. Gap junctions are also important in allowing a cell to recognize that it has bumped up against another cell; chemicals flowing from one cell to the next can signal the body to stop producing cells of a particular type.

Q How can a lack of communication between cells lead to cancer?

Normal cells generally stop dividing when they bump up against other cells, a phenomenon called **contact inhibition.** Cancer cells, however, do not have this feature. Instead, they divide continuously, eventually forming a mass of cells, called a tumor, which can cause serious health problems. Cancer cells differ from other cells in several important ways; one of these ways is that they usually have reduced numbers of gap junctions. With fewer of these "secret passageways" of communication, it may be harder for a cell to detect that it has bumped up against other cells, in which case the cell is more likely to continue dividing, possibly contributing to its eventually becoming a tumorous mass of dividing cells.

In most plants, the cells have anywhere from 1,000 to 100,000 microscopic tube-like channels, called **plasmodesmata** (*sing.* **plasmodesma**) (see Figure 3-36), connecting the cells to each other and enabling communication and transport between them. In fact, because so many of the cells are connected to one another, sharing cytoplasm and other molecules, some biologists have wondered whether we should consider a plant as just one big cell. How would you respond to that idea?

TAKE-HOME MESSAGE 3·12

In multicellular organisms, most cells are connected to other cells. The connections can form a water-tight seal between the cells (tight junctions), can hold sheets of cells together while allowing fluid to pass between the cells (desmosomes), or can function like secret passageways, allowing the movement of cytoplasm, molecules, and other signals between cells (gap junctions). In plants, plasmodesmata connect cells and enable communication and transport between them.

The Cell   Cell Membranes   Crossing the Membrane   Cell Connections

⑤ Nine important landmarks distinguish eukaryotic cells.

Cell of the voodoo lily. How many different organelles can you spot in this plant cell?

3•13

The nucleus is the cell's genetic control center.

As we have seen, all cells are surrounded by a complex plasma membrane that actively regulates what materials get into and out of a cell. Now it's time to look at the cellular contents surrounded by the plasma membrane in eukaryotic cells.

The nucleus is the largest and most prominent organelle in most eukaryotic cells. In fact, the nucleus is generally larger than any prokaryotic cell. If a cell were the size of a large lecture hall or movie theater, the nucleus would be the size of a big-rig 18-wheeler truck parked in the front rows. The nucleus has two primary functions, both related to the fact that most of the cell's DNA resides in the nucleus. First, the nucleus is the genetic control center, directing most cellular activities by controlling which molecules are produced, and in what quantity. Second, the nucleus is the storehouse for hereditary information.

Three important structural components stand out in the nucleus (**FIGURE 3-26**). First is the **nuclear membrane,** sometimes called the nuclear envelope, which surrounds the nucleus and separates it from the cytoplasm. Unlike most plasma membranes, however, the nuclear membrane consists of *two* bilayers, one on top of the other, much like the double-bagging of groceries at the market. The nuclear membrane is not a sack, though. It is perforated, covered with tiny pores made from multiple proteins embedded in the phospholipid membranes and spanning both bilayers. These pores enable large molecules to pass from the nucleus to the cytoplasm and from the cytoplasm to the nucleus.

The second prominent structure in the nucleus is the **chromatin,** a mass of long, thin fibers consisting of DNA with some proteins attached to it that keep it from getting impossibly tangled. Most of the time, as the DNA directs cellular activities, the chromatin resembles a plate of spaghetti. When it's time for cell division (a process described in detail in Chapter 6), the chromatin coils up and the threads become shorter and thicker until they become visible as chromosomes, the compacted, linear DNA molecules on which all of the organism's hereditary information is carried.

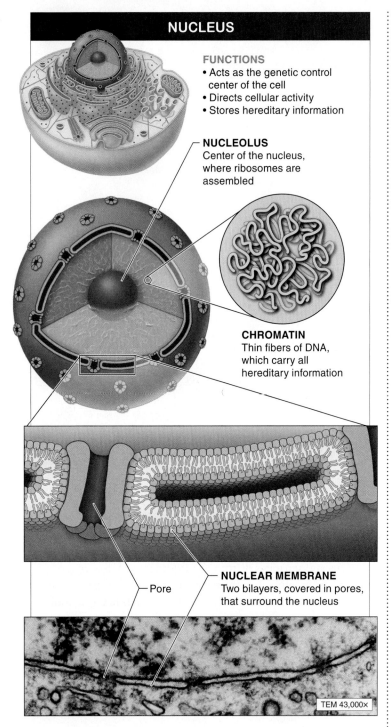

NUCLEUS

FUNCTIONS
- Acts as the genetic control center of the cell
- Directs cellular activity
- Stores hereditary information

NUCLEOLUS
Center of the nucleus, where ribosomes are assembled

CHROMATIN
Thin fibers of DNA, which carry all hereditary information

Pore

NUCLEAR MEMBRANE
Two bilayers, covered in pores, that surround the nucleus

TEM 43,000x

FIGURE 3-26 The nucleus: the cell's genetic control center.

The nucleus holds the genetic information that makes it possible to build organisms. (That's why identical twins, carrying the same genetic information, look so similar.)

A third structure in the nucleus is the **nucleolus,** an area near the center of the nucleus where subunits of the ribosomes, a critical part of the cellular machinery, are assembled. Ribosomes are like little factories that copy bits of the information stored in the DNA and use them to construct proteins, including enzymes and the proteins that make up tissues such as the bark of trees or the bone of vertebrates. The ribosomes are built in the nucleolus but pass through the nuclear pores and into the cytoplasm before starting their protein-production work.

TAKE-HOME MESSAGE 3·13

The nucleus is usually the largest and most prominent organelle in the eukaryotic cell. It directs most cellular activities by controlling which molecules are produced and in what quantity. The nucleus is also the storehouse for all hereditary information.

The Cell   Cell Membranes   Crossing the Membrane   Cell Connections

3·14

Cytoplasm and the cytoskeleton form the cell's internal environment, provide its physical support, and can generate movement.

If you imagine yourself inside a cell the size of a big lecture hall, it might come as a surprise that you can barely see that big rig of a nucleus parked in the front rows. Visibility is almost zero, not only because the room is filled with jelly-like cytoplasm, but also because there is a dense web of thick and thin, straight and branched, ropes, strings, and scaffolding, running every which way throughout the room.

This inner scaffolding of the cell, which is made from proteins, is the **cytoskeleton** (**FIGURE 3-27**). It has three chief purposes. First, it gives animal cells shape and support—making red blood cells look like little round doughnuts (without the hole in the middle) and giving neurons their very long, thread-like appearance. Plant cells are shaped primarily by their cell wall (a structure we discuss later in the chapter), but they also have a cytoskeleton. Second, the cytoskeleton controls the intracellular traffic flow, serving as a series of tracks on which a variety of organelles and molecules are guided across and around the inside of the cell. And third, because the elaborate scaffolding of the cytoskeleton is dynamic and can generate force, it gives all cells some ability to control their movement.

The cytoskeleton also plays a more direct and obvious role in helping some cells to move—whether moving the cell through its environment, or moving stuff in the environment over the cell surface. **Cilia** (*sing.* **cilium**) are short projections that often occur in large numbers on a single cell (**FIGURE 3-28**). Cilia beat

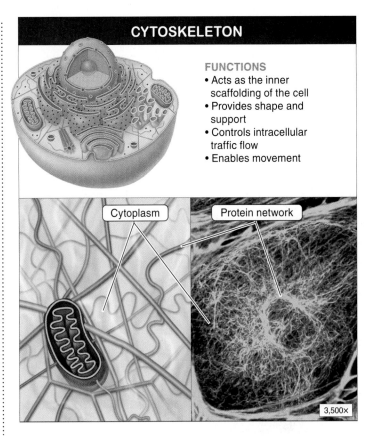

CYTOSKELETON

FUNCTIONS
- Acts as the inner scaffolding of the cell
- Provides shape and support
- Controls intracellular traffic flow
- Enables movement

Cytoplasm

Protein network

3,500×

FIGURE 3-27 **The cytoskeleton: the cell's inner scaffolding.**

CELLULAR MOVEMENT

FIGURE 3-28 **Cilia and flagella assist the cell with movement.**

With gentle beating, cilia can move fluid past cells, whereas flagella, with whip-like motions, can move the cells themselves.

swiftly, often in unison and in ways that resemble blades of grass in a field, blowing in the wind. Cilia can move fluid along and past a cell. This movement can accomplish many important tasks, including sweeping the airways to our lungs to clear them of debris (such as dust) in the air we breathe.

Flagella (*sing.* **flagellum**) are much longer than cilia. Flagella occur in many prokaryotes and single-celled eukaryotes, and many algae and plants have cells with one or more flagella. In animals, however, only one type of cell has a flagellum. But this one type of cell has a critical role in every animal species. By forming the tail of a sperm cell, a flagellum makes possible the rapid movement of the most mobile of animal cells. Some

spermicidal birth control methods prevent conception by disabling the flagellum and immobilizing the sperm cells.

TAKE-HOME MESSAGE 3·14

The inner scaffolding of the cell, which is made from proteins, is the cytoskeleton. It gives animal cells their shape and support, gives cells some ability to control their movement, and serves as a series of tracks on which organelles and molecules are guided across and around the inside of the cell.

3·15

Mitochondria are the cell's energy converters.

Cars have it easy: we generally put only a single type of fuel in them, and it is exactly the same fuel every time. With human bodies it's a different story. During some meals we put in meat and potatoes. Other times, it's juice and fruits. Still others, bread or vegetables, popcorn and gummy bears, pizza and ice cream. Yet no matter what we eat, we expect our body to utilize the energy in these various foods to power all the

reactions in our bodies that make it possible for us to breathe, move, and think. The mitochondria are the organelles that make this possible (**FIGURE 3-29**).

Mitochondria (*sing.* **mitochondrion**) are the cell's all-purpose energy converters, and they are present in virtually all plant cells, animal cells, and every other eukaryotic cell.

MITOCHONDRIA

FUNCTIONS
• Act as all-purpose energy converters
• Harvest energy to be used for cellular functions

DNA
Matrix
Outer membrane
Inner membrane
Intermembrane space

TEM 73,000×

Cells such as muscle and liver cells, which use a lot of energy, can have up to 2,000 mitochondria!

FIGURE 3-29 Mitochondria: the cell's all-purpose energy converters.

The Cell   Cell Membranes   Crossing the Membrane   Cell Connections

Although we consume a variety of foods, our mitochondria allow us to convert the energy contained in the chemical bonds of the carbohydrates, fats, and proteins in the food into carbon dioxide, water, and ATP—the molecule that is the energy source all cells use to fuel all their functions and activities. (ATP and how it works are described in detail in Chapter 4.) Because this energy conversion requires a significant amount of oxygen, organisms' mitochondria consume most of the oxygen used by each cell. In humans, for example, our mitochondria consume as much as 80% of the oxygen we breathe. They give a significant return on this investment by producing about 90% of the energy our cells need to function.

As with so many aspects of our bodies, form follows function. Cells that are not very metabolically active, such as fat cells in humans, have very few mitochondria. Other cells, such as muscle, liver, and sperm cells in animals and fast-growing root cells in plants—all of which have large energy requirements— are packed densely with mitochondria. Some of these cells have as many at 2,000 mitochondria!

To visualize the structure of a mitochondrion, imagine a plastic sandwich bag. Now take another, bigger plastic bag and stuff it inside the sandwich bag. That's the structure of mitochondria: there is a smooth outer membrane and a scrunched up inner membrane. This construction creates two separate compartments within the mitochondrion: a region outside the inner plastic bag (called the **intermembrane space**) and another region, called the **matrix,** inside the inner plastic bag. This bag-within-a-bag structure has important implications for energy conversion. And having a heavily folded inner membrane that is much larger than the outer membrane provides a huge amount of surface area on which to conduct chemical reactions. (We discuss the details of mitochondrial energy conversion and the role of mitochondrial structure in Chapter 4.)

As we learned in our earlier discussion of endosymbiosis, mitochondria may very well have existed, billions of years ago, as separate single-celled bacteria-like organisms. They are similar to bacteria in size and shape, and may have originated when symbiotic bacteria took up permanent residence within other cells. Perhaps the strongest evidence for this is that mitochondria have their own DNA. Mixed in among the approximately 3,000 proteins in each mitochondrion, there

Q We all have more DNA from one parent than the other. Who is the bigger contributor: mom or dad? Why?

are anywhere from 2 to 10 copies of its own little ring-shaped DNA. This DNA carries the instructions for making 13 important mitochondrial proteins necessary for metabolism and energy production.

We're always taught that our mothers and fathers contribute equally to our genetic composition, but this isn't quite true: the mitochondria in every one of your cells (and the DNA that comes with them) come from the mitochondria that were initially present in your mother's egg, which, when fertilized by your father's sperm, developed into you. In other words: all of your mitochondria are descended from your mother's mitochondria. The tiny sperm contributes DNA, but no cytoplasm and, hence, no mitochondria. Consequently, mitochondrial DNA is something that we inherit exclusively from our mothers. This is true not only in humans but in most multicellular eukaryotes. As the fertilized egg develops into a two-celled, then four-celled, embryo, the mitochondria split themselves by a process called fission, the same process of division and DNA duplication used by bacteria—so there are always a sufficient number of mitochondria for the newly produced cells. This similarity between mitochondria and bacteria is another characteristic that supports the theory that mitochondria were originally symbiotic bacteria.

Given the central role of mitochondria in converting the energy in food molecules into a form that is usable by cells, you won't be surprised to learn that mitochondrial malfunctions can have serious consequences. Recent research has focused on possible links between defective mitochondria and diseases characterized by fatigue and muscle pain. It seems, for instance, that many cases of "exercise intolerance"— extreme fatigue or cramps after only slight exertion—may be related to defective mitochondrial DNA.

TAKE-HOME MESSAGE 3•15

In mitochondria, which are found in virtually all eukaryotic cells, the energy contained in the chemical bonds of carbohydrate, fat, and protein molecules is converted into carbon dioxide, water, and ATP, the energy source for all cellular functions and activities. Mitochondria may have their evolutionary origins as symbiotic bacteria living inside other cells.

3·16

Lysosomes are the cell's garbage disposals.

Garbage. What does a cell do with all the garbage it generates? Mitochondria wear out after about 10 days of intensive activity, for starters. And white blood cells constantly track down and consume bacterial invaders, which they then have to dispose of. Similarly, the thousands of ongoing reactions of cellular metabolism produce many waste macromolecules that cells must digest and recycle. Many eukaryotic cells deal with this garbage by maintaining hundreds of versatile floating "garbage disposals" called lysosomes (**FIGURE 3-30**).

Q Why is Tay-Sachs disease like a strike by trash collectors?

Lysosomes are round, membrane-enclosed, acid-filled vesicles that function as garbage disposals. Lysosomes are filled with about 50 different digestive enzymes and a super-acidic fluid, a corrosive broth so powerful that if a lysosome were to burst, it would almost immediately kill the cell by rapidly digesting all of its component parts. The selection of enzymes in the lysosome represents a broad spectrum of chemicals designed for dismantling macromolecules that are no longer needed by cells or are generated as by-products of cellular metabolism. Some of the enzymes break down lipids, others

carbohydrates, others proteins, and still others nucleic acids. Consequently, lysosomes are frequently also a first step when, via phagocytosis, a cell consumes and begins digesting a particle of food or even an invading bacterium. And, ever the efficient system, the cell releases most of the component parts of molecules that are digested, such as amino acids, back into its cytoplasm, where they can be reused by the cell as raw materials.

With 50 different enzymes necessary for lysosomes to carry out their metabolic salvaging act, malfunctions sometimes occur. A common genetic disorder called Tay-Sachs disease is the result of just such a genetic mishap. In Tay-Sachs disease, an individual inherits an inability to produce a critical lipid-digesting enzyme. Even though the lysosomes cannot digest certain lipids, the cells continue to send lipids to the lysosomes, where they accumulate, undigested. This leads to a sort of lysosome constipation. The lysosome swells until it bursts and digests the whole cell or until it chokes the cell to death. Within the first few years of life, this process occurs in large numbers of cells and eventually leads to the child's death.

FIGURE 3-30 Lysosomes: digestion and recycling of the cell's waste products.

The Cell   Cell Membranes   Crossing the Membrane   Cell Connections

Although it has long been debated among biologists, it does seem that plant cells contain lysosomes, compartments with similar digestive broths that carry out the same digestive processes. Among animals, some immune system cells tend to have particularly large numbers of lysosomes, most likely because of their great need for disposing of the by-products of disease-causing bacteria.

3·17

The endoplasmic reticulum is the site where cells build proteins and disarm toxins.

The information for how to construct the molecules essential to a cell's smooth functioning and survival is stored in the DNA found in the cell's nucleus. The energy used to construct these molecules and to run cellular functions comes primarily from the mitochondria. The actual production and modification of biological molecules, however, occurs in a system of organelles called the **endomembrane system** (FIGURE 3-31). This mass of interrelated membranes spreads out from and surrounds the nucleus, forming chambers within the cell that contain their own mixture of chemicals. The endomembrane system takes up as much as one-fifth of the cell's volume and is responsible for many of the fundamental functions of the cell.

Lipids are produced within these membrane-enclosed compartments, products for export to other parts of the body are modified and packaged here, polypeptide chains are assembled into functional proteins here, and many of the toxic chemicals that find their way into our bodies—from recreational drugs to antibiotics—are broken down and neutralized in the endomembrane system.

Because the process of protein production follows a somewhat linear path, we explore the endomembrane system sequentially, beginning just outside the nucleus with the endoplasmic reticulum.

OVERVIEW OF THE ENDOMEMBRANE SYSTEM

FUNCTIONS
- Produces and modifies molecules to be exported to other parts of the organism
- Breaks down toxic chemicals and cellular by-products

Rough endoplasmic reticulum

Smooth endoplasmic reticulum

Golgi apparatus

FIGURE 3-31 The endomembrane system: the smooth endoplasmic reticulum, rough endoplasmic reticulum, and Golgi apparatus.

ROUGH ENDOPLASMIC RETICULUM

FUNCTION
• Modifies proteins that will be shipped elsewhere in the organism

Ribosomes

Rough ER is covered in ribosomes. Ribosomes are protein-making machines.

TEM 56,500x

FIGURE 3-32 The rough endoplasmic reticulum is studded with ribosomes.

Rough Endoplasmic Reticulum Perhaps the organelle with the most cumbersome name, the **rough endoplasmic reticulum,** or **rough ER** ("endoplasmic reticulum" is derived from the Greek for "within" and "anything molded" and the Latin for "small net"), is a large series of interconnected, flattened sacs that look like a stack of pancakes. These sacs are connected directly to the nuclear envelope. In most eukaryotic cells, the rough ER almost completely surrounds the nucleus (**FIGURE 3-32**). It is called "rough" because its surface is studded with little bumps. These bumps are ribosomes, the cell's protein-making machines, and generally cells with high rates of protein production have large numbers of ribosomes We cover the details of ribosome structure and protein production in Chapter 5.

The primary function of the rough ER is to fold and package proteins that will be shipped elsewhere in the organism. Poisonous frogs, for example, package their poison in rough ER of the cells where it is produced before transporting it to the poison glands on their skin. Proteins that are used within the cell itself are generally produced on free-floating ribosomes in the cytoplasm.

Smooth Endoplasmic Reticulum As its name advertises, the **smooth endoplasmic reticulum** is part of the endomembrane network that is smooth because there are no

ribosomes bound to it (**FIGURE 3-33**). Although it is connected to the rough ER, it is farther from the nucleus and, besides lacking ribosomes, differs slightly in appearance. Whereas the rough ER looks like stacks of pancakes, the smooth ER sometimes looks like a collection of branched tubes.

The smooth surface gives us the first hint that smooth ER has a different job than rough ER. Because ribosomes are absolutely essential for protein production, we can be certain that, without them, the smooth ER is not involved in folding or packaging proteins. It isn't. Instead, it synthesizes lipids such as fatty acids, phospholipids, and steroids. Exactly which lipids are produced varies throughout the organism and across plant and animal species. Inside the smooth ER of mammalian ovaries and testes, for example, the hormones estrogen and testosterone are produced. Inside the smooth ER of liver and fat cells, lipids are produced. As in the rough ER, following their production these lipids are packaged in transport vesicles and are then sent to other parts of the cell or to the plasma membrane for export.

Another critical responsibility of the smooth ER—particularly the smooth ER in human liver cells—is to help protect us from the many dangerous molecules that get into our bodies. Alcohol, antibiotics, barbiturates, amphetamines or other

The Cell　　　　Cell Membranes　　　　Crossing the Membrane　　　　Cell Connections

FUNCTIONS
- Synthesizes lipids such as fatty acids, phospholipids, and steroids
- Detoxifies molecules such as alcohol, drugs, and metabolic waste products

TEM 7,700×

Smooth ER is called "smooth" because it has no ribosomes on its surface.

Liver cells are packed with huge amounts of smooth ER, because the liver is the primary site for detoxifying harmful molecules.

FIGURE 3-33 In the smooth endoplasmic reticulum, lipids are synthesized and alcohol, antibiotics, and other drugs are detoxified.

stimulants that we may consume, along with many toxic metabolic waste products formed in our bodies, are made less harmful by the detoxifying enzymes in the smooth ER. And just as a body builder's muscles get bigger and bigger in response to weightlifting, our smooth ER proliferates in cells that are exposed to large amounts of particular drugs. This proliferation can be beneficial because it increases the capacity for detoxification. But as the cells become more and more efficient at detoxification, our tolerances to the very drugs the smooth ER is trying to destroy can increase.

As we see over and over in living organisms, form follows function. Not surprisingly, then, just as cells with high rates of protein production have large numbers of ribosomes, we find huge amounts of smooth ER in liver cells because they are the primary sites of molecular detoxification. Other cells that are packed with ER (both rough and smooth) include plasma cells in the blood that produce immune system proteins for export, and pancreas cells that secrete large amounts of digestive enzymes. On the other hand, there is relatively little smooth ER in cells such as fat-storage cells, which produce little or no protein for export.

Q How can long-term use of one drug increase your resistance to another, different drug that you have never encountered?

Chronic exposure to many drugs (from antibiotics to heroin) can induce a proliferation of smooth ER, particularly in the liver, and the smooth ER's associated detoxification enzymes. This proliferation in turn increases tolerance to the drugs, necessitating higher doses to achieve the same effect. Moreover, the increased detoxification capacity of the cells often enables them to better detoxify other compounds, even if the person has never been exposed to these chemicals. This increased detoxification capacity can lead to problems. An individual who has been exposed to large amounts of certain drugs, for example, may end up responding less well to antibiotics.

TAKE-HOME MESSAGE 3·17

The production and modification of biological molecules in eukaryotic cells occurs in a system of organelles called the endomembrane system, which includes, among other organelles, the rough and smooth endoplasmic reticulum. In rough ER, proteins that will be shipped elsewhere in the body are folded and packaged. In the smooth ER, lipids are synthesized and alcohol, antibiotics, and other drugs are detoxified.

The Golgi apparatus is the site where the cell processes products for delivery throughout the body.

Moving farther outward from the nucleus, we encounter another organelle within the endomembrane system: the Golgi (GOHL-jee) apparatus (**FIGURE 3-34**). The **Golgi apparatus** processes molecules synthesized in the cell—primarily proteins and lipids—and packages those that are destined for use elsewhere in the body. The Golgi apparatus is also a site of carbohydrate synthesis, including the complex polysaccharides found in many plasma membranes. The Golgi apparatus, which is not connected to the endoplasmic reticulum, is a flattened stack of membranes (each of which is called a Golgi body) that are *not* interconnected.

After transport vesicles bud from the endoplasmic reticulum, they move through the cytoplasm until they reach the Golgi apparatus. Here, the vesicles fuse with the Golgi apparatus membrane and dump their contents into a Golgi body. There are about four successive Golgi body chambers to be visited, and the molecules get passed from one to the next. In each Golgi body, enzymes make slight modifications to the molecule (such as the addition or removal of phosphate groups or sugars). The processing that occurs in the Golgi

apparatus often involves tagging molecules (much like adding a postal address or tracking number) to direct them to some other part of the organism. After they are processed, the molecules bud off from the Golgi apparatus in a vesicle, which then moves into the cytoplasm. If the molecules are destined for delivery and use elsewhere in the body, the transport vesicle eventually fuses with the cell's plasma membrane and dumps the molecules into the bloodstream via exocytosis.

FIGURE 3-35 shows how the various parts of the endomembrane system work to produce, modify, and package molecules within a cell.

TAKE-HOME MESSAGE 3•18

The Golgi apparatus—another organelle within the endomembrane system—processes molecules synthesized in a cell and packages those that are destined for use elsewhere in the body.

GOLGI APPARATUS

FUNCTION
• Processes and packages proteins, lipids, and other molecules for export to other parts of the organism

Transport vesicles

TEM 62,500x

FIGURE 3-34 Golgi apparatus: processing of molecules synthesized in the cell and packaging of those molecules destined for use elsewhere in the body.

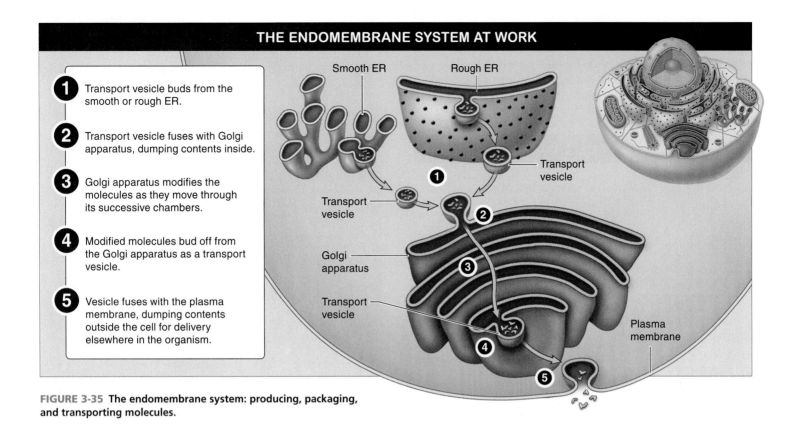

1 Transport vesicle buds from the smooth or rough ER.

2 Transport vesicle fuses with Golgi apparatus, dumping contents inside.

3 Golgi apparatus modifies the molecules as they move through its successive chambers.

4 Modified molecules bud off from the Golgi apparatus as a transport vesicle.

5 Vesicle fuses with the plasma membrane, dumping contents outside the cell for delivery elsewhere in the organism.

FIGURE 3-35 The endomembrane system: producing, packaging, and transporting molecules.

3•19

The cell wall provides additional protection and support for plant cells.

Up until now, we've discussed organelles and structures that are common to both plants and animals. Now we're going to look at several structures that are not found in all eukaryotic cells. Because plants are stuck in the ground, unable to move, they have several special needs beyond those of animals. In particular, they can't outrun or outmaneuver their competitors to get more food. Nor can they outrun organisms that want to eat them. If they want more sunshine, they have few options beyond simply growing larger or taller. And if they want to resist plant-eating animals, they must simply grow tougher outer layers.

One organelle that helps plants achieve both of these goals is the cell wall, a structure that surrounds the plasma membrane. The cell wall is made from polysaccharides in which another carbohydrate, called cellulose, is embedded. Note that although animal cells do not have cell walls, some non-plants such as archaea, bacteria, protists, and fungi also have cell walls, but the chemical composition of their cell walls differs from that found in plants.

In plants, cell wall production can be a multistage process. Initially, when the cell is still growing, a primary cell wall is laid down, along with some glue that helps adjacent cells adhere to each other. Sometime later in life, a secondary cell wall is laid down. This secondary cell wall usually contains a complex molecule called lignin, which gives the wall much of its strength and rigidity. Oddly, this strength remains even after the plant cell dies—hence wood's great value to humans as a construction material, among other things.

The cell wall is nearly 100 times thicker than the plasma membrane, and the tremendous structural strength it confers on plant cells enables some plants to grow several hundred

Nine Cell Landmarks

feet tall. Cell walls also help to make plants more water-tight—an important feature in organisms that cannot reduce water loss through evaporation by moving out of the hot sun—and provide some protection from insects and other animals that might eat them (FIGURE 3-36).

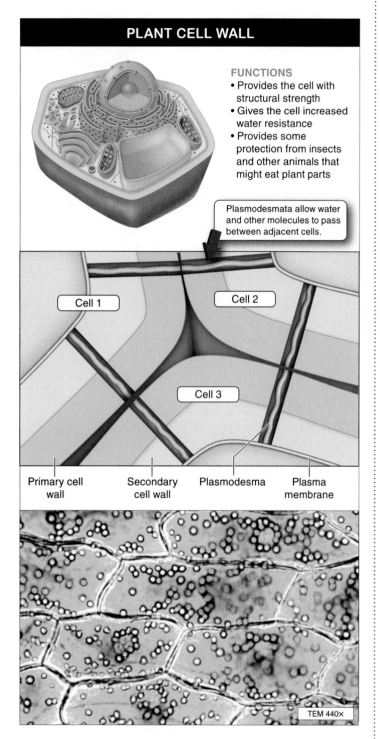

PLANT CELL WALL

FUNCTIONS
- Provides the cell with structural strength
- Gives the cell increased water resistance
- Provides some protection from insects and other animals that might eat plant parts

Plasmodesmata allow water and other molecules to pass between adjacent cells.

Cell 1
Cell 2
Cell 3

Primary cell wall | Secondary cell wall | Plasmodesma | Plasma membrane

TEM 440×

FIGURE 3-36 The plant cell wall: providing strength.

The rigid walls of the plant cell allow plants to grow tall.

Surprisingly, despite its great strength, the cell wall does not completely seal off plant cells from one another. Rather, it is porous, allowing water and solutes to reach the plasma membrane. It also has plasmodesmata, the channels connecting adjacent cells (as described in Section 3-12) and allowing the passage of some molecules between them.

TAKE-HOME MESSAGE 3·19

The cell wall is an organelle found in plants (and in some other non-animal organisms). It is made primarily from the carbohydrate cellulose and it surrounds the plasma membrane of the cell. The cell wall confers tremendous structural strength on plant cells, gives plants increased resistance to water loss, and provides some protection from insects and other animals that might eat them.

The Cell   Cell Membranes   Crossing the Membrane   Cell Connections

Vacuoles are multipurpose storage sacs for cells.

VACUOLE

FUNCTIONS
- Stores nutrients
- Retains and degrades waste products
- Accumulates poisonous materials
- Contains pigments, enabling plants to attract birds and insects that help the plant reproduce
- Provides physical support

Vacuole

TEM 5,000×

FIGURE 3-37 The vacuole: multipurpose storage.

If you look at a mature plant cell through a microscope, one organelle, the **central vacuole,** usually stands out more than all the others because it is so huge and appears empty (**FIGURE 3-37**). Although it may look like an empty sac, the central vacuole is anything but. Surrounded by a membrane, filled with fluid, and occupying from 50% to 90% of a plant cell's interior space, the central vacuole can play an important role in five different areas of plant life. (Vacuoles are also found in some other eukaryotes, including some protists, fungi, and animals, but they tend to be particularly prominent in plant cells.)

1. **Nutrient storage**—the vacuole stores hundreds of dissolved substances, including amino acids, sugars, and ions.

2. **Waste management**—the vacuole retains waste products and degrades them with digestive enzymes, much like the lysosome in animal cells.

3. **Predator deterrence**—the poisonous, nasty-tasting materials that accumulate inside the vacuoles of some plants make a powerful deterrent to animals that might try to eat parts of the plant.

4. **Sexual reproduction**—the vacuole may contain pigments that give some flowers their red, blue, purple, or other colors, enabling them to attract birds and insects that help the plant reproduce by transferring pollen.

5. **Physical support**—high concentrations of dissolved substances in the vacuole can cause water to rush into the cells through the process of osmosis. The increased fluid pressure inside the vacuole can cause the cell to enlarge a bit and push out the cell wall. This process is responsible for the pressure (called **turgor pressure**) that allows stems, flowers, and other plant parts to stand upright. The ability of non-woody plants to stand upright is due primarily to turgor pressure. Wilting is the result of a loss of turgor pressure.

TAKE-HOME MESSAGE 3•20

In plants, vacuoles can occupy most of the interior space of the cell. Vacuoles are also present in some other eukaryotic species. They function as storage spaces and play a role in nutrition, waste management, predator deterrence, reproduction, and physical support.

Chloroplasts are the plant cell's power plant.

It would be hard to choose the "most important organelle" in a cell, but if we had to, the chloroplast would be a top contender. The chloroplast, an organelle found in all plants and eukaryotic algae, is the site of photosynthesis—the conversion of light energy into the chemical energy of food molecules, with oxygen as a by-product. Because all photosynthesis in plants and algae takes place in chloroplasts, these organelles are directly or indirectly responsible for everything we eat and for the oxygen we breathe. Life on

Life on earth would be vastly different without the chloroplasts found in photosynthetic plants.

earth would be vastly different without the chloroplast (**FIGURE 3-38**).

In a green leaf, each cell has about 40–50 chloroplasts. This means there are about 500,000 chloroplasts per square millimeter of leaf surface (that's more than 200 million chloroplasts in an area the size of a small postage stamp). Chloroplasts are oval, somewhat flattened objects. On the outside, the chloroplast is encircled by two distinct layers of membranes (similar to the structure of mitochondria). The fluid in the inside compartment, called the **stroma,** contains some DNA and much protein-making machinery.

With a simple light microscope, it is possible to see little spots of green within the chloroplasts. Up close, these spots look like stacks of pancakes. Each stack consists of numerous, interconnected little flattened sacs called **thylakoids,** and it is on the membranes of the thylakoids that the light-collecting for photosynthesis occurs. (In Chapter 4, we discuss the details

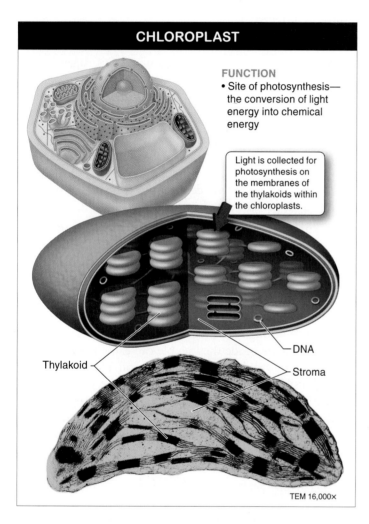

CHLOROPLAST

FUNCTION
• Site of photosynthesis—the conversion of light energy into chemical energy

Light is collected for photosynthesis on the membranes of the thylakoids within the chloroplasts.

Thylakoid

DNA

Stroma

TEM 16,000×

FIGURE 3-38 **The chloroplast: location of photosynthesis.**

The Cell | Cell Membranes | Crossing the Membrane | Cell Connections

ORGANELLE REVIEW

STRUCTURE	ANIMALS	PLANTS	FUNCTION
Nucleus	Yes	Yes	Directs cellular activity and stores hereditary information
Cytoskeleton	Yes	Yes	Provides structural shape and support and enables cellular movement
Mitochondrion	Yes	Yes	Harvests energy for cellular functions
Lysosomes	Yes	Yes	Digests and recycles cellular waste products and consumed material
Rough ER	Yes	Yes	Modifies proteins that will be shipped elsewhere in the organism
Smooth ER	Yes	Yes	Synthesizes lipids and detoxifies molecules
Golgi apparatus	Yes	Yes	Processes and packages proteins, lipids, and other molecules
Cell wall	No	Yes	Provides structural strength, protection, and increased resistance to water loss
Vacuole	Sometimes	Yes	Stores nutrients, degrades waste products, provides pigments and structural support
Chloroplast	No	Yes	Performs photosynthesis

Nine Cell Landmarks

of how chloroplasts convert the energy in sunlight into the chemical energy stored in sugar molecules.)

A peculiar feature of chloroplasts, mentioned in Section 3-3, is that they resemble photosynthetic bacteria, particularly with their circular DNA (which contains many of the genes essential for photosynthesis). Also, the dual outer membrane of the chloroplast is consistent with the idea that, long ago, a cell engulfed a photosynthetic bacterium, enveloping it with its plasma membrane in endocytosis. These features have given rise to the belief that chloroplasts might have originally been bacteria that were engulfed by a predatory cell. According to the endosymbiosis theory (see Section 3-3), the bacteria remained alive and, rather than becoming a meal, became the cell's meal ticket, providing food for the cell in exchange for protection.

FIGURE 3-39 summarizes all of the parts of a cell, in animals and plants, and their functions.

TAKE-HOME MESSAGE 3·21

The chloroplast is the organelle in plants and algae that is the site of photosynthesis—the conversion of light energy into chemical energy, with oxygen as a by-product. Chloroplasts may have originally been bacteria that were engulfed by a predatory cell by endosymbiosis.

FIGURE 3-39 Review of the structures and functions of cellular organelles.

Knowledge You Can Use

Did you know? **Drinking too much water can be dangerous!**

People have long understood the risks of becoming dehydrated, particularly during activities such as marathon running in which a great deal of fluid can be lost through sweating. Only recently, however, has it been recognized that too much water—called "water intoxication"—can also be dangerous. In 2007, a 28-year-old woman competing in a water-drinking contest died just a few hours after consuming about two gallons of water without urinating. Similarly, the death of a runner during the 2002 Boston marathon was attributed to water intoxication.

Q: **When people sweat, what do they lose?** Sweat consists of water and some dissolved solids, primarily salt. As the sweat evaporates, the body is cooled.

Q: **How do people respond to nausea and dizziness, the warning signs of dehydration?** By drinking water. Because the water they drink lacks salt and other solutes, this can lead to sodium imbalances in the body, particularly in the bloodstream and the fluid around cells (extracellular fluid).

Q: **What is osmosis?** Osmosis is the diffusion of water across a membrane toward areas where there is a greater concentration of solutes and, consequently, a lower concentration of water molecules.

Q: **If you drink lots of water (two gallons or more)—which has few or no solutes—where is it going to move once it gets into your bloodstream and extracellular fluid?** It moves by osmosis into your cells, where there is a higher concentration of solutes (primarily sodium) and a lower concentration of water molecules.

Q: **What happens to your cells if they absorb too much water?** They swell, sometimes to the point of rupturing. Such swelling and rupturing in the enclosed brain cavity can cause vomiting and confusion. Disastrously, these symptoms are often mistaken for the symptoms of dehydration and so the consumption of water is recommended, making the situation worse. Further swelling can lead to seizures, coma, and even death. (The reverse situation occurs when people drink too much seawater. Their bloodstream and extracellular fluid gain in solutes, particularly salt, so water flows out of cells by osmosis. Cell dehydration and death can occur—usually from a shrinkage of brain cells.)

What should you do? Many marathon organizers have been reducing the number of water stops they offer runners during the race and testing symptomatic runners' sodium levels in medical tents along the course. Consuming salt tablets and salty snacks helps keep blood sodium levels in a healthy range. Also, when possible, runners should weigh themselves during a race to make sure they are only replacing the fluid they have lost through sweating and are not gaining weight.

❶ What is a cell?

The most basic unit of any organism is the cell, the smallest unit of life that can function independently and perform all of the necessary activities of life, including reproducing itself. All living organisms are made up of one or more cells, and all cells arise from other, pre-existing cells. Every cell on earth is either a eukaryote or a prokaryote. Prokaryotes, which have no nucleus and are single-celled, were the first cells on earth. Eukaryotes are single-celled or multicellular organisms whose cells have a nucleus that contains linear strands of genetic material.

❷ Cell membranes are gatekeepers.

Every cell of every living organism is enclosed by a plasma membrane, a two-layered membrane that holds the contents of a cell in place and regulates what enters and leaves the cell. The plasma membrane is a fluid mosaic of proteins, lipids, and carbohydrates. Every cell in your body has a "fingerprint" made from a variety of molecules on the outside-facing surface of the cell membrane that is key to the functioning of your immune system.

❸ Molecules move across membranes in several ways.

Cells must acquire food molecules and other necessary materials from outside the cell. Similarly, metabolic waste molecules and molecules produced for use elsewhere must move out of the cell. In passive transport the molecular movement occurs spontaneously, without the input of energy. The diffusion of water across a membrane is a special type of passive transport called osmosis. Active transport is necessary if the molecules to be moved are very large or if they are being moved against their concentration gradient. It relies on proteins embedded in the plasma membrane which act like motorized revolving doors. Cells can also engulf particles with their plasma membrane in a process called endocytosis. Similarly, molecules can be moved out of a cell via exocytosis.

❹ Cells are connected and communicate with each other.

In multicellular organisms, most cells are connected to other cells. The connections can form a water-tight seal between the cells (tight junctions), can hold sheets of cells together while allowing fluid to pass between the cells (desmosomes), or can function like secret passageways between cells (gap junctions). In plants, plasmodesmata connect cells and enable communication and transport between them.

❺ Nine important landmarks distinguish eukaryotic cells.

Eukaryotic cells contain many distinct structures, each of which performs a specific life-sustaining function. The nucleus directs all protein production and is also the storehouse for all hereditary information. The cytoskeleton gives animal cells shape and support. In mitochondria, energy from macromolecules is converted into usable energy in the form of ATP. Lysosomes are like garbage disposals for dismantling waste and bacterial pathogens. The production and modification of molecules, including drug detoxification, occurs in the rough and smooth endoplasmic reticulum, and the Golgi apparatus processes molecules synthesized in the cell and packages those destined for use elsewhere in the body. In plants, the cell wall confers structural strength, herbivore protection, and resistance to water loss. Vacuoles function as storage spaces, and chloroplasts are the site of photosynthesis, the conversion of light energy into food molecules.

KEY TERMS

1. Which of the following statements about the cell theory is correct?
   a) All living organisms are made up of one or more cells.
   b) All cells arise from other, pre-existing cells
   c) All eukaryotic cells contain symbiotic prokaryotes.
   d) All prokaryotic cells contain symbiotic eukaryotes.
   e) Both a) and b) are correct.

2. Which of the following statements about prokaryotes is incorrect?
   a) Prokaryotes appeared on earth before eukaryotes.
   b) Prokaryotes have circular pieces of DNA within their nuclei.
   c) Prokaryotes contain cytoplasm.
   d) Prokaryotes contain ribosomes.
   e) Some prokaryotes can conduct photosynthesis.

3. Which of the following facts supports the claim that mitochondria developed from bacteria that, long ago, were incorporated into eukaryotic cells by the process of phagocytosis?
   a) Mitochondria have flagella for motion.
   b) Mitochondria have proteins for the synthesis of ATP.
   c) Mitochondria are the "powerhouses" of the cell.
   d) Mitochondria are small and easily transported across cell membranes.
   e) Mitochondria have their own DNA.

4. Hydrophobic molecules can pass freely through the plasma membrane, but molecules with electrical charges (such as ions and polar molecules) are impeded by the hydrophobic core. For this reason, plasma membranes can be considered:
   a) partially permeable.
   b) impermeable.
   c) hydrophobic.
   d) hydrophilic.
   e) None of these terms properly describe plasma membranes.

5. Drugs called beta-blockers do all of the following except:
   a) reduce high blood pressure.
   b) block signaling through adrenaline receptors.
   c) reduce outward symptoms of anxiety.
   d) bind to the cytoplasmic side of a receptor protein.
   e) reduce the effects of adrenaline on the heart.

6. Cellular "fingerprints":
   a) are exposed on the cytoplasmic side of the membrane.
   b) are made from cholesterol.
   c) are "erased" by the HIV virus.
   d) can help the immune system distinguish "self" from "non-self."
   e) All of the above are correct.

7. The movement of molecules across a membrane from an area of high concentration to one of low concentration is best described as:
   a) active transport.
   b) inactivated transport.
   c) passive transport.
   d) channel-mediated diffusion.
   e) electron transport.

8. The transport of water across a membrane from a solution of lower solute concentration to a solution of higher solute concentration is best describe as:
   a) osmosis.
   b) facilitated diffusion.
   c) receptor-mediated transport.
   d) active transport.
   e) general diffusion.

9. In an experiment, you measure the concentration of a polar molecule inside and outside a cell. You find that the concentration is high and gradually increasing inside the cell. You also measure the ATP concentration inside the cell and find that it is dropping. What is your best hypothesis for the process you are observing?
   a) facilitated diffusion
   b) passive transport
   c) simple diffusion
   d) active transport
   e) endocytosis

10. Which of the following allows the passage of small molecules between animal cells?
    a) nucleoli
    b) tight junctions
    c) desmosomes
    d) gap junctions
    e) black holes

11. The largest structure in a eukaryotic cell is the _____ and it is surrounded by _____ membrane(s).
    a) nucleus; one
    b) nucleus; two
    c) Golgi apparatus; one
    d) mitochondrion; two
    e) mitochondrion; one

12. The cytoskeleton:
    a) is a viscous fluid found in all cells.
    b) fills a cell's nucleus but not the other organelles.
    c) gives an animal cell shape and support, but cannot control movement.
    d) helps to coordinate intracellular movement of organelles and molecules.
    e) All of the above are correct.

13. Which of the following statements about mitochondria is correct?

    a) Mitochondria are found in both eukaryotes and prokaryotes.

    b) There tend to be more mitochondria in fat cells than in liver cells.

    c) Most plant cells contain mitochondria.

    d) Mitochondria may have originated evolutionarily as photosynthetic bacteria.

    e) All of the above are correct.

14. Which of the following organelles is not present in animal cells?

    a) lysosome

    b) Golgi apparatus

    c) rough endoplasmic reticulum

    d) mitochondrion

    e) chloroplast

15. Given that a cell's structure reflects its function, what function would you predict for a cell with a large Golgi apparatus?

    a) movement

    b) secretion of digestive enzymes

    c) transport of chemical signals

    d) rapid replication of genetic material and coordination of cell division

    e) attachment to bone tissue.

16. Cell walls:

    a) only occur in plant cells.

    b) are not completely solid, having many small pores.

    c) confer less structural support than the plasma membrane.

    d) dissolve when a plant dies.

    e) are made primarily from phospholipids.

17. Which one of the following organelles is not found in both plant and animal cells?

    a) nucleus

    b) rough endoplasmic reticulum

    c) mitochondrion

    d) smooth endoplasmic reticulum

    e) central vacuole

18. In plant cells, chloroplasts:

    a) serve the same purpose that mitochondria serve in animal cells.

    b) are the site of conversion of light energy into chemical energy.

    c) play an important role in the breakdown of plant toxins.

    d) have their own linear strands of DNA.

    e) Both a) and b) are correct..

SHORT-ANSWER QUESTION

1. Name and note the function of the organelles shown in the diagram.

See Appendix for answers. For additional study questions, go to www.prep-u.com.

ORGANELLE:
FUNCTION:

ORGANELLE:
FUNCTION:

ORGANELLE:
FUNCTION:

ORGANELLE:
FUNCTION:

ORGANELLE:
FUNCTION:

ORGANELLE:
FUNCTION:

ORGANELLE:
FUNCTION:

ORGANELLE:
FUNCTION:

ORGANELLE:
FUNCTION:

4
Energy

From the sun to you
in just two steps

❶ Energy flows from the sun and through all life on earth.

Sunlight filters through the northern California redwoods.

4•1 --

Cars that run on french fry oil? Organisms and machines need energy to work.

Imagine that you are on a long road trip. The fuel gauge of your car is nearing empty, so you pull off the highway. Instead of driving into a gas station, however, you head to a fast-food restaurant. You drive around to the back where the restaurant stores used cooking materials and fill your car's fuel tank with recycled cooking grease. You head back to the highway, ready to drive several hundred miles before needing another pit stop.

Q Humans can get energy from food. Can machines?

Fast food for your car? Yes! The idea isn't as far-fetched as it sounds. In fact, on the roads of America today many vehicles run on **biofuels,** fuels produced from plant and animal products (**FIGURE 4-1**). Most vehicles, however, run on **fossil fuels** such as gasoline. These fuels (which also include oil, natural gas, and coal) are produced from the decayed remains of ancient plants and animals modified over the course of millions of years by heat, pressure, and bacterial processes.

It turns out that biofuels, fossil fuels, and the food fuels that supply energy to most living organisms are chemically similar. This fact is not surprising because energy from the sun is the source of the energy stored in the chemical bonds between the atoms in all these fuels. Let's investigate how fuels provide energy.

When we burn gasoline, long chains of carbon and hydrogen atoms separate, releasing carbon dioxide (CO_2), water, and a lot of energy that was stored in the chemical bonds holding each gasoline molecule together. (An interesting fact: the energy released in burning one gallon of gas is equivalent to the caloric content of 15 large cheese pizzas.) In an automobile engine, some of this released energy is harnessed to push pistons, spin a crankshaft, turn wheels, and move the car.

Animal fats and the oils in many plants—such as those used to cook french fries—share an important chemical feature with

The Nature of Energy | Photosynthesis | Cellular Respiration | Alternative Pathways

FIGURE 4-1 **Biofuel technology.** Biofuels are chemically similar to fossil fuels.

gasoline. Like gasoline, these fats and oils contain chains of carbon and hydrogen atoms bound together, and just as with gasoline, breaking these bonds releases large amounts of energy (and water and CO_2). If this released energy can be captured efficiently, it, too, can be used to push pistons and turn car wheels.

Cars that run on biofuels are more than just a technological trick. The production of biofuels requires only the plant or animal source, sunlight, air, water, and a relatively short amount of time—a few months or years, depending on the source. On the other hand, the production of fossil fuels such as coal or crude oil requires plant and animal remains and millions of years. This difference gives biofuels an important advantage over fossil fuels: they are a renewable resource. For this reason, they point the way toward a future of reduced dependence on fossil fuels, the supplies of which are dwindling and the combustion of which has many harmful consequences, such as increasing global warming and the release of cancer-causing particles into the atmosphere. Are we at the point yet where all our cars can run on biofuels? Not quite: there are several significant drawbacks to the increased use of biofuels, chief among them the destruction of forests, wetlands, and other ecologically important habitats from the increased use of land to grow these fuels. This is why the search for better fuels continues.

In this chapter we explore how plant, animal, and other living "machines" run on energy stored in chemical bonds. Just as the forward motion of a car is fueled by breaking the chemical bonds in gasoline and harnessing the released energy, the activities of living organisms are fueled by breaking the chemical bonds in food molecules and harnessing the released energy.

All life depends on capturing energy from the sun and converting it into a form that living organisms can use. This

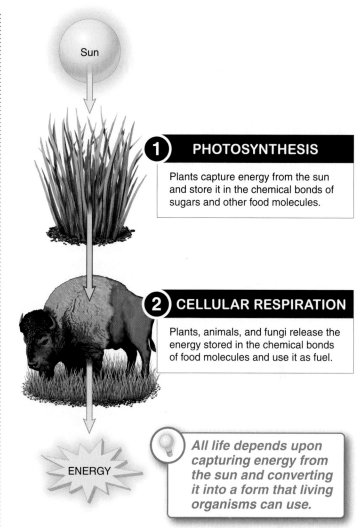

1 PHOTOSYNTHESIS

Plants capture energy from the sun and store it in the chemical bonds of sugars and other food molecules.

2 CELLULAR RESPIRATION

Plants, animals, and fungi release the energy stored in the chemical bonds of food molecules and use it as fuel.

ENERGY

All life depends upon capturing energy from the sun and converting it into a form that living organisms can use.

FIGURE 4-2 **Photosynthesis and cellular respiration.**

energy capture and conversion occurs in two important processes that mirror each other: (1) **photosynthesis,** the process by which plants capture energy from the sun and store it in the chemical bonds of sugars and other food molecules they make, and (2) **cellular respiration,** the process by which all living organisms release the energy stored in the chemical bonds of food molecules and use it to fuel their lives (**FIGURE 4-2**). The sun to you in just two steps!

TAKE-HOME MESSAGE 4·1

The sun is the source of the energy that powers all living organisms and other "machines." The energy from sunlight is stored in the chemical bonds of molecules. When these bonds are broken, energy is released regardless of whether the bond is in a molecule of food, a fossil fuel, or a biofuel like the oil in which french fries are cooked.

Energy has two forms: kinetic and potential.

"Batteries not included." For a child, those are pretty depressing words. We know that many of the toys and electronic gadgets that make our lives fun or useful (or both) need energy—usually in the form of batteries. Generating ringtones, lights, and movement requires energy. The same is true for humans, plants, and all other living organisms: they need energy for their activities, from thinking to moving to reproducing.

Energy is the capacity to do work. And work is anything that involves moving matter against an opposing force. The energy encountered in the study of living things is of two types: kinetic energy and potential energy. **Kinetic energy** is the energy of moving objects. Legs pushing bike pedals and the flapping wings of a bird are examples of kinetic energy (**FIGURE 4-3**). Heat, which results from lots of molecules moving rapidly, is

another form of kinetic energy. Because it comes from the movement of high-energy particles, light is also a form of kinetic energy—probably the most important form of kinetic energy on earth. (We explore the nature of sunlight and how it is harnessed for the work of producing food molecules when we look at photosynthesis later in this chapter.)

An object does not have to be moving to have the capacity to do work; it may have **potential energy,** which is stored energy or the capacity to do work that results from an object's location or position. Water behind a dam, for example, has potential energy. If a hole in the dam is opened, the water can flow through, and perhaps spin a water wheel or turbine. A concentration gradient, which we discussed in Chapter 3, also has potential energy: if the molecules in an area of high concentration move toward an area of lower concentration,

FIGURE 4-3 Kinetic and potential energy. Kinetic energy is the energy of motion; potential energy is energy stored in an object, such as water trapped behind a dam.

the potential energy of the gradient is converted to the kinetic energy of molecular movement, and this kinetic energy can do work. And **chemical energy,** the storage of energy in chemical bonds, is also a type of potential energy.

Because potential energy doesn't involve movement, it is a less obvious form of energy than kinetic energy. An apple has potential energy, as does any other type of food (**FIGURE 4-4**). Why? Because the chemical energy stored in the chemical bonds making up the food can be broken and the energy released during cellular respiration, enabling you to run, play, and work. We explore cellular respiration, the energy-releasing breakdown of molecules, later in this chapter. But first we need to know more about the nature of energy.

TAKE-HOME MESSAGE 4·2

Energy, the capacity to do work, comes in two forms. Kinetic energy is the energy of moving objects, while potential energy, such as chemical energy, is stored energy or the capacity to do work that results from the position or location of an object.

Food is a form of chemical energy!

Glucose
$C_6H_{12}O_6$

CHEMICAL ENERGY

A form of potential energy stored in chemical bonds

H — C
ENERGY

FIGURE 4-4 **Chemical energy.** The chemical bonds in food molecules are a form of potential energy.

4·3

As energy is captured and converted, the amount of energy available to do work decreases.

Every minute of every day—even on cloudy ones—the sun shines brightly, releasing tremendous amounts of energy. Organisms on earth cannot capture every single bit of energy released by the sun; indeed, most plants capture only about 1% of the available energy. What happens to the other 99%? This unused energy does not simply disappear. Accountants would

love to monitor the flow of energy because, as in a good accounting ledger, all of the energy numbers add up perfectly. All energy from the sun can be accounted for: some (probably less than 1%) is captured and transformed into usable chemical energy by organisms through photosynthesis (**FIGURE 4-5**). The rest of the energy from the sun is reflected back into

Energy lost in the form of

HEAT

Light energy from the sun

Chemical energy stored in plants

Energy lost in the form of

HEAT

Kinetic energy of forward motion

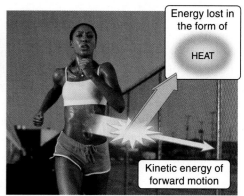

Energy lost in the form of

HEAT

Kinetic energy of forward motion

FIGURE 4-5 **Energy efficiency and heat loss.** As energy is converted to do work, some energy is released as heat.

space (probably about 30%) or absorbed by land, the oceans, and the atmosphere (about 70%), and mostly transformed into heat. Heat is not easily harnessed to do work, however, and is therefore a much less useful form of energy than the energy transformed into chemical energy in plants. The same accounting also exists on a smaller level. If you eat a bowl of rice, some portion of the chemical energy stored in the bonds of the molecules that make up the rice grains is transformed into usable energy that can fuel your cells' activities. All the rest is transformed into heat and is ultimately lost into the atmosphere.

The fact that energy can change form but never disappear is an important feature of energy in the universe, whether we are looking at the sun and the earth or a human and her rice bowl. Just as energy can never disappear or be destroyed, energy can never be created. All the energy now present in the universe has been there since the universe began, and everything that has happened since then has occurred by the transformation of one form of energy into another. In all our eating and growing, driving and sleeping, we are simply transforming energy. The study of the transformation of energy from one type to another, such as from potential energy to kinetic energy, is called **thermodynamics,** and the **first law of thermodynamics** states that energy can never be created or destroyed. It can only change from one form to another.

Because plants capture less than 1% of the sun's energy, it might seem like they are particularly inefficient. But we humans are also rather inefficient at extracting the chemical energy of plants when we eat them. These inefficiencies occur because every time energy is converted from one form to another some of the energy is always converted to heat. When a human converts the chemical energy in a plate of spaghetti into the kinetic energy of running a marathon or when a car transforms the chemical energy of gasoline into the kinetic energy of forward motion, in both cases some energy is converted to heat, the least usable form of kinetic energy. In automobiles, for example, about three-quarters of the energy in gasoline is lost as heat. Thus, for every $10 spent on gas, only about $2.50 goes toward moving the car!

The **second law of thermodynamics** states that every conversion of energy is not perfectly efficient and invariably includes the transformation of some energy into heat. Although heat is certainly a form of energy, because it is not easily harnessed to do work it is almost completely useless to living organisms for fueling their cellular activity. Put another way, the second law of thermodynamics tells us that although the quantity of energy in the universe is not changing, its quality is. Little by little, the amount of energy that is available to do work decreases. Now that we understand that organisms on earth cannot capture every single bit of energy released by the sun—and that energy conversions are inefficient—we can look at the chief energy currency of the cell: ATP.

TAKE-HOME MESSAGE 4·3

Energy is neither created nor destroyed but can change forms. Each conversion of energy is inefficient and some of the usable energy is converted to less useful heat energy.

4·4

ATP molecules are like free-floating rechargeable batteries in all living cells.

Much of the work that cells do requires energy. But even though light energy from the sun carries energy, as do molecules of sugar, fat, and protein, none of this energy can be used directly to fuel chemical reactions in organisms' cells. First it must be captured in the bonds of a molecule called **adenosine triphosphate (ATP),** a free-floating molecule found in cells that acts like a rechargeable battery that temporarily stores energy that can then be used for cellular work in plants, animals, bacteria, and all the other organisms on earth. The use of ATP solves an important timing and coordination problem for living cells: a supply of ATP guarantees that the energy required for energy-consuming reactions will be available when it's needed.

ATP is a simple molecule with three components (**FIGURE 4-6**). At the center of the ATP molecule are two of these components: a small sugar molecule attached to a molecule called adenine. But it is the third component that makes ATP so effective in carrying and storing energy for a short time: attached to the sugar and adenine is a chain of three negatively charged phosphate molecule groups (hence the "tri" in "triphosphate"). Because the bonds between these three phosphate groups must hold the groups together in the face of the three electrical charges that all repel one another, each of these bonds contains a large amount of energy and is stressed and unstable. The instability of these high-energy bonds makes the three phosphate groups like a tightly coiled

The Nature of Energy   Photosynthesis   Cellular Respiration   Alternative Pathways

FIGURE 4-6 **The structure of ATP and ADP.** When ATP ejects one of its phosphate groups, energy is released as ATP becomes ADP.

spring or a twig that is bent almost to the point of breaking. With the slightest push, one of the phosphate groups will pop off, releasing a little burst of energy that the cell can use.

It is precisely because each molecule of ATP is always on the brink of ejecting one of its phosphate groups that ATP is such an effective energy source inside a cell. As long as plenty of ATP molecules are around, they can energize the chemical reactions that make it possible for the cell to carry out reactions that require work, such as building muscle tissue, repairing a wound, or growing roots. Each time a cell expends one of its ATP molecules to pay for an energetically expensive reaction, a phosphate is broken off and energy is released. What is left is a molecule with two phosphates, called ADP (adenosine *di*phosphate), and a separate phosphate group (labeled P_i).

An organism can then use ADP, a free-floating phosphate, and an input of kinetic energy to rebuild its ATP stocks

(FIGURE 4-7). The kinetic energy is converted to potential energy when the free phosphate group attaches to the ADP molecule and makes ATP. In this manner, ATP functions like a rechargeable battery. Where does the input of energy for recharging ATP come from? When we discuss photosynthesis, we'll see that plants, algae, and some bacteria directly use light energy from the sun to make ATP from ADP and free-floating phosphate groups. Animals use the energy contained in the bonds of their food molecules. In either case, the energy is used to re-create the unstable bond in the triphosphate chain. When energy is needed, the organism can again release it by breaking the bond holding the phosphate group to the rest of the molecule. Our bodies recycle ATP molecules in this way tens of thousands of times a day.

Here's the ATP story in a nutshell. Breaking down a molecule of sugar—in a glass of orange juice, for example—leads to a miniature burst of energy in your body. The energy from the mini-explosion is put to work building the unstable high-energy bonds that attach phosphate groups to ADP molecules, creating new molecules of ATP. Later—perhaps only a fraction of a second later—when an energy-consuming reaction is needed, your cells can release the energy stored in the new ATP molecules.

FIGURE 4-7 **ATP is like a rechargeable battery.**

TAKE-HOME MESSAGE 4·4

Cells temporarily store energy in the bonds of ATP molecules. This potential energy can be converted to kinetic energy and used to fuel life-sustaining chemical reactions. At other times, inputs of kinetic energy are converted to the potential energy of the energy-rich but unstable bonds in the ATP molecule.

❷ Photosynthesis uses energy from sunlight to make food.

Sunflowers capture light energy and convert it to organic matter (including tasty seeds).

4•5

Where does plant matter come from? Photosynthesis: the big picture.

Watching a plant grow over the course of a few years can seem like watching a miracle, or at least a very subtle magic trick. Of course it's neither, but the process is nonetheless amazing. Consider that in five years a tree in a big planter can increase its weight by 150 pounds (68 kg) as it grows (**FIGURE 4-8**). Where does that 68 kg of new tree come from?

Our first guess might be the soil. Could that be it? It's easy enough to weigh the soil in the pot when first planting the tree and again 5 years later. After 5 years, though, we find that the soil in our planter has lost less than a pound, nowhere near enough to explain the massive increase in the amount of plant material. Perhaps the new growth comes from the water? Wrong again. Although the older and much larger tree holds more water in its many cells, the water provided to the plant does not come close to accounting for the increase in the dry weight of the plant.

FIGURE 4-8 Where does a plant come from?

? *When plants grow, where does the new tissue come from? From water? From the dirt? From thin air?*

5 LB. → 5 years → 155 LB.

The Nature of Energy | Photosynthesis | Cellular Respiration | Alternative Pathways

When humans grow, the new tissue comes from food we eat. When plants grow, where does the new tissue come from?

The amazing truth is that most of the new material comes from an invisible gas in the air. In the process of photosynthesis, plants capture carbon dioxide gas (CO_2) from the air and, using energy they get from sunlight, along with water and small amounts of chemicals usually found in soil, they produce solid, visible (and often tasty) sugars and other organic matter that are used to make plant structures such as leaves, roots, stems, flowers, fruits, and seeds. In the process, the plants give off oxygen (O_2), a by-product that happens to make all animal life possible.

Although plants are the most visible organisms that have evolved the ability to capture light energy and convert it to organic matter, they are not the only organisms capable of photosynthesis. Some bacteria and many other unicellular organisms are also capable of using the energy in sunlight to produce organic materials (FIGURE 4-9).

There are three inputs to the process of photosynthesis (FIGURE 4-10): light energy (from the sun), carbon dioxide (from the atmosphere), and water (from the ground). From these three inputs, the plant produces sugar and oxygen. As we will see, photosynthesis is best understood as two separate events: a "photo" segment, during which light is captured, and a "synthesis" segment, during which sugar is synthesized.

FIGURE 4-10 **Photosynthesis: the big picture.**

In the "photo" reactions, light energy is captured and temporarily stored in energy-storage molecules. During this process, water molecules split and produce oxygen. In the "synthesis" reactions, the energy in the energy-storage molecules is used to build sugar molecules from carbon dioxide in the air.

FIGURE 4-9 **Plants aren't the only photosynthesizers.** Some bacteria and single-celled organisms are capable of photosynthesis.

Cyanobacterium (magnified 660×)

Euglena (magnified 125×)

Dinoflagellate (magnified 80×)

TAKE-HOME MESSAGE 4·5

Through photosynthesis, plants use water, the energy of sunlight, and carbon dioxide gas from the air to produce sugars and other organic materials. In the process, photosynthesizing organisms also produce oxygen, which makes all animal life possible.

Photosynthesis takes place in the chloroplasts.

The best way to know which parts of a plant are photosynthetic is simply to look for the green bits. Leaves are green because the cells near the surface are packed full of **chloroplasts,** light-harvesting organelles found in plant cells, which make it possible for the plant to use the energy from sunlight to make sugars (their food) and other plant tissue (much of which animals use for food) (**FIGURE 4-11**). Other plant parts, such as stems, may also contain chloroplasts (in which case they, too, are capable of photosynthesis), but most chloroplasts are located within the cells in a plant's leaves.

Let's take a closer look at chloroplasts (**FIGURE 4-12**). The sac-shaped organelle is filled with a fluid called the **stroma.** Floating in the stroma is an elaborate system of interconnected membranous structures called **thylakoids,** which often look like stacks of pancakes. Once inside the chloroplast, you can be in one of two places: in the stroma or inside the thylakoids. The conversion of light energy to

CHLOROPLAST

TEM 16,000×

Stroma

Thylakoid

Chlorophyll

STROMA

Location of "synthesis" reaction, where chemical energy is synthesized into sugar.

THYLAKOID

Location of "photo" reaction, where light energy is converted into chemical energy.

FIGURE 4-12 The chloroplast. The chloroplast is where photosynthesis takes place in a plant.

chemical energy—the "photo" part of photosynthesis—occurs inside the thylakoids. The production of sugars—which are made in the "synthesis" part of photosynthesis—occurs within the stroma.

We examine both the "photo" and the "synthesis" processes in greater detail later in this chapter. First, however, we examine the nature of light energy and of **chlorophyll,** the special molecule found in chloroplasts that makes the capture of light energy possible.

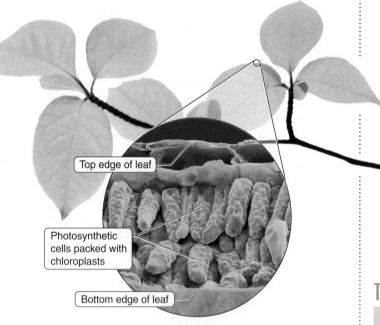

Top edge of leaf

Photosynthetic cells packed with chloroplasts

Bottom edge of leaf

FIGURE 4-11 Photosynthesis factories. Cells near the leaf's surface are packed with photosynthetic chloroplasts.

TAKE-HOME MESSAGE 4·6

In plants, photosynthesis occurs in chloroplasts, green organelles packed in cells near the plants' surfaces.

The Nature of Energy   Photosynthesis   Cellular Respiration   Alternative Pathways

Light energy travels in waves: plant pigments absorb specific wavelengths.

You can't eat sunlight. That's because sunlight is *light* energy rather than the *chemical* energy found in the bonds of food molecules. Photosynthesis is powered by **light energy,** a type of kinetic energy made up of little energy packets called **photons,** which are organized into waves. Photons can do work as they bombard surfaces such as your face (heating it) or a leaf (enabling it to build sugar from carbon dioxide and water).

Photons have various amounts of energy, and the length of the wave in which they travel corresponds to the amount of energy carried by the photon. The shorter the wavelength, the more energy the light carries. Within a ray of light, there are super-high-energy packets of photons (those with short wavelengths), relatively low-energy packets (those with longer wavelengths), and everything in between. This range, which is called the **electromagnetic spectrum,** extends from extremely short high-energy gamma rays and X rays, with wavelengths as short as 1 nanometer (a human hair is about 50,000 nanometers in diameter), to very long low-energy radio waves, with wavelengths as long as a mile (**FIGURE 4-13**).

Just as we can't hear some super-high-pitched frequencies of sound (even though many dogs can), there are some wavelengths of light that are too short or too long for us to see. The light that we can see, visible light, spans all the colors of the rainbow. Humans (and some other animals) can see colors because our eyes contain light-absorbing molecules called **pigments.** These pigments absorb wavelengths of light within the visible range. The energy in these light waves excites electrons in the pigments, which in turn stimulate nerves in our eyes. These nerves then transmit electrical signals to our brains. We perceive different wavelengths within the visible spectrum as different colors. The pigments in the human eye absorb many different wavelengths pretty well: that's why we can see so many colors. When plants use sunlight's energy to make sugar during photosynthesis, they also use the visible portion of the electromagnetic spectrum. Unlike our eyes, however, plant pigments (the energy-capturing parts of a plant) absorb and use only a portion of visible light wavelengths.

Chlorophyll is the pigment molecule in plants that absorbs light energy from the sun. Chlorophyll molecules are

FIGURE 4-13 A spectrum of energy. A ray of light emits high-energy photons, low-energy photons, and everything in between. Plants use only a fraction of the light's available energy.

embedded in the thylakoid membrane of chloroplasts, which are found primarily in plants' leaves. Just as light energy excites electrons in the pigments responsible for color vision in humans, electrons in a plant's chlorophyll can become excited by certain wavelengths of light and can capture a bit of this light energy.

Plants produce several different light-absorbing pigments (**FIGURE 4-14**). The primary photosynthetic pigment, called **chlorophyll *a*,** efficiently absorbs blue-violet and red wavelengths of light. Every other wavelength generally travels through or bounces off this pigment. Because chlorophyll *a* cannot efficiently absorb green light and instead reflects those wavelengths, our eyes and brain perceive the reflected light waves as green, and so the pigment (and the leaves in which it is found) appears green. Another pigment, **chlorophyll *b*,** is similar in structure but absorbs blue and red-orange wavelengths. Chlorophyll *b* reflects yellow-green wavelengths. A related group of pigments called **carotenoids** absorbs blue-violet and blue-green wavelengths and reflects yellow, orange, and red wavelengths.

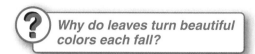

Why do leaves turn beautiful colors each fall?

FIGURE 4-14 Plant pigments. Each photosynthetic pigment absorbs and reflects specific wavelengths.

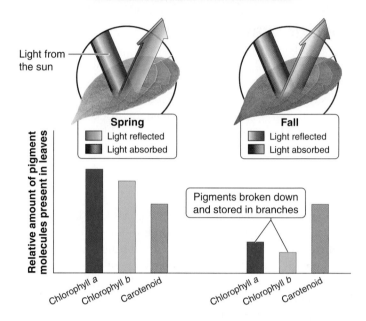

In the late summer, cooler temperatures cause some trees to prepare for the winter by shutting down chlorophyll production and reducing their photosynthesis rates, and they go into a state that resembles an animal's hibernation. Gradually, the chlorophyll *a* and *b* molecules present in the leaves are broken down and their chemical components are stored in the branches. As the amounts of chlorophyll *a* and *b* in the leaves decrease relative to the remaining carotenoids, the striking colors of the fall

Q Why do the leaves of some trees turn beautiful colors each fall?

foliage are revealed (see Figure 4-14). During the rest of the year, chlorophyll *a* and *b* are so abundant in leaves that green masks the colors of the other pigments.

TAKE-HOME MESSAGE 4·7

Photosynthesis is powered by light energy, a type of kinetic energy made of energy packets called photons. Photons hit chlorophyll and other light-absorbing molecules near the green surfaces of plants. These molecules capture the light energy and harness it to build sugar from carbon dioxide and water.

4·8

Photons cause electrons in chlorophyll to enter an excited state.

An organism can use energy from the sun only if it can convert the light energy of the sun into the chemical energy in the bonds between atoms. The most important molecule in this conversion is the pigment chlorophyll (**FIGURE 4-15**). When chlorophyll is hit by photons of certain wavelengths,

the light energy bumps an electron (e⁻) in the chlorophyll molecule to a higher energy level, an *excited* state. Upon absorbing the photon, the electron briefly gains energy, and the potential energy in the chlorophyll molecule increases.

134 CHAPTER 4 • ENERGY

The Nature of Energy   Photosynthesis   Cellular Respiration   Alternative Pathways

ENERGY MOVEMENT THROUGH CHLOROPHYLL

1 Light energy bumps an electron in the chlorophyll molecule to a higher, excited energy level.

Sun

Thylakoid

Photons

Chlorophyll

e⁻ [Higher energy state]

Potential energy increases.

e⁻ [Normal energy state]

2 The excited electron generally has one of two different fates:

Some energy is transferred to a nearby molecule, where it excites another electron.

ENERGY

or

The excited electron is transferred to a nearby molecule.

FIGURE 4-15 Capturing light energy with excited electrons. Chlorophyll electrons are excited to a higher energy state by light energy.

An electron in a photosynthetic pigment that is excited to a higher energy state generally has one of two fates: (1) the electron returns to its resting, unexcited state and, in the process, energy is released, some of which may be transferred to a nearby molecule, bumping electrons in that molecule to a higher energy state (and the rest of the energy is dissipated as heat), or (2) the excited electron itself is passed to another molecule.

The passing of electrons from molecule to molecule is one of the chief ways that energy moves through cells. Many molecules carry or accept electrons during cellular activities. All that is required is that the acceptor have a greater attraction for electrons than the molecule from which it accepts them. This receiver molecule, in turn, hands off electrons to another acceptor with an even greater attraction for them. A molecule that gains electrons always carries greater energy than it did before receiving the electron(s). For this reason, the passing of electrons from one molecule to another can be viewed as a passing of potential energy. In this way, energy moves through cells.

This transfer of electrons is one of the first steps of photosynthesis, the process that enables a plant to harness light energy from the sun efficiently and convert it to the more readily usable chemical energy. As we see later in this chapter, the dismantling of food molecules such as glucose to generate energy is also a story of breaking and rearranging chemical bonds as electrons pass from one atom or molecule to another.

Q Suppose a large meteor hit the earth. How could smoke and soot in the atmosphere wipe out life far beyond the area of direct impact?

Because any particles in the atmosphere can block the light from the sun and reduce the excitation of electrons in chlorophyll molecules, photosynthesis depends on a relatively clean atmosphere. Any reduction in the available sunlight can have serious effects on plants. Scientists believe that if a large meteor hit the earth—as one did when the dinosaurs were wiped out 65 million years ago—smoke, soot, and dust in the atmosphere could block sunlight to such an extent that plants in the region, or even possibly all of the plants on earth, could not conduct photosynthesis at high enough levels to survive. And when plants die off, all of the animals and other species that rely on them for energy die as well. All life on earth is completely dependent on the continued excitation of electrons by sunlight.

TAKE-HOME MESSAGE 4·8

When chlorophyll is hit by photons, the light energy excites an electron in the chlorophyll molecule, increasing the chlorophyll's potential energy. The excited electrons can be passed to other molecules, moving the potential energy through the cell.

Photosynthesis in detail: the energy of sunlight is captured as chemical energy.

"PHOTO" REACTION

Sunlight   Water

Chlorophyll

The "photo" reaction occurs in the thylakoids of the leaves' chloroplasts.

ATP   NADPH

Energy-storing molecules

Oxygen

Chloroplast

FIGURE 4-16 **Overview of the "photo" portion of photosynthesis.** Light energy is captured in the "photo" portion of photosynthesis. It is later used to power the building of sugar molecules.

Photosynthesis is a complex process, but our understanding of this process can be greatly aided by remembering one phrase: FOLLOW THE ELECTRONS. In the passages that follow, if you feel that you are losing focus or getting lost, just remember to think about the electrons: Where are they coming from? What are they passing through? Where are they going? And what will happen to them when they get there?

In the first part of photosynthesis, the "photo" part, sunlight hits a plant and, in a three-step process, the energy in this sunlight is ultimately captured and stored in an ATP molecule and in another molecule (called NADPH) that stores energy by accepting high-energy electrons (**FIGURE 4-16**).

The energy-capturing process occurs in a series of steps that are carried out in two **photosystems.** These photosystems are arrangements of light-catching pigments (including chlorophyll) within chloroplasts that capture energy from sunlight and transform it—first into the energy of excited electrons and ultimately into ATP and high-energy electron carriers. After these transformations, the captured energy is ready to be used to make sugar molecules in the

THE PHOTOSYSTEM (WATER-SPLITTING)

1 Light energy excites electrons in a pigment molecule (such as a chlorophyll molecule). The energy from the excited electrons is transferred to nearby pigment molecules.

2 When transferred energy excites electrons in the chlorophyll a molecule, the primary electron acceptor grabs them and sends them to the electron transport chain.

3 To replace electrons sent to the electron transport chain, water molecules are split and oxygen and hydrogen are released as by-products.

Chloroplast   Thylakoid

AREA OF DETAIL

Sun

Primary electron acceptor

To electron transport chain

Chlorophyll a molecule

Oxygen released into the atmosphere

Pigment molecule

Water

H+ H+
H+ H+

Oxygen needed for all animal life is produced here!

FIGURE 4-17 **The water-splitting photosystem.** Splitting water provides electrons for photosynthesis.

The Nature of Energy   Photosynthesis   Cellular Respiration   Alternative Pathways

Stoma open

Stoma closed

FIGURE 4-23 **Plant stomata.** Carbon dioxide enters a plant through stomata (*sing.* stoma), but water can be lost through the same openings.

Some plants, including corn and sugarcane, have evolved a process that minimizes water loss but still enables them to make sugar when the weather is hot and dry. In the process called **C4 photosynthesis,** these plants add an extra set of

steps to the usual process of photosynthesis (**FIGURE 4-24**). In these steps, the plants produce an enzyme that functions like the ultimate "CO_2-sticky tape." This enzyme has a tremendously strong attraction for carbon dioxide; it can find

C3 PHOTOSYNTHESIS

Carbon dioxide

Day

CALVIN CYCLE

Stomata open

Sugar

ADVANTAGE
• energy efficient

DISADVANTAGE
• water lost to evaporation in hot climates

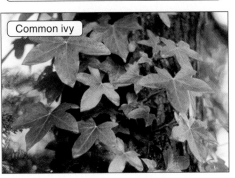

Common ivy

C4 PHOTOSYNTHESIS

Carbon dioxide

Day

ATP

CARBON UPTAKE

Stomata slightly open

CALVIN CYCLE

Sugar

ADVANTAGE
• water loss is minimized in warm climates

DISADVANTAGE
• requires more energy

Corn

CAM PHOTOSYNTHESIS

Carbon dioxide

Night

ATP

CARBON UPTAKE

Stomata open

Day

CALVIN CYCLE

Stomata closed

Sugar

ADVANTAGE
• water loss is minimized in hot climates

DISADVANTAGES
• requires more energy
• slow growth

Saguaro cactus

FIGURE 4-24 **C3, C4, and CAM photosynthesis.**

and bind carbon even when CO_2 concentration is very low. (In contrast, rubisco, the usual enzyme that plants use to pluck carbon from the atmosphere, functions poorly when CO_2 is scarce.) As a consequence, stomata can be opened just a tiny bit, and let in just a little CO_2. Having the stomata closed also reduces evaporation and conserves water for the plant.

This seems like such a good solution that we would expect all plants to use it. There is a catch, though. The extra steps in C4 photosynthesis require the plant to expend additional energy. Specifically, every time the plant generates a molecule of the "CO_2-sticky tape" enzyme, it uses one molecule of ATP. It is acceptable to pay this energy cost only when the climate is so hot and dry that the plant would otherwise have to close its stomata and completely shut down all sugar production. If the climate is mild, however, plants conducting the more energetically expensive C4 photosynthesis will be out-competed by the more efficient plants conducting standard photosynthesis (called C3 photosynthesis). Not surprisingly, we see few C4 plants in the temperate regions of the world or among photosynthetic organisms living in the oceans. In hot, dry regions, however, they are the dominant plants and displace the C3 plants wherever both occur (**FIGURE 4-25**). With global warming, many scientists expect to see a gradual expansion of the geographic ranges over which C4 plants occur and believe that non-C4 plants will be pushed farther and farther away from the equator.

A third and similar method of carbon fixation, called **CAM** (for "crassulacean acid metabolism"), is also found in hot, dry areas. In this method, used by many cacti, pineapples, and other fleshy, juicy plants, the plants close their stomata during the hot, dry days. At night, they open the stomata and let CO_2 into the leaves, where it binds temporarily to a holding molecule. During the day, when a carbon source is needed to make sugars in the Calvin cycle, the CO_2 is gradually released from the holding molecule, enabling photosynthesis to proceed while keeping the stomata closed to reduce water loss (see Figure 4-24). A disadvantage of CAM photosynthesis is that by completely closing their stomata during the day, CAM plants significantly reduce the total amount of CO_2 they can take in. As a consequence, they have much slower growth rates and cannot compete well with non-CAM plants under any conditions other than extreme dryness.

DISTRIBUTION OF PHOTOSYNTHESIS METHODS

Equator

C3 plants dominant
C4 plants dominant

FIGURE 4-25 **Map of C3 and C4 photosynthesis.**

C4 and CAM photosynthesis originally evolved because they made it possible for plants to grow better in the world's hot and dry regions. Researchers are now using these adaptations to fight world hunger. Specifically, they have introduced from corn into rice several genes that code for the C4 photosynthesis enzymes. Once in the rice, these genes increase the rice plant's ability to photosynthesize, thus leading to higher growth rates and food yields. The experiments are still in early stages, and whether the addition of C4 photosynthesis enzymes will make it possible to grow new crops on a large scale in previously inhospitable environments is not certain. Early results suggest, however, that this is a promising approach.

TAKE-HOME MESSAGE 4•11

C4 and CAM photosynthesis are evolutionary adaptations at the biochemical level that, although being more energetically expensive than regular (C3) photosynthesis, allow plants in hot, dry climates to close their stomata and conserve water without shutting down photosynthesis.

The Nature of Energy   Photosynthesis   Cellular Respiration   Alternative Pathways

③ Cellular respiration converts food molecules into ATP, a universal source of energy for living organisms.

Hamburger powered! We extract energy from all the foods we eat.

4·12

How do living organisms fuel their actions? Cellular respiration: the big picture.

Food is fuel. And all of the activities of life—growing, moving, reproducing—require fuel. Plants, most algae, and some bacteria obtain their fuel directly from the energy of sunlight, which they harness through photosynthesis. Less self-sufficient organisms, such as humans, alligators, and insects, must extract the energy they need from the food they eat.

This energy comes from photosynthetic organisms either directly (from eating plants) or indirectly (from eating animals that eat plants) (**FIGURE 4-26**).

All living organisms extract energy from the chemical bonds of molecules (which can be considered "food") through a

FIGURE 4-26 Living organisms require fuel.

143

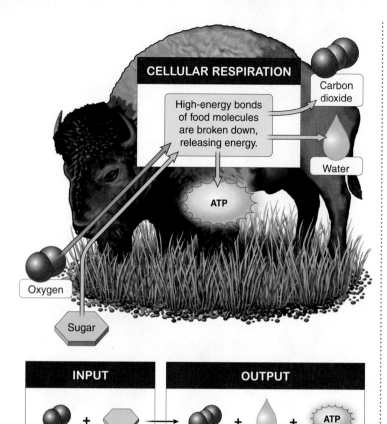

CELLULAR RESPIRATION

High-energy bonds of food molecules are broken down, releasing energy.

Carbon dioxide

Water

ATP

Oxygen

Sugar

INPUT	OUTPUT
Oxygen + Sugar	Carbon dioxide + Water + Energy (ATP)

FIGURE 4-27 Cellular respiration: the big picture.

process called cellular respiration (**FIGURE 4-27**). This process is a bit like photosynthesis in reverse. In photosynthesis, the energy of the sun is captured and used to build molecules of sugars, such as glucose. In cellular respiration, plants and animals break down the chemical bonds of sugar and other

energy-rich food molecules (such as fats and proteins) to release the energy that went into creating them. (Don't confuse cellular respiration with the act of breathing, which is also called *respiration*.) As energy is released by these reactions, cells capture it and store it in the bonds of ATP molecules. This plentiful, readily available stored energy can then be tapped as needed to fuel the work of the life-sustaining activities and processes of all living organisms.

In humans and other animals, cellular respiration starts after we eat food, digest it, absorb the nutrient molecules into the bloodstream, and deliver them to the cells of our bodies. At this point, our cells begin to extract some of the energy stored in the bonds of the food molecules. We focus here on the breakdown of glucose, but later in this chapter we'll see that the process is similar for the breakdown of fats or lipids. Ultimately, when a food molecule has been completely processed, the cell has used the food molecule's stored energy (along with oxygen) to create a large number of high-energy-storing ATP molecules (which supply energy to power the cell's activities), water, and carbon dioxide (which is exhaled into the atmosphere).

TAKE-HOME MESSAGE 4·12

Living organisms extract energy through a process called cellular respiration, in which the high-energy bonds of sugar and other energy-rich molecules are broken, releasing the energy that went into creating them. The cell captures the food molecules' stored energy in the bonds of ATP molecules. This process requires fuel molecules and oxygen and it yields ATP molecules, water, and carbon dioxide.

4·13

The first step of cellular respiration: glycolysis is the universal energy-releasing pathway.

To generate energy, fuels such as glucose and other carbohydrates and proteins and fats are broken down in three steps: (1) glycolysis, (2) the Krebs cycle, and (3) the electron transport chain. **Glycolysis** means the splitting (*lysis*) of sugar (*glyco-*) and it is the first step all organisms on the planet take in breaking down food molecules; for many single-celled organisms, this one step is sufficient to provide all of the energy they need (**FIGURE 4-28**).

As **FIGURE 4-29** illustrates, glycolysis is a sequence of chemical reactions (there are 10 in all) through which glucose is broken down to yield two molecules of a substance called **pyruvate.** Glycolysis has two distinct phases: an "uphill" preparatory phase and a "downhill" payoff phase.

Just as you sometimes have to spend money to make money, before any energy can be extracted from glucose, some

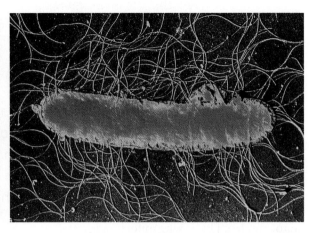

FIGURE 4-28 Plants, animals, bacteria, and all other organisms undergo glycolysis.

energy must be added to the molecule. This addition occurs during the "uphill" phase. The additional energy (which comes from ATP) makes the glucose molecule less stable and ripe for chemical breakdown. Once the glucose has been destabilized, it can be broken down chemically, and the energy stored in its bonds can be harnessed as the bonds are broken.

Three of the 10 steps in glycolysis yield energy. In two of these three, as bonds from the sugar are broken, the energy released is quickly harnessed by the attachment of phosphate groups to molecules of ADP to create energy-rich ATP molecules. In the third energy-yielding step of glycolysis, electrons originally from the glucose are transferred to NAD^+ to become the high-energy electron carrier NADH. Later (in the electron transport chain) this energy will be converted to ATP. The net result of glycolysis is that each glucose molecule is broken down into two molecules of pyruvate. During this breakdown, some of the released energy is captured in the production of energy-rich ATP molecules and molecules of the high-energy electron carrier NADH. Two molecules of water are also produced during glycolysis.

In the absence of oxygen and in many yeasts and bacteria, glycolysis is the only game in town for fueling activity. Because single-celled organisms have much lower energy needs, they can function solely on the yields of glycolysis. For many organisms (including humans), however, glycolysis is a springboard to further energy extraction. The additional energy payoffs come from the Krebs cycle and the electron transport chain.

GLYCOLYSIS

1 PREPARATORY PHASE

ATP (2)

Pᵢ

Unstable molecule prepared to be broken down

Glucose

−2 ATP

Gycolysis takes place in the cell's cytoplasm.

2 PAYOFF PHASE

Pᵢ

ATP (4)

NADH (2)

Water

Pyruvate (2)

+4 ATP +2 NADH

FIGURE 4-29 Glycolysis up close.

TAKE-HOME MESSAGE 4·13

Glycolysis is the initial phase in the process by which all living organisms harness energy from food molecules. Glycolysis occurs in a cell's cytoplasm and uses the energy released from breaking chemical bonds in food molecules to produce high-energy molecules, ATP and NADH.

The second step of cellular respiration: the Krebs cycle extracts energy from sugar.

Cells could stop extracting energy when glycolysis ends, but they rarely do because that would be like leaving most of your meal on your plate. This is because in glycolysis only a small fraction of the energy stored in sugar molecules is recovered and converted to ATP and NADH. Cells get much more of an "energy bang" for their "food buck" in the steps following glycolysis, which occur in the mitochondria. This is why mitochondria are considered ATP "factories." In the mitochondria, the molecules produced from the breakdown of glucose during glycolysis are broken down further, during two steps that are dramatically more efficient at capturing the energy in food molecules: the Krebs cycle (the subject of this section) and the electron transport chain (the subject of the next section). In breaking down the products of glycolysis, the **Krebs cycle** produces some additional molecules of ATP and, more important, captures a huge amount of chemical energy by producing high-energy electron carriers.

Before the Krebs cycle can begin, however, the end products of glycolysis—two molecules of pyruvate for every molecule of glucose used—must be modified. To start this modification process, the pyruvate molecules move from the cytoplasm into the mitochondria. Once in the mitochondria, the pyruvate molecules undergo three quick modifications that prepare them to be broken down in the Krebs cycle (**FIGURE 4-30**):

Modification 1. Each pyruvate molecule passes some of its high-energy electrons to the energy-accepting molecule NAD$^+$, building two molecules of NADH.

Modification 2. Next, a carbon atom and two oxygen atoms are removed from each pyruvate molecule and released as carbon dioxide. The CO_2 molecules diffuse out of the cell and, eventually, out of the organism. In humans, for example, these CO_2 molecules pass into the bloodstream and are transported to the lungs, from which they are eventually exhaled.

Modification 3. In the final step in the preparation for the Krebs cycle, a giant compound known as coenzyme A attaches itself to the remains of each pyruvate molecule, producing two molecules called acetyl-CoA. Each acetyl-CoA molecule is now ready to enter the Krebs cycle.

> **Q** Aerobic training can cause our bodies to produce more mitochondria in cells. Why is this beneficial?

There are eight separate steps in the Krebs cycle, but our emphasis here is on its three general outcomes (**FIGURE 4-31**):

Outcome 1. *A new molecule is formed.* Acetyl-CoA adds its two-carbon acetyl group to a molecule of the starting material of the Krebs cycle, a four-carbon molecule called oxaloacetate. This process creates a six-carbon molecule.

Outcome 2. *High-energy electron carriers (NADH) are made and carbon dioxide is exhaled.* The six-carbon molecule then gives electrons to NAD$^+$ to make the high-energy electron carrier NADH. The six-carbon molecule releases two carbon atoms along with four oxygen atoms to form two carbon dioxide molecules. This CO_2 is carried by the bloodstream to the lungs from which it is exhaled into the atmosphere.

ACETYL-CoA PRODUCTION

Glucose → Pyruvate = Molecular model of pyruvate

1 Each pyruvate molecule passes a high-energy electron to NAD$^+$, creating NADH.

2 A carbon and two oxygen atoms are released as carbon dioxide.

3 Coenzyme A attaches itself to the remaining molecule, creating acetyl-CoA.

Modifications of pyruvate take place in the cell's mitochondria.

NAD$^+$ → NADH

Carbon dioxide

+ Coenzyme A → Acetyl-CoA

to Krebs cycle

FIGURE 4-30 Preparation of pyruvate. In the mitochondria, pyruvate must be modified before it can be broken down in the Krebs cycle.

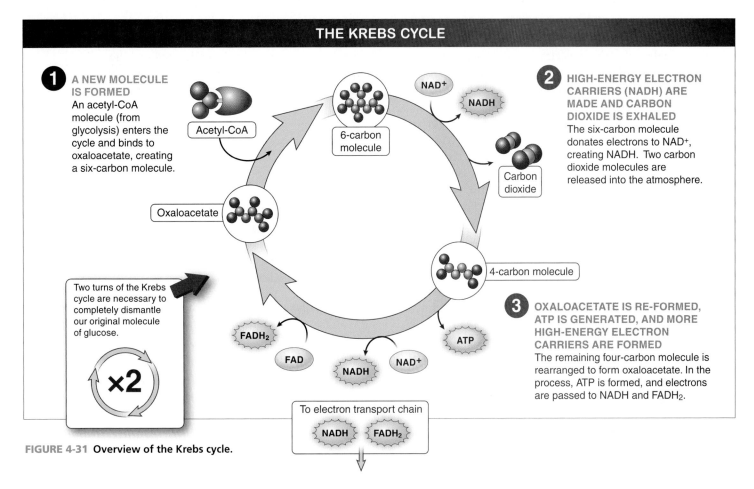

THE KREBS CYCLE

1 **A NEW MOLECULE IS FORMED**
An acetyl-CoA molecule (from glycolysis) enters the cycle and binds to oxaloacetate, creating a six-carbon molecule.

Acetyl-CoA

Oxaloacetate

6-carbon molecule

NAD+

NADH

Carbon dioxide

4-carbon molecule

2 **HIGH-ENERGY ELECTRON CARRIERS (NADH) ARE MADE AND CARBON DIOXIDE IS EXHALED**
The six-carbon molecule donates electrons to NAD+, creating NADH. Two carbon dioxide molecules are released into the atmosphere.

3 **OXALOACETATE IS RE-FORMED, ATP IS GENERATED, AND MORE HIGH-ENERGY ELECTRON CARRIERS ARE FORMED**
The remaining four-carbon molecule is rearranged to form oxaloacetate. In the process, ATP is formed, and electrons are passed to NADH and FADH₂.

Two turns of the Krebs cycle are necessary to completely dismantle our original molecule of glucose.

×2

FADH₂

FAD

NADH

NAD+

ATP

To electron transport chain

NADH   FADH₂

FIGURE 4-31 **Overview of the Krebs cycle.**

Outcome 3. *The starting material of the Krebs cycle is re-formed, ATP is generated, and more high-energy electron carriers are formed.* After the CO_2 is released, the four-carbon molecule that remains from the original pyruvate-oxaloacetate molecule formed in Outcome 1 is modified and rearranged to once again form oxaloacetate, the starting material of the Krebs cycle. In the process of this reorganization, one ATP molecule is generated and more electrons are passed to one familiar high-energy electron carrier, NADH, and a new one, FADH₂. The formation of these high-energy electron carriers increases the energy yield of the Krebs cycle. One oxaloacetate is re-formed, and the cycle is ready to break down the second molecule of acetyl-CoA. Two turns of the cycle are necessary to completely dismantle our original molecule of glucose.

We started this process of cellular respiration with a molecule of glucose, and now that we have seen the Krebs cycle in its entirety, let's trace the path of the original six carbons in that glucose molecule.

1. Glycolysis: the six-carbon starting point. Glucose is broken down into two molecules of pyruvate. No carbons are removed.

2. Preparation for the Krebs cycle: two carbons are released. Two pyruvate molecules are modified to enter the Krebs cycle and they each lose a carbon atom, in the form of two molecules of carbon dioxide.

3. Krebs cycle: the last four carbons are released. A total of four carbon molecules enter the Krebs cycle in the form of acetyl Co-A. For each turn of the Krebs cycle, two molecules of carbon dioxide are released. So the two final carbons are released into the atmosphere during the second turn of the wheel. Poof! The six carbon atoms that were originally present in our single molecule of glucose are no longer present.

In a sense, the carbon atoms that were first plucked from the atmosphere to make sugar during photosynthesis have been exhaled back into the atmosphere as six molecules of carbon dioxide. At this point, we've come full circle. In photosynthesis, carbon atoms from the air were used to build sugar molecules, which had energy stored in the bonds between their carbon, hydrogen and oxygen atoms. In cellular respiration, the energy previously stored in the bonds of the sugar is converted to molecules of ATP, NADH, and FADH₂. Carbon atoms from sugar are exhaled back into the air as CO_2 and water is produced. What happens to the high-energy electron carriers, NADH and FADH$_2$? They eventually give up their high-energy electrons to the electron transport chain. The energy released as those electrons pass through the transport chain is captured in the bonds of more ATP molecules. We explore that process in the next section.

Q Why might mitochondria malfunctions play a role in lethargy or fatigue?

As we explore the important roles of the Krebs cycle and electron transport chain in generating usable energy for organisms, it's important to note that mitochondria malfunctions have serious consequences for health. More than a hundred genetic mitochondrial disorders have been identified, all of which can lead to energy shortage, including muscle weakness, fatigue, and muscle pain. It

appears, for instance, that many cases of "exercise intolerance"—extreme fatigue or cramps after only slight exertion—may be related to inherited mutations in the mitochondrial DNA. Damage to our cellular powerhouses, not surprisingly, can create personal energy crises.

TAKE-HOME MESSAGE 4·14

A huge amount of additional energy can be harvested by cells after glycolysis. First the end product of glycolysis, pyruvate, is chemically modified. Then, in the Krebs cycle, the modified pyruvate is broken down step by step. This breakdown releases carbon molecules to the atmosphere as bonds are broken and captures some of the released energy in two ATP molecules and numerous high-energy electron carriers.

4·15

The third step of cellular respiration: ATP is built in the electron transport chain.

How do we finally get a big payoff of usable energy from our glucose molecule? Glycolysis and the Krebs cycle produce a few molecules of ATP for each molecule of glucose broken down, but it is the energy held in the high-energy electron carriers NADH and FADH$_2$ generated in these processes that ultimately generates the largest amount of usable energy in the form of ATP. In fact, almost 90% of the energy payoff from a molecule of glucose is harvested in the final step of cellular respiration when the electrons from NADH and FADH$_2$ move along the electron transport chain. This process, like the Krebs cycle, takes place in the mitochondria.

In a manner that is similar to that seen in the chloroplast during photosynthesis, mitochondria convert kinetic energy (from electrons) into potential energy (a concentration gradient of protons). Two structural features of mitochondria are essential to their impressive ability to harness energy from molecules.

Feature 1. Mitochondria have a "bag-within-a-bag" structure that makes it possible to harness the potential energy in the bonds of NADH and FADH$_2$ molecules to produce ATP (**FIGURE 4-32**). Material inside the mitochondrion can lie in one of two spaces: (1) the

FIGURE 4-32 "A bag within a bag." The structure of mitochondria makes possible their impressive ability to harness energy from food molecules.

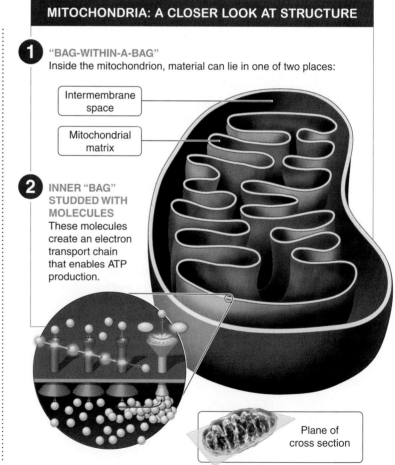

MITOCHONDRIA: A CLOSER LOOK AT STRUCTURE

1 "BAG-WITHIN-A-BAG"
Inside the mitochondrion, material can lie in one of two places:

Intermembrane space

Mitochondrial matrix

2 INNER "BAG" STUDDED WITH MOLECULES
These molecules create an electron transport chain that enables ATP production.

Plane of cross section

intermembrane space, which is outside the inner bag, or (2) the **mitochondrial matrix,** which is inside the inner bag.

With two distinct regions separated by a membrane, the mitochondrion can create higher concentrations of molecules in one area or the other. And because a concentration gradient is a form of potential energy—molecules move from the high-concentration area to the low-concentration area the way water rushes down a hill—once a gradient is created, the energy released as the gradient equalizes itself can be used to do work. In the electron transport chain, this energy is used to build the energy-rich molecule ATP.

Feature 2. The inner bag of the mitochondria is studded with molecules. These molecules, mostly electron carriers, are sequentially arranged within the inner membrane of the mitochondrion, hence their description as a "chain." This arrangement makes it possible for the molecules to hand off electrons in an orderly sequence.

Now let's explore how these features of mitochondria make it possible to harness energy from high-energy electron carriers (**FIGURE 4-33**).

> **Q** Over-the-counter NADH pills provide energy to sufferers of chronic fatigue syndrome. Why might this be?

Step 1 of the electron transport chain begins with NADH and $FADH_2$ in the mitochondrial matrix (inside the inner bag) moving to the inner mitochondrial membrane. There, the high-energy electrons they carry are transferred to molecules embedded within the membrane. After they donate their electrons, the molecules that remain, NAD^+ and FAD^+, are recycled back to the Krebs cycle.

The embedded molecules pass the electrons to the next carrier, which passes the electrons to the next, and so on. At each handoff, a bit of energy is released. Thus, as electrons move from one carrier to another through the electron transport chain, they lose energy at each handoff.

At the end of the chain (step 2), the lower-energy electrons are handed off to oxygen, which then combines with free H^+ ions in the mitochondrial fluid to form water.

As shown in step 3 of Figure 4-33, most of the energy released at each handoff from one electron carrier to another in the electron transport chain is used to pump protons (H^+ ions) from the mitochondrial matrix across the membrane and into the intermembrane space. As more and more protons are pumped across the membrane and packed in the

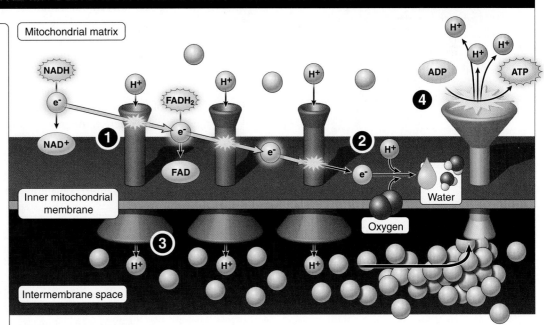

THE MITOCHONDRIAL ELECTRON TRANSPORT CHAIN

High-energy electrons are passed from the carriers NADH and FADH₂ to a series of molecules embedded in the inner mitochondrial membrane called the electron transport chain.

1 At each step in the electron transport chain's sequence of handoffs, the electrons fall to a lower energy state, releasing a little bit of energy.

2 At the end of the chain, the lower-energy electrons are handed off to oxygen, which then combines with free H^+ ions to form water.

3 The energy is used to power proton pumps, which pack hydrogen ions from the mitochondrial matrix into the intermembrane space.

4 The protons rush back into the mitochondrial matrix with great kinetic energy, which can be used to build ATP.

FIGURE 4-33 The big energy payoff. Most of the energy harvested during cellular respiration is generated by the electron transport chain in the mitochondria.

SUMMARY OF CELLULAR RESPIRATION

1 GLYCOLYSIS

Glucose

Pyruvate

ATP

Cytoplasm

Mitochondria

2 ACETYL-CoA PRODUCTION

Pyruvate

Carbon dioxide

Acetyl-CoA

3 KREBS CYCLE

Carbon dioxide

NADH   FADH₂

ATP

4 ELECTRON TRANSPORT CHAIN

e⁻

e⁻

Oxygen

Water

ATP

Each step in the breakdown of food increases the amount of usable energy that is generated!

FIGURE 4-34 The steps of cellular respiration: from glucose to usable energy.

intermembrane space, a concentration gradient is created. This gradient represents a significant source of potential energy.

If this description seems familiar, it is. In chloroplasts, during photosynthesis, great numbers of protons are pumped from the stroma outside the thylakoid sacs to the inside of the thylakoids. We likened this potential energy to the potential energy of water in an elevated tower, which can be released with great force. Similarly, in step 4 of the mitochondrial electron transport chain, the protons pumped into the intermembrane space rush back to the inner mitochondrial matrix through channels in the inner mitochondrial membrane. And as the protons pass through, the force of their flow fuels the attachment of free-floating phosphate groups to ADP to produce ATP.

In the end, the number of ATP molecules generated from the complete dismantling of one molecule of glucose is about 36, most of which are produced with the energy harnessed from high-energy electron carriers as they pass their electrons down the electron transport chain (**FIGURE 4-34**).

Given the central role of the electron transport chain in the generation of usable energy from the breakdown of food molecules, any interference in its functioning has dire consequences. And in fact, murder by cyanide poisoning, an old tradition in detective stories, is just such an interference. When cyanide gets into the mitochondria, it binds to a molecule in the electron transport chain, preventing it from accepting electrons. This halts the transfer of electrons and the pumping of protons across the mitochondrial membrane. As a consequence, the production of ATP that would occur when protons rush back across the membrane down their concentration gradient ceases. Halting the production of ATP removes a cell's energy source, starving it very quickly. For this reason, cyanide poisoning can cause death within minutes.

> **Q** *Cyanide blocks the passage of electrons to oxygen in the electron transport chain. Why does this make it a toxic poison?*

TAKE-HOME MESSAGE 4·15

The largest energy payoff of cellular respiration comes as electrons from NADH and FADH₂ produced during glycolysis and the Krebs cycle move along the electron transport chain. The electrons are passed from one carrier to another and energy is released, pumping protons into the intermembrane space. As the protons rush back to the inner mitochondrial matrix, the force of their flow fuels the production of large amounts of ATP.

The Nature of Energy   Photosynthesis   Cellular Respiration   Alternative Pathways

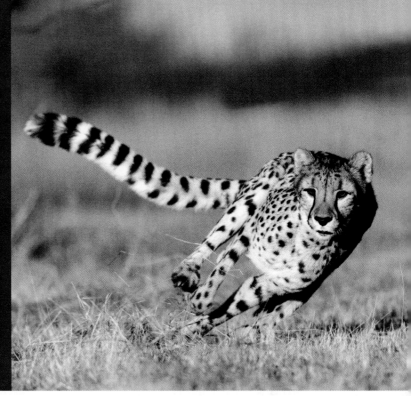

❹ There are alternative pathways to energy acquisition.

Unlike cars, living organisms (such as this cheetah) can use multiple different types of fuel.

4·16

Beer, wine, and spirits are by-products of cellular metabolism in the absence of oxygen.

Every beer brewery, the entire wine industry, and all distilleries of whiskey, vodka, tequila, and other alcoholic beverages owe their existence to microscopic yeast cells scrambling to break down their food for energy under stressful conditions. To better understand how yeast metabolism produces the alcohol we drink, it helps to investigate first what happens when humans and other animals try to metabolize energy from sugar molecules under some stressful conditions.

FIGURE 4-35 Energy production without oxygen. Organisms have a backup method for breaking down sugar when oxygen is not present.

If you run or swim as fast as you can, you soon feel a burning sensation in your muscles. What is happening? Your muscle cells are suddenly becoming very acidic. This acid buildup occurs when we demand of our bodies bursts of energy beyond that which they can sustain (**FIGURE 4-35**). (The next-day muscle soreness is not due to acid buildup, which goes away in a matter of minutes or hours; it is caused by damage to the muscle fibers. They break down a bit before growing stronger.)

With rapid, strenuous exertion, our bodies soon fall behind in delivering oxygen from the lungs to the bloodstream to the cells and finally to the mitochondria. Oxygen deficiency then limits the rate at which the mitochondria can break down fuel and produce ATP (**FIGURE 4-36**). This slowdown in ATP production occurs because the electron transport chain requires oxygen as the final acceptor of all the electrons that are generated during glycolysis and the Krebs cycle. If oxygen is in short supply, the electrons from NADH (and $FADH_2$) have nowhere to go. Consequently, the regeneration of NAD^+ (and FAD^+) in the electron transport chain is halted, leaving no recipient for the high-energy electrons harvested from the breakdown of glucose and pyruvate, and the whole process of cellular respiration can grind to a stop. Organisms don't let this interruption last long, though; most have a backup method for breaking down sugar.

151

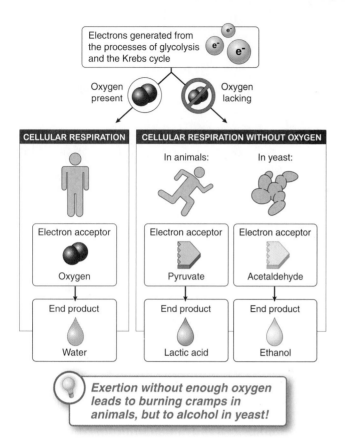

FIGURE 4-36 Energy production compared: with and without oxygen. Energy production without oxygen (anaerobic) is also called fermentation.

Among animals, there is one willing acceptor for the NADH electrons in the absence of oxygen: pyruvate, the end product of glycolysis. When pyruvate accepts the electrons, it forms lactic acid (see Figure 4-36). Once the NADH gives up its electrons, NAD$^+$ is generated, and glycolysis can continue. But as lactic acid builds up, it causes a burning feeling in our muscles. It's not an ideal situation, but if you are trying to escape a predator or are otherwise engaged in strenuous activity, the two ATP molecules generated from each glucose molecule during glycolysis are better than nothing.

Like humans, yeast normally use oxygen during their breakdown of food. And like humans, they have a backup method when oxygen is not available. But yeast make use of a different electron acceptor than do humans when oxygen isn't available, and the resulting reaction leads to the production of all drinking alcohol.

In these single-celled organisms, after glycolysis pyruvate is usually converted to a molecule called *acetaldehyde.* In the absence of oxygen, this molecule accepts the electrons released from NADH, allowing glycolysis to resume. Acetaldehyde's acceptance of NADH's electrons results in the production of **ethanol,** the molecule that gives beer, wine, and spirits their

kick. Ethanol can also be used as a fuel source because it can be combusted in much the same way as the biofuels generated from animal fats and plant oils described earlier in this chapter.

The process by which glycolysis occurs in the absence of oxygen, in which alternative molecules are used as electron acceptors, is called **fermentation.** Interestingly, although ethanol is always the alcohol produced by fermentation, the flavor of the output of fermentation depends on the type of sugar metabolized by the yeast. Fruits, vegetables, and grains all give different results. If the sugar comes from grapes, wine is produced. If the sugar comes from a germinating barley plant, beer is produced. Potatoes, on the other hand, are the sugar source usually used to produce vodka.

Because yeast prefer the more efficient process of aerobic respiration, they produce alcohol only in the absence of oxygen. That is why the fermentation tanks used in producing wine, beer, and other spirits are built specifically to keep oxygen out (**FIGURE 4-37**). Only in this circumstance will yeast be forced to use their backup respiratory pathway of fermentation.

TAKE-HOME MESSAGE 4·16

Oxygen deficiency limits the breakdown of fuel because the electron transport chain requires oxygen as the final acceptor of the electrons generated during glycolysis and the Krebs cycle. When oxygen is unavailable, yeast resort to fermentation, in which they use a different electron acceptor, pyruvate, generating in the process ethanol, the alcohol in beer, wine, and spirits.

Yeast is used in the production of beer.

FIGURE 4-37 Just a by-product of metabolism under stressful conditions. Beer, wine, and spirits are by-products of cellular metabolism in the absence of oxygen.

The Nature of Energy   Photosynthesis   Cellular Respiration   Alternative Pathways

Eating a complete diet: cells can run on protein and fat as well as on glucose.

Most automobiles can run on only one type of fuel, gasoline. If you run out of fuel and cannot get to a gas station, you are out of luck. In this chapter, we have examined the steps by which plants, animals, and other organisms use glucose as fuel. But living organisms are more flexible than autos when it comes to fuel sources. They also have more complex needs than an automobile because their fuel must also provide raw materials for growth.

Evolution has built humans and other organisms with the metabolic machinery that allows them to extract energy and other valuable chemicals from proteins, fats, and a variety of carbohydrates (**FIGURE 4-38**). For that reason, we are able to consume and efficiently utilize meals comprising various combinations of molecules.

Sugars. In the case of dietary sugars, many are polysaccharides—multiple simple sugars linked together—rather than solely the simple sugar glucose. Before they can be broken down by cellular respiration, the polysaccharides must first be separated by enzymes into glucose or related simple sugars that can be broken down by cellular respiration.

Lipids. Dietary lipids are broken down into their two constituent parts: a glycerol molecule and fatty acids. The glycerol is chemically modified into one of the molecules produced during one of the 10 steps of glycolysis. It then enters glycolysis at that step in the process and is broken down to yield energy. The fatty acids, meanwhile, are chemically modified into acetyl-CoA, at which point they enter the Krebs cycle.

Proteins. Proteins are chains of amino acids. Upon consumption they are broken down chemically into their constituent amino acids. Once that is done, each amino acid is broken down into (1) an amino group that may be used in the production of tissue or excreted in the urine and (2) a carbon compound that is converted to one of the intermediate compounds in glycolysis or the Krebs cycle, allowing the energy stored in its chemical bonds to be harnessed.

In the end, humans (indeed, all animals) are able to harvest energy from a variety of food sources beyond simple glucose. Whether a meal contains carbohydrates, lipids, proteins, or some combination thereof, the nutrients are chemically modified in some preliminary steps and are then

fed into one of the intermediate steps in glycolysis or the Krebs cycle to furnish usable energy for the organism (see Figure 4-38).

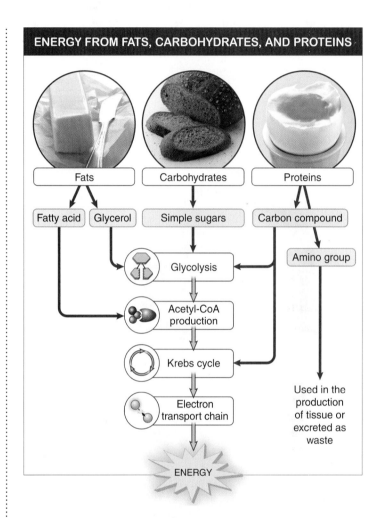

FIGURE 4-38 **"More flexible than autos."** Animals are able to harvest energy from proteins, carbohydrates, and lipids.

TAKE-HOME MESSAGE 4•17

Humans and other organisms have metabolic machinery that allows them to extract energy and other valuable chemicals from proteins, fats, and carbohydrates in addition to the simple sugar, glucose.

Knowledge You Can Use

Did you know? **If you feed and protect your flowers in a vase, they'll last longer.**

Fact: Like animals, plants require fuel for the cellular activities that keep them alive.

Q: **How do plants get the energy they need to stay alive?** Plants use **photosynthesis to harness light energy,** converting it to sugar molecules that serve as their food.

Q: **Can plants still photosynthesize once they're in a vase in your house?** Yes, but humans don't make it easy. Once cut, plants generally cannot produce sufficient sugar through photosynthesis. **Light levels in houses tend to be a bit too low, and the loss of many or most plant leaves reduces the number of chloroplasts in which photosynthesis can occur.** So they're starving to death.

Q: **Can you slow their demise?** Yes. **Plants are able to take up sugar in the vase water and use it as an energy source** for cell activities. So adding a bit of sugar is like putting fuel in their tank.

Q: **Is it that easy?** No. Putting sugar—a molecule with lots of energy stored within its chemical bonds—in the vase is like offering a free lunch. And many, many organisms are looking for a free lunch. Unfortunately, **when you add sugar, bacteria on the stems can grow rapidly, blocking the water-conducting tubes in flower stems.** This slows the flow not just of sugar, but of water as well.

Q: **What should you do?** **With the addition to vase water of both sugar and an antibacterial chemical such as chlorine bleach, you can feed and protect your flowers,** significantly increasing their longevity in vases.

Conclusion: Here's a useful summary from Martha Stewart (www .marthastewart.com): "Flowers last longer if their stems are cut under water. Use a sharp knife, and cut the stem on a slant so it can absorb as much water as possible. Next, place [the flowers] in a flask filled with 2 inches of warm water (no hotter than 110 degrees); this expedites the flow of water into the flower. Add either cut-flower food or a teaspoon of sugar and a couple drops of liquid bleach to a vase filled with cool water up to the point where the flower foliage will hit. After 5 minutes, transfer [the flowers] to the vase."

1 **Energy flows from the sun and through all life on earth.**

Energy is neither created nor destroyed but can change forms. Energy from the sun fuels all life on earth as it is converted to different forms. Each conversion is inefficient and some of the usable energy is converted to less useful heat energy. Cells temporarily store energy in the bonds of ATP molecules. This potential energy can be converted to kinetic energy and used to fuel life-sustaining chemical reactions.

2 **Photosynthesis uses energy from sunlight to make food.**

In the "photo" part of photosynthesis, chloroplasts transform light energy into the chemical energy of ATP and NADPH, while splitting water molecules and producing oxygen. In the Calvin cycle, the "synthesis" part of photosynthesis, carbon from CO_2 in the atmosphere is attached to molecules in chloroplasts to build sugars. This production of sugars consumes ATP and NADPH generated in the "photo" part of photosynthesis. C4 and CAM photosynthesis are evolutionary adaptations at the biochemical level that, although having competitive drawbacks, allow plants in hot, dry climates to close their stomata and conserve water without shutting down photosynthesis.

3 **Cellular respiration converts food molecules into ATP, a universal source of energy for living organisms.**

In cellular respiration, glucose and oxygen are converted to carbon dioxide, water, and energy that is stored in the high-energy bonds of ATP molecules. In the electron transport chain, high-energy electrons from NADH and $FADH_2$ pass from carrier to carrier, releasing energy with each step. This energy fuels the creation of a proton concentration gradient. The energy released as the protons move back down their concentration gradient is used to produce ATP. Overall, the cellular respiration of one glucose molecule yields about 36 molecules of ATP.

4 **There are alternative pathways to energy acquisition.**

Organisms are built with multiple systems for acquiring energy. They can generate ATP even when oxygen is lacking or when organic molecules other than glucose, such as lipids, proteins, and polysaccharides, are consumed.

KEY TERMS

1. Animal fats and plant oils are sometimes used as sources of fuel for automobile engines. How is energy harvested from these molecules?
   a) They contain long chains of hydrogen and carbon that, when broken, release the energy stored in the bonds linking the atoms together.
   b) They contain hydrogen and carbon tails linked by disulfide bridges that, when broken, release chemical energy.
   c) They contain multiple phosphate groups that each release energy when "liberated" from the molecule chemically.
   d) They contain long hydrophobic regions that, when mixed with water, generate explosive resistances.
   e) They contain long carbon tails and each atom has unpaired electrons that are released on exposure to extreme heat and pressure.

2. A cyclist rides her bike up a very steep hill. Pick from the following choices the statement that properly describes this example in energetic terms.
   a) Potential energy in food is converted to kinetic energy as the cyclist's muscles push her up the hill.
   b) Kinetic energy is highest when the cyclist is at the crest of the hill.
   c) The cyclist produces the most potential energy as she cruises down the hill's steep slope.
   d) Potential energy is greatest when the cyclist is at the top of the hill.
   e) Both a) and d) are correct.

3. Every time a source of energy is converted from one form to another:
   a) the potential energy of the system increases.
   b) heat is required.
   c) the second law of thermodynamics is violated.
   d) the total amount of energy in the universe is reduced by a tiny amount.
   e) some of the energy is converted to heat, which is not a very usable form of kinetic energy.

4. In your body, when energy is released from the breakdown of a molecule such as glucose:
   a) adenosine monophosphate is created.
   b) adenosine diphosphate is created.
   c) some of the energy may be harnessed by building the unstable high-energy bonds that attach phosphate groups to ADP.
   d) molecules of ATP are required to capture and absorb the heat generated by the reaction.
   e) all of the energy is lost as heat.

5. A green plant can carry out photosynthesis if given nothing more than:
   a) water, light, and carbon dioxide.
   b) water, light, and oxygen.
   c) carbon dioxide.
   d) oxygen.
   e) oxygen and carbon dioxide.

6. The actual production of sugars during photosynthesis takes place:
   a) within the outer membrane of the chloroplast.
   b) within the stroma, inside the thylakoids of the chloroplast.
   c) within the stroma, outside the thylakoid, but still inside the chloroplast.
   d) just outside the chloroplast, within the mitochondria.
   e) within the thylakoid membrane of the chloroplast.

7. The leaves of plants can be thought of as "eating" sunlight. From an energetic perspective this makes sense because:
   a) light energy, like chemical energy released when the bonds of food molecules are broken, is a type of kinetic energy.
   b) both light energy and food energy can be converted to kinetic energy without the loss of heat.
   c) the carbon-oxygen bonds within a photon of light release energy when broken by the enzymes in chloroplasts.
   d) the carbon-hydrogen bonds within a photon of light release energy when broken by the enzymes in chloroplasts.
   e) photons are linked together by hydrogen bonds, which release energy when striking the surface of a leaf.

8. A molecule of chlorophyll increases in potential energy:
   a) when it binds to a photon.
   b) when one of its electrons is boosted to a higher-energy excited state on being struck by a photon of light.
   c) when it loses an electron.
   d) only in the presence of oxygen.
   e) None of the above. The potential energy of a molecule cannot change.

9. Photosynthesizing plants rely on water:
   a) to provide the protons necessary to produce chlorophyll.
   b) to concentrate the beams of light hitting a leaf, focusing them on the reaction center.
   c) to replenish oxygen molecules that are lost during photosynthesis.
   d) to replace electrons that are excited by light energy and passed from molecule to molecule down an electron transport chain.
   e) to serve as a high-energy electron carrier.

10. During photosynthesis, which step is most responsible for a plant's acquisition of new organic material?
    a) the "building" of NADPH during the Calvin cycle
    b) the excitation of chlorophyll molecules by photons of light
    c) the "plucking" of carbon molecules from the air and fixing of them to organic molecules within the chloroplast
    d) the loss of water through evaporation
    e) the production of ATP during the light reactions

11. During C4 photosynthesis:
    a) plants use less ATP when producing sugar.
    b) plants are able to continue producing sugars even when they must close their stomata to reduce water loss on hot days.
    c) plants are able to generate water molecules to cool their leaves.
    d) plants are able to reduce water loss by producing more rubisco.
    e) plants are able to produce sugars without any input of carbon dioxide.

12. Cellular respiration is the process by which:
    a) oxygen is used to transport chemical energy throughout the body.
    b) oxygen is produced during metabolic activity.
    c) light energy is converted to kinetic energy.
    d) ATP molecules are converted to water and sugar.
    e) energy from the chemical bonds of food molecules is captured by an organism.

13. During cellular respiration, most of the energy contained within the bonds of food molecules is captured in:
    a) the conversion of the kinetic energy of food to the potential energy of ATP.
    b) the Krebs cycle and electron transport chain.
    c) digestion.
    d) glycolysis.
    e) None of the above. Energy is lost, not gained, during cellular respiration.

14. Which of the following energy-generating processes is the only one that occurs in all living organisms?
    a) the Krebs cycle
    b) glycolysis
    c) combustion
    d) photosynthesis
    e) None of the above. There are no energy-generating processes that occur in all living organisms.

15. During the Krebs cycle:
    a) the products of glycolysis are further broken down, generating additional ATP and the high-energy electron carrier NADH.
    b) high-energy electron carriers pass their energy to molecules of sugar, which store them as potential energy.
    c) the products of glycolysis are further broken down, generating additional ATP and the high-energy electron carrier NADPH.
    d) cellular respiration can continue even in the absence of oxygen.
    e) the products of glycolysis are converted to acetyl-CoA.

16. Mitochondria have a "bag within a bag" structure. This structure is necessary in order to:
    a) create two distinct regions with a concentration gradient, a form of potential energy.
    b) keep molecules of ADP in close proximity to molecules of the ATP-synthesizing enzyme.
    c) allow light-reactive accessory pigments to be embedded within the membranes.
    d) segregate the most toxic digestive enzymes from molecules of ATP and NADP$^+$.
    e) None of the above.

17. All alcoholic beverages are produced as the result of:
    a) cellular respiration by bacteria that occurs in the absence of oxygen.
    b) cellular respiration by bacteria that occurs in the absence of free electrons.
    c) cellular respiration by yeast that occurs in the absence of free electrons.
    d) cellular respiration that occurs in the absence of sugar.
    e) cellular respiration by yeast that occurs in the absence of oxygen.

18. In harvesting the chemical energy of the molecules in food:
    a) all macromolecules must first be converted to glucose.
    b) all macromolecules must first be converted to their hydrocarbon chains.
    c) all macromolecules must first be converted to glucose or another simple sugar.
    d) organisms can use sugars, lipids, and proteins.
    e) all macromolecules must first be converted to proteins or free-form amino acids.

SHORT-ANSWER QUESTIONS

1. A plant and a mouse are placed in a sealed container with a watering system. For a period of time, both will survive. Explain what the two organisms can provide for each other in order to allow them to survive temporarily.

2. It is possible to measure the rate of the "photo" reactions in plants by measuring O_2 levels produced by chloroplasts. If bicarbonate solution, a source of CO_2, is provided to the plants, the rate of the "photo" reactions increases.

    a) Explain why measuring O_2 levels provides an indication of the rate of the "photo" reactions.

    b) Being as specific as possible, describe how the plants will use the CO_2 provided by the bicarbonate solution.

    c) Propose an explanation of how increasing the CO_2 source for the plants will affect the rate of the "photo" reactions.

3. Some plants grow much more quickly than others. For a plant that is growing at a rapid rate, propose a hypothesis as to whether cellular respiration or photosynthesis is occurring at a faster rate. Justify your hypothesis.

See Appendix for answers. For additional study questions, go to www.prep-u.com.

5

DNA, Gene Expression, and Biotechnology

What is the code and how is it harnessed?

DNA microarrays measure the expression levels of many genes simultaneously.

1 DNA: What is it and what does it do?

5•1

"The DNA 200"—knowledge about DNA is increasing justice in the world.

In 1981, Julius Ruffin, a 27-year-old maintenance worker at Eastern Virginia Medical School, got onto an elevator and lost more than 20 years of his life. Several weeks earlier, a nursing student at the school had been raped by an attacker who broke into her apartment. When Ruffin got on the elevator, the student thought she recognized him as her attacker and called the police, who immediately arrested him.

Ruffin's girlfriend testified that he was with her at the time of the attack, but on the basis of the victim's eyewitness testimony, a jury found Ruffin guilty and sentenced him to life in prison. Ruffin maintained his innocence, to no avail, until 2003. At that time, the state's Division of Forensic Science performed a DNA analysis on a swab of evidence that remained from the investigation. The analysis revealed that Ruffin was not the

Q What is the most common reason why DNA analyses overturn incorrect criminal convictions?

attacker. Rather, the analysis showed that the DNA perfectly matched that of a man who was already in a Virginia prison, serving time for rape. Ruffin was freed, but only after having served more than two decades in prison (**FIGURE 5-1**).

Julius Ruffin's case is tragic, but it is not unusual. He is one of a group of 200 people in the United States who have been freed from prison as a result of DNA analyses. Called "the DNA 200," these unjustly imprisoned people spent an average of 12 years behind bars. Eighty percent had been convicted of sexual assault, 28% had been convicted of murder. In more than three-quarters of the cases, inaccurate eyewitness testimony played an important role in the guilty verdict. (Recall, from Chapter 1, the experiments that revealed the unreliability of eyewitness identification.)

What Is DNA?   Building Organisms   Genetic Damage   Biotechnology and Health

FIGURE 5-1 **Vindicated by DNA evidence.** Julius Ruffin was released from prison after two decades of wrongful incarceration.

In this chapter, we take a close look at DNA, the molecule responsible for Julius Ruffin's exoneration and the deferred justice served to the DNA 200. It is a molecule that all living organisms—people, plants, animals, bacteria, and otherwise—carry in almost every cell in their body (with just a few exceptions, such as red blood cells). Like a social security number, DNA is unique in virtually every person. In addition to being contained in our living cells, our DNA exists in what we leave behind. It is in our saliva, hair, blood, and even the dead skin cells that fall from our bodies. This is why DNA can serve as an individual identifier, a trail we all leave that is increasingly being used to ensure greater justice in our society.

The importance of DNA goes far beyond its function as an individual identifier, however. The information carried within this molecule, which is organized into individual units called genes, is among the most important of all biological knowledge. And, as witnessed by the following excerpts, it is often in the news.

> "Selfish dictators may owe their behaviour partly to their genes, according to a study that claims to have found a genetic link to ruthlessness."
> —*Nature,* April 2008

> "Whether a man has one type of gene versus another could help decide whether he's good 'husband material,' a new study suggests."
> —*Washington Post,* September 2008

By all accounts, we live in the "Age of the Gene." It's nearly impossible to open a newspaper or watch a news report these

FIGURE 5-2 **Genetics issues are in the news.**

days without being informed that yet another complex human characteristic or trait has been linked to a newly discovered gene (**FIGURE 5-2**). But what does that actually mean? What is this entity that apparently exerts influence over everything from our marriage readiness to our ruthlessness? We explore these issues and more in this chapter, beginning with a look at the structure of DNA itself and how it contains the information for producing organisms. In later sections of the chapter, we learn how modern manipulations of DNA are having far-reaching implications not just for human health but for agriculture as well.

TAKE-HOME MESSAGE 5·1

DNA is a molecule that all living organisms carry in every cell in their body. Because every person's DNA is unique and because we leave a trail of DNA behind us as we go about our lives, DNA can serve as an individual identifier. This fact is the basis of the DNA analysis that is being used increasingly to ensure greater justice in our society, such as by establishing the innocence of individuals wrongly convicted of crimes.

The DNA molecule contains instructions for the development and functioning of all living organisms.

Beginning in the 1900s and continuing through the early 1950s, a series of experiments revealed two important features of DNA. First, molecules of DNA are passed down from parent to offspring. Second, the instructions on how to create a body and control its growth and development are encoded in the DNA molecule.

Once scientists knew that DNA encoded instructions and that these instructions were passed on from one generation to the next, DNA became something of a celebrity molecule. Given that it must be able to hold the instructions for how to produce every possible type of structure in every living organism, scientists were crazed to learn all they could about it. There was a spirited race to determine the chemical structure of DNA and to understand how the molecule was assembled and shaped so that it could hold and transmit so much information.

Many formidable scientists rose to the challenge of determining the structure of DNA. American Linus Pauling had already won a Nobel prize in chemistry for his work on elucidating the structure of molecules when he began investigating the structure of DNA. Simultaneously, Maurice Wilkins and Rosalind Franklin in England devoted their research to this task as well, and produced X-ray pictures of DNA that were critical to decoding its shape. But it was Englishman Francis Crick and American James Watson, working in Cambridge, England, who happened to put all the pieces together and deduce the exact structure of DNA (**FIGURE 5-3**). As we will see later in this chapter, their discovery was more than just a description of a molecule. As soon as they figured out DNA's structure, the answers to several other thorny problems in biology—such as how DNA might be able to duplicate itself—became apparent immediately. We explore that process later in this chapter, but first we need to examine the structure of DNA in more detail.

DNA (deoxyribonucleic acid) is a **nucleic acid,** a macromolecule that stores information. It consists of individual units called **nucleotides,** which have three components: a molecule of sugar, a phosphate group (containing four oxygen atoms bound to a phosphorus atom), and a nitrogen-containing molecule called a **base.** The physical structure of DNA is frequently described as a "double helix." What exactly is a double helix? Picture a long ladder twisted around like a spiral staircase and you'll have a good idea of what a DNA molecule looks like (**FIGURE 5-4**). The molecule has two distinct strands, like the vertical sides of a ladder. These are the "backbones" of the DNA molecule, and each is made from two alternating

FIGURE 5-3 **Watson and Crick.** James Watson and Francis Crick figured out the exact structure of DNA. The first sketch of the double helix (lower right) was rendered by Crick's wife, Odile.

molecules: a sugar, then a phosphate, then a sugar, then a phosphate, and so on. The sugar is always deoxyribose and the phosphate molecule is always the same, too. It is the shapes of the backbones that cause the DNA "ladder" to twist.

The alternating sugars and phosphates hold everything in place, but they play only a supporting role. The rungs of the ladder are where things get interesting. Attached to each sugar, and protruding like half of a rung on the ladder, is one of the nitrogen-containing bases. Each sugar always has a base attached to it, and there are four different kinds of bases—adenine, thymine, cytosine, and guanine. When discussing DNA, these bases are usually referred to by their first letter: A, T, C, and G.

Both backbones of the ladder have a base protruding from each sugar. The base on one side of the ladder binds, via hydrogen bonds, to a base on the other side, and together these **base pairs** form the rungs of the ladder. They don't just pair up at random, though. Every time a C protrudes from one side, it forms hydrogen bonds with a G on the other side (and

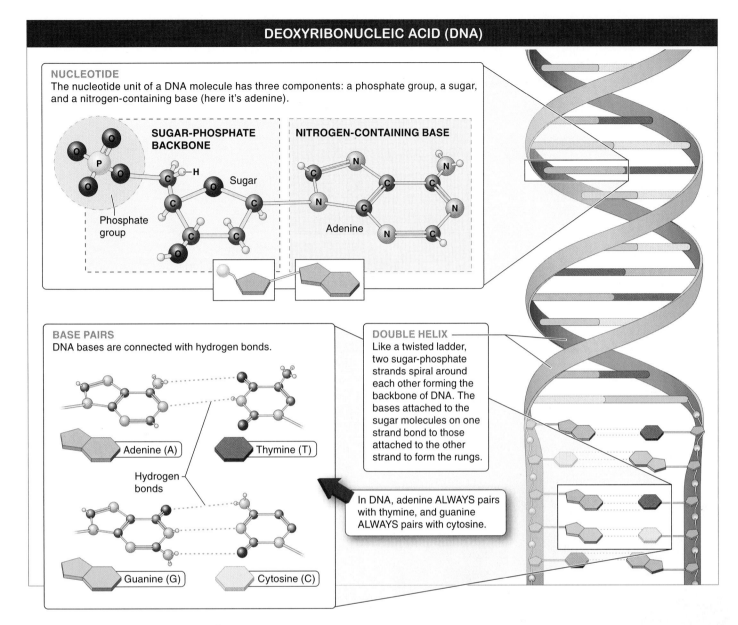

DEOXYRIBONUCLEIC ACID (DNA)

NUCLEOTIDE
The nucleotide unit of a DNA molecule has three components: a phosphate group, a sugar, and a nitrogen-containing base (here it's adenine).

SUGAR-PHOSPHATE BACKBONE

Phosphate group

Sugar

NITROGEN-CONTAINING BASE

Adenine

BASE PAIRS
DNA bases are connected with hydrogen bonds.

Adenine (A)

Thymine (T)

Hydrogen bonds

Guanine (G)

Cytosine (C)

DOUBLE HELIX
Like a twisted ladder, two sugar-phosphate strands spiral around each other forming the backbone of DNA. The bases attached to the sugar molecules on one strand bond to those attached to the other strand to form the rungs.

In DNA, adenine ALWAYS pairs with thymine, and guanine ALWAYS pairs with cytosine.

FIGURE 5-4 Overview of the structure of DNA.

vice versa: a G always bonds to a C). Similarly, every time a T protrudes from one side, it forms hydrogen bonds with an A on the other side (and vice versa). For this reason, each DNA molecule always has the same number of Gs and Cs, and the same numbers of As and Ts. Because of these base-pairing rules, it also is true that if we know the base sequence for one of the strands in a DNA molecule, we know the sequence in the other. For this reason, a DNA sequence is described by writing the sequence of bases on only one of the strands.

If our DNA molecule were really a spiraled ladder, it would be a very, very long one. One molecule of DNA can have as many as 200 million base pairs, or rungs. This length would make it difficult to squeeze the molecule into the cell. But the rungs are small and the twisting of the molecule—twists upon

twists—shortens it considerably: think about how, if you repeatedly twist your shoe lace around and around, it becomes shorter and shorter. That's what happens with DNA. Let's now investigate how DNA's structure—the rungs of the ladder, in particular—enables it to carry information.

TAKE-HOME MESSAGE 5·2

DNA is a nucleic acid, a macromolecule that stores information. It consists of individual units called nucleotides: a sugar, a phosphate group, and a nitrogen-containing base. DNA's structure resembles a twisted ladder, with the sugar and phosphate groups serving as the backbones of the molecule and base pairs serving as the rungs.

Genes are sections of DNA that contain instructions for making proteins.

One of DNA's most amazing features is that it embodies the instructions for building the cells and structures for virtually every single living organism on earth (FIGURE 5-5). Thus, DNA is like a universal language, the letters of which are the bases A, T, C, and G. (Note that the sugar-phosphate backbone only serves to hold the bases in sequence, like the binding of a book. It does not convey genetic information.)

You can think of an organism's DNA as a cookbook. Just as a cookbook contains detailed instructions on how to make a variety of foods (such as french toast, macaroni and cheese, or chocolate chip cookies), an organism's DNA carries the detailed instructions to build an organism and keep it running. And just as a book can be viewed as a sequence of letters, with the book's meaning determined by which letters are strung together and in what order, a molecule of DNA can be viewed as a sequence of bases. Letters don't have much meaning on their own, of course, but when they are put together into words and sentences, their order holds a great deal of information. Similarly, the sequence in which bases appear in a molecule of

Q Why is DNA considered the universal code for all life on earth?

an organism's DNA makes up a "code" that holds the detailed instructions for the building of the organism, whether it is a one-celled amoeba, a giant oak tree, or a biology student.

The full set of DNA present in an individual organism is called its **genome** (FIGURE 5-6). In prokaryotes, including all bacteria, the information is contained within circular pieces of DNA. In eukaryotes, including humans, this information is laid out in long linear strands of DNA in the nucleus. Rather than being one super-long DNA strand, eukaryotic DNA exists as many smaller, more manageable pieces, called **chromosomes.** Humans, for example, have three billion base pairs, divided into 23 unique pieces of DNA (and we have two copies of each piece, one from our mother and one from our father, for a total of 46 chromosomes and six billion base pairs in every cell).

The number of chromosomes varies from species to species, but is not related to how simple or complex an organism is. Corn, for example, has 10 unique chromosomes and fruit flies have only 4. Dogs and chickens have 39 different chromosomes, while goldfish put most other animal species to shame with 50 different chromosomes. As in humans, individuals in each of these species inherit one copy of each chromosome from each parent, so the total number of chromosomes in a cell is doubled— a dog, for example, has 78 chromosomes in each cell. Chromosomes vary in length, too. In humans, the longest chromosome has more than 200 million base pairs and the shortest has fewer than 50 million.

Within the long sequences of bases are relatively short sequences, on average about 3,000 bases long, that are the actual genes. "Gene" may seem like an impossibly nebulous concept because the word is often used casually in the media as if it were some magical, irresistible, and mysterious force that controls our bodies and behavior. Beyond these vague descriptions, though, the word has a literal meaning. A **gene** is a sequence of bases (strictly, as we've noted, base pairs) in a DNA molecule that carries the information necessary for producing a functional product, usually a protein or RNA molecule. Nothing more, nothing less.

Remember that a DNA molecule is like a ladder in which half of each rung is any one of the four bases— A, T, G, or C. Strung together, a segment might be read as "AAAGGCTAGGC . . ." continuing on for another

DNA provides the instructions for building virtually every organism on earth!

FIGURE 5-5 **DNA is the universal code for all life on earth.**

What Is DNA? | Building Organisms | Genetic Damage | Biotechnology and Health

FROM GENOME TO GENE

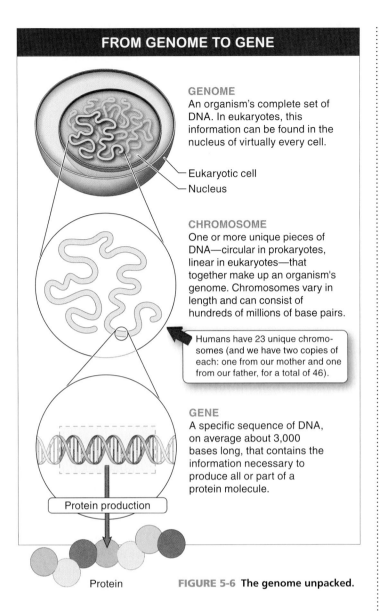

GENOME
An organism's complete set of DNA. In eukaryotes, this information can be found in the nucleus of virtually every cell.

Eukaryotic cell

Nucleus

CHROMOSOME
One or more unique pieces of DNA—circular in prokaryotes, linear in eukaryotes—that together make up an organism's genome. Chromosomes vary in length and can consist of hundreds of millions of base pairs.

Humans have 23 unique chromosomes (and we have two copies of each: one from our mother and one from our father, for a total of 46).

GENE
A specific sequence of DNA, on average about 3,000 bases long, that contains the information necessary to produce all or part of a protein molecule.

Protein production

Protein

FIGURE 5-6 The genome unpacked.

ALLELES

Within a species, individuals sometimes have alternate versions of a gene for a given protein. These instructions can result in a different version of the same trait.

ALLELE
Alternate versions of a gene that code for the same trait

ALLELE 1   ALLELE 2   ALLELE 3

Protein production   Protein production   Protein production

The protein, influenced by the environment and in some cases other genes, then produces the trait.

FIGURE 5-7 **"Different versions of the same thing."** Alleles are alternative versions of a gene.

3,000 or so bases. Returning to our cookbook analogy, this segment can be thought of as a recipe, which is also a series of letters. And just as a particular sequence of letters in a cookbook may be read and understood as the directions for baking chocolate chip cookies, the sequence of bases in DNA also carries information. Perhaps the sequence spells out part of the instructions for producing a red blood cell, for constructing the keratin that will form part of a curly strand of hair, or for assembling a chemical that alters your brain chemistry so that you exhibit a mood disorder or suicidal behavior.

Each gene is the instruction set for producing one particular molecule, usually a protein. For example, there is a gene in silk moths that codes for fibroin, the chief component of silk. And, there is a gene in humans that codes for triglyceride lipase, an enzyme that breaks down dietary fat. Within a species, individuals sometimes have slightly different instruction sets for a given protein, and these instructions can result in a different version of the same characteristic. These alternative versions of a gene that

code for the same feature are called **alleles** (**FIGURE 5-7**). And any single characteristic or feature of an organism is referred to as a **trait.** A simple hypothetical example will clarify the meaning of these terms. The color of a daisy's petals is a trait. The instructions for producing this trait are found in a gene that controls petal color. However, this gene may have many different alleles; one allele may specify the trait of red petals, another may specify white petals, and yet another may specify yellow petals. Similarly, one allele for eye color in fruit flies may carry the instructions for producing a red eye, while another, slightly different allele may have instructions for brown eyes. (Ultimately, the trait may be influenced not just by the genes an individual carries but the way those genes interact with the environment, too.)

TAKE-HOME MESSAGE 5·3

DNA is a universal language that provides the instructions for building all the structures in all living organisms. The full set of DNA an organism carries is called its genome. In prokaryotes, the DNA occurs in circular pieces. In eukaryotes, the genome is divided among smaller, linear strands of DNA called chromosomes. A gene is a sequence of bases in a DNA molecule that carries the information necessary for producing a functional product, usually a protein or RNA molecule.

Not all DNA contains instructions for making proteins.

It is debatable whether humans are the most complex species on the planet, but surely we must be more complex than an onion. "Complexity" is somewhat subjective and can be assessed in a variety of ways, such as by counting the number of different cell types in the organism. But if we measured complexity simply as the amount of DNA an organism has, we'd have to say an onion is more complex—it has more than five times as much DNA as a human (**FIGURE 5-8**)! We don't fare any better when compared with some other seemingly simple organisms, either. The salamander species *Amphiuma means,* for example, has about 25 times as much DNA as we do, and one species of amoeba—a single-celled organism—has almost 200 times as much!

Comparing the amount of DNA present in various species, in terms of both numbers of chromosomes and numbers of base pairs, reveals a paradox: there does not seem to be any relationship between the size of an organism's genome and the organism's complexity. Moreover, among species of seemingly similar complexity, there can be huge differences in genome size (among fish, for example, there is more than a threefold difference among different species of sturgeon). What is going on?

Q An onion has five times as much DNA as a human. Why doesn't that make onions more complex than us?

The description earlier in this chapter about what DNA is and how genes code for proteins is logical and tidy, but it doesn't completely explain what we observe in cells. In humans, for example, genes make up less than 5% of the DNA (**FIGURE 5-9**). In many species, the proportion of the DNA that consists of genes is even smaller. In virtually all eukaryotic species, the amount of DNA present far exceeds the amount necessary to code for all of the proteins in the organism. The fact is, a huge proportion of base sequences in DNA do not code for anything and have no obvious purpose. Some biologists even call it "junk DNA."

In what types of organisms do we find the most "junk DNA"? Bacteria and viruses tend to have very little non-coding DNA; genes make up 90% or more of their DNA. It is in the eukaryotes (with the exception of yeasts) that we see an explosion in the amount of non-coding DNA, about 25% of which occurs within genes and 75% between genes (**FIGURE 5-10**).

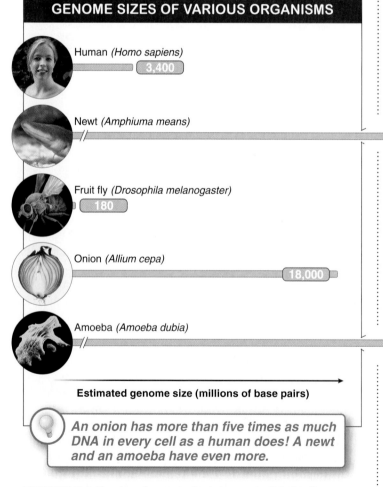

GENOME SIZES OF VARIOUS ORGANISMS

Human (*Homo sapiens*) — 3,400

Newt (*Amphiuma means*) — 84,000

Fruit fly (*Drosophila melanogaster*) — 180

Onion (*Allium cepa*) — 18,000

Amoeba (*Amoeba dubia*) — 670,000

Estimated genome size (millions of base pairs)

An onion has more than five times as much DNA in every cell as a human does! A newt and an amoeba have even more.

FIGURE 5-8 Is the size of an organism's genome related to its complexity?

Non-coding regions of DNA sometimes take the form of short sequences that are repeated thousands (or even hundreds of thousands) of times. Other times, the non-coding regions are slightly longer repeated sequences. Occasionally, the non-coding DNA consists of gene fragments, duplicate versions of genes, and pseudogenes (sequences very similar to actual genes but with a few slight alterations that make them lose their protein-coding ability). Non-coding regions occur both within genes—in which case they are called **introns**—and between genes.

PERCENTAGES OF CODING DNA FOUND IN VARIOUS ORGANISMS

Human (*Homo sapiens*) 2%

Fruit fly (*Drosophila melanogaster*) 19%

Round worm (*Caenorhabditis elegans*) 25%

Arabidopsis (*Arabidopsis thaliana*) 28%

E. coli (*Escherichia coli*) 90%

Percentage of DNA that codes for proteins

FIGURE 5-9 "Junk DNA"? The proportion of DNA that codes for proteins or RNA varies greatly among species. Which species seems to have the least "junk DNA"?

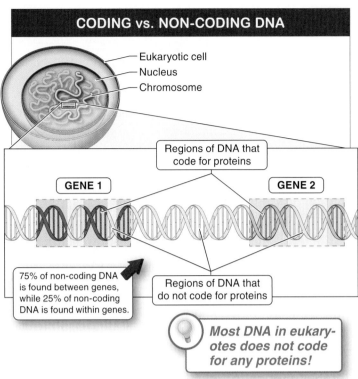

CODING vs. NON-CODING DNA

Eukaryotic cell
Nucleus
Chromosome

Regions of DNA that code for proteins

GENE 1   GENE 2

75% of non-coding DNA is found between genes, while 25% of non-coding DNA is found within genes.

Regions of DNA that do not code for proteins

Most DNA in eukaryotes does not code for any proteins!

FIGURE 5-10 Non-coding regions of DNA. These regions are found both between and within genes.

In the end, the presence of all this non-coding DNA remains a big mystery. All the extra DNA may serve some purpose. Perhaps it is a reservoir of potentially useful sequences. Or it may have some function in regulating when genes are turned on or off. Because we are not certain what non-coding DNA does, the label "junk DNA" is probably a bad description for it. The jury is still out and it is too early to assume that it has no function.

TAKE-HOME MESSAGE 5•4

Only a small fraction of the DNA in eukaryotic species is in genes, coding for proteins or RNA; the function of the rest is still a mystery.

5•5 -

How do genes work? An overview.

Just as having a recipe for chocolate chip cookies is not the same thing as having the actual cookies, having a wealth of hereditary information about how to build muscle cells or root cells or leaf cells is not sufficient to produce an organism. Think about it: every cell in an organism contains all of the information needed to manufacture every protein in its body. This means that the skin cells on your arm contain the genes

for producing proteins only found in liver cells or red blood cells or muscle tissue—but they don't produce them. Having the instructions is not the same as having the products.

The genes in strands of DNA are a storehouse of information, an instruction book, but they are only one part of the process by which an organism is built. If the genes that an organism

167

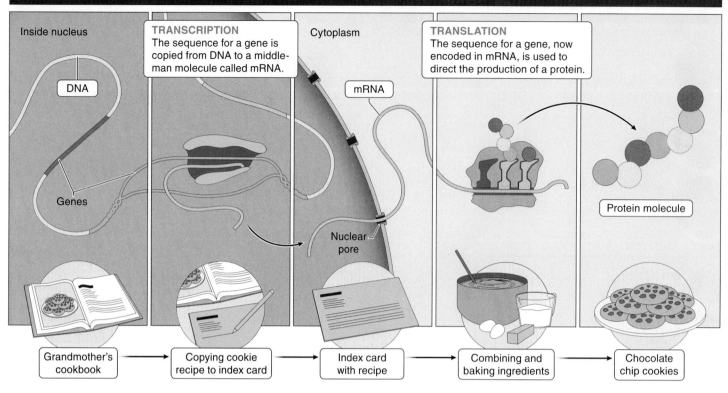

FIGURE 5-11 **Overview of the steps from gene to protein.**

carries for a particular trait—its **genotype**—are like a recipe in a cookbook, the physical manifestation of the instructions— the organism's **phenotype**—is the cookie, the macaroni and cheese, the french toast, or any of the other foods described by the recipes. And, just as you have to assemble ingredients, mix them, then bake the dough to get a cookie, there are several steps in the production of the molecules, tissues, and even behaviors that make up a phenotype.

How does a gene (a sequence of bases in a section of DNA) affect a flower's color or the shape of a nose or the texture of a dog's fur (the phenotype)? The process occurs in two main steps: **transcription,** in which a copy of a gene's base sequence is made, and **translation,** in which that copy is used to direct the production of a protein.

FIGURE 5-11 presents an overview of the processes of transcription and translation. In transcription, which in

eukaryotes occurs in the nucleus, the gene's base sequence or **code** is copied into a middleman molecule called **messenger RNA (mRNA)**. This is like copying the information for the chocolate chip cookie recipe out of the cookbook and onto an index card. In translation, the mRNA moves out of the nucleus and into the cytoplasm of the cell, where the messages encoded in the mRNA molecules are used to build proteins.

TAKE-HOME MESSAGE 5·5

The genes in strands of DNA are a storehouse of information, an instruction book. The process by which this information is used to build an organism occurs in two main steps: transcription, in which a copy of a gene's base sequence is made, and translation, in which that copy is used to direct the production of a protein.

❷ Building organisms: through gene expression, information in the DNA directs the production of the molecules that make up an organism.

Transcription and translation are a bit like copying a cookie recipe, then baking the cookies.

5•6

In transcription, the information coded in DNA is copied into mRNA.

If DNA is like a cookbook filled with recipes, transcription and translation are like cooking. In cooking, you use information about how to make chocolate chip cookies to actually produce the cookies. In an organism, the information about putting together proteins is used to build the proteins the organism needs in order to function. In this section, we examine transcription, the first step in the two-step process by which DNA regulates a cell's activity and its synthesis of proteins (see Figure 5-11). In transcription, a single copy of one specific gene within the DNA is made. Continuing our cookbook analogy, transcription is like copying a single recipe from the cookbook onto an index card. It happens in four steps (**FIGURE 5-12**).

Step 1. Recognize and bind. To start the transcription process, a large molecule, the enzyme RNA polymerase, recognizes a **promoter site,** a part of the DNA molecule that indicates the start of a gene, and, in effect, tells the RNA polymerase, "Start here." RNA polymerase binds to the DNA molecule at the promoter site and unwinds it just a bit, so that only one strand of the DNA can be read. Then, like a court reporter transcribing everything that is said in a courtroom, RNA polymerase begins to read the gene's message.

Step 2. Transcribe. As the DNA strand is processed through the RNA polymerase, the RNA polymerase builds a copy—called a "transcript"—of the gene from the DNA molecule. This copy is called messenger RNA (mRNA) because, once this copy of the gene is created, it can move elsewhere in the cell and its message can be translated into a protein. Throughout transcription, DNA is unwound ahead of the RNA polymerase so that a single strand of the DNA can be read, and is rewound after the polymerase passes.

The mRNA transcript is constructed from four different molecules called ribonucleotides (which are almost identical to DNA nucleotides, consisting of a sugar-phosphate complex with a nitrogen-containing base attached). Each

Biotechnology and Agriculture   DNA Sequences

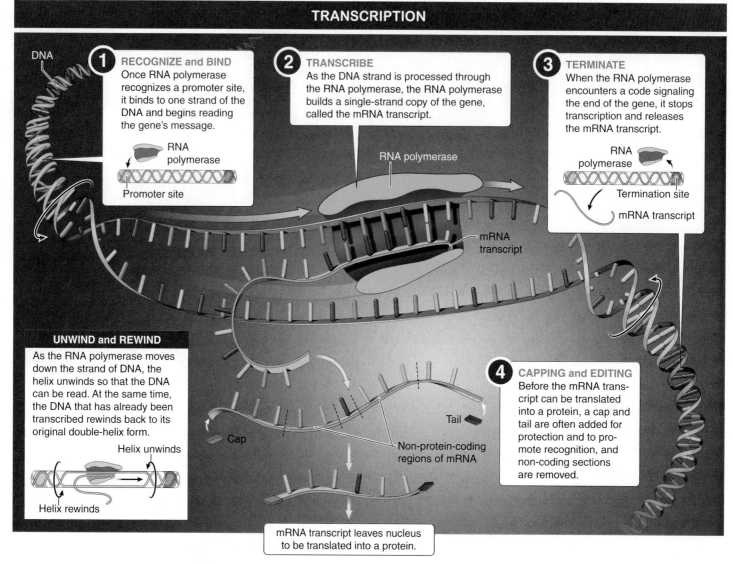

TRANSCRIPTION

1 RECOGNIZE and BIND
Once RNA polymerase recognizes a promoter site, it binds to one strand of the DNA and begins reading the gene's message.

RNA polymerase

Promoter site

DNA

2 TRANSCRIBE
As the DNA strand is processed through the RNA polymerase, the RNA polymerase builds a single-strand copy of the gene, called the mRNA transcript.

RNA polymerase

mRNA transcript

3 TERMINATE
When the RNA polymerase encounters a code signaling the end of the gene, it stops transcription and releases the mRNA transcript.

RNA polymerase

Termination site

mRNA transcript

UNWIND and REWIND
As the RNA polymerase moves down the strand of DNA, the helix unwinds so that the DNA can be read. At the same time, the DNA that has already been transcribed rewinds back to its original double-helix form.

Helix unwinds

Helix rewinds

Cap

Tail

Non-protein-coding regions of mRNA

4 CAPPING and EDITING
Before the mRNA transcript can be translated into a protein, a cap and tail are often added for protection and to promote recognition, and non-coding sections are removed.

mRNA transcript leaves nucleus to be translated into a protein.

FIGURE 5-12 Transcription: copying the base sequence of a gene. The first step in a two-step process by which DNA regulates a cell's activity and synthesis of proteins.

ribonucleotide pairs up with an exposed base on the now unwound and separated DNA, following these rules:

If the DNA strand has a thymine (T), an adenine (A) is added to the mRNA.

If the DNA strand has an adenine (A), a uracil (U) is added to the mRNA.

If the DNA strand has a guanine (G), a cytosine (C) is added to the mRNA.

If the DNA strand has a cytosine (C), a guanine (G) is added to the mRNA.

Because our court reporter (RNA polymerase) transcribes a specific sequence of DNA letters (the gene), the mRNA transcript carries this specific bit of the DNA's information. And because it is separate from the DNA, the mRNA transcript can move throughout the cell, to the places where the information is needed, while leaving the original information within the DNA.

Step 3. Terminate. When the RNA polymerase encounters a sequence of bases on the DNA at the end of the gene (called a termination sequence), it stops creating the transcript and detaches from the DNA molecule. After termination, the mRNA molecule is released as a free-floating, single-strand copy of the gene.

Step 4. Capping and editing. In prokaryotic cells, once the mRNA transcript separates from the DNA it is ready to be translated into a protein (it doesn't have a nuclear membrane to cross). In most eukaryotes, however, the transcript must first be edited in several ways. First, a cap and a tail may be added at the beginning and end of the transcript. Like a front and back cover to a book, these serve to protect the mRNA from damage and help the protein-making machinery recognize the mRNA. Second, because (as we saw earlier in the chapter)

there may have been non-coding bits of DNA transcribed into mRNA, those sections—the introns—are snipped out. Once the mRNA transcript has been edited, it is ready to leave the nucleus for the cytoplasm, where it will be translated into a protein.

TAKE-HOME MESSAGE 5·6

Transcription is the first step in the two-step process by which DNA directs the synthesis of proteins. In transcription (occurring in the nucleus in eukaryotic cells), a single copy of one specific gene in the DNA is made, in the form of a molecule of mRNA, and this moves to the cytoplasm where it can be translated into a protein.

5·7

In translation, information from DNA is used to build functional molecules.

Once the mRNA molecule moves out of the cell's nucleus and into the cytoplasm, the translation process begins. In translation, the information carried by the mRNA is read and ingredients present in the cytoplasm are used to produce a protein (see Figure 5-11). The process of translation is like the combining and baking of the ingredients listed in our chocolate chip cookie recipe to produce a cookie.

Several ingredients must be present in the cytoplasm for translation to occur. First, there must be large numbers of free amino acids floating around. Recall from Chapter 2 that amino acids are the raw materials for building proteins and an essential component of our diet. Second, there must be molecules called **ribosomal subunits,** the protein-production factories where amino acids are linked together in the proper order to produce the protein. Finally, there must also be molecules that can read the mRNA code and translate it from a sequence of bases into a protein. These molecules, called **transfer RNA (tRNA),** interpret the mRNA code and link specific base sequences on the mRNA with specific amino acids (**FIGURE 5-13**). Because tRNAs play such a central role in translation, we examine them more closely.

Human translators can hear one a word in one language and say the same word in a different language. Transfer RNA molecules function in a similar manner. Picture a two-sided

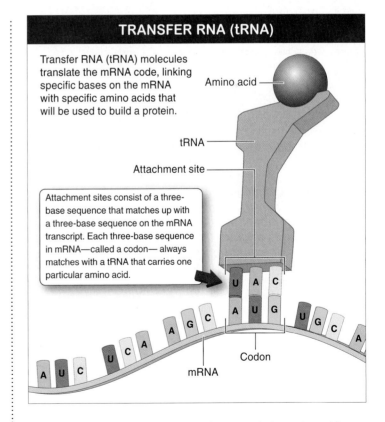

TRANSFER RNA (tRNA)

Transfer RNA (tRNA) molecules translate the mRNA code, linking specific bases on the mRNA with specific amino acids that will be used to build a protein.

Amino acid

tRNA

Attachment site

Attachment sites consist of a three-base sequence that matches up with a three-base sequence on the mRNA transcript. Each three-base sequence in mRNA—called a codon— always matches with a tRNA that carries one particular amino acid.

Codon

mRNA

FIGURE 5-13 Each transfer RNA attaches a particular amino acid to the mRNA.

TRANSLATION

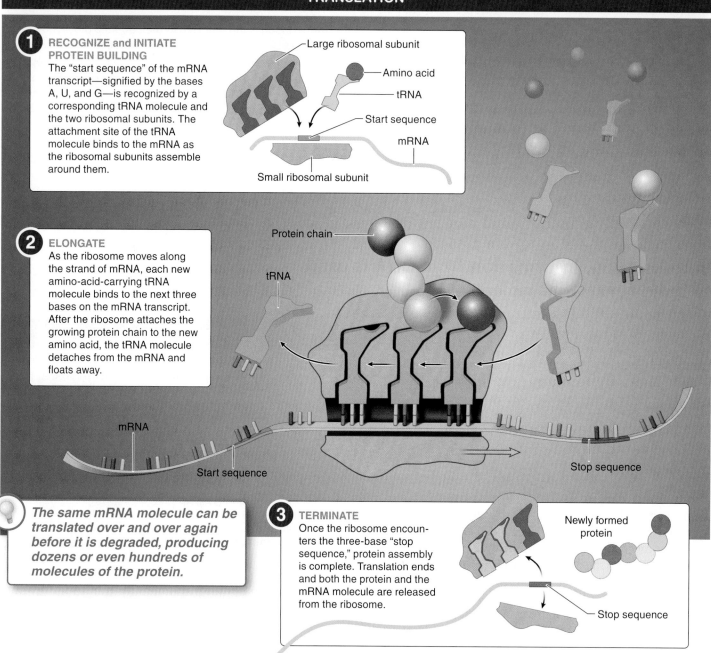

1 **RECOGNIZE and INITIATE PROTEIN BUILDING**
The "start sequence" of the mRNA transcript—signified by the bases A, U, and G—is recognized by a corresponding tRNA molecule and the two ribosomal subunits. The attachment site of the tRNA molecule binds to the mRNA as the ribosomal subunits assemble around them.

Large ribosomal subunit
Amino acid
tRNA
Start sequence
mRNA
Small ribosomal subunit

2 **ELONGATE**
As the ribosome moves along the strand of mRNA, each new amino-acid-carrying tRNA molecule binds to the next three bases on the mRNA transcript. After the ribosome attaches the growing protein chain to the new amino acid, the tRNA molecule detaches from the mRNA and floats away.

Protein chain
tRNA
mRNA
Start sequence
Stop sequence

The same mRNA molecule can be translated over and over again before it is degraded, producing dozens or even hundreds of molecules of the protein.

3 **TERMINATE**
Once the ribosome encounters the three-base "stop sequence," protein assembly is complete. Translation ends and both the protein and the mRNA molecule are released from the ribosome.

Newly formed protein
Stop sequence

FIGURE 5-14 **Translation: reading a sequence of nucleotides and producing protein.** The second step in a two-step process by which DNA regulates a cell's activity and synthesis of proteins.

What Is DNA?　　　Building Organisms　　　Genetic Damage　　　Biotechnology and Health

molecule. On one side a particular amino acid is attached. This amino acid can be any one of the 20 different amino acids that make up proteins. In other words, there are different tRNA molecules for each of the 20 different amino acid molecules, which are floating around in the cytoplasm in great numbers. On the other side of the two-sided tRNA molecule is an attachment site, consisting of a three-base sequence that matches up with a three-base sequence on the mRNA transcript. (Each three-base sequence in mRNA—called a **codon**—always matches with a tRNA that carries one particular amino acid.) This matchup enables the tRNA molecule to attach to the mRNA. Thus, the tRNA translates the code on the mRNA into an amino acid that will be used to build a protein. Translation occurs in three steps (**FIGURE 5-14**).

Step 1. Recognize and initiate protein building. Translation begins in the cell's cytoplasm when the subunits of a ribosome, essentially a two-piece protein-building factory, recognizes and assembles around a "start sequence" on the mRNA transcript. This start sequence is always the sequence A, U, and G (AUG). As the ribosomal subunits assemble themselves into a ribosome, one side of a tRNA molecule also recognizes the start sequence on the mRNA and binds to it. This initiator tRNA has the amino acid methionine bound to its other side. This will be the first amino acid in the protein that is to be produced (although occasionally, in eukaryotes, it is eventually "edited out").

Step 2. Elongate. After the mRNA start sequence (AUG), the next three bases on the mRNA specify which amino acid–carrying tRNA molecule should bind to the mRNA. If the next three bases on the mRNA transcript are GUU, for example, a tRNA molecule that recognizes that sequence, and carrying its particular amino acid, will attach to the mRNA at that point. The ribosome then facilitates the connection of this second amino acid to the first. After the amino acid carried by a tRNA molecule is attached to the preceding amino acid, the tRNA molecule detaches from the mRNA and floats away.

As this process continues, the protein grows. The next three bases on the mRNA specify the next amino acid to be added

to the first two. And the three bases after that specify the fourth amino acid, and so on. This process is called **protein synthesis** because all proteins are chains of amino acids, like beads on a string.

The mRNA continues to be "threaded" through the ribosome, with the ribosome moving down the mRNA strand reading and translating its message in little three-base chunks. Each three-base sequence specifies the next amino acid, lengthening the growing amino acid strand.

Step 3. Terminate. Eventually, the ribosome arrives at the three-base sequence on the mRNA that signals the end of translation. Once the ribosome encounters this stop sequence, the assembly of the protein is complete. Translation ends, and the amino acid strand and mRNA molecule are released from the ribosome. As it is produced, the amino acid sequence folds and bends, based on the chemical features in the amino acid side chains (as we saw in Chapter 2), acquiring a three-dimensional structure. The completed protein—such as a membrane protein or insulin or a digestive enzyme—may be used within the cell or packaged for delivery via the bloodstream to somewhere else in the body where it is needed.

Following the completion of translation, the mRNA strand may remain in the cytoplasm to serve as the template for producing another molecule of the same protein. In bacteria, an mRNA strand may last from a few seconds to more than an hour; in mammals, mRNA may last several days. Depending on how long it lasts, the same mRNA strand may be translated hundreds of times. Eventually, it is broken down by enzymes in the cytoplasm.

TAKE-HOME MESSAGE 5•7

Translation is the second step in the two-step process by which DNA directs the synthesis of proteins. In translation, the information from a gene that has been encoded in the nucleotide sequence of an mRNA is read and ingredients present in the cell's cytoplasm are used to produce a protein.

Biotechnology and Agriculture   DNA Sequences

❸ Damage to the genetic code has a variety of causes and effects.

Damage to the genetic code can interfere with normal development.

5•8

What causes a mutation and what are its effects?

Through the two-step process of transcription and translation, an organism converts the information held in its genes into the proteins necessary for life. But the process is only as good as the organism's underlying genetic information. Sometimes, however, something occurs to alter the sequence of bases in an organism's DNA. Such an alteration is called a **mutation,** and it can lead to changes in the structure and function of the proteins produced. Mutations can have a range of effects. Sometimes they result in a serious, even deadly, problem for an organism. Sometimes they have little or no detrimental effect. And, occasionally—but very rarely—they may even turn out to be beneficial to the organism.

As an example of how mutations can affect organisms, consider the case of breast cancer in humans. When two human genes, called BRCA1 and BRCA2, are functioning properly, they help to prevent breast cancer by deterring cells from dividing uncontrollably. If the DNA sequence of either of these genes is altered through mutation and the gene's normal function is lost, the person carrying the gene has a

significantly increased risk of developing breast cancer. (Because a variety of other factors, including environmental variables, are involved in development of cancer, it's impossible to know for certain whether these individuals will definitely develop breast cancer.) Currently, more than 200 different changes in the DNA sequences of these genes have been detected, each of which results in an increased risk of developing breast cancer.

Given the havoc they can cause for an organism, it's not surprising that mutations have a bad reputation. After all, because they can change the protein produced, they tend to disrupt normal processes and usually harm the individual (**FIGURE 5-15**). Fortunately, mutations are very, very rare. More surprising, though, is the paradoxical fact that mutations are essential to evolution. Those mutations that don't kill an organism, or reduce its ability to survive and reproduce, can be beneficial. Every genetic feature in every organism was, initially, the result of a mutation. In Chapter 8, we will explore the relationship between mutation and evolution.

What Is DNA?　　　　Building Organisms　　　▼　　Genetic Damage　　　Biotechnology and Health

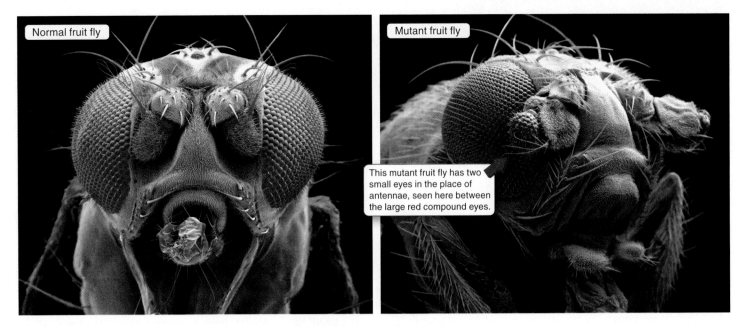

Normal fruit fly

Mutant fruit fly

This mutant fruit fly has two small eyes in the place of antennae, seen here between the large red compound eyes.

FIGURE 5-15 Wreaking havoc. Mutations can change the protein produced by the altered gene, with disastrous consequences.

It's important to note that mutations can occur in an organism's gamete-producing cells (that is, cells that produce sperm or eggs) as well as in its non-sex cells (such as skin cells or cells in the lungs). Mutations in the non-sex cells generally have bad health consequences for the person carrying them. Many forms of cancer, such as lung cancer or skin cancer, result from such mutations. On the upside, though, non-sex-cell mutations are not passed on to your children. Mutations in the sex cells (gametes), on the other hand, do not have any adverse health effects on the person carrying them, but they can be passed on to offspring, with terrible effects such as the induction of a miscarriage or the occurrence of birth defects. Individuals inheriting certain mutations from a parent—because the mutation occurred in the parent's sex cells or the parent inherited the mutation from his or her parent—can be at increased risk for certain diseases such as breast cancer or cardiovascular disease.

Mutations generally fall into two types: point mutations and chromosomal aberrations (**FIGURE 5-16**). In point mutations, one base pair is changed, whereas in chromosomal aberrations, entire sections of a chromosome are altered.

Point mutations are mutations in which one nucleotide base pair in the DNA is replaced with another, or a base pair is inserted or deleted. Insertions and deletions can be much more harmful than substitutions because the amino acid sequence of a protein is determined by reading the three-base sequence (codon) on an mRNA molecule and attaching the particular amino acid specified by that sequence. If a single base is added

or removed, the three-base groupings get thrown off and the sequence of amino acids stipulated "downstream" from that point will be all wrong—the reading frame is shifted. It's almost like putting your hands on a computer keyboard, but offset by one key to the left or right, and then typing what should be a normal sentence. It comes out as gibberish.

Chromosomal aberrations are changes to the overall organization of the genes on a chromosome. Chromosomal aberrations are like the manipulation of large chunks of text when you are working on a paper. They can involve the complete deletion of an entire section of DNA, the moving of a gene from one part of a chromosome to elsewhere on the same chromosome or to a different chromosome, or the duplication of a gene, with the new copy inserted elsewhere on the chromosome or on a different chromosome. Whatever the type of aberration, a gene's expression—the production of the protein that the gene's sequence codes for—can be altered when it is moved, as can the expression of the genes around it.

Given the potentially hazardous health consequences of mutations, it is advisable to minimize their occurrence. Can this be done? Yes and no. There are three chief causes of mutation and, although one of them is beyond our control, the other two can be significantly reduced (**FIGURE 5-17**).

1. *Spontaneous mutations.* Some mutations arise by accident as long strands of DNA are duplicating themselves—at the rate of more than a thousand bases a minute in humans—when cells are dividing (you'll read more details on this process

175

Normal DNA / mRNA / Normal protein

POINT MUTATIONS

BASE-PAIR SUBSTITUTION
DNA
mRNA
Mutated protein

BASE-PAIR INSERTION
DNA
mRNA
Mutated protein

BASE-PAIR DELETION
DNA
mRNA
Mutated protein

Insertions and deletions can be much more harmful than substitutions because they can alter the reading frame for the rest of the gene.

CHROMOSOMAL ABERRATIONS

Normal chromosome
Gene 1   Gene 2   Gene 3

GENE DELETION
Gene 1   Gene 2   Gene 3
          DELETED

GENE RELOCATION
Gene 2   Gene 3
Gene 1 RELOCATED to a different chromosome

GENE DUPLICATION
Gene 1   Gene 2   Gene 3
DUPLICATED

In point mutations, one base pair is changed, whereas in chromosomal aberrations, entire sections of a chromosome are altered.

FIGURE 5-16 Point mutations and chromosomal aberrations.

in Chapter 6). Most errors are repaired by DNA repair enzymes, but some still slip by and there's not much we can do about them.

2. *Radiation-induced mutations.* Ionizing radiation is radiation with enough energy to disrupt atomic structure—even breaking apart chromosomes—by removing tightly bound electrons. Sources of ionizing radiation include X rays and the ultraviolet (UV) rays of the sun. When you lie in the sun, for example, its ultraviolet rays can induce a mutation in

Q Why do dentists put a heavy apron over you when they X-ray your teeth?

one of your skin cells that can transform the cell into a cancer cell. This is why long-term sun exposure can contribute to the development of skin cancer. Because ionizing rays cannot pass through lead, the lead apron a doctor or dentist puts over you when you get an X ray protects your body from the ionizing radiation.

Another source of dangerous radiation is found in the core of nuclear power plants, where radioactive atoms are used and

What Is DNA?   Building Organisms   Genetic Damage   Biotechnology and Health

FIGURE 5-17 Gambling with mutation-inducing activities? You can increase or decrease your risk of mutations with your behavior: smoking, sun worshipping, and radiation (but notice the protective lead apron).

Q Why is it dangerous to be near the core of a nuclear power plant?

produced in energy-generating reactions. While it is the high energy of the radioactivity that fuels the production of usable energy, this radioactivity is potentially harmful because it can pass through your body and disrupt your DNA, causing point mutations. Radioactivity can also cause larger-scale chromosomal aberrations. With the proper safety precautions, however, nuclear power plant workers can minimize their exposure to harmful radiation.

3. *Chemical-induced mutations.* Many chemicals, such as those found in cigarette smoke and in exhaust from internal combustion engines, can also react with the atoms in DNA molecules and induce mutations.

In the next section, we examine how even tiny changes in the sequence of bases in DNA can lead to errors in protein production and profound health problems.

TAKE-HOME MESSAGE 5·8

Mutations are alterations in a single base or changes in large segments of DNA that include several genes. They are rare, but when they do occur in a gene this usually disrupts normal physiological functioning. Mutations play an important role in evolution.

5·9

Faulty genes, coding for faulty enzymes, can lead to sickness.

Isabella joins her friends in sipping wine during a dinner party. As the meal progresses, her companions become tipsy. Their conversations turn racy, their moods relaxed. They refill their glasses, reveling in a little buzz. Not so for Isabella. Before her first glass is empty, she experiences a "fast-flush" response: her face turns crimson, her heart begins to race, and her head starts to pound. Worse still, she soon feels the need to vomit.

How can people respond so differently to alcohol? It comes down to a difference in a single base pair in their DNA, a single difference that can influence dramatically a person's behavior, digestion, respiration, and general ability to function. The base-pair change leads to the production of a non-

functional enzyme, and the lack of a functional version of this enzyme leads to physical illness. Let's look at the details.

When we consume alcohol, our bodies start a two-step process to convert the alcohol molecules from their intoxicating form into innocuous molecules. Each of the two steps is made possible by a specific enzyme, whose assembly instructions are coded in the DNA of most people.

Fast-flushers like Isabella complete the first step of breaking down alcohol, but cannot complete the breakdown because they carry defective genetic instructions for making aldehyde dehydrogenase, the enzyme that makes possible the second step of the process. A poisonous substance subsequently

Biotechnology and Agriculture   DNA Sequences

accumulates, and the symptoms of the "fast-flush" reaction are due to this substance's toxic effects in the body.

Q **Why do many Asians have unpleasant experiences associated with alcohol consumption?**

Approximately half of the people living in Asia carry a non-functional form of the gene for aldehyde dehydrogenase, a mutation that may confer a greater benefit than harm. In a study of 1,300 alcoholics in Japan, not a single one was a fast-flusher, even though half of all Japanese people are fast-flushers. The minor change in the genetic code that makes alcohol consumption an unpleasant experience may be responsible for the lower incidence of alcoholism among Japanese and other Asian people.

In many other cases—perhaps in the majority of genetic diseases—the link between a particular defective DNA sequence and physical illness is equally direct. Recall from Chapter 3 the case of Tay-Sachs disease. In Tay-Sachs disease, an individual inherits genes with a mutation that causes an inability to produce a critical lipid-digesting enzyme in their lysosomes, the cellular garbage disposals. Because these organelles cannot digest certain lipids, the lipids accumulate, undigested. The lysosomes swell until they eventually choke the cell to death. This occurs in numerous cells in the first few years of life, and ultimately leads to the child's death.

Although the details differ from case to case, the overall picture is the same for many, if not most, inherited diseases. The pathway from mutation to illness includes just four short steps (**FIGURE 5-18**).

1. A mutated gene codes for a non-functioning protein, commonly an enzyme.
2. The non-functioning enzyme can't catalyze the reaction as it normally would, bringing the reaction to a halt.
3. The molecule with which the enzyme would have reacted accumulates, just as half-made products would pile up on a blocked assembly line.
4. The accumulating chemical causes sickness and/or death.

The fact that most genetic diseases involve illnesses brought about by faulty enzymes suggests some strategies for treatment. These include administering medications that contain the normal-functioning version of the enzyme. For instance, lactose-intolerant individuals can consume the enzyme lactase, which for a short while gives them the ability to digest lactose. Alternatively, lactose-intolerant individuals can reduce their consumption of lactose-containing foods to keep the chemical from accumulating, thus reducing the problems that come from its overabundance.

FROM MUTATION TO ILLNESS

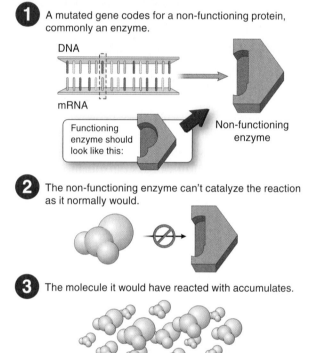

1 A mutated gene codes for a non-functioning protein, commonly an enzyme.

DNA

mRNA

Functioning enzyme should look like this:

Non-functioning enzyme

2 The non-functioning enzyme can't catalyze the reaction as it normally would.

3 The molecule it would have reacted with accumulates.

4 The accumulating chemical causes sickness (fast-flushing, for example) or death (Tay-Sachs disease, for example).

FIGURE 5-18 Faulty enzymes can interfere with metabolism.

TAKE-HOME MESSAGE 5·9

Most genetic diseases result from mutations that cause a gene to produce a non-functioning enzyme, which in turn blocks the functioning of a metabolic pathway.

What Is DNA? | Building Organisms | Genetic Damage | Biotechnology and Health

4 Biotechnology has the potential for improving human health.

A visual representation of the human genome. (Each color represents a different nucleotide.)

5·10

What is biotechnology?

Understanding a phenomenon is nice, but controlling it takes the satisfaction to an entirely new level. This may explain why there is so much excitement surrounding the field of **biotechnology,** in which organisms, cells, and their molecules are modified to achieve practical benefits. As we have seen in this chapter, the structure and function of DNA are pretty well understood. Biotechnology uses this knowledge to develop medicines for fighting viral infections, produce more nutritious food while reducing reliance on pesticides, and analyze forensic evidence as part of criminal investigations.

The modern emphasis in biotechnology is on **genetic engineering,** the manipulation of organisms' genetic material by adding, deleting, or transplanting genes from one organism to another. Such manipulation may create a better-tasting tomato or improve the treatment of a human disease.

This section introduces the most important tools and techniques used in biotechnology. We then explore the two primary areas in which biotechnology is applied: human health and agriculture. Biotech advances in human health fall into three categories: (1) producing medicines to treat diseases, (2) curing diseases, and (3) preventing diseases from

occurring in the first place. In agriculture, biotechnology is used to create more nutritious and healthful food as well as to increase the efficiency and reduce the ecological impact of farming (**FIGURE 5-19**).

FIGURE 5-19 Biotechnology in agriculture and human health.

How would you create a plant resistant to being eaten by insects? Or a colony of bacteria that can produce human insulin? Although there are many different uses of biotechnology, it employs a surprisingly small number of processes and tools. Each of these applications, for example, would use a similar sequence of steps and applications (**FIGURE 5-20**).

1. **Chop** up the DNA from a donor organism that exhibits the trait of interest.

2. **Amplify** the small amount of DNA into more useful quantities.

3. **Insert** the different DNA pieces into bacterial cells or viruses.

4. **Grow** separate colonies of the bacteria or viruses, each of which contains a different inserted piece of donor DNA.

5. **Identify** the colonies that have received the DNA containing the trait of interest.

THE 5 STEPS OF BIOTECHNOLOGY

1 CHOP up the DNA from a donor species that exhibits the trait of interest.

2 AMPLIFY the small amount of DNA into more useful quantities.

3 INSERT the different DNA pieces into bacterial cells.

4 GROW separate bacterial colonies, each containing a different inserted piece of donor DNA.

5 IDENTIFY the bacterial colonies that have received the DNA containing the gene of interest.

Not all of the steps are used in all biotechnology applications—some applications utilize only one or a few of these techniques.

FIGURE 5-20 Five important tools and techniques of most biotechnology procedures.

In other words, we can reduce decades of complex research down to five steps: Chop, Amplify, Insert, Grow, and Identify. Then we're ready to utilize the products. This may be as simple as harvesting a product of interest or may require inserting the gene into another species, possibly a crop plant or maybe a sick human. To be sure, it's easier said than done, but these steps capture the essence of most modern biotechnology. Not all of the steps are used in all biotech applications—some use only one or a few of these techniques—nonetheless, they all are mainstays of today's biotechnology world. Now let's explore each step in more detail.

Step 1. Chopping up DNA from a donor organism. To begin, researchers select an organism that has a trait they want to make use of. For example, they might want to produce human growth hormone in large amounts. Their first step would be to obtain a sample of human DNA and cut it into smaller pieces. Cutting DNA into small pieces requires the use of **restriction enzymes** (**FIGURE 5-21**).

Restriction enzymes have a single function: when they encounter DNA, they cut it up into small pieces. These enzymes evolved to protect bacteria from attack by viruses. Immediately upon encountering DNA from an invading virus, a restriction enzyme recognizes and binds to a particular sequence of four to eight bases and cuts the DNA there, thus making it impossible for the virus to reproduce within the bacterial cell. Dozens of different restriction enzymes exist, each of which recognizes a different sequence of bases and cuts the DNA molecule at a different location.

Restriction enzymes are one of the key tools in biotechnology because the first step of nearly all biotech applications involves cutting the DNA from a donor species that has a gene or genes of interest into shorter lengths (without actually cutting anywhere *within* the gene of interest).

Step 2. Amplifying DNA pieces into more useful quantities. In many situations, only a small amount of DNA will be available for analysis or for some other biotechnology use. The **polymerase chain reaction** (**PCR**) is a laboratory technique that allows a tiny piece of DNA (perhaps one that has been cut from a larger piece by restriction enzymes, or one that has been recovered from a crime scene) to be duplicated repeatedly, producing virtually unlimited amounts of that piece of DNA. PCR is especially useful in forensics, because investigators often find only small amounts of DNA at crime scenes, which must then be amplified before they can be analyzed by using the technique of DNA fingerprinting (which we discuss below).

In the first step of PCR, the DNA of interest is heated, which separates the double-stranded DNA into separate, single strands (**FIGURE 5-22**). Next, the DNA is cooled and an

What Is DNA?   Building Organisms   Genetic Damage   Biotechnology and Health

RESTRICTION ENZYMES

1 The gene of interest is located on a section of DNA from the donor species.

Gene of interest

Source DNA

2 Restriction enzymes that target a particular base-pair sequence on either side of the gene are introduced.

—Restriction enzymes—

In this example, the restriction enzymes recognize the sequence ATCGAT and cut between the first A and T.

| A | T | C | G | A | T |
| T | A | G | C | T | A |

3 The restriction enzymes bind to their target base-pair sequence and cut the strand of DNA.

4 The gene of interest has now been separated from the donor's DNA.

FIGURE 5-21 Restriction enzymes are used to isolate a gene of interest.

POLYMERASE CHAIN REACTION (PCR)

1 A solution containing an isolated segment of DNA is heated, separating the double-stranded DNA into two single strands.

Heat

2 The enzyme DNA polymerase is added along with a large number of free nucleotides, and the solution containing the segments is cooled.

DNA polymerase

Cool

Free nucleotides

3 DNA polymerase adds complementary bases to each single strand.

4 The result is two identical copies of the original segment of DNA.

This process can be repeated again and again until there are billions of identical copies of the target sequence.

FIGURE 5-22 The polymerase chain reaction can duplicate a small strand of DNA into billions of copies.

enzyme called DNA polymerase and a large number of free nucleotides are added to the DNA mixture. DNA polymerase uses one of the two strands of DNA as a template to build a complementary strand, using the free nucleotides as building blocks. This makes each of the single strands double-stranded again, resulting in two complete copies of the DNA of interest. This process of heating and cooling can be repeated

again and again until there are billions of identical copies of the target sequence. Depending on the experiment, PCR can be done so that it will only amplify a DNA fragment with a specific sequence, for example the gene for human growth hormone, which can later be separated from the original DNA. In other experiments, scientists want to amplify all the available DNA fragments so they can be surveyed later on.

Step 3. Inserting foreign DNA into the target organism. Most modern biotechnology applications involve the transfer of genes from one species to another, creating **transgenic organisms** that contain DNA from the donor species. In the human growth hormone example mentioned above, the researchers might want to transfer the human growth hormone gene into the bacterium *E. coli*, with the goal of having the bacteria produce human growth hormone.

To create a transgenic organism, researchers have to physically deliver the DNA of interest, isolated from a donor species, into the recipient organism. This delivery is often accomplished by using **plasmids,** circular pieces of DNA that can be incorporated into a bacterium's genome. This is an effective strategy because, following certain chemical treatments, researchers can construct plasmids that readily take up DNA that has been cut with restriction enzymes (**FIGURE 5-23**). Genes on the plasmid can then be expressed in the bacterial cell. The gene of interest is also replicated whenever the cell divides, so both of the new cells contain the plasmid. In some cases, rather than in plasmids, genes are incorporated into viruses, which can then be used to infect organisms and thus transfer the genes of interest to those species.

Step 4. Growing bacterial colonies that carry the DNA of interest: cloning. Once the DNA of interest has been transferred to the bacterial cell, every time the bacterium divides, it creates a **clone,** a genetically identical cell that contains the DNA of interest. The term **cloning** describes the production of genetically identical cells, organisms, or DNA molecules, a process that occurs each time a bacterium divides. As a consequence, with numerous rounds of cell division, it is possible to produce a huge number of clones, all transcribing and translating the gene of interest.

In a typical experiment, a large amount of DNA may be chopped up with restriction enzymes. Each of the pieces is incorporated into a plasmid that is introduced into the target organism, a bacterial cell. Then, all of the bacterial cells are allowed to divide repeatedly, each producing a clone of the foreign DNA fragment they carry. Together, all of the different bacterial cells containing all of the different fragments of the original DNA are called a **clone library** or a **gene library** (**FIGURE 5-24**).

Step 5. Identifying bacterial colonies that have received the gene of interest. Although a clone library contains a huge amount of potentially useful genetic information, the information is in no particular order, much like a library of books, uncatalogued and in complete disarray. If a researcher is interested in working with bacterial cells that contain the one human gene that produces human growth hormone, how can the bacteria

INSERTING DNA BY USING PLASMIDS

1 A target segment of source DNA is isolated using restriction enzymes. Using the same restriction enzyme, a single cut is made in a bacterial plasmid.

Source cell

Bacterial cell

Gene of interest

Plasmid

2 Because the same restriction enzyme was used to isolate the segment of DNA and cut the plasmid, the two share complementary base pairs and fit perfectly together.

3 The plasmid—now including the gene of interest—is inserted back into the bacterial cell. Genes on the plasmid can be expressed in the bacterial cell and are replicated whenever the cell divides.

FIGURE 5-23 A gene shuttle. Plasmids can be used to transfer DNA from one species to another.

containing that one particular piece of DNA be identified and separated from the millions of other clones in the library?

The gene of interest is usually found by using a method called **hybridization** (**FIGURE 5-25**). Hybridization requires the use of a **DNA probe,** a short sequence of single-stranded DNA that (1) contains a small part of the sequence of the gene of interest and (2) has had some of its nucleotides modified so that they carry radioactive elements. Bacteria that may carry the gene of interest are washed with a chemical that separates

What Is DNA?   Building Organisms   Genetic Damage   Biotechnology and Health

CHOP AMPLIFY INSERT **GROW** IDENTIFY

CHOP AMPLIFY INSERT GROW **IDENTIFY**

CREATING A GENE LIBRARY

1 To create a gene library, a large amount of DNA is chopped up using restriction enzymes. **2** Each piece is inserted into a plasmid that is then introduced into a bacterial cell. **3** The bacteria are allowed to divide repeatedly, each producing a clone of the foreign DNA fragment it carries. **4** Together, all of the different bacterial cells contain all of the different fragments of the original DNA.

IDENTIFYING A GENE BY USING A DNA PROBE

1 To locate a gene of interest among the millions of clones in a gene library, the bacteria are washed with a chemical that breaks down the DNA, making it single-stranded.

2 A radioactive probe is then washed over the single-stranded DNA. The probe is a short length of DNA that contains a sequence of base pairs complementary to the gene of interest.

3 The DNA probe binds to the complementary base-pair sequence found in the gene of interest and glows with radioactivity, allowing the gene to be easily identified.

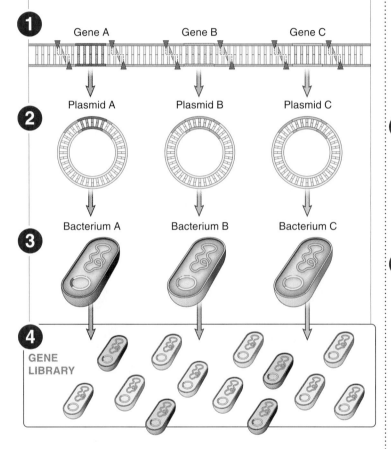

FIGURE 5-25 Finding the needle in a haystack. A DNA probe is used to locate the desired clone in a gene library.

FIGURE 5-24 DNA collection. A gene library (or clone library) is a collection of cloned DNA fragments.

the DNA strands, making single-stranded DNA. Then the radioactive probe is washed over the single-stranded DNA. Bacteria with DNA containing the sequence complementary to the probe will bind to the probe and glow with radioactivity. Once the desired clone in a library is identified, these cells can be grown in large numbers—for example, vats of *E. coli* that produce human growth hormone.

In the remainder of this chapter, we learn how these tools and techniques are being used to develop products. Keep in mind,

however, that even though biotechnology is frequently in the news and has had many important successes, the field is still in its infancy.

TAKE-HOME MESSAGE 5·10

Biotechnology is the use of technology to modify organisms, cells, and their molecules to achieve practical benefits. Modern molecular methods make it possible to cut and copy DNA from one organism and deliver it to another, not necessarily of the same species. With these methods, biotechnology has achieved some important successes in medicine, agriculture, and forensics, but the field is still in its infancy.

5·11

The treatment of diseases and production of medicines are improved with biotechnology.

You can't always get what you want. In the best of all worlds, biotechnology would *prevent* humans from ever getting debilitating diseases. Next best would be to *cure* diseases once and for all. But these noble goals are not always possible, so biotechnology often is directed at the more practical goal of *treating* diseases, usually by producing medicines more efficiently and more effectively than they can be produced with traditional methods. Biotechnology has had some notable successes in achieving this goal. The treatment of diabetes is one such success story.

Diabetes is a chronic disease in which the body cannot produce insulin, a chemical that allows cells to take up and break down sugar from the blood. Complications from diabetes can include vascular disease, kidney damage, and nerve damage. As recently as 1980, if you were one of the approximately 15 million Americans with diabetes, each day you would treat the disease by injecting yourself with insulin extracted from the pancreas of cattle or pigs that had been killed for meat. For most diabetics, the insulin injections kept the disease under control. But the traditional process of collecting insulin this way was difficult and costly. Moreover, cow and pig insulins differ slightly from human insulin in their structures, and about 5% of people with diabetes had negative reactions to the cow and pig insulin preparations.

Q Why do some bacteria produce human insulin?

Everything changed in 1982 when a 29-year-old entrepreneur, Bob Swanson, joined scientist Herbert Boyer to transform the potential of **recombinant DNA technology,** the combination of DNA from two or more sources, into a product. In doing so, they started the biotech revolution. Working with the scientist Stanley Cohen, Swanson and Boyer used restriction enzymes to snip out the human DNA sequence that codes for the production of insulin. They then inserted this sequence into the bacterium *E. coli.* After cloning the new, transgenic bacteria, the team was able to grow vats and vats of the bacteria, all of which churned out human insulin (**FIGURE 5-26**). The drug could be produced efficiently in huge quantities and made available for patients with diabetes. This was the first genetically engineered drug approved by the Food and Drug Administration and it continues to help millions of people every day.

FIGURE 5-26 Life-saving insulin. Human insulin is engineered through recombinant DNA technology.

Perhaps even more significant than providing a better source of insulin, Swanson, Boyer, and Cohen's application of biotechnology revealed a generalized process for genetic engineering. It instantly opened the door to a more effective method of producing many different medicines and treating diseases. Today, more than 1,500 companies work in the recombinant DNA technology industry, and their products generate more than $40 billion in revenues each year.

Several important achievements followed the development of insulin-producing bacteria. Here are just two examples.

1. *Human growth hormone (HGH).* Produced by the pituitary gland, human growth hormone has dramatic effects throughout the body. It stimulates protein synthesis, increases utilization of body fat for energy to fuel metabolism, and stimulates the growth of virtually every part of the body (**FIGURE 5-27**). Insufficient growth hormone production, usually due to pituitary malfunctioning, leads to dwarfism.

When treated with supplemental HGH, individuals with dwarfism experience additional growth. HGH is also used to combat weight loss in people with AIDS. Until 1994, HGH treatment was prohibitively expensive because the growth hormone could only be produced by extracting and purifying it from the pituitary glands of human cadavers. Through the creation of transgenic bacteria, using a technique similar to that used in the creation of insulin-producing bacteria, HGH can now be created in virtually unlimited supplies and made available to more people who need it.

The availability of HGH, which can increase strength and endurance, may be irresistibly tempting to some athletes—even at $7,500 for a month's supply. Recent sporting scandals suggest that the illegal use of HGH occurs frequently among elite swimmers, cyclists, and other athletes.

Q What is "blood doping"? How does it improve some athletes' performance?

FIGURE 5-27 Bulking up with a little (illegal) help. Sylvester Stallone—age 61 in this photo—used human growth hormone (along with strength training) to develop larger muscles for a film role.

2. *Erythropoietin.* Produced primarily by the kidneys, erythropoietin (also known as EPO) is a hormone that regulates the production of red blood cells. Numerous clinical conditions (nutritional deficiencies and lung disease, among others) and treatments (such as chemotherapy) can lead to anemia, a lower than normal number of red blood cells, which reduces an individual's ability to transport oxygen to tissues and cells.

Cloned in 1985, recombinant human erythropoietin (rhu-EPO) is now produced in large amounts in hamster ovaries. It is used to treat many forms of anemia. Worldwide sales of EPO exceeded $10 billion in 2004.

EPO has been at the center of several "blood doping" scandals in professional cycling. Why might a cyclist be interested in having elevated levels of this hormone? EPO increases the oxygen-carrying capacity of the blood, so some otherwise healthy athletes have used EPO to improve their athletic performance. It can be very dangerous, though. By increasing the number of red blood cells, the blood can become much thicker and this can strain the heart and increase the risk of heart attack.

Beyond these and other medicines currently produced via transgenic organisms, plans are under way to create a variety of other useful products—including potatoes that produce antibodies enabling more effective response to illnesses, and other food plants that fight tooth decay. Remember, however, that putting human genes in bacteria or other organisms makes it possible to harness the gene products to treat symptoms of diseases, but it doesn't cure or prevent the diseases. In the next section we examine the as yet less successful attempts to *cure* diseases through biotechnology.

TAKE-HOME MESSAGE 5•11

Biotechnology has led to some notable successes in treating diseases, usually by producing medicines more efficiently and more effectively than they can be produced with traditional methods.

Gene therapy: the use of biotechnology to cure diseases is having limited success.

When it comes to curing a disease by using biotechnology, there is good news and bad news. The good news is that, in the 1990s, a handful of humans with a normally fatal genetic disease called severe combined immunodeficiency disease (SCID) were completely cured through the application of biotechnology. The bad news is that these promising techniques could not be applied to other diseases.

It's not for a lack of trying: there have been more than 500 other clinical trials for **gene therapies** designed to treat or cure a variety of diseases by inserting a functional gene into an individual's cells to replace a defective version of the gene. These trials have included attempts to cure cystic fibrosis, hemophilia, and an inherited form of heart disease, among many others. But there have been no clear successes. Not one. Moreover, recent evidence has even diminished the good news about the SCID treatment, which has been suspended following the deaths of two patients from illness related to their treatment.

The case of SCID is worth examining in some detail, however, because it proves that biotechnology can indeed be used to exchange working genes for malfunctioning disease-causing genes in a way that permanently cures a disease in a human individual.

SCID is a genetic disorder in which a baby is born with an immune system unable to properly produce a type of white blood cell, leaving the infant vulnerable to most infections. The disease usually causes death within the first year of life. It is sometimes called the "bubble boy" disease because in the 1970s, a boy suffering from SCID survived for 12 years by living in a sterile, germ-free bubble (**FIGURE 5-28**).

Researchers have been able to cure the disease in some individuals by using gene therapy that is simple in concept but difficult in practice. First, the researchers removed some cells from an affected baby's bone marrow. These were **stem cells,** cells that have the ability to develop into any type of cell in the body—a feature which makes them very useful in biotechnology. In bone marrow, they normally produce white blood cells, but in individuals with SCID, a malfunctioning gene disrupts normal white blood cell production. Next, in a test tube, the bone marrow stem

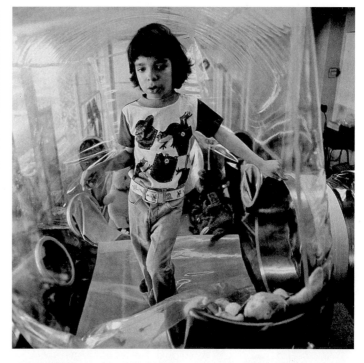

FIGURE 5-28 Protected by a bubble. This child with severe combined immunodeficiency lived 12 years in a protective bubble at Texas Children's Hospital.

cells were infected with a transgenic virus carrying the functioning gene. If all went well, the virus inserted the good gene into the DNA of the stem cells, which were then injected back into the baby's bone marrow. There, the cells could produce normal white blood cells, thereby permanently curing the disease.

Q Why has gene therapy had such a poor record of success in curing diseases?

Difficulties with gene therapy have been encountered in several different areas, usually related to the organism used to transfer the normal-functioning gene into the cells of a person with a genetic disease.

1. *Difficulty getting the working gene into the specific cells where it is needed.* For example, gene therapy for cystic fibrosis—a disease in which thick mucus accumulates in the lungs—attempts to use viruses like those responsible for the common cold to insert a normal gene into malfunctioning lung cells. Unfortunately, because the lungs have evolved to keep foreign objects out, the body is very effective at mounting immune responses against the viruses used to deliver the working gene to the lung cells.

2. *Difficulty getting the working gene into enough cells and at the right rate to have a physiological effect.* In the two deaths related to gene therapy for SCID, for example, the treatment apparently caused leukemia due to problems with the insertion of the gene disrupting normal functioning in white blood cells, triggering them to become cancerous.

3. *Problems with the transfer organism getting into unintended cells.* Viruses can cause serious side effects if they infect cells other than the target cells. Their delivery to target cells must therefore be precise.

Beyond these common problems, gene therapy faces additional difficulties because, for many diseases, the malfunctioning gene has not been identified or the disease is caused by more than one malfunctioning gene. And finally, it is important to keep in mind that gene therapy targets cells in the body other than sperm and eggs. Consequently, while the disease can, in theory, be cured in an individual, the techniques do not prevent those individuals from passing on the disease-causing gene(s) to their offspring. It's not clear what the future holds for gene therapy, but many scientists and patients remain hopeful and a great deal of research is still in progress.

TAKE-HOME MESSAGE 5•12

Gene therapy has had a poor record of success in curing human diseases. This stems primarily from technical difficulties in transferring normal-functioning genes into the cells of a person with a genetic disease.

5•13

Biotechnology can help prevent disease.

Would you want to know? Once, this was just a hypothetical question: if you carried a gene that meant you were likely to develop a particular disease later in your life, would you want to know about it? Or another question: would you want to know if there's a good chance that your future children will be born with a genetic disease? Now, for better or for worse, these are becoming real questions that we all must address. And there is more at stake than simply peace of mind. As biotechnology develops the tools to identify some of the genetic time-bombs that many of us carry, it also carries the danger that such information may become the basis for greater discrimination than we have ever known.

Intervening to prevent diseases by using biotechnology focuses on answering questions such as those posed above at three different points in time.

1. *Is a given set of parents likely to produce a baby with a genetic disease?* Many genetic diseases occur only if an individual inherits two copies of the disease-causing gene, one from each parent. This is true for Tay-Sachs disease, cystic fibrosis, and sickle-cell anemia, among others. Individuals with only a single copy of the disease-causing gene never fully show the disease, but they may pass on the disease gene to their children. Consequently, two healthy parents (that is, having no disease symptoms) may produce a child with the disease. In these cases, it can be beneficial for the parents to be screened to determine whether they carry a disease-causing copy of the gene. Such screening, combined with genetic counseling and testing of embryos following fertilization, can reduce the incidence of the disease dramatically. This has been the case with Tay-Sachs disease, for example. Since screening began in 1969, the incidence of Tay-Sachs disease has been reduced by more than 75% (**FIGURE 5-29**).

Since screening began in 1969, the incidence of Tay-Sachs disease has been reduced by more than 75%!

FIGURE 5-29 Genetic screening can determine the presence of the Tay-Sachs gene.

Possessing mutations in the genes called BRCA1 and BRCA2 (short for breast cancer 1 and breast cancer 2) can make women more susceptible to breast cancer. A genetic screening for the gene is done with a simple blood test.

2. *Will a baby be born with a genetic disease?* Once fertilization has occurred, it is possible to test an embryo or developing fetus for numerous genetic problems. Prenatal genetic screening can detect cystic fibrosis, sickle-cell anemia, Down syndrome, and other disorders. The list of conditions that can be detected is growing quickly.

To screen the fetus, doctors must sample some of the fetal cells and/or the amniotic fluid, which carries many chemicals produced by the developing embryo. This is usually done via amniocentesis or chorionic villus sampling (CVS), techniques that we explore in detail in Chapter 6. Once collected, the cells can be analyzed by a variety of means.

3. *Is an individual likely to develop a genetic disease later in life?* DNA technology can also be used to detect disease-causing

genes in individuals who are currently healthy but are at increased risk of developing an illness later. Early detection of many diseases, such as breast cancer, prostate cancer, and skin cancer, can greatly enhance the ability to treat the disease and reduce the risk of more severe illness or death.

While biotechnology may reduce suffering and the incidence of diseases, these potential benefits come with significant potential costs, including ethical dilemmas. People who have a gene that puts them at increased risk of developing a particular disease, for example, might be discriminated against, even though they are not currently sick and may never suffer from the particular disease for which they are at heightened risk. Although a federal Genetic Information Nondiscrimination Act was signed into law in 2008, the law does not cover all forms of genetic discrimination. It does not, for example, cover life insurance, disability insurance, and long-term care insurance. Insurance companies have already cancelled or denied such insurance coverage on discovering that an individual carries a gene that puts them at increased risk of disease. Additionally, parents who discover that their developing fetus will develop a painful, debilitating, or fatal disease soon after birth are confronted with the difficult question of how to proceed.

TAKE-HOME MESSAGE 5·13

Using biotechnology, many tools have been developed that enable us to identify whether a given set of parents is likely to produce a baby with a genetic disease, whether a baby is likely to be born with a genetic disease, and whether an individual carries certain disease-causing genes that may have their effect later in life. These tools can help us to reduce suffering and the incidence of diseases, but they also come with significant potential costs, particularly the risk of discrimination.

What Is DNA?   Building Organisms   Genetic Damage   Biotechnology and Health

⑤ Biotechnology is producing improvements in agriculture.

Genetically modified corn growing in a lab.

5•14

More nutritious and better foods are being produced with biotechnology.

Your breakfast cereal is probably fortified with vitamins and minerals. And for snacking you may have switched from candy bars to protein bars that have as much protein as a full chicken breast. It shouldn't come as a surprise, then, that farmers have begun using biotechnology to improve on the natural levels of vitamins, minerals, and other nutrients in the fruits, vegetables, and livestock they produce.

Using our knowledge of biology for agricultural improvements is nothing new. For thousands of years humans have been practicing a relatively crude and slow—but still very effective—form of genetic engineering by carefully selecting which plants or animals they use as the breeders for maintaining their crop or animal populations. Through this process of selecting plants or animals with certain desirable traits and breeding the organisms that have them, farmers and ranchers have improved virtually every product of agriculture, producing meatier turkeys, seedless watermelons, and big, juicy corn kernels (**FIGURE 5-30**). But what used to take many generations of breeding can now be accomplished in a fraction of the time, using recombinant DNA technology. Agriculture has entered into exciting but unknown territory.

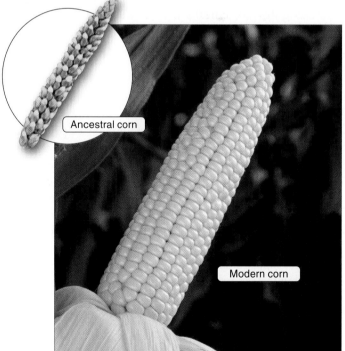

Ancestral corn

Modern corn

FIGURE 5-30 Ancestral corn and modern corn—selected for larger, juicier kernels—compared.

The process begins with the identification of a new characteristic, such as larger size or faster ripening time, that farmers would like in a particular crop. Traditionally, plant breeders would then search for an organism within the same species that had the desirable trait, breed it with their crop organisms, and hope that the offspring would express the trait in the desired way. With recombinant DNA technology, the desired trait can come from *any* species so the pool of organisms from which the trait can be taken becomes much larger. Organisms produced with recombinant DNA technology are referred to in the media as genetically modified organisms, or GMOs.

Although the rewards are potentially huge, in practice the process of creating transgenic species that are more nutritious or have other desirable traits turns out to be difficult. Nonetheless, the results so far hint at a rich and fruitful marriage of agriculture and technology.

Q How might a genetically modified plant help 500 million malnourished people?

Nutrient-Rich "Golden Rice." Almost 10% of the world's population suffers from vitamin A deficiency, which causes blindness in a quarter-million children each year and a host of other illnesses in people of all ages. These nutritional problems are especially severe in places such as southern Asia and sub-Saharan Africa, where rice is a staple of most diets. Addressing this global health issue, researchers have developed what may be the model for solving problems with biotechnology. It involves the creation of a new crop called "golden rice."

Mammals generally make vitamin A from beta-carotene, a substance found in abundance in most plants (it's what makes carrots orange). Beta-carotene is also found in rice plants, but not in the edible part of the rice grains. Researchers set out to change this by inserting into the rice genome three genes that code for the enzymes used in the production of beta-carotene: two genes from the daffodil plant and one bacterial gene (**FIGURE 5-31**). It's clear that the transplanted genes are working because the normally white rice grain takes on a golden color from the accumulated beta-carotene. The rice doesn't yet supply a full-day's requirement of vitamin A in one serving, but it does provide a significant amount. Since golden rice was first developed in 1999, new lines have been

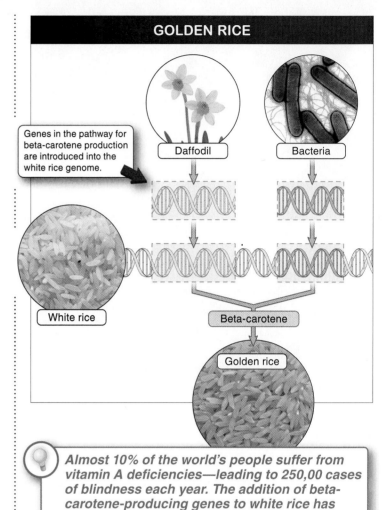

GOLDEN RICE

Genes in the pathway for beta-carotene production are introduced into the white rice genome.

Daffodil

Bacteria

White rice

Beta-carotene

Golden rice

Almost 10% of the world's people suffer from vitamin A deficiencies—leading to 250,00 cases of blindness each year. The addition of beta-carotene-producing genes to white rice has increased its vitamin A content almost 25-fold.

FIGURE 5-31 The potential to prevent blindness in 250,000 people each year. Engineering rice to prevent blindness by increasing its vitamin A content.

produced, now having only two transplanted genes, yet producing almost 25 times the vitamin A found in the original strains. Field tests of golden rice are still under way, so it is not yet being used widely. But this is viewed as one of the most promising applications yet of biotechnology.

TAKE-HOME MESSAGE 5·14

Biotechnology has led to important improvements in agriculture by using transgenic plants and animals to produce more nutritious food.

What Is DNA?   Building Organisms   Genetic Damage   Biotechnology and Health

⑤ Biotechnology is producing improvements in agriculture.

Genetically modified corn growing in a lab.

5•14

More nutritious and better foods are being produced with biotechnology.

Your breakfast cereal is probably fortified with vitamins and minerals. And for snacking you may have switched from candy bars to protein bars that have as much protein as a full chicken breast. It shouldn't come as a surprise, then, that farmers have begun using biotechnology to improve on the natural levels of vitamins, minerals, and other nutrients in the fruits, vegetables, and livestock they produce.

Using our knowledge of biology for agricultural improvements is nothing new. For thousands of years humans have been practicing a relatively crude and slow—but still very effective—form of genetic engineering by carefully selecting which plants or animals they use as the breeders for maintaining their crop or animal populations. Through this process of selecting plants or animals with certain desirable traits and breeding the organisms that have them, farmers and ranchers have improved virtually every product of agriculture, producing meatier turkeys, seedless watermelons, and big, juicy corn kernels (**FIGURE 5-30**). But what used to take many generations of breeding can now be accomplished in a fraction of the time, using recombinant DNA technology. Agriculture has entered into exciting but unknown territory.

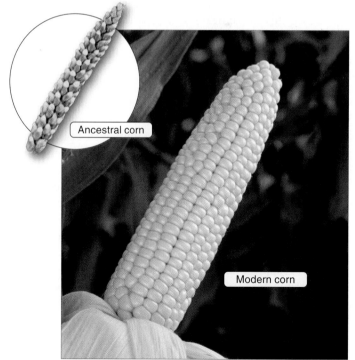

FIGURE 5-30 Ancestral corn and modern corn—selected for larger, juicier kernels—compared.

Ancestral corn

Modern corn

The process begins with the identification of a new characteristic, such as larger size or faster ripening time, that farmers would like in a particular crop. Traditionally, plant breeders would then search for an organism within the same species that had the desirable trait, breed it with their crop organisms, and hope that the offspring would express the trait in the desired way. With recombinant DNA technology, the desired trait can come from *any* species so the pool of organisms from which the trait can be taken becomes much larger. Organisms produced with recombinant DNA technology are referred to in the media as genetically modified organisms, or GMOs.

Although the rewards are potentially huge, in practice the process of creating transgenic species that are more nutritious or have other desirable traits turns out to be difficult. Nonetheless, the results so far hint at a rich and fruitful marriage of agriculture and technology.

Q How might a genetically modified plant help 500 million malnourished people?

Nutrient-Rich "Golden Rice." Almost 10% of the world's population suffers from vitamin A deficiency, which causes blindness in a quarter-million children each year and a host of other illnesses in people of all ages. These nutritional problems are especially severe in places such as southern Asia and sub-Saharan Africa, where rice is a staple of most diets. Addressing this global health issue, researchers have developed what may be the model for solving problems with biotechnology. It involves the creation of a new crop called "golden rice."

Mammals generally make vitamin A from beta-carotene, a substance found in abundance in most plants (it's what makes carrots orange). Beta-carotene is also found in rice plants, but not in the edible part of the rice grains. Researchers set out to change this by inserting into the rice genome three genes that code for the enzymes used in the production of beta-carotene: two genes from the daffodil plant and one bacterial gene (**FIGURE 5-31**). It's clear that the transplanted genes are working because the normally white rice grain takes on a golden color from the accumulated beta-carotene. The rice doesn't yet supply a full-day's requirement of vitamin A in one serving, but it does provide a significant amount. Since golden rice was first developed in 1999, new lines have been

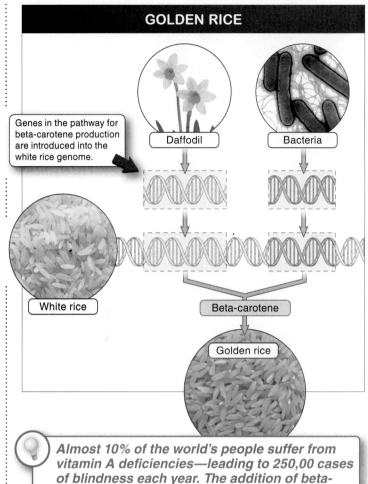

GOLDEN RICE

Genes in the pathway for beta-carotene production are introduced into the white rice genome.

Daffodil

Bacteria

White rice

Beta-carotene

Golden rice

Almost 10% of the world's people suffer from vitamin A deficiencies—leading to 250,00 cases of blindness each year. The addition of beta-carotene-producing genes to white rice has increased its vitamin A content almost 25-fold.

FIGURE 5-31 The potential to prevent blindness in 250,000 people each year. Engineering rice to prevent blindness by increasing its vitamin A content.

produced, now having only two transplanted genes, yet producing almost 25 times the vitamin A found in the original strains. Field tests of golden rice are still under way, so it is not yet being used widely. But this is viewed as one of the most promising applications yet of biotechnology.

TAKE-HOME MESSAGE 5·14

Biotechnology has led to important improvements in agriculture by using transgenic plants and animals to produce more nutritious food.

5·15

Farming can be made more efficient and eco-friendly through biotechnology.

> *Italians come to ruin most generally in three ways, women, gambling, and farming. My family chose the slowest one.*
>
> —Pope John XXIII

We are in the midst of a revolution—a green revolution—and few people are aware of it. The possibilities of biotechnology playing a role in treating, curing, and preventing human diseases may garner all the headlines, but the actual results pale in comparison with the current impact of biotechnology on agriculture. And although biotechnology can improve agriculture by making more nutritious food (as in the case of golden rice), the real revolution is the extent to which biotechnology has reduced the costs, both environmental and financial, of producing the plants and animals we eat.

The extent of biotechnology's success in the area of producing plants with built-in insecticides and genetically engineered resistance to herbicides is staggering. Currently, more than 170 million acres worldwide are planted with genetically modified crops, two-thirds of which are in the United States,

representing more than a fortyfold increase in the past 10 years. The financial benefits to farmers—at least in the short run—are so great that more and more of them are embracing the genetically modified crops.

The numbers are surprising: 45% of all corn grown in the United States—including corn grown in every single state in the continental United States—is genetically modified. Seventy-six percent of all cotton grown is genetically modified. And 85% of all soybeans grown is genetically modified (**FIGURE 5-32**). Two factors explain much of the rapid expansion and extensive adoption of genetically modified plants in U.S. agriculture. (1) Many plants have had insecticides and herbicide resistance engineered into them, which can reduce the amounts of insecticides and herbicides used in agriculture. (2) Because herbicide- and insect-resistant plants can reduce the amount of plowing required around crops to remove weeds,

Q What genetically modified foods do most people in the U.S. consume (usually without knowing it)?

GENETICALLY MODIFIED CROPS IN THE UNITED STATES

Corn

55% 45%

Cotton

24% 76%

Soybeans

15% 85%

Proportion of crops that are not genetically modified
Proportion of crops that are genetically modified

FIGURE 5-32 A significant portion of crops grown in the United States are genetically modified.

FIGURE 5-33 **Competing with us for food.** Insects can ravage crops.

a huge potential food resource for insects. Every year, about 40 million tons of corn are unmarketable as a consequence of insect damage. Increasingly, however, farmers have been enjoying greater success in their battles against insect pests, primarily through the use of genetically modified crops.

Farmers owe much of this success to soil-dwelling bacteria of the species *Bacillus thuringiensis.* These bacteria produce spores containing crystals that are highly poisonous to insects but harmless to the crop plants and to people. Within an hour of ingesting "Bt" crystals, the insects' feeding is disrupted. The toxic crystals cause pores to develop throughout the insects' digestive systems, paralyzing their gut and making them unable to feed. Within a few days the insects die from a combination of tissue damage and starvation.

genetically modified plants can reduce the costs of producing food and the loss of topsoil to erosion.

Several of the biggest successes of the application of recombinant DNA technology to agriculture are described below.

Insect Resistance Insect pests have a field day on agricultural crops (**FIGURE 5-33**). Crops that are planted at high densities and nurtured with ample water and fertilizer represent

Beginning in 1961, the toxic Bt crystals were included in the pesticides sprayed on crop plants. In 1995, however, recombinant DNA technology led to a huge reduction in the application of this toxic spray. The gene coding for the production of the Bt crystals was inserted directly into the DNA of many different crop plants, including corn, cotton, and potatoes, so that the plants themselves produced the crystals. As a consequence, it is no

Q How can genetically modified plants lead to reduced pesticide use by farmers?

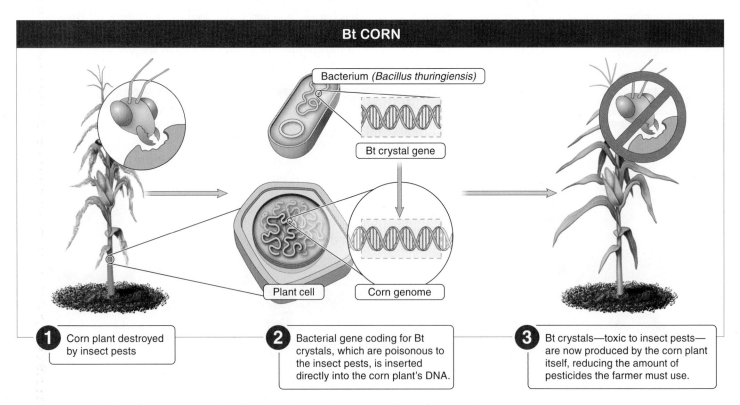

Bt CORN

Bacterium *(Bacillus thuringiensis)*

Bt crystal gene

Plant cell

Corn genome

1 Corn plant destroyed by insect pests

2 Bacterial gene coding for Bt crystals, which are poisonous to the insect pests, is inserted directly into the corn plant's DNA.

3 Bt crystals—toxic to insect pests— are now produced by the corn plant itself, reducing the amount of pesticides the farmer must use.

FIGURE 5-34 **Help from bacteria in growing disease-resistant corn.** Corn engineered to contain spores of the bacterium *Bacillus thuringiensis* (Bt) kills insect pests but does not harm humans.

What Is DNA?   Building Organisms   Genetic Damage   Biotechnology and Health

longer necessary for farmers to apply huge amounts of Bt-containing pesticides (**FIGURE 5-34**). Instead, the plants do that work themselves. The insects that try to eat the genetically modified plants ingest the toxin and soon die. There is no evidence that Bt crystals have any harmful effects on humans at all, even when they are exposed to very high levels. Because pesticides can be damaging to the environment, particularly when they flow from the farm and into the water supply as runoff, reduction in pesticide use can be beneficial for the environment.

Herbicide Resistance Bacteria have come to the aid of farmers in their fight against pests in another way, too. Consider a seemingly impossible challenge that farmers face when they attempt to kill weeds that harm crop plants by competing for light, water, and soil nutrients. Herbicides, chemicals that kill plants, can be applied to kill the weeds (**FIGURE 5-35**). These chemicals usually work by blocking the action of an enzyme that enables plants (as well as fungi and bacteria) to build three critical amino acids. Without these amino acids, the organisms soon die. The problem is that because herbicides affect *all* plants similarly, they are generally toxic to the crop plant as well as the weeds. This is where bacteria come in.

In the 1990s, researchers discovered bacteria that can resist the effect of herbicides. The gene that gives the bacteria resistance to herbicides was identified and introduced into crop plants. Integration of this gene into the plants' DNA gives them resistance to the herbicides and allows farmers to kill weeds

Herbicide-resistance genes from bacteria protect crop plants from the herbicides used to kill the weeds competing with them.

FIGURE 5-35 **Crop duster.** Herbicides like the one applied by this crop duster must kill weeds while leaving the crop unharmed.

FIGURE 5-36 **Bigger salmon.** The larger salmon carries a growth hormone gene that keeps it growing year-round rather than in the summer only.

with herbicides without harming the crop plants, greatly increasing yields.

Faster Growth and Bigger Bodies Agriculture includes the cultivation not just of plants but also of animals. Currently, researchers are developing transgenic salmon that grow significantly faster and reach a much larger size than normal salmon (**FIGURE 5-36**). These salmon carry a version of the growth hormone gene that functions year round, rather than primarily in the summer. It was isolated from a cold-water fish species, called arctic pout, and injected into the egg of a salmon with genetic engineering techniques. The super-fish can be raised much more quickly and using significantly less feed than normal salmon, reaching market size within 18 months rather than the usual 24–30 months. In taste tests, consumers cannot tell the difference between the transgenic and non-transgenic salmon.

Quite troubling, though, is the fact that researchers expected growth rate increases of about 25% but found that the genetically modified fish grew about 500% faster. Additionally, if the larger, faster-growing fish were to escape from their breeding nets back into their natural habitat—something that experts agree is inevitable—the fish might harm populations of other species because they can consume more resources and may grow too large to be consumed by their natural predators. It is unclear what the outcome would be.

TAKE-HOME MESSAGE 5·15

There has been a revolution in the extent to which biotechnology has reduced the environmental and financial costs of producing food, through the creation of herbicide-resistant and insect-resistant crops.

Fears and risks: are genetically modified foods safe?

Chickens without feathers look ridiculous (**FIGURE 5-37**). But such a genetically modified breed was developed with a valuable purpose in mind: "naked" birds are easier and less expensive to prepare for market, benefiting farmers by lowering their costs and benefiting consumers by lowering prices. Such chickens, however, turned out to be unusually vulnerable to mosquito attacks, parasites, and disease, and ultra-sensitive to sunlight. They also have difficulty mating, because the males are unable to flap their wings. Researchers currently are working to address these problems.

These chickens teach us an important lesson about genetically modified plants and animals. Although the new breed of featherless chickens was produced by traditional animal husbandry methods—the cross-breeding of two different types of chickens—rather than by recombinant DNA technology, the new breed ended up having not just the desired trait of no feathers but also some unintended and undesirable traits. Now, as more genetically modified foods are

FIGURE 5-38 **Consumer fears.** Protesters voice opposition to the use of genetically modified organisms (GMOs).

created using modern methods of recombinant DNA technology, the same risks of unintended and potentially harmful traits occurring must be weighed. For these and other reasons—some legitimate and rational, others irrational—many people have concerns about the production and consumption of genetically modified foods (**FIGURE 5-38**).

Fear 1. Organisms that we want to kill may become invincible. Weed-resistant canola plants were cultivated in Canada, making it possible for farmers to apply herbicides freely to kill the weeds but not the canola plants. But the weed-resistant canola plants accidentally spread to neighboring farms and grew out of control, because traditional herbicides could not kill them. Similarly, there is the possibility that insect pests will develop resistance to the Bt that genetically modified crops produce, which will also make these pests resistant to Bt pesticides applied to crops that are not genetically modified.

Fear 2. Organisms that we don't want to kill may be killed inadvertently. Monarch butterflies feed on milkweed plants. Recent research has demonstrated that if pollen from plants genetically modified to contain the insect-killing Bt genes accidentally lands on milkweed plants and is consumed by monarch butterflies, the butterflies can be killed, which may significantly reduce their populations. Although such an incident has not occurred outside experimental fields, it illustrates a risk that may be hard to control.

Fear 3. Genetically modified crops are not tested or regulated adequately. It is impossible to ever really know whether a new

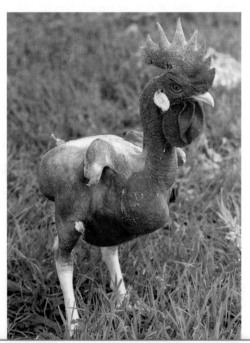

> Featherless birds are cheaper for farmers and consumers. But there are unintended consequences, including vulnerability to mosquitoes and other parasites.

FIGURE 5-37 **"Naked" birds.** The breeding of featherless chickens has benefited farmers but led to some unanticipated consequences.

What Is DNA?   Building Organisms   Genetic Damage   Biotechnology and Health

technology has been tested adequately. Still, scientists and lawmakers have been working toward an organized and responsible set of policies designed to ensure sufficient safety testing is done. For example, laboratory procedures for working with recombinant DNA have been established, and researchers have developed techniques that make it impossible for most genetically engineered organisms to survive outside the specific conditions for which they are developed.

As an example of the degree of testing used in the field of genetically engineered foods, the Monsanto Company has had its strain of herbicide-resistant soybeans evaluated and approved by 31 different regulatory agencies in 17 different countries, including, in the United States, the Department of Agriculture, the Food and Drug Administration, and the Environmental Protection Agency. In a recent report on genetically modified animals, however, an expert committee from the U.S. National Academy of Sciences warned that genetically modified organisms still pose risks that the government is unable to evaluate. Technology is moving so fast that it is difficult to even know what the new risks might be.

> And he gave it for his opinion, 'that whoever could make two ears of corn, or two blades of grass, to grow upon a spot of ground where only one grew before, would deserve better of mankind, and do more essential service to his country, than the whole race of politicians put together.'
>
> —Jonathan Swift, *Gulliver's Travels,* 1726

Fear 4. Eating genetically modified foods is dangerous. In the 1990s, a gene from Brazil nuts was used to improve the nutritional content of soybeans. The genetically modified soybeans had better nutritional content, but they also acquired some allergy-causing chemicals previously present in the Brazil nuts but not in soybeans. This outcome illustrates the risk that some unwanted features might be passed from species to species in the creation of transgenic organisms. In this case, all of the genetically modified soybeans were destroyed and this research program was suspended. To date, no evidence has appeared to suggest that consumption of any genetically modified foods is dangerous.

Fear 5. Loss of genetic diversity among crop plants is risky. As increasing numbers of farmers stop using non-genetically modified crops in favor of one or a few genetically modified strains of crops, there is a reduction in the genetic diversity of the crops. This can make them more vulnerable to environmental changes or pests. The Irish Potato Famine is an example of the value of genetic diversity in crops. In the mid-1800s, much of the population of Ireland depended on potatoes. Because most of the potato crops had been propagated from cuttings from the same plant, they were all genetically the same. When the crops were infected by a rot-causing mold, all of the potato plants were susceptible and most were wiped out, causing a famine responsible for the deaths of more than a million people.

Fear 6. Hidden costs may reduce the financial advantages of genetically modified crops. When seed companies create genetically modified seeds with crop traits desirable to farmers, the companies also engineer sterility into the seeds. As a consequence, the farmers must purchase new seeds for each generation of their crops. Such increases in the long-term costs and dependency on seed companies must be factored in by the farmers.

Among the less rational fears is that genetically modified foods are not "natural" and therefore must be harmful. This should not be a cause for concern. Smallpox, HIV, poison ivy, and cyanide, after all, are natural. The smallpox vaccine, on the other hand, is unnatural. Innumerable other valuable technological developments are equally unnatural. There simply is no value in knowing whether something is natural or unnatural when evaluating whether it is good and desirable or not.

In the end, we must compare the risks of producing genetically modified foods with the benefits. For example, farm workers will greatly benefit from spending less time applying pesticides to genetically modified, pest-resistant crops. The benefits of the workers' reduced pesticide exposure will be significantly greater in the less-developed countries, where safety regulations for pesticide use are more frequently ignored. The cost-benefit analyses will have to include the potential to reduce food costs and the ability to reduce environmental degradation from agriculture.

TAKE-HOME MESSAGE 5•16

More and more genetically modified foods are being created using modern methods of recombinant DNA technology. Some legitimate fears among the public remain, however, about the safety of these foods, given that their development relies on such new technology, and about the long-term financial advantages they offer.

6 Biotechnology today and tomorrow: DNA sequences are rich in information.

With the cloning of a sheep, a new era in biotechnology began.

5·17

DNA is an individual identifier: the uses and abuses of DNA fingerprinting.

In another time, Colin Pitchfork, a murderer and rapist, would have walked free. But in 1987, he was captured and convicted, betrayed by his DNA, and is now serving two life sentences in prison. Pitchfork's downfall began when he raped and murdered two 15-year-old high-school girls in a small village in England in the 1980s. The police thought they had their perpetrator when a man confessed, but only to the second murder. He denied any involvement with the first murder, though, which perplexed the police because the details of the two crimes strongly suggested that the same person committed both.

At the time, British biologist Alec Jeffreys made the important discovery that there were small pieces of DNA in every person's chromosomes that were tremendously variable in their base sequences. In much the same way each person has a driver's license number or social security number that differs from everyone else's, these DNA fragments are so variable that it is extremely unlikely that two people would have identical

sequences at these locations. Thus, a comparison between these regions in a person's DNA sample and in DNA-containing evidence left at a crime scene would enable police to determine whether the evidence came from that person.

Jeffreys analyzed DNA left by the murderer/rapist on the two victims and found that it did indeed come from a single person, and that that person was *not* the man who had confessed to one of the crimes. That original suspect was released and has the distinction of being the first person cleared of a crime through DNA fingerprinting. To track down the criminal, police then requested blood samples from all men in the area who were between 18 and 35 years old—a practice many viewed as an invasion of privacy—collecting and analyzing more than 5,000 blood samples. This led them to Colin Pitchfork, whose DNA matched perfectly the DNA left on both of the victims, and ultimately was the evidence responsible for his conviction. (He almost slipped through, having persuaded a friend to give a blood sample in his name.

What Is DNA?   Building Organisms   Genetic Damage   Biotechnology and Health

FIGURE 5-39 **Betrayed by his DNA.** Colin Pitchfork, the first criminal brought to justice with DNA fingerprinting.

But when the friend was overheard telling the story in a pub, police tracked down Pitchfork to get a blood sample.)

DNA fingerprinting is now used extensively in forensic investigations, in much the same way that regular fingerprints have been used for the past 100 years (**FIGURE 5-39**). But traditional fingerprinting is limited in its usefulness for many crimes because, often, no fingerprints are left behind. DNA fingerprinting, on the other hand, is not so limited because DNA samples more frequently are left behind, usually in the form of semen, blood, hair, skin, or other tissue. As a consequence, this technology has been directly responsible for bringing thousands of criminals to justice and, as we saw at the beginning of the chapter, for establishing the innocence of more than 200 people wrongly convicted of murder and other capital crimes. Let's examine how DNA fingerprinting is done, why it is such a powerful forensic tool, and why it is not foolproof.

The DNA from different humans is almost completely identical. More than 99.9% of the sequences of two individuals are the same because we are all from the same species and thus share a common evolutionary history. Even so, in a genome of three billion base pairs, even a tenth of a percent difference translates to about three million base-pair differences. These differences are responsible for the fact that all individuals have their own unique genome. (The lone exception? Identical twins, whose DNA is exactly the same.) Thus, when we are trying to evaluate whether the DNA from a crime scene matches that from a suspect, the analysis focuses on the parts of our DNA that differ. There are several dozen of these highly variable regions in the human genome.

Some of the most highly variable parts of the DNA are regions called **VNTR**s, because they contain a **v**ariable **n**umber of **t**andem **r**epeats. A VNTR region, we'll call it VNTR 1, might have a short sequence of 15 to 100 bases that repeats over and over again. Imagine that a person's genome is a book. The VNTR would be like the insertion of the word "green" over and over again on one of the pages. An individual—we'll call her Individual A—has two copies of each chromosome, one from her mother and one from her father, and the two copies of the VNTR 1 site that she carries are likely to differ from each other—maybe there are 14 repeats in one case and 3 repeats in the other. Individual B also has two copies of the VNTR 1 region, and the number of times the sequence repeats in those regions of Individual B's DNA might be 5 and 11 (**FIGURE 5-40**). Elsewhere in a person's genome are more than a dozen other VNTR regions that can be compared similarly.

Why are VNTRs so useful for forensic scientists? Because the existence of these regions means that forensic scientists don't need to look at all three billion bases in a person's entire DNA sequence. Instead, they only have to look at how many times a sequence is repeated in a suspect's two copies of VNTR 1, let's say 14 and 3 times. Then they look at how many times the VNTR 2 sequence is repeated in that person, let's say 7 and 34, and VNTR 3 might have 42 and 16 repeats. So the individual in question could be thought of as having a DNA fingerprint

Biotechnology and Agriculture   ▼   DNA Sequences

CREATING DNA FINGERPRINTS BY USING VNTRs

A DNA fingerprint is created by counting the number of times that a repetitive sequence of base pairs occurs on an individual's chromosomes. Regions of repetitive base-pair sequences are called VNTRs because they contain a **V**ariable **N**umber of **T**andem **R**epeats.

INDIVIDUAL A

	VNTR	Number of times the repeating sequence occurs
Chromosome from mother	‖‖‖‖‖‖‖	**14**
Chromosome from father	‖‖‖	**3**

14 / 3
Individual A's DNA fingerprint for this region

INDIVIDUAL B

	VNTR	
Chromosome from mother	‖‖‖‖‖	**5**
Chromosome from father	‖‖‖‖‖‖	**11**

5 / 11
Individual B's DNA fingerprint for this region

For a given region, it's the same sequence that repeats, but the number of times it repeats differs from individual to individual (and differs on the maternal and paternal copies of a chromosomes that an individual carries).

FIGURE 5-40 VNTRs in forensics. Forensic scientists use VNTRs to genetically link a person of interest with DNA-containing evidence left at a crime scene.

of 14/3, 7/34, 42/16. It's like checking whether two books are identical by looking at the first word on pages 16, 48, 123, and 200. It's a lot easier than comparing every word, but still can produce the correct answer. In the U.S., an individual's complete DNA fingerprint is generally built by sampling their DNA at more than ten different VNTR regions.

In court, DNA fingerprints are usually shown as photos of dark bands in a gel to demonstrate that the banding pattern of a suspect's DNA either does or doesn't match the banding pattern of DNA found at the crime scene (**FIGURE 5-41**).

Q What is a DNA fingerprint?

CREATING A DNA FINGERPRINT GEL

1 Using restriction enzymes, DNA samples are cut on either side of the repeating sequences. Only two VNTR regions are shown here. Actual DNA fingerprints are created by sampling up to eight regions.

FIGURE 5-41 Your unique identifier: a DNA fingerprint. DNA fingerprints are being used to match a suspect's DNA to DNA found at the scene of a crime.

How do we get from an entire genome to a simple photo of bands?

First, the DNA in an evidence sample is chopped with a restriction enzyme that makes cuts on either side of the area where the repeating sequence occurs. The DNA is then

What Is DNA?   Building Organisms   Genetic Damage   Biotechnology and Health

poured into a gel and an electrical charge is applied to induce the DNA to migrate toward the bottom of the gel. Because smaller pieces move more quickly than larger pieces, this process causes different-size pieces of DNA to separate by size. The smaller pieces—those with a small number of repeats—are found closer to the bottom; the larger pieces—those with a large number of repeats—are found toward the top. At this point, because the DNA still is not visible, the repeating DNA sequence is stained with a radioactively labeled chemical and the gel is placed in contact with X-ray film. The film records dark bands at the point in the gel to which the DNA fragments have migrated.

DNA samples from different people will form different bands, whereas different samples of DNA from one person will have exactly the same pattern of dark bands.

In the end, despite universally accepted methods, DNA fingerprinting is not foolproof, because the possibility for human error remains. Even though we should not blindly draw conclusions solely from this one type of evidence, however, DNA fingerprinting is an increasingly valuable tool for law enforcement. The FBI, for example, has reported that nearly a third of their suspects are cleared immediately by DNA testing, and that many more criminals, because of DNA fingerprinting, now plead guilty to the crimes they have committed.

TAKE-HOME MESSAGE 5·17

Comparisons of highly variable DNA regions can be used to identify tissue specimens and determine the individual from whom they came.

5·18

DNA sequences reveal evolutionary relatedness.

It's hard to get somewhere new without a map. Just as explorers of the New World created maps to get the lay of the land, genetics researchers have been working since 1990 to map the genomes of humans and other species (**FIGURE 5-42**). Today, useful drafts of these sequences have been completed for more than 180 species, including humans and various species of bacteria, yeast, plants, worms, and fruit flies. The list is expected to top 1,000 species in a few years. Now that we have all of these data and are acquiring more, the big question is, what do we do with all of this information? Progress is

being made in numerous different directions. We'll touch on just two of them.

1. *Mapping genetic landscapes.* The **Human Genome Project** is a project to decode the three billion base pairs in the human genome and to identify all the genes present in it. The project neared completion in 2000, and subsequent efforts have focused on determining some basic information, including how many genes there are, where they are, and how they work. In fact, the first big surprise of the Human Genome Project was the discovery that the number of genes in human genetic material is much smaller than expected, falling somewhere between 20,000 and 25,000. This is not much different from the number of genes in the genomes of fruit flies (13,000) or nematode worms (19,000) or the small plant *Arabidopsis* (26,000). We saw earlier that the total amount of DNA an organism carries is not related to the organism's complexity. It seems that there also is no relationship between the number of genes an organism has and its complexity, either.

2. *Building earth's family tree.* Humans crave order. To organize our cities we use address systems, to manage our digital information we use folders, subfolders, and files on our computers, and to keep track of our books we use the Library of Congress cataloging system. Not surprisingly, humans desire order when it comes to the billions of living organisms on earth. For this, we create **phylogenetic trees,** a way of grouping organisms in a hierarchical system that reflects the evolutionary history and relatedness of all

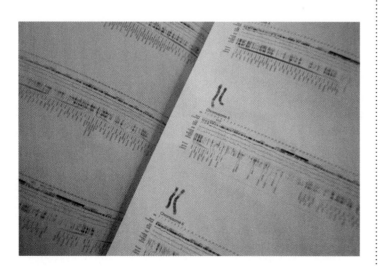

FIGURE 5-42 **The Human Genome Project: mapping the genetic landscape.**

Biotechnology and Agriculture   DNA Sequences

organisms. Phylogenetic trees are a useful organizational system that helps us understand the diversity of life on earth and the nature of how species are related to one another.

Traditionally, phylogenetic trees were built on the basis of similarities among organisms' physical structures and embryological development. We'll examine the details of how this was done in Chapter 10, but the net result was a system of six kingdoms—bacteria, archaea, protists, plants, animals, and fungi—that reflected reasonably well the consensus hypothesis about the evolutionary history of the different groups of organisms. Today, however, DNA sequences are being analyzed as a feature that can be traced over time, and this analysis is revolutionizing the construction of phylogenetic trees. The new DNA-sequence-based phylogenetic trees reflect much more accurately the branching patterns of different groups of organisms as they evolved (**FIGURE 5-43**).

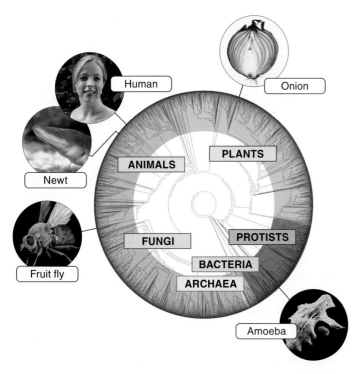

FIGURE 5-43 The tree of life. Our understanding of the evolutionary relationships among species is made more accurate through analysis and comparison of DNA sequences.

The rationale behind constructing phylogenies based on DNA sequences is straightforward. Organisms inherit DNA from their ancestors. As species evolve in their own unique ways, their DNA sequences diverge, becoming increasingly different. The more time that passes following the splitting off of one species from another, the greater the difference between their DNA sequences. Phylogenies are then constructed that place species in positions that reflect their relative genetic similarity.

Measures of overall DNA similarity between chimps and humans reveal that 96% of our base pairs are the same. (This is actually a comparison with only one chimp, a male named Clint, from the Yerkes Primate Research Center in Atlanta.)

> **Q** When we say humans and chimps are genetically 96% identical, what do we mean?

This is a greater similarity than humans share with any other species and indicates that chimps are our closest living relatives. Although most DNA doesn't code for proteins, when comparing only the sequences that code for proteins, we find that 29% of the genes code for the same exact proteins in chimps and humans.

For the purpose of comparison:

1. The difference between human DNA and mouse DNA is 60 times greater than the human–chimp difference.

2. The difference between mouse and rat DNA is 10 times greater than the human–chimp difference.

3. The difference between human and chimp DNA is 10 times greater than the average difference in DNA between two humans.

TAKE-HOME MESSAGE 5·18

Comparisons of DNA sequence similarities across species reveal their evolutionary relatedness and make it possible to construct detailed evolutionary trees.

Cloning—from organs to individuals—offers both promise and perils.

Cloning. Perhaps no scientific word more readily conjures horrifying images of the intersection of curiosity and scientific achievement. But is fear the appropriate emotion to feel about this burgeoning technology? Perhaps not.

For starters, let's clarify what the word means. "Cloning" actually refers to a variety of different techniques. To be sure, cloning can refer to the creation of new individuals that have exactly the same genome as the donor individual—a process called "whole organism cloning." That is, a clone is like an identical twin, except that it may differ in age by years or even decades. It is also possible to clone tissues (such as skin) and entire organs from an individual's cells. And, as we have already seen in this chapter, it is possible to clone genes.

Cloning took center stage in the public imagination in 1997 when Ian Wilmut, a British scientist, and his colleagues first reported that they had cloned a sheep, named Dolly. Their research was based on ideas that went back to 1938, when Hans Spemann first proposed the experiment of removing the nucleus from an unfertilized egg and replacing it with the nucleus from the cell of a different individual. Although the process used by Wilmut and his research group was difficult and inefficient, it was surprisingly simple in concept (**FIGURE 5-44**). They removed a cell from the mammary gland of a grown sheep, put its nucleus into another sheep's egg from which the nucleus had been removed, induced the egg to divide, and transplanted it into the womb of a surrogate mother sheep. Out of 272 tries, they achieved one success. But that was enough to show that the cloning of an adult was possible.

Shortly after news of Dolly's birth, teams set about cloning a variety of other species, including mice, cows, pigs, and cats (**FIGURE 5-45**). Not all of this work was driven by simple curiosity. For farmers, cloning could have real value. It can take a long time to produce animals with desirable traits from an agricultural perspective—such as increased milk production in cows. And with each successive generation of breeding, it can be difficult to maintain these traits in the population. But through the process of transgenic techniques and whole-animal cloning, large numbers of valuable animals with such traits can be produced and maintained.

FIGURE 5-44 No longer science fiction. The steps used in the cloning of Dolly the sheep.

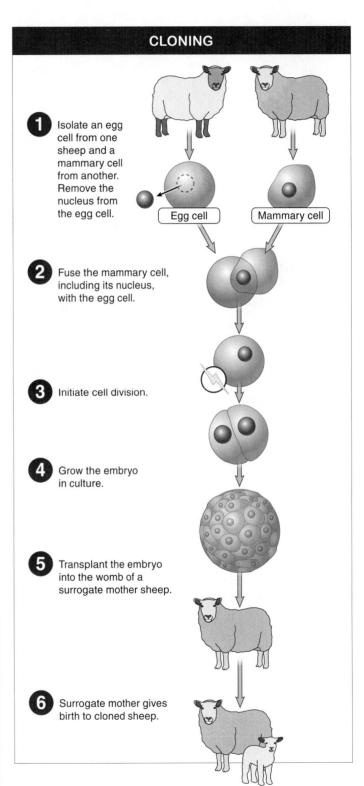

CLONING

1 Isolate an egg cell from one sheep and a mammary cell from another. Remove the nucleus from the egg cell.

Egg cell   Mammary cell

2 Fuse the mammary cell, including its nucleus, with the egg cell.

3 Initiate cell division.

4 Grow the embryo in culture.

5 Transplant the embryo into the womb of a surrogate mother sheep.

6 Surrogate mother gives birth to cloned sheep.

FIGURE 5-45 Genetically identical cloned animals. The cloning of animals can maintain desirable traits from generation to generation.

Q Are there any medical justifications for cloning?

Medical researchers, too, see much to gain from cloning. In particular, transgenic animals containing human genes—such as those discussed earlier in this chapter—can be very valuable. For instance, sheep that produce the human protein alpha-1 antitrypsin that is used to treat cystic fibrosis, or hamsters that produce human erythropoietin to treat kidney disease, are needed in large numbers. By cloning the transgenic animals, large colonies of pharmaceutical-drug-producing animals can be created. Pig clones, too, are being created with the hope of using their organs for transplantation into humans. Such pigs are unusual and useful because they have had genes inactivated that would normally cause their tissue to be rejected by the immune systems of human recipients.

Q Is it possible to clone a dinosaur? How could it be done?

Movies have raised the question of whether it is possible to clone a dinosaur (**FIGURE 5-46**). While the short answer is *yes,* that is in theory only. Several practical steps would be necessary, each of which might prove impossible. The first step would be retrieving the DNA of the dinosaur to be cloned. The most likely route would be to salvage fossilized DNA from the stomach of a mosquito, preserved since the time when dinosaurs lived. Mosquitoes from those days have actually been found, preserved in amber, a hardened, fossilized tree sap, but no ingested blood has been recovered from them. Then the dinosaur DNA would have to be extracted from the blood—assuming the mosquito had bitten a dinosaur. The biggest problem at this step is that it would be necessary to extract

What Is DNA?　　　　Building Organisms　　　　Genetic Damage　　　　Biotechnology and Health

FIGURE 5-46 Can a dinosaur be cloned?

and isolate all of the dinosaur's DNA, not just a few fragments. In the unlikely event that this could be done, this DNA would need to be inserted into the egg from a living organism, perhaps a crocodile, to create an embryo. At this point the embryo would have to be put in some sort of artificial egg where it could be incubated and fed as it grows, until it could finally hatch.

And if a dinosaur, why not a human? At this point, it is almost certain that the cloning of a human will be possible. Many people wonder, though, whether such an endeavor should be pursued. There is near unanimity among scientists that human cloning should not be attempted, and governments are struggling to develop wise regulations for this new world.

TAKE-HOME MESSAGE 5·19

Cloning of individuals has potential benefits in agriculture and medicine, but ethical questions linger.

Biotechnology and Agriculture   DNA Sequences

Knowledge You Can Use

Did you know? Mixing aspirin and alcohol can lead to metabolic interference and unexpected inebriation.

Ethanol is the form of alcohol found in cocktails, beer, and wine. When you consume alcohol, the first step in your body's metabolic breakdown of the ethanol is carried out by the enzyme alcohol dehydrogenase.

Q: What happens to ethanol molecules *not* broken down? Any ethanol molecules in the bloodstream not broken down by alcohol dehydrogenase make their way to the brain. Once there, they cause you to feel a bit inebriated—or drunk, if the amount of ethanol is sufficiently large. Aspirin has the unintended side effect of disabling alcohol dehydrogenase, thus interfering with its normal functioning.

Q: What do you think happens to someone who takes two aspirin and then drinks several alcoholic beverages? By blocking the activity of alcohol dehydrogenase, aspirin interferes with a person's ability to break down ethanol. As a consequence, the ethanol remains in the body longer and has a more pronounced effect on the brain, producing greater inebriation (26% greater, according to one study).

What can you conclude? Medications, even over-the-counter products, can have unexpected (and unintended) physiological consequences, particularly when mixed with alcohol consumption. Exercise great caution when taking them.

❶ DNA: What is it and what does it do?

DNA, the genetic material for all living organisms, is a double-helix molecule consisting of two "backbone" chains of sugar and phosphate molecules linked by pairs of nucleotide bases. Genes are sequences of bases in DNA that contain the code for the construction of protein molecules. Only a small fraction of the DNA in eukaryotes codes for genes.

❷ Building organisms: through gene expression, information in the DNA directs the production of the molecules that make up an organism.

Information in the genes is transcribed into the sequences of mRNA molecules. These mRNA sequences then direct the construction of proteins.

❸ Damage to the genetic code has a variety of causes and effects.

Mutations are rare alterations in a single base or changes in large segments of DNA that include several genes. When they occur in a gene, normal physiological functioning is usually disrupted. Mutations play an important role in evolution. Most genetic diseases result from a mutation that causes a gene to produce a non-functioning enzyme, resulting in a blocked metabolic pathway.

❹ Biotechnology has the potential for improving human health.

Biotechnology is the use of technology to modify organisms, cells, and their molecules to achieve practical benefits. With modern molecular methods, researchers can cut and copy DNA and deliver it to new organisms, not necessarily of the same species. Biotechnology has had some notable successes in treating diseases, usually by producing medicines more efficiently and more effectively than with traditional methods. Gene therapy, however, has so far had a poor record of success in curing human diseases. With biotechnology, many tools have been developed that enable us to identify individuals carrying disease-causing genes. These tools can help reduce the symptoms and incidence of diseases, but come with significant potential costs, particularly the risk of discrimination.

❺ Biotechnology is producing improvements in agriculture.

Biotechnology has led to important advances in agriculture by using transgenic plants and animals to produce more nutritious food. Even more dramatic is the extent to which biotechnology has reduced the environmental and financial costs of producing food through the creation of herbicide- and insect-resistant crops. Some legitimate fears among the public remain, however, as to the safety of genetically modified foods

❻ Biotechnology today and tomorrow: DNA sequences are rich in information.

Advances in biotechnology have led to automated methods for analyzing DNA sequences. Comparisons of highly variable DNA regions have forensic value in identifying tissue specimens and determining the individual from whom they came. Comparisons of sequence similarities across species reveal evolutionary relatedness and allow construction of detailed evolutionary trees. Cloning of individuals has potential benefits in agriculture and medicine, but ethical questions linger.

KEY TERMS

1. A person's unique DNA is carried in:
   a) muscle cells.
   b) hair.
   c saliva.
   d) skin cells.
   e) All of the above contain a person's DNA.

2. Which of the following are always the same in every DNA molecule?
   a) the sugar
   b) the base
   c) the phosphate group
   d) Only a) and b) are always the same.
   e) Only a) and c) are always the same.

3. The full set of an individual organism's DNA is called its:
   a) complement.
   b) genome.
   c) nucleosome.
   d) nucleotide.
   e) chromosome.

4. In humans, genes make up _____ of the DNA.
   a) about 75%
   b) 100%
   c) less than 5%
   d) about 10%
   e) about 50%

5. Genotype is to phenotype as:
   a) cookie is to oven.
   b) cookie is to recipe.
   c) cookbook is to cookie.
   d) recipe is to cookie.
   e) oven is to cookie.

6. To start the transcription process, a large molecule, _____, recognizes a _____.
   a) RNA polymerase; messenger RNA
   b) DNA polymerase; termination site
   c) DNA polymerase; promoter site
   d) RNA polymerase; promoter site
   e) DNA polymerase; messenger RNA

7. During transcription, at the point where the DNA strand being copied has an adenine, a(n) _____ is added to the _____.
   a) thymine; tRNA
   b) cytosine; DNA
   c) uracil; tRNA
   d) adenine; mRNA
   e) uracil; mRNA

8. There are different _____ molecules for each of the 20 different amino acids that are used in building proteins.
   a) ribosomal subunit   d) DNA
   b) tRNA   e) elongation
   c) mRNA

9. Deletions and substitutions are two types of point mutations. Which type is more likely to cause mistranslations of proteins?
   a) Substitutions, because they shift the reading frame and cause downstream amino acids to be changed.
   b) Substitutions, because one protein is substituted for another protein.
   c) Deletions, because they shift the reading frame and cause downstream amino acids to be changed.
   d) Deletions, because one protein is deleted.
   e) None of the above is correct.

10. Which of the following statements about the metabolism of ethanol (which is present in alcoholic beverages) is incorrect?
    a) Individuals who produce non-functioning aldehyde dehydrogenase exhibit "fast-flushing."
    b) The process requires two enzymes: alcohol dehydrogenase and isopropyl dehydrogenase.
    c) Individuals who are "fast-flushers" are less likely to become alcoholics.
    d Aspirin interferes with the action of alcohol dehydrogenase.
    e) All of the above are correct.

11. The polymerase chain reaction (PCR):
    a) makes it possible to create huge numbers of copies of tiny pieces of DNA.
    b) enables researchers to determine the sequence of a complementary strand of DNA when they have only single-stranded DNA.
    c) utilizes RNA polymerase to build strands of DNA.
    d) can create messenger RNA molecules from small pieces of DNA.
    e) All of the above are correct.

12. Which of the following is not a difficulty that medicine has encountered in its attempts to cure human diseases through gene therapy?
    a) The transfer organism—usually a virus—may get into unintended cells and cause disease.
    b) It is difficult to get the working gene into the specific cells where it is needed.
    c) It is difficult to get the working gene into enough cells at the right rate to have a physiological effect.
    d) For many diseases, a malfunctioning gene has not been identified.
    e) All of the above are difficulties encountered in attempts to cure human diseases through gene therapy.

13. Golden rice:
    a) grows without a husk, thereby reducing the processing required before it can be consumed.
    b) can make vitamin A without beta-carotene.
    c) could help prevent blindness in the 250,000 children who lose their sight each year because of vitamin A deficiency.
    d) supplies more vitamin A in one serving than an individual needs in a full week.
    e) is one of the most recent developments in organic farming.

14. Which of the following statements about Bt crystals is correct?
    a) They are produced by soil-dwelling bacteria of the species *Bacillus thuringiensis.*
    b) The gene coding for the production of Bt crystals has been genetically engineered into the genome of dairy cows, increasing their milk production sixfold.
    c) They are produced by the polymerase chain reaction (PCR).
    d) They are produced by most weedy species of plants.
    e) All of the above are correct.

15. A VNTR is:
    a) a gene that codes for a biochemical trait, rather than a structural trait.
    b) used in biotechnology when creating a clone.
    c) produced when a mutation occurs in a non-sex cell.
    d) a highly variable section of an individual's DNA.
    e) produced when a mutation occurs in a sperm-producing or egg-producing cell.

16. Ninety-six percent of the base-pair sequences in chimps and humans are the same. This finding indicates that:
    a) humans are more closely related to chimps than any two mice are related to each other.
    b) 96% of the genes in humans have identical counterparts in chimps.
    c) humans are more closely related to chimps than to any other species on earth.
    d) 4% of the DNA we carry contains mutations.
    e) 96% of the proteins produced by chimps is identical to proteins produced by humans.

SHORT-ANSWER QUESTIONS

1. After sequencing a molecule of DNA, you discover that 20% of the bases are cytosine. What percentage of the bases would you expect to be thymine? Explain why.

2. A cell is in the process of transcription. The single strand of DNA that is being copied has the sequence ATC GAC GGA TCC. Indicate the mRNA strand that will be synthesized from this strand, listing the correct sequence of bases. Explain why this step is important in the process of producing protein.

3. What is being "translated" in translation? What is the role of transfer RNA molecules in this step?

4. The genetic disorder known as sickle-cell disease results in an altered form of the protein hemoglobin, which ultimately can affect the structure of red blood cells. A portion of the sequence for the normal hemoglobin allele is:

    CTG ACT CCT GAG GAG AAG TCT

    Individuals with the sickle-cell disorder have the following sequence:

    CTG ACT CCT GTG GAG AAG TCT

    Identify the type of mutation and explain how it would result in an altered (in this case non-functional) protein.

5. Describe three major applications of biotechnology and their impact on human health.

See Appendix for answers. For additional study questions, go to www.prep-u.com.

6

Chromosomes and Cell Division

Continuity and variety

❶ There are different types of cell division.

Bladder cancer cells in the final stage of cell division.

6•1

Immortal cells can spell trouble: cell division in sickness and in health.

Once you are fully grown, do you have just one set of cells that live as long as you do? The answer is *no:* your cells are continually dying off and the ones that remain divide and replace the cells you've lost, in an ongoing process. But how long can this last? Can it go on forever? And does a cell even know how old it is?

Actually, a cell does have a measure of how old it is. Just as a car comes with an odometer, which keeps track of how far the car has been driven, animal cells have a sort of counter that keeps track of how many times a cell has divided. This "counter" is a section of DNA called the **telomere,** which is located at each tip of every chromosome, right next to the genes that direct the processes that keep an organism alive. The telomere functions like a protective cap at the end of the DNA. Every time a cell divides, making an exact copy of itself, its DNA divides as well. However, each time the DNA divides, the telomere gets a bit shorter. Eventually, after some

critical number of cell divisions, the telomere can become so short that additional cell divisions cause the loss of functional, essential DNA, and that means almost certain death for the cell (**FIGURE 6-1**).

At birth, the telomeres in most human cells are long enough to support about 50 cell divisions. Occasionally, however, individuals are born with telomeres that are much shorter than normal. In these people, the normal functioning of many genes is disrupted, and their cells and tissues begin to appear aged very soon after birth (**FIGURE 6-2**). As a consequence, those born with this disorder rarely live beyond the age of 13.

Just as someone might think it's a good idea to disconnect his or her car's odometer, you might think that disconnecting the cellular odometer by continually rebuilding the telomere would be an effective strategy for allowing a cell and its descendants to function for a longer time than normal. Such

▼
Types of Cell Division   Mitosis   Meiosis   Sex Differences

TELOMERES

Animal cell

Every time a cell divides, the telomere gets shorter. After a critical number of cell divisions, functional DNA is lost, which means almost certain death for the cell.

— Nucleus
— Chromosome

TELOMERE
A protective cap at the end of each chromosome

Functional DNA

↓ Cell division

↓ Cell division

↓ Cell division

Cell death

At birth, telomeres in most human cells are long enough to support about 50 cell divisions!

FIGURE 6-1 A cellular "odometer." Telomeres limit cells to a fixed number of divisions.

a line of cells would never die—so you'd think that the continually rebuilding telomere would be like a fountain of youth. It's not. We know this because there *are* some cells with a sort of disconnected odometer. These cells rebuild their telomeres after each cell division, restoring the protective cap. Unfortunately, most cells that rebuild their telomeres with each cell division present a big problem: they don't know when to stop dividing. There is a word for such cells: cancer.

Because cancer is runaway cell division, discovering a cure for cancer will necessarily involve a deep understanding of cell division. In this chapter, we explore the processes that have evolved that enable cells to divide and create new cells.

FIGURE 6-2 Just a child. Individuals with the genetic condition Hutchinson-Gilford progeria syndrome are born with shorter-than-normal telomeres and appear to age at a rapid rate. The inset shows chromosomes (in blue), with the telomeres highlighted in yellow.

❝ Cancer cells are those which have forgotten how to die.

—Harold Pinter, playwright, from the poem "Cancer Cells," 2002 **❞**

Prokaryotes have one method. It is called binary fission and it serves all of their needs. Eukaryotes, on the other hand, have two methods, mitosis and meiosis, each of which has a specific purpose. In addition to discussing what happens when cell division goes unchecked, we also will explore some of the difficulties that arise when cells wind up with too many or too few chromosomes after cell division.

TAKE-HOME MESSAGE 6·1

Within cells there is a protective section of DNA called the telomere that gets a bit shorter every time the cell divides. If a telomere becomes too short, additional cell divisions cause the loss of essential DNA and cell death. Cells that rebuild the telomere with each division can become cancerous.

211

Some chromosomes are circular, others are linear.

As a method for storing genetic information, DNA has complete market saturation. All life on earth uses it. This is pretty remarkable considering the tremendous diversity of life that exists on earth—from bacteria to plants and animals. One way in which different organisms' DNA varies is in how it is organized into chromosomes.

In most bacteria and the other prokaryotes, the genetic information is carried in a single, circular chromosome, a strand of DNA that is attached at one site to the cell membrane (**FIGURE 6-3**). Eukaryotes, on the other hand, have much more DNA than do bacteria and organize it into free-floating linear chromosomes within the nucleus. As we saw in Chapter 5, a eukaryotic organism may have as few as 2 chromosomes or as many as 1,000. For plants and animals, the number is usually between 10 and 50.

The most important part of a eukaryotic chromosome is the DNA molecule, which carries information about how to carry out the processes needed to support the life of the organism. But eukaryotic chromosomes (and some prokaryotic chromosomes) are made of more than just DNA. The long, linear strand is wrapped around proteins called **histones,** which keep the DNA from getting tangled and enable an orderly, tight, and efficient packing of the DNA inside the cell (see Figure 6-3). In the next sections we examine the behavior of prokaryotic and eukaryotic chromosomes during cell division.

TAKE-HOME MESSAGE 6•2

In most bacteria and the other prokaryotes, the genetic information is carried in a single, circular chromosome, a strand of DNA that is attached at one site to the cell membrane. Eukaryotes have much more DNA than do bacteria and organize it into free-floating linear chromosomes within the nucleus.

PROKARYOTIC AND EUKARYOTIC CHROMOSOMES

PROKARYOTIC CELL

EUKARYOTIC CELL

Chromosome

Attachment site

Chromosome

Nucleus

Histones

DNA

FIGURE 6-3 Chromosomes compared. Prokaryotic cells have a single circular chromosome attached to the cell membrane. Eukaryotic cells contain free-floating linear chromosomes within a nucleus.

Types of Cell Division   Mitosis   Meiosis   Sex Differences

Prokaryotes divide by binary fission.

When it is time for bacteria and the other prokaryotes to reproduce, they use a method called **binary fission,** which means "division in two" (**FIGURE 6-4**). This process begins with **replication,** the process by which a cell's DNA duplicates itself. Replication begins as the double-stranded DNA molecule unwinds from its coiled-up configuration. Once the strands are uncoiled, they split apart like a zipper. As the double-stranded molecule unzips, free-floating nucleotide bases attach to the bases exposed on each of the two separated, single-stranded circular molecules of DNA, matching A to T and G to C to create two identical double-stranded DNA molecules. The newly created circular chromosomes of DNA attach to the inside of the plasma membrane at a different spot than the original. The original cell, called the **parent cell,** then pinches into two new cells, called **daughter cells,** so each of the daughter cells has an exact and precise two-stranded copy of the original two-stranded circular chromosome.

In some prokaryotes, such as the *E. coli* that live in our digestive system, the complete process of binary fission can occur very quickly—often in as little as 20 minutes. Binary fission is considered **asexual reproduction** because the daughter cells inherit their DNA from a single parental cell. In **sexual reproduction,** a combination of DNA from two separate individuals is passed on to offspring, resulting in offspring that are genetically different both from one another and from their parents. We will discuss sexual reproduction in more detail later in this chapter.

TAKE-HOME MESSAGE 6•3

Bacteria divide by a type of asexual reproduction called binary fission: first the circular chromosome duplicates itself and then the cell splits into two identical new cells.

PROKARYOTIC CELL DIVISION: BINARY FISSION

Prokaryotic parent cell — Double-stranded DNA

REPLICATION
An exact copy of the cell's DNA is created.

Cell elongates and begins to pinch in two.

Daughter cells are formed.

Binary fission results in two genetically identical daughter cells.

FIGURE 6-4 "Dividing in two." Asexually reproducing prokaryotes reproduce rapidly through binary fission.

There is a time for everything in the cell cycle.

In life, we often go through phases that are defined by the focus of our interests and activities. For many years we go to school. For a while we may tend to our career. Then, for a spell, our personal life may take center stage. Most eukaryotic cells do this, too. They may spend a long period of time occupied with activities relating solely to growth of the cell and then suspend those activities as they segue into a period devoted exclusively to reproducing themselves. This alternation of activities between processes related to cell division and processes related to growth is called the **cell cycle** (FIGURE 6-5).

Before we go any further in discussing the cell cycle, we need to note an important distinction among the cells of the body. All of the cells of a multicellular eukaryotic organism can be divided into two types: **somatic cells** are the cells

forming the body of the organism; **reproductive cells** are the sex cells, or **gametes.** The processes of cell division by which somatic cells and reproductive cells are produced differ from each other. In this section, we focus on somatic cells. Later in the chapter, we examine cell division leading to the production of sex cells.

The cell cycle describes the series of phases that leads to somatic cell division. There are two main phases in the cell cycle: **interphase,** during which the cell grows and prepares to divide, and the **mitotic phase** (or **M phase**), during which first the nucleus and genetic material within the cell divides, and then the rest of the cellular contents divides. Interphase is further divided into three distinct subphases, described below. A eukaryotic somatic cell moves through the phases in precise order and, at any given moment, a somatic

EUKARYOTIC CELL CYCLE

INTERPHASE

① GAP 1
The cell's primary growth phase. Normal cellular functions (making proteins, getting rid of waste, etc.) take place.

② DNA SYNTHESIS
The cell begins preparations for division. Every chromosome creates an exact duplicate of itself in a process called replication.

③ GAP 2
Second period of growth and preparation for cell division.

④ MITOSIS
The cell's nucleus duplicates, then divides.

⑤ CYTOKINESIS
The cell then duplicates, forming two separate cells.

② DNA synthesis

INTERPHASE

① Gap 1

MITOTIC PHASE

③ Gap 2

⑤ Cytokinesis

④ Mitosis

FIGURE 6-5 **The cycle of cellular activity.** The eukaryotic cell cycle is the series of events that results in the division of somatic cells (cells that make up the body of an organism).

Types of Cell Division   Mitosis   Meiosis   Sex Differences

cell is at some point in this cycle. The phases and subphases can be summarized as follows.

Interphase

Gap 1. During this period, a cell grows and performs all cellular functions (making proteins, getting rid of waste, and so on). Cells that divide infrequently, such as most neurons and heart muscle cells, spend most of their time in the Gap 1 phase.

DNA synthesis. During this phase, the cell begins to prepare for cell division. Every chromosome creates an exact duplicate of itself by replication. Before replication, each chromosome is a long linear strand of genetic material. After replication, each chromosome has become a pair of identical long linear strands, held together near the center; this region, where the chromosomes are in contact, is called the **centromere.**

Gap 2. In this phase, the cell continues to grow and prepare for cellular division. This phase differs from Gap 1 because the genetic material has now been duplicated. Gap 2 is usually much shorter than Gap 1.

Mitotic phase

This phase begins with **mitosis,** a process in which the nucleus of the parent cell duplicates. Mitosis is generally followed by **cytokinesis,** during which the cytoplasm, organelles, and the rest of the materials in the cell duplicate and the cell separates into two daughter cells, each of which has a complete set of the parent cell's DNA and other cellular structures. (Note that each daughter cell becomes a parent cell in the next cell cycle.) The mitotic phase is usually the shortest period in the eukaryotic cell cycle.

The transition from one phase to the next is triggered by specific events. For instance, a cell moves from the Gap 1 phase (denoted G_1) to DNA synthesis only if environmental conditions outside the cell are favorable for cell division. Until conditions are favorable, the cell remains in G_1. Likewise, a cell begins mitosis only when the genetic material has been successfully duplicated. Later in this chapter we'll see that errors in this orderly, step-by-step process can lead to out-of-control cell division and cancer.

TAKE-HOME MESSAGE 6·4

Eukaryotic somatic cells alternate in a cycle between cell division and other cell activities. The cell division portion of the cycle is called the mitotic phase. The remainder of the cell cycle, called interphase, consists of two gap phases separated by a DNA synthesis phase during which the genetic material is replicated.

6·5

Cell division is preceded by replication.

In one of the great understatements in the scientific literature, James Watson and Francis Crick wrote the following sentence in their paper describing the structure of DNA: "It has not escaped our notice that the specific pairing we have postulated immediately suggests a possible copying mechanism for the genetic material." The researchers' ability to explain how DNA copied itself was a critical feature of their description of DNA's structure. After all, every single time any cell in any organism's body divides, that cell's DNA must first duplicate itself so that each of the two new cells has all of the genetic material of the original parent cell. The process of DNA duplication, as we've seen, is called replication.

What is it about DNA's structure that makes duplication so straightforward? Watson and Crick were referring to the feature of DNA called **complementarity,** the characteristic that in the double-stranded DNA molecule, the base on one strand always has the same pairing partner (called the **complementary base**) on the other strand: A pairs with T (and vice versa), and G pairs with C (and vice versa).

With this consistent pattern of pairing, when the DNA molecule separates into two strands, it is possible to perfectly reconstruct all the information on the missing half because one strand carries all the information needed to construct its complementary strand. Just before cells divide, the DNA molecule unwinds and "unzips" and each half of the unzipped molecule serves as a template on which the missing half is reconstructed. At the end of the process of reconstructing the missing halves, there are two DNA molecules, each identical to the original DNA molecule, one for each of the two new cells.

Deviation in Chromosome Numbers

The process of DNA replication occurs in two steps: unwinding and rebuilding (**FIGURE 6-6**).

1. Unwinding. Replication begins when the coiled, double-stranded DNA molecule unwinds and separates into two strands, like a zipper unzipping.

2. Rebuilding. In the rebuilding process, each of the single strands becomes a double strand as an enzyme connects the appropriate complementary base to the exposed base. If an A is exposed, a T binds to it; if a C is exposed, a G binds to it.

The end result of replication is two double-stranded DNA molecules, each identical to the cell's original double-stranded DNA molecule. DNA replication occurs in all types of cells, somatic and reproductive, prior to cell division.

A variety of mutations (such as mismatched bases and added or deleted segments) can occur during replication, but many

This artist is carefully duplicating a painting. Similarly, DNA must be duplicated prior to cell division.

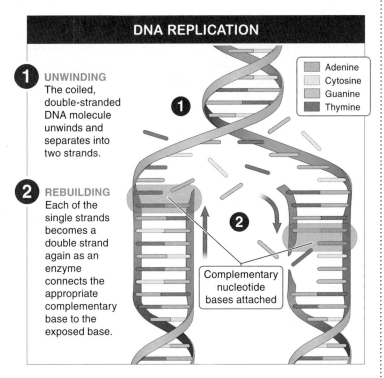

DNA REPLICATION

1 UNWINDING
The coiled, double-stranded DNA molecule unwinds and separates into two strands.

2 REBUILDING
Each of the single strands becomes a double strand again as an enzyme connects the appropriate complementary base to the exposed base.

- Adenine
- Cytosine
- Guanine
- Thymine

Complementary nucleotide bases attached

FIGURE 6-6 How DNA replicates: unwinding and rebuilding.

of these errors are caught and repaired after replication. If an error remains, however, the sequences in a replicated DNA molecule (including the genes) can be different from those in the parent molecule. A changed sequence may then produce a different mRNA sequence, which may in turn produce a different amino acid sequence, and, hence, a different protein. In this way, new genes can enter a population and be acted on by evolution. We explore the relationship between mutation and evolution in more detail in Chapter 8.

Q Errors sometimes occur when DNA duplicates itself. Why might that be a good thing?

TAKE-HOME MESSAGE 6•5

Every time a cell divides, that cell's DNA must first duplicate itself so that each of the two new cells has all the genetic material of the original parent cell. The process of DNA duplication is called replication. Errors in replication can lead to mutations, which are changes in the DNA sequence.

❷ Mitosis replaces worn-out old cells with fresh new duplicates.

For most of the cell cycle, chromosomes resemble a plate of spaghetti more than the familiar condensed, replicated, X shape seen in photos.

6•6

Most cells are not immortal: mitosis generates replacements.

Look around your dorm room. Dust is everywhere. What is it? It is primarily dead skin cells. In fact, you (and your roommates) slough off millions of dead skin cells each day—yet your skin doesn't disappear. How can that be? Obviously, your body is replacing the sloughed-off cells. The cells have simply worn out, so your body creates replacements. It does this through the process of mitosis.

Mitosis has just one purpose: to enable existing cells to generate new, genetically identical cells. There are two different reasons for this need (**FIGURE 6-7**).

1. Growth. During development, organisms get bigger. Growth happens in part through the creation of new cells. In fact, if you want to see cell division in action, one surefire

Q What is dust? Why is it your fault?

place to look is at the tip of a plant root: at a growth rate of about half an inch per day, the root is one of the fastest-growing parts of a plant.

2. Replacement. Cells must be replaced when they die. The wear and tear that come from living can physically damage cells. The daily act of shaving, for example, damages thousands of cells on a man's face. It's nothing to worry about, though. Microscopic views of human skin reveal several distinct layers, with the outermost layers—the layers under assault during shaving—made up primarily of dead cells. These cells help protect us from infection and also reduce the rate at which the underlying living cells dry out. The living cells that exist just below the layers of dead cells are being produced at a high rate by mitosis; they can also be harmed if you're not careful.

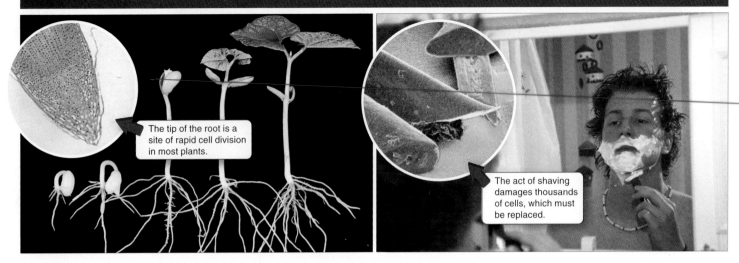

FIGURE 6-7 Reasons for mitosis. Mitosis is important in organism growth and the replacement of cells.

The tip of the root is a site of rapid cell division in most plants.

The act of shaving damages thousands of cells, which must be replaced.

Some other cells that must be replaced actually die on purpose, in a planned process of cell suicide called **apoptosis.** This seemingly counterproductive strategy is employed in parts of the body where the cells are likely to accumulate significant genetic damage over time and are therefore at high risk of becoming cancer cells (a process described later in this chapter). Cells targeted for apoptosis include many of the cells lining the

digestive tract as well as those in the liver, two locations where cells are almost constantly in contact with harmful substances.

Every day, a huge number of cells in an individual must be replaced by mitosis. In humans this number is in the billions. Nearly all the somatic cells of the body—that is, everything other than sperm- and egg-producing cells—undergo mitosis. There are a few notable exceptions. Brain cells and heart muscle cells, in particular, do not seem to divide, or, if they do divide, they do so at very, very slow rates. (It is not known why this is so.)

The rate at which mitosis occurs varies dramatically for different types of cells. The most rapid cell division occurs in the blood and in the cells lining various tissues and organs. The average red blood cell, for example, is in circulation only for about two to four months and then must be replaced. The cells lining the intestines are replaced about every three weeks. Hair follicles, too, are among the most rapidly dividing cells.

The red blood cells in your body carry oxygen to the body's organs and tissues and are replaced about every 120 days.

TAKE-HOME MESSAGE 6·6

Mitosis enables existing cells to generate new, genetically identical cells. This makes it possible for organisms to grow and to replace cells that die.

Types of Cell Division   Mitosis   Meiosis   Sex Differences

Overview: mitosis leads to duplicate cells.

For mitosis to begin, the parent cell duplicates its DNA, creating a duplicate copy of each chromosome. Once this task is completed, the other materials in the cell duplicate and the cell divides into two new duplicate cells, the daughter cells.

Mitosis occurs in just four steps. It begins with the condensing of the chromosomes, which are all tangled during interphase. Next, all of the duplicated and condensed chromosomes move to the center of the cell. Each chromosome is then pulled apart from its duplicate. In the final step, new membranes form around each complete set of chromosomes to create the new nuclei, and the cytoplasm and rest of the cell duplicate as well. Where once there was one parent cell, now there are two daughter cells (**FIGURE 6-8**).

These two human embryonic stem cells have just completed mitosis.

TAKE-HOME MESSAGE 6•7

Mitosis is the process by which cells duplicate themselves. It occurs in just four steps, following the replication of chromosomes. It leads to the production of two daughter cells from one parent cell.

MITOSIS CREATES DUPLICATE CELLS

INTERPHASE
Chromosomes are replicated during the DNA synthesis portion of interphase.

MITOSIS

1 Chromosomes condense.

2 Chromosomes line up in the middle of the cell.

3 Each chromosome is pulled apart from its duplicate.

4 New membranes form around each complete set of chromosomes and the cell divides.

FIGURE 6-8 A simplified introduction to mitosis.

Deviation in Chromosome Numbers

The details: mitosis is a four-step process.

Let's look at the process of mitosis in a bit more detail, keeping in mind that the ultimate consequence of the process is the production of two cells with identical chromosomes.

Interphase: in preparation for mitosis, the chromosomes replicate

Processes essential to cell division take place even before the mitotic phase of the cell cycle begins. During the DNA synthesis part of interphase, every chromosome replicates itself. Recall from Section 6-5 that prior to replication, each chromosome is just a long, linear strand of genetic material; after replication, each chromosome is a *pair* of identical long, linear strands, held together at the center.

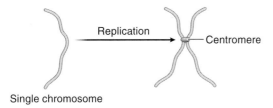

Replication → Centromere

Single chromosome

Mitosis The actual process of cell division occurs in four steps (**FIGURE 6-9**).

1. Prophase: the long, linear chromosomes that have replicated condense. When you look at a cell through a microscope, you generally won't see any chromosomes because the cells you're looking at are in some part of their interphase. At this point, the chromosomes are strung out so thin that they are not visible. Mitosis begins with prophase, when the chromosomes in the cell's nucleus become more and more tightly coiled. As they condense, they become thick enough to be seen through a light microscope.

At this point, in prophase, each chromosome looks like the letter X. This appearance is misleading for a couple of reasons. First, during all of the cell cycle except mitosis, chromosomes are uncoiled and spread out in a diffuse way like a mass of spaghetti. Because they are so stretched out, they are not dense

INTERPHASE

Nuclear membrane
Centromere
Replicated chromosome
Parent cell

INTERPHASE
• Chromosomes replicate in preparation for mitosis.

MITOSIS

Nuclear membrane
Sister chromatids
Spindle

1 PROPHASE
• Nuclear membrane breaks down.
• Sister chromatids (replicated chromosomes) condense.
• Spindle forms.

2 METAPHASE
• Sister chromatids line up at the center of the cell.

FIGURE 6-9 Mitosis: cell duplication, step by step.

enough to be visible. Second, chromosomes are not actually X-shaped. They are *linear*. Each X is really two identical linear DNA molecules: a chromosome and its identical replicated copy, joined at the centromere. (Each of the identical DNA molecules is called a **chromatid**; together, the pair are called **sister chromatids**.) The reason for the X shape in most photos of chromosomes is that the only time the chromosome is coiled tightly and thus thick enough to be seen is during cell division.

Because a lot of room is required for separation of the duplicated chromosomes, the membrane around the nucleus is dismantled and disappears near the end of this first stage of mitosis. At the same time, a structure called the **spindle** is assembled. The spindle, which is part of the cell's

Q Animal chromosomes are linear. So, in pictures, why do they look like the letter X?

cytoskeleton, can be thought of as a group of parallel threads stretching from one pole of the cell to the other. These threads (known as **spindle fibers**) pull the sister chromatids to the middle of the cells and will eventually be used to pull the chromatids apart as cell division proceeds.

2. **Metaphase: the chromosomes congregate at the cell center.** After condensing, the chromatids seem to move aimlessly around the cell, but eventually they line up at the cell's center, pulled by spindle fibers attached to the centromere. Metaphase can be considered the half-time of mitosis. At the end of metaphase, all the chromatids are lined up in an orderly fashion, straddling the center in a "single-file" congregation that is called the metaphase plate. The chromatids are at their most condensed during this part of mitosis.

3. **Anaphase: the chromatids separate and move in opposite directions.** In anaphase, the fibers attached to the centromeres begin pulling each chromatid in the sister chromatid pairs toward opposite poles of the cell. From each pair of sister chromatids, one strand of DNA is pulled

CYTOKINESIS

3 ANAPHASE
• The sister chromatid pairs are pulled apart by the spindle fibers. One full set of chromosomes goes to one side of the cell and another identical set goes to the other.

4 TELOPHASE
• The chromosomes begin to uncoil as the nuclear membrane is reassembled around them.

CYTOKINESIS
• Cytoplasm and organelles duplicate and are divided into approximately equal parts. The cell splits in two.

Deviation in Chromosome Numbers

in one direction and the other, identical strand is pulled in the opposite direction. At the end of anaphase one full set of chromosomes is at one end of the cell and another identical full set is at the other end. These chromosome sets will eventually reside in the nucleus of each of the two new daughter cells that result from the cell division.

4. Telophase: new nuclear membranes form around the two complete chromosome sets. With two full, identical sets of chromosomes collected at either end of the cell, the cell is prepared to divide into two genetically identical cells. In this last step, called telophase, the chromosomes begin to uncoil and fade from view, nuclear membranes reassemble, and the cell begins to pinch into two.

The process of mitosis, the division of a cell's nucleus into two identical sets, is generally followed by cytokinesis, during which the cell's cytoplasm is also divided into approximately equal parts, with some of the organelles going to each of the two new cells. Only when cytokinesis is complete do the two new daughter cells, each with an identical nucleus containing identical genetic material, enter interphase and begin the business of being cells.

TAKE-HOME MESSAGE 6·8

The ultimate consequence of mitosis and cytokinesis is the production of two genetically identical cells.

6·9

Cell division out of control means cancer.

Too much of a good thing can be bad. This is especially true when cell division runs amok, a situation that can lead to cancer. **Cancer** is defined as unrestrained cell growth and division and can cause serious health problems. Cancer is the second leading cause of death in the United States, responsible for more than 20% of all deaths. Only heart disease causes more deaths.

Cancer occurs when some disruption of the DNA in a normal cell interferes with the cell's ability to regulate cell division. DNA disruption can be caused by chemicals that mutate DNA or by sources of high energy such as X rays, the sun, or nuclear radiation. Cancer can even be caused by some viruses. However it begins, once a cell loses control over its cell cycle, cell division can proceed unrestrained (**FIGURE 6-10**).

Cancer cells have several features that distinguish them from normal cells; the two most significant differences are:

1. *They lose their "contact inhibition."* Most normal cells divide until they bump up against other cells or collections of cells. At that point, they stop dividing. Cancer cells, however, ignore the signal that they are at high density and continue to divide.

2. *Cancer cells can divide indefinitely.* As we saw in Section 6-1, normal cells can divide approximately 50 times. After that point, they may continue living but they lose the ability

Q What is cancer? How does it usually cause death?

to divide. Cancer cells, on the other hand, never lose their ability to divide and continue to do so indefinitely, even in the presence of conditions that normally would halt the cell cycle before cell division. (The ability of cancer cells to divide indefinitely is made possible because they rebuild their telomeres following each cell division.)

Tumors caused by excessive cell growth and division are of two very different types: benign and malignant. Benign tumors, such as many moles, are just masses of normal cells that do not spread. They can usually be removed safely without any lasting consequences. Malignant tumors, on the other hand, are the result of unrestrained growth of cancerous cells. They grow continuously and shed cells (**FIGURE 6-11**). The shedding of cancer cells from malignant tumors is how cancer spreads, a process called metastasis (meh-TASS-tuh-siss). In this process, cancer cells separate from a tumor and invade the circulatory system, which spreads them to different parts of the body where they can cause the growth of additional tumors.

How does cancer actually cause death? Somewhat surprisingly, it's not because of some chemical or genetic property of the cancer cells. It's simpler than that. As a tumor gets larger, it takes up more and more space, pressing against neighboring cells and tissue (**FIGURE 6-12**). Eventually, the tumor may block other cells and tissue from carrying out their normal functions and even kill them. This cell dysfunction or cell death can

Types of Cell Division   Mitosis   Meiosis   Sex Differences

CANCER CELLS

Cancer cells have several features that distinguish them from normal cells.

CANCER CELLS HAVE NO CONTACT INHIBITION

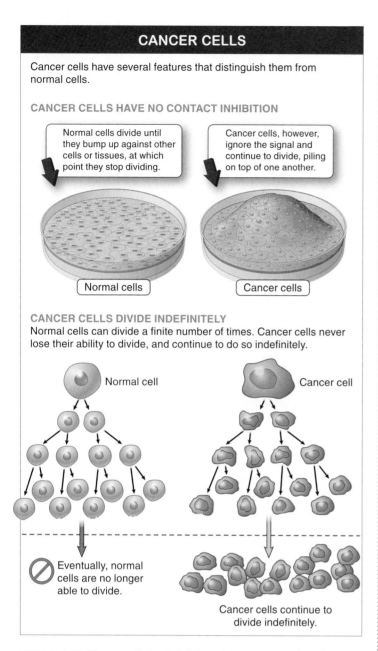

Normal cells divide until they bump up against other cells or tissues, at which point they stop dividing.

Cancer cells, however, ignore the signal and continue to divide, piling on top of one another.

Normal cells

Cancer cells

CANCER CELLS DIVIDE INDEFINITELY

Normal cells can divide a finite number of times. Cancer cells never lose their ability to divide, and continue to do so indefinitely.

Normal cell

Cancer cell

Eventually, normal cells are no longer able to divide.

Cancer cells continue to divide indefinitely.

FIGURE 6-10 Uncontrolled cell division. Cancer occurs when there is a disruption in cells' ability to regulate cell division.

have disastrous consequences when the affected normal tissue controls processes critical to life, such as breathing, heart function, or the detoxification processes in the liver.

To treat cancer, the rapidly dividing cells must be removed surgically, killed, or their division at least slowed down. Currently, the killing and slowing down are done in two ways: by chemotherapy and by radiation. In chemotherapy, drugs that interfere with cell division are administered, slowing down the growth of tumors. Because these drugs interfere with rapidly dividing cells throughout the body (not just the rapidly dividing

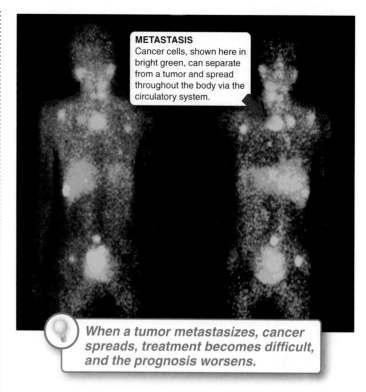

METASTASIS
Cancer cells, shown here in bright green, can separate from a tumor and spread throughout the body via the circulatory system.

When a tumor metastasizes, cancer spreads, treatment becomes difficult, and the prognosis worsens.

FIGURE 6-11 Multiple tumors. Metastasis spreads cancer cells throughout the body.

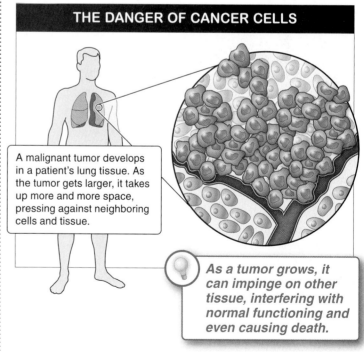

THE DANGER OF CANCER CELLS

A malignant tumor develops in a patient's lung tissue. As the tumor gets larger, it takes up more and more space, pressing against neighboring cells and tissue.

As a tumor grows, it can impinge on other tissue, interfering with normal functioning and even causing death.

FIGURE 6-12 Cancer's effects. Cancer cells harm the body by crowding and disrupting normal cells.

cancer cells), they can have terrible side effects. In particular, chemotherapy drugs disrupt normal systems that rely on the rapid and constant production of new cells. For instance, chemotherapy often causes extreme fatigue and shortness of breath because it reduces the rate at which red blood cells are produced, limiting the amount of oxygen that can be transported throughout the body. By interfering with the division of bone marrow stem cells, chemotherapy also reduces the production of platelets and white blood cells and thus increases bruising and bleeding, and increases susceptibility to infection. Another location of rapidly dividing cells commonly affected by chemotherapy and radiation is the hair follicle. Many people lose their hair when undergoing chemotherapy (it usually grows back when treatments stop).

Q Why is the treatment for cancer often considered as bad as the disease?

Like chemotherapy, radiation works by disrupting cell division. Unlike the drugs used in chemotherapy, however, which circulate throughout the entire body, radiation therapy directs high-energy radiation only at the part of the body where a tumor is located. Like chemotherapy, the radiation process is not perfect and nearby tissue is often harmed as well. The significant negative effects of chemotherapy and radiation treatment on normal tissue and the suffering this leads to has caused some patients to comment that the treatment for cancer is worse than the disease itself—especially because patients may be asymptomatic at the time of treatment (**FIGURE 6-13**).

> " Bitterness is like cancer. It eats upon the host. But anger is like fire. It burns it all clean. "
>
> —Maya Angelou, U.S. poet and writer

Is there hope for a complete cure soon? Not yet. Nonetheless, many potentially successful therapies for cancer treatment and prevention are on the horizon. Extensive research is being conducted on the mechanisms by which genes controlling the cell cycle are damaged, for example, and how such damage might be prevented or reversed.

In addition, there is a rapidly growing field of research into cancer-inhibiting chemicals that have evolved in many plant

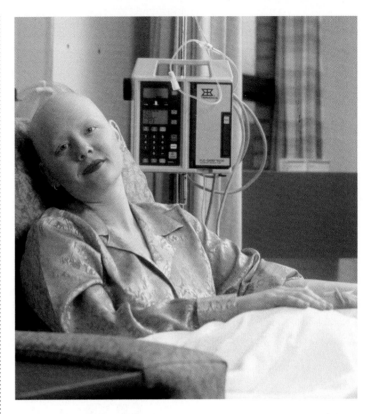

FIGURE 6-13 Cancer treatments. The treatments can be as painful and difficult as the disease.

species. One such chemical is resveratrol, which is found in grapes and peanuts. Many of these naturally occurring chemicals may be less harsh than the drugs currently used in chemotherapy. We are also becoming more conscious of the role that the environment plays in increasing cancer risk; our greater awareness is leading to changes in industrial processes and, on the individual level, to dietary or lifestyle changes. Advances in early screening, particularly in identifying cancer prior to metastasis, continue to have significant benefits as well.

TAKE-HOME MESSAGE 6·9

Cancer is unrestrained cell growth and division, leading to large masses of cells called malignant tumors that can cause serious health problems. Treatment focuses on killing or slowing down the fast-growing and dividing cells, usually using chemotherapy and/or radiation.

Types of Cell Division   Mitosis   Meiosis   Sex Differences

❸ Meiosis generates sperm and eggs and a great deal of variation.

Human gametes: sperm attempting to fertilize an egg.

6·10

Overview: sexual reproduction requires special cells made by meiosis.

There are two ways in which an organism can reproduce. Many organisms, including bacteria, fungi, and even some plants and animals, undergo asexual reproduction, in which a single parent produces identical offspring. Other organisms, including most animals and plants, undergo sexual reproduction, in which offspring are produced by the fusion of two reproductive cells in the process of **fertilization.**

Sexually reproducing organisms, humans included, have a problem to solve when it comes to cell division. When offspring are created, they carry the genetic material from two individuals. But think about the difficulties this presents. If reproductive cells were produced through mitosis, both parents would contribute a full set of genes—that is, 23 pairs of chromosomes in humans—to create a new individual, and the new offspring would inherit 46 pairs of chromosomes in all. And when that individual reproduced, if he or she contributed 46 pairs of chromosomes and his or her mate also contributed 46 pairs, their offspring would have 92 pairs of chromosomes. Where would it end? The genome would double in size every generation. That wouldn't work at all. At the very least, within a few generations cells would be so overloaded with chromosomes that they would explode.

Sexually reproducing organisms have evolved a way to avoid chromosome overload. The solution is **meiosis,** a process that enables organisms, prior to fertilization, to make special reproductive cells, the gametes, which have only half as many chromosomes as the rest of the cells in the organism's body. In other words, in anticipation of combining one individual's genome with another's, meiosis reduces each individual's genome by half in the gametes. In humans, for example, gamete cells have only one set of 23 chromosomes, rather than two sets. In genetics, the term **diploid** refers to cells that have two copies of each chromosome, and the term **haploid** refers to cells that have one copy of each chromosome. Thus, somatic cells are diploid and gametes, the cells produced in meiosis, are haploid. At fertilization, two haploid

225

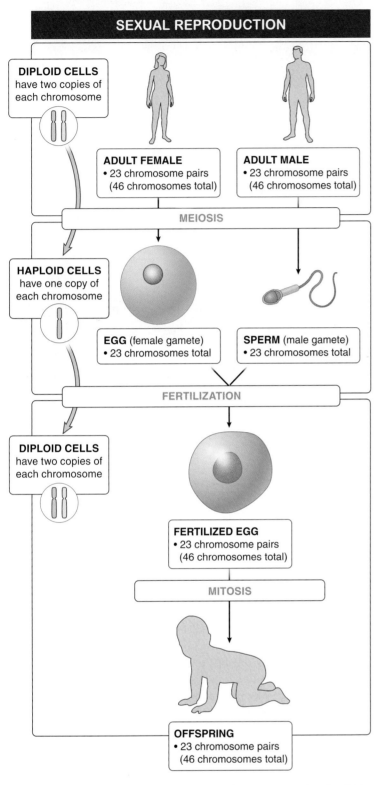

SEXUAL REPRODUCTION

DIPLOID CELLS have two copies of each chromosome

ADULT FEMALE
• 23 chromosome pairs (46 chromosomes total)

ADULT MALE
• 23 chromosome pairs (46 chromosomes total)

MEIOSIS

HAPLOID CELLS have one copy of each chromosome

EGG (female gamete)
• 23 chromosomes total

SPERM (male gamete)
• 23 chromosomes total

FERTILIZATION

DIPLOID CELLS have two copies of each chromosome

FERTILIZED EGG
• 23 chromosome pairs (46 chromosomes total)

MITOSIS

OFFSPRING
• 23 chromosome pairs (46 chromosomes total)

FIGURE 6-14 Sexual reproduction. Diploid organisms produce haploid gametes that fuse at fertilization and return to the diploid state.

cells, each with one set of 23 chromosomes, merge and create a new individual with the proper diploid human genome of 46 chromosomes. And when the time comes to reproduce, this new individual also will produce haploid gametes through meiosis that have only a single set of 23 chromosomes. With sexual reproduction, then, diploid organisms produce haploid gametes that fuse at fertilization to restore the diploid state (**FIGURE 6-14**). This process repeats perpetually and maintains a stable genome size in a species.

Although there are some variations on this pattern of alternation between the haploid and diploid states, most multicellular animals produce simple haploid cells for reproduction. And after two of those gametes come together to form a diploid fertilized egg, multiple cell divisions via mitosis again produce a diploid multicellular animal.

Meiosis achieves more than just a reduction in the amount of genetic material in gametes. As a diploid individual, you have two copies of every gene in every somatic cell: one from your mother and one from your father. When making haploid cells from cells that are diploid, an individual creates cells that have one allele for each trait rather than two. Which of these two alleles is included in each gamete cannot be predicted.

Each egg or sperm is produced with a varied combination of maternal and paternal alleles so that all of the gametes an individual makes carry a unique set of alleles. As a consequence, each offspring inherits a slightly different assemblage of that parent's alleles. All offspring resemble their parents, yet none resembles them in exactly the same way. This variation among offspring can be seen among three of the Kennedy brothers in the photo that opens this chapter (pages 208–209). Edward Kennedy is in the middle, with Robert on the left and John on the right.

In all, meiosis has two important outcomes:

1. It reduces the amount of genetic material in gametes.

2. It produces gametes that differ from one another with respect to the combinations of alleles they carry.

In the next sections, we examine the exact steps by which these outcomes occur and investigate the consequences of all the variation that meiosis and sexual reproduction generate.

TAKE-HOME MESSAGE 6·10

Meiosis is the process by which gametes are produced in sexually reproducing organisms. It results in gametes that have only half as much genetic material as the parent cell and that differ from one another with respect to the combinations of alleles they carry.

Sperm and egg are produced by meiosis: the details, step by step.

Mitosis is an all-purpose process for cell division. It occurs all over the body, all the time. Meiosis, on the other hand, is a special-purpose process. It occurs in just a single place: the **gonads** (the ovaries and testes in sexually reproducing animals). And it occurs for just a single reason: the production of gametes (reproductive cells). To better understand meiosis, let's examine in detail the steps by which a diploid cell creates haploid cells, each of which is different from the other.

Meiosis starts with a diploid cell—not just any diploid cell, but one of the specialized diploid cells, found in the gonads, that is capable of undergoing meiosis. Thus, for humans, meiosis starts with a cell that has 46 chromosomes: a maternal copy and a paternal copy of each of the 23 human chromosomes. Together, the maternal and paternal copies of a chromosome are called a **homologous pair,** or **homologues.** This means that there 23 homologous pairs in a diploid human cell (**FIGURE 6-15**).

Before meiosis can occur (and just as we saw with mitosis), all of these 46 chromosomes are duplicated. So now, instead of having just a maternal and paternal copy of chromosome 1, we have two copies of the maternal copy of chromosome 1, bound together as sister chromatids, and a pair of sister chromatids for the paternal copy of chromosome 1. This means that when meiosis begins, we have 92 (2 × 46) strands of DNA after replication. Remember that this duplication of the chromosomes occurs during the interphase portion of the cell cycle, before actual cell division has begun.

Unlike mitosis, which has only one cell division, cells undergoing meiosis divide twice. In the first division, the homologues separate. In other words, for each of the 23 chromosome pairs, the maternal sister chromatids and the paternal sister chromatids separate into two new cells. In the second division, each of the two new cells divides again, and the sister chromatids separate from each other into two newer cells. At the end of meiosis, there are four new cells, each of which has 23 strands of DNA—that is, 23 chromosomes (**FIGURE 6-16**).

HOMOLOGUES AND SISTER CHROMATIDS

Homologues are the maternal and paternal copies of a chromosome. A sister chromatid is a chromosome and its identical duplicated version held together at a centromere.

FIGURE 6-15 **Chromosome vocabulary: homologues and sister chromatids.**

MEIOSIS: REDUCING THE GENOME BY HALF

INTERPHASE
Each chromosome in a homologous pair replicates to form a sister chromatid.

Homologues
Maternal chromosome
Paternal chromosome
Diploid parent cell
Sister chromatids

MEIOSIS I
In the first division of meiosis, the homologous pairs separate.

Haploid daughter cells

MEIOSIS II
In the second division of meiosis, the sister chromatids separate. This results in four haploid cells, of which two contain a single copy of one of the original chromosomes and the other two contain a single copy of the other chromosome.

FIGURE 6-16 **Meiosis reduces the genome by half in anticipation of combining it with another genome.**

Interphase: in preparation for meiosis, the chromosomes replicate Meiosis is a one-way process: cells undergoing meiosis produce gametes that do not themselves undergo meiosis, so the cells that undergo meiosis do not have a cell cycle. Before meiosis occurs, however, preparations for cell division occur that are similar to those occurring in interphase of the mitotic cell cycle. In particular, there is a DNA synthesis phase during which every chromosome creates an exact duplicate of itself by replicating. As in mitosis, before replication each chromosome is a long, linear strand of genetic material, while after replication each chromosome is a *pair* of identical long, linear strands, held together at the centromere.

Meiosis begins following replication and occurs in eight steps (**FIGURE 6-17**).

Meiosis division I: the homologues separate

1. Prophase I: chromosomes condense and crossing over occurs. This is by far the most complex of all the phases of meiosis. It begins with all of the replicated genetic material condensing.

As the sister chromatids become shorter and thicker, the homologous chromosomes come together. That is, the maternal and paternal sets of sister chromatids for chromosome 1 come together so there are four versions of this chromosome lined up together. The maternal and paternal sets of chromosome 2, 3, 4, and so on, also line up. Under a microscope, the two pairs of sister chromatids appear as pairs of X's lying on top of each other.

At this point, the sister chromatids that are next to each other do something that makes every sperm or egg cell genetically unique: they swap little segments. In other words, when this is happening inside of you, some of the genes that you inherited from your mother get swapped onto the strand of DNA you inherited from your father, and the corresponding bit from your father is inserted into the DNA strand from your mother. This event is called **crossing over,** and it can take place at several spots on every chromosome. As a result of crossing over, every sister chromatid ends up having a unique mixture of the genetic material that you received from your two parents. Note that crossing over occurs only during the production of gametes, in meiosis. It does not occur during

| INTERPHASE | MEIOSIS DIVISION 1: HOMOLOGUES SEPARATE |

INTERPHASE
- Chromosomes replicate in preparation for meiosis.

① PROPHASE I
- Replicated chromosomes condense.
- Spindle is formed.
- Homologous pairs of sister chromatids come together and cross over.
- Nuclear membrane disintegrates.

② METAPHASE I
- Homologues move toward the center of the cell and line up.

③ ANAPHASE I
- Homologues separate and are pulled to opposite poles. Sister chromatids going to each side are a mix of maternal and paternal genetic material.

FIGURE 6-17 Meiosis: generating reproductive cells, step by step.

Types of Cell Division   Mitosis   Meiosis   Sex Differences

mitosis. Following crossing over, the nuclear membrane disintegrates.

We explore crossing over in more detail in Section 6-13. The remaining steps of meiosis are relatively straightforward.

2. Metaphase I: chromosomes all line up at the center of the cell. After crossing over has occurred, each pair of homologous chromosomes (that is, the pairs of X's lying on top of each other) moves to the center of the cell, pulled by the spindle fibers to form the arrangement called the metaphase plate. Remember that each pair of homologous chromosomes includes the maternal and paternal version of the chromosome *and* the replicated copy of each—four strands in all.

3. Anaphase I: homologues are pulled to either side of the cell. This is the beginning of the first cell division that occurs during meiosis. In this phase, the homologues are pulled apart toward opposite sides of the cell. One of the homologues (consisting of two sister chromatids) goes to the top pole, the other to the bottom. At this point, something else occurs that contributes, along with crossing over, to making all the products

of meiosis genetically unique. The maternal and paternal sister chromatids are pulled to the ends of the cell in a random fashion. As a result, the sister chromatids gathered at each pole are a mix of maternal and paternal sister chromatids. Imagine all the different combinations that can occur in a species with 23 pairs of chromosomes. For chromosome pair 1, perhaps the maternal homologue goes to the "top" pole and the paternal to the "bottom" pole. And for pair 2, perhaps the maternal homologue also goes to the top and the paternal to the bottom. But maybe for pair 3 it is the paternal homologue that goes to the top and the maternal homologue to the bottom. For each of the 23 pairs, you cannot predict which will go to the top and which will go to the bottom. There is a *huge* number of possible combinations. (Can you figure out how many? Here's a hint: if there were 2 chromosome pairs, the answer would be 4; if there were 3 chromosome pairs, the answer would be 8…)

4. Telophase I and cytokinesis: nuclear membrane reassembles around sister chromatids and two daughter cells form. After the chromatids arrive at the two poles of the cell, the nuclear membrane re-forms, then cytokinesis occurs and the cytoplasm divides, and the cell membrane pinches the cell into two daughter cells, each with its own nucleus,

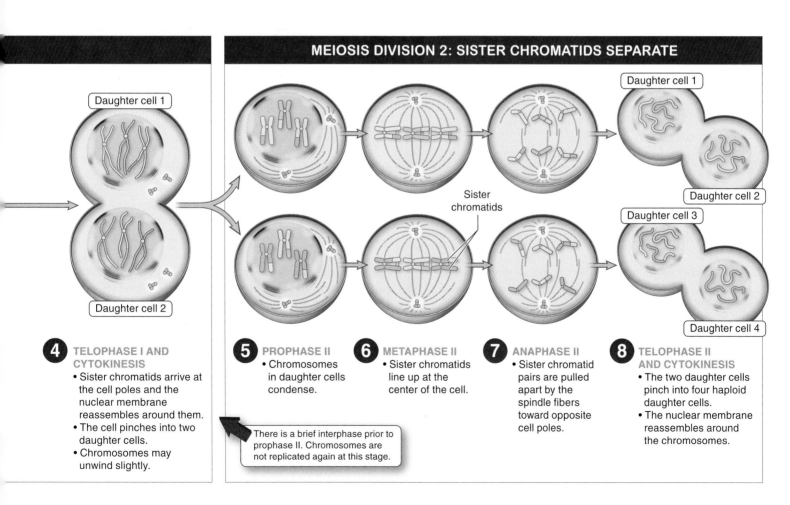

MEIOSIS DIVISION 2: SISTER CHROMATIDS SEPARATE

Daughter cell 1

Daughter cell 2

Daughter cell 1

Daughter cell 2

Daughter cell 3

Daughter cell 4

Sister chromatids

4 TELOPHASE I AND CYTOKINESIS
- Sister chromatids arrive at the cell poles and the nuclear membrane reassembles around them.
- The cell pinches into two daughter cells.
- Chromosomes may unwind slightly.

There is a brief interphase prior to prophase II. Chromosomes are not replicated again at this stage.

5 PROPHASE II
- Chromosomes in daughter cells condense.

6 METAPHASE II
- Sister chromatids line up at the center of the cell.

7 ANAPHASE II
- Sister chromatid pairs are pulled apart by the spindle fibers toward opposite cell poles.

8 TELOPHASE II AND CYTOKINESIS
- The two daughter cells pinch into four haploid daughter cells.
- The nuclear membrane reassembles around the chromosomes.

Deviation in Chromosome Numbers

containing the genetic material—two sister chromatids for each of the 23 chromosomes in humans.

Meiosis division II: separating the sister chromatids There is a brief interphase after the first division of meiosis. In some organisms, the DNA molecules (now in the form of chromatids) briefly uncoil and fade from view. In others, the second part of meiosis begins immediately. *It is important to note that in the brief interphase prior to prophase II, there is no replication of any of the chromosomes.*

5. Prophase II: chromosomes re-condense. The second part of meiosis, like the first division of meiosis, is a four-phase process. It begins with prophase II. In this phase, the genetic material in each of the two daughter cells once again coils tightly, making the chromatids visible under the microscope.

6. Metaphase II: sister chromatids line up at the center of the cell. In each of the two daughter cells, the sister chromatids (each appearing as an X) move to the center of the cell, pulled by spindle fibers attached to the centromere, where the sister chromatids are held together. The congregation of all the genetic material in the center of each daughter cell is visible as a flat metaphase plate.

7. Anaphase II: sister chromatids are pulled to opposite sides of the cell. This phase starts with 46 pieces of DNA in each cell created by meiosis I: 23 sister chromatid pairs. During this phase, the fibers attached to the centromere begin pulling each chromatid in the 23 sister chromatid pairs toward opposite poles of each daughter cell. When anaphase II is finished, each of what will become the four daughter cells (you could think of them as granddaughter cells) has one single copy of each of the 23 chromosomes.

8. Telophase II and cytokinesis: nuclear membranes reassemble and the two daughter cells pinch into four haploid gametes. Finally, the sister chromatids for all 23 chromosomes have been pulled to opposite poles. The cytoplasm then divides, the cell membrane pinches the cell into two new daughter cells, nuclear membranes begin to re-form, and the process comes to a close.

In humans, the outcome of one diploid cell undergoing meiosis is the creation of four haploid daughter cells, each with just one set of 23 chromosomes. These chromosomes contain a combination of traits from the individual's diploid set of chromosomes.

TAKE-HOME MESSAGE 6·11

Meiosis occurs only in gamete-producing cells. It is preceded by DNA replication and consists of two rounds of cellular division, one in which homologous pairs of sister chromatids separate and a second in which the sister chromatids separate. The final product of meiosis in a diploid organism is four haploid gametes.

6·12

Male and female gametes are produced in slightly different ways.

No matter how small or large they are, no matter whether they're plants or animals, there is just one way to distinguish males from females in all sexually reproducing organisms. Regardless of the species, the defining feature is always the same (and if you're thinking of anything visible to the naked eye, you're wrong). When there are two sexes—as there are in nearly every sexually reproducing plant and animal species— the females are the sex that produces the larger gamete, and the males produce the smaller, more motile, gamete (**FIGURE 6-18**). Because gametes are all produced through meiosis, it is a slight variation in how meiosis works in males and females that leads to this difference.

Q How do you distinguish a male from a female?

The size difference between the male gamete and the female gamete (sperm and eggs in animals) all comes down to the fact that sperm cells have very little cytoplasm, while eggs have a huge amount. During the production of sperm, the two divisions occur just as described in Section 6-11, resulting in four evenly sized cells that become sperm. During the production of eggs, things are a bit different. The first division occurs just as described in Section 6-11, except that in telophase I, as the cell divides, instead of forming two cells of the same size, the division is lopsided. The genetic material is evenly divided, but nearly all of the cytoplasm goes to one of the cells and almost none goes to the other. Then, in the second meiotic division of the

Types of Cell Division   Mitosis   Meiosis   Sex Differences

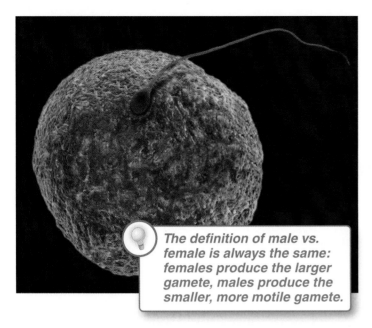

> 💡 *The definition of male vs. female is always the same: females produce the larger gamete, males produce the smaller, more motile gamete.*

FIGURE 6-18 **Egg and sperm.** In sexually reproducing organisms, the female produces the larger gamete.

larger cell, there is again an unequal division of cytoplasm. As in the first division, one of the cells gets nearly all of the cytoplasm and the other gets almost none. The net result of meiosis in the production of eggs is one large egg with lots of cytoplasm and three small cells with very little cytoplasm. These small cells degrade almost immediately and never function as gametes (**FIGURE 6-19**).

Ultimately, whether it is eggs or sperm that are being produced through meiosis, each gamete ends up with just one copy of each chromosome. That way, the fertilized egg that results from the fusion of sperm and egg carries two complete sets of chromosomes, so the developing individual will be diploid. The extra cytoplasm carried by the egg contains a large supply of nutrients and other chemical resources to help with the initial development of the organism following fertilization.

TAKE-HOME MESSAGE 6·12

In species with two sexes—including nearly every sexually reproducing plant and animal species—females are the sex that produces the larger gamete, and males produce the smaller gamete. Whether it is male or female gametes that are being produced, each gamete ends up with just one copy of each chromosome.

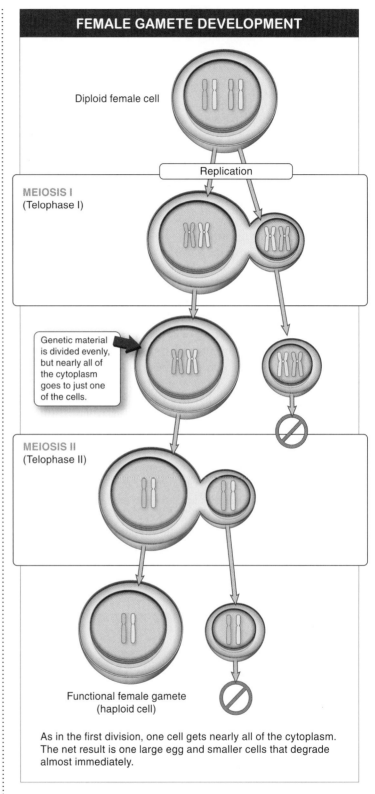

FEMALE GAMETE DEVELOPMENT

Diploid female cell

Replication

MEIOSIS I
(Telophase I)

Genetic material is divided evenly, but nearly all of the cytoplasm goes to just one of the cells.

MEIOSIS II
(Telophase II)

Functional female gamete
(haploid cell)

As in the first division, one cell gets nearly all of the cytoplasm. The net result is one large egg and smaller cells that degrade almost immediately.

FIGURE 6-19 **Unequal distribution of cytoplasm results in one large egg.**

231

6·13

Crossing over and meiosis are important sources of variation.

Genetically speaking, there are two ways to create unique individuals. The obvious way is for an organism to carry an allele that is not present in any other individuals. Alternatively—and equally successful in creating uniqueness— an individual can carry a *collection* of alleles, no single one of which is unique, that has never before occurred in the same individual. Both types of novelty introduce important variation into a population of organisms. The process of crossing over (**FIGURE 6-20**), which occurs during prophase I in meiosis, creates a significant amount of the second type of variation.

As we saw in Section 6-11, during the first prophase of meiosis, the sister chromatids of homologous chromosomes all come together. That previous sentence is a mouthful! Let's review exactly what it means. Let's look at just one of your 23 homologous pairs of chromosomes, the homologous pair of chromosome 15. When we refer to the "homologous pair" of chromosome 15, remember that this pair includes two copies of chromosome 15: one copy from your mother (which you inherited from the egg that was fertilized to create you) and one copy from your father (which you inherited from the sperm that fertilized the egg). Each chromosome in the pair carries the same genes, but because they came from different people, they don't necessarily have the same alleles.

Once the sister chromatids of the homologous chromosomes line up, regions that are close together can swap segments. A piece of one of the maternal chromatids—perhaps including the first 100 genes on the strand of DNA—may swap places with the same segment in a paternal chromatid. Elsewhere, a stretch of 20 genes in the middle may be swapped from the other maternal chromatid with one of the paternal chromatids. Every time a swap of DNA segments occurs, an identical amount of genetic material is exchanged, so all four chromatids still contain the complete set of genes that make up the chromosome. The *combination* of alleles on each chromatid, though, is now different.

Suppose there are genes relating to eye color, hair color, and height on a particular chromosome. After crossing over, the linear strands still have instructions for all of those traits. But where one chromatid previously may have had instructions for brown eyes, brown hair, and short height, it may now have some differences—perhaps brown eyes, blond hair, and tall height. All of the alleles from your parents are still carried on one DNA molecule or another. But the *combination of traits* that are linked together on a single chromatid is new. And when a gamete, let's say it's an egg, carrying a new combination of alleles is fertilized by a sperm, the developing individual will carry a completely novel set of alleles. And so without creating any new traits (perhaps yellow eyes or purple hair), crossing over still creates gametes with collections of alleles that may never have existed together. In Chapter 8, we will see that this variation is tremendously important for evolution.

TAKE-HOME MESSAGE 6·13

Although it doesn't create any new traits, crossing over during the first prophase of meiosis creates gametes with collections of traits that may never have existed before; this variation is important for evolution.

CROSSING OVER

Homologous chromosomes

Sister chromatids

Maternal copy   Paternal copy

Crossing over between the sister chromatids of the homologous chromosomes

Homologous chromosomes after the exchange of genetic information

Each of these 4 chromatids gets packaged into a haploid gamete (sperm or egg cell).

Chromatids with recombined DNA

Crossing over doesn't create new alleles but it does create new combinations of alleles on a chromatid.

FIGURE 6-20 Swapping DNA. Crossing over creates new combinations of alleles on each chromatid.

6·14

What are the costs and benefits of sexual reproduction?

There are two fundamentally different ways by which cells and organisms can reproduce. On the one hand there is mitosis and asexual reproduction via binary fission. On the other hand there is meiosis and sexual reproduction. Is one method better than the other? It depends. In fact, the more appropriate question is, what are the advantages and disadvantages of each method and under what conditions do the benefits outweigh the costs?

What are the advantages of sexual reproduction? Sexual reproduction leads to offspring that are all genetically different from one another and from either parent, through three different routes (**FIGURE 6-21**).

1. *Crossing over during the production of gametes.* As we saw, crossing over during prophase I of meiosis causes every chromosome in a gamete to carry a mixture of that individual's maternal and paternal genetic material.

2. *Shuffling and reassortment of homologues during meiosis.* Recall that as the homologues for each chromosome are pulled to opposite poles of the cell during the first division of meiosis, maternal and paternal homologues are randomly pulled to each pole. This means that there is an extremely large number of different combinations of maternal and paternal homologues that might end up in each gamete.

3. *Combining alleles from two parents at fertilization.* First and foremost, with sexual reproduction a new individual comes from the fusion of gametes from two different people. Each of these parents comes with his or her own unique set of genetic material.

Q Bacteria reproduce asexually, whereas most plants and animals reproduce sexually. Which is the better method?

The variability among the offspring produced by sexual reproduction enables populations of organisms to cope better with changes in their environment. After all, if the environment is gradually changing from one generation to the next, individuals producing many offspring increase the likelihood that one of their offspring will carry a set of genes particularly suited to the new environment. Over time, populations of sexually reproducing organisms can quickly adapt to changing environments. It's like buying lottery tickets—the more different tickets you buy, the more likely it is that one of them will be a winner.

SOURCES OF GENETIC VARIATION

There are multiple reasons why offspring are genetically different from their parents and one another.

CROSSING OVER
Crossing over during meiosis produces a mixture of maternal and paternal genetic material on each chromatid.

REASSORTMENT OF HOMOLOGUES
The homologues and sister chromatids distributed to each daughter cell during meiosis are a random mix of maternal and paternal genetic material.

ALLELES COME FROM TWO PARENTS
Each parent donates his or her own set of genetic material.

FIGURE 6-21 Crossing over and meiosis: creating many different combinations of alleles. As a result of crossing over and meiosis, each individual born from a separate fertilized egg is genetically unique.

What are the disadvantages of sexual reproduction? In the yellow dung fly (*Scathophaga stercoraria*), males sometimes wrestle each other for mating access to a female (**FIGURE 6-22**). The female awaits the outcome of the battle in a pile of dung. Occasionally, females drown in the dung pile as they wait. Dangers associated with mating are just one of the downsides to sexual reproduction. There are several others. First, when one parent reproduces, only half of its offsprings' alleles will come from that organism.

233

FIGURE 6-22 **Drowning in dung.** A downside to sexual reproduction.

Some bacteria can divide every 20 minutes!

FIGURE 6-23 **Pluses and minuses.** This asexually reproducing bacterial colony can reproduce quickly and efficiently, but because each new cell is an exact duplicate of the parent genome, there is no genetic variety to enable adaptation should the environment change.

The other half will come from the other parent. With asexual reproduction, parents produce nearly identical offspring, so there is a very efficient transfer of genetic information from one generation to the next. Also, with sexual reproduction it takes time and energy to find a partner. This is energy that asexual organisms can devote to additional reproduction. Moreover, as we see with the dung flies, sex can be a risky proposition because organisms make themselves vulnerable to predation, disease, and other calamities during reproduction. And finally, as we will see later in this chapter, the complex cellular division required for sexual reproduction offers opportunities for mistakes, sometimes leading to chromosomal disorders.

What are the advantages of asexual reproduction?
With asexual reproduction, the advantages and disadvantages are more or less reversed. Because asexual reproduction involves only a single individual, it can be fast and easy. Some bacteria can divide, forming a new generation, every 20 minutes (**FIGURE 6-23**). And for organisms in isolated habitats or when establishing new populations, asexual reproduction can be advantageous as well. Asexual reproduction is efficient, too. Offspring carry all of the genes that their parent carried—they are genetically identical. If the environment is stable, it is beneficial for organisms to produce offspring as similar to themselves as possible.

What are the disadvantages of asexual reproduction? The downside to asexual reproduction is

that the more closely an offspring's genome resembles its parent's, the less likely it is that the offspring will be suited to the environment when it changes.

In the end, we still see large numbers of species using asexual reproduction and large numbers, including the vast majority of animals and plants, that reproduce sexually. It seems that conditions favoring each occur in the great diversity of habitats of the world. That we see both sexual and asexual reproduction also highlights the recurring theme in biology that there often is more than one way to solve a problem.

TAKE-HOME MESSAGE 6·14

There are two fundamentally different ways that cells and organisms can reproduce: (1) mitosis and asexual reproduction via binary fission, and (2) meiosis and sexual reproduction. Asexual reproduction can be fast and efficient, but leads to genetically identical offspring that carry all of the genes that their parent carried. Sexual reproduction leads to offspring that are all genetically different from one another and from either parent, but takes more time and energy and can be risky.

Types of Cell Division   Mitosis   Meiosis   Sex Differences

4 There are sex differences in the chromosomes.

XX and XY: Only one of the 23 chromosome pairs determines sex in humans.

6•15

How is sex determined in humans?

In humans the sex of a baby is determined by its father. The complex sequence of events involved in sex determination is instigated by one special pair of chromosomes called the sex chromosomes. The sex chromosomes carry information that directs a growing fetus to develop as a male or as a female. An individual's sex depends on the sex chromosome inherited from his or her father.

Q Which parent determines a baby's sex? Why?

Let's take a closer look at the sex chromosomes. In humans, we noted that there are 23 pairs of chromosomes in every somatic cell. These can be divided into two different types: there is one pair of sex chromosomes and there are 22 pairs of non-sex chromosomes. The human sex chromosomes are called the **X and Y chromosomes** (FIGURE 6-24).

How do the X and Y chromosomes differ from the other chromosomes? As we described earlier, all of the genetic information is stored on the chromosomes in all cells of an organism's body. But most of this information is not sex specific—that is, if you are building an eye or a neuron

or a skin cell or a digestive enzyme, it doesn't matter whether it is for a male or for a female; the instructions are the same for both sexes. Some genetic information, however, instructs the body to develop into one sex or the other. That information—which *does* differ dramatically depending on whether it is the genetic information for building a female or a male—is found on the sex chromosomes.

An individual has two copies of all of the non-sex chromosomes. One copy is inherited from the mother, one from the father. Individuals also have two copies of the sex chromosomes, but not always two copies of the same kind. Males have one copy of the X chromosome and one copy of the Y chromosome. Females, on the other hand, don't have a Y chromosome but instead have two copies of the X chromosome.

So how does the father determine the sex of the baby? During meiosis in females, the gametes that are produced carry only one copy of each chromosome. This is true for

FIGURE 6-24 X and Y: the human sex chromosomes.

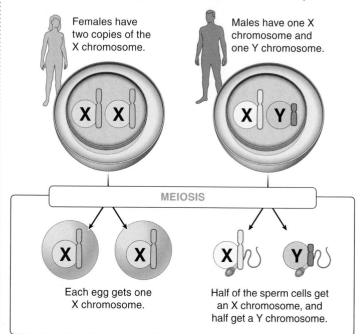

Individuals have two copies of the sex chromosomes in every cell.

Females have two copies of the X chromosome.

Males have one X chromosome and one Y chromosome.

MEIOSIS

Each egg gets one X chromosome.

Half of the sperm cells get an X chromosome, and half get a Y chromosome.

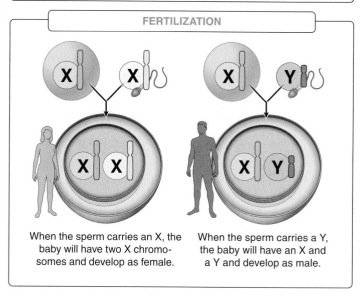

FERTILIZATION

When the sperm carries an X, the baby will have two X chromosomes and develop as female.

When the sperm carries a Y, the baby will have an X and a Y and develop as male.

FIGURE 6-25 Sex determination in humans. The sex of offspring is determined by the father.

the sex chromosomes, too. So from the two copies of the X chromosome carried by females, half of the gametes end up with a copy of one of those X chromosomes while the rest of the gametes inherit the other X chromosome. Thus, every egg has an X for its one sex chromosome. During meiosis in males, the sperm that are produced also carry one copy of each chromosome, including the sex chromosomes, but in this case, half of the sperm produced inherit the X chromosome and the other half inherit the Y chromosome. At fertilization, an egg bearing a single X chromosome (and one copy of all the non-sex chromosomes) is fertilized by a sperm bearing one copy of all the non-sex chromosomes and *either* an X chromosome or a Y chromosome. When the sperm carries an X, the baby will have two X chromosomes and will develop as a female. When the sperm carries a Y, the baby will have an X and a Y and so will develop as a male (**FIGURE 6-25**).

Q *Why can we be certain that there is no essential genetic information carried on the Y chromosome?*

Given the distribution of X and Y chromosomes into males and females, it must be true that there is no essential genetic information carried on the Y chromosome. Why?

Because females don't have a Y chromosome in any of their cells, yet they are able to develop and live normal, healthy lives. For this reason, we know that nothing on the Y chromosome is absolutely necessary.

Physically, the X and Y chromosome look very different from each other. The X chromosome is relatively large and carries a great deal of genetic information relating to a large

Types of Cell Division   Mitosis   Meiosis   Sex Differences

number of non-sex-related traits. The Y chromosome, on the other hand, is tiny and carries genetic information only about a very small number of traits. The small set of genetic instructions on the Y chromosome instructs the fetal gonads to develop as testes rather than ovaries. Once this is done, very little additional genetic input from the Y chromosome is necessary. Instead, the hormones produced by the testes or ovaries generally direct the rest of the body to develop as a male or female. In the next sections we investigate some of the ramifications of males having two different sex chromosomes and look at some of the systems for sex determination that have evolved in other groups of organisms.

6•16

The sex of offspring is determined in a variety of ways in non-human species.

For something as fundamental to a species as sex determination, you might imagine that one method evolved and all species use it. The world is more diverse than that. In most plants there aren't even distinct males and females. In corn, for example, every individual produces both male and female gametes. All earthworms and garden snails are also capable of producing both male and female gametes. Such organisms are called **hermaphrodites** (from the names of the Greek god Hermes and goddess Aphrodite) because both male and female gametes are produced by an individual. But even among the species with separate males and females there are several different methods of sex determination. The human method, in which having both an X and a Y chromosome makes you male and having two X chromosomes makes you female, is common among eukaryotes and is seen in all mammals. But it is only one of at least four different methods used by organisms for determining sex. We'll examine the three others, used by different groups, here (FIGURE 6-26).

SEX DETERMINATION IN VARIOUS OTHER SPECIES

BIRDS
In birds, the mother determines the sex in a way that is similar to how fathers determine sex in humans.

FEMALE   MALE
W Z   Z Z

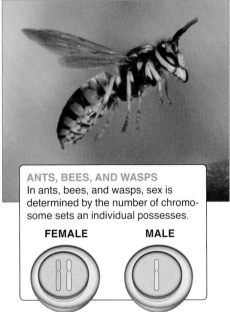

ANTS, BEES, AND WASPS
In ants, bees, and wasps, sex is determined by the number of chromosome sets an individual possesses.

FEMALE   MALE

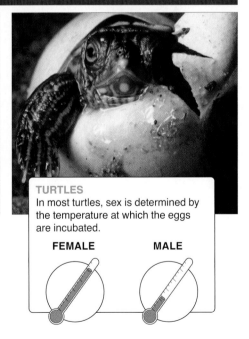

TURTLES
In most turtles, sex is determined by the temperature at which the eggs are incubated.

FEMALE   MALE

FIGURE 6-26 **Other methods of sex determination.** In some species, sex is determined by the type of chromosomes, the number of chromosomes, or incubation temperature of the eggs.

Deviation in Chromosome Numbers

Some biologists are concerned that global warming will harm turtles, like this loggerhead sea turtle. Since the sex of the animal is determined by the temperature of the incubating eggs, a warming of the nesting beaches could result in all female offspring.

Birds In birds, the mother determines the sex in a way that is similar to the way fathers determine sex in humans. Females have one copy of two different sex chromosomes, called the Z and the W chromosomes. Males, on the other hand, have only one type, carrying two copies of the Z chromosome. Consequently, the sex of bird offspring is determined by the mother rather than the father. This method is also found in some fish and butterfly species.

Ants, bees, and wasps In these insects, sex is determined by the number of chromosome sets an individual possesses.

Males are haploid, having only a single set of chromosomes, and females are diploid, carrying two full sets of chromosomes. In this unusual method, females produce haploid eggs by meiosis. They then mate with males and store the sperm in a sac. As each egg is produced, the female can allow it to be fertilized by some of the sperm she has stored, in which case the offspring will be diploid and female. Alternatively, she can lay the unfertilized egg, which can develop into a haploid, male individual. Just think, in these species, males don't have a father, yet they do have a grandfather!

Turtles In some species, sex determination is controlled by the environment rather than by the number or types of chromosomes an organism has. In most turtles, for example, offspring sex is determined by the temperature at which the eggs are kept. Eggs that are kept relatively hot during incubation become females, while the eggs incubated at slightly cooler temperatures become males. The sex of some lizards and crocodiles is also determined by the temperature at which the eggs are incubated.

TAKE-HOME MESSAGE 6·16

A variety of methods are used for sex determination across the world of plant and animal species. These include the presence or absence of sex chromosomes, the number of chromosome sets, and environmental factors such as incubation temperature.

❺ Deviations from the normal chromosome number lead to problems.

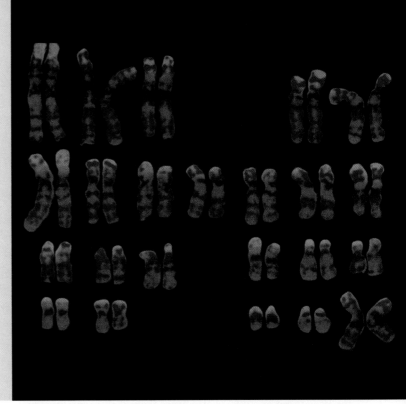

Karyotype displaying the chromosome pairs of a normal human female.

6·17

Down syndrome can be detected before birth: karyotypes reveal an individual's entire chromosome set.

It becomes riskier for women to have children as they become older. Increasingly, their gametes contain incorrect numbers of chromosomes or chromosomes that have been damaged. These problems can lead to effects that range from minor to fatal.

To find out if their offspring may have some disorders associated with incorrect numbers of chromosomes, parents can request a quick test for some common genetic problems even before their baby is born. This information is available from an analysis of an individual's **karyotype,** a visual display of a complete set of chromosomes. A karyotype can be made for adults or children, but it is most commonly done for a fetus. A karyotype is a useful diagnostic tool because it can be prepared very early in the fetus's development to assess whether there is an abnormality in the number of chromosomes or in their structure. And because it shows all of the chromosomes, even the sex chromosomes, it also reveals the sex of the individual.

Preparing a karyotype takes five steps. (1) First it is necessary to obtain some cells from the individual. (2) The cells are then encouraged to divide by culturing them in a test tube with nutrients. (3) After a few days to two weeks of cell division, the cells are treated with a chemical that stops them exactly midway through cell division—a time when the chromosomes are coiled thickly and more visible than usual. (4) Then the cells are placed on a microscope slide and a stain is added that binds to the chromosomes, making them visible. (5) And finally, the chromosomes are arranged by size and shape and displayed on a monitor or in a photograph (**FIGURE 6-27**).

The first step—obtaining cells to prepare a karyotype from— is relatively easy in an adult or child. Usually cells are collected from a small blood sample. Collecting cells from a fetus, on the other hand, poses a special problem. Two different methods for collecting cells have been developed.

1. *Amniocentesis.* This procedure can be done approximately three to four months into a pregnancy (**FIGURE 6-28**). In one quick motion (and without using anesthetic), a 3- to 4-inch needle is pushed through the abdomen, through the amniotic sac, and into the amniotic fluid that surrounds and protects the fetus. Using ultrasound for guidance, the doctor

239

FIGURE 6-27 **A human karyotype.** A visual display of a complete set of chromosomes.

FIGURE 6-28 **Amniocentesis.** Fluid surrounding a fetus (containing some of its cells) is extracted and analyzed to determine the genetic composition of the developing fetus.

aims for a small pocket of fluid as far as possible from the fetus and withdraws about 2 tablespoons of fluid. This fluid contains many cells from the fetus, which can then be used for karyotype analysis. Any chromosomal abnormalities in the fetus will be present in these cells.

2. *Chorionic villus sampling (CVS).* In this procedure, rather than sampling cells from the amniotic fluid, a small bit of tissue is removed from the **placenta,** the temporary organ that allows the transfer of gases, nutrients, and waste products between a mother and fetus. A needle is inserted either through the abdomen or through the vagina and cervix, again using ultrasound for guidance. Then a small piece of the finger-like projections from the placenta are removed via the syringe. Because the fetus and placenta both develop

from the same fertilized egg, their cells contain the same genetic composition. The chief advantage of CVS over amniocentesis is that it can be done several weeks earlier in the pregnancy, usually between the 10th and 12th weeks.

The resulting karyotype, whether from amniocentesis or CVS, will reveal whether the fetus carries an extra copy of any of the chromosomes or lacks a copy of one or more chromosomes. Of all the chromosomal disorders detected by karyotyping, Down syndrome is the most commonly observed. Named after John Langdon Down, the doctor who first described it in 1866, the syndrome is revealed by the presence of an extra copy of chromosome 21 (**FIGURE 6-29**). (For this reason, the condition that causes Down syndrome is called "trisomy 21.") Striking about 1 of every 1,000 children born, Down syndrome is

FIGURE 6-29 **Trisomy 21.** An extra copy of chromosome 21 causes Down syndrome.

Types of Cell Division   Mitosis   Meiosis   Sex Differences

NONDISJUNCTION

Nondisjunction is the unequal distribution of chromosomes during meiosis. The resulting gametes have zero or two copies of a chromosome, rather than a single copy.

METAPHASE I

— Homologues

ANAPHASE I

— Homologues do not separate.

METAPHASE II

RESULTING GAMETES

Both gametes are missing a chromatid.

Both gametes have an extra chromatid.

FIGURE 6-30 Nondisjunction: forming gametes with too many or too few chromosomes.

characterized by a suite of physical and mental characteristics that includes learning disabilities, a flat facial profile, heart defects, and increased susceptibility to respiratory difficulties.

Down syndrome and other disorders caused by a missing chromosome or an extra copy of a chromosome are a consequence of **nondisjunction,** the unequal distribution of

chromosomes during meiosis. Nondisjunction can occur at two different points in meiosis: homologues can fail to separate during meiosis I or sister chromatids can fail to separate during meiosis II (**FIGURE 6-30**). In both cases, nondisjunction results in an egg or sperm with zero or two copies of a chromosome rather than a single copy. Each of the chromosomes is approximately equally likely to fail to separate during cell division, but the ramifications of trisomy (having an extra copy of one chromosome in every cell) are greater for chromosomes with larger numbers of genes. When trisomy occurs for chromosomes with greater numbers of genes, the likelihood that the developing embryo survives to birth is reduced. Consequently, we tend to see cases of trisomy that involve only the chromosomes with the fewest genes, such as chromosomes 13, 15, 18, 21, and 22. In fact, actual observations show that trisomy 1 is never seen (all such fertilized eggs die before implantation in the uterus), trisomy 13 occurs in 1 in 20,000 newborns (and most die soon after birth), while trisomy 21 occurs in 1 in 1,000 newborns (many of whom live long lives).

As we mentioned at the beginning of this section, it becomes riskier for women to have children as they become older. Older women, for example, have more babies with Down syndrome. Why does the risk of having a baby with Down syndrome or some other disorder that results from trisomy increase with increasing age for women but not for men? As women age, their gametes tend to have more errors. The problem is that those eggs *began* meiosis near the time the woman was born and they may not complete it until she is 40 or more years old. During those decades, the cells may accumulate impairments that interfere with normal cell division. In men, the cells that are undergoing meiosis are relatively young because new sperm-producing cells are produced every couple of weeks after puberty.

Q *Why do older women have more babies with Down syndrome?*

Whereas lacking a non-sex chromosome or having an extra chromosome usually has serious consequences for health, we'll see in the next section that lacking or having an extra X or Y chromosome has consequences that are much less severe.

TAKE-HOME MESSAGE 6·17

A karyotype is a visual display of a complete set of chromosomes. A karyotype is a useful diagnostic tool because it can be prepared very early in the fetus's development to assess whether there is an abnormality in the number of chromosomes or in their structure, such as in the case of Down syndrome. Down syndrome is caused by having an extra copy of chromosome 21.

Life is possible with too many or too few sex chromosomes.

It is usually fatal to have one too many or one too few of the non-sex chromosomes. When it comes to the sex chromosomes, however, the situation is less extreme. Many individuals are born lacking one of the sex chromosomes or with an additional X or Y chromosome, and they usually survive. We can get a glimpse into the role of the sex chromosomes by looking at the physical and mental consequences of having an abnormal number of sex chromosomes. In each of the cases below, the condition is caused by nondisjunction during the production of sperm or eggs (FIGURE 6-31).

TOO MANY OR TOO FEW SEX CHROMOSOMES

GAMETE GENOTYPE	OFFSPRING GENOTYPE	SYNDROME CHARACTERISTICS
X / _	X _	**TURNER SYNDROME** • Short height • Web of skin between neck and shoulders • Underdeveloped ovaries; often sterile • Some learning difficulties
_ / X		
XX / Y	XXY	**KLINEFELTER SYNDROME** • Underdeveloped testes • Lower testosterone levels; usually infertile • Development of some female features • Long limbs and slightly taller than average
X / XY		
X / YY	XYY	**"SUPER MALES"** • Taller than average • Moderate to severe acne • Intelligence may be slightly lower than average
XX / X	XXX	**"METAFEMALES"** • May be sterile • No obvious physical or mental problems

FIGURE 6-31 **Characteristics of individuals with too many or too few sex chromosomes.**

Turner syndrome: X_ Approximately 1 in 5,000 females carry only one X chromosome and no Y chromosome, exhibiting a condition called Turner syndrome. This condition, denoted as X_ (or sometimes XO), is the only condition in humans in which a person can survive without one of a pair of chromosomes. Instead of having 46 chromosomes in every cell, these individuals have only 45. As common as Turner syndrome is, 98% of the time that this condition occurs in a fertilized egg, the egg is spontaneously aborted long before a fetus can come to term. There are both physical and mental consequences of the absence of a second sex chromosome.

- Women with Turner syndrome are usually relatively short, averaging 4 feet 8 inches in height.

- They develop a web of skin between the neck and shoulders.

- The ovaries never fully mature, so the women are almost always sterile.

- The breasts and other secondary sex characteristics develop incompletely.

- Intelligence is usually normal, but some learning difficulties are common.

Klinefelter syndrome: XXY An individual who has two X chromosomes is female; an individual with an X and a Y is male. But what happens when both of those conditions are true? An individual who carries two X chromosomes *and* a Y chromosome, a condition known as Klinefelter syndrome, develops as male. This is because, as we saw earlier, the Y chromosome carries genetic instructions that, if present, cause fetal gonads to develop as testes; if absent, the fetal gonads develop as ovaries. The extra X chromosome, however, does cause Klinefelter males to be somewhat feminized, although this effect can be reduced through treatments such as testosterone supplementation. Approximately 1 in 1,000 males have the genotype XXY, making this one of the most common genetic abnormalities in humans. Klinefelter syndrome has physical and mental consequences:

Q *If a person has two X chromosomes but also has a Y chromosome, is the individual male or female?*

- Affected individuals have testes, but they are smaller than average. Because the testes are small, levels of testosterone are low and the individuals are almost always infertile.

- They develop some female features, including reduced facial and chest hair, and some breast development.

- They have long limbs and are slightly taller than average (about 6 feet on average).

- They learn to speak at a later age than average and tend to have language impairments.

- Sometimes individuals have further additional X chromosomes, exhibiting the karyotypes XXXY or even XXXXY. These individuals also exhibit Klinefelter syndrome but more frequently have mental retardation.

Q Do human hermaphrodites exist?

A hermaphrodite is an individual with functioning male and female reproductive organs capable of producing both male and female gametes. Often it is mistakenly assumed that a person with Klinefelter syndrome must be a hermaphrodite because he has both two X chromosomes (which would usually make an individual a female) and an X and a Y chromosome (usually making the individual a male). But although hermaphroditism is common among invertebrates and occurs in some fish and other vertebrates, contrary to urban legends, human hermaphrodites do not exist. Although some men with Klinefelter syndrome may have some features of the opposite sex, they do not produce female gametes and so are not hermaphrodites.

XYY males There is no official name for the condition in which an individual has one X chromosome and two Y chromosomes, although they are sometimes referred to as "super males." This chromosomal abnormality occurs in approximately 1 in 1,000 males (and in about 1 in 325 males who are 6 feet or taller). There are no distinguishing features at birth to indicate that an individual carries an extra Y chromosome, and the vast majority live their lives normally without knowing they have an extra chromosome in every cell. Several consequences of the XYY condition have been well documented.

- XYY males are relatively tall (about 6 feet 2 inches on average).

- They tend to have moderate to severe acne.

- Their intelligence usually falls within the same range as that of XY males, although the average may be slightly lower.

Because in the United States about 20 times as many XYY males are found in prisons as in the population as a whole, there has been tremendous controversy over whether the extra Y chromosome predisposes individuals to criminal behavior. Because these observations were not part of randomized, controlled, double-blind experiments, however, they are open to numerous alternative interpretations. One large study, for example, concluded that the increased representation of XYY males in prison was more likely due to the fact that taller males and males of lower intelligence are more likely to be apprehended, regardless of which chromosomes they carry. The controversy remains unresolved, and the majority of XYY males are *not* in prison.

Although an error during meiosis in either the father *or* the mother causes Klinefelter or Turner syndrome, it is only an error in meiosis in a male that gives rise to an XYY child. Why? Because only males carry Y chromosomes. Consequently, an XYY child must receive both of his Y chromosomes from his father. This means that the father's sperm cell must have contained two copies of the Y, a result of nondisjunction during meiosis in the father.

XXX females Sometimes called "metafemales," individuals with three X chromosomes occur at a frequency of about 1 in 1,000 women. Very few studies of this condition have been completed, although initial observations suggest that some XXX females are sterile but otherwise have no obvious physical or mental problems.

TAKE-HOME MESSAGE 6·18

Although it is usually fatal to have one too many or one too few of the non-sex chromosomes, individuals born lacking one of the sex chromosomes or with an additional X or Y chromosome usually survive, though usually with physical and/or physiological problems.

Deviation in Chromosome Numbers

Knowledge You Can Use

Can you select the sex of your baby? (Would you want to?)

Some fertility clinics now promise that for about $200 they can allow a couple to choose the sex of their baby.

Q: **Given our knowledge of the behavior of the X and Y chromosomes, is it more likely that these sex-selection techniques involve manipulations of sperm or of eggs? Why?** Because all eggs have X chromosomes, the X chromosome doesn't play a role in determining the sex of a baby. Instead, the fertility clinics must somehow sort and separate sperm cells based on whether or not they carry an X or a Y chromosome (men produce approximately equal numbers of each).

Q: **What feature of sperm must be used in separating them?** This process must be based on distinguishing between sperm carrying an X chromosome and sperm carrying a Y chromosome. After the two types of sperm are distinguished, they must then be separated.

Q: **What techniques might make this possible?** One method involves determining, by weighing sperm, which cells have more DNA. The heavier sperm must be carrying the X rather than the Y chromosome because the X is so much larger. Another method is to add a fluorescent dye to sperm that temporarily attaches to DNA. Because sperm with an X chromosome have more DNA, more of the dye attaches to them and they are more fluorescent. A machine then sorts the sperm one by one, and at ovulation, insemination is performed using the sperm with the desired sex chromosome.

Q: **Does it work?** The technique is still fairly new, but initial results suggest that the procedures have a success rate of 70% to 90%.

Concerns for you to ponder.
What are the ethical concerns raised by selecting the sex of one's baby? Are there some circumstances in which such a selection would be more acceptable than in others?

❶ There are different types of cell division.

Telomeres are protective sections of DNA that get shorter every time a cell divides. If a telomere gets too short, additional cell divisions cause loss of essential DNA and cell death. Cells that rebuild telomeres with each division can become cancerous. In most prokaryotes, genetic information is carried in a single, circular DNA strand (chromosome). Eukaryotes have much more DNA, organized into linear chromosomes in the nucleus. Bacteria divide by a type of asexual reproduction, binary fission. In the cell cycle, eukaryotic somatic cells alternate between division, called the mitotic phase, and other cell activities, called interphase, consisting of two gap phases separated by a DNA synthesis phase. Every time a cell divides, its DNA must first replicate so that each of the two new cells has all the genetic material of the parent cell.

❷ Mitosis replaces worn-out old cells with fresh new duplicates.

In mitosis, cells generate new, genetically identical cells, enabling organisms to grow and to replace cells that die. Mitosis occurs in four steps, following replication of the chromosomes, to produce two genetically identical daughter cells from one parent cell. Cancer is unrestrained cell growth and division, leading to large masses of cells (tumors). Treatment focuses on killing or slowing down cell growth and division, by chemotherapy and/or radiation.

❸ Meiosis generates sperm and eggs and a great deal of variation.

Meiosis occurs only in gamete-producing cells of sexually reproducing organisms. Gametes have only half as much genetic material as the parent cell and they differ from one another in the combinations of alleles they carry. Meiosis is preceded by DNA replication and consists of two rounds of cell division: separation of homologous pairs of sister chromatids, then separation of the sister chromatids. The final product in a diploid organism is four haploid gametes. In species with two sexes, the female is the sex producing the larger gamete, while males produce the smaller gamete. Every gamete has just one copy of each chromosome. Crossing over during the first prophase of meiosis creates gametes with collections of alleles that may never have existed together before; this variation is important for evolution. Cells and organisms can reproduce by mitosis and asexual reproduction via binary fission or by meiosis and sexual reproduction.

❹ There are sex differences in the chromosomes.

In humans, the sex chromosomes carry information that directs a fetus to develop as male (if a Y chromosome is present) or female (no Y present). The sex is determined by the sex chromosome inherited from the parents. Methods for sex determination in plant and animal species include the presence or absence of sex chromosomes, the number of chromosome sets, and environmental factors such as incubation temperature.

❺ Deviations from the normal chromosome number lead to problems.

A karyotype, a visual display of a complete set of chromosomes, is a useful diagnostic tool. It can be prepared very early in fetal development to detect abnormalities in the number or structure of chromosomes. Having one too many or one too few non-sex chromosomes is usually fatal, but individuals born lacking one of the sex chromosomes or with an additional X or Y usually survive, but often with physical and/or physiological problems.

KEY TERMS

1. Which of the following statements about telomeres is incorrect?
   a) They function like a counter, keeping track of how many times a cell has divided.
   b) At birth, they are long enough to permit approximately 50 cell divisions in most cells.
   c) They are slightly shorter in prokaryotic cells than in eukaryotic cells.
   d) They function like a protective cap on chromosomes.
   e) They contain no critical genes.

2. Prokaryotic cells can divide via:
   a) mitosis.
   b) binary fission.
   c) meiosis.
   d) both mitosis and binary fission.
   e) None of the above.

3. In multicellular organisms, cells that undergo mitotic division but not meiotic division are called _____ cells.
   a) somosis
   b) skin
   c) interphase
   d) somatic
   e) germ

4. The _____ marks the break in the cell cycle between the end of mitosis and beginning of the DNA synthesis stage.
   a) mitotic phase
   b) synthesis phase
   c) Gap 2 phase
   d) Gap 1 phase
   e) None of the above.

5. Mitosis results in:
   a) daughter cells with twice as much genetic material as the parent cell and a unique collection of alleles.
   b) eight daughter cells.
   c) daughter cells with the same number and composition of chromosomes.
   d) gametes.
   e) daughter cells with the same number of chromosomes but different combinations of alleles.

6. Using a light microscope, it is easiest to see chromosomes:
   a) during mitosis and meiosis, because the condensed chromosomes are thicker and therefore more prominent.
   b) during interphase, when they are concentrated in the nucleus.
   c) in the mitochondria, because the chromosomes are circular.
   d) during asexual reproduction.
   e) during interphase, because they are uncoiled and have a more linear structure.

7. The division of the cytoplasm during cell division is referred to as:
   a) cytoplasm splicing.
   b) cytokinesis.
   c) vegetative growth.
   d) cytodivision.
   e) hybridization.

8. Which of the following statements about tumors is incorrect?
   a) Benign tumors pose less of a health risk than do malignant tumors.
   b) Malignant tumors shed cancer cells that can spread to other parts of the body.
   c) Tumors are caused by excessive cell growth and division.
   d) Cancer cells from malignant tumors can travel to other parts of the body in a process called metastasis.
   e) They contain cells with abnormally high contact inhibition.

9. During meiosis but not during mitosis:
   a) haploid gametes are produced that are all identical in their allelic composition.
   b) division of the cytoplasm occurs.
   c) chromosomes line up in the center of the cell during metaphase.
   d) genetic variation among the daughter cells is increased.
   e) two identical daughter cells are produced.

10. Sister chromatids are:
    a) the result of crossing over.
    b) identical molecules of DNA resulting from replication.
    c) homologous chromosomes.
    d) produced in meiosis but not in mitosis.
    e) single-stranded.

11. Which of the following events does not occur during prophase I of meiosis?
    a) Spindle fibers form.
    b) Homologous pairs of chromosomes align at the metaphase plate.
    c) Homologues come very close together.
    d) Crossing over occurs between sister chromatids.
    e) Chromosomes condense.

12. Which of the following is the best way to distinguish male from female?
    a) Males are larger.
    b) Males are more brightly colored.
    c) Males produce motile gametes.
    d) Males are more aggressive.
    e) All of the above statements are correct.

13. A potential disadvantage of asexual reproduction is that:
    a) it increases the time required to find a mate.
    b) it allows perpetuation of a population even when the members are isolated.
    c) it allows population size to increase rapidly.
    d) it produces genetically uniform populations.
    e) it requires additional rounds of meiosis, which is energetically much more costly than mitosis.

14. Which of the following is not a method of chromosomal sex determination that occurs in nature?
    a) In birds, the mother's sex chromosomes determine the sex of the offspring.
    b) In sea turtles, eggs laid in hot sand become females and eggs laid in cooler sand become males.
    c) In humans, the presence of a Y chromosome makes an individual male, even if he also possesses two X chromosomes.
    d) In bees, the eggs that the queen allows to be fertilized become females and the eggs she doesn't fertilize become males.
    e) All of the above are naturally occurring examples of sex determination.

15. A karyotype reveals:
    a) the shape of the spindle.
    b) 23 pairs of chromosomes.
    c) the number, shapes, and sizes of chromosomes in an individual cell.
    d) the sex chromosomes but not the non-sex chromosomes.
    e) the non-sex chromosomes but not the sex chromosomes.

16. What type of chromosomal abnormality leads to the phenotype known as Down syndrome?
    a) trisomy 21
    b) monosomy 21
    c) monosomy 22
    d) trisomy 22
    e) disomy 21

17. Nondisjunction:
    a) is the unequal division of the genetic material during cell division.
    b) occurs during meiosis but not mitosis.
    c) occurs in males but not females among mammals, and in females but not males among birds.
    d) leads to a missing chromosome or extra chromosome.
    e) Both a) and d) are correct.

18. Of all the varying types of sex chromosome abnormalities, one possibility—that of an individual having one Y but no X—has never been reported. Which of the following statements is the most likely reason that medical researchers have yet to report a person having a karyotype with one Y and no X?
    a) There is no meiotic event that could produce an egg that does not have an X chromosome.
    b) Individuals are probably born with this karyotype; however, because the phenotypic abnormalities caused are so slight, they live out their lives normally and are unaware of this genetic anomaly.
    c) The Y chromosome needs the X chromosome to be present in the cell in order for its genes to be expressed.
    d) The X chromosome contains genes that are unrelated to sex determination and are essential for life.
    e) Our current technology for karyotyping would not reveal such an abnormality.

SHORT-ANSWER QUESTIONS

1. Seventy to ninety percent of the genetic material in a sperm or egg cell made in your body could be inherited from your mother. Why isn't there a 50/50 split (half from your mother and half from your father)? Is the reverse situation (more genetic material from your father than from your mother) possible?

2. Healthy individuals may have just one sex chromosome, as long as it is an X chromosome. Why can't a person survive with a Y chromosome and no X chromosome?

3. Some organisms can switch between sexual and asexual reproduction. Under what conditions would you expect them to reproduce sexually? What conditions would be more conducive to asexual reproduction? How would you test your hypotheses under laboratory conditions?

See Appendix for answers. For additional study questions, go to www.prep-u.com.

7

Mendelian Inheritance

Family resemblance:
how traits are
inherited

1 Why do offspring resemble their parents?

Identical twins inherit the same genetic material. (These twins were not told what to do with their hands when the photographer took the photo. Note the similarities in their hand positions.)

7·1

Family resemblance: your mother and father each contribute to your genetic makeup.

Can a gene be cruel? Of course not. But if one could, consider this candidate: in humans, there is a gene for an enzyme called FMO3 (flavin-containing mono-oxygenase-3), which breaks down a chemical in our bodies that smells like rotting fish. Some unfortunate individuals inherit a defective FMO3 gene and can't break down the noxious chemical. Instead, their urine, sweat, and breath excrete it, causing them to smell like rotting fish. Worst of all, because the odor comes from within their bodies, they cannot wash it away no matter how hard they try. Called "fish odor syndrome," this disorder often causes those afflicted by it to suffer ridicule, social isolation, and depression.

For individuals born with this malady, beyond their own suffering there looms a scary question: "Will I pass this condition on to my children?" They might also wonder how

Q How can a single bad gene make you smell like a rotten fish?

they came to have the disorder, particularly if neither of their parents suffered from it.

Fortunately, most of us do not have to worry about fish odor syndrome, but we do wonder about many other inherited traits that we may or may not pass down to our children. Who do I blame for my straight hair (or curly hair or big feet)? Hearing a new baby described as having "his father's nose" and "his grandmother's lips" could make someone think that we're all cobbled together from a bunch of used parts. Where do we begin and our parents end (**FIGURE 7-1**)?

In many cases, the answers to questions about heredity are simple. Fish odor syndrome, for example, is just such a case. Recall from Chapter 6 that sexually reproducing organisms generally carry two copies of every chromosome in every cell

Family Resemblance   Genetics and Chance   Translation of Genotypes   Gene Linkage

(except their sex cells), because they inherit one copy of each chromosome from each parent. Humans, for example, have 23 pairs of chromosomes (46 individual chromosomes). At each location on the two chromosomes of a pair is the same gene: one copy from the mother and one from the father (**FIGURE 7-2**). As we noted in Chapter 5, each of the two copies of the gene is call an **allele.** The gene for FMO3 is on chromosome 1. There is a normal version—or allele—of the gene for FMO3, which most people carry, and there is a rare, defective version that is responsible for fish odor syndrome. As long as a person has at least one normal version of the FMO3 gene, he or she will produce enough of the enzyme to break down the fishy chemical. But if a person inherits two copies of the defective version of the fish odor gene, one from each parent, that person will inherit the disorder (**FIGURE 7-3**).

When it comes to having children, there is a bright side, at least. Although individuals with fish odor syndrome carry two defective copies of the allele and will pass on one set of the bad instructions in every sperm or egg cell that they make, their children won't necessarily inherit the disease. As long as the other parent supplies a normal version of the FMO3 gene, the child will not have fish odor syndrome. The unaffected children, though, will carry a silent copy of the fish odor allele,

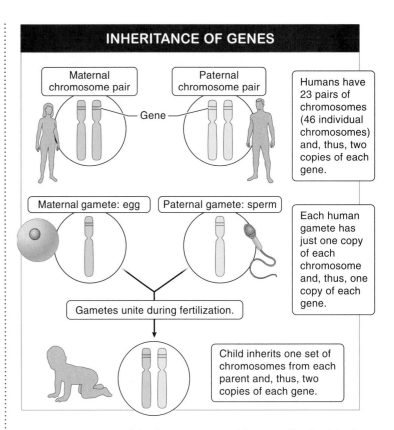

FIGURE 7-2 **How we inherit our genes.** Each human offspring inherits one maternal set of 23 chromosomes and one paternal set.

FIGURE 7-1 **Where do we begin and our parents end?** Family resemblances reveal that many traits are inherited.

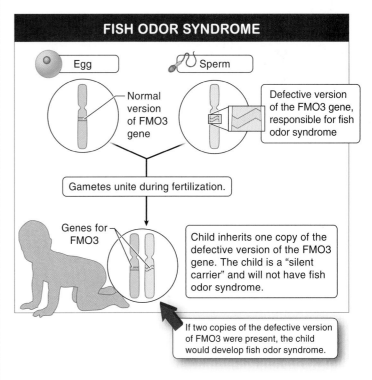

FIGURE 7-3 **Unlucky catch.** A baby inheriting two copies of the defective FMO3 gene will develop fish odor syndrome.

and if, when they have children, it comes together with a defective FMO3 allele from another person, the fish odor trait will be expressed. In this way, some alleles can exist in a population without always revealing themselves.

In this chapter, we explore how heredity works. Inheritance follows some simple rules that allow us to make sense of the patterns of family resemblance we see, and even to predict accurately how many offspring are likely to inherit particular traits—or, put another way, predict the likelihood that an offspring will inherit a trait. We'll also examine why the behavior of some traits is easy to predict while many other traits have less straightforward patterns of inheritance, yet still can be studied experimentally and yield predictable patterns.

TAKE-HOME MESSAGE 7•1

Offspring resemble their parents because they inherit genes—instruction sets for biochemical, physical, and behavioral traits, some of which are responsible for diseases—from their parents.

7•2

Some traits are controlled by a single gene.

Like father, like son. No one will ever seek a paternity test to prove that Michael Douglas is the son of Kirk Douglas. All it takes is a quick look at their legendary dimpled chins and we are sure that the men are related. We see all around us that offspring resemble their parents more than they resemble other random individuals in the population—a consequence of the passing of characteristics from parents to offspring through their genes. This is **heredity.**

> ❝ HEREDITY
>
> I AM the family face;
> Flesh perishes, I live on,
> Projecting trait and trace
> Through time to times anon,
> And leaping from place to place
> Over oblivion.
>
> The ears-heired feature that can
> In curve and voice and eye
> Despise the human span
> Of durance—that is I;
> The eternal think in man,
> That heeds no call to die.
>
> —Thomas Hardy, *Moments of Vision and Miscellaneous Verses,* 1917 ❞

Observing heredity is easy. Elucidating how it works is not. For thousands of years before the mechanisms of heredity were discovered and understood, plant and animal breeders recognized that there is a connection from parents to offspring across generations. In ancient Greece, for example, the poet Homer extolled the tremendous benefits to society that came

Once breeders recognized the existence of heredity, selective breeding—such as for increased body size in cattle—became possible.

FIGURE 7-4 **Have a cow!** Selective breeding was used to produce this steer.

from the skillful breeding of horses. The awareness of beneficial breeding practices enabled farmers to systematically create strains of crops, livestock, and even pets with desirable traits (**FIGURE 7-4**). From flocks of sheep and herds of reindeer to litters of puppies and fields of wheat, the similarity among relatives is clear. Once breeders recognized the existence of heredity, they began selecting individual plants or animals with the desired traits to breed with each other, in the hope that their offspring would also have the desirable traits. And since then, it has been possible to create a rich world of sweeter corn, loyal dogs, docile livestock, beautiful flowers, and more.

Family Resemblance   Genetics and Chance   Translation of Genotypes   Gene Linkage

EXAMPLES OF SINGLE-GENE TRAITS

Cleft chin | Non-cleft chin

Unattached earlobes | Attached earlobes

Widow's peak | Straight hairline

FIGURE 7-5 **Chins, ears, and hairlines: single-gene traits.** Some traits are determined by the instructions a person carries on one gene.

And these breeders did it all without ever really understanding how heredity works.

Practical successes were many, but people didn't understand exactly how the outcomes were achieved. They knew the result of selective breeding, but why and how it worked, or didn't always work, was a mystery. Patterns of similarity among related individuals are impossible to make sense of without an underlying understanding of *how* heredity works. This is the province of the field of genetics.

Mechanisms of heredity are like a puzzle, and they become a bit easier to grasp when we focus on aspects of family resemblances that are commonplace. Virtually everyone with a cleft chin, for instance, has a similarly dimpled parent. Everyone who has earlobes that hang from their ears rather than being attached directly to their head has at least one parent with the same feature. Everyone who has a widow's peak has a parent with the same hairline. Everyone who is farsighted has a similarly farsighted parent. The list goes on and on. In fact, there are more than 9,000 human traits that exhibit these straightforward patterns of inheritance. Traits that are determined by the instructions a person carries on one gene are called **single-gene traits** (FIGURE 7-5).

Q Are any human traits determined by a single gene?

How single-gene traits such as the cleft chin pass from parent to child is the easiest pattern of inheritance to decipher. For this reason, we first explore the mechanisms by which single-gene traits are inherited. Then we investigate how to expand our model of heritability to account for the passing on of more complex traits.

TAHE-HOME MESSAGE 7•2

More than 9,000 human traits are determined by the instructions a person carries on a single gene, and the traits exhibit straightforward patterns of inheritance.

7·3

Mendel learned about heredity by conducting experiments.

What do parents "give" their offspring that confers similarity? In other words, how is it that a physical entity is inherited—that is, how is it passed from parent to offspring? It wasn't until the mid–1800s that any real headway was made on these questions. At that time, Gregor Mendel, a monk living in what is now the Czech Republic, began some studies that not only shed light on these questions but practically answered them completely. Mendel understood the essence of the genetics puzzle and set out to piece it all together.

When Mendel turned to these questions about heredity, there were no obvious answers and the prevailing state was confusion. There were a couple of existing ideas, but they had obvious problems. One postulated that perhaps an entire, pre-made human—albeit a very, very tiny one—was contained in every sperm cell (**FIGURE 7-6**). That idea had been popular since the late 1600s. It doesn't take much imagination to see why it ran into difficulty, though, because children resemble both their mothers and their fathers, not just their fathers. Another popular idea was that offspring reflect a simple blending of their two parents' traits via the blood. But this couldn't explain how brown-eyed parents could give birth to

Q Individuals who share common ancestry are called "blood relatives," yet they don't actually share any blood. How might the phrase "blood relatives" be a reflection of early conceptions of inheritance?

blue-eyed children. Or why one tall and one short parent sometimes produced a tall child, rather than always producing children of intermediate height.

Sometimes scientific breakthroughs are made because an innovative new technique is invented or a lucky, chance observation is made. Neither played a role in Mendel's success. He didn't do anything radically new, but simply applied the tried and true methodical experimentation and scientific thinking (as described in Chapter 1) that had been a boon to chemists and physicists. Three features of Mendel's research, in particular, were critical to its success (**FIGURE 7-7**).

1. He chose a good organism to study: the garden pea. It's not that Mendel had a particular fondness for vegetables. In fact, his goals were to understand inheritance in all organisms, not just plants. Pea plants were simply the right tool. Cats and dogs or even mice wouldn't have served his purposes very well, because they would have been too hard to take care of in the large numbers—thousands and thousands of individuals—that Mendel required. Humans, too, would have made a terrible study organism. We take too long to breed (and won't produce offspring on command). Pea plants, on the other hand, have a lot going for them as experimental organisms. They are relatively easy to fertilize manually by "pollen dusting." It is easy to collect dozens or even hundreds of offspring from a single **cross,** in which male pollen is used to fertilize female eggs. And, on top of all this, pea plants are fast enough breeders that Mendel could conduct experiments that lasted for multiple generations.

2. Mendel chose to focus on easily categorized traits. For instance, shape: all peas of the variety that Mendel studied are either round or wrinkled in shape, with nothing in between. In addition, all peas are either yellow or green in color, never any intermediate shade. In all, Mendel looked at seven different traits but, for each trait, only two variants ever appeared, so he and his research assistants could easily observe and unambiguously identify them.

3. Mendel began his studies by first repeatedly breeding together similar plants until he had many distinct populations, each of which was unvarying for a particular

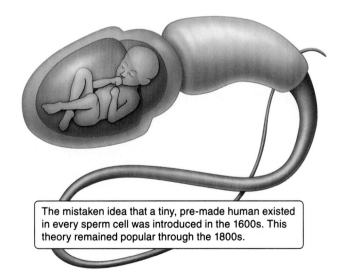

The mistaken idea that a tiny, pre-made human existed in every sperm cell was introduced in the 1600s. This theory remained popular through the 1800s.

FIGURE 7-6 The "pre-made human." Tiny humans in every sperm cell? Why would children have any resemblance to their mothers?

Family Resemblance Genetics and Chance Translation of Genotypes Gene Linkage

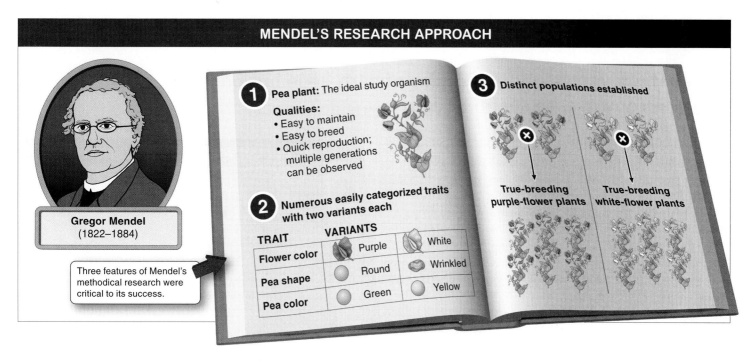

Gregor Mendel
(1822–1884)

Three features of Mendel's methodical research were critical to its success.

1 **Pea plant:** The ideal study organism

Qualities:
• Easy to maintain
• Easy to breed
• Quick reproduction; multiple generations can be observed

2 Numerous easily categorized traits with two variants each

TRAIT	VARIANTS	
Flower color	Purple	White
Pea shape	Round	Wrinkled
Pea color	Green	Yellow

3 Distinct populations established

True-breeding purple-flower plants

True-breeding white-flower plants

FIGURE 7-7 **Mendel's notebook.** Careful planning and a well-selected study organism contributed to the success of Mendel's experiments.

trait. He described these plants as **true-breeding** for that trait because they always produced offspring with the same variant of the trait as the parents. True-breeding round-pea plants always produced plants with round peas when they were crossed together. True-breeding purple-flowered pea plants always produced purple-flowered offspring, while true-breeding white-flowered pea plants always produced white-flowered offspring. It took a lot of prep work to establish these populations, but once Mendel had them, he was in a position to set up all of the different crosses that enabled him to piece together the genetics puzzle.

Once he had obtained his groups of true-breeding plants, Mendel began a straightforward process of experimentation. He crossed plants with different traits—for example, he crossed plants with green peas and plants with yellow peas—and observed large numbers of their offspring, counting the numbers of green peas and yellow peas produced. (This cross, written as "green × yellow cross," always resulted in plants with nothing but yellow peas.) Mendel devised a hypothesis that would explain his observations and generate predictions about the outcome of further crosses. Then he conducted

those crosses to see whether the predictions generated by his hypothesis were borne out.

There was a simple elegance to Mendel's work: his predictions were always in the form of simple questions that could be answered by his experiments. His crosses would come out either one way or the other. Everything was black and white. This process of performing well thought out, rigorous experiments was a radical innovation for biology. In the end, Mendel's patience, perseverance, and cleverness paid off. As we explore his results in the next section, we see that they allowed a definitive interpretation.

TAKE-HOME MESSAGE 7·3

In the mid-1800s, Gregor Mendel conducted studies that helped us understand how traits are inherited. He applied methodical experimentation and rigorous hypothesis testing, focusing on easily observed and categorized traits in garden peas.

7·4

Segregation: you've got two copies of each gene but put only one copy in each sperm or egg.

One odd and recurring result spurred Mendel to figure out the mechanism by which traits could be passed from parent to offspring. Just as brown-eyed parents can have blue-eyed children, sometimes traits that weren't present in either parent pea plant would show up in their offspring. Crosses that entailed the fertilization of plants with purple flowers by pollen from other plants with purple flowers produced *mostly* purple-flowered offspring, for instance, but *sometimes* they produced plants with white flowers. How was that possible? Where did the whiteness come from?

Here's where Mendel's meticulous and methodical experiments paid off. First, he started with some true-breeding white-flowered plants. Then he got some true-breeding purple-flowered plants. He wondered: which color wins out when a white-flowered plant is crossed with a purple-flowered plant? The answer was definitive: purple wins (**FIGURE 7-8**). All of the offspring from these crosses were purple, every time. For this reason, Mendel called the purple-flower trait **dominant,** and he considered the white-flower trait to be the **recessive** trait. In general, a dominant trait masks the effect of a recessive trait when an individual carries both the dominant and the recessive versions of the instructions for the trait.

Things got a bit more interesting when Mendel took the purple-flowered plants that came from the cross between true-breeding purple- and white-flowered plants and bred these purple offspring with each other. He found that these mixed-parentage plants were no longer true-breeding. Occasionally, they would produce white-flowered offspring. (To be exact, of the 929 plants Mendel examined, 705 had purple flowers, and 224 had white flowers.) Apparently, the directions for building white flowers—last seen in one of their grandparents—were still lurking inside the purple-flowered parent plants. The existence of traits that could disappear for a generation and then show up again was quite perplexing. Mendel devised a simple and perfect hypothesis to explain these observations (**FIGURE 7-9**).

Mendel's hypothesis for explaining this pattern of inheritance incorporated three ideas that then helped him (and now help us) make predictions about crosses he hadn't yet done.

1. Rather than passing on the trait itself, **each parent puts into every sperm or egg it makes a single set of instructions for building the trait.** Today, we call that instruction set a *gene.*

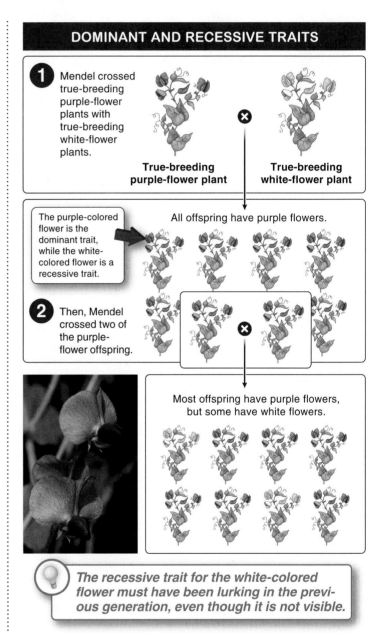

FIGURE 7-8 White or purple? By careful and repeated crosses among pea plants, Mendel determined that there were "dominant" and "recessive" traits.

2. Offspring receive two copies of the instructions for any trait. Often, both sets of instructions are identical and the offspring produce the traits according to those instructions. Other times, though, each parent contributes a slightly different set of instructions—that is, a different

Family Resemblance   Genetics and Chance   Translation of Genotypes   Gene Linkage

MENDEL'S LAW OF SEGREGATION

According to Mendel's law of segregation, only one of the two alleles for a gene is put into a gamete. At fertilization, offspring receive from each parent one allele for each gene.

Heterozygous pea plant   **Heterozygous pea plant**

Two different alleles (white, purple) for the same gene (flower color)

MEIOSIS Each gamete gets one copy of each gene.

FERTILIZATION Each fertilized egg gets two copies of each gene.

| Homozygous recessive | Heterozygous | Heterozygous | Homozygous dominant |

FIGURE 7-9 **Mendel's law of segregation.** Organisms have two copies of each gene but place only one copy into each gamete during the process of meiosis (as described in Chapter 6).

allele—for the trait. So, for example, in pea plants there are two alleles for flower color: a purple-flower allele and a white-flower allele. Each is a different allele, but because both specify instructions for producing flowers, they are both genes for flower color. Same gene (flower-color gene), different alleles (purple and white).

3. The trait observed in an individual depends on the *two* copies of the gene it inherits from its parents. When

an individual inherits the same two alleles for this gene, the individual's genotype for that gene is said to be **homozygous** and the individual shows the trait specified by the instructions embodied in those alleles. When an individual inherits a different allele from each parent, the individual's genotype for that gene is said to be **heterozygous.** Dominant and recessive alleles are defined by their action when they are in the heterozygous state: if one of the alleles is fully apparent in the individual while the other is not, the first is called the dominant allele and is said to "mask" the effect of the other allele, which is called the recessive allele.

> 'There were no questions.'
>
> —Entry in the meeting notes following Mendel's first public presentation of his ideas on how heredity works. (*No one in the audience of scientists had any idea what he was talking about. It was about 40 years before the world understood his discoveries.*)

Later, when the offspring grows up and gets ready to have its own offspring, it only gets to contribute one of its two copies of the gene to its offspring. The other parent will contribute the other allele. So, when sperm and eggs are made, each sex cell gets only one copy of the gene—as opposed to the two copies present in every other cell in the body. For a male who is heterozygous, for example, it means that half of the sperm he produces will have one of the alleles and half will have the other. The idea that, of the two copies of each gene everyone carries, only one of the two alleles gets put into each gamete is significant and important enough that it is called **Mendel's law of segregation.**

TAKE-HOME MESSAGE 7·4

Each parent puts into every sperm or egg it makes a single set of instructions for building a particular trait. This instruction set is what today we call a gene. The trait observed in an individual depends on the two copies (alleles) of the gene it inherits from its parents.

Observing an individual's phenotype is not sufficient for determining its genotype.

Things are not always as they appear. Take skin coloration, for example. Humans and many other mammals have a gene that contains the information for producing melanin, one of the chemicals responsible for giving our skin its coloring (FIGURE 7-10). Unfortunately, there is also a defective, non-functioning version of the melanin gene that is passed along through some families. An individual who inherits two copies of the defective version of the gene cannot produce pigment and has a condition known as albinism, a disorder characterized by little or no pigment in the eyes, hair, and skin. But it is impossible to tell whether a normally pigmented individual carries one of these defective alleles just by looking at his or her appearance—we would need to get a genetic analysis done to discover this information.

The inability to deduce an individual's genetic makeup through simple observation is a general problem in genetics: physical appearances don't always exactly reflect the underlying genes. A normally pigmented individual may have two copies of the pigment-producing allele or may have only one copy. In either case, the individual will look the same. The outward appearance

of an individual is called its **phenotype.** A phenotype includes features visible to the naked eye such as flashy coloration, height, or the presence of antlers. A phenotype also includes less easily visible characteristics such as the chemicals an individual produces to clot blood or digest lactose. An individual's phenotype even includes the behaviors it exhibits.

Underlying the phenotype is the **genotype.** This is an organism's genetic composition. We usually speak of an individual's genotype in reference to a particular trait. For example, an individual's genotype might be described as "homozygous for the recessive allele for albinism." Another individual's genotype for the melanin gene might be described as "heterozygous." Occasionally, the word "genotype" is also used as a way of referring to *all* of the genes an individual carries.

When an organism exhibits a recessive trait, such as albinism, we know with certainty what its genotype is. When it shows the dominant trait, on the other hand, it's impossible to distinguish whether the organism carries two copies of the dominant allele (homozygous dominant) or carries one copy of the dominant allele and one of the recessive (heterozygous). Because it's not possible to discern the genotypes of two individuals with the same phenotype just by looking at them, much of genetic analysis must make use of clever experiments and careful record-keeping to determine organisms' genotypes. Let's explore a useful method of analyzing the relationships between genotypes and phenotypes.

How do we analyze and predict the outcome of crosses? First, we must assign symbols to represent the different variants of a gene. Generally we use an upper-case letter for the dominant allele and lower-case letter for the recessive allele. In the case of the albino giraffe, we represent the albino individual as *aa,* because it must carry two copies of the recessive allele, *a.* A giraffe that is pigmented must have the genotype *AA* or *Aa.* If we don't know which of the two possible genotypes the pigmented individual is, we can write *A_,* where _ is a placeholder for the unknown second allele, whose identity we aren't certain of.

We can trace the possible outcomes of a cross between two individuals using a handy tool called the **Punnett square.** In **FIGURE 7-11** we illustrate the cross between a true-breeding

PHENOTYPE: Little or no pigment in the eyes, hair, and skin
GENOTYPE: Homozygous for the recessive allele for albinism

FIGURE 7-10 An albino deer stands out from its peers. Carrying two non-functioning versions of the gene that produces the pigment melanin causes albinism.

Family Resemblance | Genetics and Chance | Translation of Genotypes | Gene Linkage

PUNNETT SQUARE: ALBINISM

A Punnett square is a useful tool for determining the possible outcomes of a cross between two individuals.

pigmented individual, *AA*, and an albino, *aa*. Along the top of the square we list, individually, the two alleles that one of the parents produces, and along the left side of the square we list the two alleles that the other parent produces. We split up an individual's two alleles in this way because, although the individual carries two alleles, only one of the alleles is contained in each sperm or egg cell that it produces.

In the four cells of the Punnett square, we enter the genotypes of all the possible offspring resulting from our cross—a cell contains the allele given at the head of the column, which is one parent's contribution, and the allele at the left of the row, which is the other parent's contribution. The two gametes that come together at fertilization produce the genotype of the offspring. In Cross 1 illustrated in Figure 7-11, every possible offspring would be heterozygous and would be normally pigmented, because it receives a dominant allele from the pigmented parent and a recessive allele from the albino parent.

In the bottom half of Figure 7-11 (Cross 2), we trace the cross between two heterozygous individuals. Note that each parent produces two kinds of gametes, one with the dominant allele and one with the recessive allele. This cross has four possible outcomes: one-quarter of the time the offspring will be homozygous dominant (*AA*), one-quarter of the time the offspring will be homozygous recessive (*aa*), and the remaining half of the time the offspring will be heterozygous (*Aa*). Phenotypically, three-quarters of the offspring will be normally pigmented (*AA* or *Aa*) and one-quarter will be albino (*aa*).

TAKE-HOME MESSAGE 7·5

It is not always possible to determine an individual's genetic makeup, known as its genotype, by observation of the organism's outward appearance, known as its phenotype. For a particular trait, an individual may carry a recessive allele whose phenotypic effect is masked by the presence of a dominant allele. Much genetic analysis therefore makes use of clever experiments and careful record-keeping, using Punnett squares, to determine organisms' genotypes.

❷ Probability and chance play central roles in genetics.

Because of the role of chance in genetics, we cannot always predict the exact outcome of a genetic cross.

7•6

Chance is important in genetics.

Sometimes genetics is a bit like gambling. Even with perfect information, it can still be impossible to know the genetic outcome with certainty. It's like flipping a coin: you can know every last detail about the coin, but you still can't know whether the coin will land on heads or tails. The best you can do is define the probability of each possible outcome.

The rules of probability (the same ones that govern coin tosses and the rolling of dice) have a central role in genetics, for two reasons. The first is a consequence of segregation, the process Mendel described, in which each gamete an individual produces receives only one of the two copies of each gene that the individual carries in most of its other cells. As a result, it is equally likely that the haploid gamete—the sperm or egg—will include one or the other of the two alleles that the individual carries. It is impossible to know which allele it will be. The second reason probability plays a central role in genetics is that fertilization, too, is a chance event. All of the sperm or eggs produced by an individual are different from one another, and any one of those gametes may be the gamete involved in fertilization. Thus, knowing everything about the alleles a parent carries is not enough to be able to determine with certainty which alleles his or her offspring will carry.

> Chaos umpire sits,
> And by decision more embroils the fray
> By which he reigns: next him high arbiter
> Chance governs all.
>
> —John Milton, *Paradise Lost*, 1667

Let's explore how we can make predictions in games of chance, as well as in matings, based on probabilities. As with the probability of getting heads in a coin flip, any gamete produced by an individual heterozygous for a trait has a 50% probability of carrying the dominant allele and a 50% probability of carrying the recessive allele. If an individual is homozygous for a trait, 100% of his or her gametes will carry that allele. Let's use these probabilities in an example.

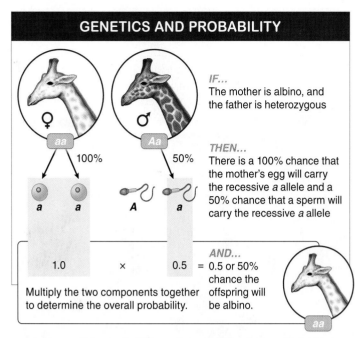

FIGURE 7-12 **Using probability to determine the chance of inheriting albinism.**

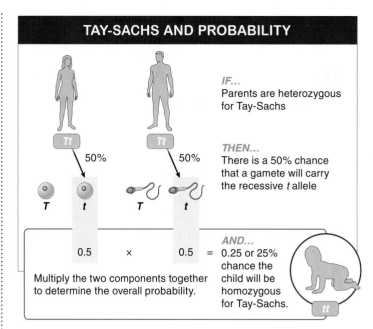

FIGURE 7-13 **The chance of inheriting Tay-Sachs disease.** The probability that a child inherits Tay-Sachs disease from two heterozygous parents is 25%.

If a male is heterozygous for albinism (*Aa*) and a female is albino (*aa*), what is the probability that their child will be homozygous for albinism (*aa*)? In order to get the *aa* outcome, two events must occur. First, the father's gamete must carry the recessive allele (*a*) and, second, the mother's gamete must carry the recessive allele (*a*). In this case, the probability of a homozygous recessive offspring is 0.5 (the probability that the father's gamete carries *a*) times 1.0 (the probability that the mother's gamete carries *a*), for a probability of 0.5. This is a general rule when determining the likelihood of a complex event occurring: if you know the probability of each component that must occur, you simply multiply all the probabilities together to get the overall probability of that complex event occurring (**FIGURE 7-12**).

Consider another example, involving Tay-Sachs disease. As described in Chapter 3, Tay-Sachs is a disease caused by malfunctioning lysosomes that do not digest cellular waste properly. It leads to death in early childhood. Tay-Sachs occurs if a child inherits two recessive alleles for the Tay-Sachs gene. If each parent is heterozygous for the Tay-Sachs gene, what is the probability that their child will have Tay-Sachs disease?

To solve this, just break down the event of the child having Tay-Sachs into two separate events: first, it must inherit a recessive allele from its heterozygous father. The probability of that is 0.5. Second, it must also inherit a recessive allele from its heterozygous mother. The probability of that is also 0.5. So the overall probability of the child having Tay-Sachs disease is 0.5 × 0.5 = 0.25, or 1 in 4 (**FIGURE 7-13**). Of course, if the couple has only one child, we can't predict with certainty whether this child will have Tay-Sachs. Mathematicians, on the other hand, might be satisfied just to be able to say that if the couple had an infinite number of children, we could be certain that one-fourth of them would have Tay-Sachs. Still, this is not much help for heterozygous couples trying to decide whether to have a child or not.

TAKE-HOME MESSAGE 7·6

Probability plays a central role in genetics. In segregation, each gamete that an individual produces receives only one of the two copies of each gene the individual carries in its other cells, but it is impossible to know which allele goes into the gamete. Chance plays a role in fertilization, too: all of the sperm or eggs produced by an individual are different from one another, and any one of those gametes may be the gamete involved in fertilization.

A test-cross enables us to figure out which alleles an individual carries.

How can you see something that is invisible to the naked eye? When it comes to genetics, this is a practical issue that has been causing problems for as long as humans have been breeding plants and animals for food or other purposes. Genes are too small to be seen, and so determining an individual's genotype requires indirect methods. Suppose you are in charge of the alligators at a zoo. Some of your alligators come from a population in which white, albino alligators have occasionally occurred, although none of your alligators are white. Because white alligators —those having two recessive pigmentation alleles, *mm* ("m" for melanin)—are popular with zoo visitors, you would like to produce some at your zoo through a mating program.

The problem is that you cannot be certain of the genotype of your alligators. They might be homozygous dominant, *MM,* or they might be heterozygous, *Mm*. In either case, their phenotype is normal coloration. How can you figure out which of these two possibilities is the genotype of a particular alligator? This is a challenge to animal breeders, but not an insurmountable one. Genes may be invisible, but their identity can be revealed by a simple tool called the **test–cross.**

In the test-cross, you take an individual exhibiting a dominant trait but whose genotype is unknown. You cross (i.e., mate) that individual with an individual that is homozygous recessive and look at the phenotypes of their offspring. In the case of your breeding program for albino alligators, you could borrow an albino alligator from another zoo and breed your unknown-genotype alligator (genotype: *M_*) with that albino alligator (genotype: *mm*). There are two possible outcomes, and they will reveal the genotype of your unknown-genotype alligator (**FIGURE 7-14**). If your alligator is homozygous dominant (*MM*), it will contribute a dominant allele, *M,* to every offspring. Even though the albino alligator will contribute the recessive allele, *m,* to all its offspring, all the offspring will be heterozygous, *Mm,* and none of them will be albino.

If, on the other hand, your unknown-genotype alligator is heterozygous, *Mm,* half of the time it will contribute a recessive allele, *m,* to the offspring. In every one of those cases, the

FIGURE 7-14   A test-cross can reveal an unknown genotype. In this test-cross, a homozygous white female alligator is bred with a normally colored male of unknown genotype. The color of their offspring will help identify whether the male is homozygous dominant or heterozygous.

TEST-CROSS: WHITE ALLIGATORS

Test-cross

FATHER
pigmented
unknown genotype
M_
(**MM** or **Mm**)

MOTHER
white
homozygous
mm

GAMETES   m   m

GAMETES   M   M   m

Mm   Mm
Mm   Mm
mm   mm

Unknown allele could be *M* or *m*.

OFFSPRING (if unknown genotype is **MM**)

	Genotype	Phenotype
	All heterozygous **Mm**	All pigmented

OFFSPRING (if unknown genotype is **Mm**)

	Genotype	Phenotype
	2/4 heterozygous **Mm**	2/4 pigmented
	2/4 homozygous recessive **mm**	2/4 white

offspring will be homozygous recessive and thus albino. So the cleverness of the test-cross is that when you cross your unknown-genotype organism with an individual showing the recessive trait, if it sometimes produces offspring with the recessive trait (and if it does, it will do so half the time, on average), its genotype must be heterozygous. If it never produces offspring with the recessive trait, it must be homozygous for the dominant allele. In order to be confident in concluding that the unknown-genotype alligator has the *MM* genotype, though, you'd have to observe as many offspring as possible. After all, even if its genotype is *Mm*, quite a few offspring in a row might be normally pigmented, with the genotype *Mm*. Eventually, however, a heterozygous individual is likely to produce an offspring with the homozygous recessive genotype, *mm*, and the albino phenotype.

TAKE-HOME MESSAGE 7·7

In a test-cross, an individual that exhibits a dominant trait but has an unknown genotype is mated with an individual that is homozygous recessive. The phenotypes of the offspring reveal whether the unknown-genotype individual is homozygous dominant (all of the offspring exhibit the dominant trait) or heterozygous (half of the offspring show the dominant trait and half show the recessive trait).

7·8

We use pedigrees to decipher and predict the inheritance patterns of genes.

People want to know things about the future, such as: what is the likelihood that I will have a child with a particular genetic disease, say hemophilia? Or what is my own risk of developing a genetic disease, such as Huntington's disease, later in my life? Geneticists who study these and other diseases want to know other things: how is a particular disease inherited? Is it recessive or dominant? Is it carried on the sex chromosomes or on one of the other chromosomes? A **pedigree** is a type of family tree that can help answer these questions.

In a pedigree, information is gathered from as many related individuals as possible across multiple generations (**FIGURE 7-15**). Starting from the bottom, each row in the chart represents a generation, listing all of the children in their order of birth and whether or not they express a particular trait. Working up the pedigree, their parents are indicated and, above them, their parents' parents, for as far back as data are available. Squares represent males and circles represent females, and these shapes are shaded to indicate that an individual exhibits the trait of interest. Sometimes the genotype (as much of it as is known) is also listed for each individual.

By analyzing which individuals manifest the trait and which do not, it may be possible to deduce the pattern of inheritance for the trait—or, at least, rule out certain patterns. For example, if an individual exhibits a trait that neither of his or her parents exhibits, the trait is recessive. For dominant traits, all affected individuals must have at least one parent who exhibits the trait. In contrast, an individual can exhibit a recessive trait even if both parents are unaffected (i.e., do not exhibit the trait). In this case, the individual's parents must be

heterozygous for that trait, each carrying one dominant and one recessive allele. Similarly, it is sometimes possible to determine whether a trait is carried on the sex chromosomes (X or Y, which we discuss in Section 7-13) or on one of the non-sex chromosomes (also called "autosomes"). Traits that are controlled by genes on the sex chromosomes are called **sex-linked traits.** Recessive sex-linked traits, for example, appear more frequently in males than females, whereas dominant sex-linked traits appear more frequently in females. These patterns may become obvious only on inspection of a large pedigree.

PEDIGREE

A pedigree is a useful tool to document a trait of interest across multiple generations of family members.

Grandfather Grandmother   Grandfather Grandmother

Aunt Uncle   Father Mother Uncle Aunt

First cousin   First cousin   Sister Me Sister

○ Female exhibiting trait of interest
○ Female not exhibiting trait
□ Male exhibiting trait of interest
□ Male not exhibiting trait

FIGURE 7-15 Family tree. A pedigree maps the occurrence of a trait in a family.

Anury is a condition in dogs and other animals in which they are born without a tail. The condition is inherited as a recessive trait.

Can you figure out the genotype of the individual labeled 1?

IF...
A dog has no tail, but his father was normal

THEN...
The normal-looking mother (1) of the tailless dog must be heterozygous.

PEDIGREE

- ⬤ Female with anury
- ◯ Female not exhibiting anury
- ◼ Male with anury
- ◻ Male not exhibiting anury

Can you now determine the probability of this puppy having anury?

FIGURE 7-16 Pedigree puzzle: tracing the occurrence of anury in a family of dogs.

An example of how pedigrees can help determine how traits are inherited is given in **FIGURE 7-16**. Anury is a condition seen in dogs and some other animals in which the animal has no tail. The pedigree reveals that anury is inherited as a recessive trait, because unaffected parents can have offspring with the disorder.

Can you figure out the genotype of the individual labeled "1"? That female must be heterozygous (*Aa*) for anury. (An individual who carries one allele for a recessive trait, and so does not exhibit the trait but can have offspring that do, is referred to as **carrier** of the trait.) The same is true for her mate. On the same pedigree, note that two individuals in the second generation have a puppy (indicated by "?"). What is the probability that this puppy has anury? Examine the pedigree carefully and see if you can come up with the answer; the next paragraph will guide you through if you get stuck.

In the cross producing the mystery puppy, the father has anury and so must be homozygous recessive; he will definitely pass on one *a* to the son. The mother's genotype can be *AA*, *Aa*, or *aA*, given that both of her parents are heterozygous. Consequently, she has a ⅔ (2 in 3) probability of carrying the *a* allele. If she does, then there is a 1 in 2 (50%) chance that she will pass it on to her offspring and he will have anury. The probability, therefore, is ⅔ × ½ = ⅓—or a 1 in 3 chance that the puppy will have anury.

As we will discuss in later sections of this chapter, some traits may not show complete dominance and many traits are also

Q Why do breeders value "pedigreed" horses and dogs so much?

influenced by the environment, so it is not always completely obvious what a trait's mode of inheritance is. In such cases, the more individuals we can include in the pedigree, the more accurate the analysis. Some human pedigrees contain thousands of individuals and stretch back six or more generations. With some other species it is possible to analyze tens of thousands of individuals per generation for a dozen or more generations. This is why plants and small insects (among other organisms) are excellent for studying inheritance patterns.

Dog and horse breeders often speak of "pedigreed" animals, a feature that adds tremendous value to the animals. This is because, with knowledge of an animal's family tree or complete lineage, it is much less likely that any genetic surprises will occur as the animal develops and reproduces.

Once you have an idea about how a trait is inherited, it is possible to make informed predictions about a couple's risk of having a child with a particular disorder. This makes earlier prenatal testing and treatment possible. It also enables the identification of individuals who are "carriers" of a recessive trait.

TAKE-HOME MESSAGE 7·8

Pedigrees help scientists, doctors, animal and plant breeders, and prospective parents determine the genes that individuals carry and the likelihood that the offspring of two individuals will exhibit a given trait.

Family Resemblance   Genetics and Chance   Translation of Genotypes   Gene Linkage

❸ The translation of genotypes into phenotypes is not a black box.

Phenotypic diversity has multiple sources and is all around us. (The photo shows schoolchildren from London.)

7·9

Incomplete dominance and codominance: the effects of both alleles in a genotype can show up in the phenotype.

As Mendel saw it, the world of genetics was straightforward and simple. We should be so lucky. Unfortunately, the world in which each trait is coded for by a single gene with two alleles—one completely dominant and one recessive—and with no environmental effects at all doesn't quite capture the complexity of the world beyond Mendel's pea plants. So, here and in the next sections, we'll build up a more complex model of how genes influence the building of bodies. We begin with the observation that, sometimes, the phenotype of heterozygous individuals differs from that of either of the homozygotes, and instead reflects the influence of both alleles rather than a clearly dominant allele.

One situation in which complete dominance is not observed is called **incomplete dominance,** in which the heterozygote appears to be intermediate between the two homozygotes. The quintessential example of incomplete dominance is the flower color of snapdragons (**FIGURE 7-17**). We can obtain true-breeding (homozygous) lines of snapdragons with red flowers, and true-breeding (homozygous) lines that produce

only white flowers. When plants from these two populations are crossed, we would expect—if one allele were dominant over the other—either all red or all white flowers. Instead, such crosses always produce plants with pink flowers. Interestingly, when we cross two plants with pink flowers, we get ¼ red-flowered plants, ½ pink-flowered plants, and ¼ white-flowered plants.

How can we interpret this cross? It seems that the plants with white flowers have the genotype $C^W C^W$ and produce no pigment. At the other extreme, the plants with red flowers have the genotype $C^R C^R$ and produce a great deal of pigment. The letter "C" refers to the fact that the gene codes for **c**olor, and the superscript "W" or "R" refers to an allele producing no pigment (**w**hite) or **r**ed pigment. We use these designations for the genotypes because it isn't clear that either white or red is dominant over the other, and so neither should be upper-case or lower-case. The pink flowers receive one of the pigment-producing C^R alleles and one of the no-pigment-producing C^W alleles, and so produce an intermediate amount

INCOMPLETE DOMINANCE: SNAPDRAGONS

Incomplete dominance occurs when a heterozygote exhibits an intermediate phenotype between the two homozygotes.

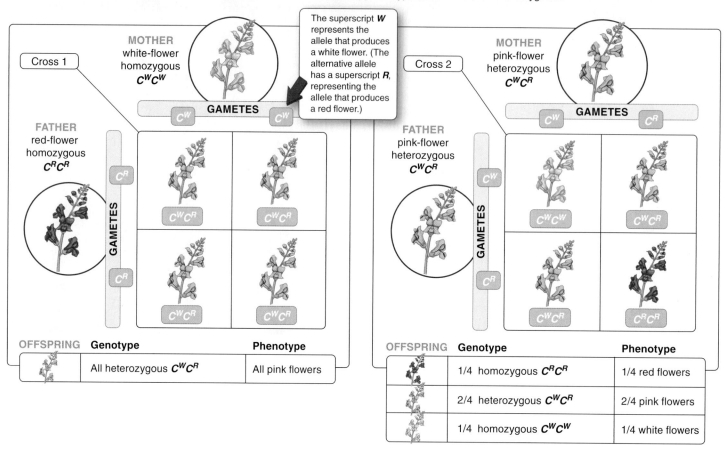

The superscript **W** represents the allele that produces a white flower. (The alternative allele has a superscript **R**, representing the allele that produces a red flower.)

Cross 1

MOTHER
white-flower homozygous
$C^W C^W$

FATHER
red-flower homozygous
$C^R C^R$

OFFSPRING	Genotype	Phenotype
	All heterozygous $C^W C^R$	All pink flowers

Cross 2

MOTHER
pink-flower heterozygous
$C^W C^R$

FATHER
pink-flower heterozygous
$C^W C^R$

OFFSPRING	Genotype	Phenotype
	1/4 homozygous $C^R C^R$	1/4 red flowers
	2/4 heterozygous $C^W C^R$	2/4 pink flowers
	1/4 homozygous $C^W C^W$	1/4 white flowers

FIGURE 7-17 Pink snapdragons demonstrate incomplete dominance. When true-breeding white and red snapdragons are crossed, offspring have pink flowers.

CODOMINANCE: SICKLE-CELL DISEASE

$Hb^A Hb^A$
Homozygote

Individual does not have sickle-cell disease.

$Hb^S Hb^A$
Heterozygote

Individual carries a defective allele and has an intermediate condition. Some cells become sickled under extreme conditions.

$Hb^S Hb^S$
Homozygote

Individual has two copies of the defective hemoglobin allele and has sickle-cell disease.

Codominance occurs when a heterozygote displays characteristics of both homozygous parents.

FIGURE 7-18 With codominance, a heterozygous individual shows features of both alleles.

Family Resemblance   Genetics and Chance   Translation of Genotypes   Gene Linkage

of pigment. Ultimately, the intensity of pigmentation just depends on the amount of pigment chemical made by the flower-color gene.

A second situation in which complete dominance is not observed is called **codominance,** in which the heterozygote displays characteristics of both homozygotes. In this situation, the alleles can be thought of as codominant because neither masks the effect of the other. An example of codominance occurs with sickle-cell disease (**FIGURE 7-18**), a condition in which individuals produce defective red blood cells that change their shape, becoming sickle-shaped, when they lose the oxygen they carry. The defective blood cells cannot effectively transport oxygen to tissues, and they accumulate in blood vessels, causing extreme pain. Individuals with sickle-cell disease suffer shortness of breath and numerous other problems that lead to a significantly reduced lifespan.

Sickle-cell disease seems, at first inspection, to be a condition seen only in individuals with two copies of the sickle-cell allele for hemoglobin, Hb^S; recall that hemoglobin is the oxygen-carrying molecule in red blood cells. Individuals with two copies of the normal hemoglobin allele, $Hb^A Hb^A$, and those who are heterozygous, $Hb^S Hb^A$, seem to be normal. However, closer observation reveals that heterozygotes—who are described as having "sickle-cell trait"—carry many red blood cells that become sickled under extreme conditions of physical exertion or low atmospheric oxygen. In other words, heterozygous individuals show an intermediate phenotype between the two homozygotes.

TAKE-HOME MESSAGE 7·9

Sometimes the effects of both alleles in a heterozygous genotype are evident in the phenotype. With incomplete dominance, a heterozygote appears to be intermediate between the two homozygotes. With codominance, a heterozygote displays characteristics of both homozygotes.

7·10

What's your blood type? Some genes have more than two alleles.

What is your blood type? It can be O, A, B, or AB. Each of these different blood types indicates something about the physical characteristics of your red blood cells and has implications for blood transfusions—both giving and receiving blood. The blood groups are also interesting from a genetic perspective, because they illustrate a case of **multiple allelism,** in which a single gene has more than two alleles. Each individual still carries only two alleles—one from the mother and one from the father. But if you survey all of the alleles present in the population, you will find more than just two alleles.

Inheritance of the ABO blood groups provides the simplest example of multiple allelism, because there are only three alleles. We can call these alleles *A, B,* and *O.* The *A* and *B* alleles are both completely dominant to *O,* so individuals are considered to have blood type A whether they have the genotype *AA* or *AO* (**FIGURE 7-19**). Similarly, an individual with the genotype *BB* or *BO* is considered to have blood type B. If you carry two copies of the *O* allele, you are blood type

MULTIPLE ALLELISM: BLOOD TYPE

Multiple allelism occurs when there are three or more alleles for a gene within a population. An individual still inherits only two alleles—one from each parent.

Gene that determines blood type

Three alleles possible:
■ *A* (dominant to *O* and codominant with *B*)
■ *B* (dominant to *O* and codominant with *A*)
□ *O* (recessive to *A* and *B*)

Type A — A A or A O
Type B — B B or B O
Type AB — A B
Type O — O O

Red blood cells have 6 different genotypes (*AA, AO, BB, BO, AB,* and *OO*). These genotypes result in 4 different phenotypes (type A, type B, type AB, and type O).

FIGURE 7-19 Multiple allelism. There are three different alleles—*A, B,* and *O*—for blood type.

BLOOD TYPE, ANTIGENS, AND ANTIBODIES

Antigens are chemicals on the surface of some cells. They act as signposts that tell the immune system whether the cell belongs in the body. Antibodies are immune system molecules that attack cells with foreign antigens.

Type A — A antigens / B antibodies

Type B — B antigens / A antibodies

Type AB — A and B antigens / Neither A nor B antibodies

Type O — Neither A nor B antigens / A and B antibodies

FIGURE 7-20 Friend or foe? Antigens are signposts that tell the immune system whether or not a cell belongs in the body.

O. The *A* and *B* alleles are codominant with each other, so the genotype *AB* gives rise to blood type AB. Consequently, with these three alleles in the population, individuals can be one of four different blood types: A, B, AB, and O.

What are the phenotypes of these alleles? An individual's blood type alleles carry instructions that direct construction of a specific set of chemicals, called *antigens,* that protrude from every red blood cell. Antigens are proteins that jut from the surface of a cell and can "turn on" its defenses against foreign invaders. The *A* allele directs the production of A antigens all over the surface of red blood cells. Similarly, the *B* allele directs

the production of B antigens on all red blood cells. The *O* allele does not code for either type of antigen. This means that individuals with blood type AB have red blood cells with both A and B antigens, while individuals with blood type O have red blood cells with a "smoother" surface because they produce no A or B antigens (**FIGURE 7-20**).

Antigens on red blood cells play a role in the body's disease-fighting immune system. Antigens are like signposts, telling the immune system whether a cell belongs in the body or not. If a red blood cell with the wrong antigens were to enter your bloodstream, your immune system would recognize it as a

THE SCIENCE BEHIND BLOOD DONATION

BLOOD TYPE	CAN DONATE TO	CAN RECEIVE FROM
Type A • Has A antigens • Produces antibodies that attack B antigens	Type A Type AB	Type A Type O
Type B • Has B antigens • Produces antibodies that attack A antigens	Type B Type AB	Type B Type O
Type AB • Has A and B antigens • Produces neither A nor B antibodies • Universal recipient	Type AB	Type A Type B Type AB Type O
Type O • Has neither A nor B antigens • Produces antibodies that attack A and B antigens • Universal donor	Type A Type B Type AB Type O	Type O

Individuals with type O blood are universal donors. Individuals with type AB are universal recipients.

FIGURE 7-21 Mapping blood compatibility. Individuals with type O blood are said to be universal donors, and individuals with type AB blood are universal recipients.

Family Resemblance   Genetics and Chance   Translation of Genotypes   Gene Linkage

foreign invader and destroy it. Such an attack is initiated by molecules in the bloodstream called *antibodies*, which attack only foreign antigens. Individuals with only A antigens on their red blood cells produce antibodies that attack B antigens. If they encounter a red blood cell with B antigens, they attack it. Such an immune response can lead to destruction of red blood cells, low blood pressure, and even death. Under normal circumstances, antibodies would not encounter a red blood cell with foreign antigens—unless, that is, red blood cells with foreign antigens are accidentally injected into the person's bloodstream in a transfusion.

Q Why are people with type O blood considered "universal donors"? Why are those with type AB considered "universal acceptors"?

Individuals with only B antigens on their red blood cells produce antibodies that attack A antigens. Individuals with blood type O, who have neither A nor B antigens on their red blood cells, produce antibodies that attack both A and B antigens. Individuals with blood type AB don't produce either type of antibody (or else they would have antibodies that attacked their own blood cells). From this information, we can deduce which blood types can be used in transfusions. This is shown in FIGURE 7-21.

Another marker on the surface of red blood cells is the Rh blood group marker. This marker results from the effects of a single gene with two alleles. Individuals' red blood cells carry the Rh cell surface marker if they have one or two copies of the dominant Rh marker allele. This is then noted in their blood type as a "+," or "positive," in addition to their ABO blood type, as in "O-positive" or "B-negative." Individuals who have two copies of the recessive allele for this gene do not have any Rh markers, and they are described as "negative," as in "O-negative" or "A-negative." If individuals who are Rh-negative are exposed to Rh-positive blood, their immune system attacks the Rh antigens as foreign invaders—an immune response that can vary from mild, and passing unnoticed, to severe, which can lead to death. Thus, the Rh marker further restricts whose blood a person can receive in a transfusion.

Beyond the ABO and the Rh marker groups, there are many, many genes with multiple alleles—a dozen or even more alleles in some cases. In fact, one gene for eye color in fruit flies has more than 1,000 different alleles!

TAKE-HOME MESSAGE 7·10

In multiple allelism, a single gene has more than two alleles. Each individual still carries only two alleles, but in the population, more than just two alleles occur. This is the case for the ABO blood groups in humans.

7·11 ----------------------------------

Multigene traits: how are continuously varying traits such as height influenced by genes?

When babies are born, the parents are often curious about how tall their child will grow to be. Old wives' tales suggest a couple of ways for predicting height. If the baby is a boy, they say to add 5 inches to the mother's height and average that with the father's height. If it is a girl, subtract 5 inches from the father's height and average that with the mother's height. Alternatively, the lore says, just take the child's height at two years of age and double it.

While these methods can be surprisingly accurate at making predictions, they don't really help us understand the underlying reason *why* height can be predicted so well. The reason these methods work is that genes play a strong role in influencing height, and so offspring do resemble their parents in measurable ways. But unlike the simple case of Mendel's

peas, where a single gene with two alleles determines the height of the plant, for humans and most animals, adult height is influenced by many different genes. Such a trait is said to be **polygenic.**

Recent research has identified two genes that help determine height in humans. Individuals with the "tall" allele for both genes are taller than those with the tall allele for only one. Individuals with one tall allele, in turn, are taller than people who do not have a tall allele for either gene. The term **additive effects** describes what happens when the effects of alleles from multiple genes all contribute to the ultimate phenotype. Traits such as skin color in humans also work this way. In the case of height genes, one of the genes is on chromosome 15 and codes for an enzyme that converts testosterone to estrogen. This enzyme influences height because estrogen helps bones to fuse at their ends and thus stop growing. In all, the two height genes identified in humans account for only about 2 inches of a person's height, but they exemplify how multiple genes can influence a single characteristic. The variety of heights from very short to very tall, with every height in between, reflects that height is a trait influenced by contributions from multiple genes as well as the environment (FIGURE 7-22).

Many other physical traits are influenced by multiple genes, including eye color and skin color in humans. Eye color, long believed to be controlled by a single gene with a dominant brown-eye allele and a recessive blue-eye allele, now seems to be the result of the interactions of at least two genes and possibly more—a situation that makes more sense, given the

Q Why might computer nerds be more likely to have autistic children?

significant continuous variation in eye color seen among adults.

Many behavioral traits are influenced by multiple genes. The developmental disorder known as autism, for example, seems to be the result of alterations in numerous—perhaps as many as 10 or even 20—different genes. Individuals with autism have difficulty interacting with others, particularly in making emotional connections. Autistic individuals also tend to have narrowly focused and repetitive interests. Autism is notoriously difficult to study, though, because its symptoms are varied, as is their intensity. It turns out that this may be because different combinations of the many genes involved in autism cause different variations of the disorder, much as different alleles for several height genes may work together to produce a variety of heights.

Interestingly, the genes responsible for autism may also influence some desirable characteristics. Behaviors that cause problems when manifest in the extreme may turn out to be beneficial when less extreme, such as unusual abilities of perception, analytical skills, and focus. This idea—called the "geek theory of autism"—was first suspected when an upsurge in cases of autism in children was noticed in areas with large numbers of high-tech workers, including Silicon Valley in California and near Cambridge, England, in the heart of the United Kingdom's high-technology industry. Perhaps alleles for genes that contribute to making individuals good at computer programming and solving complex technology problems may produce autism when an individual carries too many of them.

TAKE-HOME MESSAGE 7·11

Many traits, including continuously varying traits such as height, eye color, and skin color, are influenced by multiple genes.

POLYGENIC TRAITS

FIGURE 7-22 From many genes, one trait. Height and skin color are multigene traits.

Family Resemblance   Genetics and Chance   Translation of Genotypes   Gene Linkage

Sometimes one gene influences multiple traits.

Q *What is the benefit of "almost" having sickle-cell disease?*

Just as multiple genes can influence one trait, some individual genes can influence multiple unrelated traits, a phenomenon called **pleiotropy.** In fact, this may be true of nearly all genes. For example, earlier we described the sickle-cell allele for hemoglobin production. This allele causes sickle-cell disease when in the homozygous state, but in the heterozygous state it has the effect of conferring resistance to the parasite that causes malaria. The resistance is due to the fact that the malarial parasite—which lives in red blood cells— cannot survive well in cells that carry the defective version of the hemoglobin gene. Because individuals who are heterozygous for sickle-cell anemia have a significant number of such red blood cells, their bloodstream is just not a hospitable environment for the malarial parasite (**FIGURE 7-23**).

Another example of pleiotropy is the SRY gene. Named for "**s**ex-determining **r**egion on the **Y** chromosome," this gene causes fetal gonads to develop as testes shortly after fertilization. Following the gonads' secretion of testosterone, a cascade of other developmental changes then occurs, such as development of the internal and external male reproductive structures. Ultimately, the SRY gene is responsible for numerous behavioral characteristics as well. These are described further in Chapter 9.

TAKE-HOME MESSAGE 7·12

In pleiotropy, one gene influences multiple unrelated traits. Most, if not all, genes may be pleiotropic.

PLEIOTROPY: ONE GENE, MANY EFFECTS

Pleiotropy occurs when one gene influences multiple, different traits.

Heterozygote for the sickle-cell trait

An allele that causes cells to sickle has two effects:

1 It disrupts red blood cells' oxygen delivery.

2 It causes red blood cells to be inhospitable to malarial parasites.

Someone with no sickled cells does not have sickle-cell anemia but is susceptible to malaria.

Malarial parasite infecting a red blood cell

FIGURE 7-23 **From one gene, multiple traits.** The allele for sickle-cell disease is pleiotropic: it causes red blood cells to form an unusual, sickled shape, and it also provides resistance to malaria.

Why are more men than women color-blind? Sex-linked traits differ in their patterns of expression in males and females.

The patterns of inheritance of most traits do not differ between males and females. When a gene is on an autosome (one of the non-sex chromosomes), both males and females inherit two copies of the gene, one from their mother and one from their father. The likelihood that an individual inherits one genotype versus another does not differ between males and females.

Traits coded for by the sex chromosomes, on the other hand, have different patterns of expression in males and females. One of the most easily observed examples of this phenomenon is red-green color-blindness. On the X chromosome in humans, there is a gene that carries the instructions for producing light-sensitive proteins in the eye that, when produced normally, respond differently to red wavelengths and green wavelengths of light. These proteins make it possible to distinguish between the colors red and green. As long as an individual has at least one functioning copy of the gene carrying the instructions for production of these light-sensitive proteins, he or she produces sufficient amounts of the protein to have normal color vision.

There is a rare allele for this gene, however, that produces a non-functioning version of the protein. Having some of this non-functioning protein is not a problem as long as the person also carries another, normal version of the gene and produces some of the functioning protein. But this poses a problem for men, because they have only one X chromosome. In contrast, females inherit two X chromosomes.

Here's the problem: men only get one chance to inherit the normal version of the gene that codes for red-green color vision. It must be on the X chromosome they inherit from their mother. Women get two chances. Although a woman may inherit the defective allele from one parent, she can still inherit the normal allele from the other parent. As long as she inherits the normal gene from one parent, she will have normal color vision. As we would predict, then, the frequency of red-green color-blindness is significantly greater in males than in females (**FIGURE 7-24**). Approximately 7% to 10% of men exhibit red-green color-blindness, while fewer than 1% of women are red-green color-blind.

Q If a man is color-blind, did he inherit this condition from his mother, his father, or both parents?

While males exhibit sex-linked recessive traits more frequently than do women, the situation is reversed for sex-linked dominant traits. In these cases, because females have two chances to inherit the allele that causes the trait, they are more likely to have the allele and thus exhibit the trait than are males, who have only one chance to inherit the allele.

SEX-LINKED TRAITS: COLOR-BLINDNESS

A sex-linked trait is carried on the X chromosome. Women carry two copies of the X chromosome, while men carry an X chromosome and a Y chromosome.

X

Gene with instructions for light-sensitive proteins within the eye

Two alleles possible:
— **R** (produces functioning light-sensitive proteins)
— **r** (produces defective light-sensitive proteins)

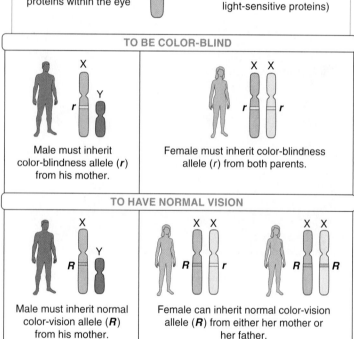

TO BE COLOR-BLIND

Male must inherit color-blindness allele (**r**) from his mother.

Female must inherit color-blindness allele (**r**) from both parents.

TO HAVE NORMAL VISION

Male must inherit normal color-vision allele (**R**) from his mother.

Female can inherit normal color-vision allele (**R**) from either her mother or her father.

FIGURE 7-24 Sex-linked traits such as color-blindness are not expressed equally among males and females.

TAKE-HOME MESSAGE 7·13

The patterns of inheritance of most traits do not differ between males and females. However, when a trait is coded for by a gene on a sex chromosome, such as color vision on the X chromosome, the pattern of expression differs for males and females.

Family Resemblance   Genetics and Chance   Translation of Genotypes   Gene Linkage

Environmental effects: identical twins are not identical.

It is a very serious warning, in boldfaced capital letters: **"PHENYLKETONURICS: CONTAINS PHENYLALANINE."** But it's not next to a skull and crossbones on a glass bottle in a chemistry lab, it's on cans of diet soda intended for human consumption (**FIGURE 7-25**). Most of us either don't notice the warning or ignore it. Still, it's always there, and to some people it's a matter of life and death. What does it mean?

> **Q** Drinking diet soda can be deadly if you carry a single bad gene. What gene is it and why is it so deadly?

At the most fundamental level, this warning is on products with phenylalanine because an organism's phenotype is a product of its genes in combination with the environment. Put another way, nature and nurture interact to produce bodies. In this case, the interaction is related to how humans metabolize a particular chemical called phenylalanine, one of the amino acids. Sometimes our bodies use phenylalanine directly to build proteins, adding it to a growing amino acid chain. At other times, phenylalanine is chemically converted into another amino acid, called tyrosine. Tyrosine is then used to build proteins and in a variety of other ways by the body.

The problem is this: at birth, some people carry two copies of a mutant version of the gene that is supposed to produce the enzyme that converts phenylalanine into tyrosine. The mutant gene produces a malfunctioning enzyme, and none of the body's phenylalanine is converted into tyrosine. Perhaps this doesn't seem as though it should be a big problem, but it is. Little by little, as these individuals consume phenylalanine, it builds up in their bodies because none of it is converted into tyrosine. Eventually, so much accumulates that it reaches toxic levels and poisons the brain, leading to mental retardation and death. The disease is called phenylketonuria, or PKU.

Here's where the warning label comes in. By limiting the amount of phenylalanine in their diet (from diet soda and other sources), individuals with the two mutant alleles for processing phenylalanine can avoid the toxic buildup of the amino acid in their brain. In essence, they modify their environment so that it contains little or no phenylalanine. And in that environment, the PKU mutant alleles are harmless. This example highlights the fact that genes by themselves do not "code" for physical characteristics. Rather, genes interact with the environment to produce physical characteristics. Unless you have information about both the genes *and* the environment, it is not usually possible to know what the phenotype will be.

A less life-threatening illustration of the interaction of genes with the environment can be seen at a pet shop. Siamese cats (as well as the Himalayan rabbit) carry genes that produce dark pigmentation. These genes interact strongly with

> **Q** Could you create a temporarily spotted Siamese cat with an ice pack? Why?

the environment and are heat-sensitive. Dark pigment is produced only in relatively cold areas of the animal's body, while warm areas remain very light in color. This is why the fur on the coldest parts of the body—the ears, paws, tail, and tip of the face—becomes the darkest, while the fur on the rest of the body remains cream-colored or white (**FIGURE 7-26**). For Siamese cats living in cold climates that spend a lot of time outside, it's interesting to notice that they become significantly darker in color during the winter months. Those that lounge indoors all winter remain lighter in color.

There are thousands of other cases in which genes' interactions with the environment influence their ultimate effects in the body. The very fact that identical twins—who inherit exactly the same set of alleles at the time of

FIGURE 7-25 Change the environment, "cure" the disease. The ingestion of phenylalanine in diet soda can be fatal to individuals with the inherited condition phenylketonuria (PKU).

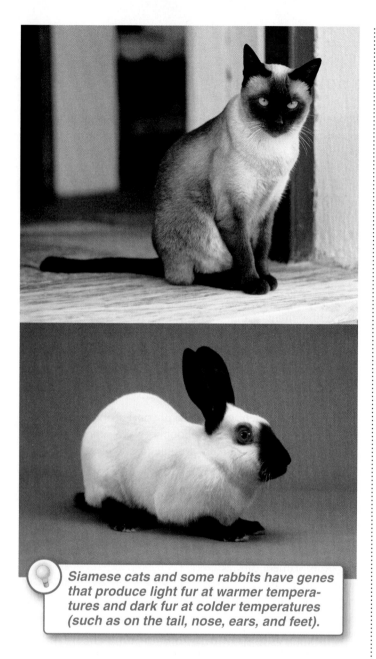

Siamese cats and some rabbits have genes that produce light fur at warmer temperatures and dark fur at colder temperatures (such as on the tail, nose, ears, and feet).

FIGURE 7-26 **Heat-sensitive fur color.** Some pigment-producing genes produce dark fur only under cold conditions. That's why these animals have darker patches of fur on their extremities.

fertilization—don't die on the same day at the same moment and from the same cause, for example, reveals that environmental variation influences the expression of genes. The scope of environmental influences ranges from traits with large and obvious environmental effects, such as body weight and its relationship with caloric intake, to traits such as eye color that are barely influenced by environmental effects, to traits with complex and subtle interactions with the environment, such as intelligence or personality.

Because of the role of environmental factors in influencing phenotypes, DNA is not like a blueprint for a house. This metaphor, though frequently used, is inappropriate, because there is nearly always a significant interaction between the genotype and the environment that influences the exact phenotype produced. The use of this metaphor might even be considered irresponsible, to the extent that it suggests that the phenotype is determined solely by the genotype. If that were the case, there would be no reason to invest in better schools, physical fitness regimens, nutritional monitoring, self-help efforts, or any other process by which individuals or societies try to improve people's lives (i.e., alter phenotypes) by enriching their environment.

Q James Watson, the co-discoverer of DNA, once wrote that when we completed the Human Genome Project, we would have "the complete genetic blueprint of man." Why was he wrong?

TAKE-HOME MESSAGE 7·14

Genotypes are not like blueprints that specify phenotypes. Phenotypes are generally a product of the genotype in combination with the environment.

Family Resemblance   Genetics and Chance   Translation of Genotypes   Gene Linkage

❹ Some genes are linked together.

Different genes influence red hair and freckles, so why are they often inherited together?

7·15

Most traits are passed on as independent features: Mendel's law of independent assortment.

Sometimes you can be right about something for the wrong reason. This happened to Gregor Mendel. He didn't know that genes were carried on chromosomes. He believed that the units of heredity were just free-floating entities within cells. Given this perspective, it made sense to him that the inheritance pattern of one trait wouldn't influence the inheritance of any other trait. He believed that all genes behaved independently.

It helps to consider an example. Earlier in this chapter we saw that the characteristic of having a dimpled chin is caused by a single dominant gene. Imagine that you had a true-breeding population of people with dimpled chins (remember, "true-breeding" for a trait means that all offspring always manifest the trait; in the case of dimples, all have the genotype *DD*). Now suppose that an individual in this population had children with someone from a population of true-breeding, non–dimple-chinned people (everyone has the genotype *dd*). In this case, all of their children would have dimpled chins, but they would be heterozygous (*Dd*), getting a dominant allele from their dimple-chinned parent and a recessive allele from the other parent. If two of those heterozygotes then had children together, they would produce three-quarters dimpled

and one-quarter non-dimpled children, with genotypes in the ratio of ¼ *DD*, ½ *Dd,* and ¼ *dd*. That is just what Mendel observed for traits in pea plants.

But what if we concurrently observed another characteristic of these people? Suppose the original true-breeding population of dimpled-chin people (*DD*) was also true-breeding for albinism (all *aa*), a disorder caused by carrying two recessive alleles for a single gene. And suppose that individuals from the normal-chinned population (*dd*) were also true-breeding for normal pigmentation (all *AA*). The question is, do the alleles an individual inherits for the dimpled-chin trait influence which alleles that individual inherits for pigmentation? And the answer is that, in most cases, they do not (FIGURE 7-27). Rather, the first cross of a dimple-chinned albino with a normal-chinned, normally pigmented individual would result in a child heterozygous for both traits, expressing each of the dominant traits. Phenotypically, this and all subsequent offspring would be dimple-chinned and normally pigmented. Then, in a mating between two of these doubly heterozygous individuals, three-quarters of their children would have the dominant trait and one-quarter would have the recessive trait,

Mendel's law of independent assortment states that one trait does not influence the inheritance of another trait.

FIGURE 7-27 **Independent assortment of genes.**

regardless of which trait you are tallying. In other words, neither trait influences the inheritance pattern for the other trait; all traits are inherited independent of each other. This is known as **Mendel's law of independent assortment.**

In the next section we'll see that, despite Mendel's correct understanding that separate traits are inherited independently, his belief that this happened because all genes just float freely around in the cell was not correct. The genes, as we now know,

are carried on chromosomes. And this fact leads occasionally to some situations in which independent assortment does *not* occur.

TAKE-HOME MESSAGE 7·15

Genes tend to behave independently, such that the inheritance pattern of one trait doesn't usually influence the inheritance of any other trait.

7·16

Red hair and freckles: genes on the same chromosome are sometimes inherited together.

Most redheads have pale skin and freckles. This simple observation is problematic for the law of independent assortment as Mendel imagined it. After all, in his law, he asserted that the

inheritance of one trait does not influence the inheritance of another. But clearly that doesn't happen in this case. Having one of the traits, such as red hair, seems to influence the presence of

Family Resemblance | Genetics and Chance | Translation of Genotypes | Gene Linkage another trait, pale skin (**FIGURE 7-28**). Strictly speaking, Mendel's second law is not true for *every* pair of traits. Sometimes the alleles for two genes are inherited together and expressed almost as a package deal. Let's investigate why that is so.

Q Why do most redheads have pale skin?

For starters, how many traits make up an individual? Of course, this depends on the species. Nonetheless, the number will almost always extend into the thousands, if not tens of thousands. In the case of humans, for example, there are about 25,000 genes in our genome. Yet we have only 23 unique chromosomes (two copies of each). Thus, genes influencing some different traits must be on the same chromosome, maybe even right next to each other. When they are close together, we say that they are **linked genes.** One 2008 study, for example, demonstrated a link between human genes that influence hair color and skin pigmentation.

Why are linked genes inherited together? To answer this, we must revisit the behavior of chromosomes during the production of gametes, discussed in Chapter 6. Imagine that you have a child. Let's focus on your gamete that took part in the fertilization to produce that child. When you made that sperm or egg by meiosis, only one of the two copies of each of your chromosomes ended up in the gamete. It may have been the one from your mother or it may have been the one from your father. In either case, all of the alleles that were on the chromosome from that one parent were passed on as a group to the child that resulted from the fertilization involving that gamete. This process continues generation after generation. The linked alleles never get split up unless, during meiosis, recombination occurs between them, moving one or more to the other chromosome in the pair so that they now become linked with the alleles on that chromosome (**FIGURE 7-29**).

> " Four be the things I'd been better without: Love, curiosity, freckles, and doubt.
>
> —Dorothy Parker, *Inventory*, 1926 "

When alleles are linked closely on the same chromosome, Mendel's second law doesn't hold true. It is very surprising—and was fortunate for Mendel—that of the seven pea traits he studied, none of them were close together on the same chromosome. For this reason they all behaved as if they weren't linked, and in his crosses of different pea plants Mendel never noticed any linked genes.

TAKE-HOME MESSAGE 7·16

Sometimes, having one trait, such as red hair, influences the presence of another trait, such as pale skin. This is because the alleles for two genes are inherited and expressed almost as a package deal when the genes are located close together on the same chromosome.

LINKED GENES

If genes are far apart on a chromosome or are on different chromosomes, the inheritance of one does not influence the inheritance of another.

FIGURE 7-29 Gene linkage. Alleles that are closely linked on the same chromosome will be passed on to offspring in one bundle.

streetBio

Knowledge You Can Use

Where did you come from? Where are you going (to dinner)? World migrations, novelty-seeking, and a single gene that influences both.

On human chromosome 11 there is a gene called D4DR. This gene carries the instructions for building a protein embedded in the membranes of many of your brain cells. The protein is a receptor for the brain chemical dopamine, which is active in several areas of the brain called "pleasure centers" and influences initiative and motivation.

Although all humans have some form of the D4DR gene, there are several different alleles that a person might carry (though only two per person, of course). Each type of allele can be described by the length of the sequence of bases that makes up its DNA: some sequences are short, others are long.

Q: Can a single gene influence your personality? Several studies have reported that individuals carrying one or two long alleles of the D4DR gene, as opposed to two short alleles, are more likely to exhibit certain personality traits, including novelty-seeking and engaging in impulsive behaviors. On personality questionnaires, several differences emerge:

People carrying two short alleles are slightly more likely to be:	People carrying one or more long alleles are slightly more likely to be:
Reflective	Impulsive
Rigid	Exploratory
Stoic	Excitable
Slow-tempered	Quick-tempered
Frugal	Extravagant

Individuals carrying the long alleles even show a preference for spicy foods!

Q: Can a single gene explain why some people are willing to migrate over long distances and others remain rooted in one location? A study of 39 populations from around the world reported a strong relationship between D4DR allele length and how likely a population was to migrate or to remain stationary. Populations in which the long alleles were more common tended to migrate long distances, while those in which the short alleles were more common tended to stay put.

Population	Frequency of the long allele
Chinese and Japanese	5%
African	16%
European	20%
North American	32%
Central American	42%
South American	69%

Q: Do you want to know which alleles you carry for the D4DR gene? As more and more companies offer personalized genetic testing, it may soon be possible for you to learn which alleles you carry for the D4DR gene. Would you want to know? What would you do with this information?

Q: Do you want your employers or insurance companies to know which alleles you carry for the D4DR gene? Carrying one or more long alleles has also been linked with increased use and abuse of alcohol and other drugs. Information about your genotype could put you at risk for genetic discrimination. Moreover, the personality differences influenced by D4DR are relatively slight, they seem to be influenced by numerous other genes, and there is a tremendous variation in personality traits that is unrelated to the D4DR gene. So it's just not clear what it means for you to have short or long D4DR allele.

What can you conclude? Possessing the long D4DR alleles does not necessarily mean you will want to go sky-diving or feel an urge to move out of the country. Conversely, possessing the short D4DR alleles may not discourage you from climbing K2. **But knowledge about the makeup of your D4DR gene might place you at risk for employment or insurance discrimination.**

1 Why do offspring resemble their parents?

Offspring resemble their parents because they inherit genes—instruction sets for biochemical, physical, and behavioral traits, some of which are responsible for diseases—from their parents. Gregor Mendel conducted studies in garden peas that helped us understand how traits are inherited, by applying methodical experimentation and rigorous hypothesis testing and focusing on easily observed and categorized traits in the pea plants. Each parent puts into every sperm or egg it makes a single set of instructions for building a particular trait. The trait observed in an individual depends on the two copies of the gene that it inherits from its parents. It is not always possible to determine an individual's genotype through observation of its phenotype. Much genetic analysis therefore makes use of clever experiments and careful record-keeping to determine organisms' genotypes.

2 Probability and chance play central roles in genetics.

Probability plays a central role in genetics. In segregation, each gamete an individual produces receives only one of the two copies of each gene the individual carries in its other cells; it is impossible to know which allele it will be. Chance plays a role in fertilization, too: all of the sperm or eggs produced by an individual are different from one another, and any one of those gametes may be the gamete involved in fertilization. Test-crosses and pedigrees help scientists, doctors, plant and animal breeders, and prospective parents determine which genes individuals carry and the likelihood that the offspring of two individuals will exhibit a given trait.

3 The translation of genotypes into phenotypes is not a black box.

Sometimes the effects of both alleles in a heterozygous genotype are seen in the phenotype. Many traits are influenced by multiple genes, and many genes may influence multiple unrelated traits. When a trait is coded for by a gene on the sex chromosomes, the pattern of expression differs for males and females. Genotypes are not like blueprints that specify phenotypes. Rather, phenotypes are a product of the genotype in combination with the environment.

4 Some genes are linked together.

Many genes tend to behave independently, such that the inheritance pattern of one trait doesn't usually influence the inheritance of any other trait. When genes are located close together on the same chromosome, however, the alleles for two genes are inherited and expressed almost as a package deal.

KEY TERMS

Check Your Knowledge

1. Most genes come in alternative forms called:
- a) alleles.
- b) heterozygotes.
- c) gametes.
- d) chromosomes.
- e) homozygotes.

2. Traits that are determined by a single gene:
- a) occur in single-celled organisms, but not in humans.
- b) are common in humans.
- c) must occur on the X chromosome, because males have only a single X and so the gene must be able to function in the absence of its homologous allele.
- d) include eye color and skin color.
- e) can have only two alleles.

3. Pea plants were well suited for Mendel's breeding experiments for all of the following reasons except:
- a) Peas exhibit variations in a number of observable characteristics, such as flower color and seed shape.
- b) Mendel and his staff could control the pollination between different pea plants.
- c) It is easy to obtain large numbers of offspring from any given cross.
- d) Many of the characteristics that vary in pea plants are not linked closely on the same chromosome.
- e) Peas have a particularly long generation time.

4. The law of segregation states that:
- a) the transmission of genetic diseases within families is always recessive.
- b) an allele on one chromosome will always segregate from an allele on a different chromosome.
- c) gametes cannot be separate and equal.
- d) the number of chromosomes in a cell is always divisible by 2.
- e) each of two alleles for a given trait segregate into different gametes.

5. In pea plants, purple flower color is dominant to white flower color. If two pea plants that are true-breeding for purple flowers are crossed, in the offspring:
- a) all of the flowers will be purple.
- b) three-quarters of the flowers will be purple and one-quarter will be white.
- c) half of the flowers will be purple and one-quarter will be white.
- d) one-quarter of the flowers will be purple and three-quarters will be white.
- e) all of the flowers will be white.

6. Pea flowers may be purple (*P*) or white (*p*). Pea seeds may be round (*R*) or wrinkled (*r*). What proportion of the offspring from a cross between purple-flowered, round-seeded individuals will have both white flowers and round seeds?
- a) $\frac{1}{16}$
- b) $\frac{1}{2}$
- c) $\frac{3}{16}$
- d) $\frac{3}{4}$
- e) $\frac{9}{16}$

7. The test-cross:
- a) makes it possible to determine the genotype of an individual of unknown genotype that exhibits the dominant version of a trait.
- b) is a cross between an individual whose genotype for a trait is not known and an individual homozygous recessive for the trait.
- c) sometimes requires the production of multiple offspring to reveal the genotype of an individual whose genotype is unknown (but who exhibits the dominant phenotype).
- d) Only a) and b) are correct.
- e) Choices a), b), and c) are correct.

8. Which of the following statements is correct regarding pedigree analysis?
- a) Darkened squares or circles always represent individuals with the trait being traced.
- b) White squares or circles always represent heterozygous individuals.
- c) Horizontal lines connect siblings.
- d) The length of the vertical lines is dependent on the relatedness between two individuals.
- e) Squares represent females, and circles represent males.

9. All of the offspring of a black hen and a white rooster are gray. The simplest explanation for this pattern of inheritance is:
- a) multiple alleles.
- b) codominance.
- c) incomplete dominance.
- d) incomplete heterozygosity.
- e) sex linkage.

10. A woman with type B blood and a man with type A blood could have children with which of the following phenotypes?
- a) AB only
- b) AB or O only
- c) A, B, or O only
- d) A or B only
- e) A, B, AB, or O

11. Which of the following traits shows a polygenic method of inheritance?
- a) flower color in snapdragons
- b) blood type in humans
- c) seed color in peas
- d) sickle-cell disease in humans
- e) skin color in humans

12. The impact of a single gene on more than one characteristic is called:
- a) incomplete dominance.
- b) environmentalism.
- c) balanced polymorphism.
- d) pleiotropy.
- e) codominance.

13. A rare, X-linked dominant condition in humans, congenital generalized hypertrichosis, is marked by excessive hair growth all over a person's body. Which of the following statements about this condition is incorrect?
    a) The son of a woman with this disease has just slightly more than a 50% chance of having this condition.
    b) The daughter of a man with this condition has a 100% chance of having this condition.
    c) The daughter of a woman with this disease has just slightly more than a 50% chance of having this condition.
    d) The son of a woman with this disorder has a 100% chance of having this condition.
    e) Males with this condition almost always have a more severe phenotype, because women usually have at least one normal allele that dilutes the effect of the mutated allele.

14. Individuals carrying two non-functioning alleles for the gene producing the enzyme that converts phenylalanine into tyrosine:
    a) will develop phenylketonuria (PKU).
    b) may or may not develop phenylketonuria, because the environment (diet, in this case) also plays a role in determining their phenotype.
    c) will develop both PKU and Tay-Sachs disease, because the genes causing the two traits are on the same chromosome.
    d) will develop both PKU and Tay-Sachs disease, because one pleiotropic gene influences both traits.
    e) do not need to worry about their phenylalanine consumption, because PKU depends on both genes and the environment.

15. Because of Mendel's law of independent assortment:
    a) individuals with red hair are more likely to have freckles.
    b) skin color and hair texture tend to be inherited together.
    c) we can deduce that genes cannot exist as free-floating entities within a cell but must be carried on chromosomes.
    d) the alleles coding for one trait do not usually influence the inheritance pattern for another trait.
    e) Both a) and b) are correct.

16. Thousands (or even tens of thousands) of different traits make up an individual. For this reason:
    a) in a species with 23 different chromosomes, some traits must be coded for by genes on the same chromosome.
    b) the environment must influence more than half of our traits.
    c) all (or nearly all) genes must be pleiotropic.
    d) knowing an individual's phenotype is not sufficient for determining the person's genotype.
    e) All of the above are correct.

SHORT-ANSWER QUESTIONS

1. Two healthy parents have two children, both of whom have a degenerative muscular disorder called SMA (spinal muscular atrophy).
   a) Explain how SMA must be inherited and what the genotype of the parents must be.
   b) The parents wish to have a third child and are contemplating adopting versus having a biological child. They seek the guidance of a genetic counselor and ask what the probability of having another child with SMA would be. What does the genetic counselor tell them? Explain.

2. Mackenzie's favorite flower is the hydrangea. She plants a large hydrangea bush in her yard and enjoys the big purple blossoms. Her sister, Anna, would like some hydrangeas for her yard as well, so they transplant a small portion of the large bush to Anna's yard. Anna is surprised when, the first time the plant in her yard blooms, the blossoms are pink instead of purple. If the genotype of this plant has not changed, what might have influenced a change in the phenotype?

3. Your friend explains to you that her father has hemophilia, a genetic disorder that affects the process of blood clotting, but her mother does not.
   a) Knowing that the inheritance pattern for this disorder is sex-linked, explain to your friend why it is more common for males to inherit a sex-linked trait.
   b) Using a Punnett square, indicate the probability that a brother of your friend would have hemophilia, assuming that her mother carries two functional copies of the gene involved in this sex-linked recessive disorder. Explain your answer.

See Appendix for answers. For additional study questions, go to www.prep-u.com.

8

Evolution and Natural Selection

Darwin's
dangerous idea

❶ Evolution is an ongoing process.

What big eyes you have! The glass wing butterfly, Cithaerias aurorina, *from Peru.*

8·1 --

We can see evolution occurring right before our eyes.

What's the longest that you've ever gone without food? Twenty-four hours? Thirty-six hours? If you've ever gone hungry that long, you probably felt as if you were going to die of starvation, but humans can survive days, even weeks, without food. In 1981, for example, 27-year-old Bobby Sands, a political prisoner in Northern Ireland, went on a hunger strike. Forsaking all food and consuming only water, he gradually deteriorated and ultimately died—after 66 days without food.

Each hour without food is even more dangerous when you're tiny. How long do you think a fruit fly can last without food? The answer is just under a day—20 hours, give or take a few (**FIGURE 8-1**). Fruit flies can't live long without food because their tiny bodies don't hold very large caloric reserves. But could you breed fruit flies that could live longer than 20 hours on average? Yes.

First, start with a population of 5,000 fruit flies. In biology, the term "population" has a specific meaning: a **population** is a group of organisms of the same species living in a particular geographic region. The region can be a small area like a test tube or a large area like a lake. In your experiment, your population occupies a cage in your laboratory.

From these 5,000 fruit flies, you choose only the "best" fruit flies to start the next generation, with "best" meaning the 20% of flies that can survive the longest without food. Choosing these long-lived, hungry flies can be accomplished in three simple steps:

1. Remove the food from the cage with the population of 5,000 flies.

2. Wait until 80% of the flies starve to death and then put a container of food into the cage.

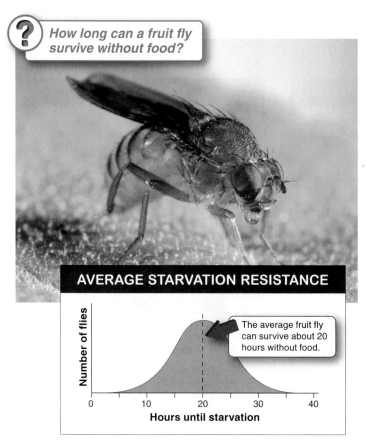

How long can a fruit fly survive without food?

AVERAGE STARVATION RESISTANCE

The average fruit fly can survive about 20 hours without food.

Number of flies

0   10   20   30   40
Hours until starvation

FIGURE 8-1 Starvation resistance in fruit flies.

3. After the surviving flies eat, they'll have the energy to reproduce. When they do, collect the eggs they lay and transfer those eggs to a new cage.

You can now start a new generation populated only by the offspring of those fruit flies able to survive without food for the longest amount of time. When these eggs hatch, do you think the flies in this new generation will live longer than 20 hours without food? When this experiment is done, the flies do indeed show increased resistance to starvation. The average fly in the new generation can live for about 23 hours without food. And again, some of the new flies survive for more than 23 hours, some for less. The surviving flies tend to be those that pack on the fat when the food is available. The new starvation-resistance curve for this second generation is similar to that of the original flies, but shifted to the right a bit so the highest part of the curve occurs at 23 hours, indicating that most flies of the new generation survive for an average of 23 hours without food.

What if you kept repeating these three steps? Start each new generation using eggs only from the fruit flies in the top 20% of starvation resistance. After five generations, would the average starvation-resistance time for a fly from your

population be even higher? In fact, it would. With each new generation, you would see a slight increase, and with the fifth generation, the population's average survival would be about 32 hours.

One benefit to conducting research on fruit flies is that they have very short lives—they reach maturity at about two weeks of age and live for one month, on average. An experiment such as this one can be continued for many generations. After 60 generations of allowing the flies that are "best" at surviving without food to reproduce, how has the population changed? Amazingly, the *average* fly in the resulting population can survive for more than 160 hours without food (**FIGURE 8-2**). In 60 generations, the flies have gone from a starvation resistance of less than a day to one of nearly a week! At the point where the food is removed, the flies in generation 60 are noticeably fatter than the flies in generation 0 were.

What happened? In a word: **evolution.** That is, there was a genetic change in the population of fruit flies living in the cage. Every fly in the generation 60 population, even the fly with the worst starvation resistance, is still more than seven times better at resisting starvation than the best fly in the original population. This evolution is the result of **natural selection.** We'll discuss natural selection in more detail later in this chapter, but, in short, it is the consequence of certain individual organisms in a population being born with characteristics that enable them to survive better and reproduce more than the offspring of other individuals in the population. In this experiment, the 20% of fruit flies that were the most starvation resistant had a huge reproductive advantage over less-resistant flies because they were the *only* flies in the population that survived to reproduce.

This experiment answers a question that is sometimes perceived as complex or controversial: does evolution occur? The answer is an unambiguous *yes.* We can watch it happen in the lab whenever we want. Recall from our discussion of the scientific method (see Chapter 1) that for an experiment's results to be valid, they must be reproducible. To be certain of these results, researchers carried out the starvation-resistance experiment described above five separate times. The results were the same every time.

In this experiment we changed starvation resistance, but what if we used dogs instead of flies and, instead of allowing the most starvation-resistant individuals to reproduce, we allowed only the smallest dogs to reproduce? Would the average body size of individual dogs decrease over time? Yes. What if the experiment were done on rabbits and only the fastest rabbits escaped death from predation by foxes? Would the average running speed in rabbits increase over time? Yes. We know

285

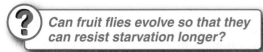

Can fruit flies evolve so that they can resist starvation longer?

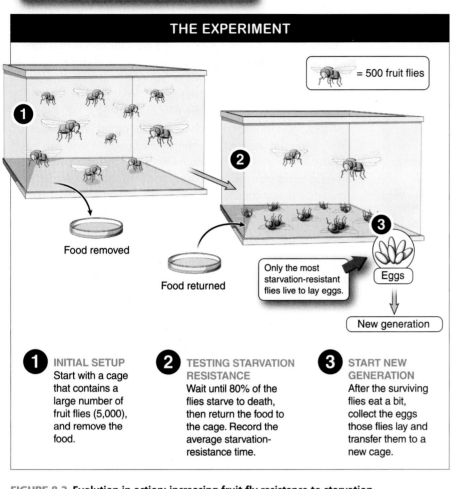

THE EXPERIMENT

= 500 fruit flies

Food removed

Food returned

Only the most starvation-resistant flies live to lay eggs.

Eggs

New generation

1 INITIAL SETUP
Start with a cage that contains a large number of fruit flies (5,000), and remove the food.

2 TESTING STARVATION RESISTANCE
Wait until 80% of the flies starve to death, then return the food to the cage. Record the average starvation-resistance time.

3 START NEW GENERATION
After the surviving flies eat a bit, collect the eggs those flies lay and transfer them to a new cage.

FIGURE 8-2 Evolution in action: increasing fruit fly resistance to starvation.

THE RESULTS

Number of flies

GENERATION 1
Average starvation resistance:
20 HR.

0   40   80   120   160   200
Hours until starvation

Number of flies

GENERATION 2
Average starvation resistance:
23 HR.

0   40   80   120   160   200
Hours until starvation

Experiment continues through 60 generations.

Number of flies

GENERATION 60
Average starvation resistance:
160 HR.

0   40   80   120   160   200
Hours until starvation

Over many generations of natural selection, the population changes! The flies now resist starvation much longer.

these answers because each of these experiments has already been done.

In this chapter we examine how evolution can occur and the type of changes it can cause in a population. We also review the five primary lines of evidence that indicate that evolution and natural selection are processes that help us to clarify all other ideas and facts in biology. Let's begin our investigation with a look at how the idea of evolution by natural selection was developed. This knowledge will help us better understand

why this idea is regarded as one of the most important ideas in human history, why it is sometimes considered a dangerous idea, and why it generates emotional debate.

TAKE-HOME MESSAGE 8·1

The characteristics of individuals in a population can change over time. We can observe such change in nature and can even cause such change to occur.

Evolution Is Ongoing   Darwin's Journey   Four Mechanisms   Natural Selection

8•4

Observing geographic similarities and differences among fossils and living plants and animals, Darwin developed a theory of evolution.

The only book Darwin brought with him on the *Beagle* was Lyell's *Principles of Geology,* which he read again and again and discussed with anyone who would listen. Intrigued by the book's premise that the earth has been, and continues to be, constantly changing, Darwin's mind was ripe for fresh ideas—particularly as he began observing seemingly inexplicable things. How could he explain finding marine fossils high in the Andes Mountains, hundreds of miles from the nearest ocean? The idea that the earth was an ever-changing planet was seeping into Darwin's mind. That idea would serve him well when, a few years into its journey, the *Beagle* stopped at the Galápagos Islands, off the northwest coast of South America.

This group of volcanic islands was home to many unusual species, from giant tortoises to extremely docile lizards that made so little effort to run away that Darwin had to avoid stepping on them. Darwin was particularly intrigued by the wide variety of birds, especially the finches, which seemed dramatically more variable than those he had seen in other locations.

Darwin noticed two important and unexpected patterns on his voyage that would be central to his discovery of a mechanism for evolution. The first involved the finches he collected and donated to the Zoological Society of London. Darwin had assumed that they were the equivalent of tall and short, curly-haired and straight-haired people. That is, he thought that all the finches were of the same species but with different physical characteristics or **traits,** such as body size, beak shape, or feather color. The staff of the Zoological Society, however, could see from the birds' physical differences that the birds were not reproductively compatible and that there were 13 unique species—a different species for every one of the Galápagos Islands that Darwin had visited. Moreover, although the birds

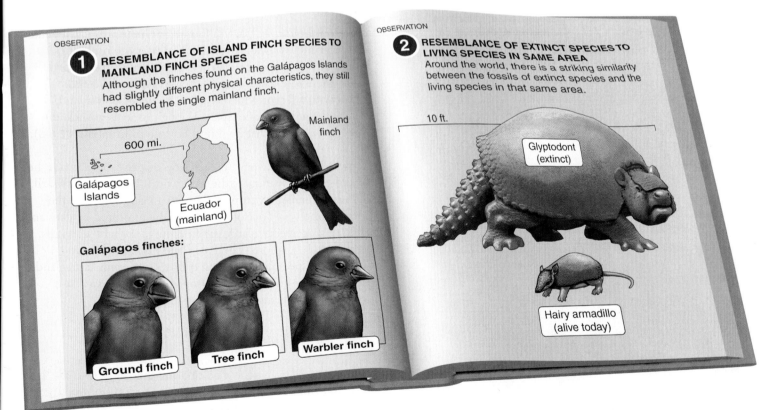

OBSERVATION

1 RESEMBLANCE OF ISLAND FINCH SPECIES TO MAINLAND FINCH SPECIES
Although the finches found on the Galápagos Islands had slightly different physical characteristics, they still resembled the single mainland finch.

600 mi.

Galápagos Islands

Ecuador (mainland)

Mainland finch

Galápagos finches:

Ground finch

Tree finch

Warbler finch

OBSERVATION

2 RESEMBLANCE OF EXTINCT SPECIES TO LIVING SPECIES IN SAME AREA
Around the world, there is a striking similarity between the fossils of extinct species and the living species in that same area.

10 ft.

Glyptodont (extinct)

Hairy armadillo (alive today)

FIGURE 8-7 Darwin observed unexpected patterns.

Evolution Is Ongoing   Darwin's Journey   Four Mechanisms   Natural Selection

2 Darwin journeyed to a new idea.

The Galápagos tortoise, Geochelone elephantopus.

8•2

Before Darwin, most people believed that all species had been created separately and were unchanging.

Charles Darwin grew up in an orderly world. When it came to humans and our place in the world, in the mid-19th century the beliefs of nearly everyone were virtually unchanged from the beliefs of people who had lived more than 2,000 years earlier. Biblical explanations were sufficient for most natural phenomena: the earth was thought to be about 6,000 years old. With the occasional exception of a flood or earthquake or volcanic eruption, the earth was believed to be mostly unchanging. People recognized that organisms existed in groups called species. (In Chapter 10 we discuss in more detail what a species is; for now, we'll just say that individual organisms in a given species can interbreed with each other but not with members of another species.) People also believed that all species, including humans, had been created at the same time and that, once created, they never changed and never died out. This was pretty much what Aristotle had believed more than two millennia earlier.

Before he left this world, Darwin had thrown into question long and dearly held beliefs about the natural world and forever changed our perspective on the origins of humans and our relationship to all other species. He didn't smash the worldview to pieces all at once, though, and he didn't do it by himself (FIGURE 8-3).

In the 1700s and 1800s, scientific thought was advancing at a rapid pace. In 1778, the respected French naturalist Georges Buffon began to shake things up by suggesting that the earth must be about 75,000 years old. He arrived at this age by estimating that 75,000 years was the minimum time required for the planet to cool from a molten state. In the 1790s, Georges Cuvier began to explore the bottoms of coal and slate mines and found fossil remains that had no obvious similarity to any living species. The implications of Cuvier's discoveries were unthinkable, since biblical accounts did not allow for species to be wiped out. Cuvier's publications documented giant fossils (including the Irish elk, the mastodon, and the giant ground sloth) that bore no resemblance to any currently living animals (FIGURE 8-4). These fossils allowed only one explanation: extinction was a fact. Troubling as this observation was for the prevailing worldview, it was only the beginning.

Evidence for Evolution

DARWIN'S INFLUENCES

Georges Buffon
(1707–1788)
Suggested that the earth was much older than previously believed.

Georges Cuvier
(1769–1832)
By documenting fossil discoveries, showed that extinction had occurred.

Jean-Baptiste Lamarck
(1744–1829)
Suggested that living species might change over time.

Charles Lyell
(1797–1875)
Argued that geological forces had gradually shaped the earth and continue to do so.

FIGURE 8-3 **Scientists who shaped Darwin's thinking.**

Not only was it starting to seem that species could disappear from the face of the earth, but the biologist Jean-Baptiste Lamarck suggested in the early 1800s that living species might change over time. Although his ideas about the mechanisms by which this change might occur were wrong—he thought that change came about through the use or disuse of features—his willingness to question previously sacred "truths" contributed to an atmosphere of unfettered scientific thought in which it was possible to challenge convention.

Perhaps the heretical ideas that most inspired Darwin were those of the geologist Charles Lyell. In his 1830 book, *Principles of Geology,* Lyell argued that geological forces had shaped the earth and were continuing to do so, producing mountains and valleys, cliffs and canyons, through gradual but relentless change. This idea that the physical features of the earth were constantly changing would most closely parallel Darwin's idea that the living species of the earth, too, were gradually—but constantly—changing.

TAKE-HOME MESSAGE 8·2

In the 18th and 19th centuries, scientists began to overturn the commonly held beliefs that the earth was only about 6,000 years old and that all species had been created separately and were unchanging. These gradual changes in scientists' beliefs helped shape Charles Darwin's thinking.

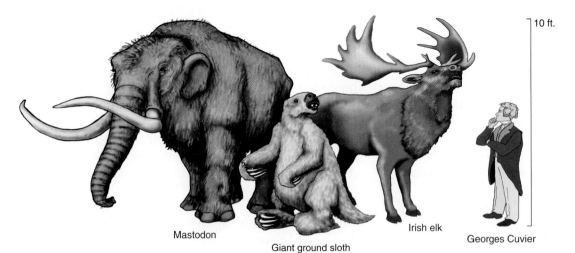

10 ft.

Mastodon

Giant ground sloth

Irish elk

Georges Cuvier

Fossils of organisms no longer found on earth means that extinction occurs!

FIGURE 8-4 **Extinction occurs.** Deep in coal mines, Cuvier discovered the fossilized remains of very large animals no longer found on earth.

8·3

A job on a 'round-the-world survey ship allowed Darwin to indulge and advance his love of nature.

Charles Darwin was born into a wealthy family in England in 1809 and his early life was unremarkable. Charles was never at the top of his class and professed to hate schoolwork. Nonetheless, at 16, he went to the University of Edinburgh to study medicine, following his father's steps. He was bored, though, and finally left at the end of his second year, when the prospect of watching gruesome surgeries (in the days before anesthesia) was more than he could bear.

Upon the abrupt ending of Charles's medical career, his father urged him to pursue the ministry and sent him to study theology at Cambridge University. Although he never felt great inspiration in his theology studies, Darwin was in heaven at Cambridge, where he could pursue his real love, the study of nature. While at Cambridge, he spent most of his time reading nature books, collecting specimens, and developing close relationships with many of his professors.

FIGURE 8-6 **Charles Darwin.** Darwin found his l[...] calling in exploration and st[...] of the natural world.

Shortly after graduation in 1831, Darwin—with significan[...] help from his botany professor—landed his dream job, a position as a "gentleman companion" for the captain of th[...] HMS *Beagle.* The *Beagle,* a 90-foot sailing vessel, was on a five-year, 'round-the-world surveying expedition (FIGURE [...] This job came as a huge relief to the young Darwin, who [...] not really interested in becoming a minister.

Once on the *Beagle,* Darwin found that he didn't actually l[...] sea travel. He was seasick for much of his time on board, [...] where he shared a room that was much tinier than a dorm[...] room. He wrote to his cousin: "I hate every wave of the ocean with a fervor which you . . . can never understand." To avoid nausea, Charles spent as much time as possible on [...] shore. It was there that he found his true calling (FIGURE 8-[...]

Darwin loved fieldwork. At each stop that the *Beagle* made, [...] would eagerly investigate the new worlds he found. In Braz[...] he was enthralled by tropical forests. In Patagonia, near the [...] southernmost part of South America, he explored beaches a[...] cliffs, finding spectacular fossils from huge extinct mammals[...] Elsewhere he explored coral reefs and barnacles, always packing up specimens to bring back to museums in Englan[...] and recording his observations for later use. He was like a schoolboy on permanent summer vacation.

TAKE-HOME MESSAGE 8·3

After initially training in medicine and theology, Charles Darwin was able to focus on studying the natural world when, in 1831, he got a job on a ship conducting a five-year 'round-the-world survey.

DARWIN'S 'ROUND-THE-WORLD VOYAGE

London, England
Atlantic Ocean
Galápagos Islands
Pacific Ocean
Indian Ocean

Galápagos Islands

FIGURE 8-5 **"Like a schoolboy on permanent summer vacation."**
In 1831, Darwin set out on the HMS *Beagle* on a five-year surveying expedition that took him around the globe.

❷ Darwin journeyed to a new idea.

The Galápagos tortoise, Geochelone elephantopus.

8·2

Before Darwin, most people believed that all species had been created separately and were unchanging.

Charles Darwin grew up in an orderly world. When it came to humans and our place in the world, in the mid–19th century the beliefs of nearly everyone were virtually unchanged from the beliefs of people who had lived more than 2,000 years earlier. Biblical explanations were sufficient for most natural phenomena: the earth was thought to be about 6,000 years old. With the occasional exception of a flood or earthquake or volcanic eruption, the earth was believed to be mostly unchanging. People recognized that organisms existed in groups called species. (In Chapter 10 we discuss in more detail what a species is; for now, we'll just say that individual organisms in a given species can interbreed with each other but not with members of another species.) People also believed that all species, including humans, had been created at the same time and that, once created, they never changed and never died out. This was pretty much what Aristotle had believed more than two millennia earlier.

Before he left this world, Darwin had thrown into question long and dearly held beliefs about the natural world and forever changed our perspective on the origins of humans and our relationship to all other species. He didn't smash the worldview to pieces all at once, though, and he didn't do it by himself (**FIGURE 8-3**).

In the 1700s and 1800s, scientific thought was advancing at a rapid pace. In 1778, the respected French naturalist Georges Buffon began to shake things up by suggesting that the earth must be about 75,000 years old. He arrived at this age by estimating that 75,000 years was the minimum time required for the planet to cool from a molten state. In the 1790s, Georges Cuvier began to explore the bottoms of coal and slate mines and found fossil remains that had no obvious similarity to any living species. The implications of Cuvier's discoveries were unthinkable, since biblical accounts did not allow for species to be wiped out. Cuvier's publications documented giant fossils (including the Irish elk, the mastodon, and the giant ground sloth) that bore no resemblance to any currently living animals (**FIGURE 8-4**). These fossils allowed only one explanation: extinction was a fact. Troubling as this observation was for the prevailing worldview, it was only the beginning.

Georges Buffon
(1707–1788)
Suggested that the earth was much older than previously believed.

Georges Cuvier
(1769–1832)
By documenting fossil discoveries, showed that extinction had occurred.

Jean-Baptiste Lamarck
(1744–1829)
Suggested that living species might change over time.

Charles Lyell
(1797–1875)
Argued that geological forces had gradually shaped the earth and continue to do so.

FIGURE 8-3 **Scientists who shaped Darwin's thinking.**

Not only was it starting to seem that species could disappear from the face of the earth, but the biologist Jean-Baptiste Lamarck suggested in the early 1800s that living species might change over time. Although his ideas about the mechanisms by which this change might occur were wrong—he thought that change came about through the use or disuse of features—his willingness to question previously sacred "truths" contributed to an atmosphere of unfettered scientific thought in which it was possible to challenge convention.

Perhaps the heretical ideas that most inspired Darwin were those of the geologist Charles Lyell. In his 1830 book, *Principles of Geology*, Lyell argued that geological forces had shaped the earth and were continuing to do so, producing mountains and valleys, cliffs and canyons, through gradual but relentless change. This idea that the physical features of the earth were constantly changing would most closely parallel Darwin's idea that the living species of the earth, too, were gradually—but constantly—changing.

TAKE-HOME MESSAGE 8·2

In the 18th and 19th centuries, scientists began to overturn the commonly held beliefs that the earth was only about 6,000 years old and that all species had been created separately and were unchanging. These gradual changes in scientists' beliefs helped shape Charles Darwin's thinking.

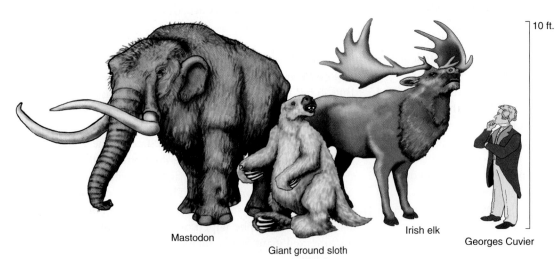

10 ft.

Mastodon

Giant ground sloth

Irish elk

Georges Cuvier

Fossils of organisms no longer found on earth means that extinction occurs!

FIGURE 8-4 **Extinction occurs.** Deep in coal mines, Cuvier discovered the fossilized remains of very large animals no longer found on earth.

Evolution Is Ongoing   Darwin's Journey   Four Mechanisms   Natural Selection

A job on a 'round-the-world survey ship allowed Darwin to indulge and advance his love of nature.

Charles Darwin was born into a wealthy family in England in 1809 and his early life was unremarkable. Charles was never at the top of his class and professed to hate schoolwork. Nonetheless, at 16, he went to the University of Edinburgh to study medicine, following his father's steps. He was bored, though, and finally left at the end of his second year, when the prospect of watching gruesome surgeries (in the days before anesthesia) was more than he could bear.

Upon the abrupt ending of Charles's medical career, his father urged him to pursue the ministry and sent him to study theology at Cambridge University. Although he never felt great inspiration in his theology studies, Darwin was in heaven at Cambridge, where he could pursue his real love, the study of nature. While at Cambridge, he spent most of his time reading nature books, collecting specimens, and developing close relationships with many of his professors.

FIGURE 8-6 **Charles Darwin.** Darwin found his life's calling in exploration and study of the natural world.

Shortly after graduation in 1831, Darwin—with significant help from his botany professor—landed his dream job, a position as a "gentleman companion" for the captain of the HMS *Beagle.* The *Beagle,* a 90-foot sailing vessel, was on a five-year, 'round-the-world surveying expedition (**FIGURE 8-5**). This job came as a huge relief to the young Darwin, who was not really interested in becoming a minister.

Once on the *Beagle,* Darwin found that he didn't actually like sea travel. He was seasick for much of his time on board, where he shared a room that was much tinier than a dorm room. He wrote to his cousin: "I hate every wave of the ocean with a fervor which you . . . can never understand." To avoid nausea, Charles spent as much time as possible on shore. It was there that he found his true calling (**FIGURE 8-6**).

Darwin loved fieldwork. At each stop that the *Beagle* made, he would eagerly investigate the new worlds he found. In Brazil, he was enthralled by tropical forests. In Patagonia, near the southernmost part of South America, he explored beaches and cliffs, finding spectacular fossils from huge extinct mammals. Elsewhere he explored coral reefs and barnacles, always packing up specimens to bring back to museums in England and recording his observations for later use. He was like a schoolboy on permanent summer vacation.

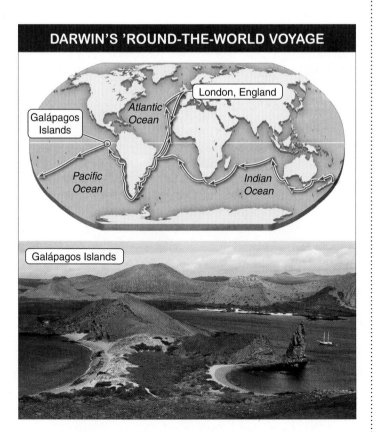

DARWIN'S 'ROUND-THE-WORLD VOYAGE

London, England
Atlantic Ocean
Galápagos Islands
Pacific Ocean
Indian Ocean

Galápagos Islands

FIGURE 8-5 **"Like a schoolboy on permanent summer vacation."** In 1831, Darwin set out on the HMS *Beagle* on a five-year surveying expedition that took him around the globe.

TAKE-HOME MESSAGE 8·3

After initially training in medicine and theology, Charles Darwin was able to focus on studying the natural world when, in 1831, he got a job on a ship conducting a five-year, 'round-the-world survey.

Observing geographic similarities and differences among fossils and living plants and animals, Darwin developed a theory of evolution.

The only book Darwin brought with him on the *Beagle* was Lyell's *Principles of Geology,* which he read again and again and discussed with anyone who would listen. Intrigued by the book's premise that the earth has been, and continues to be, constantly changing, Darwin's mind was ripe for fresh ideas—particularly as he began observing seemingly inexplicable things. How could he explain finding marine fossils high in the Andes Mountains, hundreds of miles from the nearest ocean? The idea that the earth was an ever-changing planet was seeping into Darwin's mind. That idea would serve him well when, a few years into its journey, the *Beagle* stopped at the Galápagos Islands, off the northwest coast of South America.

This group of volcanic islands was home to many unusual species, from giant tortoises to extremely docile lizards that made so little effort to run away that Darwin had to avoid stepping on them. Darwin was particularly intrigued by the wide variety of birds, especially the finches, which seemed dramatically more variable than those he had seen in other locations.

Darwin noticed two important and unexpected patterns on his voyage that would be central to his discovery of a mechanism for evolution. The first involved the finches he collected and donated to the Zoological Society of London. Darwin had assumed that they were the equivalent of tall and short, curly-haired and straight-haired people. That is, he thought that all the finches were of the same species but with different physical characteristics or **traits,** such as body size, beak shape, or feather color. The staff of the Zoological Society, however, could see from the birds' physical differences that the birds were not reproductively compatible and that there were 13 unique species—a different species for every one of the Galápagos Islands that Darwin had visited. Moreover, although the birds

OBSERVATION

1 RESEMBLANCE OF ISLAND FINCH SPECIES TO MAINLAND FINCH SPECIES
Although the finches found on the Galápagos Islands had slightly different physical characteristics, they still resembled the single mainland finch.

600 mi.

Galápagos Islands

Ecuador (mainland)

Mainland finch

Galápagos finches:

Ground finch

Tree finch

Warbler finch

OBSERVATION

2 RESEMBLANCE OF EXTINCT SPECIES TO LIVING SPECIES IN SAME AREA
Around the world, there is a striking similarity between the fossils of extinct species and the living species in that same area.

10 ft.

Glyptodont (extinct)

Hairy armadillo (alive today)

FIGURE 8-7 Darwin observed unexpected patterns.

were different species, they all resembled very closely the single species of finch living on the closest mainland, in Ecuador (**FIGURE 8-7**).

This resemblance seemed a suspicious coincidence to Darwin. Perhaps the island finches resembled the mainland species because they used to be part of the same mainland population. Over time they may have separated and diverged from the original population and gradually formed new—but similar—species. Darwin's logic was reasonable, but his idea flew in the face of all of the scientific thinking of the day.

The second important but unexpected pattern Darwin noted was that, throughout his voyage, at every location there was a striking similarity between the fossils of extinct species and the living species in that same area. In Argentina, for instance, he found some giant fossils from a group of organisms called "glyptodonts." The extinct glyptodonts looked just like armadillos—a species that still flourished in the same area. Or rather, they looked like armadillos on steroids: the average armadillo today is about the size of a house cat and weighs 10 pounds, while the glyptodonts were giants. At 10 feet long and more than 4,000 pounds, they were larger than an automobile (see Figure 8-7).

If glyptodonts had lived in South America in the past, and armadillos were currently living there, why was only one of the species still alive? And why were the glyptodont fossils found only in the same places that modern armadillos lived? Darwin deduced that glyptodonts resembled armadillos because they were their ancestors. Again, it was a logical deduction, but it contradicted the scientific dogma of the day that species were unchanging and extinction did not occur.

The wheels were turning in Darwin's head, but he didn't have a great "Eureka!" moment until several years after his voyage, when he was back in England. He was reading "for amusement" a book called *Essay on the Principle of Population* by the economist Thomas Malthus. Malthus prophesied doom and gloom for populations—including humans—based on his belief that populations had the potential to grow much faster than food supplies could. Darwin speculated that, rather than the future holding certain catastrophe for all, maybe the best individuals would "win" in the ensuing struggle for existence and the worst would "lose." If so, he suddenly saw that "favourable variations would tend to be preserved, and unfavourable ones to be destroyed." Everything was falling into place for Darwin, who wrote: "Here then I had at last got a theory by which to work."

In 1842, Darwin finally hammered out a first draft of his ideas in a 35-page paper, written in pencil, and fleshed it out over the next couple of years. He knew that his idea was important—so important that he wrote a letter to his wife instructing her, in the case of his sudden death (an odd request, given that he was only 35 years old), to give his "sketch" to a competent person, with about $1,000 and all of his books, so that this person could complete it for him.

He also had an inkling that his ideas would rock the world. In a letter to a close friend, he wrote: "At last gleams of light have come, and I am almost convinced (quite contrary to the opinion I started with) that species are not (it is like confessing to murder) immutable." But "confessing to murder" was apparently more than Darwin was ready for. Inexplicably, he put his sketch into a drawer, where it remained for 14 years.

TAKE-HOME MESSAGE 8•4

Darwin noted unexpected patterns among fossils he discovered and living organisms he observed while on the voyage of the *Beagle*. Fossils resembled but were not identical to the living organisms in the same area in which they were found. And finch species on each of the Galápagos Islands differed from each other in small but significant ways. These observations helped Darwin to develop his theory of how species might change over time.

In 1858, after decades of mulling and procrastinating, Darwin published his thoughts on natural selection.

It came in the form of a letter from a young British biologist in the throes of malaria-induced mania in Malaysia. In the letter, Alfred Russel Wallace, who had also read Malthus's book on population growth, laid out a clear description of the process of evolution by natural selection (**FIGURE 8-8**). He asked Darwin to "publish it if you think it is worthy." Crushed at having been scooped, Darwin wrote that "all my originality will be smashed." He had numerous friends among the most prominent scientists of the day, however, and they arranged for a joint presentation of Wallace's and Darwin's work to the Linnaean Society of London. As a result, both Darwin and Wallace are credited for the first description of evolution by natural selection.

Darwin then sprang into action, rapidly putting together his thoughts and observations and completing, 16 months later, a full book. In 1859, he published *Origin of Species* (its full title is *On the Origin of Species by Means of Natural Selection, or the Preservation of Favoured Races in the Struggle for Life*) (**FIGURE 8-9**).

The book was an instant hit, selling out on its first day, provoking public discussion and debate and ultimately causing a wholesale change in the scientific understanding of

WORLDVIEW BEFORE AND AFTER DARWIN

BEFORE:
• Organisms were all put on earth by a creator at the same time.
• Organisms are fixed: no additions, no subtractions.
• Earth is about 6,000 years old.
• Earth is mostly unchanging.

First published in 1859

AFTER:
• Organisms change over time.
• Some organisms have gone extinct.
• Earth is more than 6,000 years old.
• The geology of earth is not constant, but always changing.

FIGURE 8-9 Reconsidering the world. Commonly held ideas before and after Darwin.

evolution. Where once the worldview was of a young earth, populated by unchanging species all created at one time, with no additions or extinctions, now there was a new dynamic view of life on earth: species could and did change over time, and as some species split into new species, others became extinct.

Darwin's theory has proved to be among the most important and enduring contributions in all of science. It has stimulated an unprecedented diversity of theoretical and applied research programs and it has withstood repeated experimental and observational testing. With the background of Darwin's elegant idea in hand, we can now examine its details.

TAKE-HOME MESSAGE 8•5

After putting off publishing his thoughts on natural selection for more than 15 years, Darwin did so only after Alfred Russel Wallace independently came up with the same idea. The two men published a joint presentation on their ideas in 1858, and Darwin published a much more detailed treatment in *Origin of Species* in 1859, sparking wide debate and discussion of natural selection.

FIGURE 8-8 Alfred Russel Wallace. Darwin and Wallace independently identified the process of evolution by natural selection.

③ Four mechanisms can give rise to evolution.

Even with its snowy coloring, this snowshoe hare has been spotted by a cougar.

8•6

Evolution occurs when the allele frequencies in a population change.

Suppose you were put in charge of a large population of tigers in a zoo. Virtually all of them are orange and brown with black stripes. Occasionally, though, unusual all-white tigers are born. This white phenotype is the result of the possession of a rare pair of alleles that suppresses the production of most fur pigment (**FIGURE 8-10**). Because visitors flock to zoos to see the white tigers, you want to increase the proportion of your tiger population that is white. How would you go about doing that?

One possibility is that you could alter the population by trying to produce more animals with the desired white phenotype. In this case, your best strategy would be to try to breed the white tigers with each other. Over time, this would lead to the production of more and more of the white tigers you desire. And as the generations go by (this will take a while, since the generation time for tigers is about eight years), your population will include a higher proportion of white

tigers. When this happens, you will have witnessed evolution, a change in the allele frequencies of the population.

Another way that you might increase the proportion of your white tiger population would be to acquire some white tigers from another zoo, if you could find any. By directly adding white tigers to your population, your population would include a higher proportion of white tigers (you might even trade some of your orange tigers, which would further increase the proportion of your population that is white). By adding white tigers or removing orange tigers, you will again have witnessed evolution in your population.

This example illustrates that evolution doesn't involve changing the genetics or physical features of *individuals*. Individuals do not evolve. Rather, you change the proportion of white fur-pigment alleles in the *population*. An individual tiger has two alleles for any gene, and within a population, the

293

TIGER POPULATION

Allele frequencies:
- Proportion of orange fur-pigment alleles in the population
- Proportion of white fur-pigment alleles in the population

Evolution is a change in the allele frequencies of a population over time. For example, a change in the proportion of pigment alleles in the population of tigers means that evolution has occurred.

FIGURE 8-10 Evolution defined. Evolution is a change in allele frequencies in a population.

proportion of all the alleles coming from any one allele is that allele's frequency. It is helpful to think of each allele as having some "market share" of all of the alleles. In the tiger example, the white alleles initially had little market share, perhaps as small as 1%. Over time, though, they came to make up a larger and larger proportion of the total pigment alleles. And as this happened, evolution occurred.

Any time an allele's market share changes, evolution is occurring. Perhaps the most dramatic way we see this is in acts of predation. When one organism kills another, the dead animal's alleles will no longer be passed on in the population. And if certain alleles make it more likely that an animal will be killed by another—maybe they alter its coloration, or speed—or some other defense, those alleles are likely to lose market share. The converse is equally true, evolution occurs when alleles increase in frequency.

Darwin demonstrated that natural selection could be an efficient mechanism of evolution and a powerful force in adapting populations to their environment. Evolution and

natural selection, however, are not the same thing. Natural selection is one way that evolution can occur, but it is not the only agent of evolutionary change. It is one of four. They are:

1. Mutation

2. Genetic drift

3. Migration

4. Natural selection

Keeping in mind that evolution is genetic change in a population, we'll now explore each of these four forces that are capable of causing such genetic changes.

TAKE-HOME MESSAGE 8·6

Evolution is a change in allele frequencies within a population. It can occur by four different mechanisms: mutation, genetic drift, migration, and natural selection.

Mutation—a direct change in the DNA of an individual—is the ultimate source of all genetic variation.

In describing the first of the four mechanisms of evolution, it is helpful to keep in mind our precise definition of evolution: a change in the allele frequencies found in a population. **Mutation** is an alteration of the base-pair sequence of an individual's DNA, and when this alteration occurs in a section of the DNA that codes for a particular gene, the change in the DNA sequence may change the allele. Say, for example, that a mutation changes one of the two blue-eye alleles that an individual possesses into a brown-eye allele. If this mutation occurs in the sperm- or egg-producing cells it can be passed on to the next generation; the offspring may be born carrying the brown-eye allele. When this happens, the proportion of blue-eye alleles in the population, their market share, is slightly reduced and the market share of brown-eye alleles is slightly increased. Evolution has occurred.

Q *Tanning beds bombard the body with ultra-violet radiation. Can they cause mutations?*

What causes mutations? For one thing, the incredibly complex process of cell division can go awry. Duplication of the 3 *billion* base pairs of a cell's DNA and separation of the new and original versions of the DNA into two new cells, after all, are not simple processes. Mutations also can be induced by environmental phenomena. Most environmentally induced mutations occur when the DNA of cells is exposed to high-energy sources or mutation-inducing chemicals (called **mutagens**). Radioactive isotopes, for example, emit high-energy particles or photons that can disrupt DNA. This is the reason that radiation therapy is both dangerous and effective as a treatment for cancer: the high-energy source kills most of the cells in its vicinity but can also mutate the DNA in other cells. Ultraviolet radiation, even from the sun, can cause mutations as well; sunscreen serves as a barrier, blocking the harmful radiation from reaching cells and the DNA they carry.

Q *Mobile phones release radiation. Can they cause brain tumors?*

Although environmental factors influence the rate at which mutation occurs, they do not generally influence exactly *which* mutations occur. Thus, mutations are random, and whether a mutation happens or not is unrelated to how useful or harmful that mutation might be. Treatment of some agricultural pest with harmful chemicals, for example, may increase the number of mutations occurring but does not increase the likelihood that a particular mutation is beneficial or detrimental.

There are many potential mutation-causing threats in our world today, including tanning beds, for example, which cause individuals to become tan by bombarding the skin with ultraviolet rays similar to those produced by the sun. These rays induce tanning just as exposure to the sun does: by stimulating pigment-producing cells to produce more melanin, a pigment that darkens skin color. But because the ultraviolet rays are high-energy waves, they can cause mutation just as the sun's rays can. Not surprisingly, researchers have found that the greater an individual's exposure to ultraviolet rays, the greater the incidence of mutations and, as a consequence, of skin cancer.

Another potential source of mutation-causing energy is of concern to many more people: mobile phones. The scientific community is still divided on this question, however. Although mobile phones do emit electromagnetic radiation, that radiation is sufficiently low level that it is generally unable to break chemical bonds and therefore cause mutations. The majority of studies have not found any consistent causal relationship between mobile phone usage and any short- or medium-term health hazards.

Since 2004, however, several studies have suggested that further investigation is warranted. A study from 12 laboratories found evidence of DNA damage in cell cultures exposed to levels of radiation similar to those produced by mobile phones. Another study found that individuals who had used mobile phones for 10 years or more had an increased risk of brain tumors. A third study discovered an increased risk of brain tumors and noted that the increased risk was only for the side of the head on which the phone was used. Stay tuned for more information on this important question.

In addition to being an agent of evolutionary change, mutation has another important role that is relevant to all agents of evolution, including natural selection: mutation is

EVOLUTIONARY CHANGE: MUTATION

MECHANISMS OF
EVOLUTION

MUTATION
A mutation can create a new allele in an individual. When this happens, the population experiences a change in its allele frequencies and, consequently, experiences evolution.

FIGURE 8-11 Agents of evolutionary change: mutation.

DNA

Mutagen

Normal base-pair sequence

Mutated base-pair sequence

Normal protein

Mutated protein

Normal phenotype

Mutated phenotype

Despite mutation's vital role in the generation of variation, mutations almost always cause early death or lower the reproductive success of an organism.

DNA may cause the production of a gene product that has never existed before: instead of blue or brown eyes, the mutated gene might code for yellow or red. If such a new allele occurs in the sperm- and egg-producing cells and if it does not significantly reduce an individual's "fitness" (which we'll consider below), the new allele can be passed on to offspring and remain in the population. At some future time, the mutation might even confer higher fitness, in which case natural selection may cause it to increase in frequency. For this reason, mutation is critically necessary if natural selection is to occur: all variation—the raw material for natural selection—must initially come from mutation.

Despite this vital role in the generation of variation, however, nearly all mutations reduce an organism's fitness by causing its early death or by reducing its reproductive success. Suppose that you have written a 10-page paper. If you were to randomly select one letter in the paper and change it to another letter, is the change more likely to make your paper better or worse? The answer is obvious: except in *very* rare situations (can you imagine one?) the change will hurt your paper. Similarly, mutations usually change a normally functioning allele that codes for a normally functioning protein into an allele that codes for a non-functioning protein. Almost inevitably, that outcome reduces an organism's fitness. For this reason, our bodies are built in ways that protect our sperm- or egg-producing DNA with a variety of built-in error-correction mechanisms. As a result, mutations are very, very rare.

TAKE-HOME MESSAGE 8·7

Mutation is an alteration of the base-pair sequence in an individual's DNA. Such an alteration constitutes evolution if it changes an allele that the individual carries. Mutations can be caused by high-energy sources or chemicals in the environment and also can appear spontaneously. Mutation is the only way that new alleles can be created within a population, and so generates the variation on which natural selection can act.

the ultimate source of genetic variation in a population. We saw that a mutation may lead to the conversion of one allele to another that is already found within the population, as in our blue-eye and brown-eye example. More important, though, a mutation may instead create a completely novel allele that codes for the production of a new protein (**FIGURE 8-11**). That is, a change in the base-pair sequence of a person's

Genetic drift is a random change in allele frequencies in a population.

Along with mutation, another evolutionary agent is **genetic drift,** a random change in allele frequencies. Genetic drift is best illustrated with an example. Imagine that in a population there are two alleles for a particular trait, let's say a cleft chin. This is a dominant trait, so individuals with either one or two copies of the dominant allele (CC or Cc) exhibit the cleft chin (FIGURE 8-12). Now suppose that two heterozygous (Cc) people have one child. Which combination of alleles will that child receive? This is impossible to predict because it depends completely upon which sperm fertilizes which egg—the luck of the draw. If the couple's sole child inherited a recessive allele from each parent, would the *population's* allele frequencies change? Yes. After all, there is now another individual in the population, and that individual has two recessive alleles (cc). There are slightly more recessive alleles in the population. And because a change in allele frequencies

has occurred, evolution has happened. It is equally likely that this couple's only child would have received two of the dominant alleles (CC), rather than the recessive alleles. In either case, because a change in allele frequencies has occurred, evolution has happened.

The important factor that distinguishes genetic drift from natural selection is that the change in allele frequencies is not related to the alleles' influence on reproductive success. A cleft chin does not affect an individual's ability to reproduce.

This impact of genetic drift is much greater in small populations than in large populations. In a large population, it is more likely that any one couple having a child with two recessive alleles is likely to be offset by another couple having one child with two dominant alleles. When that occurs, the

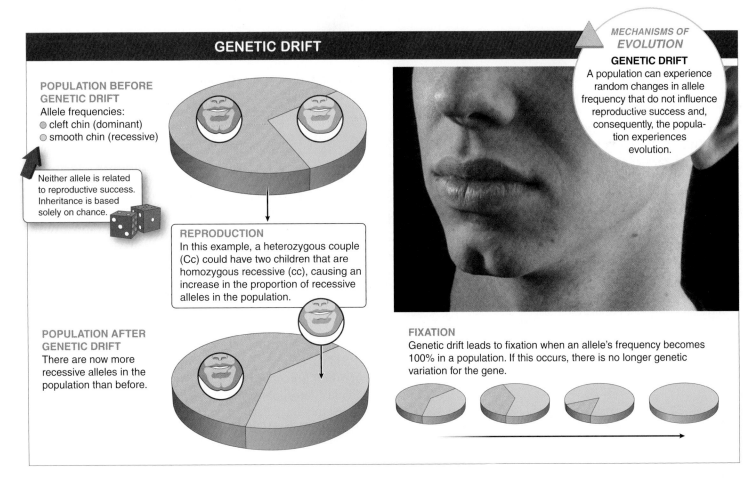

GENETIC DRIFT

POPULATION BEFORE GENETIC DRIFT
Allele frequencies:
● cleft chin (dominant)
○ smooth chin (recessive)

Neither allele is related to reproductive success. Inheritance is based solely on chance.

REPRODUCTION
In this example, a heterozygous couple (Cc) could have two children that are homozygous recessive (cc), causing an increase in the proportion of recessive alleles in the population.

POPULATION AFTER GENETIC DRIFT
There are now more recessive alleles in the population than before.

MECHANISMS OF
EVOLUTION

GENETIC DRIFT
A population can experience random changes in allele frequency that do not influence reproductive success and, consequently, the population experiences evolution.

FIXATION
Genetic drift leads to fixation when an allele's frequency becomes 100% in a population. If this occurs, there is no longer genetic variation for the gene.

FIGURE 8-12 Agents of evolutionary change: genetic drift. Genetic drift has the greatest impact in small populations.

SOURCE POPULATION
Allele frequencies:
- 5 digits per hand (recessive)
- >5 digits per hand (dominant)

A group of individuals may leave a population and become the founding members of a new, isolated population.

NEWLY FOUNDED POPULATION
The new population will be dominated by the genetic features present in the founding members.

MECHANISMS OF EVOLUTION

FOUNDER EFFECT
The founding members of a new population can have different allele frequencies than the original source population and, consequently, the new population experiences evolution.

FIGURE 8-13 One way that genetic drift occurs: founder effect.

net result is that the overall allele frequencies in the population have not changed, and evolution has not occurred.

One of the most important consequences of genetic drift is that it can lead to **fixation** for one allele of a gene in a population (see Figure 8-12). Fixation is said to occur when an allele's frequency in a population reaches 100% (and the frequency of all other alleles of that gene becomes 0%). If this happens, there is no more variability in the population for this gene; all individuals will always produce offspring carrying only that allele (until new alleles arise through mutation). For this reason, genetic drift reduces the genetic variation in a population.

Two special cases of genetic drift, the founder effect and population bottlenecks, are important in the evolution of populations.

Founder Effect   A small number of individuals may leave a population and become the founding members of a new, isolated population. The founder population may have different allele frequencies than the original "source" population, particularly if the founders are a small sample. If this new population has different allele frequencies, there has

Q Why are Amish people more likely than other people to have extra fingers and toes?

been evolution. Because the founding members of the new population will give rise to all subsequent individuals, the new population will be dominated by the genetic features that happened to be present in those founding fathers and mothers. This type of genetic drift is called the **founder effect.**

The Amish population in the United States is believed to have been established by a small number of founders who happened to carry the allele for *polydactyly*—the condition of having extra fingers and toes. As a consequence, today this trait, while rare, occurs much more frequently among the Amish than among the rest of the U.S. population (**FIGURE 8-13**).

Population Bottlenecks   Occasionally, a famine, disease, or rapid environmental change may cause the deaths of a large proportion (sometimes as many as 90% or more) of the individuals in a population. Because the population is quickly reduced to a small fraction of its original size, this reduction is called a **bottleneck.** If the catastrophe is equally likely to strike any member of the population, the remaining members are essentially a random, small sample of the original population. For this reason, the remaining population may not possess the same allele frequencies as the original population.

Evolution Is Ongoing   Darwin's Journey   Four Mechanisms   Natural Selection

BOTTLENECK EFFECT
The surviving members of a catastrophic event can have different allele frequencies than the source population and, consequently, the new population experiences evolution.

Occasionally, famine or disease or rapid environmental change may cause the deaths of a large, random proportion of the individuals in a population.

SOURCE POPULATION

EXTREME AND RAPID ENVIRONMENTAL CHANGE

NEW POPULATION
The new population will be dominated by the genetic features present in the surviving members.

FIGURE 8-14 Another way that genetic drift occurs: bottleneck effect.

All cheetahs living today can trace their ancestry back to a dozen or so individuals that happened to survive a population bottleneck about 10,000 years ago!

Thus, the consequence of such a population bottleneck would be evolution through genetic drift (**FIGURE 8-14**).

Just such a population bottleneck occurred with cheetahs near the end of the last ice age, about 10,000 years ago. Although the cause is unknown—possibly environmental cataclysm or human hunting pressures—it appears that nearly all cheetahs died. And although the population has rebounded, all cheetahs living today can trace their ancestry back to a dozen or so lucky individuals that survived the bottleneck. As a result of this past instance of evolution by genetic drift, there is almost no genetic variation left in the current population of cheetahs. (And, in fact, a cheetah will accept a skin graft from any other cheetah, much as identical twins will from each other.)

TAKE-HOME MESSAGE 8·8

Genetic drift is a random change in allele frequencies within a population, unrelated to the alleles' influence on reproductive success. Genetic drift is a significant agent of evolutionary change primarily in small populations.

Migration into or out of a population may change allele frequencies.

The third agent of evolutionary change is **migration.** Migration, also called **gene flow,** is the movement of some individuals of a species from one population to another (**FIGURE 8-15**). This movement from population to population within a species distinguishes migration from the founder effect, in which individuals migrate to a new habitat, previously unpopulated by that species. If migrating individuals can survive and reproduce in the new population, and if they also carry a different proportion of alleles than the individuals in their new home, then the recipient population experiences a change in allele frequencies and, consequently, experiences

evolution. And because alleles are simultaneously lost from the population that the migrants left, that population, too, will experience a change in its allele frequencies.

> " I was taught that the human brain was the crowning glory of evolution so far, but I think it's a very poor scheme for survival.
>
> —Kurt Vonnegut, Jr., American writer "

Gene flow between two populations is influenced by the mobility of the organisms. Because they usually are more mobile than plants, animals tend to have greater migratory potential than plants (although pollen and seeds can sometimes be transported great distances). Gene flow can also be influenced by a variety of barriers, such as mountains or rivers. And, as we'll see in Chapter 16, human activities, too, can dramatically influence gene flow, whether it is a new highway restricting movement of organisms or boats, trains, and planes that drastically increase the migratory potential of organisms.

MIGRATION (GENE FLOW)

MECHANISMS OF EVOLUTION

MIGRATION
After a group of individuals migrates from one population to another, both populations can experience a change in their allele frequencies and, consequently, experience evolution.

1 BEFORE MIGRATION
Two populations of the same species exist in separate locations. In this example, they are separated by a mountain range.

Population 1   Population 2

2 MIGRATION
A group of individuals from Population 1 migrates over the mountain range.

Population 1   Population 2

3 AFTER MIGRATION
The migrating individuals are able to survive and reproduce in the new population.

Population 1   Population 2

TAKE-HOME MESSAGE 8·9

Migration, or gene flow, leads to a change in allele frequencies in a population as individuals move into or out of the population.

FIGURE 8-15 Agents of evolutionary change: migration (gene flow).

Evolution Is Ongoing   Darwin's Journey   Four Mechanisms   Natural Selection

When three simple conditions are satisfied, evolution by natural selection is occurring.

The fourth agent of evolutionary change is natural selection. This is the mechanism that Darwin identified in *The Origin of Species,* in which he noted that three conditions are necessary for natural selection to occur:

1. There must be variation for the particular trait within a population.

2. That variation must be inheritable (that is, it must be capable of being passed on from parents to their offspring).

3. Individuals with one version of the trait must produce more offspring than those with a different version of the trait.

Let's examine these conditions more closely.

Condition 1: Variation for a Trait   Close your eyes and imagine a dog. What does the dog look like? Chances are, if 50 people were asked this question, we would get descriptions of 50 different dogs (**FIGURE 8-16**): Chihuahuas, Great Danes, sheepdogs, greyhounds, spaniels, and more. Some are big, some are small. Some have short hair, some long. They vary in just about every way that you can imagine. Similarly, if 50 people were to imagine a human face, a similarly broad range of images would pop into their heads. Variation is all around us. Beyond making the world an interesting place to live in, variation serves another purpose: it is the raw material on which evolution feeds.

Variation is not limited to physical features such as fur color and texture or face shape and eye color. Organisms vary in

physiological and biochemical ways, too. Some people can quickly and efficiently metabolize alcohol, for example. Others find themselves violently ill soon after sipping a glass of wine. The same goes for digesting milk. Similarly, we vary in our susceptibility to poison ivy or diseases such as malaria. Behavioral variation—from temperament to learning abilities to interpersonal skills—is dramatic and widespread, too. So impressed was Darwin with the variation he observed throughout the world, he devoted the first two chapters of *The Origin of Species* to a discussion of variation in nature and among domesticated animals. Darwin knew that the variation he saw all around him was an essential component of evolution by natural selection. He considered it the first of three conditions necessary for natural selection.

Condition 2: Heritability   The second condition that Darwin identified as necessary for natural selection was a no more complex discovery than the first: for natural selection to act on a trait within a population, offspring must inherit the trait from their parents. Although inheritance was poorly understood in Darwin's time, it was not hard to see that, for many traits, offspring look more like their parents than like some other random individual in the population (**FIGURE 8-17**). Animal breeders had long known that the fastest horses generally gave birth to the fastest horses. Farmers, too, understood that the plants with the highest productivity generally produced seeds from which highly productive plants grew. And everyone knew that children resembled their parents, from their appearance to their behavior to their temperament. It was enough to know that

FIGURE 8-16 Necessary conditions for natural selection: 1. Variation for a trait.

Evidence for Evolution

FIGURE 8-17 Necessary conditions for natural selection: 2. Heritability. Goldie Hawn and daughter Kate Hudson resemble each other.

this similarity between offspring and parents existed—it was not necessary to understand how it occurred or to be able to quantify just how great the similarity was. We call the transmission of traits from parents to their children through genetic information **inheritance** or **heritability.**

Condition 3: Differential Reproductive Success

It would be nice to say that Darwin made a stunning and insightful discovery for the third of the three conditions necessary for natural selection to occur, but he didn't. Rather, he derived the third condition for natural selection from three fairly simple observations that he made:

1. More organisms are born than can survive.

2. Organisms are continually struggling for existence.

3. Some organisms are more likely to win this struggle and survive and reproduce. The struggle does not always involve direct physical contact, but in a world of scarce, limited resources, finding food or shelter is a zero-sum game: if one organism is feasting, another is more likely to be starving.

This three-part observation (which goes by the complex name **differential reproductive success**) led Darwin to his third condition for natural selection: from all the variation existing in a population, individuals with traits most suited to reproduction in their environment generally leave more offspring than do individuals with other traits (**FIGURE 8-18**). For example, in the experiments selecting for fruit flies that can survive for a long time without food, we saw that the fruit flies that inherit the ability to pack on the fat when food is available end up leaving more offspring than those inheriting a poor ability to pad their little fruit fly frames with fat deposits. The portly fruit flies have greater reproductive success than other individuals.

Q Most agricultural pests evolve resistance to pesticides. How does this happen?

That's it. Natural selection—certainly one of the most influential and far-reaching ideas in the history of science—occurs when three basic conditions are met (**FIGURE 8-19**):

1. Variation for a trait

2. Heritability of that trait

3. Differential reproductive success based on that trait

When these three conditions are satisfied, evolution by natural selection is occurring. It's nothing more and nothing less; no mysterious black box is required. Over time, the traits that lead some organisms to have greater reproductive success than others will increase in frequency in a population while traits that reduce reproductive success will fade away.

When it comes to some traits, the reason they confer greater reproductive success is that they make the individual exhibiting them more attractive to the opposite sex. Such traits—including the brightly colored tail feathers seen in male peacocks, the large antlers seen in male red deer, and a variety of other "ornaments" that increase an individual's status or appeal—because they satisfy the three conditions for natural selection, increase in frequency just the same, in a phenomenon referred to as **sexual selection.**

Another way of looking at natural selection is to focus not on the winners (the individuals who are producing more offspring), but on the

The tiniest dog in a litter has reduced differential reproductive success. Its more robust siblings prevent access to the food it needs to grow and thrive.

FIGURE 8-18 Necessary conditions for natural selection: 3. Differential reproductive success.

Evolution Is Ongoing   Darwin's Journey   Four Mechanisms   Natural Selection

EVOLUTION BY NATURAL SELECTION: A SUMMARY

1 **VARIATION FOR A TRAIT**
Different traits are present in individuals of the same species.

+

2 **HERITABILITY**
Traits are passed on from parents to their children.

+

3 **DIFFERENTIAL REPRODUCTIVE SUCCESS**
In a population, individuals with traits most suited to reproduction in their environment generally leave more offspring than individuals with other traits.

MECHANISMS OF
EVOLUTION

NATURAL SELECTION
When these three conditions are satisfied, the population's allele frequencies change and, consequently, evolution by natural selection occurs.

FIGURE 8-19 **Agents of evolutionary change: natural selection.**

NATURAL SELECTION IN NATURE

1 **VARIATION FOR A TRAIT**
Running speed in rabbits can vary from one individual to the next.

Speed ⟶

2 **HERITABILITY**
The trait of running speed is passed on from parents to their offspring.

3 **DIFFERENTIAL REPRODUCTIVE SUCCESS**
In a population, rabbits with slower running speeds are eaten by the fox and their traits are not passed on to the next generation.

losers. Natural selection can be turned on its head and viewed as the elimination of some heritable traits from a population. If you carry a trait that makes you a slower-running rabbit, for example, you are more likely to be eaten by the fox (**FIGURE 8-20**). If running speed is a heritable trait (and it is), the next generation in a population contains fewer slow rabbits. Over time, the population is changed by natural selection. It evolves.

One of Darwin's contemporaries, Thomas Huxley, supposedly cursed himself when he first read *The Origin of Species,* saying that he couldn't believe that he didn't figure it out on his own. Each of the three basic conditions is indeed simple and obvious. The brilliant deduction, though, was to put the three together and appreciate the consequences.

TAKE-HOME MESSAGE 8·10

Natural selection is a mechanism of evolution that occurs when there is heritable variation for a trait and individuals with one version of the trait have greater reproductive success than do individuals with a different version of the trait. It can also be thought of as the elimination from a population of alleles that reduce the reproductive rate of individuals carrying those alleles relative to the reproductive rate of individuals who do not carry them.

FIGURE 8-20 **Removing the losers.** Natural selection can be thought of as the elimination from a population of traits that confer poor reproductive success.

Evidence for Evolution

4 Through natural selection, populations of organisms can become adapted to their environment.

Adapted to the hot, dry desert: A camel caravan in the Sahara.

8·11

Traits causing some individuals to have more offspring than others become more prevalent in the population.

"Survival of the fittest." This is perhaps the most famous phrase from Darwin's most famous book. Unfortunately, it can be a confusing and even misleading phrase. Taken literally, it seems to be a circular phrase: those organisms that survive *must* be the fittest if being fit is defined as the ability to survive. But we will see that when using the word "fitness" in discussing evolution, it has little to do with an organism's ability to survive or to its physical strength or health. Rather, fitness has everything to do with an organism's reproductive success.

Here's an interesting side note: the phrase "survival of the fittest" was coined not by Darwin but by Herbert Spencer, an influential sociologist and philosopher. Moreover, the phrase did not appear in *Origin of Species* when Darwin first published it. It wasn't until the fifth edition, 10 years later, that he used the phrase in describing natural selection.

Before we see how fitness affects natural selection and a population's adaptation to its environment, let's define fitness in a precise way. **Fitness** is a measure of the relative amount of reproduction of an individual with a particular phenotype, as compared with the reproductive output of individuals with alternative phenotypes. The idea is much easier to understand with an example. Suppose there are two fruit flies. One fly carries the genes for a version of a trait that allows it to survive a long time without food. The other fly has the genes for a different version of the trait that allows it to survive only a short while without food. Which fly has the greater fitness? If the environment is one in which there are long periods of time without food, such as in the experiment described at the beginning of the chapter, the fly that can live a long time without food is likely to produce more offspring than the other fly and so over the course of its life has greater fitness. The alleles carried by an individual with high fitness will

Evolution Is Ongoing   Darwin's Journey   Four Mechanisms   ▼   Natural Selection increase their market share in a population over time and the population will evolve.

There are three important elements to an organism's fitness.

1. An individual's fitness is measured relative to other genotypes or phenotypes in the population. Those traits that confer on an individual the highest fitness will generally increase their market share in a population, and their increase will always come at the expense of the market share of alternative traits that confer lower fitness.

2. Fitness depends on the specific environment in which the organism lives. The fitness value of having one trait versus another depends on the environment an organism finds itself in. A sand-colored mouse living in a beach habitat will be more fit than a chocolate-colored mouse. But that same sand-colored mouse will practically call out to potential predators when living in the darker brush away from the beach. An organism's fitness, although genetically based, is not fixed in stone and unchanging—it can change over time and across habitats (**FIGURE 8-21**).

3. Fitness depends on an organism's reproductive success compared with other organisms in the population. If you carry an allele that gives you the trait of surviving for 200 years, but that allele also causes you to be sterile and incapable of producing offspring, your fitness is zero; that allele will never be passed down to future generations and its market share will soon be zero. On the other hand, if you inherit an allele that gives you a trait that causes you to die at half the age of everyone else, but also causes you to have twice as many offspring as the average while you are alive, your fitness is increased. It is reproductive success that is most important in determining

Q *"Survival of the fittest" is a misnomer. Why?*

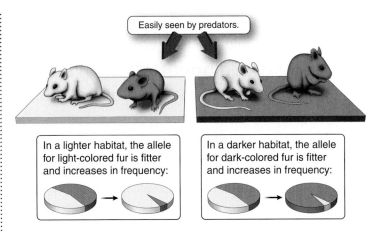

FIGURE 8-21 An organism's fitness depends on the environment in which it lives.

whether particular traits increase their market share.

"Survival of the fittest" is a misleading phrase since it is the individuals that have the greatest *reproductive output* that are the most fit in any population. It becomes a more meaningful phrase if we consider it as a description of the fact that the alleles that increase an individual's fitness will "survive" in a population more than the alleles that decrease an individual's fitness.

TAKE-HOME MESSAGE 8·11

Fitness is a measure of the relative amount of reproduction of an individual with a particular phenotype, as compared with the reproductive output of individuals with alternative phenotypes. An individual's fitness can vary, depending on the environment in which the individual lives.

Organisms in a population can become better matched to their environment through natural selection.

If you put a group of humans on the moon, will they flourish? If you take a shark from the ocean and put it in your swimming pool, will it survive? In both cases, the answer is *no*. Organisms are rarely successful when put into novel environments. And the stranger the new environment, the less likely it is that the transplanted organism will survive. Why is that?

As Darwin noted over and over again during his travels, the organisms that possess traits that allow them to better exploit the environment in which they live will tend to produce more offspring than the organisms with alternative traits. With passing generations, a population will be made up of more and more of these fitter organisms. And, as a consequence, organisms will tend to be increasingly well matched or adapted to their environment.

Adaptation refers both to the process by which organisms become better matched to their environment *and* to the specific features that make an organism more fit. Examples of adaptations abound. Bats have an extremely accurate type of hearing (called echolocation) for navigating and finding food, even in complete darkness. Porcupine quills make porcupines almost impervious to predation (**FIGURE 8-22**). Mosquitoes produce strong chemicals that prevent blood from clotting, so that they can extract blood from other organisms.

FIGURE 8-22 Adaptations increase fitness. Quills are an adaptation that almost completely eliminates the risk of predation for porcupines.

TAKE-HOME MESSAGE 8•12

Adaptation—the process by which organisms become better matched to their environment and the specific features that make an organism more fit—occurs as a result of natural selection.

Natural selection does not lead to perfect organisms.

In Lewis Carroll's *Through the Looking-Glass,* the Red Queen tells Alice that "in this place it takes all the running you can do, to keep in the same place." She might have been speaking about the process of evolution by natural selection. After all, if the least fit individuals are continuously weeded out of the population, we might logically conclude that, eventually, fitness will reach a maximum and all organisms in all populations will be perfectly adapted to their environment. But this never happens. That is where the Red Queen's wisdom comes in.

Consider one of the many clearly documented cases of evolution: that of the beak size of Galápagos finches. Over the course of a multidecade study, the biologists Rosemary and Peter Grant closely monitored the average size of the finches' beaks. They found that the average beak size within a population fluctuated according to the food supply. During dry years—when the finches had to eat large, hard seeds—bigger, stronger beaks became the norm. During wet years, smaller-beaked birds were more successful since there was a surplus of small, softer seeds.

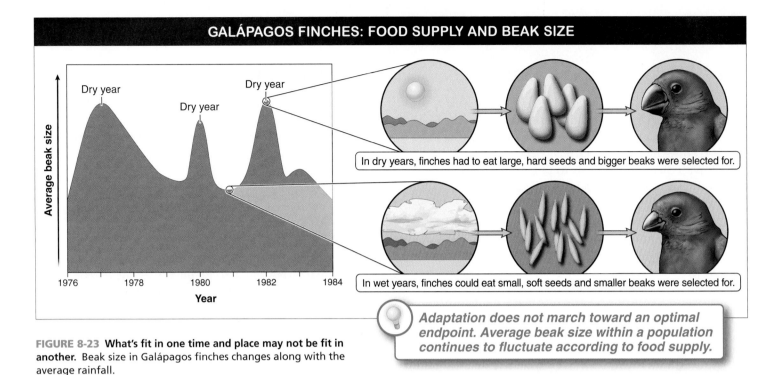

GALÁPAGOS FINCHES: FOOD SUPPLY AND BEAK SIZE

In dry years, finches had to eat large, hard seeds and bigger beaks were selected for.

In wet years, finches could eat small, soft seeds and smaller beaks were selected for.

FIGURE 8-23 What's fit in one time and place may not be fit in another. Beak size in Galápagos finches changes along with the average rainfall.

Adaptation does not march toward an optimal endpoint. Average beak size within a population continues to fluctuate according to food supply.

The ever-changing "average" finch beak illustrates that adaptation does not simply march toward some optimal endpoint (**FIGURE 8-23**). Evolution in general, and natural selection specifically, do not guide organisms toward "better-ness" or perfection. Natural selection is simply a process by which, in each generation, the alleles that cause organisms to have the traits that make them most fit in that environment tend to increase in frequency. If the environment changes, the alleles causing the traits favored by natural selection may change, too.

Q Why doesn't natural selection lead to the production of perfect organisms?

In the next to last paragraph of *Origin of Species,* Darwin wrote: "We may look with some confidence to a secure future of great length. And as natural selection works solely by and for the good of each being, all corporeal [physical] and mental endowments will tend to progress towards perfection." In this passage, he overlooks several different factors that prevent populations from progressing inevitably toward perfection:

1. Environments change quickly. Natural selection may be too slow to adapt the organisms in a population to the constantly moving target that is the environment.

2. Variation is needed as the raw material of selection—remember, it is the first necessary condition for natural selection to occur. If a mutation creating a new, "perfect" version of a gene never occurs, the individuals within a population will never be perfectly adapted. Why, for instance, are there no mammals with wheeled appendages? That might be a great trait, but until the genes for it exist, natural selection cannot increase their market share in the population.

3. There may be multiple different alleles for a trait, each causing an individual to have the same fitness. In this case, each allele represents an equally fit "solution" to the environmental challenges.

TAKE-HOME MESSAGE 8·13

Natural selection does not lead to organisms perfectly adapted to their environment because (1) environments can change more quickly than natural selection can adapt organisms to them; (2) all possible alleles are not produced by mutation; (3) there is not always a single optimum adaptation for a given environment.

8·14

Artificial selection is a special case of natural selection.

In practice, plant and animal breeders understood natural selection before Darwin, they just didn't know that they understood it. Farmers bred crops for maximum yield, and dog, horse, and pigeon fanciers selectively bred the animals with their favorite traits to produce more and more of the offspring with more and more exaggerated versions of the trait.

The artificial selection used by animal breeders and farmers is also natural selection because the three conditions are satisfied, even though the differential reproductive success is being determined by humans and not by nature. Apple growers, for example, use artificial selection to produce the wide variety of apples available: green, yellow, and red, tart and sweet, large and small. What is important is that it is still differential

reproductive success, and the results are no different. It was a stroke of genius for Darwin to recognize that the same process farmers were using to develop new and better crop varieties was occurring naturally in every population on earth and always had been.

TAKE-HOME MESSAGE 8·14

Animal breeders and farmers are making use of natural selection when they modify their animals and crops, because the three conditions for natural selection are satisfied. Since the differential reproductive success is determined by humans and not by nature, this type of natural selection is also called artificial selection.

8·15

Natural selection can change the traits in a population in several ways.

Certain traits are easily categorized—some people have blue eyes and others have brown eyes, just as some tigers have white fur and some have orange fur. Other traits, such as height in humans, are influenced by many genes and environmental factors so that a continuous range of phenotypes occurs (FIGURE 8-24). Whether a given trait is influenced by one gene or by a complex interaction of many genes and the environment, it can be subject to natural selection and be changed in any of several ways.

Directional Selection   In **directional selection,** individuals with one extreme from the range of variation in the population have higher fitness. Milk production in cows is an example. There is a lot of variation in milk production from cow to cow. As you might expect, farmers select for breeding those cows with the highest milk production and have done so for many decades. The result of such selection is not surprising: between 1920 and 1945, average milk production increased by about 50% in the United States (FIGURE 8-25).

TRAIT CATEGORIZATION

SPECIFIC CATEGORY
Some traits, such as chin shape, can be easily placed into one category or the other.

or   Cleft chin   Smooth chin

CONTINUOUS RANGE
Other traits, such as height, may fall into a wide range of values.

Height (feet)

Whether a given trait is influenced by one gene or a complex interaction of many genes and the environment, it can be subject to natural selection and be changed in any of several ways.

FIGURE 8-24 **Some traits fall into clear categories, others range continuously.**

Evolution Is Ongoing   Darwin's Journey   Four Mechanisms   Natural Selection

FIGURE 8-25 **More milk?** Cows have been selected for their ability to produce more and more milk.

Q Turkeys on poultry farms have such large breast muscles that they can't get close enough to each other to mate. How can such a trait evolve?

naturally or through artificial selection by breeders.) In subsequent generations, the average value for the milk-production trait increases. Note, however, that the variation for the trait decreases a bit, too, because the alleles contributing to one of the phenotype extremes are eventually removed from the population. Although the experiment has never been done, we can confidently predict that directional selection for *reduced* milk production on these dairy farms would have worked equally well.

Hundreds of experiments in nature and laboratories demonstrate the power of directional selection. In fact, it is one of the marvels of laboratory selection and animal breeding that nearly any trait they choose to exaggerate via directional selection responds dramatically—sometimes with absurd results (**FIGURE 8-26**). Turkeys, for instance, have been selected for increased size of their breast muscles, which makes them more valuable to farmers. Directional selection has been so successful that the birds' breast muscles are now so large that it has become impossible for them to mate. All turkeys on large-scale, industrial poultry farms must now reproduce through artificial insemination!

Those at the other end of the range, the cows that produce the smallest amounts of milk, have reduced fitness, since the farmers do not allow them to reproduce and therefore contribute their "less milk" alleles to the subsequent generation. (This is true whether the selection occurs

---

DIRECTIONAL SELECTION

SELECTIVE PRESSURE
Farmers select against turkeys with smaller breast muscles because they are less valuable.

Population before selection

Evolved population

Change in average breast size

Proportion of individuals in population

Smaller — Larger

Relative size of breast muscles in farmed turkeys

FIGURE 8-26 **Patterns of natural selection: directional selection.**

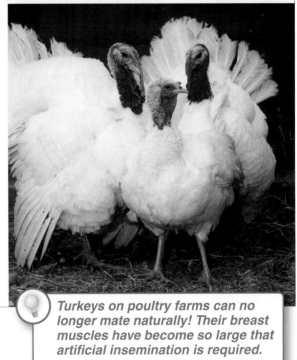

Q Turkeys on poultry farms can no longer mate naturally! Their breast muscles have become so large that artificial insemination is required.

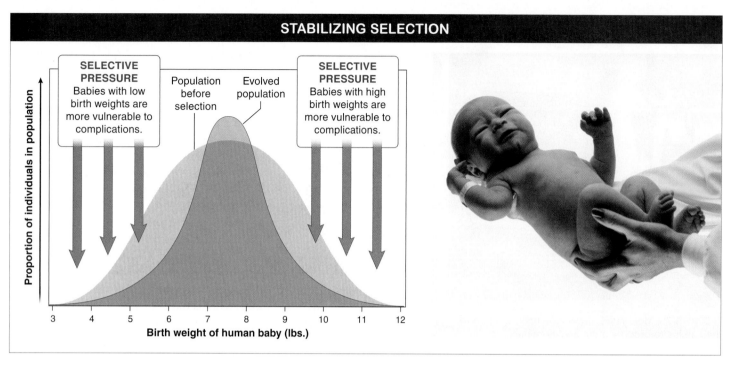

FIGURE 8-27 **Patterns of natural selection: stabilizing selection.**

Stabilizing Selection How much did you weigh at birth? Was it more than 10 pounds? More than 11? Was it close to the 22½ pounds that Christina Samane's child weighed when born in South Africa in 1982? Or was it 5 pounds or less?

Unlike directional selection, for which there is increased fitness at one extreme and reduced fitness at the other, **stabilizing selection** is said to occur when individuals with intermediate phenotypes are the most fit. The death rate

FIGURE 8-28 **Patterns of natural selection: disruptive selection.**

Evolution Is Ongoing   Darwin's Journey   Four Mechanisms   Natural Selection

among babies, for example, is lowest between 7 and 8 pounds, rising among both lighter and heavier babies (**FIGURE 8-27**). This outcome has kept the average weight of a baby constant over many generations, but at the same time the variation in birth weight has decreased as stabilizing selection has reduced the market share of those genes associated with high and low birth weights.

Modern technologies, including Caesarean deliveries and premature-birth wards, allow many babies to live who would not have survived without such technology. This has the unintended consequence of reducing the selection against any alleles causing those traits, reducing the rate of their removal from the population.

Disruptive Selection   There is a third kind of natural selection, in which individuals with extreme phenotypes experience the highest fitness and those with intermediate phenotypes have the lowest. Although examples of this type of selection, called **disruptive selection,** are rare in nature, the results are not surprising. Among some species of fish—

Q How is medical technology undoing the work of natural selection in optimizing the number of babies with normal birth weights?

the Coho salmon, for instance—only the largest males acquire good territories. They generally enjoy relatively high reproductive success. While the intermediate-size fish regularly get run out of the good territories, some tiny males are able to sneak in and fertilize eggs before the territory owner detects their presence. Consequently, we see an increase in the frequency of small and large fish with a reduction in the frequency of medium-size fish (**FIGURE 8-28**).

TAKE-HOME MESSAGE 8·15

Acting on multigene traits for which populations show a large range of phenotypes, natural selection can change populations in several ways, including directional selection, in which the average value for the trait increases or decreases; stabilizing selection, in which the average value of a trait remains the same while extreme versions of the trait are selected against; and disruptive selection, in which individuals with extreme phenotypes have the highest fitness.

8·16

Natural selection can cause the evolution of complex traits and behaviors.

We have seen that natural selection can change allele frequencies and modify the frequency with which simple traits like fur color or turkey-breast size appear in a population. But what about complex traits, such as behaviors, that involve numerous physiological and neurological systems? For instance, can natural selection improve maze-running ability in rats?

The short answer is *yes*. Remember, evolution by natural selection is occurring, changing the allele frequencies for traits whenever (1) there is variation for the trait, (2) that variation is heritable, and (3) there is differential reproductive success based on that trait. These conditions can easily be satisfied for complex traits, including behaviors.

In 1954, to address this question, William Thompson trained a group of rats to run through a maze for a food reward (**FIGURE 8-29**). There was a huge amount of

Q How can a wing evolve if 1% of a wing doesn't help an organism fly or glide at all?

variation in the rats' abilities: some rats learned much more quickly than others how to run the maze. Thompson then selectively bred the fast learners (called "maze-bright" rats) with each other and the slow learners (the "maze-dull" rats) with each other. Over several generations, he developed two separate populations: rats descended from a line of fast maze learners and rats descended from a line of slow maze learners. After only six generations, the maze-dull rats made twice as many errors as the maze-bright rats before mastering the maze, while the bright rats were adept at solving complex mazes that would give many humans some difficulty. Fifty years later, it still is unclear which actual genes are responsible for maze-running behavior, yet the selection experiment still demonstrates a strong genetic component to the behavior.

Natural selection can also produce complex traits in unexpected, roundabout ways. One vexing case involves the question of how natural selection could ever produce a

Can natural selection improve complex traits such as maze-running ability in rats?

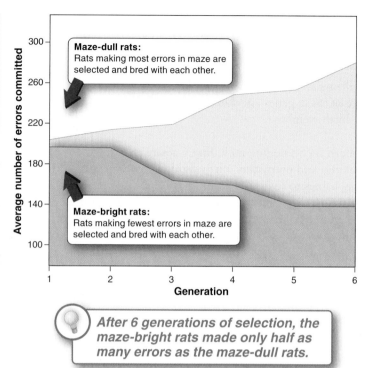

Maze-dull rats: Rats making most errors in maze are selected and bred with each other.

Maze-bright rats: Rats making fewest errors in maze are selected and bred with each other.

After 6 generations of selection, the maze-bright rats made only half as many errors as the maze-dull rats.

FIGURE 8-29 **"Not too complex . . ."** Natural selection can produce and alter complex traits, including behaviors.

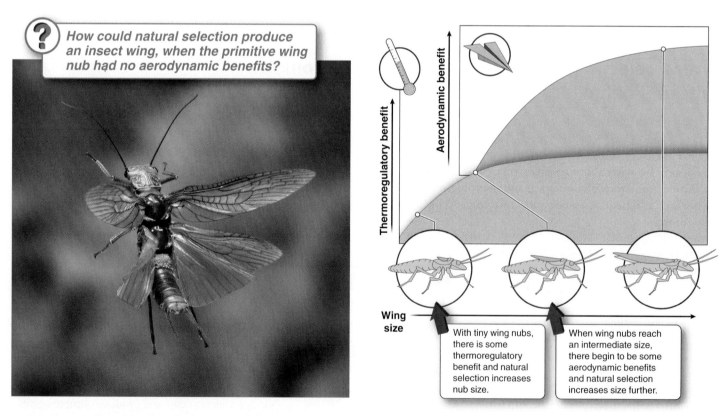

How could natural selection produce an insect wing, when the primitive wing nub had no aerodynamic benefits?

With tiny wing nubs, there is some thermoregulatory benefit and natural selection increases nub size.

When wing nubs reach an intermediate size, there begin to be some aerodynamic benefits and natural selection increases size further.

FIGURE 8-30 **"Not necessarily a wing . . ."** Traits selected for one function may be co-opted by natural selection for a new function.

Evolution Is Ongoing   Darwin's Journey   Four Mechanisms   Natural Selection

complex organ such as a fly wing, when 1% or 2% of a wing—that is, an incomplete structure—doesn't help an insect to fly. In other words, while it is clear how natural selection can preserve and increase traits' frequencies in populations, how do these soon-to-be-useful traits increase during the early stages if they don't increase the organism fitness at all (**FIGURE 8-30**)?

> All things must change
> To something new, to something strange.
>
> —Henry Wadsworth Longfellow

The key to answering this question is that 1% of a wing doesn't actually need to function as a wing at all to increase an individual's fitness. Often, structures are enhanced or elaborated on by natural selection because they enhance fitness by serving some other purpose. In the case of little nubs or "almost-wings" on an insect, experiments using models of insects demonstrated that, as expected, the nubs confer no benefit at all when it comes to flying. (They don't even help flies keep their orientation during a "controlled fall.") The incipient wings do help the insects address a completely different problem, though. They allow much more efficient temperature control, allowing an insect to gain heat from the environment when the insect is cold and to dissipate heat when the insect is hot. Experiments on heat-control efficiency, in fact, show that as small nubs become more and more pronounced, they are more and more effective, probably conferring increased fitness on the individual in which they occur—but only up to a point.

Eventually, the thermoregulatory benefit stops increasing, even if the nub length continues to increase. But it is right around this point that the proto-wing starts to confer some

Wings may have originally evolved because they enabled an animal to absorb or release heat, thereby regulating body temperature more effectively.

aerodynamic benefits (see Figure 8-30). Consequently, natural selection may continue to increase the length of this "almost-wing," but now the fitness increase is due to a wholly different effect. Such functional shifts explain the evolution of numerous complex structures that we see today and may be common in the evolutionary process.

TAKE-HOME MESSAGE 8·16

Natural selection can change allele frequencies for genes involving complex physiological processes and behaviors. Sometimes a trait that has been selected for one function is later modified to serve a completely different function.

⑤ The evidence for the occurrence of evolution is overwhelming.

The well-camouflaged veiled chameleon sits and waits for insect prey to come near.

8•17

The fossil record documents the process of natural selection.

> ❝ It is indeed remarkable that this theory [evolution] has been progressively accepted by researchers, following a series of discoveries in various fields of knowledge. The convergence, neither sought nor fabricated, of the results of work that was conducted independently is in itself a significant argument in favor of this theory. ❞
>
> —Pope John Paul II, 1996

In the nearly 150 years since Darwin first published *The Origin of Species,* thousands of experiments involving his theory of evolution by natural selection have been conducted, both in the laboratory and in natural habitats. A wide range of modern methodologies has also been developed, all contributing to a much deeper understanding of the process of evolution. This ongoing accumulation of evidence overwhelmingly supports the basic premise that Darwin put forward, while filling in many of the gaps that frustrated him.

Here we review the five primary lines of evidence demonstrating the occurrence of evolution:

1. **The fossil record**—physical evidence of organisms that lived in the past

2. **Biogeography**—patterns in the geographic distribution of living organisms

3. **Comparative anatomy and embryology**—growth, development, and body structures of major groups of organisms

4. **Molecular biology**—the examination of life at the level of individual molecules

5. **Laboratory and field experiments**—implementation of the scientific method to observe and study evolutionary mechanisms

Evolution Is Ongoing   Darwin's Journey   Four Mechanisms   Natural Selection

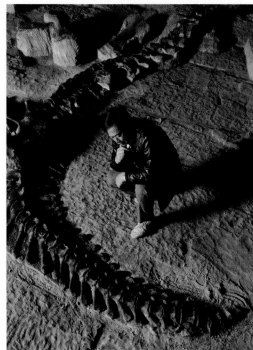

FIGURE 8-31 **Evidence for evolution: the fossil record.** Fossils can be used to reconstruct the appearance of organisms that lived long ago.

The first of the five lines of evidence is the fossil record. Although it has been central to much documentation of the occurrence of evolution, it is a very incomplete record. After all, the soft parts of an organism almost always decay rapidly and completely after death. And there are only a few unique environments (such as tree resin, tar pits, and the bottom of deep lakes) in which the forces of decomposition are reduced so that an organism's hard parts, including bones, teeth, and shells, can be preserved for thousands or even millions of years. These remains, called **fossils,** can be used to reconstruct what organisms must have looked like long ago. Such reconstructions often provide a clear record of evolutionary change (**FIGURE 8-31**).

The use of **radiometric dating** helps in further painting a picture of organisms' evolutionary history by telling us the age of the rock in which a fossil has been found (**FIGURE 8-32**). In Darwin's time, it was assumed that the deeper down in the earth a fossil was found, the older it was. Radiometric dating goes a step further, making it possible to determine not just the *relative* age of fossils, but also their *absolute* age. This is accomplished by evaluating the amounts of certain radioactive isotopes present in fossils. Radioactive isotopes in a rock begin breaking down into more stable compounds as soon as the rock is formed, and they do so at a constant rate. Nothing can alter this. By measuring the relative amounts of the radioactive isotope and the leftover decay product in the rock where a fossil is found, the age of the rock and thus of the fossil can be calculated.

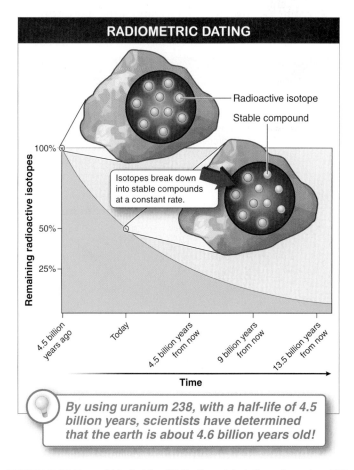

FIGURE 8-32 **How old is that fossil?** Radiometric dating of uranium 238 helps to determine the age of rocks and the fossils in them.

Evidence for Evolution

Radiometric dating confirms that the earth is very old. Rocks more than 3.5 billion years old have been found on all of the earth's continents, with the oldest so far found in northwestern Canada. By using the radioactive isotope uranium 238, with a half-life of 4.5 billion years, researchers have determined that the earth is about 4.6 billion years old and that the earliest organisms appeared at least 3.8 billion years ago. Radiometric dating also makes it possible to put the fossil record in order. By dating all the fossils discovered in one locale, paleontologists can learn if and how the organisms were related to each other and how groups of organisms changed over time.

Paleontologists must deal with the fact, however, that fossil formation is an exceedingly unlikely event, and when it does occur it represents only those organisms that happened to live in that particular area and that also had physical structures that could leave fossils. For this reason, the fossil record is annoyingly incomplete. Entire groups of organisms have left no fossil record at all and some others have numerous gaps. Still, fossils have been found linking all of the major groups of vertebrates.

The evolutionary history of horses is among the most well preserved in the fossil record. First appearing in North

America about 55 million years ago, horses then radiated around the world, with more recent fossils appearing in Eurasia and Africa. These fossils exhibit distinct adaptations to those environments. Later, about 1.5 million years ago, much of the horse diversity—including all North American horse species—disappeared, leaving only a single remaining genus, or group of species, called *Equus*. Because there is now only one horse genus on earth, it is tempting to imagine a simple linear path from modern horses straight back through their 55-million-year evolutionary history. But that's just not how evolution works. In reality, there have been numerous branches of horses that have split off over evolutionary time, flourished for millions of years, and only recently gone extinct. What we see living today is only a single branch of a greatly branched evolutionary tree (**FIGURE 8-33**).

The fossil record provides another valuable piece of evidence for the process of natural selection in the form of "missing links." These are fossils that demonstrate a link between groups of species believed to have shared a common ancestor. One such "missing link" is *Tiktaalik* (**FIGURE 8-34**). First found in northern Canada and estimated to be 375 million years old, *Tiktaalik* fossils seem to represent a transitional phase between fish and land animals. These creatures, like fish, had gills, scales,

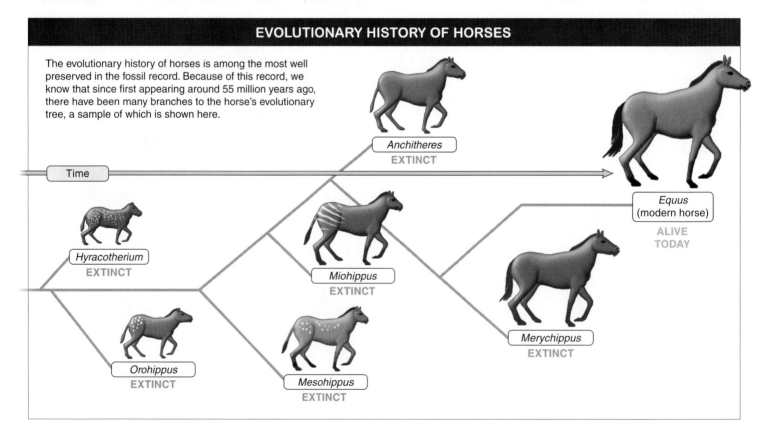

EVOLUTIONARY HISTORY OF HORSES

The evolutionary history of horses is among the most well preserved in the fossil record. Because of this record, we know that since first appearing around 55 million years ago, there have been many branches to the horse's evolutionary tree, a sample of which is shown here.

Time

Anchitheres
EXTINCT

Equus
(modern horse)
ALIVE TODAY

Hyracotherium
EXTINCT

Miohippus
EXTINCT

Merychippus
EXTINCT

Orohippus
EXTINCT

Mesohippus
EXTINCT

FIGURE 8-33 An evolutionary family tree. The branching evolutionary tree of the horse.

Evolution Is Ongoing   Darwin's Journey   Four Mechanisms   Natural Selection

TIKTAALIK: "THE MISSING LINK"

Lobe-finned fish   Tiktaalik   Early amphibian

Tiktaalik fossils have gills, scales, and fins, like fish, but arm-like joints resembling those in amphibians.

FIGURE 8-34 A missing link? *Tiktaalik* seems to be a transitional phase between fish and land animals.

and fins, but they also had arm-like joints in their fins and could drag their bodies across land in much the same way as a marine mammal such as a seal.

TAKE-HOME MESSAGE 8·17

Radiometric dating confirms that the earth is very old and makes it possible to determine the age of fossils. Analysis of fossil remains enables biologists to reconstruct what organisms looked like long ago, learn how organisms were related to each other, and understand how groups of organisms evolved over time.

8·18

Geographic patterns of species distributions reflect their evolutionary histories.

The study of the distribution patterns of living organisms around the world is called **biogeography.** This is the second line of evidence that helps us to see that evolution occurs and to better understand the process. The patterns of biogeography that Darwin and many subsequent researchers have noticed provide strong evidence that evolutionary forces are responsible for these patterns. Species often more closely resemble other species that live less than a hundred miles away but in radically different habitats than they do species that live thousands of miles away in nearly identical habitats. On Hawaii, for example, nearly every bird is some sort of modified honeycreeper (a small, finch-like bird). There are seed-eating honeycreepers, woodpecker honeycreepers, and curved-bill nectar-feeding honeycreepers. There are even parrot honeycreepers (**FIGURE 8-35**).

BIOGEOGRAPHY: HAWAIIAN HONEYCREEPERS

The honeycreepers of Hawaii have adapted to a wide range of habitats, yet still closely resemble a finch-like shared ancestor found nearly 2,000 miles away.

Mainland finch (probable shared ancestor)   'Akeke'e honeycreeper   Maui Parrotbill honeycreeper   'I'iwi honeycreeper

FIGURE 8-35 Evidence for evolution: biogeography. The Hawaiian honeycreepers resemble a common ancestor from mainland North America, but all have unique features.

When it comes to species distributions, history matters. Species were not designed from scratch to fill a particular niche. Rather, whatever arrived first—usually a nearby species—took up numerous different lifestyles in numerous different habitats, and the populations ultimately adapted to and evolved in each environment. In Hawaii, it seems that a finch-like descendant of the honeycreepers arrived 4–5 million years ago and rapidly evolved into a large number of diverse species. The same process has occurred and continues to occur in all locales, not just on islands.

Large isolated habitats also have interesting biogeographic patterns. Australia and Madagascar are filled with unique organisms that are clearly not closely related to organisms elsewhere. In Australia, for example, marsupial species, rather than placental mammals, fill all of the usual roles. There are marsupial "wolves," marsupial "mice," marsupial "squirrels," and marsupial "anteaters" (**FIGURE 8-36**). They physically resemble their placental counterparts for most traits, but molecular analysis shows that they are actually more closely related to each other, sharing a common marsupial ancestor. Their relatedness to each other is also revealed by similarities in their reproduction: females give birth to offspring at a relatively early state of development and the offspring finish their development in a pouch. The presence of marsupials in Australia does not simply mean that marsupials are better adapted than placentals to the Australian habitat. When placental organisms have been transplanted to Australia they do just fine, often thriving to the point of endangering the native species.

A good designer would use the best designs over and over again wherever they might fit. Biogeographic patterns such as those seen in honeycreepers and in the marsupials of Australia illustrate that evolution, unlike a good designer, is more of a tinkerer. Evolution takes whatever populations are at hand in a particular location then gradually changes their traits, and the species become better adapted to the habitats they occupy.

TAKE-HOME MESSAGE 8·18

Observing geographic patterns of species distributions, particularly noting similarities and differences among species living close together but in very different habitats and among species living in similar habitats but located far from one another, helps us to understand the evolutionary histories of populations.

BIOGEOGRAPHY: AUSTRALIAN MARSUPIALS AND THEIR PLACENTAL COUNTERPARTS

AUSTRALIAN MARSUPIALS

Sugar glider   Numbat   Tasmanian wolf

PLACENTAL COUNTERPARTS

Gray squirrel   Giant anteater   Gray wolf

Though less related to each other than you are to a shrew, these marsupials and their placental counterparts have come to resemble each other as natural selection has adapted them to similar habitats.

FIGURE 8-36 Evidence for evolution: biogeography. Many Australian marsupials resemble placental counterparts, though they are not closely related.

Evolution Is Ongoing   Darwin's Journey   Four Mechanisms   Natural Selection

Comparative anatomy and embryology reveal common evolutionary origins.

If you observe any vertebrate embryo, you will see that it passes through a stage in which it has little gill pouches on the sides of the neck. It will also pass through a stage in which it possesses a long bony tail. This is true whether it is a human embryo or that of a turtle or a chicken or a shark. These gill pouches disappear before birth in all but the fishes. Similarly, we don't actually find humans with tails. Why do they exist during an embryo's development? Such common embryological stages indicate that the organisms share a common ancestor, from which all have been modified (**FIGURE 8-37**). Study of these developmental stages and the adult body forms of organisms provides our third line of evidence, helping us to see that evolution occurs and to better understand the process.

Among adults, several features of anatomy reveal the ghost of evolution in action. We find, for example, that many related organisms show unusual similarities that can be explained only through evolutionary relatedness. The forelimbs of mammals are used for a variety of functions in bats, porpoises, horses, and humans (**FIGURE 8-38**). If each had been designed specifically for the uses necessary to that species, we would expect dramatically different designs. And yet in each of these species, we see the same bones—modified extensively—that betray the fact that they share a common ancestor. These features are called **homologous structures.**

HOMOLOGOUS STRUCTURES: EMBRYO STAGE

Gill pouches
Bony tail

Shark

Turtle

Human

Chicken

FIGURE 8-37 Evidence for evolution: embryology. Structures derived from common ancestry can be seen in embryos.

HOMOLOGOUS BONE STRUCTURES

Humerus
Radius
Ulna
Phalanges
Metacarpals
Carpals

Human

Horse

Bat

Porpoise

FIGURE 8-38 Evidence for evolution: comparative anatomy. Homologous bone structures among some mammals.

The similarities in the bone structure of the forelimbs of mammals demonstrate common ancestry.

> **Q** The human appendix appears to serve no function. Why are we all born with one?

At the extreme, homologous structures sometimes come to have little or no function at all. Such evolutionary leftovers, called **vestigial structures,** exist because they had value ancestrally. Some vestigial structures in mammals include the molars that continue to grow in vampire bats, even though these bats consume a completely liquid diet (**FIGURE 8-39**); eye sockets (with no eyes) in some populations of cave-dwelling fish; and pelvic bones in whales that are attached to nothing (but serve as an important attachment point for leg bones in nearly all other mammals). Even humans may have a vestigial organ—the appendix. It is greatly enlarged in our relatives the great apes, in whom it hosts cellulose-digesting bacteria that aid in breaking down the plants in their diet, but in humans it appears to serve no purpose. (Recently, however, it has been suggested that the appendix may play a role in the immune system.)

Not all organisms with adaptations that appear similar actually share ancestors. We see flying mammals (bats) and flying insects (locusts) (**FIGURE 8-40**). Similarly, dolphins and penguins live in similar habitats and have flippers that help

> **CONVERGENT EVOLUTION**

> *Different starting materials come to perform the same function through convergent evolution.*

FIGURE 8-40 Evidence for evolution: convergent evolution and analogous structures.

them to swim. In both examples, however, all of the analogous structures developed from different original structures. Natural selection—in a process called **convergent evolution**—uses the different starting materials available (such as a flipper or a forelimb) and modifies them until they serve similar purposes, much as we saw in the marsupial and placental mammals in Figure 8-36.

TAKE-HOME MESSAGE 8·19

Similarities in the anatomy of different groups of organisms and in their physical appearance as they proceed through their development can reveal common evolutionary origins.

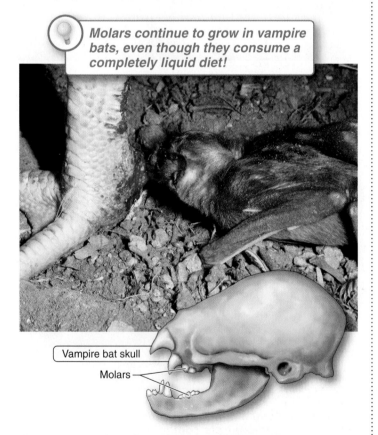

> *Molars continue to grow in vampire bats, even though they consume a completely liquid diet!*

Vampire bat skull

Molars

FIGURE 8-39 Evidence for evolution: vestigial structures.

Evolution Is Ongoing   Darwin's Journey   Four Mechanisms   Natural Selection

Molecular biology reveals that common genetic sequences link all life forms.

The development of new technologies for deciphering and comparing the genetic code provides our fourth line of evidence that evolution occurs. First and foremost is the fact that all living organisms share the same genetic code. From microscopic bacteria to flowering plants to insects and primates, the molecular instructions for building organisms and transmitting hereditary information are the same: four simple bases, arranged in an almost unlimited variety of sequences. This discovery strongly suggests that all living organisms are related.

When we examine the similarity of DNA among related individuals within a species, we find that they share a greater proportion of their DNA than do unrelated individuals. This is not unexpected; you and your siblings got all of your DNA from the same two parents, while you and your cousins each got half of your DNA from the same two grandparents. The more distantly you and another individual are related, the more your DNA differs.

We can measure the DNA similarity between two species by comparing their DNA sequences for individual genes. For example, let's look at the gene that codes for the amino acids used to build the hemoglobin molecule. In vertebrate animals, hemoglobin is found inside red blood cells, where its function is to carry oxygen throughout the body. It is made of two chains of amino acids, the alpha chain and the beta chain. In humans, the beta chain has 146 amino acids. In rhesus monkeys, this beta chain is nearly identical: of the 146 amino acids, 138 are the same as those found in human hemoglobin, and only 8 are different. In dogs the sequence is still similar, but a bit less so, with 32 different amino acids. When we look at non-mammals such as birds, we see about 45 amino acid differences, and between humans and lamprey eels (still vertebrates, but it has been more than 500 million years since we shared a common ancestor), there are 125 amino acid differences.

The differences in the amino acid structure of the beta hemoglobin chain (and remember that this structure is governed by an allele or alleles of a particular gene) seem to indicate that humans have more recently shared a common ancestor with rhesus monkeys than with dogs. And that we have more recently shared an ancestor with dogs than with birds or lampreys. These findings are just as we would expect, based on estimates of evolutionary relatedness made from comparative anatomy and embryology as well as those based on the fossil record. It is as if there is a molecular clock that is

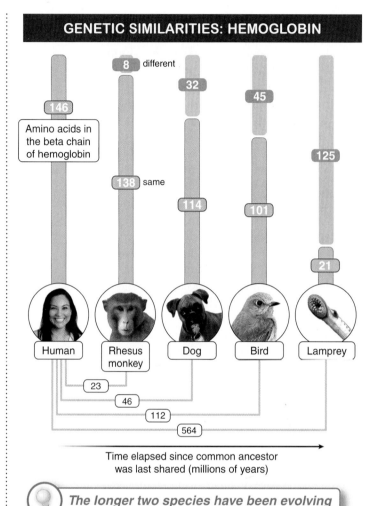

GENETIC SIMILARITIES: HEMOGLOBIN

8 different

Amino acids in the beta chain of hemoglobin

138 same

Human | Rhesus monkey | Dog | Bird | Lamprey

Time elapsed since common ancestor was last shared (millions of years)

The longer two species have been evolving on their own, the greater the number of genetic differences that accumulate.

FIGURE 8-41 An evolutionary clock? Genetic similarities (and differences) demonstrate species relatedness.

ticking. The longer two species have been evolving on their own, the greater the number of changes in amino acid sequences—or "ticks of the clock" (**FIGURE 8-41**).

TAKE-HOME MESSAGE 8 • 20

All living organisms share the same genetic code. The degree of similarity in the DNA of different species can reveal how closely related they are and the amount of time that has passed since they last shared a common ancestor.

8·21

Laboratory and field experiments enable us to watch evolution in progress.

A fifth line of evidence for the occurrence of evolution comes from multigeneration experiments and observations. Until recently, people thought that evolution was too slow a process to be observed in action. By choosing the right species—preferably organisms with very short life spans—and designing careful experiments, however, it turns out that it is actually possible to observe and measure evolution as it is happening.

In one clever study, researchers studied populations of grass on golf courses—a habitat where lawnmower blades represent a significant source of mortality. They realized that although all of the grass on golf courses was the same species, on the putting greens it was cut very frequently, whereas on the fairways it was cut only occasionally, and in the rough it was almost never cut at all (FIGURE 8-42). Over the course of only a few years, significant changes occurred in these different grass populations. The grass plants on the greens came to be short lived, with rapid development to reproductive age and very high seed output. For the plants in these populations, life was short and reproduction came quickly. (For those in which it did not come quickly, it did not come at all; hence their lack of representation in the population.) Plants on the fairways had slightly slower development and reduced seed output, while those in the rough had the slowest development and the lowest seed output of all. When plants from each of the habitats were collected and grown in greenhouses, the dramatic differences in growth, development, and life span remained, confirming that there had been changes in the frequency of the various alleles controlling the traits of life span and reproductive output in the three populations; in other words, evolution had occurred.

A more disturbing line of evidence for the occurrence of evolution in nature comes from the evolution of antibiotic-resistant strains of bacteria that cause illness in humans. For example, in the 1940s when penicillin was first used as a treatment for bacterial infections, it was uniformly effective in killing *Staphylococcus aureus*. Today, more than 90% of isolated *S. aureus* strains are resistant to penicillin (FIGURE 8-43). Because penicillin has become such a pervasive toxin in the environment of *Staphylococcus,* natural selection has led to an increase in the frequency of the alleles that make these strains resistant. As a consequence, humans are increasingly at risk for becoming infected with *Staphylococcus* and getting diseases such as pneumonia and meningitis. In the 1960s, the antibiotics methicillin and oxacillin were developed. Like penicillin, they initially had nearly complete effectiveness

against *Staphylococcus.* Today, though, nearly a third of staph infections are resistant to these antibiotics. The meaning of such unintentional natural selection "experiments" is clear and consistent: evolution is occurring all around us.

And finally, let's return to the spectacularly starvation-resistant flies introduced at the beginning of this chapter. In them we saw an unambiguous demonstration that natural selection can

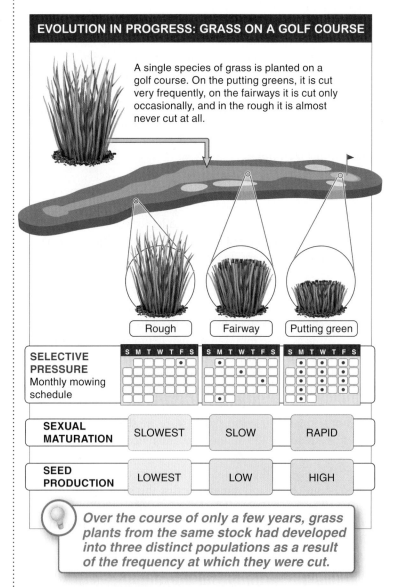

FIGURE 8-42 **Evolution in progress: grasses.** Different mowing patterns can cause evolution in the grass on a golf course.

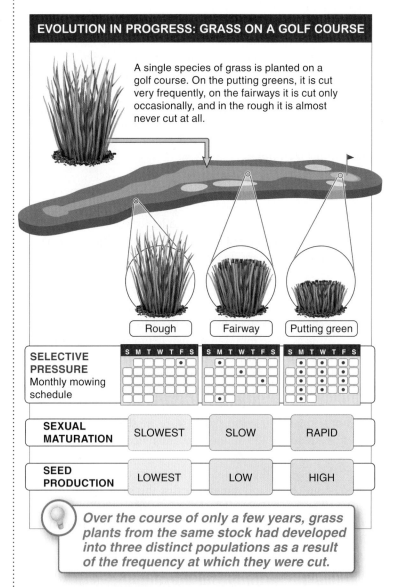

EVOLUTION OF ANTIBIOTIC RESISTANCE

When first used as medicine in the 1940s, penicillin was uniformly effective in killing the bacterium *Staphylococcus aureus*. Today, natural selection has led to an increase in antibiotic-resistant alleles, and humans are increasingly at risk from untreatable *Staphylococcus* infections.

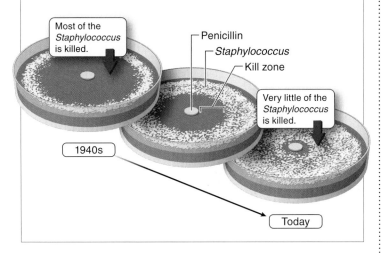

Most of the *Staphylococcus* is killed.

Penicillin

Staphylococcus

Kill zone

Very little of the *Staphylococcus* is killed.

1940s

Today

FIGURE 8-43 Evolution in progress: disease-causing bacteria. Antibiotic resistance has evolved in *Staphylococcus*.

occur, that it has the potential to produce dramatic changes in physical features and physiological processes, and that it can bring about these changes very quickly. And, perhaps most important, we saw that replicating the same evolutionary process over and over repeatedly produced the same predictable results. In the laboratory, the farm, the doctor's office, the bathroom sink, the deserts, streams, and forests, evolution is occurring.

Reflecting on both the process and products of evolution and natural selection, in the final paragraph of *The Origin of Species* Darwin eloquently wrote: "There is grandeur in this view of life . . . and that . . . from so simple a beginning endless forms most beautiful and most wonderful have been, and are being, evolved."

TAKE-HOME MESSAGE 8·21

Multiply replicated, controlled laboratory selection experiments and long-term field studies of natural populations—including observations on antibiotic-resistant strains of disease-causing bacteria—enable us to watch and measure evolution as it occurs.

323

Evidence for Evolution

Knowledge You Can Use

Did you know? **Antibacterial soaps may be dangerous. And it may be your fault.**

Fact: The antimicrobial agent, called triclosan, in most antibacterial soaps takes several minutes to kill bacteria.

Q: How long do you wash your hands? On average, **people spend less than a minute washing their hands.**

Q: If you're not killing all of the bacteria, which ones are you killing? Some bacteria are easily killed by the antimicrobial chemical in soaps. Others take longer. In the first minute you're killing only the weak bacteria (and not all of the 800,000 to 1,000,000 bacteria residing on your hands).

Q: **After one minute, what is left? After a minute of hand-washing, only the strong bacteria—those that happen to carry triclosan-resistant genes—remain and continue reproducing.** (And the next time you wash your hands, you may pick up those super-strength drug-resistant bacteria from the sink area.) **You're causing evolution to happen.** (And you may not want to.)

Q: **What do you make of this? What can you do?** The two simplest alternatives are to wash your hands for much longer or to stop using soaps with antibacterial agents in them.

1 **Evolution is an ongoing process.**

The characteristics of individuals in a population can change over time. We can observe such change in nature and can even cause such change to occur.

2 **Darwin journeyed to a new idea.**

Following years of observations on worldwide patterns of the living organisms and fossils of plant and animal species, Charles Darwin developed a theory of evolution by natural selection that explained how populations of species can change over time.

3 **Four mechanisms can give rise to evolution.**

Evolution—a change in the allele frequencies in a population—can occur via genetic drift, migration, mutation, or natural selection.

4 **Through natural selection, populations of organisms can become adapted to their environment.**

When there is variation for a trait, and the variation is heritable, and there is differential reproductive success based on that trait, evolution by natural selection is occurring. Fitness is an individual's relative reproductive output. The alleles that yield traits that confer the highest fitness on the individuals carrying them will increase their market share in a population over time, but this process does not necessarily lead to "perfect" organisms in an environment. Artificial selection, in which humans determine which individuals will have the highest reproductive success, is a type of natural selection. Directional, stabilizing, and disruptive selection can alter the average value of a trait and/or the variation for a trait in a population. Natural selection also can produce and modify complex traits, including behaviors. This modification may occur as traits selected for one purpose later confer increased fitness for a different purpose.

5 **The evidence for the occurrence of evolution is overwhelming.**

Many overwhelming lines of evidence document the occurrence of evolution and point to the central and unifying role of evolution by natural selection in helping us to better understand all other ideas and facts in biology.

KEY TERMS

1. The average time to death from starvation in a fruit fly is about 20 hours. Selecting for increased starvation resistance in fruit flies:
   a) has no effect because starvation resistance is a not a trait that influences fruit fly fitness.
   b) has little effect because ongoing mutation continuously reduces starvation resistance, counteracting any benefits from selection.
   c) cannot increase their survival time because there is no genetic variation for this trait.
   d) has no effect because starvation resistance is too complex a trait, dependent on the effects of too many genes.
   e) can produce populations in which the average time to death from starvation is 160 hours.

2. Georges Cuvier's discovery of fossils of Irish elk and giant ground sloths:
   a) demonstrated that extinction must occur.
   b) was possible because of Buffon's determination that the earth was more than 6,000 years old.
   c) was possible only following Darwin's publication of *The Origin of Species.*
   d) was made in deep ocean trenches.
   e) suggested that species were immutable.

3. While on the voyage of the HMS *Beagle,* Darwin:
   a) wrote *The Origin of Species.*
   b) nurtured his love of studying nature, exploring plant and animal diversity, and collecting fossils.
   c) discovered a love for sea travel.
   d) studied Malthus's book *Essay on the Principle of Population.*
   e) corresponded extensively with Alfred Russel Wallace about their ideas on evolution.

4. Which one of the following statements best describes the difference between artificial and natural selection?
   a) Natural selection is limited to physical traits and artificial selection is not.
   b) Artificial selection has produced many of the most delicious food items for humans; natural selection has not.
   c) Natural selection acts without the input of humans; artificial selection requires human input.
   d) Charles Darwin understood natural selection but was unaware of artificial selection in his time.
   e) Natural selection works on all species; artificial selection works only on laboratory-raised species.

5. Which of the following statements about Charles Darwin is incorrect?
   a) He spent five years traveling the world observing living organisms and collecting fossils.
   b) He was under constant pressure from his father to make something of himself.
   c) He dropped out of medical school.
   d) He was enthusiastic about unleashing his theory of natural selection on the world as soon as he thought of it.
   e) He and Alfred Russel Wallace independently came up with the theory of evolution by natural selection.

6. Evolution occurs:
   a) only when the environment is changing.
   b) only through natural selection.
   c) almost entirely because of directional selection.
   d) only via natural selection, genetic drift, migration, or mutation.
   e) by altering physical traits but not behavioral traits.

7. Which of the following statements about mutations is incorrect?
   a) Mutations are almost always random with respect to the needs of the organism.
   b) A mutation is any change in an organism's DNA.
   c) Most mutations are harmful or neutral to the organism in which they occur.
   d) The origin of genetic variation is mutation.
   e) All of these statements are correct.

8. The chief concern among conservation biologists trying to protect small populations is:
   a) to preserve genetic diversity.
   b) that mutation rates are much higher in small populations.
   c) that natural selection can only operate in large populations.
   d) to maximize rates of allele fixation.
   e) to reduce the duration of genetic bottlenecks.

9. When a group of individuals colonizes a new habitat, the event is likely to be an evolutionary event because:
   a) members of a small population have reduced rates of mating.
   b) gene flow increases.
   c) mutations are more common in novel environments.
   d) new environments tend to be inhospitable, reducing survival there.
   e) small founding populations are rarely genetically representative of the initial population.

10. To establish that evolution by natural selection is operating in a population, one must demonstrate variability for a trait, heritability of that trait, differential reproductive success based on that trait, and:
    a) increased complexity of the organism.
    b) random mating.
    c) progress.
    d) continuous change in the environment.
    e) nothing else.

11. "Survival of the fittest" may be a misleading phrase to describe the process of evolution by natural selection because:
    a) it is impossible to determine the fittest individuals in nature.
    b) survival matters less to natural selection than reproductive success does.
    c) natural variation in a population is generally too great to be influenced by differential survival.
    d) fitness has little to do with natural selection.
    e) reproductive success on its own does not necessarily guarantee evolution.

12. Adaptation:
   a) refers both to the process by which organisms become better matched to their environment and to the features of an organism that make it more fit than other individuals.
   b) cannot occur in environments influenced by humans.
   c) is possible only when there is no mutation.
   d) is responsible for the fact that porcupines are at unusually high risk of predation.
   e) occurs for physical traits but not behaviors.

13. Adaptations shaped by natural selection:
   a) are magnified and enhanced through genetic drift.
   b) are unlikely to be present in humans living in industrial societies.
   c) may be out of date, having been shaped in the past under conditions that differed from those in the present.
   d) represent perfect solutions to the problems posed by nature.
   e) are continuously modified so that they are always fitted to the environment in which an organism lives.

14. Artificial selection was used on corn to produce a single strain of corn with increased growth rates and greater resistance to a fungus. Although farmers have continued to select for these traits, the productivity of this strain is no longer increasing. This suggests that:
   a) the population size has been decreasing.
   b) all or most of the natural variation for these traits has been eliminated.
   c) gene migration is a major evolutionary agent in corn.
   d) long-term disruptive selection may lead to speciation.
   e) artificial selection is not as strong as natural selection.

15. In a population in which a trait is exposed to stabilizing selection over time:
   a) neither the average nor the variation for the trait changes.
   b) both the average value and variation for the trait increase.
   c) the average value increases or decreases and the variation for the trait decreases.
   d) the average value for the trait stays approximately the same and the variation for the trait decreases.
   e) the average value for the trait stays approximately the same and the variation for the trait increases.

16. Maze-running behavior in rats:
   a) is too complex a trait to be influenced by natural selection.
   b) is a heritable trait.
   c) is not influenced by natural selection because it does not occur in rats' natural environment.
   d) shows no variation.
   e) is influenced primarily by mutation in the laboratory.

17. A fossil is defined most broadly as:
   a) the preserved pieces of hard parts (e.g., shell or bone) of extinct animals.
   b) any preserved remnant or trace of an organism from the past.
   c) the preserved bones of vertebrates.
   d) a piece of an organism that has turned into rock.
   e) the process of preservation of intact animal bodies.

18. Which of the following are correct about marsupial mammals?
   a) fill many niches in Australia that are occupied by placental mammals in other parts of the world.
   b) are less fit than placental mammals.
   c) have gone extinct as a result of the greater fitness of placental mammals.
   d) are more closely related to each other than they are to placental mammals.
   e) Both a) and d) are correct.

19. Convergent evolution can occur only when two species:
   a) have a recent common ancestor.
   b) live in the same geographic area.
   c) are separated by a barrier such as a new river.
   d) evolve under similar selective forces.
   e) are both unpalatable to predators.

20. What can be concluded about comparing differences in molecular biology between different species?
   a) Extremely different species are fundamentally unrelated in any way.
   b) Only DNA sequences can be used to compare relatedness between species.
   c) Birds are more closely related to humans than dogs are.
   d) Genetic similarities and differences demonstrate species relatedness.
   e) The longer two species have been evolving on their own, the fewer the genetic differences between them.

21. Evolution:
   a) occurs too slowly to be observed in nature.
   b) can occur in the wild but not in the laboratory.
   c) is responsible for the increased occurrence of antibiotic-resistant bacteria.
   d) does not occur in human-occupied habitats.
   e) None of these statements is correct.

SHORT-ANSWER QUESTIONS

1. Distinguish between evolution and natural selection. (Restrict your answer to 30 words or fewer.)

2. What is genetic drift? Why is it a more potent agent of evolution in small populations rather than large populations?

3. In *The Origin of Species,* Charles Darwin wrote: "We may look with some confidence to a secure future of great length. And as natural selection works solely by and for the good of each being, all corporeal and mental endowments will tend to progress towards perfection." Describe three reasons why he is wrong.

4. How does the increasing frequency of antibiotic-resistant strains of bacteria represent an example of the occurrence evolution?

See Appendix for answers. For additional study questions, go to www.prep-u.com.

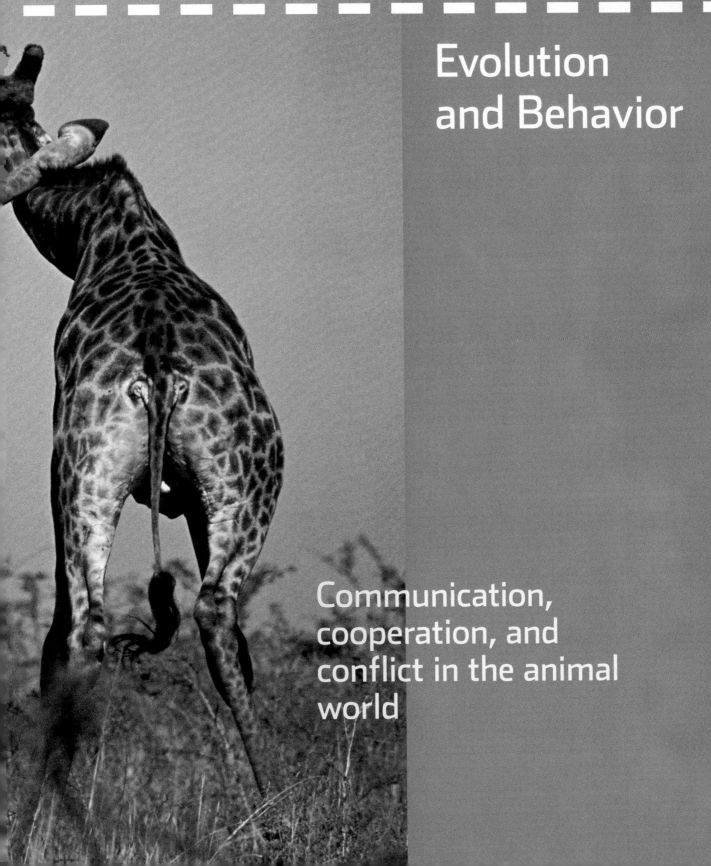

9

Evolution
and Behavior

Communication,
cooperation, and
conflict in the animal
world

1 Behaviors are traits that can evolve.

This chimpanzee has learned to use a twig as a tool to capture termites.

9·1

Behavior has adaptive value, just like other traits.

Make a mental list of your favorite things to eat. Perhaps donuts or hot fudge sundaes top your list. Or maybe pizza, or cheeseburgers and french fries. One thing is almost certain about the food preferences of any human: the list is filled with calorie-rich substances. Put another way, when it comes to eating, humans show an almost complete aversion to dirt or pebbles or other substances from which no energy can be extracted (**FIGURE 9-1**).

In controlled studies, humans demonstrate a clear and consistent preference for sweet and fatty foods. Specifically, when people are presented with food consisting of different mixtures, their preferences for the food increase as the sugar concentration in the food increases—up to a point, after which adding more sugar makes the food less preferable. But the preference becomes stronger and stronger as the food's fat content increases. In short, we prefer sweet (but not too sweet) foods that are packed with as much fat as possible (**FIGURE 9-2**).

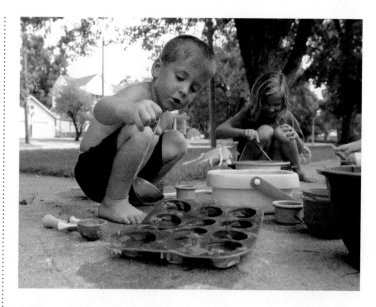

FIGURE 9-1 Mud pie for dinner? Humans show an aversion to eating dirt (and other substances from which we cannot extract nutrition).

Behaviors Evolve   Cooperation and Altruism   Sexual Conflict   Communication

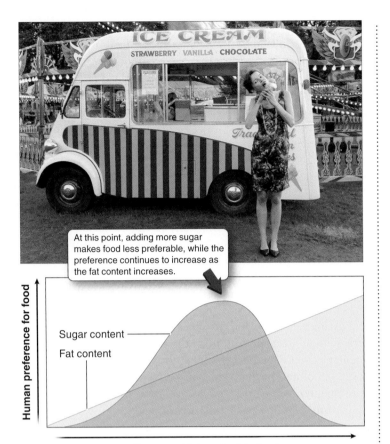

At this point, adding more sugar makes food less preferable, while the preference continues to increase as the fat content increases.

Human preference for food

Sugar content

Fat content

Sugar and fat content in food

 In controlled experiments, humans exhibit preferences for sugary foods, high in fat—and the more fat, the stronger the preference.

FIGURE 9-2 **Fatty and sweet.** Across cultures, humans prefer sweet, but not too sweet, foods that are packed with fat.

Watch birds in a field and you'll see that they, too, have preferences. A starling, for example, does not eat everything in its path. Rather, it walks through the grass, probing the ground and only occasionally picking up something to eat, such as a beetle larva. Shore crabs, too, have definite preferences. When presented with mussels of different sizes, shore crabs prefer the mussels that provide the greatest amount of energy relative to the amount of energy a crab must expend to open the mussel. The largest mussels take too long to open and so are less preferred; the smallest mussels are easily opened but don't provide enough caloric content to make the effort worthwhile (**FIGURE 9-3**). Thus, the research shows, shore crabs prefer medium-sized mussels.

Humans, starlings, shore crabs, and all other animals have taste preferences for the same reason: animals (including humans), unlike plants, cannot create their own food, so they must consume materials from which they can extract the most energy and acquire essential nutrients. If they don't, they die. The feeding choices of animals affect the number of offspring they can produce, and so taste preferences directly influence the evolutionary fitness of organisms. As a consequence, natural selection can shape feeding behaviors just as it can bring about changes in physical structures. We may believe that some things simply taste good while others do not, but these taste sensitivities and preferences are a direct result of the actions of molecules within our taste buds. And those molecules are produced by a large number of genes that have been shaped by natural selection.

Before we continue, let's define exactly what we mean by "behavior." **Behavior** encompasses any and all of the actions performed by an organism, often in response to its environment or to the actions of another organism. Feeding behavior is only one of many behaviors influenced by natural selection. Maternal care is another. In mice, an allele occasionally arises through mutation that causes a mother to show reduced maternal care. She leaves her pups unattended and neglects feeding them. Not surprisingly, this allele never increases in frequency in the mouse population, because most of the pups that inherit it fail to survive long enough to pass it

ENERGY EFFICIENT FEEDING BEHAVIOR HAS EVOLVED

The feeding behavior of shore crabs has evolved so that they preferentially choose mussels that provide the most energy relative to the effort it takes to open the shell.

FIGURE 9-3 **Efficient eater.** A shore crab feeds on a mussel.

on to their own offspring. The only reason this behavior ever occurs is that the mutation causing it occasionally occurs.

Another behavior influenced by natural selection is the singing behavior of songbirds. Among warblers, males sing complex and loud songs. They aren't singing for our enjoyment, though. They are singing to impress female warblers. Females prefer males with more complicated songs, apparently because only the healthiest, most well-nourished males are able to produce the most complex songs. Thus, in warblers, males' singing behavior has a strong effect on their evolutionary fitness.

The scope of animal behavior is vast and varied. Even a brief listing of well-studied topics in animal behavior is impressively long:

Conflict, aggression, and territoriality

Cooperation, alliance building, and sociality

Competing for food and avoiding predation

Migration and navigation

Behavioral control of body temperature

Courtship and mate choice

Pair bonding and fidelity

Breeding and parental behavior

Communication

Learning and tool use

In Chapter 8, we learned about the evolutionary origins of many *physical traits*. We did not investigate each and every physical trait that occurs in organisms, however. It was sufficient to look at a few examples, such as birth weight in babies, milk production in cows, and fur color in tigers, to understand the general process by which natural selection acts and how it can lead to the evolution of organisms adapted to their environments. Similarly, in this chapter, although we explore the evolution of *behavioral traits,* we will not investigate each and every behavioral trait in organisms. Rather, we will explore three broad areas of animal behavior—cooperation and conflict, communication, and mating and parenting—and focus on why those behaviors evolved. That is, we'll be answering this question: how do these behaviors contribute to an organism's fitness, its survival and reproduction?

Our approach to the study of evolution and behavior parallels our approach to the study of the evolution of non-behavioral traits. Recall from the discussion of natural selection in the last chapter that when a heritable trait increases an individual's reproductive success relative to that of other individuals, that trait tends to increase in frequency in the population. This is Darwin's mechanism for evolution by natural selection and, as we saw in Chapter 8, its effects are evident all around us, from the fancy ornamentation of male peacocks' feathers produced by sexual selection to the cryptic coloration that camouflages so many organisms. In this chapter, we see that *behavior* is just as much a part of an organism's phenotype as is its anatomical structures, and that behavior is produced and shaped by natural selection.

TAKE-HOME MESSAGE 9•1

Behavior encompasses any and all of the actions performed by an organism. When a heritable trait increases an individual's reproductive success relative to that of other individuals, that trait tends to increase in frequency in the population. Behavior is as much a part of an organism's phenotype as is an anatomical structure, and as such can be produced and shaped by natural selection.

9•2

Some behaviors are innate.

In his study of pea plants, Gregor Mendel described a single gene for plant height that caused a plant to be either tall or short. The production of a trait like plant height, however, is not completely genetically determined. Certain environmental conditions, such as the type of soil and the availability of water, nutrients, and sunlight, also have a role to play. Nearly all physical traits of all organisms are the products not only of genes, but also of environmental conditions. When it comes to the production of behaviors, the environment also plays an important role.

The degree to which a behavior depends on the environment for expression, however, varies a great deal. At one extreme are behaviors—called **instincts** or **innate behaviors**—that don't require any environmental input to develop. Innate behaviors are present in all individuals in a population and do not vary

Behaviors Evolve   Cooperation and Altruism   Sexual Conflict   Communication

During non-breeding season

During breeding season

In geese, the sight of an egg outside the nest triggers a fixed action pattern: the goose uses a side-to-side egg-retrieval movement all the way back to the nest, even if the egg is taken away during the process.

During breeding season, a red belly on any other male stickleback fish triggers an aggressive response.

FIGURE 9-4 No learning required. Egg-retrieval behavior in geese and the aggressive behavior of male sticklebacks are fixed action patterns.

much from one individual to another or over an individual's life span. An example of innate behavior is a **fixed action pattern.** Triggered under certain conditions, this is a sequence of behaviors that requires no learning, does not vary, and, once started, runs to completion. Examples of fixed action patterns include:

Egg-retrieval in geese. When a goose spots an egg outside its nest, a fixed action pattern is triggered. The goose gets out of the nest and rolls the egg back, using a side-to-side motion, keeping the egg tucked underneath its bill. Once started, a goose continues the side-to-side egg-retrieval movement all the way back to the nest, even if the egg is taken away during the process (**FIGURE 9-4**).

Aggressive displays and attacks by stickleback fish. The bellies of male stickleback fish turn bright red when the breeding

season arrives. During this time, the sight of a red belly on any other stickleback triggers aggressive behavior in a male stickleback. In fact, the males become aggressive at the sight of anything remotely resembling a red belly. One researcher even noticed that his sticklebacks performed their aggressive displays every day when a red mail truck drove by his window.

TAKE-HOME MESSAGE 9·2

The degree to which a behavior depends on the environment for expression varies a great deal. At one extreme are instincts, innate behaviors present in all individuals in a population that do not vary much from one individual to another or over an individual's life span. Innate behaviors don't require any environmental input to develop.

9·3

Some behaviors must be learned (and some are learned more easily than others).

In contrast to innate behaviors are those behaviors that are influenced to a much greater degree by the individual's environment. Requiring some degree of **learning,** these behaviors are altered and modified over time in response to past experiences. There is tremendous variation among behaviors that require learning: some come relatively easily and are learned by most individuals in a population, while other behaviors are less easily learned.

Consider a trait common to most primates, including humans: fear of snakes. In the wild, rhesus monkeys are afraid of snakes. The sight of a snake causes the monkeys to engage in fear-related responses, including making alarm calls and rapidly moving away from the snake.

Observations of monkeys in captivity, however, reveal that they aren't born with a fear of snakes. Captive monkeys will reach over a plastic model of a snake to get a peanut. Experiments also show that monkeys learn to fear snakes if they see another monkey terrified at the sight of a snake (**FIGURE 9-5**). Even if they see only a videotape of another monkey expressing fear at the sight of a snake, monkeys immediately and permanently become afraid of snakes. They will no longer reach over a snake model to get a peanut, even when they are very hungry. Instead, they scream and move as far away from the artificial snake as possible. Studies on humans show an equally easily learned fear of snakes. Rather than shrieking and cowering in the corner of a cage, however, we respond with sweaty palms and an increased heart rate.

Q *Why is it so much easier for an infant to learn a complex language than for a college student to learn biology?*

Behaviors that are learned easily and by all (or nearly all) individuals are called **prepared learning.** In addition to the snake-fearing behavior of monkeys, examples of prepared learning abound. The acquisition of language in humans is a dramatic example. Most children don't talk until they are a year old, but by the age of three they understand most rules of sentence construction, and the average six-year-old who is a native English-speaker already has a vocabulary of about 13,000 words. These skills are impressive, particularly

PREPARED LEARNING

BEFORE EXPOSURE TO FEAR OF SNAKES
A captive monkey will reach over a plastic snake for food.
Monkey has no fear of snakes.

EXPOSURE TO FEAR OF SNAKES
A captive monkey views another monkey expressing fear at the sight of a snake.
In real life **or** On video

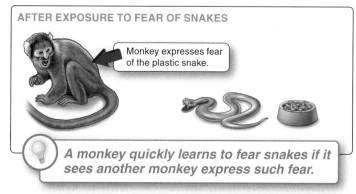

AFTER EXPOSURE TO FEAR OF SNAKES
Monkey expresses fear of the plastic snake.

A monkey quickly learns to fear snakes if it sees another monkey express such fear.

FIGURE 9-5 Learning to be afraid. The monkey can easily learn to fear snakes by observing fear in other monkeys. This is a type of prepared learning.

given that, at this age, children are generally not competent in reading, writing, and fine motor coordination.

Examples of prepared learning also underscore the fact that organisms don't learn everything with equal ease. For example, researchers demonstrated how easily monkeys learned to fear snakes, but in a related experiment, they demonstrated that the

Behaviors Evolve   Cooperation and Altruism   Sexual Conflict   Communication

monkeys were not prepared to learn everything with equal ease. They altered the videotape that previously showed a monkey expressing fear on encountering a snake. On the modified tape, a monkey was shown having the same fear reaction in response to a flower or a toy rabbit. The captive monkeys—which had never seen flowers or rabbits or snakes—did not learn to fear flowers or rabbits. Only snakes.

Q Human babies quickly and easily develop a fear of snakes. Yet they don't easily develop a fear of guns. Why?

These observations point to an evolutionary basis for the acquisition of certain behaviors. It seems that organisms are well-prepared to learn behaviors that have been important to their ancestors' survival and reproductive success over the course of their evolutionary history, and are less prepared to learn behaviors irrelevant to their evolutionary success. Consider, for example, that although human babies quickly and easily develop a fear of snakes, they don't easily develop a fear of guns. Given that more than 30,000 people are killed by guns in the United States each year, while fewer than three dozen people are killed by snakes, it would seem that we ought to be very afraid of guns and relatively unconcerned about snakes. But we are built in just the opposite way (**FIGURE 9-6**). The likely explanation is that evolution can be slow in producing populations that are adapted to their environments—that is, evolutionary change cannot always keep up with a rapidly (in evolutionary terms) changing environment.

Like all genes, any genes involved in behaviors have been handed down to us from our ancestors. Snakes caused many human deaths over the course of our evolutionary history. In contrast, guns didn't kill a single person until very recently on

FIGURE 9-6 Instincts out of date. A human infant is born with a fear of snakes.

the evolutionary time scale. Accordingly, we still fear our ancient enemy, the snake, and have no instinctual response to novel threats, regardless of how deadly.

TAKE-HOME MESSAGE 9·3

In contrast to innate behaviors are those behaviors that are influenced more by the individual's environment, requiring some learning, and are often altered and modified over time in response to past experiences. Organisms are well-prepared to learn behaviors that have been important to the reproductive success of their ancestors, and less prepared to learn behaviors irrelevant to their evolutionary success.

Complex-appearing behaviors don't require complex thought in order to evolve.

Why do humans and other animals have sex? Is it because they are thinking, robot-like: "Must maximize reproductive success. Must maximize reproductive success"? Of course not. Yet they do go about their lives behaving as if they were consciously trying to maximize their reproductive success.

Q *Animals don't consciously try to maximize their reproductive success, yet they behave as if they do. How does this happen?*

In actuality, natural selection produces organisms that exhibit relatively simple behaviors in response to certain situations or environmental conditions—behaviors that to us may seem complex and sophisticated. Let's look at an example. An animal that experiences pleasure when it has sex has the incentive to seek out additional opportunities to experience that pleasure. As long as reproductive success is enhanced in the process, it is not necessary (from the reproductive-success standpoint) for the animal to deliberately seek that outcome. In other words, natural selection doesn't have to produce animals consciously trying to maximize reproductive success. It only needs to produce animals that behave in a way that actually results in reproductive success, and this outcome will occur if organisms experience a pleasurable sensation by having sex. It's like an evolutionary short cut. Behaviors that lead to a specific outcome that increases the animal's relative reproductive success will be favored by natural selection.

We can investigate experimentally whether natural selection works in this way by tricking animals. Consider the example of egg-retrieval in geese described above. When an egg falls out of the nest, a goose retrieves it in exactly the same way every time. Is a goose able to do this because it vigilantly keeps tabs on all of its eggs? This seemed to be the case, until researchers decided to trick some geese. Researchers learned that putting *any* object that remotely resembles an egg near a goose's nest triggers the retrieval process. Geese will retrieve beer cans, door knobs, and a variety of other objects that will not increase the animal's reproductive success. Moreover, if multiple egg-like items are just outside the nest, the goose will retrieve the largest item first.

To test the limits of the retrieval behavior, researchers began putting larger and larger models of eggs outside the nests. When given a choice between an actual goose egg and an artificial egg the size of a basketball, the goose always tried to retrieve the giant egg. It seems that geese aren't actually keeping tabs on their eggs. Rather, they are following a rule of thumb dictating that they retrieve any nearby egg-like objects, and preferentially retrieve the larger objects first (**FIGURE 9-7**). Thus, natural selection can produce organisms that exhibit silly and clearly

RETRIEVING BEHAVIOR IN GEESE

NATURALLY OCCURRING FIXED ACTION PATTERN

By following simple rules that have evolved, an animal can exhibit complex-appearing behaviors.

When an egg rolls out of a nest, a goose retrieves it in exactly the same way every time.

EXPERIMENTAL FIXED ACTION PATTERN

A goose retrieves any object that remotely resembles its eggs.

A goose retrieves the largest egg-like object first instead of the actual goose eggs.

 By tricking an animal, setting up a situation that would never occur in nature, we can reveal that it is following simple behavioral rules.

FIGURE 9-7 Programmed to retrieve. The goose will retrieve any egg-like object outside its nest.

Behaviors Evolve   Cooperation and Altruism   Sexual Conflict   Communication

maladaptive (i.e., fitness-decreasing) behaviors in experiments—the goose trying to retrieve a basketball-sized egg—if, in nature, individuals' performance of those simple actions nearly always leads to fitness-increasing behaviors.

Humans, too, may behave in ways that bolster reproductive success without their awareness. One way involves the avoidance of incest. Because sexual activity among close relatives such as siblings leads to a higher proportion of offspring with genetic defects—the closely related parents may have inherited a disease-causing allele from the same family member and then pass it to their offspring—humans and many other species have evolved an aversion to incest. The kibbutzim in Israel are large, communal systems of child-rearing and education. All of the children in a kibbutz are reared together, interacting constantly and sharing most of their daily experiences from infancy through maturity. Somewhat unexpectedly, there are no records of any children raised in the same kibbutz from infancy who marry each other as adults—even in the face of social encouragement. It seems that

Q Unrelated people who grow up together from infancy on the same kibbutz never marry each other. Why?

communal living during infancy and childhood triggers an incest aversion, the same aversion that is generally triggered in biological siblings living together in typical families.

Taking an evolutionary approach to the study of the behavior of animals, including humans, is not new. In 1859, Charles Darwin wrote in *On the Origin of Species:* "In the distant future I see open fields for far more important researches. Psychology will be based on a new foundation."

And in the past several decades, Darwin's prophecy has increasingly proved true. In a synthesis of evolutionary biology and psychology, researchers taking an approach called "evolutionary psychology" have begun to view the human brain and human behaviors, including emotions, as traits produced by natural selection, selected as a result of their positive effects on survival and reproduction. So, too, has the evolutionary approach to studying behavior begun to influence the field of economics. The 2002 Nobel prize for economics was awarded to researchers who used insights from evolutionary biology and psychology in their analyses of human decision making.

In the rest of this chapter, working from the understanding that organisms' traits, including their behaviors, have evolved by natural selection, we examine the question of why organisms behave as they do. We begin with an exploration of selfishness and cooperation in animals, investigating in particular how natural selection has produced organisms that engage in apparent acts of altruism.

Four children change for bed after baths at the Kibbutz Baram's children's house in Israel. Children live in the children's house and are cared for by all community members.

TAKE-HOME MESSAGE 9•4

If an animal's behavior in natural situations usually increases its relative reproductive success, the behavior will be favored by natural selection. The natural selection of such behaviors does not require the organism to consciously try to maximize its reproductive success.

❷ Cooperation, selfishness, and altruism can be better understood with an evolutionary approach.

Placing herself at risk, a female ground squirrel alerts the colony of approaching danger.

9·5

A general theory of "kindness" helps explain the evolution of apparent acts of altruism.

If we look closely, we see many behaviors in the animal world that *appear* to be **altruistic behaviors**—that is, they seem to be behaviors that come at a cost to the individual performing them while benefiting a recipient. When discussing altruism, we define costs and benefits in terms of their contribution to an individual's fitness.

Take the case of the Australian social spider (*Diaea ergandros*). After giving birth to about 50 hungry spiderlings, the mother's body slowly liquefies into a nutritious fluid that the newborn spiders consume. Over the course of about 40 days, her offspring literally eat her alive; they ultimately kill their mother, but start their lives well-nourished. The cost to the mother is huge—she is unable to produce any more offspring—but there is a clear benefit to the many recipients (**FIGURE 9-8**).

Such altruistic-appearing behavior is common enough in the natural world that it puzzled Darwin. Natural selection, he believed, generally works to produce selfish behavior. The alleles that cause the individual carrying them to have the

greatest reproductive success should increase their market share—that is, should increase as a proportion of all the alleles for that gene within the population—in any given generation. At the same time, alleles that cause the individual carrying them to help increase other individuals' reproductive success should decrease their market share. Put another way, natural selection should never produce altruistic behavior. Darwin worried that if the apparent instances of altruism were indeed truly altruistic, they would prove fatal to his theory.

As it turns out, Darwin's theory is safe. Virtually all of the apparent acts of altruism in the animal kingdom prove, on closer inspection, to be not truly altruistic; instead, they have evolved as a consequence of either **kin selection** or **reciprocal altruism.**

Kindness toward close relatives: kin selection. Kin selection can lead to the evolution of apparently altruistic behavior toward close relatives. Suppose that, for one gene, you carry allele *K*, which causes you to behave in a way that

Behaviors Evolve   Cooperation and Altruism   Sexual Conflict   Communication

This mother spider is actually being selfish—not altruistic—when she lets her offspring eat her alive.

FIGURE 9-8 Parental care to the extreme. The female *Diaea ergandros* spider feeds her offspring with her own body.

increases the fitness of a close relative of yours, while decreasing your own fitness. Allele *K* might actually increase its market share in the population, because you and your relatives tend to share many of the same alleles—including allele *K*. Thus the increased fitness of the relatives you help might compensate for your own reduced fitness, because allele *K* will, overall, increase its frequency in the population when you help your relatives increase their reproductive output.

Kindness toward unrelated individuals: reciprocal altruism. Reciprocal altruism can lead to the evolution of apparently altruistic behavior toward unrelated individuals. Suppose that, for another gene, you carry allele *R*, which causes you to help individuals to whom you are not related. This help increases their fitness and decreases yours in the process. Allele *R* might still increase its market share in the population, if the individuals whom you help become more likely to return the favor and help you at some point in the future, increasing your fitness.

Seen in this light, both kin selection and reciprocal altruism can lead to the evolution of behaviors that are *apparently*

altruistic but, in actuality, are beneficial—from an evolutionary perspective—to the individuals engaging in the behaviors.

In the next two sections, we explore kin selection and reciprocal altruism in more detail. We also investigate some of the many testable predictions about when acts of apparent kindness should occur, whom they should occur between, and how we could increase or decrease the frequency of their occurrence through a variety of modifications to the environment.

> It is not from the benevolence of the butcher, the brewer, or the baker that we expect our dinner, but from their regard to their own interest. We address ourselves, not to their humanity, but to their self-love, and never talk to them of our own necessities, but of their advantages.
>
> —Adam Smith, *An Inquiry into the Nature and Causes of the Wealth of Nations,* 1776

Occasionally, we see individuals engaging in behavior that *is* genuinely altruistic. We will examine how and why certain environmental situations cause individuals to behave in a way that decreases their fitness and, as such, is evolutionarily maladaptive. It is also important to note that even as genes play a central role in both fostering cooperation and spurring conflict, one of the hallmarks of humans is our ability to override the impulses that often push us toward selfishness. This capacity for self-control sets us apart from the rest of the animal kingdom and is responsible for much of the rich diversity in human behavior that we see around us.

TAKE-HOME MESSAGE 9·5

Many behaviors in the animal world appear altruistic. In almost all cases, the apparent acts of altruism are not truly altruistic and have evolved as a consequence of either kin selection or reciprocal altruism and, from an evolutionary perspective, are beneficial to the individual engaging in the behavior.

Apparent altruism toward relatives can evolve through kin selection.

The grasslands of the western United States are home to the Belding's ground squirrels. The squirrels form large colonies of hundreds or even thousands of individuals. They feed mostly on plants, and they live in underground burrows. Because the colonies are so large, they attract many birds of prey. When a bird comes to the colony, it succeeds in killing a squirrel about 10% of the time. Squirrels have a system for reducing predation risk that resembles a neighborhood watch program. When an aerial predator approaches, it is common for a squirrel standing on top of a burrow to produce a loud whistle-like call that serves as an alarm, warning squirrels of the impending danger and reducing their risk of death. On hearing an alarm call, squirrels quickly take cover in their burrows. Making an alarm call is a very dangerous activity: about half the time that an alarm call is made, the squirrel making the alarm call is killed by the predator (**FIGURE 9-9**).

Some squirrels see predators and make alarm calls, while other squirrels see predators and keep their mouths shut. This raises the question: why would any squirrel make an alarm call? And given that some do, which squirrels are most likely to engage in this altruistic-appearing behavior? It certainly isn't random: about 80% of the squirrels making alarm calls are female. Moreover, older females are five times more likely than young females to make alarm calls.

These differences between males and females, and between young and old, reveal that alarm calling is about protecting relatives. The more kin an individual is likely to have, the more likely that individual is to sound the alarm. Because males travel long distances to live in new colonies shortly after reaching maturity, most adult males don't live near their parents, siblings, or any other relatives besides their own offspring. Females, on the other hand, remain near the area where they were born and are likely to have many close relatives nearby. Older females, then, are likely to have the largest number of relatives.

The biologist W. D. Hamilton expressed the idea of "kin selection," that one individual assisting another could compensate for its own decrease in fitness if it helped a close

KIN SELECTION

An individual Belding's ground squirrel will produce a whistle-like call to warn members in its colony that a predator is nearby. Eighty percent of squirrels making the alarm calls are older females, because they have more relatives to protect in the colony.

Predator

Making an alarm call seems like a bad idea—a calling squirrel frequently is killed in the process—but because she protects her relatives, the alarm-calling behavior is favored by natural selection.

FIGURE 9-9 Protecting relatives by making an alarm call. At great risk to herself, a female Belding's ground squirrel will make an alarm call likely to save her family members from a predator.

Behaviors Evolve   Cooperation and Altruism   Sexual Conflict   Communication

INAPPROPRIATE ALARM CALLING IN UNNATURAL SITUATIONS

Researchers transplanted an older female Belding's ground squirrel from her original colony to a new colony, where she had no relatives. Even though this older female is not related to any individuals in the colony, she behaves as if she were and is likely to make an alarm call.

Predator

💡 *As with the egg-retrieval experiments in geese, by tricking an animal, setting up a situation that would never occur in nature, we can reveal that the animal is following simple behavioral rules.*

FIGURE 9-10 **Following simple behavioral rules.** A transplanted female Belding's ground squirrel that is not related to the squirrel colony behaves as though she is.

relative in a way that increased the relative's fitness. After all, he realized, the recipient of the aid was likely to have at least some genes in common with the altruistic individual. And the more genes they shared in common (i.e., the more closely the individuals were related), the more likely it was that the alleles being propagated by the recipient of the altruistic-like behavior were the same alleles found in the altruistic individual.

Thus, altruistic-appearing behavior will most likely occur when the benefits to close relatives are greater than the cost to the individual performing the behavior. That is, the more closely related two individuals are, the more likely they are to act altruistically toward each other. It seems that individuals are acting selflessly/altruistically, but they are really acting in their genes' best interests. (The converse is also true: the less closely related two individuals are, the more likely they are to experience conflict.)

Belding's ground squirrels probably don't know exactly how closely they are related to the squirrels around them, so it is likely that natural selection has led to an evolutionary short-cut that allows them to behave as if they did know. In a clever experiment that tested this idea, researchers trapped adult female squirrels and relocated individuals to distant ground squirrel colonies in which they had no close relatives. If the

transplanted females were able to determine exactly how closely they were related to the squirrels around them, they would not make alarm calls. To do so would put the caller at risk, but would not benefit her genes. What the researchers found, however, was that transplanted females were just as likely to sound the alarm for the colony of strangers as they would be if they had been left on their home territories, surrounded by close relatives (**FIGURE 9-10**). It seems that female ground squirrels have evolved to follow a simple rule that says "if I am an older female, behave as if I have many close relatives around me."

Another example of apparent altruistic behavior toward kin is seen in how people give their money and property away when they die. Kin selection theory gives rise to the prediction that people will bequeath their assets based on their degree of relatedness to others. In an analysis of 1,000 randomly chosen wills, researchers tested two hypotheses: (1) that individuals would leave more of their estates to genetic kin, including their parents, children, and siblings (or spouses, who, the researchers assumed, would distribute the assets to the dead person's genetic kin), and (2) that closer relatives would be left greater amounts than more distant relatives. Both predictions were supported by the analysis: 92% of assets divvied up in the wills were given to genetic kin. And of this 92%, 46% was given to those individuals most closely related to the dead person (siblings and children), while only 8% was given to half-siblings, grandchildren, and nieces and nephews, and about 1% was given to cousins or more distant relations (**FIGURE 9-11**).

Q Who are people most likely to bequeath money to in their will?

KIN SELECTION: DIVIDING WEALTH

Individuals are more likely to leave their estates to genetic kin, which includes parents, children, and siblings, and to spouses (who ultimately distribute the assets to genetic kin). Closer relatives receive more of the estate than distant relatives.

ESTATE DISTRIBUTION

● Genetic kin or spouse: 92%
● Non-relatives: 8%

- Siblings and children 46%
- Spouses 37%
- Non-relatives 8%
- Cousins and distant relatives 1%
- Half-siblings, grandchildren, nieces, and nephews 8%

FIGURE 9-11 **Reading the will.** How likely are you to inherit money from a family member?

Based on the idea of kin selection, it is necessary to redefine an individual's fitness. An individual's fitness is not just measured by his or her total reproductive output (this is called **direct fitness**). Fitness also includes the reproductive output that individuals bring about through their seemingly altruistic behaviors toward their close kin. This is called an individual's **indirect fitness.** The sum of an individual's indirect and direct fitness is called **inclusive fitness.**

Q James Joyce wrote: "Whatever else is unsure in this stinking dunghill of a world, a mother's love is not." According to the concept of kin selection, he may not be completely right. Why?

No two individuals—with the exception of identical twins—are genetically identical. And because different individuals do not share all of the same alleles, we expect that they should experience some conflict. One disturbing example of such conflict takes place between a pregnant woman and her developing fetus.

A mother and fetus differ on the question of how much food—doled out as nutrients in the blood flow across the placenta—the fetus ought to get. There is a point at which it is in the mother's best interests to reduce the amount of glucose and other nutrients given to the fetus. Her genes gain if she saves some of her resources for future fetuses, to whom she will be related equally as to the present fetus.

Now consider the fetus's "point of view." Future siblings will carry some but not all of the same genes as the fetus. Consequently, the fetus does not necessarily benefit from sacrificing nutritional intake for the benefit of future siblings. In a sense, it is a sibling rivalry that starts before the sibling is even conceived! The conflict results in a physiological battle throughout pregnancy. The fetus produces chemicals that increase the diameter of the mother's blood vessels, thus increasing the amount of sugar delivered to the fetus. In response, the mother produces insulin, a chemical that has exactly the opposite effect, causing a reduction in the amount of sugar in the bloodstream available to the fetus. In some mothers this conflict causes gestational diabetes, an inability to properly regulate her blood sugar levels—a condition that disappears as soon as the baby is born. In all pregnancies, though, the conflict escalates until the mother is producing a thousand times the normal amount of insulin.

Gestational diabetes might also be the result of conflict between the fetus's alleles that come from its father and those that come from its mother. The parents may have different interests concerning how much to invest in the current offspring versus how much to save for future offspring.

The idea of kin selection gives rise to another prediction about conflict among relatives and non-relatives. Specifically, it generates the prediction that children in step-families will be injured or killed more frequently than those living with their biological families. And in fact, evidence from numerous studies across many different cultures, including evaluation of 20,000 reports from the American Humane Association in the United States, supports what has been called "the Cinderella syndrome." These findings included, for example, an estimate that the risk of abuse for a pre-schooler is about 1 in 3,000 for a child living with two biological parents (with whom he or she shares considerable genetic relatedness) and 40 in 3,000 for a child living with a step-parent (with whom he or she has no genetic relatedness). This difference in risk for children living in a home with a step-parent versus those living with their biological parents remains even when socioeconomic factors are taken into account.

Q Is a child living with one or more step-parents at greater risk of abuse than a child of the same age living with his or her biological parents?

Although the effect has been noted in multiple other cultures, the finding that living with a step-parent represents a significant risk factor for an individual has met with shock and has been criticized. Nonetheless, it is essential to note that while these data are unambiguous, abuse is still exceedingly rare—with homicide rates of approximately 0.06% per year among children 0–2 years of age living with a step-parent and 0.001% for those living with biological parents. Put simply, in the overwhelming majority of step-families no abuse occurs—a testament that "counter-Darwinian" behavior is also an important part of human nature. Ongoing research continues to explore a range of explanations for the effect. As would be expected, infanticide by unrelated "step-parent" males is also common in other species, such as lions. Males, upon taking over a pride, will usually kill the cubs that are under about nine months of age (although females defend their young aggressively), causing the females to become reproductively ready sooner than they would have if they had continued nursing their cubs.

TAKE-HOME MESSAGE 9·6

Kin selection describes apparently altruistic behavior in which an individual that assists its genetic relative compensates for its own decrease in direct fitness by helping increase the relative's fitness and, consequently, its own inclusive fitness.

Behaviors Evolve   Cooperation and Altruism   Sexual Conflict   Communication

Apparent altruism toward unrelated individuals can evolve through reciprocal altruism.

It is ironic that studies of altruism in the world reveal that natural selection has primarily produced selfish behavior among animals. And, as we saw in the previous section, when behavior appears altruistic, this is frequently because individuals are helping kin and, by doing so, are promoting the reproduction of the copies of the genes (i.e., the alleles) they carry that are also carried by their close relatives. Does any apparently altruistic behavior toward non-relatives (what humans might call "friendship") occur in the animal world? Yes, but there's not much of it. In this section we explore the conditions that give rise to reciprocal altruism, how it may have arisen, and why it is so common among humans but so rare among most other animal species.

We start by examining one species with well-documented altruistic-appearing behavior: the vampire bat. Vampire bats live in social groups of 8–12 adults, roosting primarily in caves and hollow trees. They feed by landing on large mammals such as cattle, horses, and pigs, piercing the skin with their razor-sharp teeth, and drinking the blood that flows from the wound. A chemical in their saliva keeps the blood from clotting, and thus allows the bats to feed for a longer period of time. Because of their small body size (about the size of your thumb) and their very high metabolic rate, vampire bats must consume almost their entire body weight in blood each night. If they go for more than about 60 hours without finding a meal, they are likely to die from starvation. Here's where the apparent altruism comes in: a bat that has not found food and is close to death will beg food from a bat that has recently eaten. In many cases, the bat that has just eaten will regurgitate some of the blood it has consumed into the mouth of the hungry bat, saving it from starvation. This act obviously has very high benefit for the recipient of the blood, but it comes at a cost to the sharing bat, which loses some of the caloric content of a meal it has just obtained (**FIGURE 9-12**).

Kin selection is responsible for some of the blood sharing (females often regurgitate blood for their own offspring), but, in many cases, individuals give blood to unrelated individuals. How might this behavior have arisen? To answer the question, it is important to note three other features of vampire bats. First, they are able to recognize more than a hundred distinct individuals. Second, bats that receive blood donations from non-relatives reciprocate significantly more

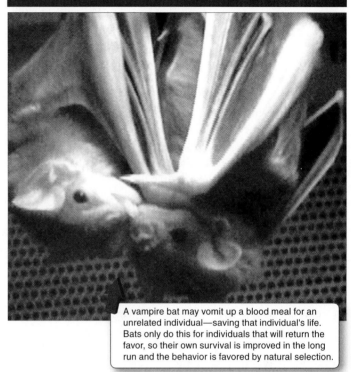

RECIPROCAL ALTRUISM IN VAMPIRE BATS

A vampire bat may vomit up a blood meal for an unrelated individual—saving that individual's life. Bats only do this for individuals that will return the favor, so their own survival is improved in the long run and the behavior is favored by natural selection.

FIGURE 9-12 The gift of life. A starving vampire bat receives blood from another, well-fed bat. The recipient of the blood will return the favor one day.

than average with the bats that have shared blood with them. And third, bats that are not familiar with each other (and do not have a history of helping each other) generally do not regurgitate for each other.

One method proposed to explain the evolution of this apparent altruism is that the bats giving blood to other bats in need are repaid the favor when they are in need of blood. In other words, the act only seems to be selfless, when in actuality it is selfish. With such reciprocal altruism, both individuals (at different encounters) give up something of *relatively* low value in exchange for getting something of great value at a later time when they need it most. In other words, they are storing goodwill in another individual, in much the same way that a person might put money in a bank for a rainy day. In both cases, individuals are protected from some of the world's uncertainties.

Q We're all born with a spare kidney, yet virtually no one donates theirs to one of the thousands of non-relatives in need. Why?

Imagine that you are a parent and your child is in need of a kidney transplant to survive. Would you consider donating a kidney to save your son or daughter? Most people do agree to give a kidney in this situation, and this type of donation is an illustration of kin selection. Your action may indirectly improve your fitness (and also help someone you cherish.) Now, imagine a circumstance where you are asked to donate a kidney to someone you have never met. Would you?

You might be surprised to know that kidney donation to a stranger is an extremely uncommon occurrence; a consideration of the conditions under which reciprocal altruism evolves helps to explain why. Reciprocal altruism can evolve if certain conditions are met:

1. Repeated interactions among individuals, with opportunities to be both the donor and the recipient of altruistic-appearing acts

2. Benefits to the recipient that are significantly greater than the costs to the donor

3. The ability to recognize and punish cheaters, individuals that are recipients of altruistic-appearing acts but do not return the favor

In the absence of these conditions, selfishness is expected to be the norm among unrelated individuals. In the presence of all three conditions, on the other hand, reciprocity masquerading as altruism is likely to occur, as it does with blood sharing among the vampire bats.

The three conditions required for the evolution of reciprocal altruism are not satisfied in many animal species, which may be why altruistic-appearing behavior among unrelated individuals is rare. But there are some exceptions. The cleaner fish, for example, is a small fish that is allowed to swim right into the mouth of a grouper, a much larger fish, where it cleans the grouper of parasites by eating them. Stomach analyses reveal that groupers almost never eat the cleaner fish, opting to pass up this opportunity for a meal. And so, while the grouper is cleaned of parasites, the cleaner fish gets to eat a meal of parasites. In this case, it is easy to rule out kin selection as the reason for the altruistic-appearing behavior; the fish are from different species and can't possibly be close relatives. It is difficult to determine with certainty how this relationship got started, but many mutually beneficial relationships such as this

Q Why are humans among the few species to have friendships?

RECIPROCAL ALTRUISM IN GROOMING BEHAVIOR

Time that could be spent finding food? Individuals that groom others are more likely to receive assistance if they ever are in danger—and so the grooming behavior is favored by natural selection.

FIGURE 9-13 You scratch my back, I'll scratch yours. One vervet monkey will groom another, but expects assistance in return.

seem to develop from unavoidable antagonistic encounters, particularly when the environment is stressful. In other words, when the conditions are right, behaviors can evolve that help organisms live more harmoniously with each other—when doing so improves their reproductive success.

Reciprocal altruism has also been documented in vervet monkeys. Monkeys that would otherwise be searching for food or, perhaps, a mate spend time grooming other monkeys (mostly by picking parasites from their fur). In return for their grooming behavior, the monkeys are more likely to receive assistance in response to their solicitations for aid (**FIGURE 9-13**).

As rare as it is in other species, reciprocal altruism is very common among humans. Friendship, among other human relationships, is built on reciprocity and is almost universal. The opportunity for friendship may be enhanced by our long life span and our ability to recognize thousands of faces and keep track of cheaters. These features are essential in individuals engaging in reciprocal altruism, because an individual becomes very vulnerable when he or she acts in an altruistic manner toward an unrelated individual. The risk is that the altruism will not be repaid, in which case the cheater enjoys greater fitness than the altruist.

While cooperators, those who repay altruistic-appearing behavior, have evolutionary advantages over loners, this advantage disappears if the cooperators are the givers all or most of the time, never, or rarely, getting anything in return. Thus, it pays to avoid cheaters, individuals who accept altruistic behavior without repaying it. The importance of keeping track of cheaters and of identifying good potential reciprocity partners (i.e., other cooperators) may be why humans are so interested in social information and seemingly trivial gossip. As with many of the complex behaviors we have explored, humans probably use some rules of thumb when making decisions about reciprocal altruism. This might include keeping track of *all* available social information to best determine any promising individuals with whom to engage in reciprocal altruism. This includes information about health, generosity, social status, and reproduction. Researchers have even hypothesized that the evolution of maintaining an interest in social information, while valuable in most contexts, may now have maladaptive manifestations. We may find ourselves interested in and distracted by the social life of Julia Roberts, Tom Cruise, or other individuals whom we will most likely never even meet.

TAKE-HOME MESSAGE 9·7

In reciprocal altruism, an individual engages in an altruistic-appearing act toward another individual. Although giving up something of value, the actor does so only when likely to get something of value at a later time. Reciprocal altruism occurs only if individuals have repeated interactions and can recognize and punish cheaters, conditions satisfied in humans but in few other species.

9·8

In an "alien" environment, behaviors produced by natural selection may no longer be adaptive.

A transplanted female squirrel makes an alarm call, putting herself in grave danger to help out dozens of squirrels that share none of her genes. A person in the United States donates money to a fundraising drive to benefit refugees in Africa. Or someone finds it difficult to maintain her body weight, craving high-fat food even as she knows it may increase her risk of heart disease and shorten her life.

These situations have something in common. In each, genetically influenced traits, behaviors favored by natural selection over evolutionary time, cause individuals to behave in a way that *reduces* their fitness. But how can that be? How can natural selection lead to fitness-reducing behaviors?

To better understand these situations, it is helpful to return to our discussion of fitness, in Chapter 8. Recall that fitness not only is an individual's reproductive success, relative to that of other individuals in the population, but also depends on the specific environment in which the organism lives. An adult female Belding's ground squirrel that makes alarm calls generally increases her inclusive fitness by doing so, because all around her are individuals who share her genes. For this reason, such behavior has been favored by natural selection and has increased in frequency in the population.

The world in which alarm calling evolved did not include biologists with pickup trucks who trapped female squirrels, drove them long distances, and released them into colonies of unrelated individuals. (In a world where long-distance transplantation was common, alarm calling would not have evolved.) So, when a female squirrel suddenly finds herself in that world, her evolved behaviors can no longer be expected to be adaptive—no more than a human could be expected to survive on the moon. But an individual's genes and all the behaviors they influence cannot change overnight. Adaptation to a new environment takes time, and the more quickly an environment changes, the more likely it is that the evolved behaviors of a population will no longer be appropriate.

Let's return to the donations to refugees on another continent. To understand how natural selection could lead to this behavior, it is necessary to understand a bit about human evolutionary history. We know from archaeological deposits that for more than 2 million years, our ancestors lived as hunter-gatherers in small groups of a few hundred people, at most. Their success depended on joint efforts against predators and prey; being "nice" paid off when the prospect of hunting alone or sleeping outside the camp meant almost certain death. The loners died, so we are descended from those who could work well with others.

It is only recently, the blink of an eye in evolutionary time, that humans invented agriculture, industrialization, and the means of food production and distribution. As a result of these changes, population group size has increased dramatically; on a given day, you may see 10 or even 100 *times* as many people as a hunter-gatherer ancestor might have seen. Additionally, if you were a hunter-gatherer, every one of the people in your group would be a regular part of your life. You would have many opportunities to help them and, in turn, to be helped by them. Reciprocity paid. And we evolved so that "altruistic" acts gave us pleasure, stimulating parts of our brain in ways that made us want to repeat such actions.

It is with this brain that you approach the issue of refugees halfway across the world, or a homeless person somewhere in the United States. It is almost certain that you will not have repeated interactions with any of those people. Nor will you ever be in a position to be helped by them. From an evolutionary perspective, your action almost certainly will not increase your fitness. But in the world humans evolved in, such altruistic-appearing behavior would most likely have been reciprocated at some future time, and thus your instincts guide you to, and reward you for, your kindness (**FIGURE 9-14**). It feels right.

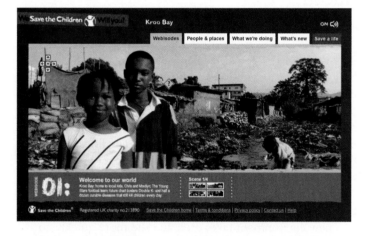

FIGURE 9-14 **Charitable acts can give us pleasure.** In the small hunter-gatherer groups in which humans evolved, altruistic behaviors would be reciprocated. Today, the pleasure we feel in response to such behaviors remains—even though the favor may never be returned.

In addition to small population sizes, our ancestors' hunter-gatherer world was characterized by unpredictable food sources that could not be stored for long periods of time. In response to this environment, humans evolved to have great appetites and taste preferences for calorie-laden foods. Fitness increased with food consumption and the storage of fat, helping to protect individuals during the times of starvation that occurred with regularity. From an evolutionary perspective, today's world is truly an alien environment: most of us have easy access to unlimited amounts of food, including fat-laden fast foods. Because humans have not yet adapted to this environment of plentiful food, our instincts to consume large amounts are strong, making it difficult for most people to control their body weight.

From battles against obesity to charitable contributions to alarm calling in transplanted Belding's ground squirrels, when organisms of any species find themselves in an environment that differs from the environment to which they are evolutionarily adapted, we expect (and see) behaviors that appear to be (and are) not evolutionarily adaptive. Still, understanding the process of natural selection can help us make sense of these seemingly nonsensical behaviors. It is important not to lose sight of the fact that just because a behavior such as making a donation to help people on a faraway continent is not adaptive from an evolutionary perspective, this aspect of the behavior is completely irrelevant when you're deciding whether or not it is something you want to do. When it comes to ethical decision making, there is nothing inherently good or important about the behaviors that natural selection has favored.

Can you think of other ways in which the modern industrial environment differs from the environment to which humans are adapted? How do these differences lead to situations in which our behavior, like the behavior of the transplanted Belding's ground squirrel, may not be adaptive from an evolutionary perspective?

TAKE-HOME MESSAGE 9·8

When organisms are in environments that differ from the environment to which they are adapted, the behaviors they exhibit are not necessarily evolutionarily adaptive.

Behaviors Evolve   Cooperation and Altruism   Sexual Conflict   Communication

Selfish genes win out over group selection.

Kin selection and reciprocal altruism can evolve in a population of animals, as we have seen in the previous three sections, and examples of both abound in nature. As a consequence, casual observers of nature frequently see individuals acting in ways that appear altruistic, even though, as we have seen, these individuals are truly acting—from the perspective of evolutionary fitness—in their own selfish interests. This nearly universal selfishness raises the question of whether evolution ever leads to behaviors that are good for the species or population but detrimental to the individual exhibiting the behavior, a process called **group selection.**

It might seem that evolution would favor individuals that behave in a manner that benefits the group, even if it comes at a cost to the individual's reproductive success. But this does not happen. Behaviors that reduce an individual's reproductive output (relative to that of other individuals in the population) are not likely to evolve. Natural selection favors individuals producing more offspring relative to those producing fewer offspring.

Let's look at an example. Imagine that a new allele appears in a population (perhaps through mutation) that causes the individual carrying the allele to double its reproductive output, even though this might spell doom for the species as individuals over-use their resources. An individual carrying this "selfish" allele will pass it to more offspring than an individual carrying the alternative allele, coding for the production of fewer offspring, will pass its allele—and this will lead to excessive, "selfish" consumption of the species' resources. And the selfish offspring also will pass on the selfish allele at a higher rate than the alternative allele is passed on. This scenario may lead to extinction of the species, yet it still occurs. With its market share perpetually increasing, the selfish allele eventually will be favored by evolution and will predominate (**FIGURE 9-15**).

Just because a behavior (determined by an "unselfish" allele) leads to a better outcome for the group doesn't mean that natural selection will favor that behavior. In the end, natural

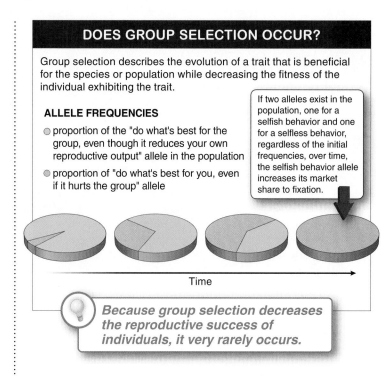

DOES GROUP SELECTION OCCUR?

Group selection describes the evolution of a trait that is beneficial for the species or population while decreasing the fitness of the individual exhibiting the trait.

ALLELE FREQUENCIES

- proportion of the "do what's best for the group, even though it reduces your own reproductive output" allele in the population
- proportion of "do what's best for you, even if it hurts the group" allele

If two alleles exist in the population, one for a selfish behavior and one for a selfless behavior, regardless of the initial frequencies, over time, the selfish behavior allele increases its market share to fixation.

Time

Because group selection decreases the reproductive success of individuals, it very rarely occurs.

FIGURE 9-15 Can a "selfless" gene increase in frequency in a population?

selection generally causes increases in the alleles that benefit the *individual* carrying them, even when this comes at the expense of the group. In some special situations, it is possible for natural selection to lead to group selection. But the stringent conditions necessary for this to occur are so rarely found in nature that we almost never see it.

TAKE-HOME MESSAGE 9•9

Behaviors that are good for the species or population but detrimental to the individual exhibiting such behavior are not generally produced in a population under natural conditions.

❸ Sexual conflict can result from disparities in reproductive investment by males and females.

The wandering albatross extends its wings during a courtship display as a potential mate looks on.

9·10

There are big differences in how much males and females must invest in reproduction.

As we've seen in this chapter so far, many behaviors have evolved that influence the ways in which animals interact with each other, reducing conflict and sometimes leading to cooperation. One aspect of animal life that necessarily involves interaction among individuals—for sexually reproducing species—is reproduction, from courtship and mating to parental investment and the forming and breaking of ties between two animals. In this and the next few sections, we will explore these behaviors and the factors that influence them.

How many babies can a woman produce over her lifetime? Is the number larger or smaller than the maximum number of babies a man can produce? The answer to the first question is probably higher than you would guess: a Russian woman had 69 children (in 27 pregnancies). But even this high number is greatly exceeded by the 888 offspring produced by one man (the Emperor of Morocco from 1672 to 1727). This large difference between the number of offspring that can be produced by females and males is less surprising. Among other species of mammals, the pattern is consistent: a male elephant seal can produce 100 offspring, while the maximum produced by a female over her lifetime is 8; a male red deer can produce 24 offspring, while the maximum produced by a female is 14.

Here we examine the physical differences between males and females and how they lead to the differences in male and female sexual behavior that have evolved.

The very definition of "male" and "female" hinges on a physical difference between the sexes. Recall from Chapter 6 that in species with two distinct sexes, a **female** is defined as the sex that produces the larger gamete, while a **male** produces the smaller gamete (**FIGURE 9-16**). At conception, the mother's material and energetic contribution to the offspring, the energy she will expend in the growth, feeding, and care of offspring—her **reproductive investment**—exceeds the father's. This is true for all animals, whether mammals, birds, insects, or sharks. (It is also true for plants.) Not only are female gametes larger, they tend to be relatively immobile and are produced in smaller numbers. Male gametes, on the other hand, while smaller, are more plentiful and very motile.

This discrepancy in size and quantity between the sperm produced by a male and the eggs produced by a female may seem trivial, but it sets the stage for evolutionary developments that magnify this initial difference in reproductive investment. For starters, the difference in the number of gametes that males

| Behaviors Evolve | Cooperation and Altruism | Sexual Conflict | Communication |

At conception, the mother's energetic contribution to the offspring, her reproductive investment, exceeds the father's.

FIGURE 9-16 **Many sperm, just one egg.**

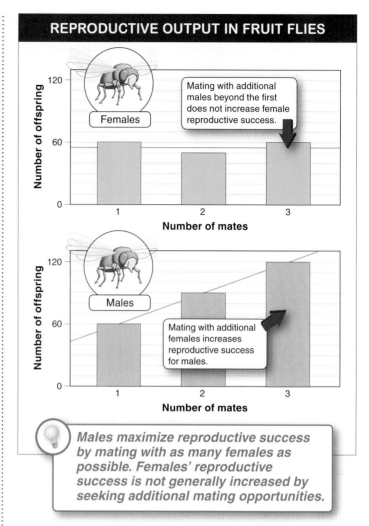

REPRODUCTIVE OUTPUT IN FRUIT FLIES

Females

Mating with additional males beyond the first does not increase female reproductive success.

Males

Mating with additional females increases reproductive success for males.

Males maximize reproductive success by mating with as many females as possible. Females' reproductive success is not generally increased by seeking additional mating opportunities.

FIGURE 9-17 **Maximizing reproductive success in fruit flies.**

and females can produce means that males have the potential to produce many, many more offspring than females. Put another way, a male's **total reproductive output,** the lifetime number of offspring he can produce, tends to increase as the number of females he is able to fertilize increases. A female, on the other hand, does not generally increase her reproductive success by mating with additional males beyond the first (**FIGURE 9-17**).

Q Why do males usually compete for females rather than the opposite?

Because additional matings usually lead to greater increases in reproduction (and fitness) for males relative to females, selective pressure has resulted in the evolution of some differences in male and female reproductive behavior. For males, the most effective way to maximize reproductive success is often to find and gain access to mating opportunities with additional females. For females, an effective way to maximize reproductive success is often to put more effort into parenting their offspring and less effort into seeking mates.

Two physical differences that can exist between males and females are particularly important when it comes to reproduction. First, in species with internal fertilization, including most mammals, fertilization takes place in the female. The offspring also grow and develop within the female's body. Because of internal fertilization and development, the amount of energy invested in reproduction by females—their reproductive investment—is much greater than the investment by males. It also limits a female's reproductive output; she can only be pregnant once at a time. A second important physical difference between females and males occurs among the mammals: lactation occurs in females and not in males (with a

very small number of exceptions). In these species then, nurturing during both pregnancy and lactation can only be accomplished by the female. This difference in reproductive investment has led to the evolution of some very different reproductive behaviors between males and females.

Although the gamete size difference is consistent across all animals (i.e., the egg is always bigger and energetically more costly to make than the sperm), the physical differences between the sexes in the early nurturing (i.e., gestation and feeding) of offspring can vary considerably across animal species. In some cases, the early investment in reproduction is much greater for females than for males, but male and female investments become more nearly equal after fertilization. In birds, for example, after fertilization but prior to emergence of the chicks, much of the development of the fertilized egg is external: the female lays an egg, but either the male or the female can protect and incubate the developing embryo. Further, birds do not lactate. Once hatched, the chicks must be fed—a task that can be done by both parents.

The female mammal invests much energy in the production and care of offspring.

Amphibian eggs are left to develop on their own; there is very little maternal or paternal reproductive investment after fertilization.

Among most bird species, males and females have an equal reproductive investment.

FIGURE 9-18 **Male and female reproductive investment differs across species.** Female investment is greatest when offspring develop internally and are fed by lactation after birth. In animals whose offspring develop externally, males or females can provide the care.

Because incubation of the eggs and early feeding can be done by either parent, the early investment in reproduction in birds does not necessarily differ so dramatically between males and females as it does in mammals. In many bird species, the maximum lifetime reproductive output of males and females is similar. In the kittiwake gull, for example, the largest number of offspring produced by a male is 26, while for a female the number is 28. External fertilization in fish and amphibian species further reduces the reproductive investment of the female—she does not spend any energy as the fertilized eggs begin developing into embryos (**FIGURE 9-18**).

We have seen, then, that in species where fertilization occurs inside the female body, there is an unequal energetic investment in reproduction, and as a result the female expends more energy in growth and care of offspring (much more care for mammals; less so for birds). Another profound consequence of internal fertilization is that a male cannot be 100% certain that any offspring a female produces are his progeny. Because it is possible for a female to mate with multiple males, any of whom

could be the father, male mammals and birds will always have some degree of **paternity uncertainty.**

In the next few sections, we'll continue to explore how the physical differences between males and females in reproductive investment, along with paternity uncertainty, have led to the evolution of differences in male and female reproductive behavior.

TAKE-HOME MESSAGE 9•10

In mammals and many other types of animals, there are important physical differences between males and females relating to reproduction. Fertilization usually takes place in the female. Lactation occurs only in the female. And in species where fertilization occurs inside the female, males cannot be certain that offspring are their progeny. These physical differences have led to the evolution of differences in male and female reproductive behavior.

9•11

Males and females are vulnerable at different stages of the reproductive exchange.

Suppose you are on your college campus and a person of the opposite sex comes up to you and says, "Hi. I have been noticing you around campus. I find you very attractive. Would you go out with me tonight?" What percentage of men would answer *yes*? What percentage of women would answer *yes*? In a study conducted at Florida State University in 1978 and 1982,

this encounter was set up for both males and females, and the percentage answering yes was 50% for both males and females.

Now imagine a different version of the encounter. In this case, the question was: "Would you have sex with me tonight?" Among the men asked, 75% said *yes*; among the women, not a

Behaviors Evolve   Cooperation and Altruism   Sexual Conflict   Communication

FIGURE 9-19 **The choosier sex.** Males and females differ in their attitude toward mating opportunities. For the female more than the male, the choice of the wrong mate could have expensive consequences.

single one said *yes*. It is tempting to interpret such results as a consequence solely of the culture of Western society, characterized by significant differences in the expectations and tolerances of male and female sexual conduct. However, cultural expectations don't adequately explain the consistency of the pattern found across Western and non-Western societies: men and women differ in their approach to sexuality.

Gender roles with regard to sexual willingness are greatly influenced by socialization, and they can vary cross-culturally. Studies of male-female differences in selectiveness about sexual partners, however, drawing on a wide range of industrial and preliterate human cultures—including many as far removed from Western influence as possible, such as the Trobriand Islanders of Melanesia in the South Pacific—most commonly reveal a difference in men's and women's willingness to have sex. Why might this difference exist?

Humans, like nearly all mammals, are characterized by greater initial reproductive investment by females. As we saw in the college sex study, women are more hesitant to take a mating opportunity, while men are much more willing. For females, at the point of mating, the cost of a poor choice can have significant consequences, from an evolutionary perspective— pregnancy and lactation, with offspring from a low-quality male or a male who deserts her and does not provide any parental investment or access to valuable resources (FIGURE 9-19). For a male, the consequences of a poor choice are less dire—little beyond the time and energy involved in mating.

Two differences in the sexual behavior of males and females have evolved:

1. The sex with the greater energetic investment in reproduction will be more discriminating when it comes to mating.

2. Members of the sex with less energetic investment in reproduction will compete among themselves for access to the higher-investing sex.

A dramatic illustration of how a high reproductive investment leads to the evolution of choosiness in mating

behavior comes from the insect world. When bush crickets mate, the male loses about a quarter of its body weight contributing a massive ejaculate, which the female then uses for energy (FIGURE 9-20). (It can represent up to one-tenth of her lifetime caloric intake.)

Not surprisingly, male crickets are very choosy when selecting a mate. Their contribution during sex would be the equivalent of nearly 50 pounds of semen in humans, a much greater up-front investment than usual. As a consequence, male bush crickets reject small females that would produce relatively few offspring. (Females, you won't be surprised to learn, spend a great deal of effort courting males.)

So far, we have been talking about the consequences of reproductive investment. We see how the sex with the greater reproductive investment must be choosy. This isn't to say that paternal investment will not or should not occur, however. In many species, particularly those in which the offspring are not

 When bush crickets mate, the male loses about a quarter of his body weight contributing a massive ejaculate that the female uses for energy. Why has this led to the evolution of choosiness among male bush crickets?

FIGURE 9-20 **A costly decision.** Because of the investment the male bush cricket is required to make when fertilizing a female, he chooses a mate very carefully.

351

well-developed at birth (including humans), offspring survival is significantly increased with paternal investment. As a consequence, this benefit frequently outweighs the evolutionary risks associated with paternity uncertainty. The female's reproductive success is vulnerable because a poor choice of mate could be disastrous. Males are not made vulnerable through their mating choices, because the matings are of little energetic consequence. However, there is a point at which males *are* vulnerable in the reproductive exchange.

The point of greatest vulnerability for males comes when they provide parental care to offspring. Due to paternity uncertainty, there is some chance that the male may be investing in offspring that are not his own. This is an action that has significant evolutionary costs: rather than increasing his own fitness, he is increasing the fitness of another male. This isn't to say that paternal investment will not or should not occur, however. In many species, particularly those in which the offspring are not well-developed at birth (including humans), offspring survival is significantly increased with paternal investment. As a conse-quence, this benefit frequently outweighs the evolutionary risks associated with paternity uncertainty. Females, conversely, are not at all vulnerable at the point of providing parental care to offspring, because a female can be completely certain that the offspring she gives birth to are her own.

In the next two sections, we explore the evolution of reproduc-tive behaviors that help males and females reduce their vulnera-bility and maximize their reproductive success. First we explore competition and courtship tactics used in selecting and securing a mate. And then we investigate the various strategies of "mate guarding" employed by males to increase their paternity certainty.

As we explore some general patterns of reproductive behavior among males and females, it is important to keep two critical points in mind:

1. There is tremendous variability among different species in male and female behaviors. For example, the use of DNA fingerprinting to identify the parents of bird offspring has lead to some surprising revelations. In significantly more cases than researchers originally predicted, offspring have different fathers than observers believed. It seems that things are not always as they appear to observing biologists. As a consequence, much of the new research calls into question some long-held assumptions about the behavioral consequences of physical differences in reproductive investment among birds versus mammals. There are general behavioral patterns among birds and among animals, but it is increasingly clear that these are not universal features of their biology, and a reasonable skepticism is important when identi-fying and interpreting broad trends across large groups of species.

2. Also, throughout history there have been many cases of people using observations and scientific findings to justify a wide variety of discriminatory thoughts and behaviors—for example, that if among mammals males "naturally" tend to be less faithful to their mates, such observations reduce individual responsibility for behaviors among humans. Such thinking ignores the tremendous variation in behavior among species and the power of cultural norms and socialization to influence and shape human behavior, while encouraging the inappropriate assumption that biology can supply meaningful insights into morality or ethical decision making.

TAKE-HOME MESSAGE 9•11

Differing patterns of investment in reproduction make males and females vulnerable to exploitation at different stages of the reproductive process. This has contributed to the evolution of differences in the sexual behavior of males and females. The sex with greater energetic investment in reproduction is more discriminating about mates, and members of the sex with less energetic investment in reproduction compete among themselves for access to the higher-investing sex.

9•12

Tactics for getting a mate: competition and courtship can help males and females secure reproductive success.

Throughout the animal kingdom, there is tremendous variation in the relative investment in reproduction among males versus females. A common pattern, however, is that females make greater investment in producing and raising offspring, while the male contribution is less. Accordingly, in these cases, males generally increase their reproductive

success by mating with many females and have evolved to compete among themselves to get the opportunity to mate. Females, conversely, do not increase their reproductive success through extra matings. Instead, they increase their reproductive success by caring for their offspring and being choosy about selecting a mate.

| Behaviors Evolve | Cooperation and Altruism | Sexual Conflict | Communication |

Female choosiness (and the male–male competition that it leads to) tends to increase the likelihood that a female will select only those males that have plentiful resources or relatively high-quality genes, either of which is beneficial to the female, causing her to produce more offspring or better offspring—where "better" may mean increased disease resistance or simply the possession of physical traits that will be found attractive by future mates. Female choosiness is manifested by four general rules.

1. Mate only after subjecting a male to courtship rituals. In many bird species, including the western grebe, females require the male to perform an elaborate and time-consuming courtship dance before she will mate with him. For the grebe, this courtship dance involves fancy dives into water, graceful hovering, various head movements, and flamboyant twists and turns. The courtship process can go on for several days. But if the male passes the time-consuming audition, he can generally be counted on to stick around to see a brood through hatching and early care (**FIGURE 9-21**).

Among bower birds, a courting male builds a small thatched structure, at which he makes displays that can involve complex dance steps and dramatic poses, while courting a female. The female will walk around the structure and inspect it. Only if she deems it satisfactory will she mate with the male. Males that are rejected by multiple females will usually tear down their bower and rebuild it from scratch to try again to attract a female.

2. Mate only with a male who controls valuable resources. Territorial defense is a common means by which males compete for access to females. Among arctic ground squirrels, for example, a female chooses a mate based, in part, on the territory he defends, which is where she will reside after mating. With greater quality and quantity of resources in

FACTORS IN MATE SELECTION

COURTSHIP RITUALS
A female grebe requires the male to perform a courtship dance before she will mate with him.

GIFTS UP FRONT
A female hanging fly will not mate with a male unless he brings her a large offering of food.

CONTROL OF VALUABLE RESOURCES
A female arctic squirrel is attracted to a male squirrel that controls the best territory.

GOOD LOOKS
A female peacock is attracted to a male with the most beautiful tail feathers.

FIGURE 9-21 Four factors that influence a female's choice of mate.

his territory, a male is better able to attract females, whose reproductive success can be increased if the territory is rich in resources.

> It is a truth universally acknowledged, that a single man in possession of a good fortune must be in want of a wife.
>
> —Jane Austen, *Pride and Prejudice*, 1813

3. Mate only with a male who contributes a large parental investment up front. Better than a believable pledge to commit resources to future offspring is an actual, on-the-spot exchange in which a female requires a male to give her his parental investment up front, in the form of resources that will help her maximize her reproductive success. In the hanging fly, for example, a female will not mate with a male unless he brings her a big piece of food—usually a dead insect. If a male offers an acceptable food item or other token, called a **nuptial gift,** as part of his courtship behavior, she will mate with him (see Figure 9-21). The larger the food item, the longer she will mate; and the more she eats, the larger the number of eggs that she will lay. After about 20 minutes of mating, though, when a male has transferred all of the sperm that he can, he is likely to break off the mating and take back whatever remains of the "gift," which he may use to try to attract another mate. Nuptial feeding is common among birds and insects.

4. Mate only with a male that has a valuable physical attribute. Male-male competition for the chance to mate with females can also take a more literal form: actual physical contests. Across the animal kingdom, from dung beetles to hippopotamuses weighing more than 5,000 pounds, male-male contests determine the dominance rankings of males. Females then mate primarily with the highest-ranking males.

In a similar process, rather than choosing the best-fighting or largest males, females sometimes base their choice on some physical attribute, such as antler size in red deer, the bright red chest feathers of frigate birds, or the elaborate tail feathers of the male peacock (see Figure 9-21). In each case, the physical feature serves as an indicator to females of the relative quality of the male, possibly because the feature is correlated with the male's health.

With so many examples of male-male fighting as part of the courtship rituals that have evolved for attracting a mate, it is reasonable to ask: why is it so rare for females to fight? And why do females generally not have to advertise their health with flashy feathers or other ornamentation? The answer is that as long as females are making the greater investment in reproduction, nearly any male will mate with them. Consequently, there is nothing more to be gained by trying to outcompete other females or otherwise attract the attention of males.

Q *Why do so few women get into barroom brawls?*

Moreover, among humans, social and cultural values also have powerful influences over mating behavior, complicating interpretation. We are not lumbering robots, destined to follow some genetic program. It's important to take into consideration that researchers have noted some subtle manifestations of female health and fertility, including waist-to-hip ratios and patterns of facial and body symmetry, in many different species. There is a rich and complex world of mating tactics, many of which we do not fully understand.

TAKE-HOME MESSAGE 9·12

As a consequence of male-female differences in initial reproductive investment, males tend to increase their reproductive success by mating with many females and have evolved to compete among themselves to get the opportunity to mate. Females increase their reproductive success not through extra matings but rather by caring for their offspring and being choosy about selecting a mate.

9·13

Tactics for keeping a mate: mate guarding can protect a male's reproductive investment.

If a male simply abandons a female after mating and searches for other mating opportunities, rather than making any investment in the potential offspring, he has no risk of investing further energy in offspring that are not his. This is one way to minimize the potential costs associated with paternity uncertainty. If you don't play, you can't lose.

This strategy, however, is not necessarily the most effective way for a male to maximize his reproductive success. If a male has a hundred or even a thousand matings but no offspring survive, that behavior is not evolutionarily successful. Consequently, in species for which offspring's survival can be enhanced with greater parental investment, there is an incentive for a male to

Behaviors Evolve   Cooperation and Altruism   Sexual Conflict   Communication

provide some parental care, even though such behavior makes him vulnerable to paternity uncertainty.

Q Why do so few females guard their mates as aggressively as males do?

In situations in which males provide parental care, it is common for the male to reduce his vulnerability through some form of **mate guarding.** Mate guarding is a consequence of (and an attempt to reduce) paternity uncertainty. As long as the offspring emerge from the female's body, she can be certain that they contain her genes. In contrast, a male inhabits a "danger zone" that lasts as long as the female is fertile. If she mates with any other males during this time, the offspring she produces may not be his. If he is going to help raise the offspring, he benefits by minimizing his risk in the danger zone. It is during this period that mate guarding is particularly common.

Mate guarding methods range from the simple to the macabre. If a male wants to ensure that a female does not mate with another male, why bother to stop mating at all? In many species, males take this approach to reduce their risk in the danger zone. Among house flies, even though the male has completed the transfer of sperm to the female in 10 minutes of copulation, he does not separate from her for a full hour. Moths go even further and continue to mate for a full 24 hours. And in the extreme case of this strategy, certain frog species continue individual bouts of mating for several months (**FIGURE 9-22**). If humans mated for a similar

Some frog species continue individual bouts of mating for several months continuously. If humans mated for a similar percentage of our lives, a single round of intercourse would last almost ten years!

FIGURE 9-22 Preventing paternity uncertainty. A prolonged period of mating prevents the female from accessing other males, assuring the male of reproductive success.

EXTREME MATE GUARDING

The male black widow spider ensures his paternity certainty by breaking off his sexual organ inside the female after mating—precluding her mating again. (He also is eaten by her in many cases, giving her a resource to help produce more and healthier offspring.)

FIGURE 9-23 A reasonable trade-off? The male black widow spider ensures his paternity by an extreme form of mate guarding.

percentage of their lives, they'd spend almost 10 years on a single round of intercourse.

In a slightly subtler form of mate guarding that occurs in reptiles, insects, and many mammalian species, males block the passage of sperm into the female by producing a copulatory plug. Formed in the female reproductive tract from coagulated sperm and mucus, copulatory plugs can be very effective. Male garter snakes that encounter a female snake with a copulatory plug, for example, do not attempt to court or mate with her, treating her instead as if she were not available.

A much more extreme form of mate guarding occurs in the black widow spider: the male breaks off his sexual organ inside the female, preventing her from ever mating again. Interestingly, when the act is completed, the female usually kills and eats the male (**FIGURE 9-23**). In sealing his mate's reproductive tract, the male assures himself of fathering the offspring, and in consuming her mate's nutrient-filled body, the female gets resources that help her produce the offspring.

TAKE-HOME MESSAGE 9·13

Because paternity certainty is low in animals with internal fertilization, males that provide parental care are vulnerable to loss of their investment. As a consequence, mate guarding and other tactics have evolved to increase paternity certainty for species in which males contribute to parental care.

355

Monogamy versus polygamy: mating behaviors can vary across human and animal cultures.

As we continue our tour of animal mating behavior, we turn again to the elephant seal. During the breeding season, these animals appear on islands off the coast of northern California. In December of each year, the males begin to compete with each other for possession of the beach. Through bloody fights, they establish dominance hierarchies, with the biggest males—which are 13 feet long and weigh more than 2 tons—generally winning control of the beach.

In mid-January, the females arrive and are ready to mate. They congregate in large groups on just a few prime beaches. Because the females stick close together, the biggest males, who control the prime beaches, can dominate the other males in the competition for sexual access to females (**FIGURE 9-24**). In one study that observed 115 males, the 5 highest-ranking males fathered 85% of the offspring. While nearly every female will mate and produce offspring, the majority of males never get the chance to mate during the 10–20 years of their life.

The elephant seals' mating pattern exemplifies **polygamy,** a system in which some individuals attract multiple mates while other individuals attract none. Polygamous mating systems can be subdivided into **polygyny** and **polyandry.** In polygyny, individual males mate with multiple females; in polyandry, individual females mate with multiple males. Polygamy can be contrasted with **monogamy,** in which most individuals mate and remain with just one other individual. Polygamy and monogamy are two types of **mating systems,** which describe the patterns of mating behavior in a species. In this section, we explore the features of environments and species that influence mating systems and survey the range of mating systems observed in nature.

As we have seen, throughout the animal world, males tend to make less parental investment than females. As a consequence, females are choosy about whom they mate with and males compete for access to mating opportunities.

POLYGAMY IN ELEPHANT SEALS

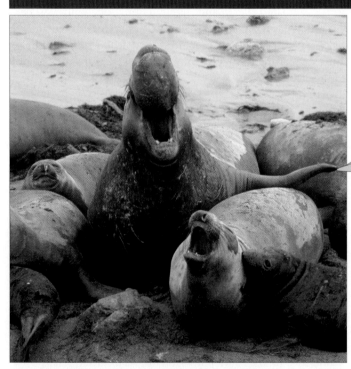

In a polygamous mating system, some individuals have multiple mates, while others have few or none.

DOMINANT MALE ELEPHANT SEAL

LOW-RANKING MALE ELEPHANT SEALS

OFFSPRING

FIGURE 9-24 **King of the beach.** The biggest, best-fighting male elephant seals control the beach and mate with as many females as possible.

In an extremely polygynous mating system, such as in elephant seals, the vast majority of males have no reproductive success at all.

Behaviors Evolve   Cooperation and Altruism   Sexual Conflict   Communication

Not surprisingly, multiple females often end up selecting the same male—usually a male on a territory rich in resources or a male with unusually pronounced physical features, such as antler size. Such female selection is the reason for the fancy ornamentation we see in male peacocks. Although, the male peacock's tail is like a giant bull's-eye to predators, the number of eyespots on the tail is directly related to how well he can attract a mate: below 140 eyespots, he gets no mates; at 150, he gets two to three mates on average; and with 160 eyespots he gets six or more mates (see Figure 9-21). Given their greater parental investment, virtually *all* females are able to attract a mate, regardless of whether they are the most fit, high-status female or the least.

Mating systems are not as easy to define as the elephant seal example might lead us to believe. Three issues, in particular, complicate the task. First, there are often differences between animals' mating behavior and their bonding behavior. That is, it may seem that a male and female have formed a **pair bond**—in which they spend a high proportion of their time together, often over many years, sharing a nest or other "home" and contributing equally to parental care of offspring in what appears to be a monogamous relationship. Closer inspection (often including DNA analysis of the offspring) sometimes reveals, however, that the male and/or the female may be mating with other individuals in the population, and perhaps the mating system is better described as a variation on polygamy. A second difficulty in defining a species' mating system arises because the mating system may vary within the species. That is, some individuals may be monogamous, while others are polygamous. And the mating system may even change over the course of an individual's life. A third difficulty is that males and females often differ in their mating behavior. In the elephant seals described above, it could be said that the females are all mating monogamously, while the males are polygamous.

Within almost any large group of related animal species, there will be a variety of species with each type of mating system, so it is not usually possible to make generalizations. And in a recent surprising discovery, researchers were able to alter the behavior of males of a polygynous species of meadow voles by inserting into their brain a single gene from a related but monogamous species of prairie voles. With the new gene, the formerly polygynous males showed a preference for their current partner over other females, and their ability to form a long-term bond with their partner increased.

Examination of birds and mammals in general, however, reveals one sharp split. As we discussed in Section 9-10, the vast majority of female mammals have greater parental

investment than male mammals. In birds, females and males have a more equal parental investment.

Does the difference in parental investment patterns in birds and mammals lead to different mating systems? Yes. In mammals, polygyny is the most common mating system across all large groups, from rodents to primates. Polygyny is a consequence of the significant female investment and lesser male investment. Males can generally benefit more by seeking additional mating opportunities, leading to male-male competition. In birds, the relatively equal parental investment by males and females has led to much less polygyny. Among the approximately 10,000 species of birds we know about, more than 90% appear to be monogamous (**FIGURE 9-25**).

And what of humans? Across a variety of cultures, males consistently have greater variance in reproductive success than females—that is, some males have very high reproductive success while many others have little or none. **FIGURE 9-26**, for example, presents data from a study of the Xavante Indians of Brazil. The *average* number of children does not differ for men and women, but some men have very large numbers (as many as 23!) and many have none. There is significantly less

> **Q** Are humans monogamous or polygamous?

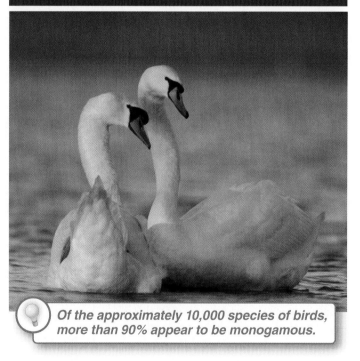

MONOGAMOUS BIRDS

Of the approximately 10,000 species of birds, more than 90% appear to be monogamous.

FIGURE 9-25 Let's stay together. Parental investment that is roughly equal often leads to monogamous mating behavior in birds.

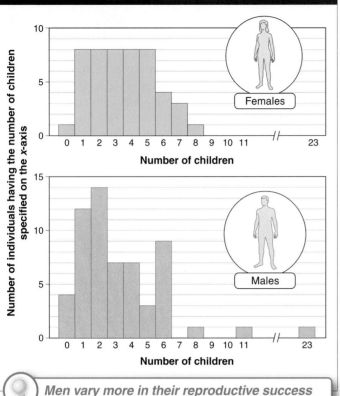

VARIANCE IN THE REPRODUCTIVE SUCCESS OF XAVANTE INDIANS

Females

Males

Men vary more in their reproductive success than women because nearly all women have access to a mate while some men are outcompeted for access to women. This male-female difference is characteristic of polygyny.

FIGURE 9-26 Larger male variance in reproductive success. Studies of the Xavante Indians of Brazil show that the range in number of offspring over their lifetime is much greater for men than for women.

variability in reproductive success from one woman to another. Such differences in the variance in reproductive success among males and females, which can be significantly influenced by socialization and other cultural forces, are generally associated with polygynous mating systems, but the difference between the sexes found in the Brazilian study is quite small when compared with many other mammalian species, such as elephant seals, and is close to that seen in populations with a monogamous mating system. Humans, consequently, seem to have a mating system close to, but not completely, monogamous. Consistent with the Brazilian study is the fact that the majority of societies in the world permit a man to have more than one wife.

TAKE-HOME MESSAGE 9•14

Mating systems—monogamy, polygyny, and polyandry—describe the variation in number of mates and the reproductive success of males and females. They are influenced by the relative amounts of parental investment by males and females.

9•15

Sexual dimorphism is an indicator of a population's mating behavior.

Male elephant seals are three to four times the size of female elephant seals. In contrast, the males and females of most bird species are the same size. In these species, even expert bird-watchers often cannot distinguish between the sexes, except when the female bird is carrying eggs (FIGURE 9-27). When the sexes of a species differ in size or appearance, this is called **sexual dimorphism.** Why do species have such dramatic differences in the degree to which males and females resemble each other in size and appearance?

Q It's almost impossible to distinguish males from females in most bird species. Why does this simple observation tell us a huge amount about their sexual behavior, including that they are monogamous?

Body size is an important clue to behavior. We have seen how male elephant seals have a winner-take-all tournament for control of the beach, and as the tournament winner, a male has access to mating opportunities. Because the largest individual will have the most offspring, there is selection for larger and larger body size. The smaller males will have relatively low reproductive success. There is no selection for larger body size among female elephant seals, because females of any size can mate. Hence the dramatic size difference between the sexes.

SEXUAL DIMORPHISM

With sexual dimorphism, the sexes differ in size or appearance.

BEHAVIORS ASSOCIATED WITH SEXUAL DIMORPHISM
- One parent invests more in caring for the offspring.
- Mating system tends toward polygamy.
- One sex (usually females) is choosier when selecting a mate.
- One sex (usually males) competes for access to mating opportunities with the other sex.

SEXUAL MONOMORPHISM

With sexual monomorphism, the sexes are indistinguishable.

BEHAVIORS ASSOCIATED WITH SEXUAL MONOMORPHISM
- Both parents invest (approximately) equally in caring for the offspring.
- Mating system tends toward monogamy.
- Both sexes are equally choosy when selecting a mate.

FIGURE 9-27 Size says a lot. Species with large males and small females are very likely to be polygamous. Species with similarly sized males and females are likely to be monogamous.

In addition to body size, coloration is also a clue to behavior. Because females of some polygynous species choose the males with the brightest or flashiest coloration, rather than the largest males, male–male competition sometimes results in differences in physical appearance between the sexes instead of (or in addition to) differences in size.

Q Men are bigger than women. What does that tell us about our evolutionary history of monogamy versus polygamy?

What happens when there is little male-male competition for mates? Among bird species, males can provide significant investment in offspring. Eggs have to be incubated, and many chicks emerge in a very poorly developed condition and it takes two parents to raise them. So a pair tends to stay together for a season or longer. If each female can only pair up with one male, there really isn't much competition for mates. And without male-male competition, there is little

selection for increased size. As a consequence, there is little sexual dimorphism among most bird species.

In turn, we can predict a great deal about the parental practices of a species just by looking at a picture of a male and a female and examining the ratio of body sizes. If the two are dramatically different, as in elephant seals, you can be confident that the smaller sex is doing most of the care of the offspring and that the species is more likely to be polygamous than monogamous.

TAKE-HOME MESSAGE 9·15

Differences in the level of competition among individuals of each sex for access to mating opportunities can lead to the evolution of male-female differences in body size and other aspects of appearance. In polygynous species, this results in larger males that are easily distinguished from females visually. In monogamous species, there are few such differences between males and females.

❹ Communication and the design of signals evolve.

Now hear this! Two common European toads communicate.

9•16

Animal communication and language abilities evolve.

The challenge of communication is central to the evolution of animal behaviors, whether they relate to cooperation, conflict, or the many interactions inherent to selecting a mate. An animal must be able to convey information about its status and intentions to other individuals. And equally important is that animals must be able to interpret and assess the information they get from other individuals, so as to reduce the likelihood that they are tricked or taken advantage of. In this section and the next, we explore the evolution of a variety of methods of animal communication and the ways in which communication can influence fitness and the evolution of other behaviors.

Not all animals need to open their mouths when they have something important to communicate. When the female silkworm moth is ready to mate, for example, she releases a potent chemical called bombykol into the air. As molecules of the chemical come in contact with the bushy antennae of a male silkworm moth—who might be as far away as three miles—he recognizes the mate attractant and responds by flying upwind, looking for the source of the chemical. He flies in a zigzag pattern until he detects the bombykol again, then

he re-orients himself toward the chemical and continues to fly toward the source until he zeroes in on the female and they mate (**FIGURE 9-28**).

This is just one example of **communication,** an action or signal on the part of one organism that alters the behavior of another organism. But a huge range of animal behaviors require communication, including defending territory, alerting individuals to threatening predators, announcing one's readiness for mating, establishing one's position in a dominance hierarchy, and caring for offspring.

Animals use many different types of communication signals. Three types are most common.

1. Chemical. Molecules released by an individual into the environment that trigger behavioral responses in other individuals are called **pheromones.** They include the airborne mate attractant of the silkworm moth, territory markers such as those found in dog urine, trail markers used by ants, and even chemicals released by humans. An example in humans

Behaviors Evolve   Cooperation and Altruism   Sexual Conflict   Communication

is chemicals that influence the length of the menstrual cycle in women and are responsible for the synchronization of menstrual cycles that occurs in women living in group housing situations, such as in dormitories and prisons.

2. Acoustical. Sounds that trigger behavioral responses are abundant in the natural world. These include the alarm calls of Belding's ground squirrels described earlier, the complex songs of birds, whales, frogs, and crickets vying for mates, and the territorial howling of wolves.

3. Visual. Individuals often display visual signals of threat, dominance, or health and vigor. Examples include the male baboon's baring of his teeth and the vivid tail feathers of the male peacock.

With increasingly complex forms of communication, increasingly complex information can be conveyed. One extreme case is the honeybee **waggle dance.** When a honeybee scout returns to the hive after successfully locating a source of food, she performs a set of maneuvers on a honeycomb that resemble a figure 8, while wiggling her abdomen. Based on the specific angle at which the scout dances (relative to the sun's position in the sky) and the duration of her dance, the bees that observe the dance are able to leave the hive and quickly locate the food source.

The complexity and power of the waggle dance raises an important question: at what point does communication become **language?** Language is a very specific type of communication in which arbitrary symbols represent concepts, and a system of rules, called grammar, dictates the way the symbols can be manipulated to communicate and express ideas (**FIGURE 9-29**).

It's not always easy to identify language. Consider one type of communication in vervet monkeys. Vervet monkeys live in groups, and when an individual sees a predator coming, it makes an alarm call to warn the others. Unlike in most other alarm-calling species, the vervet monkey's alarm call differs depending on whether the predator is a hawk, a snake, or a leopard. And the response of other vervet monkeys to hearing the alarm call also differs and is appropriate to the type of predator approaching. The snake alarm causes individuals to stand on their toes and look down at the ground. The leopard alarm causes individuals to run for the nearest tree and quickly climb it. And the hawk alarm causes vervet monkeys to look to the sky.

Does alarm calling in vervet monkeys or the honeybee waggle dance constitute language? This is debatable. Is the form of American Sign Language taught to chimpanzees and gorillas a language? (See Figure 9-29.) There are opinions on both sides of this question. Interestingly, the great apes have shown a capacity for creating new signs, expressing abstract ideas, and referring to things that are physically distant or distant in time—important features of human language. One thing that is certain, however, is that human language is near the most complex end of the communication continuum.

FORMS OF ANIMAL COMMUNICATION

CHEMICAL COMMUNICATION

AUDITORY COMMUNICATION

VISUAL COMMUNICATION

FIGURE 9-28 Chemical, auditory, and visual communication. Many animal behaviors require communication: the ability to convey and receive information from other animals.

FIGURE 9-29 Dancing, signing, and speaking. Animals have evolved a variety of ways to convey complex information to others.

The evolution of language influences many of the other behaviors discussed throughout this chapter. The evolution of reciprocal altruism, for example, may be influenced as language makes it easier for individuals to convey both their needs and their resources to other individuals. Similarly, in courtship and the maintenance of reproductive relationships, language gives individuals a tool for conveying complex information relating to the resources and value they may bring to the interaction.

TAKE-HOME MESSAGE 9·16

Methods of communication—chemical, acoustical, and visual—have evolved among animal species, enabling them to convey information about a variety of features of their condition and situation. These abilities influence fitness and the evolution of virtually all other behaviors.

9·17

Honest signals reduce deception.

Among Natterjack toads, found in northern Europe, males produce booming calls to attract females. Because females desire large males and size determines the volume of the calls, females will push their way through murky swamps until they reach the loudest croaker. They are drawn by calls that can be heard a mile away and are often louder than the legal noise limit for a car engine (**FIGURE 9-30**).

It is not surprising that females desiring the largest possible male should select their mates based on the volume of the call. The Natterjack's call is an **honest signal,** a signal that cannot be faked and that is given when both the individual making the signal and the individual responding to it have the same interests. An honest signal is one that carries the most accurate information about an individual or situation, and animals that respond to signals of any sort have evolved to value most highly

those signals that cannot be faked. The Natterjack's call, for example, cannot be faked by a small Natterjack toad. And so there is evolutionary pressure for animals to both produce and respond to honest signals, because these behaviors enable them to maximize their fitness by selecting the best possible mate.

When one animal can increase its fitness by deceiving another, deception can also be expected to evolve. An example might be baby birds begging for food from their parent. If an allele that causes a chick to exaggerate its need for food leads to faster growth and better health for that individual, the allele is likely to increase in frequency in the population.

Communication and signaling, therefore, are features of populations that are continually evolving. We would expect this evolutionary "arms race" between honest signals and deception

Behaviors Evolve   Cooperation and Altruism   Sexual Conflict   Communication

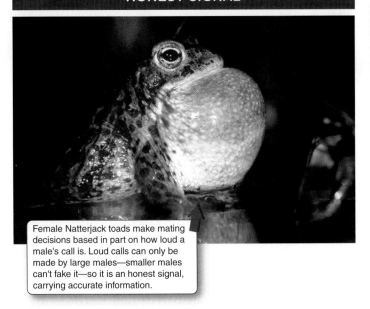

HONEST SIGNAL

Female Natterjack toads make mating decisions based in part on how loud a male's call is. Loud calls can only be made by large males—smaller males can't fake it—so it is an honest signal, carrying accurate information.

FIGURE 9-30 Loud and true. The female Natterjack toad is attracted to the biggest male with the loudest call—an "honest signal," because there is no way for a small toad to fake the call volume of a larger toad.

to lead, over time, to both increasingly unambiguous signals and ever more sophisticated patterns of deception.

We have only touched the surface of the rich field of animal behavior here. Still, we've seen how our understanding of the behavior of animals, including humans, has been greatly expanded by applying scientific thinking, an experimental approach, and careful consideration of the process by which the evolutionary mechanism of natural selection shapes populations. All of this has been made possible through the subtle but important shift in thinking that we highlighted at the beginning of this chapter: an organism's phenotype is not limited to its physical traits but includes the organism's behaviors, as well. And consequently, behaviors respond to selective pressures and evolve.

TAKE-HOME MESSAGE 9·17

Animals have evolved to rely primarily on signals that cannot easily be faked, in order to gain the maximum amount of information from them.

Knowledge You Can Use

How to win friends and influence people.

As we have seen, in the animal world it is very, very rare to find individuals engaging in costly behaviors that benefit unrelated individuals; with few exceptions, we see little that resembles animal "friendship."

Humans are a rare exception—every day, in myriad ways, we see friendly behaviors. Just as we put money in the bank for a rainy day, we buffer ourselves from the world's uncertainties by storing goodwill in our neighbors. Cooperators have evolutionary advantages over loners.

Q: Why is "kindness" a risky behavior evolutionarily? From an evolutionary perspective, we remain vulnerable. With each costly act we perform for unrelated individuals, strangers and friends alike, there is the risk that our efforts will be in vain, our energies lost. For that reason, in our evolution as reciprocal altruists, we acquired a cautiousness, a hesitancy to stick our necks out.

Q: How can we help others feel less vulnerable and more willing to cooperate? Our ability to override the impulses that often push us toward selfishness is one of the hallmarks of being human. Still, there is no reason to put people in situations where they may be better off rejecting friendship. In each interaction there is an individual on the other side who is also feeling vulnerable. By taking steps, often absurdly simple, to address the unconscious vulnerabilities that others feel, we can increase the likelihood that they choose cooperation, reciprocity, and friendship.

- **Learn and use other people's names.** It tells them you recognize them specifically and, consequently, that you understand exactly who it is that you "owe." Smile and make eye contact—even when attempting to change lanes in traffic. These gestures feel like the beginning of a relationship, thus stimulating favor-granting instincts.

- **Embrace etiquette:** acknowledge your debts to others. Be effusive and public in your thanks for kindnesses done to you. Send thank-you cards.

- **Take the first step in reciprocity.** Human cooperation is so tied to reciprocal exchange that even tiny gestures of good faith can play an important role in building relationships. Whenever possible, give gifts. Even small ones. Even when they are not required. Especially when they are not required.

- **Develop a good reputation:** for generosity, for loyalty, for remembering and acknowledging kindnesses done to you.

Can you think of other steps?

1 Behaviors are traits that can evolve.

Behavior encompasses any and all of the actions performed by an organism. When a heritable trait increases an individual's reproductive success relative to that of other individuals, that trait tends to increase in frequency in the population. Behavior is as much a part of an organism's phenotype as is its anatomical structures, and as such can be produced and shaped by natural selection. Innate behaviors are present in all individuals in a population and do not vary much from one individual to another or over the life span of an individual. Organisms are well-prepared to learn behaviors that have been important to their reproductive success over the course of their evolutionary history and less prepared to learn behaviors irrelevant to their evolutionary success. It is not necessary that an organism consciously tries to maximize its reproductive success in order for a behavior to evolve.

2 Cooperation, selfishness, and altruism can be better understood with an evolutionary approach.

Many behaviors in the animal world appear altruistic. In almost all cases, the apparent acts of altruism are not truly altruistic and have evolved as a consequence of either kin selection or reciprocal altruism and, from an evolutionary perspective, are beneficial to the individual engaging in the behavior. Kin selection describes apparently altruistic behavior in which an individual assisting a genetic relative compensates for its own decrease in direct fitness by helping increase the relative's fitness and, consequently, its own inclusive fitness. Reciprocal altruism describes the situation in which an individual engages in an altruistic-appearing act toward another individual but is likely to get something of value at a later time. Behaviors that are good for the species or population but detrimental to the individual exhibiting the behavior are not generally produced under natural conditions.

3 Sexual conflict can result from disparities in reproductive investment by males and females.

In mammals and many other animal species, there are important physical differences between males and females relating to reproduction. These physical differences have led to the evolution of differences in male and female reproductive behaviors. Males tend to increase their reproductive success by mating with many females and have evolved to compete among themselves to get the opportunity to mate. Females do not increase their reproductive success through extra matings, but rather by caring for their offspring and being choosy when selecting a mate. Mate guarding and other tactics to increase paternity certainty evolve when males provide parental care. Mating systems describe the variation in mate number and reproductive success of males and females. Differences in the level of competition among the individuals of each sex for access to mating opportunities can lead to the evolution of male-female differences in body size and other aspects of appearance.

4 Communication and the design of signals evolve.

Methods of communication—chemical, acoustical, and visual—have evolved in animal species, enabling individuals to convey information about a variety of features of their condition and situation. These abilities influence fitness and the evolution of virtually all other behaviors. Animals have evolved to rely primarily on signals that cannot easily be faked, in order to gain the maximum amount of information from them.

KEY TERMS

1. Why do human taste preferences exist?
   a) They are culturally generated; consumers are influenced by suggestive advertising.
   b) Human feeding choices influence energy intake and, consequently, fitness.
   c) Humans can extract energy from a variety of non-food sources, and so taste preferences cause us to focus on foods that are more plentiful in our environment.
   d) Fats actually do taste better than sugars.
   e) Vitamins and minerals give food unusual tastes, which causes us to seek them out.

2. From an evolutionary perspective, behavior can best be viewed as:
   a) a trait that arises by learning and not by natural selection.
   b) non-heritable.
   c) a trait subject to drift and mutation, but not natural selection.
   d) part of the phenotype.
   e) All of the above are correct.

3. Why is it so much easier for an infant to learn a complex language than for a college student to learn biology?
   a) Language involves memorization only, without the need for understanding rules, while biology involves both.
   b) Language is a feature with great evolutionary relevance for humans.
   c) Learning biology was not a behavior with evolutionary relevance for humans.
   d) Biology involves much more vocabulary than learning a complex language.
   e) Both b) and c) are correct.

4. Babies in the United States quickly and easily develop a fear of snakes. Yet they don't easily develop a fear of guns. Why?
   a) Humans cannot develop fears of inanimate objects.
   b) Evolution can be slow in producing populations that are adapted to their environments.
   c) Babies are more likely to encounter snakes than guns as they develop in the United States.
   d) Fewer individuals are killed by guns than by snakes in the United States each year.
   e) All of the above are correct.

5. In Belding's ground squirrels, why are females much more likely than males to engage in altruistic behavior by sounding alarm calls?
   a) Belding's ground squirrels have a sex ratio that is biased toward females.
   b) Females invest more in foraging and food storage, so they are more likely to lose their lives or their food if a predator attacks.
   c) Belding's ground squirrels have a sex ratio that is biased toward males.
   d) Females tend to remain in the area where they were born, so the females that call are warning their own kin.
   e) Males forage alone, so their alarm calls are useless.

6. Gestational diabetes is thought to be the consequence of:
   a) pregnant women decreasing their average daily amount of activity.
   b) a mother withholding investment in future offspring in order to invest more in the current pregnancy.
   c) a mother consuming too much sugar during gestation.
   d) conflict between the mother and fetus with respect to how much food the fetus should be given; the mother is equally related to the fetus and to any future offspring, while the fetus is not equally related to itself and any subsequent siblings.
   e) physiological constraints on the amount of insulin a mother can provide for the fetus.

7. Vampire bats:
   a) sometimes regurgitate blood into the mouth of another bat that is close to starving, but the likelihood is a function of whether the individuals are genetically related.
   b) are unusual in that they are one of the few animal species that exhibit kin selection.
   c) sometimes regurgitate blood into the mouth of an unrelated bat that is close to starving.
   d) exhibit reciprocal altruism but not kin selection.
   e) There are no such things as vampire bats; they're found only in a Dracula novel.

8. Altruistic behavior in animals may be a result of kin selection, a theory maintaining that:
   a) genes promote the survival of copies of themselves when behaviors by animals that possess those genes assist other animals that share those genes.
   b) aggression within sexes increases the survival and reproduction of the fittest individuals.
   c) companionship is advantageous to animals because in the future they can recognize those that have helped them and provide help to those individuals.
   d) aggression between sexes increases the survival and reproduction of the fittest individuals.
   e) companionship is advantageous to animals because in the future they can recognize those that have helped them and request help once again.

9. All of the following are necessary conditions for reciprocal altruism to evolve in a species except:
   a) the ability to recognize different individuals.
   b) the ability to punish cheaters who do not reciprocate.
   c) repeated interactions with the same individuals.
   d) at least one of the sexes must not disperse, so that some individuals always live near their kin.
   e) None of the above is necessary for the evolution of reciprocal altruism.

10. In a situation in which males guard eggs and care for the young without help from the female, which of the following statements would most likely be correct?
    a) Males are large and more brightly colored in order to attract the very best females.
    b) Males and females are equally brightly colored, but males court females aggressively.
    c) The population is monogamous with no sexual dimorphism.
    d) A single male controls a harem of females to which he has exclusive reproductive access.
    e) Females are more brightly colored than males and court males aggressively.

11. In mammals, as well as many other species, males generally compete for females. The best explanation for this phenomenon is:
    a) males are more aggressive.
    b) males, on average, have higher fitness.
    c) females have a higher parental investment.
    d) males are choosy.
    e) females are better looking.

12. Mate guarding is a reproductive tactic that functions to:
    a) reduce paternity uncertainty.
    b) increase the female's investment in offspring.
    c) reduce the male's reproductive investment.
    d) reduce the female's fitness.
    e) increase the number of mates to which a male has access.

13. Relative to birds, more mammalian species are:
    a) polygynous.
    b) monogamous.
    c) polyandrous.
    d) hermaphroditic.
    e) sexually monomorphic.

14. In a species such as pigeons, in which males are almost indistinguishable in appearance from females, the most likely mating system is:
    a) monomorphism.
    b monogamy.
    c) polygyny.
    d) polyandry.
    e) It is impossible to predict the mating system with only this information.

15. If you find a species of fish in which males are much more brightly colored and larger than females, what might you infer about their mating system?
    a) The degree of sexual dimorphism does not give any information about the mating system.
    b) They are simultaneous hermaphrodites.
    c) They exhibit parallel monogamy.
    d) They are serially monogamous.
    e) They are polygynous.

16. Polygynous species:
    a) usually employ external fertilization.
    b) are usually sexually dimorphic, with males larger and more highly ornamented.
    c) are usually sexually dimorphic, with females larger and more highly ornamented.
    d) usually have males and females that are physically indistinguishable.
    e) are more commonly found among birds than among mammals.

SHORT-ANSWER QUESTION

1. Female moorhens are larger and more aggressive than males. They also compete among themselves for access to the smaller, fatter males. Which sex do you think provides more parental care? Explain your answer.

See Appendix for answers. For additional study questions, go to www.prep-u.com.

10

The Origin and Diversification of Life on Earth

Understanding
biodiversity

❶ Life on earth most likely originated from non-living materials.

Life on earth arose in an inhospitable environment. (Shown here: the formation of volcanic rocks.)

10·1

Complex organic molecules arise in non-living environments.

In the beginning, there was nothing. Now there is something. That, in a nutshell, describes one of the most important, yet difficult to resolve, questions in science: how did life on earth begin? We open our investigation of this question by describing a few things that we do know and exploring how they help us speculate about the things we do not yet know. But before we start, let's define what we mean by "life." Here we give just a basic definition: **life** is defined by the ability to replicate and by the presence of some sort of metabolic activity. In the next section we expand on this definition and explore the transition from non-living to living.

Earth formed about 4.5 billion years ago from clouds of dust and gases left over after the formation of the sun. The planet was initially super-hot, and as it very gradually cooled, a crust formed at the surface and condensing water formed the oceans. This probably took several hundred million years. The oldest rocks, found in Greenland, are about 3.8 billion years old. And the earliest life forms appeared not long after these rocks were formed: fossilized bacteria-like cells have been found in rocks

CONDITIONS ON EARTH AT THE TIME LIFE BEGAN

The early atmosphere differed from that of today, and contained large amounts of carbon dioxide, nitrogen, methane, ammonia, hydrogen, and hydrogen sulfide.

Small organic molecules eventually formed, providing the building blocks of life.

FIGURE 10-1 Darwin's "warm little pond." The first life on earth tolerated an atmosphere without oxygen.

that are 3.4 billion years old. (Note that 0.4 billion years—or 400 million years—is "not long" in geological terms.)

From that initial point at earth's formation, there is now tremendous **biodiversity**—variety and variability among all genes, species, and ecosystems—on the planet. In this chapter we explore in more detail how this biodiversity might have come to be, and how we name groups of organisms and determine the relatedness of these groups to each another. We begin by returning to the question of the origin of life on earth.

How did these first organisms arise? Some have suggested that life may have originated elsewhere in the universe and traveled to earth, possibly carried on a meteor. There is little evidence for this idea, however, and it is hotly debated whether microbes could even survive the multi-million-year trip to earth in the cold vacuum of space with no protection from ultraviolet and other forms of radiation. Experimental data have been unable to answer this question definitively. Most scientists believe, instead, that life originated on earth, probably in several distinct phases.

Phase 1: The formation of small molecules containing carbon and hydrogen

The conditions on earth around the time of the origin of life were very different from those of today. In particular, chemical analyses of old rocks reveal that no oxygen gas was present. The atmosphere included large amounts of carbon dioxide, nitrogen, methane, ammonia, hydrogen, and hydrogen sulfide. Most of these molecules were produced by volcanic eruptions. It was in this environment that some locations on earth probably served as the cradle of life, or what Darwin called the "warm little pond" (**FIGURE 10-1**).

Critical to the origin of life was the formation of small molecules containing carbon and hydrogen. Because of their chemical structures, these molecules bond very easily and in many ways, causing them to have a huge variety of forms and functions that make them indispensable for the chemical processes of life. There are several plausible scenarios for how the creation of small organic molecules may have occurred at this time. The most likely comes from some simple but revealing four-step experiments done in 1953 by a 23-year-old student named Stanley Miller and his advisor, Harold Urey (**FIGURE 10-2**).

1. They created a model of the "warm little pond" and their best estimate of earth's early atmosphere: a flask of water with H_2, CH_4 (methane), and NH_3 (ammonia).

2. They subjected their mini-world to sparks, to simulate lighting.

3. And they cooled the atmosphere so that any compounds formed in it would rain back down into the water.

4. Then they waited, and examined the contents of the water to see what happened.

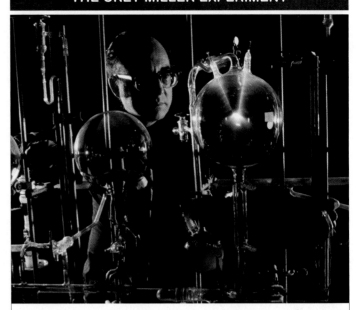

THE UREY-MILLER EXPERIMENT

In 1953, Stanley Miller and Harold Urey developed a simple four-step experiment that demonstrated how complex organic molecules could have arisen in earth's early environment.

1 They created a model of the chemicals present in the "warm little pond" and atmosphere early in earth's history: H_2, CH_4 (methane), and NH_3 (ammonia).

2 The atmosphere was subjected to sparks, to simulate lightning.

3 The atmosphere was cooled so that any compounds in it would rain back down into the water.

4 They examined the water, looking for organic molecules.

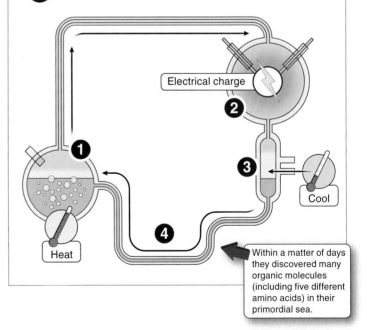

Within a matter of days they discovered many organic molecules (including five different amino acids) in their primordial sea.

FIGURE 10-2 A promising first step. The Urey-Miller experiment generated organic molecules from hydrogen, methane, and ammonia.

Overview of Diversity

They didn't have to wait long to get exciting results. Within a matter of days—not millions of years or even a few months—they discovered many organic molecules, including five different amino acids, in their primordial sea. (Using more sensitive equipment, re-analyses in 2008 of the residues from their original experiments revealed that all 20 amino acids present in living organisms were present in the residues.) This was the first demonstration that complex organic molecules could have arisen in earth's early environment. The Urey-Miller experiments are still far from proof that life arose from non-life, and questions remain, such as whether the environment they assumed to exist on the early earth was reasonable. Without oxygen gas, for example, earth's atmosphere would not have had an ozone layer protecting the early environment from the harmful ultraviolet radiation that breaks down methane and ammonia. Still, the experiments are a promising first step, suggesting that complex organic compounds—including amino acids, the primary constituents of proteins and an essential component of living systems—could have been produced from inorganic chemicals and lightning energy in the primitive environment of earth.

In the next section, we consider the final two phases by which these small organic molecules may have given rise to living organisms.

TAKE-HOME MESSAGE 10·1

Under conditions similar to those on early earth, small organic molecules form, and these molecules have some chemical properties of life.

10·2

Cells and self-replicating systems evolved together to create the first life.

After the generation of numerous organic molecules such as amino acids, the second phase in the generation of life from non-life was probably the assembly of these building block molecules into self-replicating, information-containing molecules. This is where things get a bit more speculative. It's complicated enough to generate a complex organic molecule, but it's a whole lot more complicated to generate an organic molecule that can replicate itself. Researchers believe that to get to the replication phase, enzymes were required.

Phase 2: The formation of self-replicating, information-containing molecules

Recently, researchers discovered a molecule that could unexpectedly function as an enzyme that links together nucleic acids (such as DNA or RNA). The molecule wasn't a protein but was, instead, a molecule of the nucleic acid RNA, which stores information just as DNA does. The discovery that RNA can do what proteins do—catalyze reactions necessary for replication—is notable because it means that this single, relatively simple molecule could have been a self-replicating system and a precursor to cellular life. These findings have given rise to the so-called **RNA world hypothesis,** which proposes that the world may have been filled with RNA-based life before it became filled with the DNA-based life we see today.

These self-replicating molecules raise an important question: when exactly was the threshold between living and non-living crossed? In the RNA world, self-replicating nucleic acid molecules carried the information of how to replicate *and* served as the machinery to actually carry out the replication. Is that enough? At this point, it is reasonable to ask, again: what exactly is "life"? How do we define it?

Fossils of 3.4-billion-year-old cells have been found in rocks from South Africa. These cells appear to be prokaryotic cells,

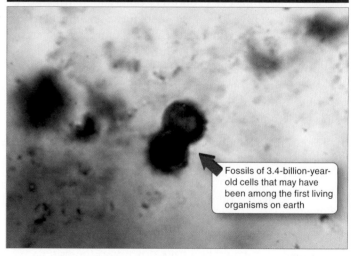

THE EARLIEST LIFE ON EARTH?

Fossils of 3.4-billion-year-old cells that may have been among the first living organisms on earth

FIGURE 10-3 Ancient prokaryotic cells. These fossilized cells may have been among the first living organisms on earth.

similar to living bacterial cells, with no nucleus, no organelles, and a circular strand of genetic information (**FIGURE 10-3**). They even look as if they are in the process of dividing. Were these the first living organisms on earth? And were they descendants of earlier, self-replicating molecules of RNA? Because no earlier fossils have been found, it is difficult to answer these questions.

Among humans, it's obvious whether someone is living or non-living. But this distinction becomes difficult to make when discussing simpler collections of molecules, and the border between living and non-living can be surprisingly fuzzy. Although no perfectly acceptable, precise definition exists, most scientists agree with the general definition given at the start of this chapter—that life is defined by two characteristics: the ability to replicate and the ability to carry out some sort of metabolism. An organism has metabolic activity when it can acquire certain molecules and use them in controlled reactions that maintain the chemical and energetic conditions necessary for its life. By this definition, following the first two phases described above, the self-replicating RNA molecules were right on the border, satisfying the first condition for life (replicating) but not the second (carrying out metabolism). We now explore the critical third phase in the generation of life from non-life: the development of a membrane that separated these self-replicating small molecules from their surroundings, thus compartmentalizing them into cells and making metabolic activity possible.

Phase 3: The development of a membrane, enabling metabolism and creating the first cells

One way that self-replicating molecules could acquire chemicals and use them in the controlled reactions of metabolism was by packaging the molecules within membranes. As we saw in Chapter 3, cell membranes are semi-permeable barriers that separate the inside of the cell from its external environment. Membranes make numerous aspects of metabolism possible. In particular, they make it possible for chemicals inside the cell to be at higher concentrations than they are outside the cell. Differences in chemical concentrations inside and outside a cell are essential to most life-supporting reactions.

So, if we could combine a self-replicating molecule and some random metabolic chemicals into a unit, surrounded by a membrane, life would be possible. But how could this have happened initially? Some evidence suggests that the first cells may simply have come together spontaneously. Specifically, researchers have found that mixtures of phospholipids placed in water or salt solutions tend to spontaneously form small spherical units that resemble living cells. These units may even "sprout" new buds at their surface, appearing to divide. Because these cell-like units don't pass on any genetic

The compartmentalization of self-replicating, information-containing molecules into microspheres may have been an important stage as the first living cells arose.

FIGURE 10-4 Are microspheres a key stage in the origin of life?

material, however, they cannot be considered to be alive. But if at some point in the past, units like these incorporated some self-replicating molecules inside, maybe by forming around them, such **microspheres** may have been important in the third phase in the generation of life from non-life: the compartmentalization of self-replicating, information-containing molecules into cells. If this did occur, the final step in the creation of something from nothing would be complete (**FIGURE 10-4**).

The exact process by which life on earth originated is still uncertain, and research continues. What is clear, however, is that life now abounds in great diversity. In the next sections, we begin to explore that diversity and how it is generated. We start with an investigation of what a species is and how individual species split and create additional species.

TAKE-HOME MESSAGE 10·2

The earliest life on earth, which resembled bacteria, appeared about 3.5 billion years ago, not long after the earth was formed. No one is certain how life arose, but evidence supports the idea that self-replicating molecules— possibly RNAs—may have formed in earth's early environment and later acquired or developed membranes, enabling them to replicate and making metabolism possible, the two conditions that define life.

Overview of Diversity

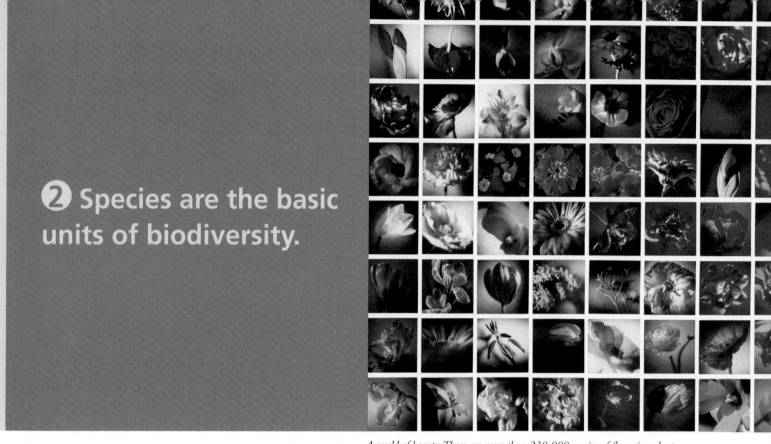

❷ Species are the basic units of biodiversity.

A world of beauty. There are more than 230,000 species of flowering plants.

10·3

What is a species?

A cat and a mouse are different things. Oprah Winfrey and Bill Gates, on the other hand, are different *versions* of the same thing. That is, although they are different individuals, both are humans. We generally distinguish between different kinds of living organisms: a rose versus a daisy, a wasp versus a fly, a snake versus a frog. Similarly, we lump some organisms into the same group and recognize them as different versions of the same thing: red roses and yellow roses, Chihuahuas and Dalmatians. How do we know when to classify two individuals as members of different groups and when to classify them as two individuals that are members of the same group (FIGURE 10-5)? And how do we classify two individuals that are somewhat similar but

 Do similar-looking organisms belong to the same or different species? How do we know?

FIGURE 10-5 **The same but not the same.** We recognize Bill Gates and Oprah Winfrey as members of the same species, but can you tell whether the rose and the tulip or the wood mouse and the common brown rat are from the same species? (Hint: they're not.)

Origin of Life ▼ What Are Species? Evolutionary Trees Macroevolution seem too different to classify as the same thing, such as a mouse and a rat, or a goat and a sheep?

Biologists use the word **species** to label different kinds of organisms. According to the **biological species concept,** species are populations of organisms that interbreed, or could possibly interbreed, with each other under natural conditions, and that cannot interbreed with organisms outside their own group (**FIGURE 10-6**).

Notice that the biological species concept completely ignores physical appearance when defining a species and instead emphasizes **reproductive isolation,** the inability of individuals from two populations to produce fertile offspring with each other, thereby making it impossible for gene exchange between the populations to occur.

Let's clarify two important features of the biological species concept. First, it says that members of a species are either actually interbreeding or *could possibly* interbreed. This emphasis means that just because two individuals are physically separated, they aren't necessarily in different species. A person living in the United States and a person living in Iceland, for example, may not be able to mate because of the distance between them. But if they were brought to the same location, they could mate if they wanted to, so we do not consider them to be reproductively isolated. Second, our definition refers to "natural" conditions. This distinction is important because occasionally, in captivity, individuals may interbreed that would not interbreed in the wild, such as lions and tigers (**FIGURE 10-7**).

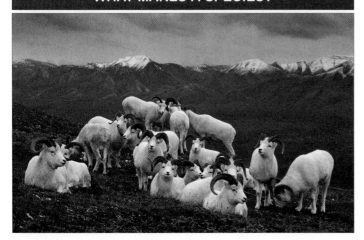

WHAT MAKES A SPECIES?

SPECIES ARE
• populations of organisms that interbreed with each other,
• or could possibly breed, under natural conditions,
• and are reproductively isolated from other such groups.

FIGURE 10-6 **An interbreeding population.** This herd of Dall sheep lives in the remote regions of Denali National Park, Alaska.

Even if they can be produced in captivity, organisms that do not naturally occur in the wild—such as the "liger"—are not considered species.

FIGURE 10-7 **Interbreeding is not enough.** This "liger" is the offspring of a lion and a tiger, two species that will interbreed, but only in captivity.

There are two types of barriers that prevent individuals of different species from reproducing: prezygotic and postzygotic barriers (**FIGURE 10-8**). (Remember, an egg that has been fertilized by a sperm cell is a *zygote.*)

Prezygotic barriers make it impossible for individuals to mate with each other or, if they can mate, make it impossible for the male's reproductive cell to fertilize the female's reproductive cell. These barriers include situations in which the members of the two species have different courtship rituals or have sufficient physical differences that they are unable to mate. In some cases, individuals from two different species can and do mate, but physical or biochemical factors prevent the male gamete from fertilizing the female gamete.

Postzygotic barriers occur after fertilization and generally prevent the production of fertile offspring from individuals of two different species—such offspring are called **hybrids.** These barriers are responsible for the production of hybrid individuals that either do not survive long after fertilization or, if they do survive, are infertile or have reduced fertility. Mules, for example, are the hybrid offspring of horses and donkeys, and although they can survive, they cannot breed with each other or produce offspring.

At first glance, the biological species concept seems to make it possible to determine unambiguously whether individuals belong

PREZYGOTIC BARRIERS
• Individuals are physically unable to mate with each other.
  OR
• If individuals are able to mate, the male's reproductive cell is unable to fertilize the female's reproductive cell.

POSTZYGOTIC BARRIERS
• Matings produce hybrid individuals that do not survive long after fertilization.
  OR
• If hybrid offspring survive, they are infertile or have reduced fertility.

FIGURE 10-8   Barriers to reproduction. With postzygotic barriers to reproduction, even if fertilization does occur, the animal (such as the mule, on the right) is usually sterile.

to the same or different species. For plants and animals this is usually true. For many other organisms, however, it is impossible to apply the biological species concept. We examine those species in Section 10-5, and explore some alternative methods of identifying a species. Even though it is not perfect, however, the biological species concept gives us an important tool for conceptualizing and categorizing the tremendous diversity of organisms found on earth, and it is used by most biologists.

TAKE-HOME MESSAGE 10•3

Species are generally defined as populations of individuals that interbreed with each other, or could possibly interbreed, and that cannot interbreed with organisms outside their own group. This concept can be applied easily to most plants and animals, but not to many other types of organisms.

10•4

How do we name species?

Keeping track of a large group of anything requires an organizational system. Many libraries, for example, catalog their books by using the Dewey Decimal System, which organizes books into 10 different classes, further subdivides each class into 10 divisions, and each division into 10 sections. On your computer, you store your electronic files within virtual folders and subfolders, which can be organized by name, size, date, or type.

With the huge number of species on earth, such a classification system is particularly important. Biologists use the system developed by the Swedish biologist Carolus

Linnaeus in the mid-1700s and published in his book *Systema Naturae* ("System of Nature").

Here's how it works (**FIGURE 10-9**). Every species is given a scientific name that consists of two parts, a **genus** (*pl.* genera) and a **specific epithet.** Linnaeus gave humans the name *Homo sapiens,* meaning "wise man." *Homo* is the genus and *sapiens* is the specific epithet. (The genus is capitalized, and the genus and specific epithet are both italicized.) The redwood tree has the name *Sequoia sempervirens.* The strength of Linnaeus's system is that it is a hierarchical system—that is, each element of the system falls under a single element in the level just above it.

Origin of Life   What Are Species?   Evolutionary Trees   Macroevolution

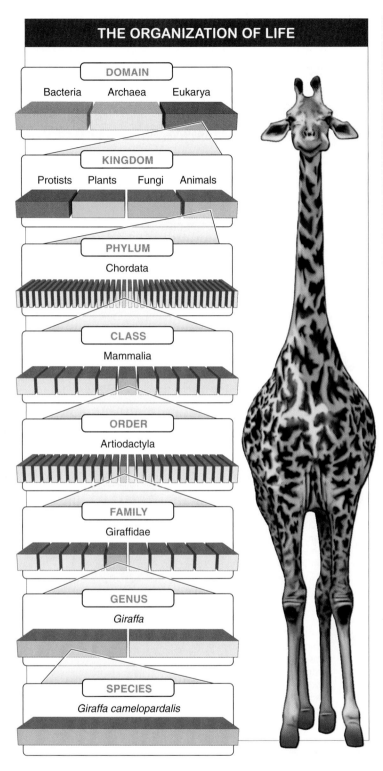

THE ORGANIZATION OF LIFE

DOMAIN
Bacteria   Archaea   Eukarya

KINGDOM
Protists   Plants   Fungi   Animals

PHYLUM
Chordata

CLASS
Mammalia

ORDER
Artiodactyla

FAMILY
Giraffidae

GENUS
Giraffa

SPECIES
Giraffa camelopardalis

FIGURE 10-9 Name that giraffe. Each species on earth, like *Giraffa camelopardalis,* is given a scientific name and is categorized according to hierarchical groups.

In Linnaeus's system, the species is the narrowest classification for an organism. The specific epithet for every species within a genus is unique, but many different species may be placed within the same genus. Similarly, many genera are grouped within a broader group, called a **family.** And many families are grouped within an **order.** Orders are grouped within a **class.** Classes are grouped within a **phylum.** And at the highest level—though this has changed today, as noted below—all phyla were classified under one of three **kingdoms:** the animal kingdom, the plant kingdom, or the "mineral kingdom."

Today, many of the species classifications that Linnaeus described have been revised, and some of his designations, such as the "mineral kingdom," have been left out. Also, all of the kingdoms are now classified under an even higher order of classification, the **domain.** But Linnaeus's basic hierarchical structure remains, and all life on earth is still named using this system, with all organisms belonging to one of the three domains.

When new species are discovered, they are given names based on Linnaeus's system. The higher-level groups into which a new species should be classified are generally clear—that is, it's usually obvious whether something is in the animal or plant kingdom, or whether it is in the mammalian or amphibian class. But when scientists are assigning a specific epithet, they frequently have a little fun. In recent years, for example, an amphibian fossil was named *Eucritta melanolimnetes,* which translates roughly into "creature from the black lagoon." And in honor of Elvis Presley, a wasp species was named *Preseucoila imallshookupis.* A variety of other celebrities also have had species named after them, including the Beatles (several trilobite species), Mick Jagger (a mollusk species and a trilobite species), Kate Winslett (a beetle species), Steven Spielberg (a dinosaur species), and Greta Garbo (a wasp).

Plants and animals often are referred to by common names, which are different from their "official" names in the Linnaean system and are based on similarities in appearance. These common names, however, can cause confusion. "Fish" for example is used as part of the names of jellyfish, crayfish, and silverfish, none of which are closely related to the vertebrate group of fishes that includes salmon and tuna. Later in this chapter we'll see how an important goal of modern classification and the naming of organisms is to link an organism's classification more closely with its evolutionary history than with its physical characteristics.

TAHE-HOME MESSAGE 10•4

Each species on earth is given a unique name, using a hierarchical system of classification. Every species falls into one of three domains.

Overview of Diversity

Species are not always easily defined.

Biologists, like all humans, can be biased. When investigating the natural world, for example, they often focus on plants and animals, to the exclusion of the rest of the earth's rich bio-diversity. This gets them into trouble when it comes to an idea such as the biological species concept. While the biological species concept is remarkably useful when describing most plants and animals, it falls short of representing a universal and defini-tive way of distinguishing many life forms (**FIGURE 10-10**).

Difficulties in classifying asexual species The biological species concept defines species as groups of interbreeding individuals. But this is a useless distinction for the asexual species of the world. Recall from Chapter 6 that asexual reproduction, common among single-celled organisms (including all bacteria), many plants, and some animals, is a form of reproduction that doesn't involve fertilization or even two individuals. Rather, the cell (or cells) of an individual simply divides, creating new individuals. Asexually reproducing organisms don't have a partner or need a partner, they just divide. Because asexual reproduction does not involve any interbreeding, the concept of reproductive isolation is no longer meaningful and it might seem that every individual should be considered a separate species. Clearly, that's not a helpful rule to follow.

Difficulties in classifying fossil species When classifying fossil species, differences in the size and shape of fossil bones from different individuals can never reveal definitively whether there was reproductive isolation between those individuals. This makes it impossible to apply the biological species concept.

Difficulties in determining when one species has changed into another Based on fossils, it seems that modern-day humans, *Homo sapiens,* evolved from a related species called *Homo erectus* about 250,000–400,000 years ago. This seems reasonable, until you consider that your parents—who are in the species *Homo sapiens*—were born to your *Homo sapiens* grandparents, who were born to your *Homo sapiens* great-grandparents, and so on. If humans evolved from *Homo erectus,* at what *exact* point did *H. erectus* turn into *H. sapiens*? It may not be possible to identify the exact point at which this change occurred.

Difficulties in classifying ring species Living in central Asia are some small, insect-eating songbirds called greenish warblers, which are unable to live at the higher elevations of the Tibetan mountain range. Because of this limitation, the warblers

FIGURE 10-10 A useful concept that can't always be easily applied.

THE BIOLOGICAL SPECIES CONCEPT DOESN'T ALWAYS WORK

The biological species concept is remarkably useful when describ-ing most plants and animals, but it doesn't work for distinguishing all life forms.

1 CLASSIFYING ASEXUAL SPECIES
Asexual reproduction does not involve interbreeding, so the concept of reproductive isolation is no longer meaningful.

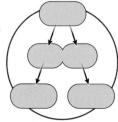

2 CLASSIFYING FOSSIL SPECIES
Differences in size and shape of fossil bones cannot reveal whether there was reproductive isolation between the individuals from whom the bones came.

3 DETERMINING WHEN ONE SPECIES HAS CHANGED INTO ANOTHER
There is rarely a definitive moment marking the transition from one species to another.

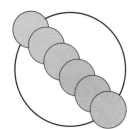

4 CLASSIFYING RING SPECIES
Two non-interbreeding popula-tions may be connected to each other by gene flow through another population, so there is no exact point where one species stops and the other begins.

5 CLASSIFYING HYBRIDIZING SPECIES
Hybridization—the interbreeding of closely related species—sometimes occurs and produces fertile offspring, suggesting that the borders between the species are not clear cut.

live in a ring around the mountain range. At the southern end of the ring, in northwest India, the warblers interbreed with each other. In more northern locations, along either side of the mountain range, the warbler population is split. On each side, the warblers interbreed, but warblers from one side do not interbreed with those from the other, because the mountain range separates them. Where the two "side" populations of warblers meet up again at the northernmost end of the ring, in the forests of Siberia, they can no longer interbreed.

What has happened? Gradual changes in the warblers on each side of the mountain range have accumulated so that the two populations that meet up in Siberia are sufficiently different, physically and behaviorally, that they have become reproductively incompatible. But because the two non-interbreeding populations in the north are connected by gene flow through other populations farther south, there is no exact point at which one species stops and the other begins. So where do you draw the line? The greenish warblers are just one example of more than 20 such **ring species** that have been described.

Difficulties in classifying hybridizing species

Increasingly, **hybridization**—the interbreeding of closely related species—has been observed among plant species and among animal species. This phenomenon fits with the biological species concept as long as postzygotic barriers have evolved, so that the hybrids are weak and unable to reproduce. But in some cases, such as among butterflies in the genus *Heliconius,* the hybrids have high survival rates *and* are fertile, whether interbreeding with other hybrids or with individuals from either of the parental species. This suggests that the borders between the species are not clear cut.

All of these shortcomings have prompted the development of several alternative approaches to defining what a species is.

Q Chihuahuas and Great Danes generally can't mate. Does that mean they are different species?

These alternatives tend to focus on aspects of organisms other than reproductive isolation as defining features. The most commonly used alternative is the **morphological species concept,** which characterizes species based on physical features such as body size and shape. Although the choice of which features to use is subjective, an important feature of the morphological species concept is that it can be used effectively to classify asexual species. And because it doesn't require knowledge of whether individuals can actually interbreed, the morphological species concept is a bit easier to use than the biological species concept when observing organisms in the wild.

When it comes to species definitions, although the biological species concept is the most widely used and can be applied without difficulty to most plants and animals, we should not expect one size to fit all, and there will probably never be a universally applicable definition of what a species is. From asexual species to ring species to hybridizing species, the diversity of the natural world is simply too great to fit into neat, completely defined and distinct little boxes. Nonetheless, scientists can generally use a species definition that is satisfactory for a particular situation.

TAKE-HOME MESSAGE 10·5

The biological species concept is useful when describing most plants and animals, but it falls short of representing a universal and definitive way of distinguishing many life forms. Difficulties arise when trying to classify asexual species, fossil species, species arising over long periods of time, ring species, and hybridizing species. In these cases, alternative approaches to defining species can be used.

10·6

How do new species arise?

Biologists don't really have a clue about how many species there are on earth. Estimates of the number vary tremendously, from a low of 5 million to a high of 100 million. Biologists do know, however, the process by which all species arose.

The process of **speciation,** in which one species splits into two distinct species, occurs in two phases and requires more than just evolutionary change in a population. The first phase of

speciation is *reproductive isolation* (which we've already discussed), through which two populations come to have independent evolutionary fates. The second phase of speciation is *genetic divergence,* in which two populations evolving as separate entities accumulate physical and behavioral differences over time as they become adapted differently to features of their separate environments, including to different predators and types and abundances of food available.

The initial reproductive isolation necessary for speciation often comes about, as we've seen, when two populations are geographically separated. Although this is an effective and common way for speciation to occur, speciation can occur with or without it.

Speciation with geographic isolation: allopatric speciation

Suppose one population of squirrels is split into two separate populations because the climate in its habitat grows wetter and a river forms and splits the habitat in two. Because the squirrels cannot cross the river, the population on one side is reproductively isolated from the population on the other side. Over time, the two populations have different evolutionary paths as they adapt to particular features of their habitats, which may differ on either side of the river. Eventually, the two populations may genetically diverge enough that if the river separating them disappeared and the populations came back into contact, squirrels from the two groups might no longer be able to interbreed. Two species of antelope ground squirrels formed on the north and south rims of the Grand Canyon as a result of this type of speciation, known as **allopatric speciation** (**FIGURE 10-11**).

An example of allopatric speciation can also be seen in the various species of finches found in the Galápagos Islands. Individual finches from the nearest mainland, now Ecuador, in South America, originally colonized one or more of the islands. (And later, additional islands may have been colonized by birds from the mainland or from previously colonized islands.) But because the islands are far apart from one another, the finches tended not to travel between islands and these colonization events are rare, so the populations remained reproductively isolated from one another. Consequently, 14 different species of finches have evolved in the Galápagos Islands, each of which specializes in eating certain of the wide range of insects, buds, and seeds that occur on the islands (**FIGURE 10-12**). Only one species of finch is found in Ecuador.

The barrier doesn't have to be an expanse of water. A forming glacier could split a population into two or more isolated populations. Or a drop in the water level of a lake might expose strips of land that divide the lake into separate, smaller bodies of water, separating one large population of fish into two distinct populations. In each case, the result is the same: geographic isolation that enforces reproductive isolation.

Q *Could you create a new species in the laboratory? How would you do it?*

Researchers can easily create new species in the lab, using an analogous strategy. In one experiment, a single population of fruit flies (*Drosophila pseudoobscura*) was divided into two. The two populations were

ALLOPATRIC SPECIATION

Allopatric speciation occurs when a geographic barrier causes one group of individuals in a population to be reproductively isolated from another group.

1 INITIAL POPULATION

2 REPRODUCTIVE ISOLATION
Suppose a river forms through the squirrels' habitat, separating the population. Because they cannot cross the river, they are reproductively isolated.

3 GENETIC DIVERGENCE
Over time, the populations on either side diverge enough genetically that they are no longer able to interbreed.

Harris's antelope ground squirrel

White-tailed antelope ground squirrel

FIGURE 10-11 Geographic isolation can result in genetic divergence. Two species of antelope ground squirrel formed on the north and south rims of the Grand Canyon.

then maintained on different diets—one population received food made from the sugar maltose, while the other population was fed a starch-based food. After only eight generations of reproductive isolation and adaptation to their differing

Origin of Life   What Are Species?   Evolutionary Trees   Macroevolution

Large cactus finch

Vampire finch

Large ground finch

Woodpecker finch

Small tree finch

FIGURE 10-12   Five of the 14 Galápagos Island finch species. Each species of finch specializes in eating the insects, buds, and/or seeds found in its island habitat.

Mainland finch

> *Due to allopatric speciation, fourteen different species of finches have evolved in the Galápagos Islands. Only one species of finch is found on the nearest mainland.*

nutritional environments, the populations had diverged sufficiently to form separate species; when the populations were mixed, the fruit flies from one population would no longer interbreed with flies from the other population.

Speciation without geographic isolation: sympatric speciation

Speciation can also occur among populations that overlap geographically. This type of speciation is called **sympatric speciation.** Among animals, it is rare for populations of the same animal to become reproductively isolated when they coexist in the same area, so this method of speciation is relatively uncommon. But it *is* common among plants, and it occurs in one of two ways (**FIGURE 10-13**).

During cell division in plants (both in reproductive cells and in other cells of the plant body), an error sometimes occurs in which the chromosomes are duplicated but a cell does not divide. This creates a new cell that may then grow into an individual with twice as many sets of chromosomes as the parent from which it came. The new individual may have four sets of chromosomes, for example, while the original individual had two sets. This doubling of the number of sets of chromosomes is called **polyploidy.** The individual with four sets can no longer interbreed with individuals having only two sets of chromosomes, because their offspring would have three sets (two sets from the parent that had four sets, and one

set from the parent that had two sets), which could not divide evenly during cell division. The individual with four sets can, however, propagate through self-fertilization or by mating with other individuals that have four sets. As a consequence, the individuals with four sets of chromosomes have achieved instant reproductive isolation from the original population and are therefore considered a new species. Although rare in animals, speciation by polyploidy has occurred several times among some species of tree frogs.

A much more common method of speciation through polyploidy occurs when plants from different but closely related species interbreed, forming a hybrid. The hybrid may not be able to interbreed with either of the parental species but may be able to grow and propagate itself asexually—as many plants can. Subsequent errors in cell division and repeated hybridization, however, can ultimately produce fertile individuals with chromosomes from multiple different species (see Figure 10-13). This method of speciation, called **allopolyploidy,** has led to the production of a large number of important crop plants, including wheat, bananas, potatoes, and coffee.

When reproductive separation occurs, for whatever reason, the new populations are isolated from each other. But because the biological species concept considers individuals from populations that could *potentially* interbreed as still belonging

381

SYMPATRIC SPECIATION

Sympatric speciation results in the reproductive isolation of populations that coexist in the same area. Two scenarios, common in plants, lead to this method of speciation.

POLYPLOIDY

1 During cell division, an error occurs in which chromosomes are duplicated but the cell does not divide. This creates a gamete with twice as many sets of chromosomes as a gamete of the parent from which it came.

2 A gamete with two sets of chromosomes cannot produce offspring by fertilizing a gamete with one set of chromosomes.

3 A gamete with two sets of chromosomes can, however, produce offspring by fertilizing another gamete with two sets of chromosomes, producing an individual with four sets of chromosomes.

Parent cell

MEIOSIS

Gametes

Gamete   Gamete

Gamete   Gamete

FERTILIZATION

The new individual has achieved instant reproductive isolation from the original population and, therefore, is considered a new species.

New individual

ALLOPOLYPLOIDY

1 Two plants from different but closely related species interbreed, forming a hybrid.

Species 1   Species 2

FERTILIZATION

2 The hybrid may no longer be able to interbreed with either of the parental species.

Species 1   Hybrid   Species 2

ASEXUAL REPRODUCTION

3 The hybrid may, however, be able to propagate itself asexually—as many plants can.

The hybrid individual has achieved reproductive isolation from the original parental populations and, therefore, is considered a new species.

Hybrid

FIGURE 10-13 Speciation without geographic isolation. Plants may genetically diverge through sympatric speciation.

to the same species, speciation is not complete until sufficient differences have evolved in the two populations that they can no longer interbreed even if they do come in contact. It is in isolation from each other, however, that two populations generally diverge into separate species—that is, the prezygotic and/or postzygotic reproductive isolating mechanisms evolve that form the biological barriers to genetic exchange between the two new species.

Reproductive separation can occur relatively quickly—as long as it takes for a new river (or freeway) to divide one large population into two. But the genetic divergence that causes

true reproductive isolation—no successful interbreeding even when individuals come into contact with each other—can take a very long time, sometimes thousands of years. For this reason, speciation can be difficult to study and observe.

TAKE-HOME MESSAGE 10·6

Speciation is the process by which one species splits into two distinct species that are reproductively isolated. It can occur by polyploidy or by a combination of reproductive isolation and genetic divergence.

❸ Evolutionary trees help us conceptualize and categorize biodiversity.

The diversity of life on earth can be thought of as branching like a tree.

10•7

The history of life can be imagined as a tree.

Although Carolus Linnaeus's method for naming species was an important step in categorizing and cataloguing earth's biodiversity, his underlying assumption about where species came from was wrong. He believed, as did all biologists of his day, that all species had been created at the same time and were unchanging. For this reason, his classification of organisms into hierarchical groups was based on nothing more than his own evaluation of how physically similar various organisms appeared.

A hundred years later, however, Charles Darwin proposed and documented that species could, in fact, change and give rise to new species. With Darwin, the classification of species acquired a new goal and a more important function. In *The Origin of Species* Darwin wrote: "Our classifications will come to be, as far as they can be so made, genealogies." That is, Darwin proposed that the classifications of organisms would resemble family trees that link generations of parents and offspring over long periods of time. With these words, Darwin was the first to link classification with evolution.

Classification such as that used by Linnaeus involved placing organisms within groups as a function of their apparent similarity with each other. The modern incarnation of

Darwin's vision of classification is called **systematics** and has the broader goal of reconstructing the **phylogeny,** or evolutionary history, of organisms. That is, through systematics, all species—even extinct species—are named and arranged in a manner that indicates the common ancestors they share and the points at which they diverged from each other. The common-ancestor points at which species diverge are called **nodes.** A complete phylogeny of all organisms is like a family tree for all species, past and present.

A phylogenetic tree not only shows the relationships among organisms but also presents a hypothesis about the evolutionary history of species. The hypothesis represented by a phylogenetic tree is an unfolding of a story (**FIGURE 10-14**). At the beginning of life on earth there was the first living organism, one that could replicate itself and that had a metabolism. Then a **speciation event** occurred. After this event, the population of the first living organism split into independent evolutionary lineages. The phylogenetic tree had its first branch, and there was now biodiversity. Over hundreds of millions of years, speciation events continued to occur, and today the tree has branches with millions (or possibly tens of millions) of tips that represent all the species on earth. Moving up a branch of any evolutionary tree from its trunk toward its

383

FIGURE 10-14 **Mapping common ancestors.** Evolutionary trees reveal the evolutionary history of every species and the sequence of speciation events that gave rise to them.

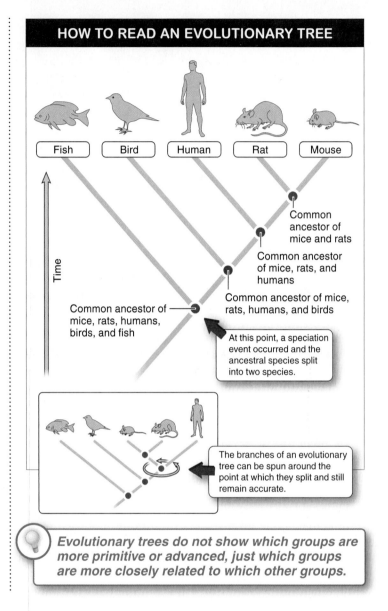

HOW TO READ AN EVOLUTIONARY TREE

Fish | Bird | Human | Rat | Mouse

Time

Common ancestor of mice and rats

Common ancestor of mice, rats, and humans

Common ancestor of mice, rats, humans, and birds

Common ancestor of mice, rats, humans, birds, and fish

At this point, a speciation event occurred and the ancestral species split into two species.

The branches of an evolutionary tree can be spun around the point at which they split and still remain accurate.

Evolutionary trees do not show which groups are more primitive or advanced, just which groups are more closely related to which other groups.

tips, we can see when groups split, with each branching point representing a speciation event. And it doesn't stop here—even today, speciation continues to add new branches all the time.

It is important to remember that phylogenies are hypotheses and, like any hypotheses, they are subject to revision and modification. In the next section we explore how phylogenetic trees are constructed and look at the rich information they convey about the history of life on earth. We also investigate the modern methods of molecular systematics and see how data from DNA sequences are illuminating phylogenies with greater accuracy than ever before. Based on this new molecular and DNA evidence, biologists are revising many earlier phylogenies that were based on observations of organisms' physical features.

TAKE-HOME MESSAGE 10·7

The history of life can be visualized as a tree; tracing from the branches back toward the trunk follows the pathway of descendants back to their ancestors. The tree reveals the evolutionary history of all species and the sequence of speciation events that gave rise to them.

10·8

Evolutionary trees show ancestor-descendant relationships.

Most human cultures and religions have a sacred "Tree of Life." The bodhi tree that the Buddha sat beneath for seven years to achieve the ultimate truth is the Tree of Enlightenment; the apple tree in the Garden of Eden is the Tree of Knowledge. The history of the relationship between all organisms that constitute life on earth is another tree of life. Like most trees, evolutionary trees have a trunk and branches. But these parts represent ancestor-descendant relationships that link living organisms with all life that has ever existed on

earth. Although the evolutionary tree of life can be thought of as one giant tree, as a practical matter biologists often study only particular branches of the tree. These branches can be illustrated as big trees or little trees and can be expanded to include whatever organisms the biologist is studying: the tree might, for example, include all animals or just the rodents.

When drawing an evolutionary tree, it does not matter on which side you put a particular group (see Figure 10-14). Any

Origin of Life | What Are Species? | Evolutionary Trees | Macroevolution branches can be spun around the point at which they split (the node) as if they were a pinwheel. This pinwheel effect means that you cannot assume that rats are more evolutionarily advanced than mice, or vice versa. They are equally advanced in the sense that both groups derived from the same speciation event.

> **Q** Are humans more advanced, evolutionarily, than cockroaches? Can bacteria be considered "lower" organisms?

What are the implications of this? Evolutionary trees tell us many things, but one thing they do not tell us is which groups are most primitive and which are most advanced. This property of phylogenetic trees can serve to undermine some of our most sacred beliefs, such as the one that humans are the pinnacle of evolution. Many trees can be drawn to support this idea, including Figure 10-14. But notice that if you rotate this tree around any one of its nodes, you can get a number of different trees.

What trees do tell us is which groups are most closely related to which other groups. One of the most interesting revelations of "tree thinking" is that—despite appearances— fungi, such as mushrooms, yeasts, and molds, are more closely related to animals than to plants. By analyzing the genetic composition of plants, fungi, and animals, biologists have been able to determine that the relationship between the three groups is that shown in **FIGURE 10-15**. We investigate this surprising fact in Chapter 11.

> " We all should know that diversity makes for a rich tapestry, and we must understand that all the threads of the tapestry are equal in value no matter what their color.
>
> —Maya Angelou, in *Wouldn't Take Nothing for My Journey Now,* 1993 "

Biologists use the term **monophyletic** to describe a group in which all of the individuals are more closely related to each other than to any individuals outside that group. Monophyletic groups are determined by looking at the nodes of the trees. Animals and fungi, taken together, compose a monophyletic group because they share a more recent common ancestor (designated by node A in

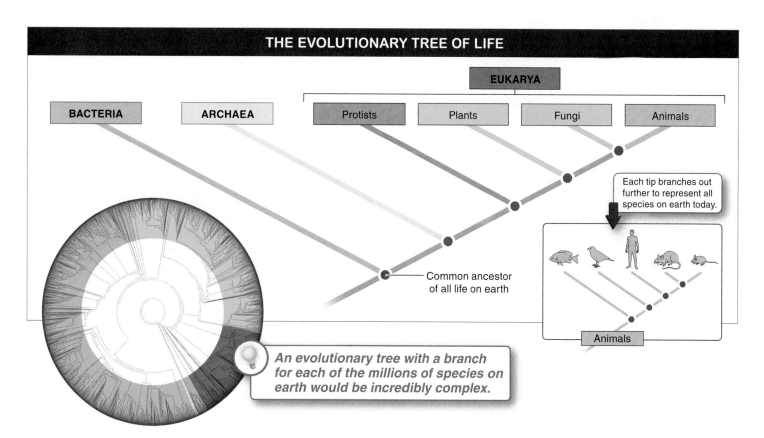

FIGURE 10-15   A growing tree. The evolutionary tree of life has branches with millions of tips representing all species on earth. As speciation events occur, new branches are added to the tree.

MONOPHYLETIC GROUPS

MONOPHYLETIC GROUP
A group in which all of the individuals are more closely related to each other than to any individuals outside that group.

| Animals | Fungi | Plants |

Ⓐ

Ⓑ

Animals and fungi together compose a monophyletic group: they all share a single common ancestor and the two groups contain all of the descendants of that ancestor.

Plants and fungi together do not make a monophyletic group: although they all share a common ancestor, some descendants of that ancestor are in another group (the animals). Plants, fungi, and animals, all taken together, do make a monophyletic group.

FIGURE 10-16 Members of a monophyletic group share a common ancestor and the group contains all of the descendants of that ancestor.

FIGURE 10-16) than either group shares with plants. Plants and fungi, taken together, do not compose a monophyletic group, because their common ancestor (at node B) is also shared by animals. But plants, fungi, and animals, all taken

together, *do* compose a monophyletic group, by virtue of all three sharing the common ancestor at node B.

Constructing evolutionary trees by comparing similarities and differences between organisms

Reading an evolutionary tree reveals which groups are most closely related and approximately how long ago they shared a common ancestor. But how are evolutionary trees—which might hypothesize historical events, such as the cat-dog split, that happened 60 million years ago—constructed in the first place?

Until recently, these trees were assembled by looking carefully at numerous physical features of species and generating tables that compared these features across the species. FIGURE 10-17 is a simple example of such a table. With only three traits, this table shows a clear split between the characteristics of the lion and the hyena, on the one hand, and those of the wolf and the bear, on the other. For most of the 20th century, biologists classifying organisms would often use 50 or more traits to generate a tree.

Then, beginning in the 1980s, biologists began using molecular sequences rather than physical traits to generate evolutionary trees. The rationale for this approach is that organisms inherit DNA from their ancestors and so, as species diverge, their DNA sequences also diverge, becoming increasingly different. As more time passes following the

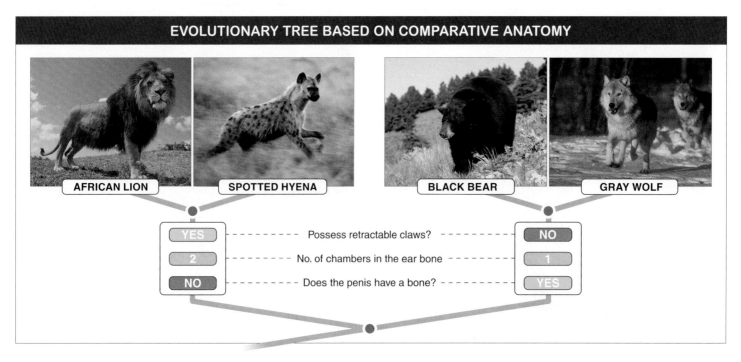

EVOLUTIONARY TREE BASED ON COMPARATIVE ANATOMY

AFRICAN LION | **SPOTTED HYENA** | **BLACK BEAR** | **GRAY WOLF**

YES	Possess retractable claws?	NO
2	No. of chambers in the ear bone	1
NO	Does the penis have a bone?	YES

FIGURE 10-17 Looking for clues in body structures. Before DNA sequencing, physical features of species were used to determine evolutionary relatedness.

Origin of Life | What Are Species? | Evolutionary Trees | Macroevolution splitting of one species into two, the differences in their DNA sequences become greater. By comparing how similar the DNA sequences are between two groups, it is possible to estimate how long ago they shared a common ancestor (FIGURE 10-18).

In theory, the construction of evolutionary trees using DNA sequences is not really different from using physical traits. It is just a way to use many, many more traits in the comparison—each DNA base pair can be thought of as a "trait" being compared.

DNA-BASED EVOLUTIONARY TREES

By comparing how similar the DNA sequences are between two groups, it is possible to estimate how long it has been since they shared a common ancestor.

Difference in DNA base pairs

Species 1 and 2 are more similar to each other than they are to either species 3 or 4. This is because they probably share a more recent common ancestor.

Species 1   Species 2   Species 3   Species 4

A
T
C
G

The more similar their DNA sequences, the more closely two species are related.

FIGURE 10-18 DNA sequences reveal evolutionary relatedness.

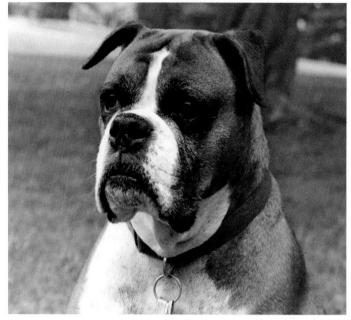

The genome of Tasha, a boxer from upstate New York, was sequenced in 2004. Knowing the makeup of a dog's DNA helps researchers understand where genes responsible for specific traits or defects are located. Knowing the sequence of an animal's DNA allows comparisons with human DNA, improving understanding of both humans and animals.

Although most evolutionary trees are now produced by using molecular sequence comparisons, when it comes to actually naming species we still use Linnaeus's system, assigning a species name and fitting it within the other categories such as kingdoms, phyla, classes, orders, and families. This can be difficult to do because there is no objective way to decide how to assign groups above the level of species—for example, whether multiple genera should be grouped in the same or different families, or whether multiple families should be grouped in the same or different orders.

TAKE-HOME MESSAGE 10·8

Evolutionary trees constructed by biologists are hypotheses about the ancestor-descendant relationships among species. The trees represent an attempt to describe which groups are most closely related to which other groups based on physical features, usually DNA sequences.

Overview of Diversity

Similar structures don't always reveal common ancestry.

Bats and many insects have wings; whales and sharks have fins. One of the goals of evolutionary biology is to figure out how the vast diversity of species and their characteristics evolved. Do bats have wings because they are most closely related to insects and inherited their wings from the common, winged ancestor they shared with insects, or did insects' wings and bats' wings evolve independently? As the evolutionary tree in **FIGURE 10-19** makes clear, insects' wings and bats' wings evolved independently. Bats did not inherit their wings from insects. Wings are an adaptation that arose separately on more than one occasion.

 Did bats inherit their wings from insects or did they evolve independently?

REPEATED EVOLUTION OF THE WING

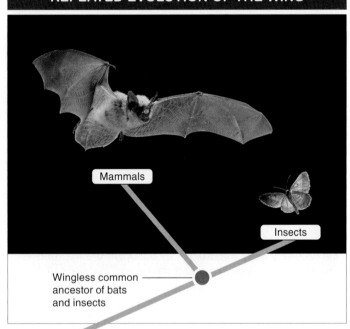

Mammals

Insects

Wingless common ancestor of bats and insects

Bats and insects share a wingless common ancestor, proving that they evolved wings independently.

FIGURE 10-19 What do wings tell us? Because wings are an adaptation that has evolved independently many times in populations living in similar environments under similar selective pressures, the presence of wings does not necessarily indicate evolutionary relatedness.

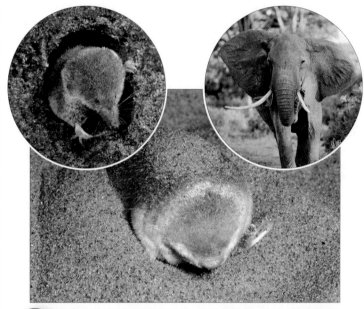

The African golden mole resembles a shrew, but DNA sequence comparisons reveal that it is more closely related to an elephant.

FIGURE 10-20 Looks can be deceiving. Which two of these three animals are more closely related? Can you tell just by looking?

The mapping of species' characteristics onto phylogenetic trees provides us with the story of evolution. Before the 1980s, biologists had always tried to decode the process of evolution by comparing physical features of organisms, such as the presence or absence of wings. Then, with the advent of methods for comparing DNA sequences in the 1980s, tracing the history of life on earth became a more rigorous science. Let's look at a case that illuminates why the original methods were weaker (**FIGURE 10-20**).

Initially, biologists thought that African golden moles belonged in the order known as insectivores, which includes shrews, hedgehogs, and other moles. This belief seemed reasonable because the moles have many characteristics in common with these animals: they are small, they have long, narrow snouts, their eyes are tiny, and they live in underground burrows. Biologists thought that this group of characteristics evolved just once, and that every species in the insectivore order possessed these characteristics because they inherited them from a common ancestor.

388   CHAPTER 10 • THE ORIGIN AND DIVERSIFICATION OF LIFE ON EARTH

Origin of Life   What Are Species?   Evolutionary Trees   Macroevolution

Recent evidence from DNA sequencing, however, surprised everyone who studies these animals. The DNA evidence reveals that the African golden moles are actually more closely related to elephants than to the insectivores, including all of the other mole species! Why do they look so similar to the insectivores? Because of a phenomenon called **convergent evolution,** which occurs when populations of organisms live in similar environments and so experience similar selective forces. **Analogous traits** are characteristics (such as bat wings and insect wings) that are the same because they were produced by convergent evolution, not because they descended from a common structure in a shared ancestor.

Features that are inherited from a common ancestor are called **homologous features.** All mammals have hair because they inherited that trait from a common ancestor. Similarly, among all reptiles, having a mouth is a homologous trait because all reptiles share a common ancestor who passed on the trait to them.

> **Q** Humans have several fused verte-brae at the bottom of the spinal cord that look like a tiny, internal "tail." Are we evolving a tail or losing a tail?

Analogous features are problematic when constructing evolutionary trees, because they are the result of natural selection rather than common ancestry. As a consequence, these traits should not be considered a sign of relatedness between the organisms that have them, and they should not be reflected as a sign of common ancestry when constructing an evolutionary tree. But, how do we know whether traits are homologous or analogous? Through DNA analysis. DNA sequences do not become similar during convergent evolution, so molecular phylogenies cannot be fooled. This is an important reason why molecular-based evolutionary trees are preferable to trees based on physical features.

Although analogous traits and the process of convergent evolution are not helpful in constructing evolutionary trees, that doesn't mean they are useless to biologists. Far from it. Convergence provides some of the best evidence for the power of natural selection. For example, from the 19th to the 20th century in industrialized parts of Europe, a large number of distinct butterfly and moth species became more and more darkly colored. This change was not the result of their sharing a common ancestor, but instead came about because industrialization had caused an accumulation of dark soot on the bark of many trees. This soot accumulation caused light-colored moths and butterflies to stand out against the background when they landed on trees, making them an easier target for predators than dark-colored individuals. As a consequence, natural selection led to a rapid evolution of body color, with many species converging on a darker coloration at the same time.

These peppered moths are the same species, Biston betularia. *The dark coloration of the moth on the left acts as camouflage when it sits on the bark of trees blackened by industrial soot. The moth on the right is adapted to an environment free from soot, and its salt and pepper coloration perfectly blends into the bark of a tree.*

TAKE-HOME MESSAGE 10·9

Evolutionary trees are best constructed by comparing DNA sequences among organisms, because convergent evolution can cause distantly related organisms to appear much more closely related but it doesn't increase their DNA sequence similarity.

4 Macroevolution gives rise to great diversity.

More than 400 Anolis *lizard species exist, including this colorful lizard* (Anolis grahami)*, which lives in trees.*

10·10

Macroevolution is evolution above the species level.

When water runs over rocks, it wears them away. The process is simple and slow, yet it is powerful enough to have created the Grand Canyon. To be sure, water running over rocks does not always make a Grand Canyon. Nonetheless, no additional physical processes are necessary. The process of evolution has a lot in common with a stream of water running over rocks: in the short term, it produces small changes in a population, yet the accumulation of these changes can be "canyon-esque" (**FIGURE 10-21**). Let's consider some examples.

The production of 200-ton dinosaurs from rabbit-size reptile ancestors. The diversification of flowering plants from a single species into more than 230,000 species. The emergence of animals that live on land. These large-scale examples, which are the products of evolutionary change involving the origins of entirely new groups of organisms, are referred to as **macroevolution.** These examples can be contrasted with the increase in milk production in cows during the first half of the 20th century, or the gradual change in the average beak size of birds with changing patterns of rainfall, phenomena involving changes in allele frequencies in a population that are referred to as **microevolution** (**FIGURE 10-22**).

No additional processes necessary! Just as a stream of water running over rocks can create the Grand Canyon, accumulated microevolutionary changes can lead to dramatic macroevolutionary changes.

FIGURE 10-21 From a trickle of water to the Grand Canyon. The Grand Canyon is a product of small changes to the landscape over millions of years.

EVOLUTION: MICRO vs. MACRO

Evolution is one thing only: a change in allele frequencies within a population. But over time, these changes can lead to new species and groups of species that vary tremendously.

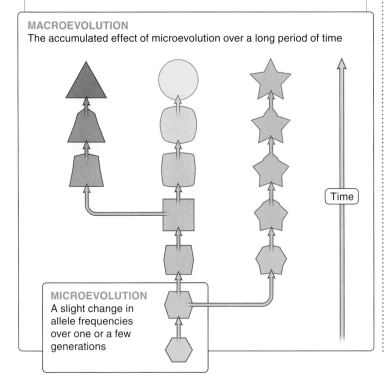

MACROEVOLUTION
The accumulated effect of microevolution over a long period of time

Time

MICROEVOLUTION
A slight change in allele frequencies over one or a few generations

FIGURE 10-22 The scale of evolution. Microevolutionary changes in allele frequencies in a population over time can lead to macroevolution, changes on a grand scale, including vast diversification of species.

These micro and macro events might seem like two very different processes, but they are not. Evolution, whether at the micro or macro level, is one thing only: a change in allele frequencies over time. In the short term, over one or a few generations, evolution can appear as a slight and gradual change within a species. But over a longer period of time, the accumulated effects of this process, acting continuously and combined with reproductive isolation of populations, can lead to the dramatic phenomena described as macroevolution. Just as a trickle of water, given enough time, produced the Grand Canyon, so evolution has created the endless forms of diversity on earth. In a sense, microevolution is the process and macroevolution is the result.

TAKE-HOME MESSAGE 10·10

The process of evolution—changes in allele frequencies within a population—in conjunction with reproductive isolation is sufficient to produce speciation, diversification, and the rich diversity of life on earth.

10·11

The pace of evolution is not constant.

If you listened in on some scientists debating their research, you might hear one side describing "evolution by jerks" and the other side making light of "evolution by creeps." They're not gearing up for a fight out in the school yard, though. They're just describing two examples of the pace of evolution.

The traditional model of how evolutionary change occurred was that used in the previous section to describe microevolution. Populations changed slowly but surely, gradually accumulating sufficient genetic differences for speciation—hence the phrase "evolution by creeps." Spurred on by findings from the fossil record that do not always support this view, however, researchers have come to believe that evolution may often occur in a different way: brief periods of rapid evolutionary change immediately after speciation, followed by long periods with relatively little change—hence "evolution by jerks." This newer view of the

pace of evolution, in which rapid periods of evolutionary change are punctuated by longer periods with little change, is called **punctuated equilibrium** (**FIGURE 10-23**).

In nature, we can find examples of both gradual change and the irregular pattern of punctuated equilibrium. It's not a situation in which one version is right and one is wrong. For many groups of organisms, including many mollusks and mammals, many fossils from a particular species exist and reveal that it seems to have persisted with little change for very long periods of time. Then, a large number of mollusk or mammal fossils appear in the fossil record that are similar enough to the older fossils that they appear to be in the same evolutionary lineage, yet there are no fossils representing the gradual transition between the newer and older fossils. For other groups, such transitional fossils do exist. Very complete sets of fossils reveal the transitional sequence from fox-like

The pace of evolution varies for different species. Some species have evolved gradually over time, while others spend vast amounts of time with little change.

GRADUAL CHANGE
Evolution by creeps: The pace of evolution occurs gradually in incremental steps.

Time

PUNCTUATED EQUILIBRIUM
Evolution by jerks: Rapid periods of evolutionary change are punctuated by longer periods with little change.

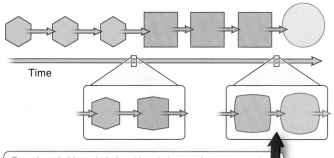

Time

Even though this period of rapid evolutionary change may only cover 1% of the species' evolutionary history, it still may cover hundreds or thousands of generations. This could take tens of thousands of years in a primate or a matter of months in bacteria.

FIGURE 10-23 **The pace of evolution can be rapid or slow.**

ancient terrestrial mammals to modern whales, for example. And fairly complete sequences of fossils allow the tracing of modern horses back to similar small, fox-like mammals.

Punctuated equilibrium has erroneously been portrayed as a "problem" for evolutionary biology, for two primary reasons.

1. It is mistakenly believed that the rapid evolutionary changes of punctuated equilibrium are the result of some new mechanism of evolution—thus seeming to undermine evolutionary theory. In actuality, although the punctuated change appears lightning-fast on a geological time scale, the per-generation rates of change necessary to produce it are relatively slow and not at all beyond the rates of change that have been measured in evolving populations. As an example, a fossil species may appear unchanged for most of its 3- to 5-million-year history, save for a "rapid" burst during the first 50,000–100,000 years after the species appears (following a

speciation event). Even though the period of change represents only about 1% of that species' evolutionary history, it still covers tens of thousands of generations—a more than ample period for even the most dramatic of morphological changes.

2. The existence of "gaps" in the fossil record seems to imply that the transitional species did not exist—again seeming to contradict evolutionary theory. To the contrary, though, punctuated equilibrium should generate exactly the patterns found. Consider an example. In a large population, speciation is most likely to occur at the edges of the species' range. At the edges, small populations are more likely to be separated and geographically isolated from the rest of the population. The population is also likely to experience the most extreme selective pressures at the edges of its range, where the habitat differs most from the rest of its range and may represent the edge of the species' ecological tolerance. In this border region, speciation and rapid evolutionary change are most likely to occur. But because the individuals there represent only a tiny fraction of all the individuals in the population, they will make up only a tiny fraction of the fossil record. And this is in addition to the fact that the probability of an organism (or part of an organism) becoming a fossil is very low to start with.

In the end, there is no single rate at which evolution occurs across all species. Some species have clearly changed gradually, over time, for their entire evolutionary history. Others have spent vast periods of time with little change at all. The coelacanth is a species of fish, for example, that seems to have undergone almost no physical changes over more than 300 million years. For any organisms, the rate of evolutionary change depends on the selective forces acting on the population. Strong and directional forces may act over long periods of time and lead to rapid change, while stabilizing selection may lead to very little change.

Darwin foreshadowed the discussion about the pace of evolution in *The Origin of Species,* writing, "the periods during which species have undergone modification, though long as measured in years, have probably been short in comparison with the periods during which they retain the same form."

TAKE-HOME MESSAGE 10•11

The pace at which evolution occurs can be rapid or very slow. In some cases, the fossil record reveals rapid periods of evolutionary change punctuated by longer periods with little change. In other cases, species may change at a more gradual but consistent pace.

392   CHAPTER 10 • THE ORIGIN AND DIVERSIFICATION OF LIFE ON EARTH

Origin of Life   What Are Species?   Evolutionary Trees   Macroevolution

10·12

Adaptive radiations are times of extreme diversification.

Even the greatest of success stories often owe something to a bit of luck. As mammals, we owe more than a little of our success to luck. Flash back to 65 million years ago. Our mammalian ancestors had their place among the organisms on earth, but it didn't remotely resemble our dominant position today. Our ancestors were rodent-size, insect-eating creatures, relegated to a nocturnal existence. It was the dinosaurs' time, and the giant reptiles dominated the earth.

All that changed in an instant when the earth was struck near what is now the eastern part of Mexico (the Yucatán Peninsula) by an asteroid about 6 miles (10 km) in diameter. This caused the environmental conditions on earth to change quickly and dramatically. We'll explore the details of this catastrophic event in the next section, but one outcome was definitive: all of the dinosaur species were wiped out. Our mammalian ancestors, on the other hand, survived and found themselves living on a planet where most of their competitors had suddenly disappeared. We were in the right place at the right time.

What followed was an explosive expansion of mammalian species. In a brief period of time, a small number of species diversified into a much larger number of species, able to live in a wide diversity of habitats. Called an **adaptive radiation,** such a large and rapid diversification has occurred many times throughout history (**FIGURE 10-24**).

Three different phenomena tend to trigger adaptive radiations. After one of these events, surviving species find themselves in locations where they suddenly have access to plentiful new resources.

1. *Mass extinction events.* With the disappearance of the dinosaurs, a world of "opportunities" opened up for the mammals. Where previously the dinosaurs had prevented mammals from utilizing resources, mammals suddenly had few competitors. Not surprisingly, the number of mammalian species increased from a very small number—perhaps a few hundred—to about 130 genera, with more than 4,000 species overall, including primates, bats, horses, rodents, and the first completely aquatic mammals. This happened in about 10 million years, barely the blink of an eye by geological standards. Following other large-scale extinctions, numerous other groups that suddenly lost most of their competitors experienced similar adaptive radiations.

ADAPTIVE RADIATION

Adaptive radiation occurs when a small number of species diversify into a larger number of species. Three phenomena tend to trigger adaptive radiation.

1 **MASS EXTINCTION EVENTS**
With their competition suddenly eliminated, remaining species can rapidly diversify.

2 **COLONIZATION EVENTS**
Moving to a new location with new resources (and possibly fewer competitors), colonizers can rapidly diversify.

3 **EVOLUTIONARY INNOVATIONS**
With the evolution of an innovative feature that increases fitness, a species can rapidly diversify.

FIGURE 10-24 **A rapid diversification of species.**

---

2. *Colonization events.* In a rare event, one or a few birds or small insects will fly off from a mainland and end up on a distant island group, such as Hawaii or the Galápagos Islands. Once they make it there, they tend to find a large number of opportunities for adaptation and diversification. In the Galápagos, as we learned in Section 10-6, 14 finch species evolved from the single species of finch found on the nearest mainland, Ecuador, 600 miles away. In Hawaii, there are several hundred species of fruit flies, all believed to have evolved from one species that colonized the islands and experienced an adaptive radiation. Perhaps a few flies were blown there by a storm, or maybe they were carried there, stuck in the feathers of a bird.

3. *Evolutionary innovations.* In the world of computers, software developers are always looking for the "killer app"—the killer application, a new type of software so useful that it immediately leads to huge success, opening up a large new niche in the software market or greatly expanding an already-existing niche. The first spreadsheet, email program, and web browser were all killer apps. In nature, evolution sometimes produces killer apps, too. These are innovations such as the

Overview of Diversity

wings and rigid outer skeleton that appeared in insects and helped them diversify into the most successful group of animals, with more than 800,000 species today (more than a hundred times the number of mammalian species). The flower is an innovation that propelled the flowering plants to an explosion of diversity and evolutionary success relative to the non-flowering plants, such as ferns and pine trees. Today, about 9 out of 10 plant species are flowering plants.

10·13

There have been several mass extinctions on earth.

Forests keep disappearing, rivers dry up, wild life's become extinct, the climate's ruined and the land grows poorer and uglier every day.

—Anton Chekhov, *Uncle Vanya,* 1899

If the past is a guide to the future, we know this: no species lasts forever. Speciation is always producing new species, but **extinction,** the complete loss of all individuals in a species population, takes them away. Extinction, which is always occurring, is a risk faced by all species.

For any given time in earth's history, it is possible to estimate the rate of extinctions at that time, and the evidence reveals that these rates are far from constant (**FIGURE 10-25**). Although the particular details differ in most cases, extinctions generally fall into one of two categories: "background" extinctions or mass extinctions.

Background extinction is the term used to describe the extinctions that occur at lower rates during periods other than periods of mass extinction. Background extinctions occur mostly as the result of natural selection. Competition with other species, for example, may reduce a species' size or the range over which it can roam or grow. Or a species might be too slow to adapt to gradually changing environmental conditions, and becomes extinct as its individuals die off.

Mass extinctions are periods during which a large number of species on earth become extinct over a relatively short period of time. There have been at least five mass extinctions on earth and, during each of these extinctions, 50% or more of the animal species living at the time became extinct (see the red parts of Figure 10-25).

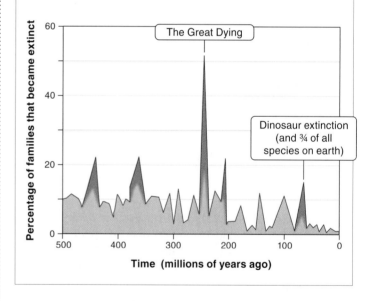

TWO CATEGORIES OF EXTINCTIONS

Extinctions generally fall into two categories:

● **MASS EXTINCTIONS**
A large number of species become extinct over a short period of time due to extraordinary and sudden environmental change.

● **BACKGROUND EXTINCTIONS**
These extinctions occur at lower rates during times other than mass extinctions.

FIGURE 10-25 **Extinction never sleeps.**

There is a fundamental difference between background and mass extinctions that goes beyond differences in rates. They have different causes. Mass extinctions are due to extraordinary and sudden changes to the environment (such as the asteroid that brought about the dinosaurs' extinction). As a consequence, nothing more than bad luck is responsible

394 CHAPTER 10 • THE ORIGIN AND DIVERSIFICATION OF LIFE ON EARTH

Origin of Life   What Are Species?   Evolutionary Trees   Macroevolution

for the extinction of species during mass extinctions; fit and unfit individuals alike perish.

Of the five mass extinctions during the past 500 million years, the most recent is also the best understood. Sixty-five million years ago, an almost unimaginable catastrophe occurred. As discussed in the previous section, a massive asteroid smashed into the Caribbean near the Yucatán Peninsula of Mexico (**FIGURE 10-26**). The impact left a crater more than 100 miles

Global climate change! Worldwide mass extinction! These catastrophic events set the stage for explosive speciation of the remaining groups, including the mammals.

FIGURE 10-26 **Catastrophic collision.** Sixty-five million years ago, an asteroid 6 miles in diameter smashed into earth near the Yucatán Peninsula of Mexico and wiped out three-fourths of all species on earth, including the dinosaurs.

wide, and probably created an enormous fireball that caused fires worldwide, followed by a cloud of dust and debris that blocked all sunlight from the earth and disturbed the climate worldwide for months. There may also have been other asteroid impacts around this same time. In the aftermath of this catastrophe, about 75% of all species on earth were wiped out, including the dinosaurs, a tremendously successful group that had been thriving for 150 million years.

As bad as the dinosaur-exterminating asteroid event was, from a biodiversity perspective it is not the worst catastrophe in earth's history. That distinction falls to a mass extinction that took place 250 million years ago, called the Great Dying (see Figure 10-25). Although the cause is not clear—hypotheses include an asteroid impact, continental drift, a supernova (a star explosion), or extreme volcanic eruptions—more than 95% of all marine life became extinct, along with almost 75% of all terrestrial vertebrates. The causes of the three other mass extinctions also are poorly understood, but the evidence in the fossil record of the tremendous upheavals is clear.

It is interesting to note that although, from time to time, sudden and extreme events lead to mass extinctions that wipe out a large proportion of the species on earth, in every case these mass extinctions are followed by a time of explosive speciation in the groups that survive. An important question now being debated is whether we are currently in the midst of a human-caused mass extinction event. We will explore whether this is true—and why—in Chapter 16.

TAKE-HOME MESSAGE 10·13

As new species are being created, others are lost through extinction, which may be a consequence of natural selection or large, sudden changes in the environment. Mass extinctions are periods during which a large number of species on earth become extinct over a short period of time. These events are usually followed by periods of unusually rapid adaptive radiation and diversification of the remaining species.

❺ An overview of the diversity of life on earth: organisms are divided into three domains.

Purple-ring top snails (Calliostoma annulatum) *on a kelp bulb.*

10·14

All living organisms are classified into one of three groups.

Learning about biological diversity can be humbling. At the very least, it makes it harder to believe that humans are particularly special. We are not at the center of earth's "family tree." Nor are we at its peak. We are simply another branch. A newer branch, and definitely connected to the same trunk as all the other branches, but a bit small and off to one side, just the same.

When Linnaeus first put together his system of classification, he saw a clear and obvious split: all living organisms were either plant or animal. Plants could not move and they made their own food. Animals could move but could not make their own food. So in Linnaeus's original classification, all organisms were put in either the animal kingdom or the plant kingdom (his third, "mineral" kingdom, now abandoned, only included non-living matter).

With the refinement of microscopes and subsequent discovery of the rich world of **microbes**—microscopic organisms—the two-kingdom system was inadequate. Where did the microbes belong? Some could move, but many of those could also make their own food, seeming to put them somewhere between plants and animals. And the problems didn't stop with the microbes: mushrooms and molds, among other organisms originally categorized as plants, didn't move but they didn't make their own food either—they digested the decaying plant and animal material around them.

The two-kingdom system gave way in the 1960s to a five-kingdom system. At its core, the new system was a division based on the distinction between *prokaryotic* cells (those without nuclei) and *eukaryotic* cells (those with nuclei). The prokaryotes were put in one kingdom, the monera, and the only residents in this kingdom were the bacteria: single-celled organisms with no nucleus, no organelles, and genetic material in the form of a circular strand of DNA. The eukaryotes—having a nucleus, compartmentalized organelles, and individual, linear pieces of DNA—were divided into four separate kingdoms: plants, animals, fungi, and **protists,** a group that includes most of the single-celled eukaryotes, such as algae.

The classification of organisms took a huge leap forward in the 1970s and 1980s, and the five-kingdom system had to be discarded. Until that point, organisms had been classified primarily based on their appearance. But because the ultimate goal had changed to reconstructing phylogenetic trees that reflected the evolutionary history of earth's diversity, Carl Woese, an American biologist, and his colleagues began classifying organisms by examining the nucleotide sequences they carried.

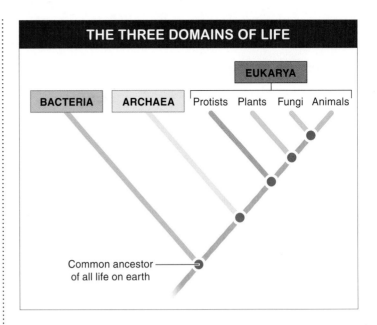

FIGURE 10-27 **All living organisms are classified into one of three groups.**

Q All living organisms have probably evolved from a single common ancestor. How can the existence of one particular molecule support this claim?

Woese assumed that the more similar the genetic sequences were between two species, the more closely related they were, and he built phylogenetic trees accordingly. The only way Woese could compare the evolutionary relatedness of all the organisms present on earth today was by examining one molecule that was found in *all* living organisms and looking at the degree to which it differed from species to species. He discovered a perfect candidate for this role: a molecule called ribosomal RNA, which helps translate genes into proteins (see Chapter 5). Ribosomal RNA has the same function in all organisms on earth, almost certainly because it comes from a common ancestor. Over time, however, its genetic sequence (i.e., the DNA that codes for the ribosomal RNA) has changed a bit. Tracking these changes makes it possible to reconstruct the process of diversification and change that has taken place.

The trees that Woese's genetic sequence data generated had some big surprises. First and foremost, the sequences revealed that the biggest division in the diversity of life on earth was not between plants and animals. It wasn't even between prokaryotes and eukaryotes. The new trees revealed instead that the diversity among microbes was dramatically greater than ever imagined—particularly because of the discovery of a completely new group of prokaryotes called **archaea,** which thrive in some of the most extreme environments on earth and differ greatly from bacteria. The tree of life was revised to show three primary branches, called domains: the bacteria, the archaea, and the eukarya (**FIGURE 10-27**).

Woese put the domains above the kingdom level in Linnaeus's system. In Woese's new system, which is the most widely accepted classification scheme today, the bacteria and archaea domains each have one kingdom and the eukarya domain has four. Because both bacteria and

Q Are viruses alive?

archaea are microscopic, it can be hard to believe that the two domains are as different from each other as either domain is from the eukarya. Each of the three domains is monophyletic, however, meaning that each contains species that share a common ancestor and includes all descendants of that ancestor. Close inspection even reveals that the archaea are more closely related to the eukarya than they are to the bacteria.

The three-domain, six-kingdom approach is not perfect and is still subject to revision. Within the eukarya, for example, it seems that single-celled protists are so diverse that they should be split into multiple distinct kingdoms. Also problematic is the fact that bacteria sometimes engage in **horizontal gene transfer,** which means that, rather than passing genes simply from "parent" to "offspring," they transfer genetic material directly into another species. This process complicates the attempt to determine phylogenies based on sequence data, because it creates situations in which two organisms might have a genetic sequence that is similar not because they share a common ancestor, but as a result of a direct transfer of the sequence from one species to another.

Additionally, a fourth group of incredibly diverse and important biological entities, the **viruses,** is not even included in the tree of life, because they are not considered to be living organisms. Viruses can replicate, but can only have metabolic activity by taking over the metabolic processes of another organism. Their lack of metabolic activity puts viruses just outside the definition of life we use in this book, but some scientists view viruses as living.

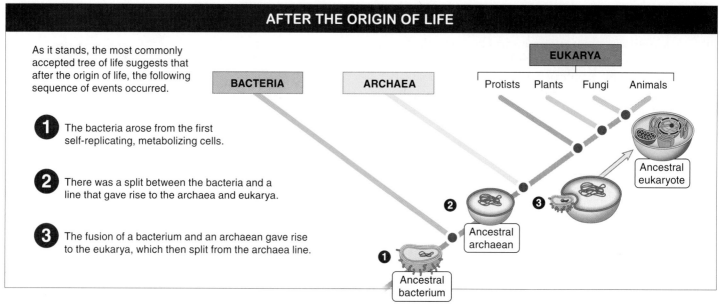

As it stands, the most commonly accepted tree of life suggests that after the origin of life, the following sequence of events occurred.

EUKARYA

BACTERIA   ARCHAEA   Protists Plants Fungi Animals

1 The bacteria arose from the first self-replicating, metabolizing cells.

2 There was a split between the bacteria and a line that gave rise to the archaea and eukarya.

3 The fusion of a bacterium and an archaean gave rise to the eukarya, which then split from the archaea line.

Ancestral eukaryote

Ancestral archaean

Ancestral bacterium

FIGURE 10-28 From self-replicating metabolizing cells to complex organisms. The biggest branches on the tree of life separate the three domains.

As it stands currently, the most commonly accepted tree of life suggests that, after the origin of life, the following sequence of events occurred (**FIGURE 10-28**):

1. The bacteria arose from the first self-replicating, metabolizing cells.

2. There was a split between the bacteria and a line that gave rise to the archaea and eukarya.

3. The fusion of a bacterium and an archaean gave rise to the eukarya, which then split from the archaea line.

The next three sections introduce each of the three domains, surveying the broad diversity that has evolved within each domain and the common features that link all the members of each. Chapters 11, 12, and 13 cover the domains in greater detail.

TAKE-HOME MESSAGE 10·14

All life on earth can be divided into three domains—bacteria, archaea, and eukarya—which reflect species' evolutionary relatedness to each other. Plants and animals are just two of the four kingdoms in the eukarya domain, encompassing only a small fraction of the domain's diversity.

10·15

The bacteria domain has tremendous biological diversity.

Morning breath is stinky. And learning the cause of the offensive smell may make the situation even worse. You see, when you wake up, your mouth contains huge amounts of bacterial waste products. Perhaps the only consolation for this situation is that it gives us a glimpse into just how diverse and resourceful bacteria are.

Q Why is morning breath so stinky?

At any given time, there are several hundred species of bacteria in your mouth—mostly residing on your tongue—all competing for the resources you put there (**FIGURE 10-29**). Some of the bacteria are aerobic, requiring oxygen for their metabolism, and others are anaerobic. At night, because the flow

of saliva slows down and the oxygen content of your mouth decreases, the anaerobic bacteria start to get the upper hand in terms of growth and reproduction. These bacteria metabolize food bits in your mouth, plaque on your teeth and gums, and dead cells from the lining of your mouth, breaking down proteins in these materials to use as their energy source. As they do so, waste products accumulate. Because proteins are made from amino acids, some of which contain the smelly chemical sulfur, their breakdown leads to the bad smell.

In the morning, once you wake up, you breathe more and produce more saliva, both of which increase the oxygen level in

FIGURE 10-29 **Bacteria thrive on your tongue and in your mouth.**

your mouth. This tips the battle for space and food back in favor of the aerobic bacteria. Because aerobic bacteria prefer carbohydrates as their energy source and because carbohydrates don't contain sulfur, as the aerobic bacteria start to outcompete the anaerobic bacteria, the sulfur smell goes away. The aerobic bacteria are, of course, also filling your mouth with waste products, it's just that their waste products don't smell as bad.

On a small scale, your mouth reveals some of the tremendous biological versatility of the bacteria: hundreds of species can live in a tiny area—a teaspoon of soil, for example, is home to more than a billion bacteria—they can thrive in a variety of unexpected habitats, they can utilize a variety of food sources, and they can survive and thrive with or without oxygen.

Bacteria have a biomass greater than that of all the plants and animals on earth. They thrive in a variety of habitats including arctic ice, volcanic vents, and the human intestine.

FIGURE 10-30 **We're outnumbered.** Bacteria are everywhere and greatly outnumber eukarya.

Looking around the world, we find that the diversity in the bacteria domain is even greater. A survey of the numbers shows the clear dominance of bacteria on our planet. By any measure, it is their planet: the biomass of bacteria (if they were all collected, dried out, and weighed) exceeds that of all the plants and animals on earth. Bacteria live in soil, air, water, arctic ice, and volcanic vents (**FIGURE 10-30**). Many can even make their own food, utilizing light from the sun or harnessing energy from chemicals such as ammonia.

While they differ in many ways, the bacteria are a monophyletic group, sharing a common ancestor. For this reason, they all have a few features in common. All bacteria are single-celled organisms with no nucleus or organelles, with one or more circular molecules of DNA as their genetic material, and using several methods of exchanging genetic information. Because they are asexual—they reproduce without a partner, just by dividing—the biological species concept cannot be applied to bacteria when classifying them into narrower categories. Thus bacteria are classified on the basis of physical appearance or, preferably, genetic sequences.

It is rather unfair to introduce the bacteria by describing how they can cause bad breath, given that many people already think of bacteria as illness-causing organisms. While bacteria are responsible for many diseases, including strep throat, cholera, syphilis, pneumonia, botulism, anthrax, leprosy, and tuberculosis, disease-causing bacteria are only a small fraction of the domain, and bacteria seem to get less credit for their many positive effects on our lives. Consider that bacteria (*E. coli*) living in your gut help your body digest the food you eat and, in the process, make certain vitamins your body needs. Other bacteria (actinomycetes) produce antibiotics such as streptomycin. Still other bacteria live symbiotically with plants as small fertilizer factories, converting nitrogen into a form that is usable by the plant. Bacteria also give taste to many foods, from sour cream to cheese, yogurt, and sourdough bread. Increasingly, bacteria are used in biotechnology —from those that can metabolize crude oil and help in the cleanup of spills to transgenic bacteria used in the production of insulin and other medical products, as described in Chapter 5.

We explore the great diversity of the bacteria in more detail in Chapter 13.

TAKE-HOME MESSAGE 10·15

All bacteria share a common ancestor and have a few features in common. They are prokaryotic, asexual, single-celled organisms with no nucleus or organelles, with one or more circular molecules of DNA as their genetic material, and using several methods of exchanging genetic information. Bacteria have evolved a broad diversity of metabolic and reproductive abilities relative to the eukarya.

The archaea domain includes many species living in extreme environments.

In a bubbling hot spring in Yellowstone National Park, the temperatures range from the 212° F (100° C) of boiling water down to a relatively cool 165° F (74° C) at the surface. It would seem a most inhospitable place for life. Yet researchers have found 38 different species of archaea thriving there. Perhaps even more surprising, the genetic differences among these species are more than double the genetic differences between plants and animals.

In the freezing waters of Antarctica, too, archaea abound. More than a third of the organisms in the Antarctic surface waters are archaea. In swamps, completely devoid of oxygen, and in the extremely salty water of the Dead Sea, the story is the same: where once it was assumed that no life could survive, the archaea not only exist but thrive and diversify (FIGURE 10-31).

If biologists ever find themselves thinking that we have a handle on the breadth of life that is possible on earth, they would be wise to remember the archaea. Until relatively recently, our perception of life on earth was that there were bacteria and there were

Q Will life be found elsewhere in the universe? How does the discovery of archaea alter the likelihood of this?

eukaryotes, such as the plants and animals. But in the past several decades, as researchers have explored some of the most unlikely of habitats, they have found archaea thriving. There are entire worlds of life on earth that we never even imagined.

Analyses of genetic sequences indicate that the archaea and the bacteria diverged about 3 billion years ago. Although it is likely that some genetic exchanges continued to occur between them, they have evolved along largely independent paths ever since. Approximately 2.5 billion years ago, the eukarya split off from the archaea. The archaea are still grouped in one kingdom within the domain archaea, but we have no idea how many species exist. Given that archaea are the dominant microbes in the deep seas, it may very well be that these organisms—which we were completely ignorant of until recent decades—are the most common organisms on earth. It is still too early to tell.

We do know that, like the bacteria, all archaea are single-celled prokaryotes. For that reason, under a microscope they look very similar to bacteria. Several physical features distinguish them from the bacteria, however. Archaea have cell walls made from materials that are different from those found in the cell walls of bacteria and eukaryotes. Specifically, the archaean cell walls contain polysaccharides not found in either bacteria or eukaryotes. The archaea also have cell membranes, ribosomes, and some enzymes similar to those found in the eukarya.

The archaea exhibit tremendous diversity and are often divided into five groups based on their physiological features:

1. Thermophiles ("heat lovers"), which live in very hot places

2. Halophiles ("salt lovers"), which live in very salty places

3. High- and low-pH-tolerant archaea

4. High-pressure-tolerant archaea, found as deep as 4,000 meters (about 2.5 miles) below the ocean surface, where the pressure is almost 6,000 pounds

FIGURE 10-31 **Archaea can thrive in the most inhospitable-seeming places.** Archaea live in a diversity of environments; they resemble bacteria but are actually more closely related to eukarya.

per square inch (compared with an air pressure of less than 15 pounds per square inch at sea level)

5. Methanogens, which are anaerobic, methane-producing archaea

There also seem to be large numbers of archaea living in relatively moderate environments that are also commonly home to bacteria. We explore the great diversity of the domain archaea in more detail in Chapter 13.

TAKE-HOME MESSAGE 10·16

Archaea, many of which are adapted to life in extreme environments, physically resemble bacteria but are more closely related to eukarya. Because they thrive in many habitats that humans have not yet studied well, including the deepest seas and oceans, they may turn out to be much more common than currently believed.

10·17

The eukarya domain consists of four kingdoms: plants, animals, fungi, and protists.

After exploring the archaea and bacteria domains, turning to the eukarya feels like coming home. We are most familiar with this domain: all of the living organisms that we can see are eukarya. Plants, mushrooms, slime molds, insects, fish, birds, and, of course, mammals, including humans. There are three kingdoms of eukarya that can be seen with the naked eye: plants, animals, and fungi. All are made up from eukaryotic cells—they have a membrane-enclosed nucleus—and each kingdom is almost entirely multicellular.

A fourth kingdom, called protista (protists), which is most often invisible to the naked eye, is a sort of grab bag that includes a wide range of mostly single-celled eukaryotic organisms, including amoebas, paramecia, and algae. Discovery of new species of protists, as with the other microscopic organisms on earth (the bacteria and archaea), continues at very high rate. It is likely that the protista will turn out to be the most diverse of all the eukarya kingdoms, or it may even be split into multiple kingdoms.

The eukarya split from the archaea about 2.5 billion years ago (**FIGURE 10-32**). At the time, the eukarya were single-celled and probably resembled modern protists more than any other

modern eukarya. The split may have occurred when symbiotic bacteria became incorporated within an ancestor of the eukaryotes, resulting in what would become the mitochondria. Later, approximately 1.5 billion years ago, a second important symbiosis between bacteria and eukarya resulted in what would become chloroplasts (see Chapter 3).

Because they are so much easier to see than bacteria and archaea, a disproportionate number of the named species on earth are in the domain eukarya. In fact, of the 1.5 million named species, the majority are eukarya, with about half being insects. This is more a result of the interests and biases of biologists than a reflection of the relative numbers of actual species in the world. We explore the great diversity of the eukarya in Chapters 11 (on animals) and 12 (on plants and fungi).

TAKE-HOME MESSAGE 10·17

All living organisms that we can see with the naked eye (and many that are too small to be seen) are eukarya, including all plants, animals, and fungi. The eukarya are unique among the three domains in having cells with organelles.

THE EUKARYA

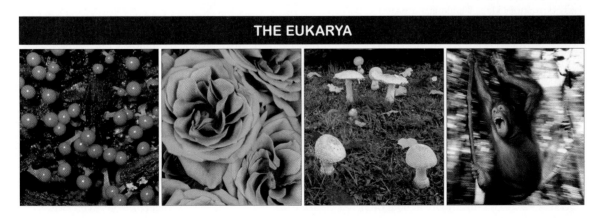

FIGURE 10-32 **All shapes and sizes.** The slime mold, passion flower, fungi, and an orangutan reflect the diversity of eukaryotic organisms.

401

Knowledge You Can Use

Do racial differences exist on a genetic level?

In a fraction of a second we are aware of someone's gender, size, and age—and race. But what is race? Does race have biological meaning? And can humans be categorized into groups that can scientifically be called races?

Q: What is a race, biologically speaking? The biological species concept hinges on a straightforward question: can two individuals interbreed under natural conditions? The concept of a race, conversely, is not straightforward. There is no universally agreed upon definition of race. Because systematics has the goal of reconstructing the evolutionary relationships of organisms, however, biologists have proposed that races are groups of interbreeding populations phylogenetically distinguishable, but reproductively compatible with each other

It is important to note that religious, cultural, ethnic, and geographic groups (among others) are commonly referred to as races. These racial distinctions, though often poorly defined, are a deeply entrenched aspect of many legal, social, and political issues today. Moreover, regardless of whether these socially constructed groupings reflect underlying genetic differences, they continue to be a real part of our lives. But the question we raise here is whether any of these uses of "race" reflect meaningful genetic divisions.

Q: What do racial groupings reflect? Are blacks and whites genetically different? The answer is obviously yes; black-skin alleles differ from white-skin alleles. Moreover, the prevalence of some genetic diseases varies by race and ethnicity. Sickle cell anemia, for example, is relatively common among Africans and Southeast Asians because the genes that cause it also improve a person's resistance to malaria, a disease prevalent in Africa and Southeast Asia.

These two observations—racial differences in skin color and patterns of disease prevalence—reveal one important problem with racial groupings. The genetic resistance to malaria, and the consequent higher risk of sickle cell anemia, is shared by Africans to the south of the Mediterranean and Europeans to the north. Africans from the southernmost part of the continent, on the other hand, have no greater risk of sickle cell anemia than do the Japanese, because malaria is similarly rare in both of their homelands. Should we use this trait rather than skin color

when determining a person's race? If we do, the southern Africans more closely resemble the Japanese than they do the North Africans, and so must be grouped differently.

Even if we decide to base race definitions on the traits we can see, how do we even decide who is white and who is black (or who is Basque and who is Native American)? Walking down Main Street, USA, this may seem like a simple observation—"they" look alike. But if you instead travel from tropical Africa up through Egypt into the Middle East and further north, it is impossible to discern where one race ends and the next begins. In a world in which people fall at every point along the continuum, grouping becomes arbitrary. Thus it may be more useful to ask a slightly different question.

Q: How much genetic difference exists between major geographic groups, besides the genes that are relevant to skin color? DNA-sequence comparisons among races reveal an important feature of the genetic variation in the human species. There is little rhyme or reason to how this variation is parceled out from one person to the next, and *similarity in skin color between two people does not tell you about overall genetic similarity.* There is, for example, huge variation in blood type among Africans: some are type O, some AB, others A or B. But the same goes for Asians and Turks, Russians and Spaniards. Yet we do not group people according to blood type. The same goes for variation in many other physiological and biochemical traits. But again, we do not group people according to these variations.

What is the future of race? Racial groupings such as "Hispanic" or "black or African American" or "Asian" do not reflect consistent genetic differences among the groups. Because no human race is completely isolated from all the others, human racial groups have not developed consistent genetic differences across a large number of genes. Still, even as the U.S. Census Bureau has defined race as not primarily biological or genetic, race remains an important part of individual social, cultural, and ethnic identity for many, and its use is mandated in the official collecting and presenting of data for federal agencies. For these reasons, it will remain a powerful social construct, even as scientific consensus emerges that racial groupings do not reflect meaningful genetic distinctions. Interestingly, with our extensive movements around the globe and frequent intermarriage between groups, it is likely that the existing genetic differences between races will shrink in the future.

1 Life on earth most likely originated from non-living materials.

Under conditions similar to those on early earth, small organic molecules form that have some chemical properties of life. Eventually, self-replicating molecules—possibly RNA—may have acquired or developed membranes, enabling them to replicate and making metabolism possible, the two conditions that define life. However it came about, the earliest life, which resembled bacteria, appeared about 3.5 billion years ago, not long after the earth was formed.

2 Species are the basic units of biodiversity.

Species are distinct biological entities, named using a hierarchical system of classification. Reproductive isolation and genetic divergence together lead to speciation. The distinction between two species is not easily defined for all types of organisms.

3 Evolutionary trees help us conceptualize and categorize biodiversity.

The history of life can be visualized as a tree; tracing from the branches back toward the trunk follows the pathway of descendants back to ancestors. The tree reveals the evolutionary history of every species and the sequence of speciation events that gave rise to them. Evolutionary trees are best constructed by comparing genetic similarity among organisms, because convergent evolution can cause distantly related organisms to appear much more closely related.

4 Macroevolution gives rise to great diversity.

The process of evolution—changes in allele frequencies within a population—in conjunction with reproductive isolation is sufficient to produce speciation, diversification, and the rich diversity of life on earth. The pace at which this occurs can be rapid or very slow. As new species are being created, others are lost through extinction, which may be a consequence of natural selection or large, sudden changes in the environment. Mass extinctions are usually followed by periods of unusually rapid adaptive radiation and diversification of the remaining groups or organisms.

5 An overview of the diversity of life on earth: organisms are divided into three domains.

All life on earth can be divided into three domains, which reflect species' phylogeny, their evolutionary relatedness to each other. Bacteria have evolved a broad diversity of metabolic and reproductive abilities relative to eukarya. The eukarya are unique in having cells with organelles. Archaea, many of which are adapted to life in extreme environments, physically resemble bacteria but are more closely related to eukarya.

KEY TERMS

CHECK YOUR KNOWLEDGE --

1. In a set of classic experiments performed in the early 1950s, Urey and Miller subjected an experimental system composed of H_2, CH_4 (methane), and NH_3 (ammonia) to electrical sparks. A few days later, they found _____ in their system.
   a) amino acids
   b) DNA
   c) microspheres
   d) cells
   e) RNA

2. What kind of molecule is thought to have been the first genetic material?
   a) protein
   b) DNA
   c) carbohydrate
   d) RNA
   e) microsphere

3. According to the biological species concept, species are natural populations of organisms that have the potential to interbreed and are _____ isolated from other such populations.
   a) behaviorally
   b) prezygotically
   c) postzygotically
   d) geographically
   e) reproductively

4. The classification rank that includes genera but not orders is:
   a) species.   d) kingdom.
   b) domain.   e) family.
   c) class.

5. Populations of *Larus* gulls around the North Pole show an unusual pattern of reproductive isolation: each population is able to interbreed with its neighboring populations, but populations separated by larger geographic distances are not able to interbreed. *Larus* gulls are an example of a(n) _____ species.
   a) ring   d) Escher
   b) polyploid   e) linked
   c) circular

6. In animals, it is believed that the most common mode of speciation is:
   a) autopolyploidy.
   b) chromosomal.
   c) directional.
   d) allopatric.
   e) sympatric.

7. Polyploidy:
   a) arises only when there is an error in meiosis resulting in diploid gametes instead of haploid gametes.
   b) is a common method of sympatric speciation for plants and animals.
   c) arises when sympatric speciation causes plants to have more sets of chromosomes than their parent plants.
   d) is an increased number of sets of chromosomes.
   e) always results in sympatric speciation.

8. In the plant kingdom, all of the species are descended from a single common ancestor. In terms of phylogeny, what type of tree of life is this?
   a) monophyletic   d) sympatric
   b) uniphyletic   e) allopatric
   c) punctuated

9. Phylogenetic trees should be viewed as:
   a) true genealogical relationships of species.
   b) the result of vertical, but never horizontal, gene transfer.
   c) intellectual exercises, not to be interpreted literally.
   d) representations of allopatric speciation events.
   e) hypotheses regarding evolutionary relationships of groups of organisms.

10. The difference between microevolution and macroevolution is that:
    a) they take place on different time scales.
    b) macroevolution occurs with physical structures, whereas microevolution occurs with physiological traits.
    c) microevolution involves changes to individuals within a population, whereas macroevolution involves changes to all of the individuals within a population.
    d) microevolution has been proven, whereas macroevolution is very speculative.
    e) microevolution occurs in prokaryotes, whereas macroevolution takes place among eukaryotes.

11. The idea of punctuated equilibrium challenges which component of Darwin's theory of evolution?
    a) steady change
    b) gradualism
    c) species stasis
    d) Both a) and b) are correct.
    e) None of the above is correct.

12. Which of the following scenarios would best facilitate adaptive radiation?
    a) A population of birds native to an island archipelago is forced to relocate to the mainland by a storm.
    b) A population of cheetahs goes through an event in which all genetic diversity in the population is wiped out.
    c) Darker-colored moths have a selective advantage over lighter-colored moths due to industrial soot on trees.
    d) A population of birds becomes stranded on an island archipelago.
    e) All of the above would facilitate adaptive radiation equally.

13. The mass extinction that occurred on earth 65 million years ago was immediately followed by:
    a) the rise of archaea.
    b) the emergence of the first non-photosynthetic organisms.
    c) the rise of the reptiles, including the dinosaurs.
    d) an increase in atmospheric oxygen levels.
    e) the rapid divergence and radiation of modern mammals.

14. Prokaryotes are classified into _____ domain(s).

a) 1   d) 4
b) 2   e) 5
c) 3

15. Protists are all alike in that all are:

a) bacterial.   d) prokaryotic.
b) photosynthetic.   e) archaean.
c) eukaryotic.

16. Which of the following domains share the most recent common ancestor?

a) archaea and eukarya
b) bacteria and eukarya
c) archaea and bacteria
d) None of the above; all three domains evolved from different ancestors.
e) None of the above; all three domains are equally related to each other.

17. Which of the following groups would be placed nearest the fungi in an evolutionary tree based on DNA sequences?

a) plants   d) archaea
b) bacteria   e) protists
c) animals

SHORT-ANSWER QUESTIONS

1. Explain the difference between defining a species according to the biological species concept and according to the morphological species concept. Include an example of why one method might be more appropriate than the other for certain situations.

2. Evolutionary trees are important in helping to explain the history of life. Outline the difference between how evolutionary trees were constructed in the 20th century and how they are constructed today. Explain what types of information these trees can and cannot convey.

3. The shape of the skull, as well as the shape and size of teeth, often reflects the method of feeding in mammals. A skull with a more rounded jaw and shorter, narrower teeth indicates a mammal that is well adapted to pulling bunches of leaves off trees and grinding them up. A skull with a more square jaw indicates a mammal adapted to grazing and cropping vegetation at the ground level. Explain how the adaptations of long, flat teeth and a more square jaw or muzzle could have triggered adaptive radiation in horses.

See Appendix for answers. For additional study questions, go to www.prep-u.com.

11

Animal
Diversification

Visibility in motion

❶ Animals are just one branch of the eukarya.

A kangaroo rests in the sun at Koala Sanctuary, Brisbane, Australia.

11·1

What is an animal?

Looking at a kangaroo, it's obvious that it is an animal. The same goes for a mosquito, earthworm, or jellyfish. It gets a bit more difficult, though, if you look at a sponge. Is it an animal? Or is it a plant? Or is it something else? This is when it is helpful to have some guidelines for identifying and organizing the biodiversity we can see on earth.

For the sponge, the answer is *yes,* it is classified as an animal. Here's why. When defining what is or is not an animal, we look for certain features that are common to all animals. Surprisingly, considering the enormous differences among even the few animals mentioned above, just three characteristics are generally sufficient to define **animals** (**FIGURE 11-1**).

1. All animals eat other organisms. That is, rather than manufacturing their own food, as plants and some bacteria do through photosynthesis, animals consume plants, bacteria, other animals, fungi, or some combination of these.

And because the best way to catch other organisms is to go to them or chase after them, that brings us to the second characteristic that sets animals apart from plants.

2. All animals move—*at least at some stage of their life cycle.* That qualifying phrase is important because some animals move only during a period called the "larval stage" that comes early in their lives, and when they turn into adults they fasten themselves to a surface. If you visit a rocky seashore, you can find two of these animals—mussels and barnacles—living on rocks in the tidal zone. Organisms that are fastened in place, such as adult mussels and barnacles, are said to be **sessile,** but even animals that are sessile as adults moved when they were larvae.

3. All animals are multicellular, and they generally have body parts that are specialized for different activities. For example, most sense organs (the eyes, ears, nose, tongue, antennae, whiskers, and so on) are at an animal's front end, where

408   CHAPTER 11 • ANIMAL DIVERSIFICATION

▼

What Is an Animal?   Vertebrates and the Move onto Land   Terrestrial Vertebrates   Invertebrates

1 ANIMALS EAT OTHER ORGANISMS
All animals acquire energy by consuming other organisms.

2 ANIMALS MOVE
All animals have the ability to move—at least at some stage of their life cycle.

3 ANIMALS ARE MULTICELLULAR
Animals consist of multiple cells and have body parts that are specialized for different activities.

FIGURE 11-1 **Characteristics of an animal.** The cup coral (*left*) is eating a young octopus; a herd of spotted deer move through a forest; the animal with multiple appendages is a mantis.

it detects prey. Once an animal detects prey, the animal attempts to capture the prey, and the animal consumes it.

There are a few other general features of animals—most reproduce sexually, for example—but these three are the easiest criteria to use in determining whether or not you're looking at an animal.

TAKE-HOME MESSAGE 11·1

Animals are organisms that share three characteristics: all of them eat other organisms, all of them can move during at least one stage of their development, and all of them are multicellular.

11·2

Four key distinctions divide the animals.

More than two-thirds of the almost two million species on earth that have been identified are animals. In general, we see that closely related animals have more similar body forms than distantly related animals (although convergent evolution can obscure this; see Section 10-9). But with such a large number of animals to organize and identify, it is useful to divide them into a few large categories, based on their evolutionary relatedness.

Analyses of DNA and RNA sequences have helped biologists identify four key distinctions, into which we can group all of the animals. Each of these distinctions hinges on a particular adaptation and on the question of whether or not an animal descends from an ancestor with that adaptation (**FIGURE 11-2**).

1. Does the animal have defined tissues, with specialized cells? In the sponges, there is no distinctive tissue. Rather, the organisms are just aggregations of similar cells, with little coordination of activities between the cells. This is in contrast to the animals with defined tissues, such as humans, which have highly specialized cells such as skin cells, muscle cells, and sensory cells.

2. Does the animal develop with radial symmetry or bilateral symmetry? Among the animals with defined tissues, there is an important distinction between those with **radial symmetry** and those with **bilateral symmetry.** Radial symmetry describes organisms with a body structured like a pie. It is possible to make multiple slices, all going through

409

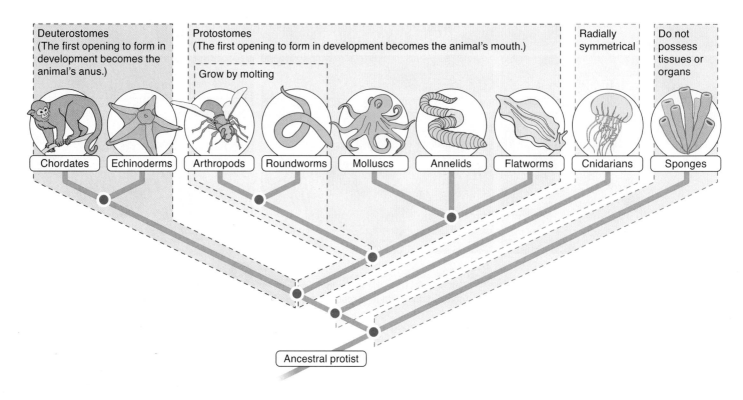

FIGURE 11-2 Innovations in body plan. The nine groups of animals in this figure represent just 25% of all animal phyla, but contain 96% of all animal species. Each of these groups will be explored in this chapter

the center, that divide the organism into identical pieces. This differs from bilateral symmetry, in which organisms have a left and right side, which are mirror images. Radially symmetrical organisms tend to be slow-moving or free-floating, while bilaterally symmetrical organisms tend to be faster-moving and more adept at searching for food and avoiding predators.

3. During development, does the animal's gut develop from front to back or back to front? Among the bilaterally symmetrical animals with defined tissues, there is an important distinction—at least with respect to determining the evolutionary lineage to which they belong—in how their early development occurs. Early in the evolution of animals—about 630 million years ago—a major split occurred that allows us to separate the bilaterally symmetrical animals with defined tissues into two distinct lineages: the **protostomes** (which can be translated as "mouth first") and the **deuterostomes** ("mouth second"). These names refer to the way the gut develops—from front to back or from back to front. In protostomes the gut develops from front to back, so the first opening that forms is the mouth of the adult animal, whereas in deuterostomes the gut develops from back to front and the second opening becomes the mouth. Although it's something you can see only when closely observing animals' embryonic development, the protostome/deuterostome split is the

most basic division of animals, in terms of revealing their evolutionary relatedness to one other.

4. Does growth occur by molting or by adding continuously to the skeletal elements? Among the bilaterally symmetrical animals that are protostomes, there is an important distinction between animals that molt—shedding an exoskeleton (a hard outer layer) and replacing it with a larger one at regular intervals during their life—and animals that grow by adding to the size of their skeletal elements, in a more continuous manner than occurs in the molting animals.

While all of the animals share a common ancestor, an ancestral protist, the four distinctions above help us identify and organize all of the approximately 36 living phyla of animals.

TAKE-HOME MESSAGE 11·2

The animals probably originated from an ancestral protist. Four key distinctions divide the extant animals: (1) tissues or not, (2) radial or bilateral symmetry, (3) protostome or deuterostome development, and (4) growth through molting or through continuous skeletal enlargement.

410 CHAPTER 11 • ANIMAL DIVERSIFICATION

What Is an Animal?   Vertebrates and the Move onto Land   Terrestrial Vertebrates   Invertebrates

Everything that is not extinct is evolutionarily successful.

"Never use the word higher or lower." This is a note that Charles Darwin wrote to himself in the margin as he was reading a book that proposed a theory of cosmic and biological evolution. It was an important insight, but one that can be difficult to adhere to—particularly as we turn our attention to the animals.

What is "success," evolutionarily speaking? There are two possible states for a species. It can be extant, meaning that it currently exists. Or it can be extinct. From an evolutionary perspective, every living species is successful. The fact that a species exists means that it is able to do the three things that all organisms must do: find food, escape predators, and reproduce. Earthworms and tigers live very different lives, and the characteristics that work well for one species would not work at all for the other, but the earthworm is as successful at being a worm as the tiger is at being a very large cat (**FIGURE 11-3**).

Throughout Darwin's writings, rather than viewing one species as higher or lower, as more or less advanced or primitive, he emphasized adaptation among the species he described and their relationship to their environment. One species was never better or worse than another. Rather, each species was differentiated from the others, showing specializations that adapted individuals of that species to the particular niche in which they occurred. Different environments posed different challenges. All extant species, as evidenced by their existence and persistence, are able to overcome those challenges.

In most biology textbooks, the tour through the animal kingdom begins with animals having a simple body plan, such as sponges, jellyfish, and worms, continues through the more complex invertebrates, the molluscs and arthropods, and culminates in the complexity of the vertebrates (mammals, amphibians, reptiles [including birds], and fishes). But because phylogenetic trees do not provide any information about the relative success of any species, there is no reason to begin our tour with the simplest animals, building until we reach humans at the "crown of creation."

So, our tour begins with the vertebrates. We chose this group because humans are vertebrates and vertebrates are the animals we are most familiar with. If we followed that logic to its conclusion, though, we would start by discussing humans. But we decided to begin with jawless fishes and move across to humans, because that sequence makes the evolution of certain organs and body parts, such as lungs, legs, and the amniotic egg, easier to understand.

As we survey the many different kinds of animal, we will see that even as each extant species represents an evolutionary success, certain adaptations have led to the rapid and extensive diversification of particular groups of species. Paying attention to these adaptations, we will investigate nine of the most successful phyla (shown in Figure 11.2) and the evolutionary innovations that contributed to their success. Although the nine groups represent only a quarter of all animal phyla, they actually account for 96% of the animal species extant today.

> From an evolutionary perspective, every living species is successful.

FIGURE 11-3 Two equally successful organisms: the earthworm and the tiger.

TAKE-HOME MESSAGE 11•3

From an evolutionary perspective, all extant species are successful. Among the higher levels of taxonomic grouping, however, some groups are represented by more species than others. Of the 36 animal phyla, 9 phyla account for more than 96% of all described animal species.

411

❷ Vertebrates are animals with a backbone.

Animals in motion: these X-ray images show tuatara walking.

11·4

All vertebrates are members of the phylum Chordata.

All vertebrates are deuterostome animals, so they have defined tissues and are bilaterally symmetrical. Further, they are all part of a phylum called Chordata (the chordates), whose members are defined by four distinct body structures (**FIGURE 11-4**).

1. The **notochord,** a rod of tissue extending from the head to the tail, is the structure that gives chordates their name. The notochord stiffens the body when muscles contract during locomotion (movement). Primitive chordates retain the notochord throughout life, but in advanced chordates, such as vertebrates, the notochord is present only in early embryos and is replaced by the vertebral column (backbone).

2. A **dorsal hollow nerve cord** extends from head to tail. In vertebrates, this nerve cord forms the central nervous system, which consists of the spinal cord and the brain. Other kinds of animals (worms, insects, and so on) also have a nerve cord, but it lies in the lower portion of the front (ventral) part of the body and is solid instead of hollow.

3. The earliest chordates were aquatic and passed water through slits in the pharyngeal region (the area between the back of the mouth and the top of the throat) for breathing and feeding. **Pharyngeal slits** are present in the embryos of all chordates, even in the embryos of humans, which lost their gills far back in evolutionary time.

4. A **post-anal tail** is another characteristic of chordates. "Post-anal" means that the tail extends back beyond the end of the trunk. (The posterior end of the digestive system is the anus, and this structure marks the end of the trunk.) The anus of worms, insects, and other non-chordate animals lies at the extreme hind end of the body. Some vertebrates, including humans, have a tail, but only for a brief period of time during embryonic development.

The phylum Chordata contains three subphyla: the tunicates (Urochordata; with about 2,000 species), lancelets (Cephalochordata; with about 20 species), and vertebrates (Vertebrata; with about 56,000 species). All three of the subphyla have the four characteristic chordate structures for at

THE CHORDATES

COMMON CHARACTERISTICS

All chordates possess four common body structures, although in many chordates, these structures are only present during specific life stages.
• Notochord
• Dorsal hollow nerve cord
• Pharyngeal slits
• Post-anal tail

MEMBERS INCLUDE
• Tunicates (about 2,000 species)
• Lancelets (about 20 species)
• Vertebrates (about 56,000 species)

FIGURE 11-4 **Understanding the chordates.** Shown here (*from left*) are Pacific double saddle butterfly fish, a golden lion tamarin, a thorny devil lizard, and compound ascidians (tunicates).

least a portion of their lives, but they look very different from one another (**FIGURE 11-5**).

Tunicates are marine animals, and the adults, which are about the size of your thumb, look like balls of brownish green jelly. You can find them attached to docks and mooring lines of boats. It is the free-swimming larvae of tunicates that reveal their chordate characteristics.

Lancelets are slender, eel-like animals, about the length of your little finger, that live in coastal waters. Unlike tunicates, adult lancelets have all of the chordate characteristics. Lancelets (and

GROUPS OF CHORDATES

TUNICATES
• Filter-feeding marine animals
• Can often be found attached to docks
• Larvae are free-swimming

LANCELETS
• Filter-feeding marine animals
• Live in coastal waters

VERTEBRATES
• The most diverse subphylum of chordates
• Range in size from the tiny hummingbird to the enormous blue whale
• Can be found in most of the habitats on earth

FIGURE 11-5 **Representing the subphyla of chordates: (*left to right*) singing sea squirt, lancelet, and hummingbird.**

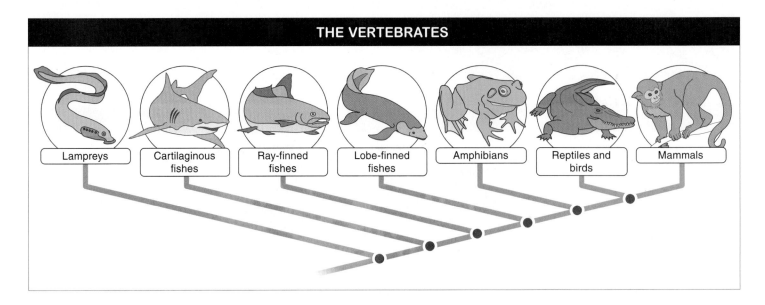

FIGURE 11-6 Vertebrate phylogeny.

tunicates) are filter-feeders: cilia draw a current of water through the mouth into the pharynx and out through the pharyngeal slits, and microscopic food items are trapped by the bars between the slits.

The capybara is a semi-aquatic vertebrate found in South America in forested areas near bodies of water, such as lakes, rivers, swamps, ponds, and marshes. It is the largest of all living species of rodents, and can weigh as much as 150 pounds.

Vertebrates are the most diverse subphylum of chordates (**FIGURE 11-6**). The size range of vertebrates extends from tiny hummingbirds and shrews that weigh only a tenth of an ounce (about 3 grams) to the blue whale, which weighs in at more than 130 tons (120,000 kilograms). Vertebrates move by swimming, burrowing, crawling, walking, running, and flying, and they have invaded all of the habitats on earth, from the depths of the sea to the tops of mountains, and from rain forests to deserts and the ice at the Poles. But despite all these differences, vertebrates still have the four chordate characteristics at some stage of their lives.

TAKE-HOME MESSAGE 11·4

All chordates have four characteristic structures: a notochord, a dorsal hollow nerve cord, pharyngeal gill slits, and a post-anal tail. The three subphyla of chordates are superficially very different, but all are united by possessing these four characteristics—although some of the subphyla reveal these characteristics only during specific life stages.

414   CHAPTER 11 • ANIMAL DIVERSIFICATION

What Is an Animal?   Vertebrates and the Move onto Land   Terrestrial Vertebrates   Invertebrates

The evolution of jaws and fins gave rise to the vast diversity of vertebrate species.

Vertebrates eat other organisms, and they need a mouth to do so. But a mouth is just an opening through which food passes—jaws help an animal get the food into its mouth and chew it. The earliest vertebrates were fish-like animals that lived more than 500 million years ago and did not have jaws. Two kinds of jawless vertebrates still exist: the lampreys (with 41 species) and the hagfishes (with 43 species). Lampreys are parasitic; they attach to other fishes by creating suction with a circular oral disk that is studded with sharp spines. First, the spines scrape an opening through the scales and skin. Then, a protein produced by the lamprey keeps the wound from closing, and the lamprey feeds on the blood and body fluids that seep out. Hagfishes feed on dead animals, using two spiny dental plates that they embed in their prey.

Q Why are jaws needed?

Lampreys and hagfishes do the two things that animals do—feed on other organisms and use locomotion to reach their food—but they are about as basic as it is possible to be. And because they do not have jaws, they cannot seize prey or chew food before swallowing. It was the evolution of fins and jaws that set the stage for the evolutionary explosion of diversity of vertebrates.

Fish swim the same way that lampreys and hagfishes swim—that is, by bending the body from side to side to create an S-shaped wave that moves from the head of the fish toward its tail. A typical fish has a total of seven fins serving different purposes: to drive the fish forward, to minimize rolling from side to side, and for steering and stopping. The combined effects of the seven fins allow the fish to swim rapidly in a straight line when it is in open water or to weave its way through a dense stand of plants or corals, with great maneuverability.

The evolution of fins was paralleled by the evolution of jaws, because the two kinds of structure work together. Fins get you to the organism you are going to eat, and jaws capture and kill it (**FIGURE 11-7**).

THE EVOLUTION OF JAWS AND FINS

FISHES WITHOUT JAWS OR PAIRED FINS
• Tail fin propels organism through water
• Feed by attaching mouthparts to prey

FISHES WITH JAWS AND PAIRED FINS
• Paired fins provide controlled movement through water
• Jaws allow for seizing and chewing prey

FIGURE 11-7 The lamprey and the shark. The lamprey (*left*) has an oral disk that it uses for sucking in a meal. The shark has a powerful jaw lined with sharp teeth, and pairs of fins that allow it to move rapidly through the water and attack prey.

 The development of fins and jaws set the stage for the evolutionary explosion of vertebrate diversity.

JAWED FISHES

CARTILAGINOUS FISHES
• Characterized by a skeleton made completely from cartilage
• Contain about 880 species, including sharks and rays

RAY-FINNED FISHES
• Characterized by having rigid bones and fins lined with hardened rays
• Possess a swim bladder, which aids in flotation
• Contain about 27,000 species, including almost everything you think of as "fish," from salmon to goldfish and guppies

LOBE-FINNED FISHES
• Characterized by having two pairs of sturdy fins on the underside of their body
• Contain six species of lungfish and two species of coelacanths

FIGURE 11-8 Fishes with jaws: cartilaginous fishes, ray-finned fishes, and lobe-finned fishes. Shown are stingray, rainbow trout, and Australian lungfish.

The development of jaws and fins allowed great diversification in vertebrate body shapes and composition. The **cartilaginous fishes,** with about 880 species, include sharks and rays (**FIGURE 11-8**). The cartilaginous fishes are characterized by a skeleton made completely from cartilage, the same tissue that gives your nose its shape. The largest group of jawed fish is also the most diverse group of vertebrates, the **ray-finned fishes,** with about 27,000 species, which includes almost everything you think of as "fish," from salmon to goldfish. Ray-finned fishes have rigid bones and a mouth at the narrow tip, or apex, of the body. They get their name from the fact that their fins are lined with hardened rays.

An important evolutionary development in the ray-finned fishes was the swim bladder, a gas-filled organ that keeps the fish from sinking. This bladder later evolved into the lung. The cartilaginous fishes, because they have no swim bladder, must constantly move around in the water or they will sink. A third group of jawed fishes, the **lobe-finned fishes,** are represented by just eight species. These fishes have sturdy fins on the underside of their body (see Figure 11-8), and as we will see in the next section, their features were useful in initiating the move onto land.

TAKE-HOME MESSAGE 11·5

The development of two structures—fins and jaws—set the stage for the enormous diversity of modern vertebrates.

416 CHAPTER 11 • ANIMAL DIVERSIFICATION

What Is an Animal?   Vertebrates and the Move onto Land   Terrestrial Vertebrates   Invertebrates

The movement onto land required lungs, a rigid backbone, four legs, and eggs that resist drying.

Movement from water onto land required four major evolutionary innovations: lungs, a backbone, four legs, and eggs that won't dry out. All four of these characteristics are necessary for a land animal, and all four evolved in the immediate ancestors of terrestrial vertebrates, which were predatory fish that lived in shallow water.

Q Where did legs and lungs come from?

As we've noted, lungs developed from the swim bladder found in ray-finned fishes. All of the terrestrial (land-dwelling) vertebrates—amphibians, reptiles (including birds), and mammals— are descendants of the lobe-finned fishes that lived during the Devonian period, some 400 million years ago. At this time, the lobe-finned fishes had lungs and the rudiments of limbs, and they were still fully aquatic and living near the shore. The four sturdy fins on the underside of their body helped them move through shallow water, and lungs allowed them to breathe air when the oxygen concentration in the warm, stagnant water was low. Thus, these fishes already had some of the basic characteristics of terrestrial animals—they could walk using their fins, and they could get oxygen from air via their lungs instead of from water via their gills. The jointed bones in the fins of lobe-finned fishes don't look much like your arms and legs, but they are evolutionarily homologous structures.

To move onto land, a vertebrate needs more than just legs and lungs; it needs structural support to resist the pull of gravity. Each vertebra of a terrestrial vertebrate has projections that interlock with projections from the vertebra ahead of it and the vertebra behind it. These interlocking projections prevent the backbone from sagging under the pull of gravity, and the body weight is transmitted through the limbs to the ground.

The last innovation necessary to move onto land is an egg that resists drying out (**FIGURE 11-9**). When eggs are deposited on land, they are exposed to air and lose water by evaporation. The eggs of terrestrial animals need a waterproof covering—a membrane and eggshell—to prevent them from drying out before they hatch. The appearance of shelled eggs about 380 million years ago marks the final step in the evolution of entirely terrestrial vertebrates, the groups that evolved into mammals and reptiles (including birds). These eggs have a water-tight membrane that keeps the embryo surrounded by a bath of fluid.

FROM WATER TO LAND

The transition of vertebrates from life in water to life on land required overcoming three main obstacles. Four major evolutionary innovations allowed for this transition.

1 **PROBLEM: RESPIRATION**
Aquatic animals use gills to acquire dissolved oxygen from water. The transition onto land required the ability to breathe air.

SOLUTION: LUNGS
Gas exchange was transferred from gills to lungs, which evolved from the swim bladder found in ray-finned fishes.

Ray-finned fish

Early terrestrial vertebrate

2 **PROBLEM: GRAVITY**
The transition onto land required structural support to resist the pull of gravity.

SOLUTIONS: LIMBS and MODIFIED VERTEBRAE
Limbs evolved from the jointed fins found on the underside of lobe-finned fishes.

Vertebrae were modified to transmit the body weight through the limbs to the ground.

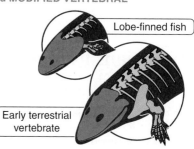

Lobe-finned fish

Early terrestrial vertebrate

3 **PROBLEM: EGG DESICCATION**
The transition onto land required an egg that resisted drying out when exposed to air.

SOLUTION: AMNIOTIC EGG
Terrestrial animals developed a waterproof eggshell, which prevents eggs from drying out before they hatch.

FIGURE 11-9 How did vertebrates make the transition from life in water to life on land?

TAKE-HOME MESSAGE 11•6

In the transition of vertebrates from life in water to life on land, fins were modified into limbs, vertebrae were modified to transmit the body weight through the limbs to the ground, and the site of gas exchange was transferred from gills to lungs. The only entirely new feature to appear in the early development of terrestrial vertebrates was an egg that resisted drying out.

❸ All terrestrial vertebrates are tetrapods.

A flap-necked chameleon crosses a salt flat in Botswana.

11·7

Amphibians live a double life.

The ancestors of all vertebrates that live on land had both lungs and four legs. Some modern vertebrates (such as snakes and whales) have evolved so that their limbs are reduced to a few shrunken and unused bones. Whales (and some snakes) have returned to living in water. And some salamanders have lost

their lungs and breathe through the skin, but these are recent changes. Taxonomically, all terrestrial vertebrates are **tetrapods** (*tetra* = four; *poda* = feet). And it is the presence or absence of the amniotic egg that allows us to distinguish two large groups of terrestrial vertebrates: the non–amniotes (amphibians, or

---

THE AMPHIBIAN LIFE CYCLE

Amphibians, such as frogs, toads, and salamanders, are terrestrial vertebrates (tetrapods) with non-amniotic eggs. Most species live on land as adults, but develop in water.

EGGS
Amphibians have non-amniotic eggs, which must be laid in water to prevent desiccation.

LARVAE
Amphibians spend their larval stage underwater and undergo metamorphosis to develop legs and lungs.

ADULTS
Only the adults are true land animals; however, most of the species in this group stay close to water to lay their eggs.

FIGURE 11-10 **From egg, to larva, to adult: the amphibian life cycle.**

418 CHAPTER 11 • ANIMAL DIVERSIFICATION

What Is an Animal?   Vertebrates and the Move onto Land   Terrestrial Vertebrates   Invertebrates

tetrapods without desiccation-proof amniotic eggs) and the **amniotes** (reptiles and mammals, tetrapods with amniotic eggs) that are the subjects of this and the next three sections.

The first terrestrial vertebrates were **amphibians,** from the Greek word *amphibios,* meaning "living a double life." Only the adults are true land animals—the young of most amphibian species remain aquatic (**FIGURE 11-10**). Most of the 6,000 species in this group are still significantly tied to life in the water, because they must always be near water to lay their eggs, which are simple structures, not unlike fish eggs. Frogs and toads make up the vast majority of amphibians (about 5,400 species, with salamanders comprising another 550 species, and a group of legless burrowing amphibians, called caecilians, comprising 170).

The larval stage of the frog, the tadpole, lives in the water, lacks legs, and eats algae. During metamorphosis, the tadpole develops

legs, lungs, and a digestive system fit for its adult life as a carnivore. Adult frogs have thin moist skin through which much of their gas (oxygen and carbon dioxide) exchange with the air takes place.

The last two decades have seen a stunning decline in amphibian populations throughout the world, attributable to a combination of causes: climate change, habitat degradation, fungal diseases, and increased pollution.

TAKE-HOME MESSAGE 11·7

Amphibians are terrestrial vertebrates (tetrapods), but most species still lay eggs in water and the eggs hatch into aquatic larvae (in frogs: tadpoles).

11·8

Birds are reptiles in which feathers evolved.

Soon after amniotic vertebrates appeared, two different evolutionary lineages began to diverge (**FIGURE 11-11**). One of these lineages is the mammals, the group to which humans belong. The other lineage is the one we have been calling "reptiles (including birds)." That is an awkward term, but the evolutionary tree (see Figure 11-11) shows why it is necessary. Birds (about 9,700 species) are merely one branch of the reptile lineage that

also includes snakes and lizards (about 8,000 species), turtles (about 300 species), crocodiles and alligators (23 species), the New Zealand tuatara (2 species), as well as the extinct dinosaurs.

The characteristics that hold the bird-crocodile-dinosaur group together are mostly similarities in bones (especially the bones of the skull and legs) and their DNA sequences.

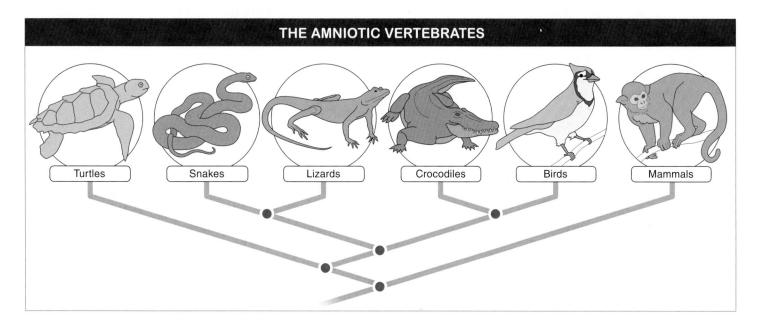

FIGURE 11-11 Amniotic vertebrate phylogeny.

More Invertebrate Diversity

REPTILES
- Skin is covered in scales.
- Body temperature is controlled by external conditions, such as air temperature (exothermic).
- Include snakes and lizards (about 8,000 species), turtles (about 300 species), crocodiles and alligators (23 species), and tuatara (2 species)

BIRDS
- Skin is covered in feathers, enabling flight and providing insulation.
- Body temperature is maintained by heat generated from cellular respiration (endothermic).
- Include about 9,700 species

FIGURE 11-12 Comparison of reptiles and birds. A Jackson's chamelon (*left*) and a scarlet macaw.

Reptiles are amniotes, and thus have amniotic eggs. You are familiar with the hard-shelled eggs of birds, and some lizard, snake, and turtle eggs also have hard shells. Other species in these groups have eggs with a parchment-like shell.

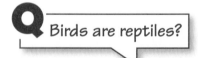

Most people find it confusing that birds are grouped with turtles, lizards and snakes, and alligators and crocodiles, and that confusion is easy to understand (**FIGURE 11-12**). After all, birds are covered by feathers and can fly, whereas the other reptiles have bare skin and do not fly. And there is an important physiological difference as well: birds are **endotherms**—meaning that they use the heat produced by their cellular respiration to raise their body temperature above air temperature. The other reptiles are **ectotherms**—they bask in the sun to raise their body temperature, and seek the shade when the air is too warm.

How can it be that animals as different as crocodiles and hawks are closely related? During the Mesozoic era, from about 250 million years ago to about 65 million years ago, dinosaurs were the dominant terrestrial vertebrates. The fossil record shows a remarkably clear series of forms that bridge the transition between bare-skinned dinosaurs and feathered birds—birds are, in fact, just one group of dinosaurs.

The most apparent difference between reptiles (including most dinosaurs) and birds is the presence of feathers in birds. The fossil record reveals that feathers actually evolved before birds, appearing in many reptile species, and that the appearance of feathers probably had nothing to do with flight. In fact, feathers evolved in many different kinds of dinosaur, and feathered dinosaurs continued to live for millions of years after birds appeared.

Q What were the first feathers used for?

Among reptiles, many different species had feathers. A small, agile dinosaur called *Sinosauropteryx*, for example, living about 120 million years ago, had feathers that were simple spiky filaments on the neck, back, and tail, along with shorter filaments covering the body. These filaments were probably brightly colored and used by male *Sinosauropteryx* for courtship displays to females, as well as for aggressive displays to other males. *Sinosauropteryx* and many of the other feathered dinosaurs were latecomers, however. The pigeon-sized *Archaeopteryx*, which was among the first bird species, had already, by 147 million years ago, developed feathers that it used to fly; by 140 million years ago, the skies were filled with birds.

If *Archaeopteryx* and a host of other early birds already had the specialized feathers needed for flight, why did later dinosaurs

What Is an Animal? Vertebrates and the Move onto Land Terrestrial Vertebrates Invertebrates

> Originally, colorful feathers were used for behavioral displays. In modern birds, feathers provide insulation and aid in flight.

FIGURE 11-13 *Sinosauropteryx:* an early feathered, flightless dinosaur.

still have primitive feathers? The answer is that feathers have many functions, of which flight is only one. *Sinosauropteryx,* as we saw above, probably used their feathers for courtship or

territorial displays, and they may also have used them as insulation (**FIGURE 11-13**).

Another big difference between birds and reptiles is that, as we noted earlier, birds use internally generated heat to maintain a high body temperature, while the reptiles rely on the sun to heat their bodies. This evolutionary change may have been related to feathers. As feathers evolved for display and flight, they may also have provided some insulation. This insulation may then have enabled the evolution of high rates of cellular respiration and the maintenance of a high and constant body temperature.

TAKE-HOME MESSAGE 11·8

Birds are just one branch of the reptile lineage but possess feathers and can generate body heat. The complex anatomical and physiological systems that we see in living animals, such as feathers and endothermy in birds, are the products of hundreds of millions of years of step-by-step changes that began with simple structures. Feathers were originally colorful structures used for behavioral displays, and their current uses as insulation and in flight developed later.

11·9

Mammals are animals that have hair and produce milk.

Long before the first birds appeared, the mammals evolved from the reptiles, originally as small, nocturnal insect-eaters. These early mammals remained small and, judging from the scarce fossil record, they were never in great abundance while the dinosaurs were around. The early mammals had several features in common, however, that are still present in all mammals today. Two important features are **hair**—dead cells filled with the protein keratin—which serves as an insulator, and **mammary glands** in female mammals, which enable the nursing of young with nutritious, calorie-rich milk.

As the earliest mammals evolved, there was a gradual transition from bodies with short legs that projected out horizontally from the trunk, like the legs of an alligator, to bodies with long legs held vertically beneath the trunk, like those of a dog (**FIGURE 11-14**). These anatomical changes would have allowed the earliest mammals to run faster and farther to capture prey.

FIGURE 11-14 How did endothermy evolve in early mammals?

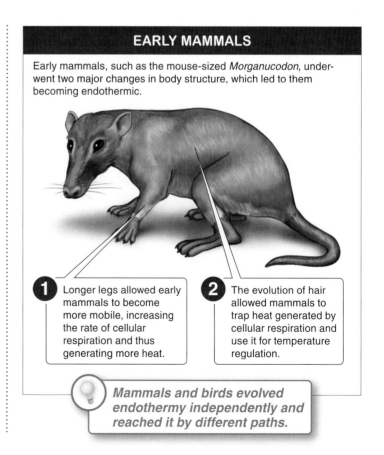

EARLY MAMMALS

Early mammals, such as the mouse-sized *Morganucodon*, underwent two major changes in body structure, which led to them becoming endothermic.

1 Longer legs allowed early mammals to become more mobile, increasing the rate of cellular respiration and thus generating more heat.

2 The evolution of hair allowed mammals to trap heat generated by cellular respiration and use it for temperature regulation.

> Mammals and birds evolved endothermy independently and reached it by different paths.

Mammals are endothermic vertebrates that have hair and produce milk for their young.

MONOTREMES
• Females lay eggs.
• Females produce milk, but do not have nipples—babies suck milk from the hairs on their mother's chest.
• Only five species survive—the platypus and four species of spiny animals called echidnas.

MARSUPIALS
• Females give birth after a short period of development.
• Females of most species have a pouch where the young complete their development.
• About 300 species survive, including kangaroos, koalas, wallabies, and possums.

PLACENTAL MAMMALS
• Females have a placenta that provides oxygen and nutrients to embryos in the uterus.
• About 4,500 species survive.

FIGURE 11-15 Three groups of mammals: monotremes, marsupials, and placentals. Shown here are a platypus, a kangaroo, and a bonobo.

But an increase in muscle activity for running had to be accompanied by an increase in the rate of cellular respiration to provide the energy to keep the leg muscles working. This increase in metabolism, along with the evolution of hair, would have allowed the early mammals to trap the heat produced and use it for temperature regulation, setting the stage for endothermy (but through a different sequence of events than in the birds).

One feature common to most mammals—**viviparity,** or giving birth to babies ("live birth") rather than laying eggs—is not a defining mammalian characteristic. This is because it isn't common to all mammals. **Monotremes** retain the ancestral condition of laying eggs. Monotremes do produce milk, but they do not have nipples—the babies suck milk from the skin and hairs on their mother's chest.

Today, only five species of monotremes survive—the platypus and four species of spiny animals called echidnas. The platypus lives in streams and rivers in eastern Australia. It uses its broad, leathery, electrosensitive bill to probe under rocks and sunken logs for prey such as aquatic insect larvae and crayfish, which it finds by sensing the electrical activity of their muscles as they try to hide. The echidnas—three species in Australia and one in New Guinea—also have leathery electrosensitive beaks.

The remaining two lineages of mammals, the **marsupials** and the **placentals,** are both viviparous, but the newborn young of marsupials and placentals are quite different (**FIGURE 11-15**). Marsupials are called the "pouched mammals" because the females of most marsupial species have a pouch where the young complete their development, following a short period of embryonic life in the uterus.

Placental mammals (including humans) take their name from the *placenta,* which is the structure responsible for the transfer of nutrients, respiratory gases, and metabolic waste products between the mother and the developing fetus. Dense capillary beds in the placenta send finger-like projections into the wall of the uterus, where they are surrounded by blood from the mother's circulation.

TAKE-HOME MESSAGE 11·9

The presence of hair and lactation are defining characteristics of mammals. Monotremes are egg-laying mammals. Marsupial mammals give birth after a short period of intrauterine development and have a pouch in which the baby completes its development. Placental mammals have a placenta that provides oxygen and nutrients to embryos that undergo a longer development in the uterus.

422 CHAPTER 11 • ANIMAL DIVERSIFICATION

What Is an Animal?   Vertebrates and the Move onto Land   Terrestrial Vertebrates   Invertebrates

Humans tried out different life styles.

The evolutionary lineage to which humans belong, the primates, originated about 55 million years ago. We can get an idea of what the ancestral primate looked like from the modern species of tree-living mammals called prosimians—a group that includes the lemurs. Many of the anatomical characteristics of humans and the other primates can be traced to our arboreal origin. Our forward-directed eyes and binocular vision that allow us to judge distances accurately, our shoulder and elbow joints that allow our arms to rotate, and the retention of 10 fingers and 10 toes that allow us to grasp objects—all these are traits we inherited from our arboreal ancestors (**FIGURE 11-16**).

Humans are part of the anthropoid lineage of primates, which includes the New World and Old World monkeys (both of which have tails) and the apes (which lack tails). Among the apes, gibbons and orangutans live in pairs or alone, while gorillas and chimpanzees live in social groups that consist of one or more adult males and several females, the ancestral social structure for human societies. Within the apes, genetic and anatomical characteristics show that chimpanzees are our closest living relatives. Human and chimpanzee genes are very similar—their base sequences differ by about 1%, and one-third of human and chimpanzee genes are identical (**FIGURE 11-17**).

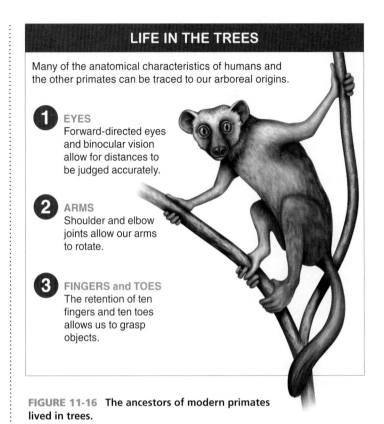

LIFE IN THE TREES

Many of the anatomical characteristics of humans and the other primates can be traced to our arboreal origins.

1 **EYES**
Forward-directed eyes and binocular vision allow for distances to be judged accurately.

2 **ARMS**
Shoulder and elbow joints allow our arms to rotate.

3 **FINGERS and TOES**
The retention of ten fingers and ten toes allows us to grasp objects.

FIGURE 11-16 **The ancestors of modern primates lived in trees.**

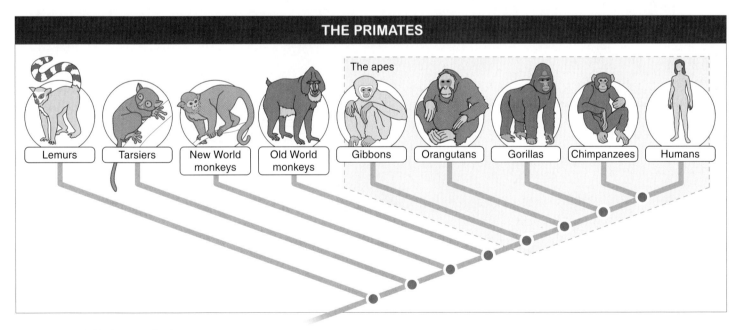

THE PRIMATES

The apes

Lemurs | Tarsiers | New World monkeys | Old World monkeys | Gibbons | Orangutans | Gorillas | Chimpanzees | Humans

FIGURE 11-17 **Phylogeny of primates.**

More Invertebrate Diversity

The amount of genetic difference between humans and chimpanzees indicates that the chimpanzee and modern human lineages separated only 5 or 6 million years ago. Humans differ from chimpanzees in three major anatomical characteristics: humans are bipedal (we normally walk on two legs, whereas chimpanzees usually walk on four legs), humans are bigger than chimpanzees, and the human brain is about three times the size of the chimpanzee brain.

When we trace the appearance of these human characteristics through the fossil record, we find that humans did not gradually become bipedal, big, and brainy. Instead, the three characteristics evolved one-by-one. Bipedality evolved first, then brain volume increased, and finally an increase in body size was accompanied by a further increase in the size of the brain.

The fossil record of human evolution looks like a bush with three episodes of branching that produced clusters of species. The first branching episode, which occurred between 3.5 and 4 million years ago, resulted from changes in the jaws and teeth associated with different diets. The second branching episode, about 2 million years ago, was associated with

Q What are the advantages of walking on two feet rather than four?

increased brain size and probably with the first use of tools. And a final branching event, half a million years ago, followed an increase in body size.

Shifting from walking on four legs to walking on two legs requires changes in several parts of the skeleton. The evolution of bipedality was not a simple process, but it was the first change in the human lineage and it seems to have set the stage for all the changes that followed. The primary advantage of bipedal locomotion for early humans was probably energetic efficiency. Bipedal locomotion at walking speed uses less energy than quadrupedal locomotion. Efficient walking locomotion thus could have been a valuable adaptation as the human lineage was moving out of the forest onto the savannas, while the chimpanzee lineage remained in the forest.

About 3.5 or 4 million years ago, several bipedal groups, the australopithecines, appeared (**FIGURE 11-18**). They were no larger than chimpanzees and they had the same brain volume, just a bit larger than a pint jar (350–400 cubic centimeters). The fossil known as Lucy was the first of many australopithecines to be discovered. An adult female, Lucy was only 3 feet (about

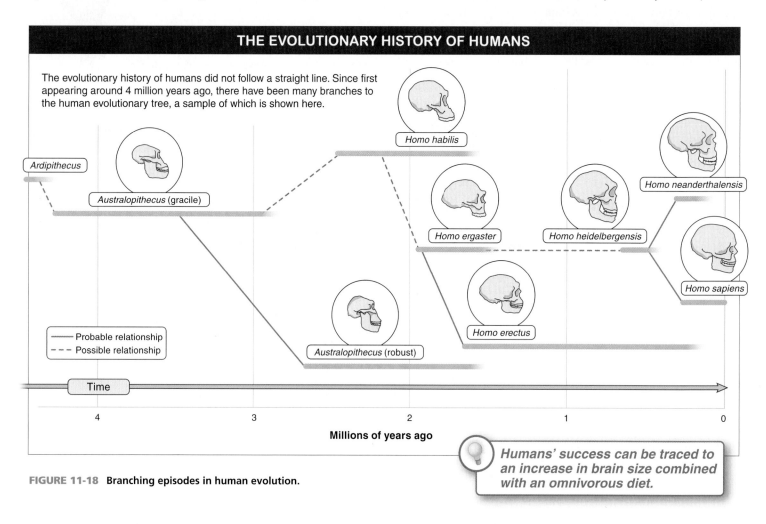

THE EVOLUTIONARY HISTORY OF HUMANS

The evolutionary history of humans did not follow a straight line. Since first appearing around 4 million years ago, there have been many branches to the human evolutionary tree, a sample of which is shown here.

Ardipithecus

Australopithecus (gracile)

Homo habilis

Homo ergaster

Homo heidelbergensis

Homo neanderthalensis

Homo sapiens

—— Probable relationship
- - - Possible relationship

Homo erectus

Australopithecus (robust)

Time

Millions of years ago

4   3   2   1   0

FIGURE 11-18 Branching episodes in human evolution.

Humans' success can be traced to an increase in brain size combined with an omnivorous diet.

What Is an Animal?   Vertebrates and the Move onto Land   Terrestrial Vertebrates   Invertebrates

1 meter) tall and probably weighed no more than 60 pounds (a bit less than 30 kilograms). The australopithecine group to which Lucy belonged lived in grassy habitats with scattered trees; their hands and feet retained the curved finger and toe bones that are characteristic of arboreal animals.

Probably, Lucy and the other closely related species of australopithecines foraged on the ground for food and climbed into trees to escape predators and perhaps at night to sleep. With jaws and teeth much like the jaws and teeth of chimpanzees, their diet was probably also like that of chimpanzees, a mixture of leaves, soft fruits, and nuts.

Following the appearance of bipedal primates, a second branching episode in human evolution originated among the australopithecines and produced the earliest species in our own genus, *Homo*. These earliest humans had brain volumes about twice those of chimpanzees, but little increase in body size. The species in this radiation had markedly smaller teeth than their australopithecine ancestors, a change that might indicate they had started to use tools instead of their teeth for the initial preparation of food. Stone tools are found in the same deposits as the fossils of *Homo habilis,* and this species may have been the first to use tools. Up to this time, all of human evolution had taken place in the southern and eastern parts of Africa, but about 2 million years ago, one branch, represented by *Homo erectus,* migrated out of Africa and spread through eastern Europe and Asia, while a second branch, represented by *Homo ergaster,* remained in Africa.

A third branching episode in human evolution (see Figure 11-18) gave rise to several species of advanced humans, including Neandertals (*Homo neanderthalensis*) and our own species (*Homo sapiens*). This radiation coincided with an increase in body size to approximately the height and weight of modern humans and an increase in brain volume to nearly twice that of earlier ancestors. In addition, the body form of the species that evolved during this branching episode looked like that of modern humans, with shorter arms and longer legs.

The evolutionary lineage to which *H. sapiens* belongs became bipedal, started using tools, and developed larger brains and larger bodies. These features, along with an omnivorous diet, turned out to be a winning strategy—other lineages closely related to that of *H. sapiens* became extinct while our lineage survived.

TAKE-HOME MESSAGE 11·10

Humans' forward-looking eyes, hands and feet with 10 fingers and 10 toes, and shoulder and elbow joints that allow the arms to rotate are characteristics retained from our arboreal ancestors. The early ancestors of humans left the trees and took up life on the ground, where they walked on two legs. Our success can be traced to an increase in brain size combined with an omnivorous diet.

11·11

How did we get here? The last 100,000 years of human evolution.

A bit more than 100,000 years ago, a new species of human evolved in Africa—the first modern *Homo sapiens*. One hundred thousand years is a ballpark figure based on a combination of molecular and fossil evidence. At that time, a small group—probably no more than 100—of these modern humans left Africa and ultimately spread across the earth.

Mitochondrial DNA shows that the initial human migration out of Africa followed three major pathways (FIGURE 11-19). One lineage turned west after leaving Africa and spread into Europe. A second group turned southeast and spread into southern Asia and through the Indo-Australia Archipelago to

HUMAN MIGRATION OUT OF AFRICA

About 100,000 years ago a small group of modern humans left Africa, and ultimately spread across the earth. Mitochondrial DNA shows that this migration followed three major pathways.

FIGURE 11-19 *Homo sapiens* on the move: three migratory paths out of Africa.

Australia. The third group moved northeast, populating northern Asia. This is the group that crossed the Bering Straits bridge about 15,000 years ago and spread southward through the Americas.

There was something about the world those *Homo sapiens* moved into as they left Africa that we would find bizarre—*they were not alone.* Populations of at least three other species of humans (i.e., species in the genus *Homo*) were already present in some of the areas that modern humans were moving into. That situation is totally different from the world we know today, in which we are the only species of human. What would it be like to live with species that were as similar to us as dogs are to wolves, coyotes, and jackals? How would you interact with an individual who was clearly human, but not human in exactly the same way that you are?

This is the situation that the first modern *Homo sapiens* encountered when they left Africa. Neandertals (*Homo neanderthalensis*) had already spread across Europe and the Middle East. Neandertals were about the same size as modern humans, but they were more robust and muscular. Fossils of Neandertals often include bones that had been broken and healed. These injuries provide two types of information about Neandertals.

1. Neandertals must have lived in organized groups that included a social support system, because these fossils reveal serious injuries. The injured individuals would have been incapacitated for days or even weeks until they healed, and family or clan members must have cared for them during this period. Ritual burials of Neandertals have also been found, and these provide additional support for the hypothesis that they lived in organized social groups.

2. The pattern of injuries found in Neandertal skeletons is distinctive—the only modern counterpart is found in professional rodeo bull and bronco riders, people who come in very close contact with large, angry animals. Neandertals probably hunted large mammals (bison, mammoths, wooly rhinoceroses) with short spears that were used for close-up jabbing instead of being thrown from a safe distance.

The *Homo sapiens* who migrated through the Indo-Australian Archipelago encountered two species of humans: *Homo erectus* and *Homo floresiensis. Homo erectus* had migrated to Asia at least a million years earlier, and they were still present on the island of Java when modern humans arrived about 50,000 years ago. Although *H. erectus* was the same size as modern humans, it had a smaller brain—an average brain volume of just over a quart (about 1,000 cubic centimeters) for *H. erectus* compared

with a quart and a half (about 1,400 cubic centimeters) for *H. sapiens*. Nonetheless, *H. erectus* had technological skills. They may have been the first humans to use fire (the evidence on this is not definitive) and they almost certainly built boats that allowed them to move along coasts and from island to island.

> **Q** Were two species of humans ever alive at the same time? If so, what happened?

In 2003, stone tools and fossils of a species of human only 3 feet (about 1 meter) tall and with a brain volume of just over a pint (about 350 cubic centimeters) were discovered on Flores Island, which lies east of Java. The paleontologists who discovered this species gave it the scientific name *Homo floresiensis* ("Flores man"), but the world press promptly called it the Hobbit because of its tiny size. At the time that the dwarf human *H. floresiensis* lived, Flores Island was also home to a dwarf elephant, which *H. floresiensis* probably ate, and a giant monitor lizard that probably ate *H. floresiensis.*

All three of the other humans disappeared after modern *Homo sapiens* spread into the areas where they occurred. Neandertals persisted until about 30,000 years ago, *Homo erectus* became extinct about 27,000 years ago, and *Homo floresiensis* disappeared about 12,000 years ago. Why did these three species vanish? Broadly speaking, there are two possible explanations. They might have interbred with *H. sapiens,* in which case modern humans would carry genes from the three recently extinct species. No evidence supports that possibility, however, and it is more likely that *H. sapiens* exterminated the other species, either by monopolizing access to food and living space or by killing them in battles for food and space.

Now we are the only extant species of human, a situation that is unique in the evolutionary history of humans. Ever since the first branching event about 4 million years ago, multiple species of closely related species have coexisted. We are the first species of human to be alone.

TAKE-HOME MESSAGE 11·11

Modern humans (*Homo sapiens*) evolved in Africa between 200,000 and 100,000 years ago, and all living humans are descended from that evolutionary radiation. About 100,000 years ago, a small group of modern humans moved out of Africa, and the descendants of this group ultimately populated Europe, Asia, and the Americas. Three other species of humans were living at this time. They all became extinct between 30,000 and 12,000 years ago, after modern humans had spread into the areas where they were living.

4 Invertebrates are animals without a backbone.

A brightly colored shield bug sits on a leaf.

11·12

Invertebrates are the largest and most diverse group of animals.

Up to this point, we have been discussing the evolution of deuterostomes, primarily one particular group within the phylum Chordata, the vertebrates. Although the vertebrates may be the most well understood animals, they are nowhere near the most diverse or abundant. The animal group that takes the prize for diversity and abundance is the invertebrates—defined simply as animal species without a backbone.

Although the invertebrate grouping has a convenient and easy-to-apply definition, the invertebrates are not a monophyletic group. There is one large group of invertebrates that, like the vertebrates, is made up of deuterostomes. Called echinoderms—and including sea stars (starfish), sea urchins, and sand dollars—this phylum of animals shows the same back-to-front gut development as the vertebrates, and these animals are the chordates' closest relatives. All of the other invertebrates are protostomes. Taken together, the invertebrates include more than 96% of the animal kingdom (**FIGURE 11-20**).

The invertebrate protostomes include eight separate evolutionary lineages with an enormous diversity of body forms, body sizes, habitats, and behaviors. Worms are

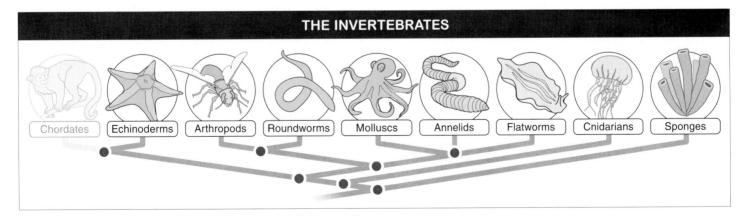

THE INVERTEBRATES

Chordates — Echinoderms — Arthropods — Roundworms — Molluscs — Annelids — Flatworms — Cnidarians — Sponges

FIGURE 11-20 Phylogeny of invertebrates.

invertebrates, and they are long and flexible with a soft body covering, but molluscs (clams, oysters, and their relatives) are also invertebrates and they are round or oval and some are encased in rigid shells. Sponges are the most sedentary and plant-like of the invertebrates, and octopuses and squids are as active and exploratory as any vertebrate.

Some invertebrates (such as worms) have soft bodies, others (snails, for example) have a shell, and still others (including insects and crustaceans such as lobsters) have a rigid external covering called an **exoskeleton.**

No invertebrate can be considered to represent the group as a whole, but the phylum Arthropoda (the **arthropods**) has the largest number of species—about 75% of the animals on earth are arthropods. This phylum includes the most familiar invertebrates: insects, spiders, scorpions, lobsters, crabs, shrimp, and their relatives. We begin by examining the vertebrates' closest relatives, the deuterostome echinoderms, followed by a tour of the protostome invertebrates.

TAKE-HOME MESSAGE 11·12

Invertebrates, defined as animals without a backbone, are the largest and most diverse group of animals, comprising 96% of all living animal species. The invertebrates are not a monophyletic group, however, and include protostomes and deuterostomes.

11·13

Echinoderms are vertebrates' invertebrate relatives: sea stars, sea urchins, and sand dollars.

Appearances can be deceiving—take the sea star, for example. Considering the great diversity among the invertebrates—including insects, crabs, worms, and other bilaterally symmetrical organisms that live on land, many with an obvious head region, eyes, and mouth—it is a bit surprising that the sea star and other members of the phylum Echinodermata are, as deuterostomes, more closely related to humans than they are to any other invertebrate groups. They stand as an example of how classifying organisms based on evolutionary relatedness rather than physical similarity can reveal a great deal about how the force of natural selection, adapting organisms to their environment, can have dramatic effects on body form.

The echinoderms include about 6,000 species of marine animals, most enclosed by a hard skeleton that lies beneath a spiny skin (**FIGURE 11-21**). Adult sea stars (starfish and brittle stars) are the most recognizable echinoderms, because they have five or more appendages evenly distributed around the circumference of their bodies. This is an example of radial symmetry, as described earlier, and radially symmetrical animals have no front or back ends. They move with equal ease in any direction, and their sense organs are distributed around their circumference, because any side of the animal can be the leading edge. Adult sea urchins and sand dollars don't have projecting arms, but they are also radially symmetrical.

Although radial symmetry is characteristic of adult echinoderms, their larvae are bilaterally symmetrical, having a left and right side of the body. Bilaterally symmetrical animals generally move in one direction, with their sense organs grouped at the front end of the body. Echinoderms evolved from bilaterally symmetrical ancestors, and radial symmetry is an evolutionary specialization associated with their mode of locomotion and feeding specializations. The larvae of echinoderms have anatomical characteristics in common with the larvae of primitive chordates, and as we've noted, echinoderms and chordates are each other's closest relatives.

Echinoderms creep on little tube feet that are extensions of an internal system of water-filled tubes that radiate throughout the body. The undersides of the arms of a sea star and the bodies of sea urchins and sand dollars are carpeted with tube feet that extend and contract in waves, grasping and releasing the substrate (the surface on which they move) as the animal glides along. Although each tube foot is tiny, there are thousands of them, and the combined force they exert allows a sea star to pull the two shells of a clam or mussel apart.

And then comes something really unexpected. The sea star can push its stomach out through its mouth and insert it into the open shell of the clam or mussel. The stomach then secretes digestive enzymes that break down the tissue of the

THE ECHINODERMS

COMMON CHARACTERISTICS
• Enclosed by a hard skeleton under spiny skin
• Larvae are bilaterally symmetrical and share some anatomical features with chordates.
• Adults are radially symmetrical.
• Undersides are covered with tube feet that aid in locomotion and grasping.

MEMBERS INCLUDE
• Sea stars (about 1,600 species)
• Sea urchins and sand dollars (about 940 species)
• Sea cucumbers (about 1,100 species)

FIGURE 11-21 What is an echinoderm? Examples of echinoderms (*from top left, clockwise*) are a sand dollar, a club sea urchin, and a blue sea star.

prey, and the soup that results is absorbed by the sea star. Sea urchins feed on algae, scraping them loose from rocks with sharp tooth-like surfaces made of calcium, and sand dollars capture floating particles of algae and other organic matter by trapping them in streams of mucus that are moved toward their mouth by beating cilia.

TAKE-HOME MESSAGE 11·13

Echinoderms are, like chordates, deuterostomes, but they lack a backbone. Their aquatic larvae are bilaterally symmetrical and share some anatomical features with chordates, but adult echinoderms are radially symmetrical.

11·14

An external skeleton and metamorphosis produced the greatest adaptive radiation ever.

When it comes to biodiversity, the insects take the cake. They outnumber all other forms of life in species diversity. The flies, alone, comprise 90,000 species; the bees, wasps, and ants number 150,000 species; and the butterflies and moths have 120,000 species. But even those numbers are dwarfed by the beetles, with 320,000 species! To put it in perspective, there are more species of beetles than there are species of plants, and nearly six times as many species of beetles as there are species of vertebrates. In this section, we'll investigate why insects have become so diverse.

The protostome phylum of arthropods (Arthropoda) contains invertebrates that have segmented bodies (i.e., bodies with a

distinct head, thorax, and abdomen), an exoskeleton made of a stiff carbohydrate called chitin, and legs with joints (think of all the joints in a lobster's claw). In all, the phylum contains almost 1 million recognized species, or approximately 75% of all animal species, and as many as 60% of *all* the species on earth (**FIGURE 11-22**). If diversification is a measure of evolutionary success, the arthropods are unmatched. One characteristic of insects has been central to their success: the way they cope with the change in body size as they grow.

Baby animals are smaller than adults—that's true of any kind of animal, and it can be a problem because a tiny baby cannot do the things that an adult does. In particular, babies cannot feed

THE ARTHROPODS

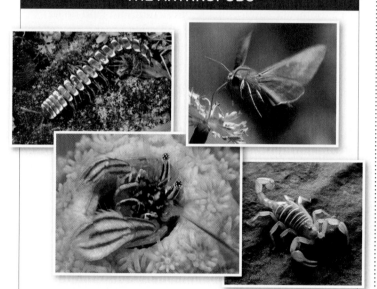

COMMON CHARACTERISTICS
• Segmented body with a distinct head, thorax, and abdomen
• Exoskeleton made of chitin
• Jointed appendages

MEMBERS INCLUDE
• Insects (over 750,000 species)
• Arachnids (about 60,000 species)
• Crustaceans (about 52,000 species)
• Millipedes and centipedes (about 10,000 species)

> *Arthropods make up about 75% of the animal species on earth.*

FIGURE 11-22 **Exploring the arthropods.** Shown here (*from top left, clockwise*) are a blue millipede, an elephant hawk moth, a scorpion sitting in the dust, and a hermit crab peeking out from its home in hard coral.

 Q *Mammals get bigger and bigger the more they eat. Why don't insects?*

themselves. Mammals solve that problem by supplying milk to infants and providing parental care until the young animal has grown large enough to feed and take care of itself. Many insects have a different way of solving the problem—their life history is divided into three completely different stages (**FIGURE 11-23**).

1. An egg hatches into a **larva,** which looks completely different from an adult. A caterpillar is the larval stage of a

butterfly or moth, and a grub is the larval stage of a beetle. A larva's job is to eat and grow large enough to enter the next life stage, the pupa.

2. A **pupa** is covered by a case, and inside that covering the body structures of the larva are broken down to molecules and then reassembled into the adult form. This rebuilding process is called **metamorphosis,** from two Greek words meaning "change of form."

3. When the **adult** form hatches from the pupa, it is at its full size. Insects do not grow after they emerge from their pupae—for this reason, some adults do not even eat. Rather than eating and growing, the job of an adult insect is reproduction.

Metamorphosis is the key to this life history pattern. In the first step of metamorphosis, the larva encloses itself in a case. Inside the case, the insect activates the genes that code for the adult body form. Proteins that made up the larva are broken down to amino acids, which are used to synthesize adult proteins. The entire process requires several days, and by the end of the pupal stage the adult insect is curled tightly inside the pupal case. The pupal case then splits and the adult

COMPLETE METAMORPHOSIS

Complete metamorphosis is the division of an organism's life history into three completely different stages.

LARVA
An egg hatches into a larva, which eats to grow large enough to enter into the pupal stage.

PUPA
The larva encloses itself in a case, where its body structures are broken down into molecules that are reassembled into the adult form.

ADULT
The adult emerges from the pupa, and its primary job is to reproduce.

> *The separation of life stages has contributed to the enormous ecological diversity of insects.*

FIGURE 11-23 **From larva to adult monarch butterfly.** Some insects have a life history divided into unique stages, each with a distinct purpose.

What Is an Animal? | Vertebrates and the Move onto Land | Terrestrial Vertebrates | Invertebrates squirms out. The wings of butterflies and moths are tightly folded inside the pupa, and they require time to expand and stiffen before the adult is ready to fly.

"Complete metamorphosis," involving the three stages, allows the larva and adult to act as if they were different species, optimized for very specialized tasks. Caterpillars eat leaves, for example, and spend their entire larval period on a single plant. In contrast, butterflies feed on nectar that they collect by flying from flower to flower. As a larva, the organism must feed and grow. As an adult, it reproduces. The genes that control the larval body form are different from the genes that determine the adult form, so natural selection can act on the larval and adult stages independently.

Not all insects undergo the dramatic change (complete metamorphosis) from larva to adult that we see in beetles and butterflies. Juveniles of some primitive insects, such as grasshoppers and cockroaches, look and behave very much like the adults. These juveniles, which are called nymphs, do not have wings or reproductive organs, but they live in the same habitats as adults and eat the same foods.

About 83% of insect species have complete metamorphosis, and this separation of the life stages has helped insects diversify into nearly 20 times as many named species as vertebrates, and at least that many more species of insects have yet to be named.

TAKE-HOME MESSAGE 11•14

The arthropods are protostome invertebrates, including the insects. The life cycle of most insects includes a larval stage that is devoted to feeding and growth and an adult stage in which the insect reproduces. This separation of life stages has contributed to the enormous ecological diversity of insects and has produced remarkable specializations among the nearly one million species of insects that have been named so far.

11•15

Other arthropods include arachnids, crustaceans, millipedes, and centipedes.

The arthropod phylum is so large that even if we excluded all the insects, more species still would remain than are present in any other phylum. In this section we survey the other major groups of arthropods: the arachnids, crustaceans, and millipedes and centipedes (**FIGURE 11-24**).

Arachnids Arachnids are land-dwelling arthropods that include spiders, scorpions, mites, and ticks. Arachnids usually have four pairs of walking legs and a specialized feeding apparatus, such as the fangs of spiders and the grinding mouthparts of scorpions. Arachnids (like insects) are distinguished from other arthropods by the positioning of their legs: they have legs only on the middle section of their body (the thorax), whereas other arthropods have legs on the posterior section of their body (the abdomen) as well as on the thorax.

Of the 60,000 species of arachnids, spiders and scorpions are the ones you are most likely to see—and to fear. But their reputation as fear-worthy animals may not be fully deserved. Most species are small, harmless, and often beneficial, because they eat many insects that are pests in agriculture or in homes.

Arachnids are predators. Many spiders construct webs by spinning fibers of silk (which consists of interconnected protein molecules). Orb-weaving spiders construct intricate webs that trap (and sometimes even lure) flying insects, and they place their web across the routes followed by flying insects. A spider rests inconspicuously until an insect gets trapped in the web, then runs out to wrap the prey in silk threads.

When a spider captures prey, it uses its fangs to inject venom that contains two types of enzymes. The first set of enzymes disrupts the prey's nervous system and paralyzes it, and then other enzymes dissolve its internal organs. When the prey's innards have liquefied, the spider sucks out its contents. (Maybe their reputation *is* deserved.)

The venom of most spiders, however, usually is harmless to animals as large as humans, but two North American spiders *are* dangerous. The black widow spider is the better known, and is easily identified by a red hourglass marking on the bottom surface of the female. The brown recluse spider lacks distinctive markings, so it is harder to identify, but it is more dangerous. Both species are likely to move indoors, constructing webs in little-used spaces, such as in closets and cabinets and behind furniture that stands against a wall. Brown recluse bites are notorious for creating open wounds that can last for months or years and sometimes require repeated skin grafts to close.

More Invertebrate Diversity

INSECTS
- Three pairs of walking legs
- Legs are located on the thorax
- Life cycle consists of separate life stages

ARACHNIDS
- Usually four pairs of walking legs
- Legs are located on the thorax
- Specialized mouthparts
- Predators

CRUSTACEANS
- Many pairs of legs
- Usually five pairs of appendages that extend from head
- Mostly aquatic

MILLIPEDES AND CENTIPEDES
- Many pairs of legs (one pair per segment)
- Long, segmented body

FIGURE 11-24 Arthropod diversity: insects, arachnids, crustaceans, and millipedes and centipedes. Shown here (*from left to right*) are a shield bug, a banded garden spider, a red swamp crayfish, and a giant desert centipede.

Scorpions look dangerous. With their large claws (which are modified legs) and curved tails, with the stinger prominently displayed, they clearly say, "Don't mess with me!" But being stung by most scorpions is no worse than being stung by a bee. Scorpions are hunters, searching at night for insects. Scorpions seize prey with their claws; the sting is not used unless the prey struggles. Some of the scorpion's mouth structures are covered with spines that grind the prey, while glands secrete enzymes that liquefy it. The scorpion sucks up the soup and discards the hard parts of the prey.

Crustaceans If arachnids are the most feared arthropods, the aquatic crustaceans—lobsters, crayfish, crabs, and shrimp—may be regarded as the most desirable arthropods, at least at meal time. In the United States, the top three fishery crops, by their annual market value, are shrimp, crabs, and lobsters. There are about 52,000 species of crustaceans, and although crustaceans can look very different—from sessile barnacles to mobile, tiny krill—they all have five pairs of appendages extending from their heads. Three of these appendages are used for feeding and two are antennae that sense the environment.

Crustaceans have many pairs of legs, and their legs are modified for many purposes. Shrimp and barnacles (which are essentially shrimp that spend their adult lives fastened to a rock and enclosed in a shell) have legs with comb-like projections that they use to capture tiny plankton, and free-living crustaceans (shrimp, crabs, lobsters) have modified limbs in the abdominal region that are used to hold eggs or newly hatched young.

Although most crustaceans are aquatic, the inconspicuous wood lice (also known as pill bugs, sow bugs, or roly-polies) are terrestrial forms that you can find when you turn over a rock in your garden. Wood lice are herbivores and play an important role in recycling dead plant material.

Millipedes and centipedes The names *millipede* and *centipede* translate to "a thousand feet" and "a hundred feet" and refer to the enormous numbers of legs on these long, skinny animals. They don't really have a thousand—or even a hundred—feet, but millipedes do have more legs than centipedes, and that is the easiest way to tell them apart. Millipedes and centipedes have long, segmented bodies that seem almost worm-like, but their jointed legs and hard exoskeletons are characteristics of arthropods. Both live among fallen leaves, especially in forests where the leaf litter layer is usually cool and moist. Millipedes feed on decaying plant material, while centipedes are predators that use fangs equipped with venom to kill insects and even small mammals. Unlike spiders and scorpions, centipedes use their jaws to tear prey into pieces that are small enough to swallow.

TAKE-HOME MESSAGE 11·15

Spiders and scorpions are predatory arthropods that eat insects and occasionally small vertebrates. Lobsters, crabs, shrimps, and barnacles are predatory marine crustaceans. Centipedes are predators with fangs that inject venom, while millipedes are herbivores that feed on dead plant material.

What Is an Animal?   Vertebrates and the Move onto Land   Terrestrial Vertebrates   Invertebrates

11·16

Most molluscs live in shells.

Molluscs (phylum Mollusca) are a large group of animals—nearly 100,000 named species and many more that have not yet been named—that live in the ocean, in fresh water, and on land. There are many familiar molluscs, including clams, snails, scallops, oysters, mussels, squids, and octopuses. They are so diverse that it is difficult to describe any single defining characteristic. The position of molluscs within the animal phylogeny reflects the fact that they have true tissue, are bilaterally symmetrical, and are protostomes that do not molt. Three features are common to many molluscs, however: a shell that protects the soft body, a mantle (the tissue that secretes calcium carbonate to form the shell), and a sandpaper-like tongue structure, called the radula, that is used during feeding (**FIGURE 11-25**).

There are three major groups of molluscs, sharing most of the common features but with varied body plans: **gastropods, bivalve molluscs,** and **cephalopods.**

Gastropods Snails and slugs are gastropod molluscs. They get their name because *gastropod* means "belly foot" and they have an expanded foot on the bottom of their body that allows them to climb a vertical surface as easily as they glide across a horizontal one. Snails have a one-piece, curled shell, and slugs are snails that have reduced the shell to a tiny remnant that is not even visible because it is covered by the mantle. Found in both aquatic and terrestrial environments, snails and slugs account for three-quarters of all molluscs.

The snail's shell is its primary protection against predators, but because both terrestrial slugs and sea slugs have very little shell material, other defense mechanisms are necessary. Slime is what a terrestrial slug relies on—when a slug is attacked, it secretes slime that sticks to the predator. Worse still, anything that touches the slime on the predator sticks to it, so a bird that attacks a slug quickly finds that its beak and face are covered by pieces of dead leaves, clods of soil, and twigs. Sea slugs have their own methods of defense. Some species synthesize toxic chemicals, and others, which feed on sponges, recycle sponge toxins into their own slime. One of the most remarkable uses of another animal's defense equipment is found among sea slugs that eat sea anemones. Anemones are protected by stinging cells, and the sea slugs that eat anemones transfer some of those cells from the anemones to their own skin, enabling them to sting other creatures.

Bivalve molluscs Clams, scallops, oysters, and mussels have a pair of shells that clamp together, and these animals are the

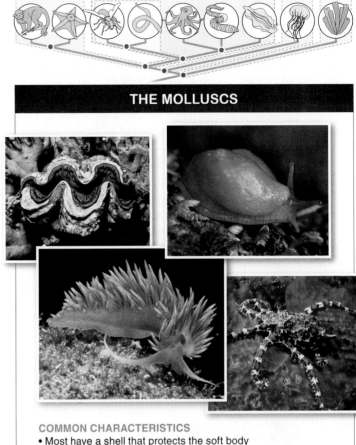

THE MOLLUSCS

COMMON CHARACTERISTICS
- Most have a shell that protects the soft body
- Mantle (tissue that forms the shell)
- Radula (sandpaper-like tongue structure used during feeding)

MEMBERS INCLUDE
- Gastropods (about 35,000 species)
- Bivalves (about 8,000 species)
- Cephalopods (about 700 species)

FIGURE 11-25 Molluscs: an overview. Some examples of molluscs (*from top left, clockwise*) are a giant clam, a banana slug, a Caribbean reef octopus, and a nudibranch.

bivalve molluscs. There are about 8,000 species of bivalves and most of them live in the ocean, although there are some freshwater species. Clams spend their lives buried in mud or sand, scallops live on the sea floor, and oysters and mussels fasten themselves to underwater objects, such as rocks, the pilings that support piers, and the hulls of boats. All bivalves are filter-feeders that draw a current of water through a tube called the "incurrent siphon," across their gills, where tiny food particles are captured, and out through the "excurrent siphon."

When a grain of sand is trapped in the shell of a bivalve, the mantle may secrete layer after layer of a shell-like protein material that covers the sand grain and forms the iridescent gem called a pearl. Oysters are the best known source of pearls, but clams and other bivalve molluscs also form pearls.

Cephalopods The third major group of molluscs is the cephalopods (**FIGURE 11-26**). It includes 6 species of nautilus and about 600 species of squids and octopuses. The nautilus has an external shell that resembles a snail shell, except that it is not twisted. Squids have very small shells that are covered by the mantle, and octopuses have lost the shell entirely. Their tentacles appear to grow directly from the head, which explains the name "cephalopod" (translated as "head-footed").

The most obvious feature of cephalopods may be their tentacles, which they use to walk and to capture prey. Squids have eight short tentacles called arms and two long sucker-bearing tentacles that are used to capture prey. After a slow approach to its unsuspecting prey, a squid propels its tentacles forward with astonishing speed—accelerations of more than 800 feet per second per second (about 250 meters per second per second) have been measured, the equivalent of your car accelerating from 0 to 6,000 mph in 10 seconds! The suckers on the tentacles adhere to the prey and draw it back toward the squid, where the arms take over, turning and manipulating the prey as it is consumed by the squid's sharp beak.

Many species of squid are more than 3 feet (about 1 meter) long, and the giant squids reach lengths of 40 feet (about 13 meters). When a squid is in a hurry, it swims tail-end first, using jet propulsion. Water is expelled through the siphon at the tentacle end of the body, shooting the animal backward. Octopuses, which can have tentacles that spread 12 feet (about 4 meters), are bottom dwellers, living in coral reefs and rocky coasts. Because they have completely lost the shell, octopuses can squeeze through astonishingly small openings, and their ability to escape from even carefully covered tanks is a perpetual challenge for the keepers of zoos and aquaria. Keepers are accustomed to arriving in the morning and finding an octopus wandering around the room.

In the next section we further explore the cephalopods—do some of their skills of manual dexterity make them the smartest of the invertebrates?

TAKE-HOME MESSAGE 11•16

Molluscs are protostome invertebrates that do not molt. Including the snails and slugs, clams and oysters, and squids and octopuses, the molluscs are the second most diverse phylum of animals. Most feature a shell for protection, a mantle of tissue that wraps around their body, and a specialized tongue called a radula.

GROUPS OF CEPHALOPODS

SQUIDS
- 8 short tentacles and 2 long sucker-bearing tentacles
- Free-swimming
- Contains about 300 species

OCTOPUSES
- 8 short tentacles called arms
- Bottom dwellers, living in coral reefs and rocky coasts
- Contains about 300 species

NAUTILUSES
- Chambered shell used for protection
- Free-swimming
- Contains 6 species

FIGURE 11-26 Cephalopod diversity. Shown here (*from left*) an oval squid, big red octopus, and chambered nautilus.

434 CHAPTER 11 • ANIMAL DIVERSIFICATION

What Is an Animal?   Vertebrates and the Move onto Land   Terrestrial Vertebrates   Invertebrates

Are some animals smarter than others?

Q Are octopuses smart?

Squids and octopuses are wide-ranging predators that must locate, capture, and subdue other animals. This is a far more active life style than that of bivalves or snails, and it requires quick movements and rapid responses to stimuli. All of these features are conspicuous elements of the behavior of an octopus as it moves through a coral reef, inserting a tentacle into every opening. In fact, as an expert multi-tasker, an octopus can capture a prey item with one of its tentacles, use a second tentacle to hold something else, and still keep the remaining six tentacles in motion. In addition, octopuses are very good at manipulating things with their tentacles. It doesn't take a captive octopus long to learn how to screw the top off a glass jar to reach a fish that's swimming inside—after all, that's just a minor variation on twisting a clam to open it (**FIGURE 11-27**).

Because of their manipulative skills and easily observed exploratory behavior, octopuses are sometimes featured on TV shows and in magazine articles, where they are often described as the smartest invertebrates, and their intelligence compared to that of mammals. But this is the sort of claim that must be viewed skeptically.

What is intelligence, after all? The concept of intelligence loses its meaning when animals that live in completely different worlds and respond to entirely different stimuli are compared. We can say that octopuses are very good at doing the things that octopuses need to do—searching for prey, capturing it, and manipulating it, all within their marine environment, for example. But is following a complex sequence of actions the same thing as intelligence? Spiders, for example, construct elaborate webs that are strategically placed in the flight paths of insects; they detect the impact of an insect with the web and rush out to wrap the insect in silk before it can escape. Does that require more or less intelligence than opening a glass jar to eat a fish? And what about the red squirrels of northern Europe, which hide more than 3,000 acorns, cones, and nuts each fall, ultimately recovering more than 80% of their hidden nuggets of food over the course of the winter—can a human manage that? An octopus can't.

These questions are analogous to the question asked earlier in the chapter about what makes a species evolutionarily successful. Recall that, from an evolutionary perspective, any

? *Octopuses quickly learn to open jars. Does that make them smart?*

FIGURE 11-27 An octopus attempts to open a jar.

non-extinct species must be considered a success. Similarly, the very concept of "intelligence" may actually be relevant only to humans, to the extent that intelligence is defined as the ability to do some problem-solving task valued by humans. Applying the question of intelligence to non-human species may just be an assessment of how well they can accomplish tasks. For this reason, questions of comparative animal intelligence are not generally useful. Rather, animals' abilities should be considered as evolutionary responses to particular selective pressures imposed by their environment, and the fact that a species currently survives should be taken as an indication of some mastery of its particular niche.

TAKE-HOME MESSAGE 11·17

The predatory behavior of octopuses involves exploration and manipulation behaviors often considered as intelligent by humans, but the concept of intelligence cannot be applied objectively to other species, which have evolved in response to the selective forces relevant only to their own particular niche.

A Lion's Mane jellyfish swims in the Atlantic waters off the coast of Nova Scotia, Canada.

11·18

Segmented worms, flatworms, and roundworms come in all shapes and sizes.

It's a good thing that biologists don't run bait shops—if they did, a simple trip to buy worms for fishing would be way too complicated. The name "worm" is applied to a long, skinny animal without a backbone, but you can find animals fitting that description in eight different phyla. You will never encounter most of them, and that's a good thing, because many are parasites that cause really unpleasant diseases. We consider here just three phyla—**segmented worms** (phylum Annelida), **flatworms** (phylum Platyhelminthes) and **roundworms** (phylum Nematoda)—that illustrate the diversity of the animals we call worms.

Segmented worms Segmented worms, also called annelids, number about 13,000 species and are easy to recognize by the grooves running around their bodies that mark the divisions between segments—if you've seen an earthworm, you are familiar with those grooves. The segmented worms are protostomes with defined tissues and do not molt. They are organized into three different groups: marine polychaetes (pronounced POL-ly-keets), terrestrial oligochaetes, and leeches (**FIGURE 11-28**).

Polychaetes are marine worms, living on the sea floor. *Polychaete* means "many bristles," and the combination of

segments and bristles makes polychaetes easy to recognize. Some species burrow through the mud and extract organic material from it. Tube worms use sand grains or limestone to make a tube in which they live, with just their waving tentacles exposed. Small particles of food are trapped in mucus on the tentacles and transported to the worm's mouth.

Earthworms, which belong to the group called *oligochaetes* ("few bristles"), are the annelid worms you are most likely to have seen. The night crawler is a typical earthworm. It gets its name from its habit of emerging from its burrow on rainy nights to crawl across the surface of the ground. More than 4,000 species of earthworms have been named, ranging in size from less than half an inch (about a centimeter) long to the giant Gippsland earthworm, 7–10 feet (2–3 meters) long!

Earthworms are bulk-feeders—that is, as an earthworm burrows, it consumes particles of soil and organic material. The organic material is digested as it passes through the worm's gut, and the fecal material plus the inorganic part of the soil are excreted as feces, called castings. Earthworm castings are valued by gardeners as soil supplements, and an earthworm can produce its own weight in castings every day. Because they mix the soil components, creating a more even mixture of nutrients,

What Is an Animal?   Vertebrates and the Move onto Land   Terrestrial Vertebrates   Invertebrates

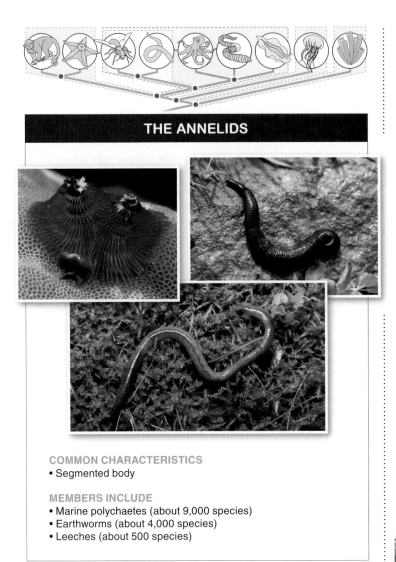

THE ANNELIDS

COMMON CHARACTERISTICS
• Segmented body

MEMBERS INCLUDE
• Marine polychaetes (about 9,000 species)
• Earthworms (about 4,000 species)
• Leeches (about 500 species)

FIGURE 11-28 Annelid diversity. Shown here (*from top left, clockwise*) are a polychaete worm, a leech, and an earthworm.

and expedite the breakdown of organic materials in soil, making the nutrients available for plants, the economic value of earthworms to agriculture is enormous. A government study published in 2008 estimated the economic value of earthworms in Ireland alone to be more than $1 billion per year.

Have you ever waded in a pond or swamp and found a long, dark brown worm clinging to your leg when you emerged? If so, you have met the third major group of annelids, the *leeches.* Probably you pulled the leech off (not an easy thing to do, because leeches are both slippery and stretchy) and threw it as far as you could. If you had looked at it carefully, though, you would have noticed that, like earthworms, leeches have segmented bodies.

If you have had this close encounter with a leech, you'll also have noticed that the wound that it made on your leg continued to ooze blood for several hours. That's because the saliva of blood-sucking leeches contains an anticoagulant substance that prevents blood from clotting. If the leech is undisturbed while it is feeding, blood keeps flowing and the leech inflates like a small balloon.

Not all leeches are blood-suckers—in fact, more than half the species of leeches are predators. The horse leech, for example, which reaches a length of 8 inches (about 20 centimeters), feeds on smaller annelid worms, snails, and aquatic insect larvae.

Roundworms The roundworms are protostomes with defined tissues and, unlike the segmented worms and flatworms, they grow by molting. They are probably the most abundant animals on earth—a spoonful of garden soil contains several thousand individuals, and some species produce more than 200,000 eggs every day. More than 90,000 species of roundworms have been named, and there may be five times as many species that have not yet been identified (**FIGURE 11-29**)

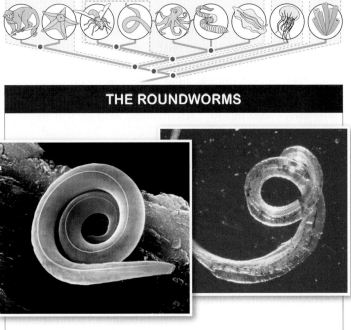

THE ROUNDWORMS

COMMON CHARACTERISTICS
• Long, narrow unsegmented body
• Bilaterally symmetrical
• Surrounded by a strong, flexible cuticle
• Must molt in order to grow larger

MEMBERS INCLUDE
• About 90,000 species (scientists estimate there may be five times as many species that have not yet been identified)

FIGURE 11-29 Roundworm diversity. A nematode worm (*left*) and a whipworm.

437

Nematodes are soil-dwelling roundworms that live in the roots of plants, causing damage or even death of the plant. About 15,000 species of roundworms are parasites of vertebrate animals, and roundworms are responsible for a large number of human diseases. Most roundworms are transmitted by fecal contamination of soil or food.

Tiny parasitic roundworms called *filariae* are responsible for several tropical diseases, including some that have an enormous economic and social impact in certain parts of the world. Elephantiasis, for example, is a disease in which filariae transmitted by the bite of a mosquito block the lymph ducts so that fluid accumulates in the limbs or scrotum, causing grotesque swelling. These filariae occur in India, Africa, South Asia, Pacifica, and tropical America.

Flatworms Like the segmented worms, but unlike the roundworms, the flatworms are protostomes that do not molt (**FIGURE 11-30**). Flatworms include more than 20,000 species and can be parasites (the flukes and the tapeworms) or brilliantly colored, free-living aquatic creatures.

Flatworms have well-defined head and tail regions, with clusters of light-sensitive cells called eyespots. Although many flatworms have a digestive system, one feature of free-living flatworms (but not the segmented or roundworms) that many humans find distasteful is that their gut has only one opening, requiring them to consume food and eliminate undigested food through the same opening. Tapeworms, by contrast, completely lack a digestive system; these parasitic worms utilize nutrients from their host by living in the gut and absorbing nutrients directly through their body wall. Most flatworms are hermaphroditic, each individual worm producing both male and female gametes, and they engage in both sexual and asexual reproduction.

All 5,000 species of tapeworms are parasites, and most have a two-stage life cycle that is split between two different host species. For example, the common tapeworm that infects dogs spends the other half of its life cycle in fleas, while the human pork tapeworm splits its life cycle between pigs and humans. Tapeworms have long, flat bodies made up of repeated segments, and each segment is a reproductive unit. Mature tapeworms spread by breaking off segments, which are then shed in the feces of an infected individual.

THE FLATWORMS

COMMON CHARACTERISTICS
- Well-defined head and tail regions
- Hermaphroditic and can engage in both sexual and asexual reproduction
- Some have a single opening in the body, which serves as a mouth and an anus

MEMBERS INCLUDE
- Tapeworms
- Flukes

FIGURE 11-30 Overview of the flatworms. A liver fluke (*left*) and a tapeworm of the mammalian intestine.

TAKE-HOME MESSAGE 11·18

Worms are found in several different phyla and are not a monophyletic group. All are bilaterally symmetrical protostomes with defined tissues. The segmented worms (annelids) and flatworms do not molt, while the roundworms do. Earthworms are annelids that play an important role in recycling dead plant material. Many roundworms are parasites of plants or animals and are responsible for several widespread human diseases. Flatworms include parasitic flukes and tapeworms, many of which infect humans.

What Is an Animal?   Vertebrates and the Move onto Land   Terrestrial Vertebrates   Invertebrates

Jellyfish and other cnidarians are among the most poisonous animals in the world.

If you feel a sting when you are swimming in the ocean, you have met a jellyfish. If you are lucky, it will be one of the thousands of species that feed on tiny floating organisms called plankton, because these species of jellyfish have only mild stings. If it is a species that feeds on fish or shrimp, however, you are definitely unlucky, because these jellyfish have powerful stings that can cause extreme pain or even death.

The jellyfish, along with sea anemones and corals, belong to the phylum Cnidaria (pronounced nigh-DARE-ee-ah), the cnidarians. This phylum includes about 11,000 species, all of which have defined tissues and radial symmetry and are generally simpler than the bilaterally symmetrical invertebrates we have discussed so far (**FIGURE 11-31**).

There are two types of cnidarian bodies: a sessile "polyp" and a free-floating "medusa." In some species, individuals spend part of their life cycle as a polyp and part as a medusa. Other species occur only as medusas or, as in corals and sea anemones, only as polyps. All cnidarians can reproduce both sexually and asexually.

Cnidarians are carnivores, using their tentacles to capture and feed on a wide variety of marine organisms, from protists all the way to fish and shellfish. Their method of capturing prey relies on a structure unique to the cnidarians—a stinging cell called a cnidocyst. All cnidarians have tentacles that are armed with rows of stinging cnidocysts. Each cnidocyst has a coiled thread with barbs inside and a "trigger" on the outside. When something comes in contact with the trigger, perhaps a prey item (or perhaps your leg as you swim in the ocean), the coiled thread is ejected and, like a harpoon, can penetrate the prey, often injecting toxin. The Portuguese man-o'-war is one of the dangerously poisonous species of cnidarians. Being stung by a Portuguese man-o'-war is painful (extremely painful if the unlucky swimmer becomes entangled in the tentacles and receives hundreds of stings), but these encounters are rarely fatal. Although all cnidarians possess cnidocysts, there is great diversity among the three major groups of cnidarians.

Jellyfish Jellyfish range tremendously in size. The Asian giant jellyfish is more than 6 feet (about 2 meters) across and weighs nearly 500 pounds (more than 200 kilograms). When swarms of these giant jellyfish appear, they clog the nets of fishing boats with a glutinous mass of toxic stingers. At the opposite

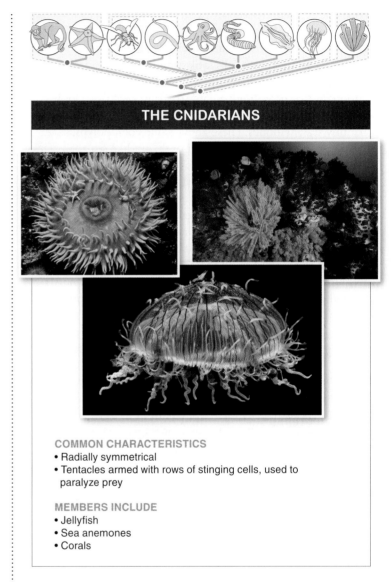

THE CNIDARIANS

COMMON CHARACTERISTICS
• Radially symmetrical
• Tentacles armed with rows of stinging cells, used to paralyze prey

MEMBERS INCLUDE
• Jellyfish
• Sea anemones
• Corals

FIGURE 11-31 Understanding cnidarians. Shown here (*from top left, clockwise*) are a giant green sea anemone, a portion of coral reef, and an aptly named flower hat jellyfish.

end of the size scale, the irukandji jellyfish is about the size of a hen's egg, but it is so deadly that when just five irukandji were sighted in Hervey Bay, Queensland, Australia, this completely halted the filming of a major Hollywood movie, which had to be completed in the safety of a studio.

Sea anemones As their name suggests, sea anemones look rather like flowers, and some of the colorful species are popular additions to saltwater aquaria. The polyp body form of a sea anemone resembles an upside-down medusa form of a jellyfish—the tentacles with their stinging cells are at the top of the body, surrounding the mouth. Sea anemones have a larval stage that swims freely and then settles on a rock and metamorphoses into the adult form. Even as adults, many sea anemones can crawl a few inches a day, and the wandering sea anemone is often found floating in seaweed.

Corals The corals live as small polyps in large colonial groups. At first glance, corals look even less like jellyfish than sea anemones do—because they secrete a chemical, calcium carbonate, that gives them a hard structure, but if you look inside the hard external structure you'll find animals that look like tiny sea anemones. These animals, the coral polyps, have stinging tentacles surrounding a mouth, just like sea anemones; they can use the tentacles to catch small fish and plankton. Corals grow in a wide variety of shapes—from free-standing spherical or branched forms to crusts that grow on rocks or on other corals. A coral reef creates a complex three-dimensional environment that is home to a greater diversity of species than any other marine habitat. The Great Barrier Reef, located in the Coral Sea off the coast of Queensland, extends more than 1,500 miles (about 2,600 kilometers) from north to south. It is the largest biological structure in the world and is easily visible from the International Space Station.

Q How is global warming affecting the coral reefs of the world?

The effects of global warming are already being seen in coral reefs. Although coral polyps can catch prey using their cnidocytes, they obtain most of their nutrition from algae, called zooxanthellae, that live symbiotically within the polyps. These algae use the carbon dioxide produced by the polyps to conduct photosynthesis. Strangely, when coral polyps are too hot they expel their zooxanthellae. The zooxanthellae are responsible for the colors

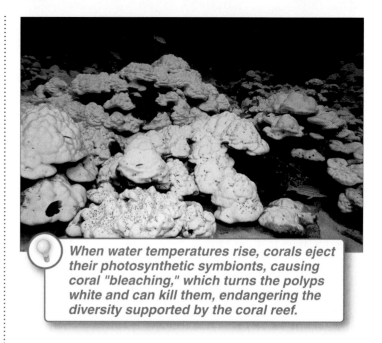

When water temperatures rise, corals eject their photosynthetic symbionts, causing coral "bleaching," which turns the polyps white and can kill them, endangering the diversity supported by the coral reef.

FIGURE 11-32 Coral bleaching.

of many corals, so when coral polyps expel their zooxanthellae, the coral turns white, a phenomenon called coral bleaching. The zooxanthellae return to the polyps when the water cools, but repeated bleaching events lead to the death of the coral polyps. The world's oceans are already warming, and coral bleaching events are becoming more and more frequent (**FIGURE 11-32**).

TAKE-HOME MESSAGE 11·19

Jellyfish, sea anemones, and corals are radially symmetrical animals with defined tissues, in the phylum Cnidaria. All cnidarians are carnivores and use specialized stinging cells located in their tentacles to capture prey.

11·20

Sponges are animals that lack tissues and organs.

Sponges (phylum Porifera), though they scarcely resemble most of the organisms we think of as animals, are indeed animals, too (**FIGURE 11-33**). They have no specialized tissues or organs. Most lack any symmetry at all. Despite their simplicity, however, sponges are remarkably efficient at gathering food. You can think of the anatomy of a sponge as a

cylinder with holes in it. Epidermal cells cover the outside of the sponge, and collar cells cover the inside. The middle layer is merely a gel with fibers that stiffen the sponge's body.

Each collar cell has a long, whip-like flagellum, and the beating of the flagella creates a current that carries water upward and

THE SPONGES

COMMON CHARACTERISTICS
- No tissues or organs
- Body consists of a hollow tube with pores in its wall
- Feed by pumping water, along with bacteria, algae, and small particles of organic material, through their pore cells
- Free-swimming larvae
- Sessile as adults

MEMBERS INCLUDE
- About 5,000 species

FIGURE 11-33 An overview of the sponges. A brown tube sponge (*left*) and a purple sponge.

out through the opening at the top of the sponge. That water is replaced by an inward flow of water through pore cells on the sides of the sponge, and the incoming water carries bacteria and algal cells, other microscopic organisms, and small particles of organic material into the sponge. The volume of water moving through a sponge daily can be as much as 20,000 times the volume of the sponge.

The collars of the collar cells are covered with sticky mucus that traps food particles. Cilia move the trapped food particles to the cell body, where they are absorbed by endocytosis. Sponges have no circulatory systems to move nutrients from one place to another. Instead, a special kind of cell, an amoebocyte cell, picks up food particles from the collar cells

and moves around through the sponge, transferring food particles to the epidermal cells. Amoebocytes can move freely because sponges lack tight junctions (see Section 3-12) between their cells.

Sponges are hermaphrodites, each individual containing both male and female reproductive organs, but they produce only one kind of gamete (eggs or sperm) at a time. Sponges that are acting as males release a cloud of sperm that swim to other sponges and fertilize their eggs. The eggs develop into larvae, which are released into the water and then drift for a few days before settling on a rock or coral outcrop and developing into a new sponge.

Sponges also reproduce asexually—small sponges bud from the outside of the sponge and eventually break off, settle to the bottom, and grow into new sponges. Even a piece accidentally broken off a sponge by wave action or a collision with a passing fish will grow into a new sponge. Most remarkable is the ability of sponge cells to reassemble. Picture this experiment. You put a living sponge into a food blender and puree it, then strain the liquid through a fine sieve to remove all the chunks, leaving a suspension of individual cells, which you then dump into an aquarium. Within a day you will see small clumps of sponge forming as the individual cells move around and attach to each other when they meet, and within a week a new sponge will have formed. For an even more dramatic demonstration, you can repeat the experiment with two species of sponge of different colors. Not only will the sponges reassemble, but the cells from each species will assemble only with other cells of the same species, so you will have two sponges, each with only its correct cells.

Because sponges are superficially the most "plant-like" of the animals, they make a fitting transition to the subject of our next chapter: the diversity of plants. Although less numerous than animal species, plants are complex organisms that live on land and in water and are found on nearly every surface of the earth.

TAKE-HOME MESSAGE 11·20

Sponges are among the simplest of the lineages of animals. A sponge consists of a hollow tube with pores in its wall; it has no tissues or organs. Sponges reproduce asexually (by budding) and sexually by producing eggs and sperm. The fertilized eggs grow into free-swimming larvae that settle and metamorphose into sessile adult sponges.

Knowledge You Can Use

The box jellyfish: nature at its nastiest? When it comes to acquiring food, we expect strong selection for effective strategies.

Natural selection has produced some pretty lethal structures, from the razor-sharp teeth of the piranha, to the hypodermic, poison-injecting fangs of a viper, to the toxic sting of a hornet. But these animals all have an aura of danger and we are not surprised by their lethality. Everyone's got to eat, and their methods for catching prey are impressively effective.

Q: **Can a gentle-looking animal with simple body plan be a lethal predator?** Quietly swimming in the ocean are some animals that reach a whole new level of lethality. One of nature's most serene-appearing animals, the box jellyfish, has no teeth, no muscles, and no warning hiss. Throughout the waters north of Australia and extending east to the beaches of Hawaii, these animals simply contract and relax their bodies, gently propelling themselves forward.

Q: **What is the fastest cellular process in nature?** If you find yourself sharing the water with a box jellyfish, you may get a deadly demonstration of nature's fastest cellular process. Along the lengths of their tentacles, these jellyfish have high densities of the stinging cells known as cnidocytes. But these aren't just any stingers. Each cell has a coiled, barbed hair within it. The slightest contact with a trigger on the cell releases the stinger, like a dart. Because the stinger is held at such high pressure, when released, it is ejected 10–20 micrometers in less than a millionth of a second. In a study using a high-speed camera, the acceleration of the dart was measured at over one million times the force of gravity (for comparison, during a rocket launch, astronauts experience about five times the force of gravity). The cnidocyte dart can strike prey with the force of a bullet, easily penetrating the shell of a crustacean (or the skin of a human).

Q: **What does the sting do to a human?** Feeling the bullet-like strikes of the hundreds of cnidocytes on a single tentacle isn't even the worst of it. The toxin injected when the cnidocyte fires is among the top 10 most toxic animal products on earth. The pain of a box jellyfish sting is so excruciating that swimmers may go into shock or experience heart failure and drown before they can reach shore. More than a hundred people have been killed by box jellyfish stings. Even a non-fatal sting can produce severe scarring and pain that persists for weeks.

The best treatment is prevention? In the disastrous event of an encounter with a box jellyfish, what can you do? Immediately pouring vinegar on the sting and any still-attached tentacles is extremely helpful, disabling any cnidocytes that have not yet fired (though doing nothing for the pain). Many lifeguard stations on Australian beaches keep vinegar on hand for such occasions. But because little else can be done, the best strategy is to avoid swimming at beaches where box jellyfish have been sighted. Stand on the shore instead and safely appreciate one of natural selection's most deviously dangerous creations.

1 Animals are just one branch of the eukarya.

Animals are multicellular organisms that feed on other organisms and can move during at least one stage of their life. Animals probably originated from an ancestral protist. Four key distinctions divide the extant animals: tissues or not, radial or bilateral symmetry, protostome or deuterostome development, and growth through molting or without molting.

2 Vertebrates are animals with a backbone.

All chordates have four characteristic structures: a notochord, a dorsal hollow nerve cord, pharyngeal gill slits, and a post-anal tail. The development of fins and jaws set the stage for the enormous diversity of modern vertebrates. The transition of vertebrates from life in water to life on land was enabled by four developments: fins modified into limbs, vertebrae modified to transmit body weight through the limbs to the ground, transfer of the site of gas exchange from gills to lungs, and an egg resistant to drying out.

3 All terrestrial vertebrates are tetrapods.

Amphibians are terrestrial vertebrates (tetrapods), but most species still lay eggs in water and the eggs hatch into aquatic larvae. Feathers were originally colorful structures used for behavioral displays in reptiles. Their current uses as insulation and in flight developed later. Hair and lactation are defining characteristics of mammals. The evolutionary success of humans can be traced to an increase in brain size combined with an omnivorous diet. Modern humans evolved in Africa 200,000–100,000 years ago, and all living humans are descended from that evolutionary radiation.

4 Invertebrates are animals without a backbone.

Invertebrates, defined as animals without a backbone, are the largest and most diverse group of animals, comprising 96% of all living species of animals. The invertebrates are not a monophyletic group, however, and include protostomes and deuterostomes. Echinoderms, like chordates, are deuterostomes. The arthropods, protostome invertebrates, are the most diverse phylum on earth, with nearly 1 million species of named insects alone. Molluscs are protostome invertebrates that do not molt and include the snails and slugs, clams and oysters, and squids and octopuses. The concept of intelligence cannot be applied objectively to compare species, which have evolved in response to the selective forces relevant only to their own particular environments.

5 Worms, jellies, and sponges are examples of invertebrate diversity.

Worms fall into several different phyla. All are bilaterally symmetrical protostomes with defined tissues. The segmented worms (annelids) and flatworms do not molt; the roundworms do. Jellyfish, sea anemones, and corals are radially symmetrical animals with defined tissues, in the cnidarian phylum. All cnidarians are carnivores and use specialized stinging cells located in tentacles to capture prey. Sponges, lacking specialized tissues, organs, and symmetry, are the simplest lineage of animals.

KEY TERMS

adult, p. 430
amniote, p. 419
amphibian, p. 419
animal, p. 408
arthropod, p. 428
bilateral symmetry, p. 409
bivalve mollusc, p. 433
cartilaginous fishes, p. 416
cephalopod, p. 433

deuterostome, p. 410
dorsal hollow nerve cord, p. 412
ectotherm, p. 420
endotherm, p. 420
exoskeleton, p. 423
flatworm, p. 436
gastropod, p. 433
hair, p. 421
larva, p. 430

lobe-finned fishes, p. 416
mammary glands, p. 421
marsupial, p. 422
metamorphosis, p. 430
monotreme, p. 422
notochord, p. 412
pharyngeal slits, p. 412
placental, p. 422

post-anal tail, p. 412
protostome, p. 410
pupa, p. 430
radial symmetry, p. 409
ray-finned fish, p. 416
roundworm, p. 436
segmented worm, p. 436
sessile, p. 408
tetrapod, p. 418
viviparity, p. 422

Check Your Knowledge

1. Which of the following is not a characteristic of all animals?
   a) They are able to move at some point in their life.
   b) They get their energy by eating other organisms.
   c) They are multicellular.
   d) They are sexually reproducing at some point in their life cycle.
   e) All of the above are characteristics of all animals.

2. The lineage that first separated from the common ancestor of all animals, and retains many of those primitive features to this day, includes which of the following modern organisms?
   a) the grasshopper   d) the earthworm
   b) the jellyfish   e) the sea turtle
   c) the sponge

3. Which of the following extant species is the most evolutionarily successful?
   a) the house mouse (*Mus domesticus*)
   b) the housefly (*Musca domestica*)
   c) the Western gorilla (*Gorilla gorilla*)
   d) irukandji, the most toxic box jellyfish (*Carukia barnesi*)
   e) All of the above species are equally successful because they are not extinct.

4. Which of the following are chordates?
   a) fish
   b) humans
   c) frogs
   d) All of the above are chordates.
   e) Only a) and c) are chordates.

5. The two most important evolutionary innovations in vertebrates, which resulted in their eventual domination among the large animals, were:
   a) air-breathing lungs and a more efficient heart.
   b) opposable thumbs.
   c) jaws and amniotic eggs.
   d) the ability to walk or fly.
   e) air-breathing lungs and endothermy.

6. The molluscan mantle is used primarily for:
   a) feeding.
   b) gas exchange.
   c) producing the shell.
   d) excretion.
   e) reproduction.

7. Which of these animals is a tetrapod that does not produce amniotic eggs?
   a) salamander
   b) human
   c) monkey
   d) elephant
   e) python

8. Why is the amniotic egg considered a key evolutionary innovation?
   a) It prohibits external fertilization, thereby facilitating the evolutionary innovation of internal fertilization.
   b) It has an unbreakable shell.
   c) It greatly increases the likelihood of survival of the eggs in a terrestrial environment.
   d) It enables eggs to float in an aquatic medium.
   e) It extends the time of embryonic development.

9. Which came first, the chicken or the egg?
   a) The chicken, because the amniotic egg did not evolve until after the first chicken appeared.
   b) The egg, because the amniotic egg evolved well before the first birds.
   c) The chicken, because during speciation the adult stage always precedes the juvenile stage.
   d) The egg, because the chicken is not a real species.
   e) It is impossible to determine, because eggs leave no fossils.

10. Which of the following is not a reptile?
    a) sparrow   d) turtle
    b) snake   e) dinosaur
    c) frog

11. Marsupials and which of the following groups combine to make a monophyletic assemblage?
    a) birds
    b) carnivores
    c) primates
    d) monotremes
    e) placental mammals

12. According to the fossil record, the first humans appeared approximately _____ years ago.
    a) 6 million
    b) 6,000
    c) 190 million
    d) 100,000
    e) 1 million

13. Which of the following insect groups has the most species currently named by scientists?
    a) bees, wasps, and ants
    b) beetles
    c) grasshoppers and crickets
    d) flies
    e) butterflies and moths

14. Which is the only animal phylum to have over one million described species?
    a) chordates
    b) arthropods
    c) flatworms
    d) nematodes
    e) molluscs

15. The phylum Arthropoda includes all of the following kinds of animals except:
    a) snails.
    b) crabs.
    c) crayfish.
    d) butterflies.
    e) scorpions.

16. Which of the following traits is unique to arthropods?
    a) a ventral nerve cord
    b) an exoskeleton
    c) segmentation
    d) a large ratio of brain size to body size
    e) wheeled appendages

17. In cnidarians, cnidocytes are primarily used for:
    a) creation of water flow across the body wall.
    b) formation of free-living medusas.
    c) secretion of digestive enzymes.
    d) prey capture and defense.
    e) muscular contraction during movement.

18. Sponges are sessile, meaning that they:
    a) reproduce asexually.
    b) are parasitic and depend on their host for a constant supply of nutrients.
    c) have exoskeletons that they must shed as they grow.
    d) live within shells they find on the ocean floor.
    e) live attached to a solid structure and do not move around.

SHORT-ANSWER QUESTIONS

1. The cheetah is an amazingly fast cat that hunts its prey on the African savannas, sprinting for short distances at speeds approaching 70 miles per hour. The sloth in the Amazon rain forests of South America is a very slow creature that hangs from a tree by all four limbs and eats vegetation. How can it be explained, in evolutionary terms, that the cheetah can sprint at high speeds, while the sloth moves almost in slow motion? What modifications might occur to the cheetah's ability to run if it were to occupy the same niche as the sloth, assuming the cheetah has no natural enemies there?

2. Explain why an ectotherm, such as a snake, requires only an occasional meal, while an endotherm, such as a mouse, requires frequent meals to survive?

3. A caring pet owner brings her boa constrictor to the veterinarian because it has been more sluggish than normal and appears to be ill. The veterinarian is concerned when she sees that the owner has placed a thick sweater, specially knitted, on the snake. Why might the sweater worry the veterinarian?

4. You have a friend who confuses a terrestrial slug with a segmented worm. Describe some of the characteristics of the slug that you could point out to him that are not found in any segmented worms.

See Appendix for answers. For additional study questions, go to www.prep-u.com.

12

Plant and Fungi Diversification

Where did all the plants and fungi come from?

❶ Plants are just one branch of the eukarya.

The primitive land plant Psilotum nudum, *growing in Hawaii Volcanoes National Park, Hawaii.*

12·1

What makes you a plant?

An animal can move from place to place to get food or water, to avoid being eaten, or to find a mate. It's harder for a plant. It must do all the things an animal does—obtain food and water, protect itself from predators, and reproduce—*but* it can't move a centimeter in the process. Wherever a seed puts down roots, that's where the plant stays anchored for its entire life.

So plants are organisms that are fixed in place, but that's only one aspect of being a plant. A **plant** is a multicellular eukaryote that produces its own food by photosynthesis and has an embryo that develops within the protected environment of the female parent (**FIGURE 12-1**). Plants vary in size from less than 0.04 inch (1 millimeter) to 380 feet (117 meters) tall, and most plants live on land.

WHAT IS A PLANT?

PLANTS CREATE THEIR OWN FOOD
Plants carry out photosynthesis, using energy from sunlight to convert carbon dioxide and water into sugar.

PLANTS ARE SESSILE AND TERRESTRIAL
Plants are anchored in place at their bases.

PLANTS ARE MULTICELLULAR
Plants consist of multiple cells and have structures that are specialized for different functions.

FIGURE 12-1 Defining characteristics of a plant.

Plants Are a Branch of Eukarya   Mosses, Ferns, and Their Relatives   The Advent of the Seed   Flowering Plants

There are other organisms on earth that are multicellular and photosynthetic eukaryotes. These include many species of algae, for example the brown algae commonly known as seaweed, and include the closest relatives of land plants. They differ from plants, however, in that they live only in water or on very moist land surfaces. This is in sharp contrast to most plants, which can live even in deserts.

As we have said, most plants make their own food—that is, they carry out photosynthesis, using energy from sunlight to convert carbon dioxide and water to sugar, and then convert some of the sugar to starch. However, a plant can't live on carbohydrate alone. A plant needs nitrogen to build proteins, phosphorus to make ATP, and salts to create concentration gradients between the inside and outside of cells. Plants use **roots,** the part of a plant below ground, to obtain these needed substances from the soil. Above ground, plants have a **shoot** that consists of a stem and leaves. The stem is a structure that supports the main photosynthetic organ of a plant: its leaves.

When we think of plants, we often think of them as "chlorophyll-containing," but some plants have no chlorophyll—their ancestors had chlorophyll, but they have lost almost all of it over evolutionary time—and can't carry out photosynthesis. Instead, they live as parasites that steal nutrients from other plants. Dodder is an example of a parasitic plant that almost completely lacks chlorophyll and gets its sugar from the host plant it grows on (**FIGURE 12-2**). You can probably find dodder growing on plants in a vacant lot or at a roadside near where you live.

Plants that carry out photosynthesis need sunlight, and that creates a challenge for some plants. Think about plants growing

Dodder has evolved to perform reduced levels of photosynthesis. As a result, it lives as a parasite and steals nutrients from the plant on which it grows.

FIGURE 12-2 **Dodder: a parasitic, non-photosynthetic plant.**

Q If a plant growing in the shade can't move to a new, sunnier location, how can it reach the available sunlight?

on the floor of a forest. It's mostly shady under the tree canopy, with just a few sunny spots where light penetrates the tree branches overhead. If a seed sprouts (germinates) in the shade, the young plant needs to somehow get to the nearest sunny spot. Plants can do that, of course, but not by moving. Instead, they grow toward the light (**FIGURE 12-3**). You may have seen a plant called "lucky bamboo" (it's not really a bamboo) for sale in a grocery or drug store. It catches your eye because it grows in spirals or loops. The growers have produced these shapes by moving the light source every few days, forcing the plant to bend to follow the light. If you buy a lucky bamboo, you must continue rotating the light source when you get it home or it will start to grow straight, like any other plant.

STATIONARY LIVING CREATES CHALLENGES

OBTAINING FOOD
Because plants can't move to reach sunlight, they bend in place and grow toward light.

Diploid ferns   Haploid fern

FINDING A MATE
Male and female plants can't meet to reproduce, so they have developed ways of getting the male gamete to the female gamete, including alternating haploid and diploid life stages, and using other organisms to transport the gametes.

RESISTING PREDATION
Plants can't run from predators, so they have developed adaptations such as thorns to defend themselves.

FIGURE 12-3 **Plants must overcome the constraint of being immobile.**

FIGURE 12-4 Phylogeny of the plants.

Sex can also be a challenge for an organism that is anchored in place by its roots. Not surprisingly, plants have a sex life that is very different from that of animals. Consider humans as a typical animal. From the moment an egg is fertilized by a sperm, humans have the diploid number of chromosomes—that is, two sets of chromosomes, one set from each parent. Only the egg and sperm are haploid—an egg or a sperm has only one set of chromosomes. Humans and other animals have no haploid stage in their lives (other than their gametes), but many plants do. The haploid stage is the one that allows the male and female plants to reproduce, even though they will never meet because they can't move. Some plants enlist the help of animals to carry the male gamete to the female gamete.

Resisting predators is another challenge for organisms that cannot move. Running away is how most animals react to a predator, but plants have found other ways to defend themselves. Everyone is familiar with thorns, which are an anatomical defense plants use to avoid predation (see Figure 12-3). Plants also use chemicals to deter predators, as many people learn when they develop an itchy rash from poison ivy.

The earliest land plants were the first multicellular organisms to live on land; these simple, non-vascular plants had no

vessels to transport water and nutrients (**FIGURE 12-4**). The subsequent evolution of land plants was a series of radiations of forms with characteristics that made them increasingly independent of water. The evolutionary tree of plants shows these stages clearly: first the development of vessels to conduct water from the soil through the plant (the vascular plants), then seeds that provide nutrients to get the next generation off to a good start (the gymnosperms), and finally flowers that allow plants to entice or trick insects and birds to spread the plant's male gametes (the angiosperms). We'll look at each of these plant adaptations in this chapter. We also explore the ecologically important kingdom of fungi. Although fungi are not plants, they are very closely associated with plants.

TAKE-HOME MESSAGE 12·1

Plants are multicellular organisms that spend most of their lives anchored in one place by their roots. The inability of plants to move shapes the way they obtain food, reproduce, and protect themselves from predation. The evolutionary history of plants reveals the progressive appearance of characteristics that allow plants to succeed on land despite their inability to move.

Plants Are a Branch of Eukarya   Mosses, Ferns, and Their Relatives   The Advent of the Seed   Flowering Plants

❷ The first plants had neither roots nor seeds: mosses, ferns, and their relatives.

Close view of a filmy fern in Westland National Park, New Zealand.

12•2

Colonizing land brings new opportunities and new challenges for plants.

The aquatic ancestors of land plants were green algae. Like plants, green algae are multicellular, photosynthetic eukaryotes, but green algae live only in water or on very moist land surfaces. As water dwellers, green algae do not require specialized structures to obtain water and nutrients; water simply enters their cells by osmosis, and the nutrients they require are in solution in the water that surrounds them. Some green algae, such as sea lettuce and stoneworts, even look like plants—and there is enough uncertainty among taxonomists that some classify the green algae within the plant kingdom. Others, however, classify green algae within the protist kingdom and consider some green algae that look like slime on rocks—organisms called coleochaetes (pronounced KOH-lee-oh-keets) to be the closest relatives of plants (FIGURE 12-5). An individual coleochaete is about the size of a pinhead and is only one-cell thick. Coleochaetes can withstand exposure to air, so they survive when the water level in a lake falls and leaves them high and dry. This resistance to drying was the first evolutionary step that plants took as they moved from water to land.

Coleochaetes, a type of green algae, are the closest relative to plants. They can withstand exposure to air when water levels fall.

💡 *Ancestors of plants began the transition from water to land with the evolution of resistance to drying.*

FIGURE 12-5 **Plants' closest relatives: a green alga called coleochaete.**

Plant and Animal Relationships   Fungi

When plants emerged onto land, they faced the same two challenges that terrestrial animals faced 25 million years later.

1 **PROBLEM:** GRAVITY
SOLUTION: The earliest plants grew very close to the ground, as mosses do today, in order to resist the pull of gravity.

2 **PROBLEM:** DESICCATION
SOLUTION: Plants developed an outer waxy layer called a cuticle that covers their entire surface.

FIGURE 12-6 **Leaving the water.** With the transition to land, plants had to develop ways to resist the pull of gravity and—a more immediate need—to resist drying out.

The first land plants appeared about 475 million years ago. They weren't impressive-looking, just some patches of low-growing green stems at the water's edge. They did not have any of the structures we associate with plants today—no roots, leaves, or flowers. But from an evolutionary perspective, those early plants were enormously important because, until terrestrial plants evolved, there was nothing on land for other land organisms to eat. Thus, the first land plants not only set the stage for the tremendous diversity of plant life that we know today, but also paved the way for the evolution and diversification of land animals.

When plants emerged onto land, they faced the same two challenges that were to confront the first terrestrial animals, some 25 million years later: supporting themselves against the pull of gravity and reducing evaporation so they didn't dry

out. The second of these problems was the more urgent. The earliest plants did not have to grow upward—they could creep along the ground—but they *did* have to avoid drying out (**FIGURE 12-6**). The material that protects land plants from drying is a waxy layer called the **cuticle** that covers the leaves and stem of the plant. The cuticle is what makes leaves shiny, and all terrestrial plants have a cuticle.

TAKE-HOME MESSAGE 12·2

The first land plants were small, had no leaves, roots, or flowers, and could grow only at the water's edge. Nonetheless, these inconspicuous little plants set the stage for the enormous diversity of terrestrial plants and animals on earth today.

12·3

Mosses and other non-vascular plants lack vessels for transporting nutrients and water.

The earliest land plants were low-growing for a reason—they had no structures that could transport water and nutrients from the soil upward into the plant. The only way that these substances could move was by diffusing from one cell into an adjacent cell, and then into the next cell, and so on. Diffusion

is a slow process, and plants that rely on diffusion can grow only a few centimeters tall.

Despite the limitations of diffusion, three groups of plants, all known as **bryophytes,** still use diffusion to move substances,

Plants Are a Branch of Eukarya   Mosses, Ferns, and Their Relatives   The Advent of the Seed   Flowering Plants

rather than having any sort of "circulatory system." They are the liverworts, hornworts, and mosses. Liverworts and hornworts are small (less than an inch in height), simple plants that grow in moist and shady places and resemble flattened moss. Mosses are the plants you are most likely to be familiar with, and more than 12,000 species of mosses grow in habitats extending from arctic and alpine regions to the tropics. These bryophyte plants are referred to as **non-vascular** because they do not have vessels to transport water and food. Water and nutrients are absorbed by projections from the outermost layer of cells that penetrate a few micrometers into the soil. Because these projections are so short, non-vascular plants must either live in places where the soil is always moist or become dormant when the soil surface dries out (**FIGURE 12-7**).

THE NON-VASCULAR PLANTS

COMMON CHARACTERISTICS
• Distribute water and nutrients throughout plant by diffusion
• Release haploid spores, which grow and produce gametes
• Life cycle with multicellular haploid and diploid phases

MEMBERS INCLUDE
• Mosses (about 12,000 species)
• Liverworts (about 8,000 species)
• Hornworts (about 100 species)

FIGURE 12-7 Overview of the non-vascular plants. The plants shown here (*from left, clockwise*) are moss, liverwort, and hornwort.

In order to adapt to land, the non-vascular plants had to develop a method of reproduction that protected the plant embryo from drying out and provided it with a source of nutrients. The innovation that made this possible was a life cycle of alternating haploid and diploid generations, in which the diploid embryo is protected by the haploid female parent. This "alternation of generations" in plants is radically different from the life cycle of human and most other animals. In animals, a diploid organism (the organism you see) produces haploid gametes (which are single-celled and remain that way, never visible to the naked eye). The haploid gametes, at fertilization, produce a new, diploid cell that becomes a multicellular embryo, eventually growing until we can see it. And then, once we have the adult animal, the process starts all over again.

The life cycle of a moss provides a useful example of how the alternation of generations works. When you look at a spongy mass of moss (or any other non-vascular plant), what you see is the *haploid* part of the life cycle (**FIGURE 12-8**). The adult moss plants are haploid plants—that is, all the cells have only one set of chromosomes. There are male and female moss plants that have male and female reproductive structures located among the feathery leaves at the tips of the stems. Water collects here during rainstorms, allowing the sperm to "swim" (or be splashed) from the male structures to fertilize eggs in the female structures. Once the egg is fertilized in the female structure, a diploid zygote is formed, and it divides to become an embryo. The embryo is sheltered within the female reproductive structure while it continues to divide by mitosis and to mature. The female reproductive structure provides water and nutrients for the growing embryo, and both the embryo and the reproductive structure enlarge. Eventually, the female structure elongates to such a degree that it breaks in two and forms a capsule extending over the top of the plant like a raised fist. Inside the capsule, haploid spores are formed by meiosis. A **spore** is a single cell, containing DNA, RNA, and a few proteins. When a capsule ruptures, it releases hundreds of spores. The spores that land in moist, sheltered spots grow into new (haploid) male or female moss plants.

Some non-vascular plants are economically or ecologically important. "Peat moss" is the name given to *Sphagnum* moss when it is harvested and sold as a soil enhancer for gardening. Peat bogs are found in many parts of the world and consist of partly decayed moss. Peat harvested from bogs is dried and burned as fuel in areas where trees are scarce, and peat bogs—such as those on the coasts of Malaysia and Indonesia—are important in flood control, because they can absorb enormous amounts of water during the monsoon season and release it over a period of months. Another commercial application: mounds of burning peat are used to dry the barley used to

MOSS LIFE CYCLE

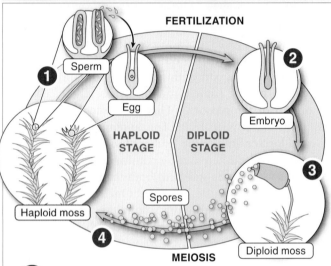

FERTILIZATION

Sperm

Egg

HAPLOID STAGE

DIPLOID STAGE

Embryo

Spores

Haploid moss

Diploid moss

MEIOSIS

1 During rainstorms, sperm swim from male reproductive structures to female reproductive structures, where they fertilize the egg.

2 A diploid embryo forms and develops into an adult diploid moss.

3 The diploid moss develops a capsule, which bursts and releases haploid spores.

4 A spore lands on moist soil and grows into an adult haploid moss.

FIGURE 12-8 **Alternation of generations in mosses.** The spongy part of the moss, at ground level in the photograph, is the haploid form. Hovering above, on stalks, are the spore capsules, the diploid form.

produce Scotch whisky. This gives Scotch its distinctive smoky taste (**FIGURE 12-9**).

Unlikely as it seems, non-vascular plants can grow in deserts, and mosses are important as components of the biological crust that holds desert soils in place. This crust is composed of mosses, lichens, and cyanobacteria. It cements the soil particles together and allows the soil to resist wind erosion, but the crust is

Non-vascular plants such as peat moss have many important economic and ecological uses, from flood control to gardening to the production of Scotch whiskey (shown here).

FIGURE 12-9 **Uses of peat.** Peat can be dried and then burned as fuel to heat homes. In traditional Scotch whisky production, the burning of peat during the distillation process provides the smoky flavor.

extremely fragile and very slow to regenerate. When people or cattle walk over the crust, they break it into small pieces that cannot resist wind, and wind erosion can then strip a meter or more of soil from the unprotected ground surface in a few decades, leaving tree trunks supported in mid-air by their roots.

TAKE-HOME MESSAGE 12·3

Non-vascular plants—mosses, liverworts, and hornworts—have scarcely evolved beyond the stage of the earliest land plants. They lack roots and vessels to move water and nutrients from the soil into the plant, and they reproduce with spores that form when a sperm from a male reproductive structure "swims" through a drop of rainwater to fertilize the egg in a female reproductive structure.

The evolution of vascular tissue made large plants possible.

What a difference some tubes make. That's all vascular tissue is—a sort of infrastructure of tubes that begins in the roots and extends up the stem of a plant and out to the farthest tips of its leaves. The evolution of vascular tissue allowed early land plants to transport water and nutrients faster and more

effectively than the cell-to-cell diffusion that non-vascular plants must rely on, and their roots penetrate far enough into the soil to reach moisture even when the soil surface is dry. Also, roots that reach deep into the soil provide the support that a plant needs to grow upward without falling over. As a consequence, the **vascular plants** can grow taller than non-vascular plants and are more successful in areas where the surface of the ground dries out between rainstorms (**FIGURE 12-10**).

Ferns are the most familiar of the primitive vascular plants. During the Carboniferous period, which extended from 360 to 300 million years ago, ferns were a major component of the huge swamp forests that eventually formed the coal deposits we are mining now. A related group of plants, the horsetails, like their Carboniferous ancestors, grow in wet habitats where the oxygen concentration around the roots is low because the spaces between soil particles are filled with water (rather than air). To adapt to this wet environment, horsetails developed hollow stems that allow oxygen from the air to diffuse down to the roots. The thin leaves of horsetails have a single nutrient- and water-carrying vessel extending from the base to the tip.

THE VASCULAR SEEDLESS PLANTS

COMMON CHARACTERISTICS
• Distribute water and nutrients throughout plant with a "circulatory system" of vascular tissue
• Release haploid spores, dispersed by the wind, which grow and produce gametes
• Life cycle (unlike in animals) with free-living, multicellular haploid and diploid phases

MEMBERS INCLUDE
• Ferns (about 12,000 species)
• Horsetails (about 15 species)

FIGURE 12-10 Snapshot of the vascular plants: ferns and horsetails. These are the groups of vascular plants that don't produce seeds. The plants shown here (*from left, clockwise*) are Christmas fern, horsetails, and ferns in a forest.

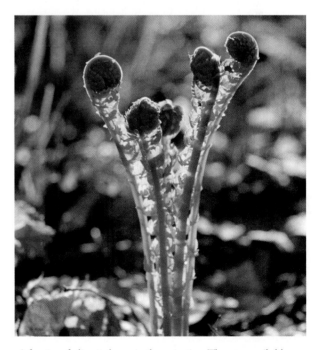

A fern is unfurling its leaves in the springtime. The young coiled leaves are called fiddleheads.

FERN LIFE CYCLE

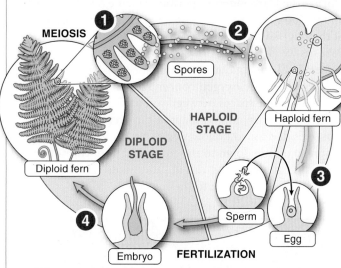

MEIOSIS

Spores

HAPLOID STAGE

Haploid fern

DIPLOID STAGE

Diploid fern

Sperm

Egg

Embryo   **FERTILIZATION**

1 Adult ferns release haploid spores, which are carried by the wind to a new location.

2 A spore lands on moist soil and grows into a haploid fern called a prothallus.

3 During rainstorms, sperm swim from male reproductive structures to female reproductive structures, where they fertilize the egg.

4 A diploid embryo forms and continues to grow into an adult diploid fern.

FIGURE 12-11 Haploid and diploid life stages in ferns. The photograph shows the sporangia, the spore-producing bodies, located on the underside of most fern leaves—on the diploid plant. The diploid and haploid plants in the fern life cycle are very different in appearance.

Simply having vessels, as horsetails do, is an important evolutionary innovation, but ferns took an additional step. In addition to the central vessel in each leaflet, the leaves of ferns have vessels branching from the central vessel to the edges of the leaflet. This arrangement places a channel for the movement of water and nutrients close to each cell in the leaf.

Like non-vascular plants, horsetails and most ferns reproduce with spores. At the top of the central hollow stem of the horsetail is a conical structure where the haploid spores are produced and released. In many ferns, the structures in which the spores are produced are called **sporangia** (*sing.* sporangium), and these are located on the underside of the leaves. Some species have sporangia that branch from the fronds or grow from the base of the plant. Because horsetails and ferns are taller than non-vascular plants, the spores can be blown by the wind when they are released, and they may settle some distance from the parent plant. This increased dispersal ability was an important adaptation—the non-vascular plants are so low-growing that wind can't play much of a role in moving their spores.

A spore that lands on moist soil grows into a tiny heart-shaped structure called a **prothallus,** which is the free-living haploid life stage of a fern (**FIGURE 12-11**). The prothallus produces the haploid gametes: some cells produce eggs (with one set of chromosomes) and others produce sperm (also with one set of chromosomes). Sperm "swim" through drops of rainwater to fertilize an egg, and the fertilized egg (a diploid zygote with two sets of chromosomes) grows into an adult fern.

TAKE-HOME MESSAGE 12·4

Like a circulatory system, vessels are an effective way to carry water and nutrients up from the soil to the leaves, and the first vascular plants—including the earliest ferns and horsetails—were able to grow much taller than their non-vascular predecessors.

Plants Are a Branch of Eukarya   Mosses, Ferns, and Their Relatives   The Advent of the Seed   Flowering Plants

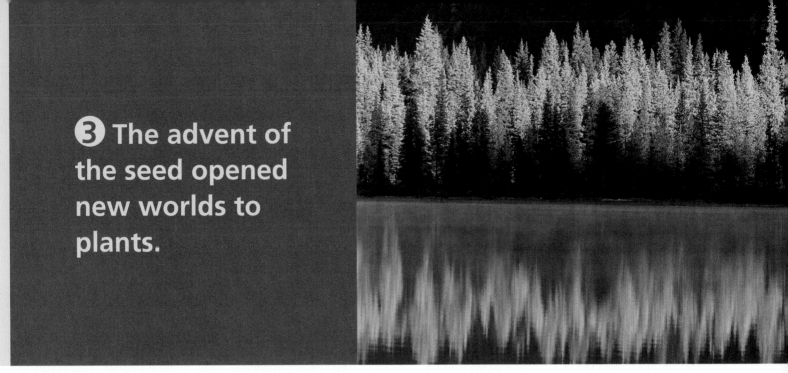

3 The advent of the seed opened new worlds to plants.

Forests surrounding Mistaya Lake in Banff National Park, Alberta, Canada.

12·5

What is a seed?

As we've seen, the evolutionary development of vascular tissue allowed plants to grow large. Plants could now colonize areas where the soil at the surface of the ground was not always wet, because their roots could penetrate the soil to reach water and nutrients. The next big innovation in plant evolution was the **seed,** an embryonic plant with its own

supply of water and nutrients encased within a protective coating (**FIGURE 12-12**).

Seeds are goodbye gifts that adult plants give to their offspring—they contain the materials needed to begin a new life. Unlike spores, which are single cells that contain only DNA, RNA, and

---

SEEDS: STRUCTURE AND GROWTH

STRUCTURE
Fertilization produces a diploid seed, which contains a multicellular embryo and a store of carbohydrate (endosperm) to fuel its initial growth.

Embryo
Protective coating
Endosperm

GROWTH
A seedling draws energy from the endosperm while it extends its leaves upward to begin photosynthesis and its roots downward into the soil to reach water and nutrients.

FIGURE 12-12   What is a seed? A seed is a package that contains a multicellular embryo and a store of carbohydrate, the endosperm, to fuel its initial growth. (This makes many seeds desirable as food sources—shown here are seeds for sale at a market in Jordan.)

a few proteins, seeds contain both a multicellular embryo and a store of carbohydrate, called **endosperm,** that can fuel the seed's initial growth. A seedling draws energy from the endosperm while it extends its leaves upward to begin photosynthesis and its roots downward into the soil to reach water and nutrients. There are two modern groups of seed-producing plants: **gymnosperms** (including pines, firs, and redwoods) and **angiosperms** (all of the flowering plants and trees).

Q How are seeds formed?

Seed plants have a very small life stage that produces haploid gametes (sperm and egg). This is analogous to an individual human, which produces haploid gametes. Botanists call this haploid plant form the **gametophyte.** **Pollen grains** and **ovules** are the male and female gametophytes, respectively, of seed plants. In brief (we expand on this later in the chapter), a haploid female gamete (egg) forms inside the ovule. When a pollen grain lands near the ovule, it produces a pollen tube that grows into the ovule. Sperm from the pollen grain move through the pollen tube into the ovule and fertilize the egg. The external layer of the ovule then forms the seed coat.

Another of the challenges that plants face is dispersing their seeds. The seed stage is the only opportunity most plants have to send their offspring away from home, and seeds and seed pods have many ways to do this. These range from the forceful send-off of exploding seed pods, to seeds that hitch a ride in or on passing animals, to those that are so small and light that they can float in water or almost fly (think of dandelions)—not to mention the use of fruits in seed dispersal. As we will see later in the chapter, many methods for dispersing seeds have evolved.

TAKE-HOME MESSAGE 12·5

Seeds are the way that plants give their offspring a good start in life and get them to leave home. A seed contains a multicellular embryo plus a store of carbohydrate and other nutrients. Seeds are distributed by wind, by attaching themselves to animals, or by floating through air or water.

12·6

With the evolution of the seed, gymnosperms became the dominant plants on earth.

Gymnosperms include four major groups: the conifers, cycads, gnetophytes, and ginkgo (just one species) (**FIGURE 12-13**). All of the 900 or more species of gymnosperms are seed-bearing plants that produce ovules on the edge of a

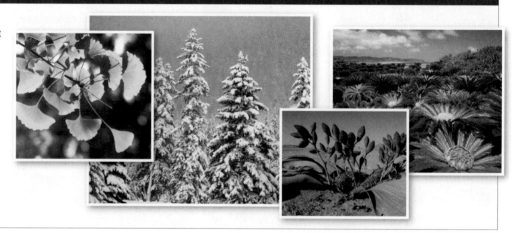

THE GYMNOSPERMS

COMMON CHARACTERISTICS
- Distribute water and nutrients throughout plant with a "circulatory system" of vascular tissue
- Have needle-like leaves
- Reproductive structures called cones produce the gametes
- Fertilization produces seeds

MEMBERS INCLUDE
- Conifers (about 600 species)
- Cycads (about 300 species)
- Gnetophytes (about 65 species)
- Ginkgos (1 species)

FIGURE 12-13 Overview of the gymnosperms: non-flowering plants with seeds. Shown (*from left*) are ginkgo, Douglas fir, welwitschia, and cycad.

Plants Are a Branch of Eukarya | Mosses, Ferns, and Their Relatives | ▼ The Advent of the Seed | Flowering Plants

GROUPS OF GYMNOSPERMS

CONIFERS
- Most commonly found in colder temperate and sometimes drier regions of the world
- Important source of timber
- Include pines, spruces, firs, cedars, hemlocks, yews, larches, cypresses

CYCADS
- Slow-growing gymnosperms of tropical and subtropical regions
- Most resemble palm trees
- Several species are facing extinction in the wild.

GNETOPHYTES
- Composed of 3 groups: *Gnetum, Ephedra,* and *Welwitschia*
- Most gnetophyte species are *Ephedra,* a shrub-like plant sometimes used as a herbal remedy for respiratory ailments.

GINKGO
- *Ginkgo biloba* is the only remaining species.
- Distinctive fan-shaped leaves
- The outer covering of the seeds emits a foul odor.

FIGURE 12-14 Major groups of gymnosperms: conifers, cycads, gnetophytes, and ginkgos.

cone-like structure. If you were a dinosaur living in the middle of the Mesozoic era, 160 million years ago, the forests you walked through would consist entirely of gymnosperms—pine trees, redwoods, cycads, and their relatives. Flowering plants (angiosperms) would not appear for another 35 million years. Indeed, when it comes to population size and the range of habitats in which they can thrive, pine trees and their relatives are among the most evolutionarily successful groups of plants on earth, growing on every ice-free continent and extending from sea level to the tree line on mountains (**FIGURE 12-14**).

Pines, spruces, firs, redwoods and their relatives—collectively, the conifers—are the most familiar gymnosperms, at least to residents of the temperate regions, and they all have needle-like leaves—but that is not true of all gymnosperms. The ginkgo has distinctive fan-shaped leaves that are nearly identical in size and shape to fossils of ginkgo leaves (see Figure 12-14). Cycads have palm-like fronds with many small leaflets, and these, too, are nearly identical to fossils from the Mesozoic.

Q *Why does tree pollen coat the windshield of cars parked outside in the spring?*

The reproductive structures of gymnosperms—the cones— are male or female. The pine cones you are familiar with are the female cones, and these produce the ovules and, eventually, the seeds (**FIGURE 12-15**). The male cones are smaller and release pollen that is blown

CONES

MALE CONES
The male cone releases pollen grains that require wind to reach a female cone.

FEMALE CONE
The female cone has ovules on the protruding scales. They produce seeds when fertilized by pollen.

 Wind dispersal of pollen is highly inefficient. For each grain that fertilizes an egg, billions of pollen grains are wasted. Nonetheless, pines have been using wind pollination successfully for more than 200 million years.

FIGURE 12-15 Cones are the reproductive structures of gymnosperms.

Plant and Animal Relationships   Fungi

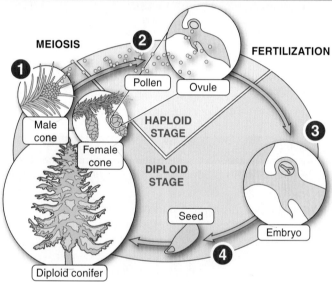

1 Male cones release pollen grains that are dispersed by the wind to ovules found beneath the scales of female cones.

2 Pollen grains release sperm that fertilize an egg within the ovule.

3 Fertilization creates a diploid embryo that matures into a seed.

4 Eventually, the seed is released from the female cone, and grows into an adult tree.

The evolution of seeds by gymnosperms eliminates the free-living haploid life stage seen in mosses and ferns.

FIGURE 12-16 Assisted by the wind: the life cycle of the gymnosperms.

by the wind, and some of it reaches the ovules, which lie beneath the protruding scales of the female cones. The quantity of pollen released by conifers is beyond imagination—the air in a pine forest becomes hazy and every surface becomes coated with the yellow pollen. Wind dispersal is clearly an inefficient method of **pollination** (getting the pollen to the vicinity of the ovule): billions of pollen grains are wasted for each grain that lands on a female cone and produces sperm to fertilize an egg in an ovule. Nonetheless, this "brute force" method of ensuring fertilization works: pines have been using wind pollination successfully for more than 200 million years.

When the pollen arrives at the female cone, a pollen tube forms and transports the haploid sperm to the ovule, where fertilization occurs and a diploid embryo begins to grow. The embryo develops slowly within the female cone over the course of many months, until the scales of the cone open and release the seed, ready to sprout into a new plant (**FIGURE 12-16**). The evolution of seeds in gymnosperms almost completely eliminated the haploid life stage that we see in mosses and ferns. Non-vascular plants spend most of their lives in the haploid stage of their cycle, and ferns produce spores that grow into free-living haploid plants. With the appearance of gymnosperms, plants developed a life history with no free-living haploid stage.

This evolutionary change in life history probably reflects the different degrees of evolutionary fitness of haploid versus diploid organisms. A haploid organism has just one copy of each chromosome, which means it has just one copy of each allele. If the haploid organism carries a defective allele for a critically important gene, the organism is doomed because it has only that one, defective version of the gene. In contrast, diploid organisms have two sets of chromosomes and thus two copies of each allele. If one copy is defective, the second copy can function as a backup, enabling the plant to produce the gene product. As a result, mutations are much less likely to be lethal for diploid organisms than for haploid organisms.

TAKE-HOME MESSAGE 12·6

Gymnosperms (pine trees and their relatives) were the earliest plants to produce seeds, and this mode of reproduction offers advantages over the spores of earlier plants. Seeds gave gymnosperms the boost they needed to become the dominant plants of the early and middle Mesozoic era. Gymnosperms depend on wind to carry their pollen—not an efficient method of dispersal. Conifers protect the developing seeds in the female cone.

Plants Are a Branch of Eukarya   Mosses, Ferns, and Their Relatives   The Advent of the Seed   Flowering Plants

Conifers include the tallest and longest-living trees.

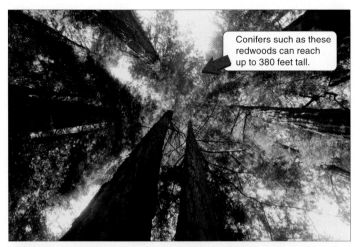

Conifers such as these redwoods can reach up to 380 feet tall.

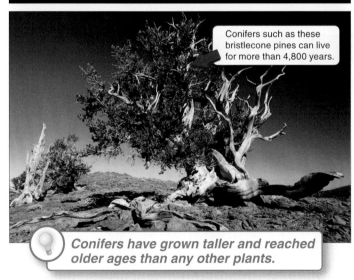

Conifers such as these bristlecone pines can live for more than 4,800 years.

Conifers have grown taller and reached older ages than any other plants.

FIGURE 12-17 Towering and ancient gymnosperms.

In addition to being abundant, diverse, and widespread, the cone-bearing trees—specifically, the conifers—include both the tallest and the oldest living organisms on earth. The four tallest trees in the world are conifers: a coast redwood that is 380 feet (113 meters) tall, a Douglas fir and a Sitka spruce, each 318 feet (97 meters), and a Sierra redwood at 311 feet (95 meters). The oldest tree trunk belongs to a Great Basin bristlecone pine (**FIGURE 12-17**). The current record holder, the Methuselah tree, has lived for more than 4,800 years, and a slightly older tree was cut down in 1964. There's more to a tree than its trunk, though, and trunks can be replaced. That is what a Norway spruce in Sweden has been doing: the current trunk is about 600 years old, but the roots are 9,550 years old. This tree has persisted by sending up a new trunk each time the old one dies. But not all conifers are big; there also are miniature species of conifers, such as the shore pine, which can be just 20 centimeters tall.

Trees can grow large and live to great ages because woody plants can be exceptionally strong and resistant to attack by herbivores. A cross section of a tree trunk shows the structural characteristics that allow a tree to stand erect and transport water to leaves more than 100 meters above the ground. Rigidity is provided primarily by the heartwood, which is a core of dead tissue that contains complexly cross-linked molecules.

Bark covers the outside of the tree trunk and branches, and the outer layers of bark are dead tissue that can be shed without damage to the tree, protecting the living tissue from attack by plant-eating insects. Part of the success of conifers can also be traced to their ability to defend themselves by exuding a sticky pine pitch that can engulf and smother insects.

TAKE-HOME MESSAGE 12•7

Conifers are the success stories among gymnosperms, with more species and a larger geographic range than all of their relatives combined. Rigidity, an exterior layer of bark, and the ability to exude sticky pitch protect conifers, allowing them to grow taller and reach older ages than any other plants.

Plant and Animal Relationships   Fungi

4 Flowering plants are the most diverse and successful plants.

Bleeding heart flowers, surrounded by neat, heart-shaped leaves.

12·8

Angiosperms are the dominant plants today.

The appearance of flowering plants (angiosperms) about 135 million years ago set the stage for the botanical world we know today, with flowering trees, flowering bushes, and all the grasses and herbaceous (non-woody) plants we see around us. The vast majority of plants on earth are flowering plants in the angiosperm group. Many of the early flowering plants would look familiar to us, and angiosperms dominate the plant world now, with some 250,000 species, compared with approximately 1,000 species of gymnosperms (**FIGURE 12-18**).

THE ANGIOSPERMS

COMMON CHARACTERISTICS
• Distribute water and nutrients throughout plant with a "circulatory system" of vascular tissue
• Produce flowers, which produce gametes
• Seeds are enclosed within an ovule

MEMBERS INCLUDE
• Flowering trees, bushes, herbs, and grasses (about 250,000 species)

FIGURE 12-18 Snapshot of the angiosperms. The angiosperms shown here (*from left*) are water lily, apple tree, Gewürztraminer grapes, and rhododendron in a forest.

Plants Are a Branch of Eukarya | Mosses, Ferns, and Their Relatives | The Advent of the Seed | Flowering Plants

Flowers come in a bewildering variety of sizes, shapes, and colors, but they all have similar structures: a supporting stem with modified leaves—the flashy petals and the sepals, which are a (usually) green wrapping that encloses the flower while it is in bud. Most plants combine the male and female reproductive structures in the same flower. The male structure is the **stamen** and includes the **anther,** which produces the pollen, and its supporting stalk, the **filament.** The female reproductive structure is the **carpel.** It has an enclosed chamber, called the **ovary,** at its base, which contains one or more ovules in which eggs develop; a stalk (the style), extending from the ovary; and a sticky tip (the stigma) (**FIGURE 12-19**).

Pollination, in angiosperms, is the transfer of pollen from the male reproductive structures to the female reproductive structures of a flower, sometimes the same flower, and sometimes a flower on a different individual. Pollination is a multi-step process in which a pollen grain sticks to the stigma, forms a tube that grows until it reaches the ovule, and thus provides a route for sperm to travel down the tube to fertilize the egg. In the next section, we explore the tremendously rich and varied methods by which plants get the male gametes to the female gametes.

Although most flowers contain both male and female structures, several thousand species of plants produce flowers

The lilies in this photo clearly show the carpel, the female reproductive structure in the center of the flower, surrounded by the male stamens, each topped with a brown anther.

that are only male or only female. Sometimes, male and female flowers are borne on different individual plants, but in quite a few plants, the same plant has some male flowers and some female flowers, often side by side. Maize (corn) is a familiar plant with male and female flowers in different places on the same plant, although you might not recognize the reproductive structures as flowers. The tassel at the top of the plant is the male flower, and the ear is formed by female flowers.

TAKE-HOME MESSAGE 12·8

Flowering plants appeared in the Cretaceous period, about 100 million years ago, and diversified rapidly to become the dominant plants in the modern world. A flower houses a plant's reproductive structures. Most flowers have both male and female structures, but some species have flowers with only male or only female reproductive structures.

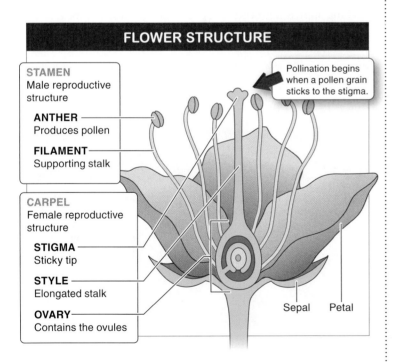

FLOWER STRUCTURE

Pollination begins when a pollen grain sticks to the stigma.

STAMEN
Male reproductive structure

ANTHER
Produces pollen

FILAMENT
Supporting stalk

CARPEL
Female reproductive structure

STIGMA
Sticky tip

STYLE
Elongated stalk

OVARY
Contains the ovules

Sepal   Petal

FIGURE 12-19 A flower houses a plant's reproductive structures.

Plant and Animal Relationships   Fungi

A flower is nothing without a pollinator.

Flowers may play only a tiny role in the sex lives of humans—did you give or receive any last Valentine's Day?—but they play the lead part in the sex lives of most plants. Here we examine how it is possible for individual plants, rooted in the ground in different locations, to bring their gametes together in sexual reproduction.

Picture a flower. What does it look like? Not everyone forms the same image in his or her mind, because flowers differ hugely from one to another, and we all have our favorites. There are giant sunflowers and colorful roses and some rotten-smelling orchids and lilies, just to name a few. The label "flower" would seem to indicate little beyond the fact that flowers are usually brightly colored, frequently have a prominent odor, and generally sit at the end of a plant's branches. Closer inspection, however, reveals that, with very few exceptions, all flowers have the same four fundamental structures introduced in Section 12-8—sepals, petals, stamens, and carpels.

Fertilization occurs when the male gamete merges with the female gamete. When it comes to plants, that's a lot easier said than done. Before fertilization can occur, the male gamete has to get to where the female gamete is. The first step toward this is for a pollen grain to make the journey to the stigma. This is the step referred to as pollination.

A small number of angiosperm species achieve pollination by simply releasing their pollen into the wind, as do the gymnosperms, or into water—on the slim chance that through luck (random movement in air or water), some of the pollen will land on the female reproductive organs of another plant. Given the really minuscule probability of any one pollen grain actually doing that, wind- and water-pollinated plants respond in the only reasonable way: they produce tremendous amounts of pollen. As we saw earlier, although this method is not particularly efficient, it works. Most angiosperms, though, have a different way of moving pollen from the anthers of one flower to the stigma of another: they use animals to carry it (**FIGURE 12-20**).

Q **Why are flowers so flashy?**

To ensure that animals will be willing to visit a flower, picking up and delivering their pollen cargo, two different and clever strategies for achieving pollination have evolved among the flowering plants.

1. *Trickery:* The plant deceives some animals into carrying its pollen from one plant to another. Among the tricksters are some orchid species that achieve pollination by producing flowers that resemble female wasps. The mimicry is so good that male wasps mount the flower and attempt to fly off and mate with it. Because the flower is attached to the plant, however, the male wasp resembles a cowboy riding a bucking bronco, twirling wildly on the flower and repeatedly whacking his head against the strategically located anthers. In the process, he doesn't end up having any reproductive success, but he does succeed in getting pollen stuck all over his head and body. That is not enough for the plant to achieve pollination—but it's a start. If that male wasp gets fooled again by a flower on another orchid, when he mounts that flower and tries to mate, he will inadvertently deposit some of the pollen from his body onto the also strategically placed stigma of that flower. In the end, the wasp looks a bit foolish and does not gain from his actions, but the orchids have used an effective system of pollination.

2. *Bribery:* The plant bribes some animals to carry its pollen from one plant to another. This more common strategy for achieving pollination is less selfish than trickery. Rather than tricking animals into carrying pollen, the plant offers something of value to the animal. For this bribery method of pollination to work, the plant must produce (a) a sticky pollen, (b) a flower that catches the attention of the pollinator, and, most important, (c) something of value to the pollinator. The payoff for the pollinator can be food, such as nutritious nectar rich in sugars

Angiosperms have developed a way to transfer pollen efficiently from the anthers of one flower to the stigma of another: get an animal to carry it!

FIGURE 12-20 **Delivering precious cargo (inadvertently).** The bee in this photo is completely covered in yellow pollen grains.

COLORS AND PATTERNS
WHITE: Nocturnal pollinators, such as moths and bats
BRIGHT: Visually oriented, diurnal pollinators, such as birds, butterflies, and bees

FLOWER STRUCTURE
TUBE: Pollinators with long tongues, such as moths
INTRICATE/CLOSED: Pollinators such as bees

ODORS
SWEET: Pollinators with a good sense of smell, such as moths, butterflies, and bees
STINKY: Pollinators, such as flies, looking for rotten meat on which to lay eggs
NO ODOR: Pollinators with a poor sense of smell, such as birds

NECTAR
ABUNDANT: Pollinators with high energy needs, such as bees, birds, and butterflies
NO NECTAR: Pollinators, such as flies, looking for a place to lay eggs, or such as beetles, looking for petals, pollen, and other parts to eat

FIGURE 12-21 Evolving together: plants and pollinators. You can often determine the type of animal that pollinates a flower just by examining the features of the flower.

and amino acids, or perhaps a safe, hospitable location for an insect to lay its eggs.

The variety of flower structures is tremendous. They differ in shape, color, smell, time of day they are open, whether or not they produce nectar, and whether or not their pollen is edible. And just as the variety of flower types is wide, so, too, is the variety of pollinators: birds (mostly hummingbirds), bees, flies, beetles, butterflies, moths, and even some mammals (mostly bats) (**FIGURE 12-21**). In each case, there has been strong coevolution between the plants and their pollinators: the plants get more and more effective at attracting the pollinators and deterring other species from visiting the flower, while the pollinators get more and more effective at exploiting the resources offered by the plants. Because of this strong coevolution between the plant and animal species, for most flowers we can determine, just by examining the features of the flower, the type of animal that will pollinate it.

Pollination is just one step toward fertilization. Fertilization itself doesn't happen until the male and female sex cells meet and fuse. This can only occur after the male gamete is taken to the female gamete by a pollen tube, which grows from the stigma, where the pollen grain lands, down into the ovary. We explore this process in the next section.

TAKE-HOME MESSAGE 12·9

Angiosperms have found a way to transfer pollen efficiently from the anthers of one flower to the stigma of another—let an animal carry it. Flowers are conspicuous structures that advertise their presence with colors, shapes, patterns, movements, and odors. Using these devices, plants are able to trick or bribe animals into transporting male gametes to female gametes, where fertilization can occur.

Plant and Animal Relationships   Fungi

Angiosperms improve seeds with double fertilization.

The process of fertilization—the fusion of two gametes to form a zygote—*begins* when a pollen grain lands on a stigma of a flower, but that is far from the end. In angiosperms, the process is called **double fertilization** because there are two separate fusions of male nuclei (each carrying a complete set of the organism's genetic material) from the pollen grain with female nuclei in an ovule. As we will see, double fertilization is a more efficient system than fertilization in gymnosperms because whenever an embryo is produced at fertilization (and only then), so, too, is a more substantial, ready-made food source.

When we left off in the previous section, a pollen grain—which will deliver the haploid male gamete—had just arrived at the stigma, in the female reproductive structure. The pollen grain forms a pollen tube that extends downward through the stigma and style to the ovary, and ultimately enters an ovule. The pollen grain forms two haploid sperm by mitosis, both of which move down the pollen tube (**FIGURE 12-22**).

While the pollen tube is growing, a haploid spore in the ovule divides mitotically to form a total of seven haploid cells, but only four of these participate in fertilization and formation of a seed. The remaining three are degenerated. One of the five cells is the female gamete (the egg), and this remains near the place where the pollen tube will be guided into the ovule by two of the other haploid cells, which aid in fertilization. The remaining haploid cells of the four participating in fertilization move into the middle of the ovule. So, when the pollen tube reaches the ovule, one of the two sperm fuses with the egg to form a zygote. The other sperm fuses with the two nuclei in the middle of the ovule to form the endosperm, which has *three sets* of chromosomes (i.e., is triploid). The process is called double fertilization because two sperm enter the ovule and combine with haploid female cells in two separate events, forming (1) a zygote (with two sets of chromosomes) and (2) an endosperm (with three sets of chromosomes).

The final steps in producing a seed occur as the diploid zygote cell undergoes multiple mitotic divisions to form an embryo, while the triploid cells multiply mitotically to produce the endosperm that will provide nutritional support for the seedling through its initial growth stages. The outer layers of the ovule form the seed coat that will protect the seed until it sprouts. The enclosure of the seed within the ovule is a distinction between angiosperms and gymnosperms. The seeds of a gymnosperm are unenclosed and are sometimes referred to as "naked."

FIGURE 12-22 Double fertilization fortifies the seeds of angiosperms.

ANGIOSPERM LIFE CYCLE

1. Male anthers release haploid pollen grains that are delivered to the stigma of another flower.

2. A pollen grain produces a tube that extends through the stigma to the ovary. Within the ovule of the ovary, a spore divides into seven haploid cells. One cell becomes the egg, another cell—with two nuclei—will form the endosperm following fertilization by a sperm cell.

3. Two sperm are released by the pollen grain. In a process called double fertilization, one sperm fuses with the egg to form a zygote, while the other sperm fuses with the two nuclei of the endosperm-forming cell.

4. The zygote and the endosperm continue to develop within the ovule, forming a seed that will eventually be released and grow into a mature plant.

What is the advantage of such a complex method of fertilization? Gymnosperms do not do this—a single sperm fuses with an egg to form a zygote in an ovule that contains hundreds of other female cells that are used as a source of nutrients for the resulting embryo.

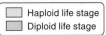

Q What advantage does double fertilization give to angiosperms?

There are two important advantages to the angiosperm method.

1. Double fertilization initiates formation of endosperm only when an egg is fertilized, whereas gymnosperms use hundreds of haploid female cells to make the endosperm *before* fertilization occurs. Waiting to be sure that an egg is fertilized is a good strategy, because making endosperm is a large energy investment for a plant. Gymnosperms invest that energy up front, and nutrients in ovules that are not fertilized are wasted because those "seeds" do not contain an embryo. In contrast, angiosperms do not waste energy forming endosperm in ovules that will not contain embryos.

2. Angiosperms can produce smaller gametes than gymnosperms, because the large energetic reserves will be produced only *after* fertilization occurs. The small size of the male and female gametes of angiosperms ensures that seeds are produced quickly. **FIGURE 12-23** shows that as different reproductive strategies have evolved in plants, their gametes have become progressively smaller. Rapid production of seeds allows angiosperms to grow as annual plants (i.e., plants that complete their life cycle from sprouting to seed production in one growing season), which is something gymnosperms cannot do.

Outbreeding, the combination of haploid cells from two different individuals, produces offspring with greater genetic diversity (i.e., carrying a greater diversity of alleles) than offspring that result from inbreeding, the combination of a male and female gamete from the same individual. Plants have a variety of ways to ensure that only sperm from another individual will fertilize the female gamete. Plants that have male and female reproductive structures in different flowers or even on different individuals increase the chance that a pollen grain landing on a stigma comes from a different plant. But many flowers with both male and female parts also have mechanisms to prevent self-fertilization. For example, the anthers may mature before the stigma, so the stigma is not ready to receive pollen when the anthers are active. And when the stigma becomes functional, the anthers are no longer producing pollen. Other plants use a molecular recognition system to prevent inbreeding: proteins on the surface of the stigma will not allow pollen from the same individual to form a pollen tube.

Q Do flowers with both male and female structures fertilize themselves?

HAPLOID AND DIPLOID LIFE STAGES

Haploid life stage
Diploid life stage

NON-VASCULAR PLANTS
The majority of the life cycle is spent in the haploid stage.

VASCULAR SEEDLESS PLANTS
The haploid and diploid stages are both multicellular and physically independent from one another.

GYMNOSPERMS
The evolution of seeds in gymnosperms almost completely eliminates the prominent haploid stage seen in mosses and ferns.

ANGIOSPERMS
Haploid gametes are further reduced in size, enabling more rapid seed production.

💡 *As plants have developed different reproductive strategies, they have progressed from having a prominent haploid stage of life to simply having haploid gametes.*

FIGURE 12-23 Overview of the haploid and diploid stages of plant life cycles.

TAKE-HOME MESSAGE 12·10

Angiosperms undergo a process of double fertilization, which ensures that a plant does not invest energy in forming endosperm for an ovule that has not been fertilized. Angiosperms have also developed methods to reduce the occurrence of self-fertilization, thereby ensuring greater genetic variation among offspring than through inbreeding.

Plant and Animal Relationships   Fungi

5 Plants and animals have a love-hate relationship.

Fruits like these peaches are structures that contain fertilized seeds.

12·11

Fleshy fruits are bribes that flowering plants pay animals to disperse their seeds.

Leaving home is an inevitable part of growing up. But as difficult as it may be, imagine you had to move away but didn't have a car or moving van—or even legs. How would you do it? This is yet another issue facing plants. Ingenious methods have evolved for transporting the male gametes to the female sex organs for fertilization, but that still results in all of the fertilized eggs, and all the new offspring they develop into, living right on the female plant—or in nearly the exact same physical location, if they manage to fall off the parent plant. This is a problem, because the new offspring must leave home. Otherwise, parent and offspring will end up competing for the same light, space, soil nutrients, and other resources.

The answer to this problem is the fruit, a structure that aids in dispersing seeds—the reproductive packets made up of the embryo, some food reserves, and a hard coat. The fruit usually develops from the ovary (and sometimes from nearby tissue), right around the seeds, which develop from the ovules. This should cause you to look at fields of flowers differently: given that the ovary is part of every flower, all of the flowers you see, after they are pollinated and fertilized, will turn into fruits. Or conversely: every fruit you eat used to be a flower.

Pea pods, sunflower seeds, corn kernels, and hazelnuts are fruits. These fruits are "dry fruits," and the seeds they contain are transported by wind or water or by sticking to an animal (**FIGURE 12-24**). Many angiosperms, though, make a **fleshy fruit** that consists of the ovary plus some additional parts of the flower. For example, the core of an apple is the ovary and the flesh of the apple is derived from adjacent parts of the flower. Blueberries, watermelons, oranges, tomatoes, and peaches are other examples of fleshy fruits.

The fleshy part of a fruit is often larger than the ovary, and a plant invests a lot of energy in producing fleshy fruit. The payoff comes when an animal eats the fruit and then defecates the seeds at a location far removed from the parent plant. In other words, fleshy fruit is the bait that some flowering plants use to get animals to disperse their seeds.

When you look at fruits, you can see several characteristics that help this system to work:

• Fruits are colorful—red is the most common color for fruits, followed by yellow and orange. All of those colors contrast with green foliage and make fruits conspicuous.

HITCHING A RIDE
Seed pods have spines or projections that attach them to passing animals.

FLYING AND FLOATING
The structure of the seed allows it to be carried away from the parent plant by wind or water.

PROVIDING A FOOD SOURCE
Fleshy fruit is a form of bait that lures an animal to eat the seed and carry it far from the parent plant before eliminating it.

FIGURE 12-24 Methods of seed dispersal.

- Fruits taste good—plants dump sugars into fruits, and their sweetness appeals to many animals.

- Fruit is good for animals—many birds incorporate the red and yellow colors of fruits into their own color patterns, and colorful male birds are appealing to female birds. Thus, male birds that eat fruit are likely to produce a lot of offspring who also eat fruit—a good deal for the plant.

Q *Can seeds still sprout after being eaten by an animal?*

Of course, there's one more requirement to make fruit a successful way for a plant to disperse its seeds: the seeds must survive passage through the animal's digestive tract and be able to germinate when they come out. That's not a problem—seeds are fully viable when they emerge. It's easy enough to test this yourself. The next time you eat fresh tomatoes, check the toilet about 24 hours later. The seeds that have passed through your digestive system are still intact. If you want to test their viability, that's easy—recover a few, rinse them off, and put them on a wet paper towel in a sealed plastic bag. If you want to do a controlled experiment, you can take some seeds directly from a tomato and put them in another bag. Put both bags in a warm, dark place, and check them in a couple of days. You'll find that both sets of seeds have sprouted. (Or, you can just take our word for it.)

Seeds not only survive being eaten by animals—some seeds will not germinate *unless* they are eaten. For some plants, the pulp that surrounds the seeds in the fruit inhibits germination and it must be digested away before the seed will sprout. The seeds of other plants have chemicals in the seed coats that must be removed by the acidity of an animal's stomach before the seeds will germinate. And there's another plus to this system: the seeds are deposited with a bit of manure that provides nutrients for the seedlings.

TAKE-HOME MESSAGE 12·11

Following pollination and fertilization, plants often use the assistance of animals to disperse their fruits, which contain the fertilized seeds, depositing them at a new location where the seedlings can grow. Fruits are made up from the ovary and, occasionally, some surrounding tissue.

Unable to escape, plants must resist predation in other ways.

The interactions of plants and animals are not simple. On the one hand, plants depend on animals to pollinate their flowers and disperse their seeds. On the other hand, many kinds of animals eat plants, and plants are vulnerable because they can't run away. If you have ever been poised above a dunking tank as your friends throw softballs at a target, hoping to dump you into the water, you know how vulnerable a plant feels.

Because plants can't run away from plant-eating animals, a host of defensive devices have evolved in plants, which fall into two categories: anatomical structures, such as thorns, and chemical compounds, including hallucinogens (**FIGURE 12-25**).

Spines, prickles, and thorns are a common way to discourage herbivores, and some of them are impressively large. The acacias, a group of plants that grow as large bushes or small trees, are notoriously spiny. In these plants, the deterrent effects of spines are sometimes enhanced by the presence of ants that live in the swollen bases of the spines and fight off other animals, from insects to mammals, including humans, that would feed on their host acacia. Spines on palm fronds carry pathogenic bacteria that infect the wounds that the spines create, providing a long-lasting reminder that it's not a good idea to try to eat a palm frond. In most plants, the surfaces of leaves and stems have microscopic hairs called trichomes. Most trichomes protect plants against insect herbivores by exuding a sticky fluid that traps the insect, but stinging plants such as nettles use larger trichomes to inject a potent fluid into any animal that brushes against a leaf.

Many chemical defenses are substances that make plants bitter or otherwise unpalatable, but some plants play hardball by producing chemicals that affect the nervous or reproductive systems of animals that eat them. Locoweeds, which occur all over western North America, contain substances called alkaloids that can cause permanent damage to the nervous system. When cattle and horses eat locoweed, they become lethargic and stop feeding. Other defensive chemicals are hallucinogens. Tetrahydrocannabinol, or THC, is found in the leaves and buds of marijuana plants, and the disorientation it produces causes mammals to stop eating.

FIGURE 12-25 Spines, sticky traps, and toxic compounds. Plants have developed a range of defenses against predation.

PLANT DEFENSES

Because plants can't run from plant-eating animals, they have developed many defensive features.

ANATOMICAL STRUCTURES
Some plants have spines, spikes, and thorns that deter predators. Some trees have thick layers of bark that are shed in order to get rid of attacking insects.

STICKY TRAPS
Conifers exude pitch, a sticky substance that can engulf and smother attacking insects.

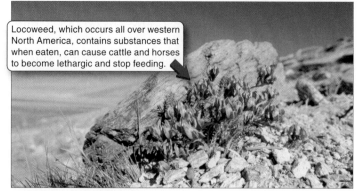

Locoweed, which occurs all over western North America, contains substances that when eaten, can cause cattle and horses to become lethargic and stop feeding.

CHEMICAL COMPOUNDS
Some plants synthesize chemicals that induce physiological and behavioral changes in the animals eating them.

Plants Are a Branch of Eukarya   Mosses, Ferns, and Their Relatives   The Advent of the Seed   Flowering Plants

Many of the chemicals that plants produce to protect themselves have physiological effects, and for thousands of years humans have used plants for medicinal purposes. The aboriginal peoples of North America and Eurasia, for example, knew that extracts of the bark from willows and aspens could relieve pain. Salicin is the compound in the bark that relieves pain, and we still use acetylsalicylic acid (aspirin) for pain relief. Opium poppies produce opium, a more powerful pain reliever; foxglove flowers contain digitalin, a compound that slows the heart rate; and the ipecacuanha tree produces ipecac, a compound that causes vomiting.

The term "bioprospecting" is applied to the search for plants with medicinal uses, and representatives of pharmaceutical companies trudge through rain forests and deserts to find compounds with medicinal applications. The rewards from such a discovery can be great. For example, artemisin from Chinese sweet wormwood is effective in treating malarial infections that are resistant to conventional pharmaceutical compounds.

Although plants can't move, they do send messages. They release volatile chemicals when they are attacked by insects—a volatile chemical is one that floats in the air—and the chemicals can warn nearby plants of the impending threat or can call protective insects to their aid.

When plants are nibbled by insects, they release the chemical methyl jasmonate (MJ), which is carried with the breeze to nearby plants. When these plants detect the presence of MJ, they ramp up production of defensive chemicals so that they are prepared if the insects move onto them. Insects can also detect MJ, and insects that prey on other insects are attracted to plants emitting the chemical, because it signals the presence of prey. Release of MJ by lima beans and apple trees that are being attacked by spider mites summons predatory mites. When corn and cotton plants are attacked by caterpillars, release of MJ summons parasitic wasps that deposit their eggs inside the caterpillars. When the eggs hatch, the wasp larvae eat the caterpillars from the inside out.

About 900 species of plants turn the tables and eat insects. They use the protein in their insect meals to supplement the nitrogen compounds they get from the soil. Insectivorous plants are most common in boggy areas, because boggy soil often has low nitrogen concentrations. (FIGURE 12-26). Pitcher plants are an example—their name comes from conspicuous pitcher-shaped structures that capture insects. It's easy for an insect to enter the open top of a pitcher, but hard to get out, because the inner walls are lined with hair-like structures that

Some plants living in nitrogen-deficient soil have turned the tables, becoming predators on insects.

FIGURE 12-26 Carnivorous plants feed on insects.

point downward, forcing the insect into a pool of liquid. Some pitcher plants merely absorb the nitrogen that is released when the trapped insects decay, but other species secrete digestive enzymes into the pool. These enzymes, which are basically the same as the enzymes in your stomach, digest the insects and make the nitrogen available more rapidly.

TAKE-HOME MESSAGE 12·12

Plants have a wide range of defenses against herbivorous animals: from physical defenses such as thorns to chemicals that have complex effects on animals' physiology. Plants respond to insect attack by synthesizing chemicals that make the plant less palatable. Some plants living in soil that is deficient in nitrogen have switched roles, preying on insects.

6 Fungi and plants are partners but not relatives.

An elegant purple fan: a view of the underside of the fungus amethyst deceiver (Laccaria amethystea).

12·13

Fungi are closer to animals than they are to plants.

Think about your toes for a minute. How do they feel? Do they itch? Maybe there's a burning sensation between the toes? Hopefully, they feel fine, because itching and burning are the symptoms of athlete's foot. One study found that 15% of all people in the United Kingdom had athlete's foot, and almost everyone has it at some time during his or her life. The reason we bring up this unpleasant topic is that athlete's foot is a fungal infection of the skin, and it is the encounter with a fungus that many people remember most clearly.

Fungi make up their own monophyletic kingdom within the eukarya domain. Most fungi are multicellular, sessile decomposers (FIGURE 12-27). Although they were originally thought to be plants lacking chlorophyll, it turns out that they

WHAT IS A FUNGUS?

FUNGI ARE DECOMPOSERS OR SYMBIONTS
Fungi acquire energy by breaking down the tissues of dead organisms or by absorbing nutrients from living organisms.

FUNGI ARE SESSILE
Fungi are anchored to the organic material on which they feed.

FUNGI HAVE CELL WALLS MADE OF CHITIN
Fungi have cell walls made of chitin, a chemical that is also important in producing the exoskeleton of insects.

FIGURE 12-27 Defining characteristics of fungi.

Plants Are a Branch of Eukarya   Mosses, Ferns, and Their Relatives   The Advent of the Seed   Flowering Plants

THE FUNGI

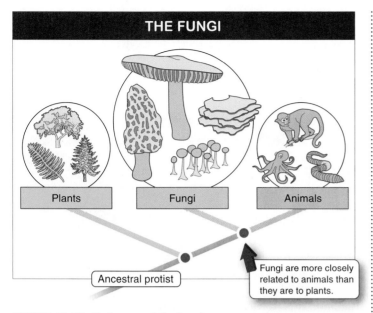

Plants Fungi Animals

Ancestral protist

Fungi are more closely related to animals than they are to plants.

FIGURE 12-28 **Phylogeny of the fungi.**

have little in common with plants. In fact, DNA sequence comparisons reveal that fungi are more closely related to animals than they are to plants (**FIGURE 12-28**). As eukaryotes, fungi have all the basic cellular components you would expect to find: nuclei, mitochondria, an endomembrane system, and a cytoskeleton. They also have cell walls, but the cell walls, instead of including cellulose, as in plants, are made of the carbohydrate chitin, a chemical also important in producing the exoskeleton of insects.

The most commonly encountered fungi are mushrooms, molds, and yeasts (the only single-celled fungi) (**FIGURE 12-29**). The fungus that causes athlete's foot is multicellular and consists of thread-like structures made up of long strings of cells called **hyphae** (pronounced HIGH-fee). Because of the thinness and length of the hyphae (*sing.* hypha), the fungal cells have an enormous surface area. This means they are very good at taking up nutrients from your skin, but it also makes them very vulnerable to drying out. The fungus responsible for athlete's foot grows best in moist places—like on the skin between your toes.

Unless you actually have an active athlete's foot infection right now, your most recent encounter with a fungus (or, at least, with evidence that a fungus was once present) was most likely when you last ate bread. Bread rises because yeast cells are mixed into the dough, which is then kept in a warm place while the yeast consumes sugar and produces carbon dioxide through fermentation. The carbon dioxide released by the yeast makes the dough swell into the shape we associate with a loaf of bread.

TAKE-HOME MESSAGE 12·13

Fungi are eukaryotes with the same internal cellular elements other eukaryotes have—and one distinctive feature: a cell wall formed from the carbohydrate chitin. Some fungi, the yeasts, live as individual cells; most of the other fungi are multicellular.

DIVERSITY OF FUNGI

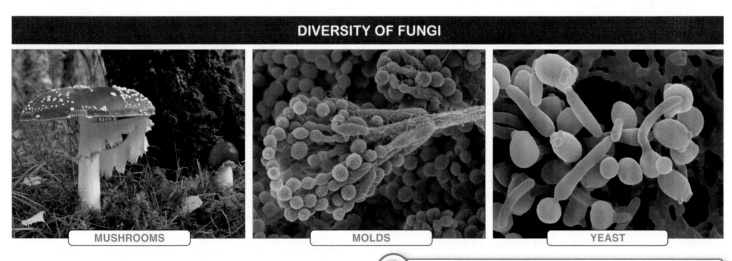

MUSHROOMS MOLDS YEAST

FIGURE 12-29 **The fungi that humans know best.** The mushroom shown here is a fly agaric; the mold is *Penicillium chrysogenum*; and the yeast is *Candida albicans*.

There are more than 1.5 million fungal species and they are found everywhere on earth: from the tropics to the arctic, on land and in the water.

473

12·14

Fungi have common structures, but exploit an enormous diversity of habitats.

As we've seen, most fungi are multicellular and are composed of long strings of barely visible, thread-like hyphal cells. The hyphae interconnect to form a mass of tissue called a **mycelium,** and this is the form in which a fungus spends most of its time, usually underground or in a decaying tree or log. Unless you dug down to reach the mycelium, you would not know a fungus was there.

The structure that most people associate with fungi is the mushroom. But a mushroom is just a temporary reproductive structure (or "fruiting body"), part of a complex reproductive cycle in some fungi (**FIGURE 12-30**):

1. Underground, genetically distinct haploid hyphae fuse. But the *nuclei* from these cells *don't* fuse, so instead of being diploid, they are "dikaryotic," meaning that each cell in the hyphae has two nuclei.

2. The dikaryotic hyphal state can last for years, with the dikaryotic mycelium growing and spreading.

3. At some point, a mushroom may form. It is produced from tightly packed dikaryotic hyphae.

4. In some cells in the mushroom, the haploid nuclei in dikaryotic hyphae fuse, putting the mushroom into a diploid state.

5. The diploid cells in the mushroom undergo meiosis, producing huge numbers of haploid spores (up to a billion in a single mushroom!), which are dispersed by wind or water or on the bodies of animals.

6. When the spores land in a hospitable place, they start growing as haploid hyphae and the cycle begins anew.

Fungi have an unusual and effective method of getting nutrition. Unlike humans, they digest their food outside their "body." While growing underground, hyphae secrete strong enzymes that break down the organic molecules around them. The hyphae then absorb those nutrients. How effective are they at absorbing nutrients and growing? Here's a better question: what is the largest living organism in the world? Not surprisingly (considering where we are asking the question), it is a fungus—specifically, a yellow honey mushroom fungus that covers an area of about 4 square miles

FIGURE 12-30 The three-stage reproductive cycle of a fungus.

FUNGUS LIFE CYCLE

1. Underground, distinct haploid hyphae fuse to form dikaryotic hyphae, in which single cells have two distinct nuclei.

2. The dikaryotic hypha state can last for years, as the dikaryotic mycelium spreads throughout an area.

3. At some point, a reproductive structure—the mushroom—forms. It is produced from tightly packed dikaryotic hyphae.

4. In some cells within the mushroom, the haploid nuclei in dikaryotic hyphae fuse, becoming diploid.

5. The diploid cells within the mushroom then undergo meiosis, producing haploid spores.

6. When the spores land in a hospitable place, they start growing as haploid hyphae and the cycle is complete.

Plants Are a Branch of Eukarya   Mosses, Ferns, and Their Relatives   The Advent of the Seed   Flowering Plants

(nearly 10 square kilometers) in eastern Oregon (**FIGURE 12-31**). This fungus is estimated to be at least 2,400 years old, and it may be more than 8,000 years old, which would place it in the category of the oldest living organisms. Huge as it is, though, the fungus is hard to detect because the mycelium is underground. The fungus appears on the surface only as mushrooms (edible, but not particularly tasty) and as hyphae in dying and dead trees. Indeed, no one even knew that this enormous fungus existed until 2000, when Oregon foresters sought an explanation for an epidemic of dead and dying trees in the Malheur National Forest.

Many mushrooms are gastronomic delicacies, and commercially grown portobello and shiitake mushrooms command high prices. The white button mushroom that you buy in plastic containers when you are making pizza is a species of *Agaricus*. Truffles are the underground reproductive structures of a fungus called *Tuber* that grows in forests. Truffles are not much to look at—they resemble small, dirt-covered potatoes—but the most prized variety, the white truffle, sells for $1,750 to $3,500 per pound. Because truffles grow underground, there is nothing on the surface to indicate their presence, and truffle hunters use trained pigs or dogs to locate them by scent.

Fungi can grow in many different habitats because, as **decomposers,** all they need for their nutrient supply is some

Molds can contaminate poorly ventilated buildings, causing a variety of irritating and potentially dangerous health problems.

FIGURE 12-32 Cleaning out toxic mold. A building overrun with mold spores may cause its occupants to become ill.

sort of organic material that they can break down. Fungi play an enormously important ecological role in speeding the decay of organic material in forests. They don't need light, so they can grow underground or inside dead trees and logs. There, the hyphae release enzymes to digest organic material and then absorb the resulting nutrient-rich fluid.

Fungi can also thrive in poorly ventilated spaces in buildings (**FIGURE 12-32**). Molds are multicellular fungi that are responsible for many cases of "sick-building syndrome"—a situation in which people living or working in a building experience a variety of unpleasant symptoms. Burning or watering eyes, a runny nose, and itchy skin are allergic reactions to the proteins in fungal spores, whereas more severe effects, such as cancers and miscarriages, are probably produced by toxins released when the fruiting bodies of the fungi disintegrate. Curing a sick building can be so expensive that in some cases, the entire building is destroyed and rebuilt.

If you break down the tissues of dead organisms you are a decomposer, and if you break down the tissues of living organisms you are a parasite. Fungi are equally happy in either role. Fungi are happy to grow in and on people, and "mycosis" is a general term for a disease that is caused by a fungus. The fungi that cause athlete's foot, and the related fungi that infect other parts of the body (causing problems with names like jock itch, beard itch, scalp itch, ringworm, and toenail fungus), get their nutrition by digesting some of the organic molecules of your body! And all of these fungal diseases are quite contagious because they are spread by spores that can linger on moist surfaces and in clothing, so

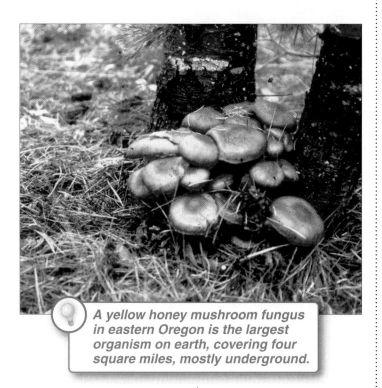

A yellow honey mushroom fungus in eastern Oregon is the largest organism on earth, covering four square miles, mostly underground.

FIGURE 12-31 The largest organism in the world: a yellow honey mushroom fungus. Shown here is just a small cluster of the fungus's temporary reproductive structures—what we call "mushrooms."

they tend to be a problem in places like dormitories, gymnasiums, and fitness centers.

Fungi also form mutually beneficial relationships with chlorophyll-containing bacteria and/or algae. These two-way or three-way partnerships are called **lichens,** and they grow on surfaces such as tree trunks and rocks. The fungus is fed by its photosynthetic partner and helps it absorb water and nutrients. In addition, the fungus provides nutrients for its partners by excreting enzymes that dissolve organic material and acid that dissolves rock.

12·15

Most plants have fungal symbionts.

If you examine the roots of a plant with a microscope, you will find round structures and fibers closely associated with the fine rootlets and root hairs. These are root fungi, or **mycorrhizae** (pronounced my-ko-RYE-zay) (**FIGURE 12-33**). Associations between roots and fungi are ancient—they have been found in 400-million-year-old fossils of early land plants—and nearly all species of modern plants have them. Some mycorrhizae, called ectomycorrhizae, have hyphae that press closely against the outer side of the walls of root hair cells. The endomycorrhizae, by contrast, actually send hyphae through the root cell walls and into the space between the cell wall and the plasma membrane. The association between mycorrhizal fungi and plants is beneficial to both partners. The fungus extracts phosphorus and nitrogen from the soil and releases them inside the roots of the plant, and the additional nutrients that the plant receives from the fungus allow it to grow faster and larger. The mycorrhizal fungus benefits by drawing sugar from the plant, which it uses to support its own cellular respiration.

Not all plants play fair, however. Some take nutrients from the fungus and give nothing in return. One of the best known of these mycorrhizal parasites is the ghost pipe, which grows in forests in the northern hemisphere (**FIGURE 12-34**). The ghost pipe is an angiosperm related to heaths and heathers. The plant's name comes from its ghostly white appearance, which makes it visible even at night when only moonlight shines on the forest floor. The roots of the ghost pipe have mycorrhizal fungi associated with them, and those same mycorrhizae are also connected to the roots of trees. The ghost pipe obtains all its nutrients from or through the

FIGURE 12-33 The association between mycorrhizal fungi and plants is beneficial to both.

MYCORRHIZAE

Mycorrhizal fungi grow in association with the roots of plants, receiving sugar from the plant and transferring nitrogen and phosphorus from the soil to the plant.

ECTOMYCORRHIZAE
Hyphae press closely against the outer side of the cell walls of the root hairs.

Hyphae

Cell wall   Root hair

ENDOMYCORRHIZAE
Hyphae grow through cell walls into the space between the cell wall and plasma membrane of root hair cells.

Hyphae

Plasma membrane

Plants Are a Branch of Eukarya   Mosses, Ferns, and Their Relatives   The Advent of the Seed   Flowering Plants

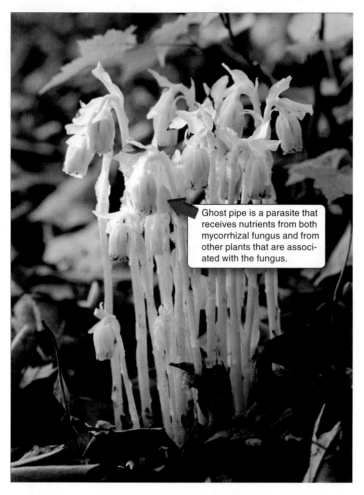

Ghost pipe is a parasite that receives nutrients from both mycorrhizal fungus and from other plants that are associated with the fungus.

FIGURE 12-34 Ghost pipe: a plant that parasitizes mycorrhizae for nutrients.

mycorrhizae—nitrogen and phosphorus are provided by the mycorrhizal fungus, and sugar travels from tree roots to the ghost pipes via the mycorrhizae.

The significance of mycorrhizae is appreciated in the horticultural industry. Lumber companies have found that they have greater success when replanting areas they have clear-cut if the conifer seedlings are first grown in soil that has been inoculated with mycorrhizal fungi. Home gardeners can buy potting soil with mycorrhizal spores or buy spores compressed into tablets that can be placed in the soil when seedlings are transplanted.

Some plants manipulate mycorrhizae to gain a competitive advantage over other plants—garlic mustard is an example. Garlic mustard was brought to North America by European colonists, because they liked using it to add a tangy flavor to salads. Initially it was planted in kitchen gardens, but, inevitably, it escaped into the wild where it has proved itself to be an extremely successful—and unusually diabolical—invasive species. Like many weedy plants, garlic mustard can thrive without mycorrhizal fungi, but garlic mustard goes further than other weeds. It excretes compounds from its roots that interfere with the partnership between mycorrhizal fungi and local native plants. Because the native plants *do* need mycorrhizal fungi to prosper, the destruction of the fungi by garlic mustard weakens the plants. This chemical warfare is so successful that garlic mustard has wiped out entire populations of native woodland plants and is threatening to damage the trees that make up the forest itself.

TAKE-HOME MESSAGE 12·15

Plants and fungi have a close and mutually beneficial association in mycorrhizae. Mycorrhizal fungi grow in intimate association with the roots of most plants, receiving sugar from the plant and transferring nitrogen and phosphorus to the plant.

Knowledge You Can Use

Yams: nature's fertility food? **Do you know any twins? Have you ever wondered how common twins are?**

In the United States and Europe, about 12 in every 1,000 births are twin births. In Asia the number is slightly lower, at 8 sets of twins per 1,000 births. Elsewhere, the rates are similar—except for southwest Africa, particularly in Nigeria, a place that has been called "The Land of Twins." There, the rate of twin births among the Yoruba people is more than four times that in the United States, with about 50 pairs of twins per 1,000 births. (Triplet births are unusually common, too, occurring 16 times more frequently in Nigeria than the United States.)

Q: **Why are so many twins born in Nigeria? Is it something in the water?** Nope. But that's not too far off. It's the diet of the Yoruba people. A staple in their diet is the white yam, a starchy vegetable that looks a bit like a potato; many people eat yams several times each day.

Q: **What's in the yams?** Some preliminary studies suggest that an estrogen-related compound in yams is responsible for stimulating the ovaries, increasing the likelihood, in any given month, that more than one egg will be released.

Q: **How can we be sure it's the yams?** At this point, no one is certain, and the question awaits a good application of the scientific method. In one intriguing laboratory study, rats fed a diet rich in yams *doubled* their litter size. As yet, however, a definitive link between yams and increased rates of twin births in humans has not been demonstrated.

So, will it work for you? It's unclear. Anecdotes abound about women having twins after purposely increasing their yam consumption. But a randomized, controlled, double-blind study has yet to be conducted. How would you set up and analyze a study like that?

1 Plants are just one branch of the eukarya.

Plants are multicellular organisms that spend most of their lives anchored to one place by their roots. Plants' inability to move shapes the way they obtain food, reproduce, and protect themselves from predation. The evolutionary history of plants reveals the progressive appearance of characteristics that allow plants to succeed on land, despite their inability to move.

2 The first plants had neither roots nor seeds: mosses, ferns, and their relatives.

The first land plants were small, had no leaves, roots, or flowers, and could grow only near water. Non-vascular plants—mosses, liverworts, and hornworts—have scarcely evolved beyond the earliest land plants. They lack roots and vessels to move water and nutrients from the soil into the plant. Vessels are an effective way to carry water and nutrients from the soil to the leaves, and the first vascular plants—the earliest ferns, horsetails, and some related forms—were able to grow much taller than their non-vascular predecessors.

3 The advent of the seed opened new worlds to plants.

Seeds are the way plants give their offspring a good start in life and get them to leave home. A seed contains a multicellular embryo and a store of nutrients. Seeds are distributed by air, water, or animals. Gymnosperms were the earliest plants to produce seeds, which offer advantages over the spores of earlier plants, but they depend on wind to carry their pollen. Conifers are the most successful gymnosperms, with more species and a larger geographic range than all of their relatives combined.

4 Flowering plants are the most diverse and successful plants.

Flowering plants appeared in the Cretaceous period, about 100 million years ago, and diversified rapidly to become the dominant plants in the modern world. A flower houses a plant's reproductive structures and advertises their presence with colors, patterns, and odors, which it uses to trick or bribe animals into transporting male gametes to female gametes, where fertilization can occur. Double fertilization in angiosperms ensures that a plant does not invest energy in forming endosperm for an ovule that has not been fertilized.

5 Plants and animals have a love-hate relationship.

Overcoming one big challenge of being rooted in the ground, plants often use the assistance of animals to disperse their fruits, which contain fertilized seeds, depositing them at a new location where the seedlings can grow. Fruits are made up from the ovary and, occasionally, some surrounding tissue. Overcoming a second big challenge posed by immobility, plants have a wide range of defenses against herbivorous animals, from physical defenses such as thorns to chemicals with complex effects on animals' physiology. Some plants living in nitrogen-deficient soil have switched roles, preying on insects.

6 Fungi and plants are partners but not relatives.

Fungi are eukaryotes with a cell wall made of chitin. Some fungi (yeasts) live as individual cells, but most are multicellular. Fungi are decomposers, and all they need to thrive is organic material to consume and a moist environment. They can attain enormous sizes. Fungi have complex life cycles, with sexual and asexual phases, and the parts of fungi that are most often visible are their temporary spore-producing bodies. Plants and fungi have a close and mutually beneficial association. Mycorrhizal fungi grow in intimate association with the roots of most plants, receiving sugar from the plant and transferring nitrogen and phosphorus to the plant.

KEY TERMS

CHECK YOUR KNOWLEDGE --

1. In the plant kingdom, all the different species are descended from a single common ancestor. In terms of phylogeny, what type of tree is this?
   a) monophyletic
   b) unbranched
   c) biphyletic
   d) uniphyletic
   e) branched

2. Which of the following is a significant new challenge that plants faced when they moved from their aquatic environment onto the land?
   a) light availability
   b) desiccation
   c) nutrient availability
   d) predation
   e) the need for osmotic regulation

3. Which is the best brief description of the vascular system of the very first terrestrial plants?
   a) The first plants developed specialized vessel cells that conducted water.
   b) The first plants did not possess a vascular system.
   c) The first plants had a very basic vascular system with a simple method of internal transport.
   d) The first plants had only long, needle-like leaves, from which water could evaporate easily.
   e) None of the above answers is correct.

4. Which of the following statements about ferns is incorrect?
   a) Their sporophyte is dominant.
   b) They require liquid water for fertilization.
   c) Their seeds are dispersed via the wind.
   d) They have vascular tissue for distributing water and nutrients throughout the plant.
   e) Their spores are contained in sporangia.

5. Mosses and ferns differ in their reproductive strategies from gymnosperms and angiosperms in which of the following ways?
   a) Mosses and ferns rely on liquid water for fertilization, whereas angiosperms and gymnosperms do not need liquid water for fertilization.
   b) Mosses and ferns have much larger seeds than do angiosperms and gymnosperms.
   c) Mosses and ferns use wind pollination, whereas angiosperms and gymnosperms use insects for pollination.
   d) Mosses and ferns are primarily diploid in their adult (reproductive) form, whereas gymnosperms and angiosperms are primarily haploid.
   e) Mosses and ferns are primarily haploid in their adult form, whereas gymnosperms and angiosperms are primarily diploid.

6. Which of the following is characteristic of gymnosperms?
   a) They are more diverse than the angiosperms.
   b) They are wind-pollinated.
   c) The gametophyte generation is dominant.
   d) They are water-pollinated.
   e) All of the above are correct.

7. In terms of their adaptation to living on land, how are reptiles similar to the seed plants?
   a) Both reptiles and seed plants became completely independent of water.
   b) Reptiles eat plants.
   c) Seed plants and reptiles have developed structures such as cuticles and impermeable skin to minimize desiccation.
   d) Reptiles and seed plants have developed structures that house their gametes and protect them from the surrounding environment.
   e) Both c) and d) are correct.

8. Angiosperms and gymnosperms differ from each other in that:
   a) angiosperms have vascular systems, but gymnosperms do not.
   b) angiosperms produce seeds, but gymnosperms do not.
   c) angiosperms tend to rely on animal pollinators, whereas gymnosperms tend to rely on wind pollination.
   d) angiosperms use pollen, whereas gymnosperms use cones.
   e) angiosperms have a single fertilization process, whereas gymnosperms use double fertilization.

9. Anthers and stigmas can be found on:
   a) bryophytes.   d) gymnosperms.
   b) fungi.   e) all of the above.
   c) angiosperms.

10. Which of the following is not an example of a group of angiosperms?
    a) cacti   d) grasses
    b) cherry trees   e) orchids
    c) pine trees

11. Over the evolutionary history of plants:
    a) the sporophyte has become smaller, though more independent.
    b) the gametophyte and sporophyte have grown increasingly independent of each other.
    c) there has been a trend toward gametophyte dominance.
    d) there has been a trend toward gametophyte dependence.
    e) the gametophyte has become larger, though more dependent.

12. Which of the following comparisons and contrasts between fungi and plants is incorrect?
    a) Fungi cannot photosynthesize, but plants can.
    b) Both fungi and plants use chitin as a structural stabilizer.
    c) Fungi are heterotrophs, but plants are not.
    d) Both fungi and plants have cell walls.
    e) Both fungi and plants have a sexual stage in their reproductive cycle.

13. Dispersal of fungal spores is typically done by:
    a) movement of cilia.
    b) insects.
    c) hummingbirds.
    d) wind.
    e) movement of flagella.

14. You are taking a hike down a forest trail and see the familiar sight of a mushroom on the ground. This visible portion of a fungal body is the structure also referred to as the:
    a) hypha.
    b) fruiting body.
    c) thallus.
    d) spore sac.
    e) mycelium.

15. In most cases, the relationship between roots and fungi in mycorrhizae can best be described as:
    a) mycelium.
    b) competition.
    c) trickery.
    d) symbiosis.
    e) parasitism.

SHORT-ANSWER QUESTIONS

1. What distinguishes a vascular plant from a non-vascular plant? How do these differences affect their geographic distribution?

2. How do fungi benefit plants? What benefit do the fungi derive from this relationship?

3. How does development of a fleshy fruit benefit plants?

See Appendix for answers. For additional study questions, go to www.prep-u.com.

13

Evolution and Diversity Among the Microbes

Bacteria, archaea,
protists, and viruses:
the unseen world

❶ There are microbes in all three domains.

This strain of E. coli *produces a toxin that causes extreme gastrointestinal discomfort.*

13•1

Microbes are the simplest, but most successful organisms on earth.

Humans are large organisms, and being large comes with a lot of baggage that we don't usually think about. We need a skeletal system to support our weight against the pull of gravity, and a respiratory system to take in oxygen and get rid of carbon dioxide. We need a circulatory system to move oxygen and carbon dioxide and other molecules around our bodies, and a digestive system that includes a mouth to take in food and teeth to grind the food into small pieces that can be broken down by enzymes in our stomach and intestines. We even need a nervous system so our brain knows what distant parts of our body are doing.

Q How can a microbe function when its body is just a single cell?

Microbes—the most abundant organisms on earth—don't have skeletal, respiratory, circulatory, digestive, or nervous systems because they are too small to need them; in fact, "microbe"

RELATIVE SIZES OF MICROBES

Although almost all microbes are invisible to the naked eye, they vary tremendously in size. Here, the relative sizes of several microbes are shown proportionally, using everyday objects.

| Influenza virus 100 nm | Escherichia coli 2 μm (2,000 nm) | Euglena gracilis 50 μm (50,000 nm) | Amoeba dubia 1 mm (1,000,000 nm) |
| Head of a pin | Paper clip | Bicycle tire | Oak tree |

FIGURE 13-1 The most abundant organisms on earth are too small to see. If an influenza virus were the size of a pinhead, an amoeba would be the size of an oak tree.

Microbes Are in All Domains   Bacteria   Archaea   Protists

is a combination of Greek words that means "small life," and the "small" part of the name fits them well. Take an amoeba, for example—it has a volume about a million billion (10^{15}) times smaller than a human. When you are that tiny, the force of gravity becomes trivial, so the amoeba needs no skeleton to support it. Plenty of oxygen diffuses inward across its cell membrane, and carbon dioxide diffuses outward, so an amoeba does not need a respiratory system; and because every part of its interior is close to the body surface, it doesn't need a circulatory system to transport gases. An amoeba eats by enclosing food items in a piece of its cell membrane and digests the food with the same enzymes it uses to recycle its own proteins, lipids, and carbohydrates, so it does not need a digestive system. And no part of an amoeba is far enough from another part to require a specialized nervous system for communication.

Most microbes are actually even smaller than an amoeba: a typical bacterium or archaean is about one thousand million billion (10^{18}) times smaller than a human, and an influenza virus is about one thousand billion trillion (10^{24}) times smaller than you are (**FIGURE 13-1**).

They may be invisible, but microbes could never be considered unsuccessful just because of their (lack of) size. To the contrary, they are more successful than humans in almost every imaginable way.

Q How would you decide whether a group of organisms can be considered successful?

Microbes are genetically diverse. More than 500,000 kinds of microbes have been identified by their unique nucleotide sequences, and further studies will almost certainly distinguish millions of additional species of microbes.

Microbes can live almost anywhere and eat almost anything. As you read these lines, more than 400 species of microbes are thriving in your intestinal tract, 500 more species thrive in your mouth, and nearly 200 species call your skin home (**FIGURE 13-2**). The microbes that live in you and on you eat mostly what you eat—some of the bacteria in your mouth and intestine compete with you, trying to digest your food before you can, and the others use the waste products you release after you have broken down the food. Others feed on the leftovers released by the breakdown of your cells during the normal process of cell renewal.

Living conditions in the human body are relatively moderate. Other microbes inhabit some of the toughest environments on earth—in the almost boiling water of hot springs, at depths a mile below the earth's surface, and more than a mile deep in the oceans where hydrothermal vents emit water at 400° C (750° F).

Microbes are abundant. Surface seawater contains more than 100,000 bacterial cells per milliliter, and diatoms (protists in the

MICROBES CAN LIVE ALMOST ANYWHERE

Human intestine

Hot spring

Hydrothermal vent

Escherichia coli

Thermus aquaticus

Methanopyrus kandleri

💡 *Microbes live in nearly every kind of environment, including water at temperatures of up to 750° F!*

FIGURE 13-2 Microbes are everywhere on earth.

Viruses

Total number of cells in the human body (trillions)

100 — Human cells

80

60

40 — Microbial cells that live in and on you

20

0

FIGURE 13-3 Microbe majority. Your body has more microbial cells than human cells.

Everywhere yet unseen. *Microbes outnumber all other species and make up most living matter, including this Malaysian rain forest.*

eukarya domain) are as abundant as bacteria. These densities translate to about 8,000 million billion trillion (8×10^{30}) individuals of just these two kinds of microbes in the world's oceans. Your own body is a testament to the abundance of microbes: it contains about 100 trillion cells, but only one-tenth of those cells are actually human cells—the remaining 90 trillion cells are the microbes that live in and on you (**FIGURE 13-3**). You're a minority in your own body.

TAKE-HOME MESSAGE 13·1

Microbes are very small, simple organisms, but they do everything that larger, multicellular organisms do. They can live anywhere, from moderate to extreme environments. There are millions of different kinds of microbes on earth, in enormous numbers.

Microbes Are in All Domains   Bacteria   Archaea   Protists

Not all microbes are evolutionarily related.

"Microbes" is an appropriately descriptive name, but it's sloppy. We could use it to point to any one of many different types of organisms too small to see without magnification. In Chapters 3 and 10 we introduced two very different kinds of microbes: bacteria and viruses. In this chapter we explore bacteria and viruses in more detail, and also include two more kinds of microbes—the protists and the archaea. The amoeba, a kind of protist, may be familiar to you, but the archaea are probably very unfamiliar to you. That's not surprising, because archaea were only recently recognized as a distinct group of microbes. The group doesn't even have a common name (unless you like "the group formerly known as archebacteria"). We'll just call them archaea.

The sloppiness of the name "microbes" stems from the fact that not all microbes are evolutionarily related. As a consequence, it is hard to make generalizations about them as we can for other organisms. We can say of all animals, for example, that they have the ability to move, if only during one portion of their life cycle. Microbes, on the other hand, are grouped together simply because they are small, not because they all share a recent common ancestor. In fact, microbes occur in all three domains of life—bacteria, archaea, and eukarya—and so the various types of microbes could not be more widely separated (**FIGURE 13-4**). In this chapter we focus on the tiny organisms from each of these domains. The microbes in the domains bacteria and archaea are prokaryotic, although archaea have some characteristics that are like prokaryotes and others that are like eukaryotes. Protists are the microbe members of domain eukarya. And viruses, another type of microbe, are not classified into any domain at all, because they are only at the borderline of life.

TAKE-HOME MESSAGE 13•2

Microbes are grouped together only because they are small, not because of evolutionary relatedness. They occur in all three domains of life, and include the viruses, which are not included in any of the domains.

MICROBES IN THE TREE OF LIFE

DOMAIN BACTERIA

E. coli

DOMAIN ARCHAEA

Halobacteria

DOMAIN EUKARYA

Amoeba

Microbes are found in all three domains of life—and beyond.

Viruses aren't truly living organisms and they are not classified into any of the three domains.

Bacteriophage virus

FIGURE 13-4 Microbes are highly diverse. Microbes are found in every domain of life.

Viruses

❷ Bacteria may be the most diverse of all organisms.

Staphylococcus aureus *(green) on the surface of the small intestine.*

13·3

What are bacteria?

A bacterium is the simplest organism you can imagine—and in many ways, it is the most efficient. A bacterium has a cell envelope consisting of a plasma membrane and, in most cases, a cell wall—that's what it needs to maintain conditions inside that are different from conditions outside. Inside the cell envelope it has cytoplasm—the liquid that fills all kinds of cells (including your own cells). Proteins in the cytoplasm carry out essential functions, such as digesting molecules of food and transferring the energy gained to ATP. DNA in the cytoplasm carries the instructions for making those proteins, and messenger RNAs carry this information to ribosomes, where the proteins are synthesized. That's it—everything an organism needs with no extra baggage.

Bacteria may be classified by their shape: some are spherical cells (known as the cocci), some are rod-shaped (the bacilli), and others are spiral-shaped (the spirilli) (**FIGURE 13-5**). Bacteria usually reproduce by binary fission, and in a few hours, a single cell can form a culture containing thousands of cells.

As a bacterial cell divides, the number of cells doubles every generation, producing a colony of cells, each of which is a

clone of the original cell. Colonies of different species of bacteria look different. The familiar human intestinal bacterium *Escherichia coli,* for example, forms beige or gray colonies that have smooth margins and a shiny mucus-like covering. Species of *Proteus,* which are often responsible for spoiling food because they can grow at refrigerator temperatures, form colonies with a surface that looks like a contour map.

Microbiologists can often identify bacteria simply by looking at the colors and shapes of their colonies. They can get additional information by examining a single cell under a microscope, but living bacterial cells are transparent, so you can't see them with an ordinary microscope unless they have been dyed. In 1884, Hans Christian Gram, a Danish microbiologist, described a method of dying the cell walls of bacteria to make them visible under a microscope, and a **Gram stain** is still the first test microbiologists use when they are identifying an unknown bacterium (**FIGURE 13-6**).

Gram-positive bacteria are colored purple by the stain, because they have a thick layer of a glycoprotein called

Microbes Are in All Domains   Bacteria   Archaea   Protists

BACTERIA: CLASSIFICATION AND STRUCTURE

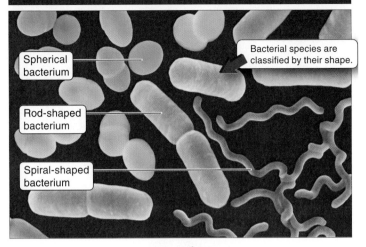

Spherical bacterium

Rod-shaped bacterium

Spiral-shaped bacterium

Bacterial species are classified by their shape.

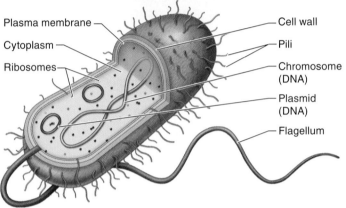

Plasma membrane

Cytoplasm

Ribosomes

Cell wall

Pili

Chromosome (DNA)

Plasmid (DNA)

Flagellum

FIGURE 13-5 Bacteria basics. Bacteria are single-celled organisms that lack a nucleus. They are usually classified by shape.

peptidoglycan on the outside of the cell wall. In *Gram-negative bacteria,* the layer of peptidoglycan lies beneath a membrane and is not stained by the dye. Penicillin is effective in treating infections by Gram-positive bacteria because it interferes with the formation of peptidoglycan cross-links. Penicillin is less effective on Gram-negative bacteria because it does not pass through the outer membrane that covers the peptidoglycan layer.

Being able to distinguish these two groups of bacteria is a big help to microbiologists trying to identify a bacterium, but the peptidoglycan is there because it is important for the bacterial cells—not to make life easier for microbiologists. The extensive interlocking bonds of the long peptidoglycan molecules provide strength to the cell envelope. Many bacteria also have a **capsule** that lies outside the cell wall. This capsule may restrict the movement of water out of the cell and allow bacteria to live in dry places, such as on the surface of your skin. In other cases, the capsule may be important in allowing the bacteria to bind to solid surfaces such as rocks or to attach to human cells.

TAKE-HOME MESSAGE 13·3

Bacteria are efficient single-celled organisms, with an envelope surrounding the cytoplasm, which contains the DNA (they have no nuclei and no intracellular organelles). Bacterial cells undergo binary fission, and a single cell can grow into a colony of cells.

METHODS OF IDENTIFYING BACTERIA

Some bacteria can be identified by looking at the colors and shapes of their colonies.

Gram-positive bacterium

Gram-negative bacterium

Gram staining can help identify unknown bacteria species based on the properties of their cell walls.

FIGURE 13-6 Bacterial IDs. Bacteria are often identified by the appearance of an entire colony or of the cells' response to Gram staining.

13·4

Bacterial growth and reproduction is fast and efficient.

The time it takes for a bacterium to reproduce can be very short; most bacteria have generation times between 1 and 3 hours, and some are even shorter. *Escherichia coli,* for example, has a generation time of 20 minutes in optimal conditions, and in just over 3 hours a single *E. coli* could give rise to a population of 20 billion cells.

Bacteria carry genetic information in two structures: the genes that provide instructions for all of the basic life processes of a bacterium are usually located in a circular DNA molecule, the bacterial chromosome. Most bacteria have just one chromosome, but some have more than one. A bacterial chromosome is organized more efficiently than a eukaryotic chromosome, in two ways. First, in bacteria, the genes that code for proteins with related functions—enzymes that play a role in a pathway that breaks down food for energy, for example—are often arranged right next to each other on the chromosome. This makes it possible to efficiently control the transcription of all the genes together.

Additionally, unlike in eukarya, almost all the DNA in a bacterial chromosome codes for proteins, so bacteria do not use time and energy transcribing mRNA that will not be translated. As we saw in Chapter 5, as much as 90% of the DNA in the chromosomes of eukaryotes does not code for genes and is edited from mRNA after it has been transcribed and before it is translated into protein.

Besides their main chromosome, many bacteria have additional genetic information—circular DNA molecules called **plasmids** that carry genes for specific functions. These include *metabolic plasmids,* with genes enabling bacteria to break down specific substances, such as toxic chemicals; *resistance plasmids,* with genes enabling bacteria to resist the effects of antibiotics; and *virulence plasmids,* with genes that control how sick an infectious bacterium makes its victim. The strain of *E. coli* that occasionally sickens patrons of fast-food restaurants carries a virulence plasmid that magnifies the effects of a gene for a sometimes-lethal toxin. (*E. coli* without that virulence plasmid are normal components of the bacterial community of the human intestine.)

When a bacteria cell divides, it creates two new daughter cells, each carrying the genetic information that was present in

Q Would it be useful to be able to transfer genetic information from one adult human to another?

the chromosome of the mother cell. Thus, binary fission transmits genetic information from one bacterial generation to the next (**FIGURE 13-7**). However, bacteria can also transfer genetic information laterally—in other words, not just to their "offspring," but to other individuals *within* the same generation—through any of three different processes: conjugation, transduction, and transformation.

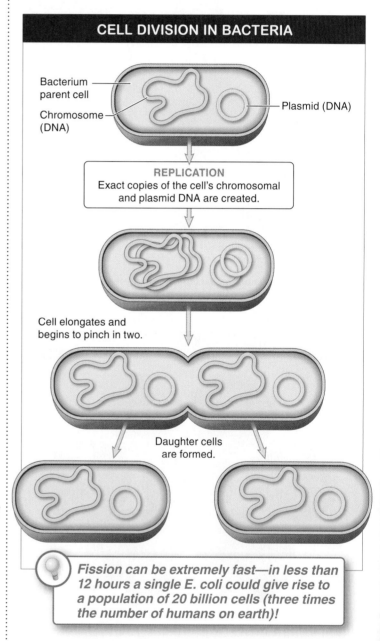

CELL DIVISION IN BACTERIA

Bacterium parent cell

Chromosome (DNA)

Plasmid (DNA)

REPLICATION
Exact copies of the cell's chromosomal and plasmid DNA are created.

Cell elongates and begins to pinch in two.

Daughter cells are formed.

Fission can be extremely fast—in less than 12 hours a single E. coli could give rise to a population of 20 billion cells (three times the number of humans on earth)!

FIGURE 13-7 Binary fission. This asexual cell division method is used by prokaryotes.

Microbes Are in All Domains · Bacteria · Archaea · Protists

METHODS OF GENETIC EXCHANGE IN BACTERIA WITHIN THE SAME GENERATION

CONJUGATION
A bacterium transfers a copy of some or all of its DNA (from the main chromosome or a plasmid) to another bacterium, giving the second bacterium genetic information it did not have before.

Donor bacterium

Recipient bacterium

TRANSDUCTION
A virus containing pieces of bacterial DNA that it inadvertently picked up from its previous host infects a bacterial cell, and passes along new bacterial genes to the bacterium.

Virus

Recipient bacterium

TRANSFORMATION
After a bacterial cell bursts open, short lengths of DNA can be taken up by a living bacterial cell and inserted into its own chromosome, potentially adding genes that it did not have before.

DNA fragments

Recipient bacterium

FIGURE 13-8 Lateral transfer of genetic information: conjugation, transduction, and transformation.

Conjugation is the process by which one bacterium transfers a copy of some of its genetic information to another bacterium—even when the two bacteria are different species. It's very much like plugging your iPod into a computer to transfer songs from the computer to the iPod. This is not reproduction, because you start with one iPod and one computer and that's what you have when you finish. But the iPod now contains songs that it did not have before. Plasmid transfer does exactly that for a bacterium—you still have just two bacterial cells, but the newly acquired plasmid has given the recipient bacterium genetic information it did not have before. These genes could enable the bacterium to make an enzyme that allows it to metabolize a new chemical or to defend itself against a new antibiotic (**FIGURE 13-8**).

Transduction occurs when a kind of virus called a bacteriophage (see bacteriophage photo, Figure 13-4) infects a bacterial cell. The virus reproduces inside the bacterial cell, and sometimes, inadvertently, the new virus particles are filled with pieces of bacterial DNA in addition to or instead of the viral DNA. When these viruses are released and infect new bacterial cells, the bacterial DNA can be inserted into the host bacterium's chromosome, passing new bacterial genes to that bacterium.

Transformation is the process by which bacterial cells scavenge DNA from their environment. This DNA comes from other bacterial cells that have burst open, releasing their cellular contents. The circular chromosomes break into short lengths of DNA, which can then be taken up by living bacterial cells and inserted into their own chromosomes, potentially adding genes they did not originally have.

TAKE-HOME MESSAGE 13·4

Bacteria grow rapidly. They have efficiently organized chromosomes—genes are organized in groups with related functions and virtually all the DNA codes for proteins. Bacteria sometimes carry genes for specialized traits on small DNA molecules called plasmids, which can be transferred from one bacterial cell to another by conjugation. DNA can also be transferred laterally between bacterial cells by transduction or transformation.

Many bacteria are beneficial.

Do you like the taste of yogurt? You can thank bacteria for that—*Lactobacillus acidophilus* and several other species of bacteria are added to milk to create yogurt. As the bacterial cells use the milk sugar for energy, the by-product is lactic acid, which reacts with the milk proteins to produce the

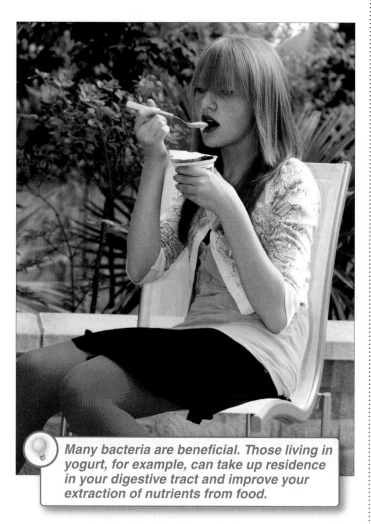

Many bacteria are beneficial. Those living in yogurt, for example, can take up residence in your digestive tract and improve your extraction of nutrients from food.

FIGURE 13-9 Yogurt contains beneficial bacteria.

characteristic taste and texture of yogurt. If you buy a brand of yogurt that is labeled as containing "live cultures," you are consuming living bacterial cells as you eat the yogurt (**FIGURE 13-9**). Bacteria are also used to produce many other foods, such as cheeses, and bacteria (along with yeasts) are used in the production of beer, wine, and vinegar. Industrial microbiology is a multi-billion-dollar industry.

But you can thank bacteria for more than just tasty snacks—you owe your life to bacteria. As we've seen, hundreds of species of bacteria grow in and on your body; these microbes are called your "normal flora." The normal flora take up every spot on your body that a disease-causing bacterium could adhere to, and they consume every potential source of nutrition, making it difficult for a disease-causing bacterium to gain a foothold. Thus, maintaining a robust population of these benign bacteria is your first line of defense against infection by harmful bacteria.

Probiotic therapy is a method of treating infections by deliberately introducing benign bacteria in numbers large enough to swamp the harmful forms. *Lactobacillus acidophilus,* which is a normal inhabitant of the human body, is used to treat gastrointestinal upsets, such as traveler's diarrhea, and urinary tract infections. In addition to growing so vigorously that it crowds out harmful bacteria, *L. acidophilus* releases lactic acid, which interferes with the growth of other bacteria and prevents them from adhering to the walls of the urinary tract and bladder.

TAKE-HOME MESSAGE 13•5

Your body fights bacteria with bacteria. A disease-causing bacterium must colonize your body before it can make you sick, and your body is already covered with harmless bacteria. If the population of harmless bacteria is dense enough, it will preclude invading bacteria from gaining a foothold.

Microbes Are in All Domains   Bacteria   Archaea   Protists

Metabolic diversity among the bacteria is extreme.

One important feature making bacterial diversity possible is that bacteria can metabolize almost anything. Some of them can even use energy from light to make their own food. Microbiologists place bacteria into "trophic" (feeding) categories that reflect their metabolic specialization.

Chemical organic feeders (**chemoorganotrophs**) are bacteria that consume organic molecules, such as carbohydrates. You probably see the products of organic feeders every time you take a shower—they are responsible for the pink deposits on the shower curtain and the floor of the shower (**FIGURE 13-10**). Most of the bacteria that live in and on your body are also organic feeders. Some compete with you to metabolize the food you eat. Others digest things you can't digest or can't eat.

Chemical inorganic feeders (**chemolithotrophs,** meaning "rock feeders") are able to use inorganic molecules such as ammonia, hydrogen sulfide, hydrogen, and iron as sources of energy. The most common inorganic feeders are the iron bacteria responsible for the brown stains that form on plumbing fixtures in regions where tap water contains high levels of iron. Sulfur bacteria are associated with iron bacteria, and are responsible for the slimy black deposits that you will probably

find if you lift the stopper out of the drain in your bathroom sink. It's hard to understand how bacteria can live in that environment, because it contains little that we would consider food; but in iron compounds and other inorganic molecules, these bacteria utilize a completely different type of "food."

On a larger scale, inorganic feeders are responsible for acid mine drainage. The ore that is being mined usually makes up only a small part of the total amount of rock that is removed from a mine. The portion that does not contain ore is discarded on the ground surface in piles called "tailings." This material is often rich in minerals such as pyrites (iron sulfides). It's not of any nutritional value to us, but inorganic feeders can gain energy by oxidizing these minerals and, in the process, they release compounds that combine with rainwater to produce strong acids, such as sulfuric acid. When this acidic water drains into streams, it can kill fish and aquatic plants and insects.

"Light self-feeders" (**photoautotrophs**)—bacteria that use the energy from sunlight—contain chlorophyll and use the light energy to convert carbon dioxide to glucose by photosynthesis. The floating mats of gooey green material that you see in roadside ditches are a type of photoautotroph

METABOLIC DIVERSITY AMONG BACTERIA

CHEMOORGANOTROPHS
• Feed on organic molecules

CHEMOLITHOTROPHS
• Feed on inorganic molecules

PHOTOAUTOTROPHS
• Use energy from sunlight to produce glucose via photosynthesis

FIGURE 13-10 **Resourceful feeders.** Bacteria can metabolize almost anything.

You may view them as "gross." But bacteria thrive in so many places because they are resourceful, able to extract "food" from a very broad range of sources, including organic molecules on your shower curtain, inorganic molecules in your pipes, and even using energy from the sun.

called cyanobacteria. The cyanobacteria living today closely resemble the first photosynthetic organisms that appeared on earth about 2.6 billion years ago. That date was the start of a major global change, because until then the earth's atmosphere contained no free oxygen—instead, air consisted almost entirely of nitrogen and carbon dioxide. Cyanobacteria were the first organisms that could use solar energy to build organic compounds from carbon dioxide, and in the process break down water molecules to release free oxygen. The accumulation of oxygen released by cyanobacteria is called the **Oxygen Revolution.** Oxygen—

which humans depend on—now makes up about 21% of the volume of air, and cyanobacteria still release important quantities of oxygen into the atmosphere.

TAKE-HOME MESSAGE 13•6

Some bacteria eat organic molecules, some eat minerals, and still others carry out photosynthesis. About 2.6 billion years ago, the photosynthesizing bacteria were responsible for the first appearance of free oxygen in the earth's atmosphere.

13•7

Bacteria cause many human diseases.

The number of **pathogenic** (disease-causing) bacteria is tiny compared with the total number of species of bacteria, but some pathogens kill millions of people annually, despite advances in medicine and sanitation. Some bacteria are always pathogenic (such as those that produce cholera, plague, and tuberculosis), but others (the ones responsible for acne, strep throat, scarlet fever, and "flesh-eating" necrotizing fasciitis) are normal parts of the communities of bacteria that live in or on humans. These bacteria in our "normal flora" become pathogenic only under special circumstances.

The great cholera epidemic that devastated London in 1854 became a milestone in epidemiology (the study of the occurrence of outbreaks of disease), when Dr. John Snow identified the Broad Street pump as the source of infection by mapping the pattern of deaths from cholera (**FIGURE 13-11**). When Dr. Snow persuaded the authorities to shut down the pump by removing the handle, new cases of cholera in the area dropped sharply.

Cholera epidemics are a continuing threat, especially in areas without effective sanitation. Epidemics of cholera are occurring now in parts of Iraq where the sewage systems have been destroyed. The strains of cholera found in these areas are especially potent; as the bacteria multiply inside their hosts, they cause severe diarrhea with a massive loss of water. Victims of these severe strains of cholera are incapacitated and soon die, but before they die they release billions of cholera bacteria in diarrhea—an effective strategy for the cholera bacteria, because they contaminate water used for drinking or bathing and rapidly infect new victims.

Streptococcus pyogenes is a normal part of the bacterial community of your nose and mouth. Normally it is harmless, and many people pass their entire lives without

Death from cholera
Street pump

By mapping the pattern of deaths from cholera, Dr. John Snow was able to identify the Broad Street water pump as the source of the infection.

FIGURE 13-11 Contaminated! Water from the Broad Street pump in London was the source of a cholera outbreak that killed more than 500 people.

experiencing any of the diseases it can cause. When the population of *S. pyogenes* is not held in check by competition with other members of the bacterial community, however, it can become a pathogen. Strep throat (more formally known as "streptococcal pharyngitis") is the most common disease caused by *S. pyogenes;* it produces a severe sore throat and a

Microbes Are in All Domains   Bacteria   Archaea   Protists

Streptococcus pyogenes *is usually harmless, but some strains are responsible for strep throat, scarlet fever, and necrotizing fasciitis (flesh-eating bacteria).*

FIGURE 13-12 Strep throat is the most common disease caused by *S. pyogenes.*

distinctive rash of red abscesses with white pus at the top of the throat (**FIGURE 13-12**). Some strains of *S. pyogenes* produce a toxin that is released into the bloodstream and produces a red rash that spreads across the skin—the disease called scarlet fever. Most threatening is an infection caused by strains of *S. pyogenes* that have a toxin that allows them to enter body tissues, where they produce necrotizing fasciitis. These strains of *S. pyogenes* are known by their well-earned name, "the flesh-eating bacteria."

TAKE-HOME MESSAGE 13·7

Some bacteria always cause disease and others do no harm except under certain conditions. For example, *Streptococcus pyogenes* can be harmless, but under some conditions releases toxins that are responsible for strep throat, scarlet fever, and necrotizing fasciitis (caused by the flesh-eating strains).

13·8

Bacteria evolve drug resistance quickly.

Penicillin was the first antibiotic to be manufactured and used widely in the battle against illness-inducing bacteria. It came into use during the Second World War and caused a revolution in the care of the wounded. But antibiotic-resistant infections soon appeared, and the number of resistant bacteria has increased rapidly ever since (**FIGURE 13-13**). Now, 60 years after the first use of antibiotics, many bacteria are resistant to many, or even to most, antibiotics. In the United States, more people now die of *Staphylococcus aureus* infections that are resistant to many different antibiotics (MRSA infections) than die of HIV/ AIDS. A few years ago, antibiotic-resistant staph infections were acquired only in hospitals, but now these infections have spread from hospitals to the community at large. Instead of contracting antibiotic-resistant staph infections in hospitals, patients (or their visitors) are now bringing the resistant bacteria into the hospital with them!

Microbes live everywhere they can, and compete for the best places to attach themselves and the richest sources of food to eat. This competition takes many forms: rapid growth to crowd competitors out of a living space, superior ability to take in nutrients to starve competitors, and—this is the key to both the benefits of antibiotics and the problem of antibiotic resistance for humans—production of chemicals that kill other microbes,

Q *Where do antibiotics come from, and why do they so quickly lose their effectiveness?*

or at least stop them from growing. Antibiotics are produced by microbes to help them compete with other microbes, and most of the antibiotics we use today are derived from microbes.

Bacteria and other microbes have developed a variety of ways to resist antibiotics. For example, some bacteria pump an antibiotic out of their cells as fast as it enters, so it never reaches a lethal concentration inside the bacterial cell. Other bacteria have proteins that bind to the antibiotic molecule and block its lethal effect. Still others carry that approach a step further—they have enzymes that break down the antibiotic molecules, which are then used as fuel to help the bacteria grow faster! Many of the genes that code for resistance are on plasmids. This means that a bacterium carrying a resistance gene can transmit the gene to other bacteria by conjugation; there's no need to wait for multiple generations of natural selection.

Q *Why is it essential to take every dose of an antibiotic prescribed by a doctor?*

When an antibiotic is taken as prescribed—that is, at the times specified on the label and until all the pills have been consumed—the population of target bacteria is greatly reduced. All of the target bacteria that

ANTIBIOTIC RESISTANCE IN BACTERIA

Patient A and Patient B both have an infection caused by a harmful strain of bacteria.

○ Harmful bacteria
○ Harmless bacteria

Patient A

Patient B

Both patients are prescribed an antibiotic to treat the infection, which reduces the initial number of harmful bacteria.

Patient A

Patient B

Patient A continues to take the antibiotic as prescribed until all of the pills have been consumed.

Patient B stops taking the antibiotic before finishing all of the prescribed amount.

Patient A

Patient B

The target bacteria population is greatly reduced and the growth of the remaining bacteria will be held in check by competition with other types of bacteria.

The many bacterial cells still alive are the most resistant to the antibiotic. They are the founders of a new population, and the next time Patient B takes the drug, it will be ineffective.

FIGURE 13-13 **Evolving resistance.** Sixty years after the appearance of antibiotics, many bacteria are resistant to many, or even most, antibiotics.

remain are resistant to the antibiotic, but there are not very many of them. The growth of these resistant bacteria will be held in check by competition with other types of bacteria. But, if you stop taking the antibiotic before you have finished all of the prescription, many of the target bacterial cells will remain alive, including the ones that are most resistant to the antibiotic. These resistant cells will be the founders of a new population of bacteria in your body, so the next time you take that drug it will be ineffective. Even worse, taking antibiotics when they are not needed—to treat a viral infection, for

Agriculture in the United States uses about 25 million pounds of antibiotics each year—about eight times more than is used for all human medicine!

FIGURE 13-14 **Antibiotics are used in agriculture.** Livestock are given antibiotics to prevent diseases that are easily spread in crowded living conditions. The practice can cause the evolution of bacteria that are resistant to antibiotics.

example—selects for resistant bacteria without providing any benefit, because antibiotics have no effect on viruses.

The use of antibiotics in agriculture is another reason for the spread of antibiotic resistance. Low concentrations of antibiotics are routinely added to the feed for cattle, hogs, chickens, and turkeys. This can be beneficial in the short-term, promoting growth and minimizing disease in the crowded conditions of commercial meat and milk production. But in the long run it can have disastrous consequences, as the practice can lead to selection for bacteria resistant to the antibiotics. The antibiotics can also pass through the food chain to humans (**FIGURE 13-14**). Data gathered by the Union of Concerned Scientists indicate that agriculture in the United States uses about 25 million pounds of antibiotics each year—about eight times more than is used for all human medicine!

TAKE-HOME MESSAGE 13·8

Microbes routinely evolve resistance to antibiotics. The genes that allow bacteria to combat antibiotics are located on plasmids, and plasmid transfer allows an antibiotic-resistant bacterium to pass that resistance to other bacteria. Excessive use of antibiotics in medicine and agriculture has made several of the most important pathogenic bacteria resistant to every known antibiotic, and infections caused by these bacteria are nearly impossible to treat.

Microbes Are in All Domains   Bacteria   Archaea   Protists

❸ Archaea exploit some of the most extreme habitats.

A steamy pool in Rotorua, North Island, New Zealand, heated by volcanic vents and home to heat-tolerant archaea.

13·9

Archaea are profoundly different from bacteria.

When you look at the photographs of archaea in **FIGURE 13-15**, you may find it hard to believe that they are even a little bit different from bacteria, let alone *profoundly* different. And you would be right to wonder, because the differences between bacteria and archaea are not externally visible. Both live as single cells or colonies of cells, both are surrounded by a plasma membrane, and both have species with flagella that twirl like propellers. In fact, biologists considered the archaea to be

bacteria until the 1970s. Only then did comparisons of DNA reveal that archaea are as different from bacteria as humans are.

Those studies of archaean nucleotide sequences stimulated other comparisons, which identified additional differences among bacteria, archaea, and eukarya. For example, the chemical compositions of the plasma membranes, cell walls, and flagella of archaea are qualitatively different from those of bacteria. And beyond the large DNA sequence difference and differences in the plasma membrane, cell walls, and flagella, a third difference helps explain why archaea are placed between bacteria and eukarya on the tree of life (see Figure 13-4): eukarya have a distinct cell nucleus that is separated from the cytoplasm by a nuclear membrane, whereas bacteria and archaea do not have either a nucleus or a nuclear membrane. Having a nucleus provides protection for the chromosomes and allows the cell to control what molecules interact with its DNA.

Archaea look very much like bacteria. But closer inspection—of their physiology, biochemistry, and DNA—reveals them to be profoundly different from all bacteria.

FIGURE 13-15 Appearances are deceiving.

TAKE-HOME MESSAGE 13·9

Archaea show a set of characteristics that places them between bacteria and eukaryotes on the tree of life. Archaea and bacteria may look similar, but they possess large and significant differences in their DNA sequences, as well as differences in their plasma membranes, cell walls, and flagella. Furthermore, neither archaea nor bacteria resemble eukarya in one key way. Eukarya alone possess a distinct cell nucleus and nuclear membrane.

Archaea thrive in habitats too extreme for most other organisms.

Archaea are famous for their ability to live in places where life would seem to be impossible, such as in water around hydrothermal vents that emerge from the seafloor that would be boiling if it were at sea level. At 2,000 meters below the surface, the water pressure reaches 200 atmospheres, or almost 3,000 pounds per square inch, and the 400° C water cannot boil. Yet the archaea thrive there. Most organisms die at temperatures between 40° and 50° C, because their protein molecules are denatured, so the ability of archaea to survive at temperatures above 100° C is truly remarkable.

Equally impressive is the ability of archaea to live in water as acidic as pH 0 or as salty as a saturated NaCl solution, and to metabolize an extraordinary range of compounds—including sulfur, iron, and hydrogen gas—to obtain energy (FIGURE 13-16). Some bacteria also can tolerate extreme physical and chemical conditions, and the organisms that can live in these conditions, both bacteria and archaea, are called **extremophiles** ("lovers of extreme conditions").

Archaea in your intestine break down a chemical bond found in beans (a bond humans cannot break)...but the process generates gas, which can cause discomfort as it tries to escape.

FIGURE 13-17 **Helping you digest tough bonds.** Intestinal archaea help you digest some chemical bonds in beans.

But not all archaea are extremophiles. Archaea live everywhere that bacteria do, and many of them live in places you would find perfectly comfortable yourself. In fact, you are home to many archaea and, on occasion, some of them make their presence all too obvious. Beans are notorious for their tendency to produce gas in the intestine, but the beans themselves are not really to blame—archaea are the culprits. Beans contain a couple of carbohydrates with chemical bonds that are not broken down very well by any human enzymes (FIGURE 13-17). Methane-producing archaea, on the other hand, have no such difficulties, producing an enzyme that targets those bonds. As a result, it is the archaea in your intestine that digest most of these carbohydrates. But in the process of breaking these bonds, they produce gases that, as they escape the digestive system, can cause considerable distress.

TAKE-HOME MESSAGE 13·10

Archaea can tolerate extreme physical and chemical conditions that are impossible for most other living organisms, but some live in moderate conditions and even in the human intestine.

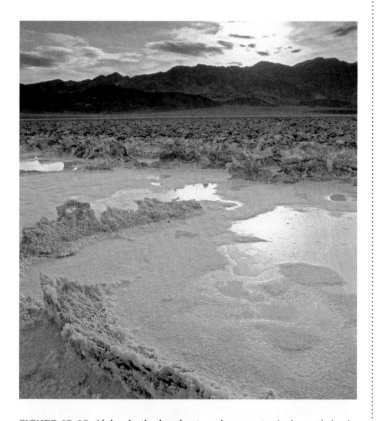

FIGURE 13-16 **Living in the harshest environments.** Archaea thrive in many extreme environments, such as this salt pond.

Microbes Are in All Domains   Bacteria   Archaea   Protists

Much archaean diversity has yet to be discovered.

How do you grow an organism in your lab that requires 200 atmospheres of pressure and a water temperature of 100° C or higher, and eats a gas that is toxic to humans? It's not easy, and only about one-quarter of the identified archaean species have been cultured in the lab. We know that the other species exist, because their transfer RNA has been identified in samples taken from the environment, and biologists are confident that thousands, perhaps millions, of species of archaea await discovery.

Extremophile archaea may have important applications in bioengineering and environmental remediation. For example, *Thermus aquaticus,* an extremophile archaean that lives in hot springs where the water is nearly boiling, produces an enzyme important in making PCR possible (see Chapter 5), and this enzyme is now sold for laboratory use under the name *Taq*

polymerase. Far from being destroyed at high temperature, *Taq* polymerase works best at 75° C, a temperature at which most human enzymes would simply fall apart.

> " He was a killer, a thing that preyed, living on the things that lived, unaided, alone, by virtue of his own strength and prowess, surviving triumphantly in a hostile environment where only the strong survive. "
>
> —Jack London, *Call of the Wild,* 1903

Bioengineers and biotechnologists believe that extremophile archaea and bacteria that thrive at high temperatures and pressures and metabolize toxic substances have enormous potential for industries that have to carry out activities under those conditions. Recent experiments have demonstrated the ability of some archaea to efficiently degrade hydrocarbons, making it possible for them to be used in the removal of sludge that accumulates in oil refinery tanks and, potentially, in contaminated environments such as oil slicks (**FIGURE 13-18**). Other archaea show promise in clearing mineral deposits from pipes in the cooling systems of power plants.

FIGURE 13-18 Practical applications for archaea. The metabolic activity of archaea has potential for use in the cleanup of oil spills.

TAKE-HOME MESSAGE 13 • 11

Archaea are hard to study because many require extreme heat or pressure to grow, and these conditions are not easy to provide in a laboratory. But the ability of archaea to grow in such extreme conditions makes them potentially valuable for industrial and environmental purposes.

❹ Protists are single-celled eukaryotes.

This diatom, a species of Actinocyclus, *is a single-celled protist with a cell wall made of supportive silica.*

13·12

The first eukaryotes were protists.

For the first 2 billion years of life on earth, organisms were really, really small—bacteria and archaea less than 10 m across, one-eighth the diameter of a human hair. But in rocks about 1.9 billion years old, we find fossils of new kinds of organisms that are 10 times larger (**FIGURE 13-19**). These are a group of organisms called acritarchs (a name that can be translated as "confusing old things"). They were the first eukaryotes.

The larger size is the first thing you notice about these fossil cells, but the internal changes in the cells are even more significant because they are the basis for the success of the entire eukaryote lineage, including humans. For the first time in the history of life, cells had internal structures that carried out specific functions. These structures, the cellular organelles, perform the specialized activities that make eukaryotic cells more complex than prokaryotes.

The nucleus is an evolutionary innovation that appeared for the first time in protists. The nucleus is a region inside the cell that is enclosed by its own plasma membrane. The nuclear membrane was probably formed by the fusion of folds of

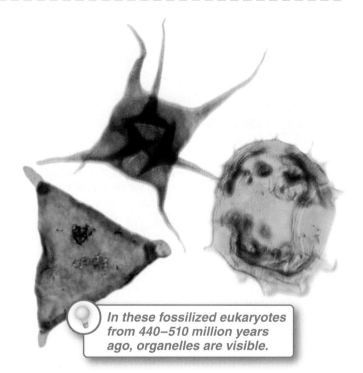

In these fossilized eukaryotes from 440–510 million years ago, organelles are visible.

FIGURE 13-19 Ancient protists. These acritarch fossils represent the first eukaryotes on earth. The oldest fossils of this type were found in rock 1.9 billion years old.

Microbes Are in All Domains   Bacteria   Archaea   Protists

plasma membrane that surrounded the cell. We know that many modern prokaryotes have complex infoldings of their plasma membranes—these increase the total surface area of the cell and allow better exchange of material between the cell and the external environment (see Figure 3-7 on page 84)—and infoldings of this type probably fused around the DNA, creating a membrane-enclosed compartment with DNA in it. Thus, the cell nucleus was formed, separated by a double membrane from the cytoplasm of the cell.

Subsequently, these first nuclear membranes developed two specializations: they incorporated proteins that controlled the movement of molecules into and out of the nucleus, and they extended outward from the nucleus to form a folded membrane called the endoplasmic reticulum. The endoplasmic reticulum is the part of a eukaryotic cell where some proteins are assembled. Further development of internal membranes produced the Golgi apparatus, where newly synthesized proteins are given some final processing steps, and sac-like structures called lysosomes, which contain enzymes that break down damaged molecules. Finally, a lineage of protists took in a guest—a bacterial cell that subsequently became specialized as the mitochondrion, the organelle that produces most of the ATP synthesized by a cell.

TAKE-HOME MESSAGE 13·12

The nucleus is an evolutionary innovation that appeared for the first time in protists. Early protists took in a bacterial cell that subsequently became specialized as the mitochondrion, an organelle in eukaryotic cells that produces ATP.

13·13

There are animal-like protists, fungus-like protists, and plant-like protists.

Six major lineages of protists have been named, but even that classification does not capture the huge diversity of the group, and some of the best-known protists do not fit into any of the six lineages. Protists include forms that are very much like animals, others that seem a lot like fungi, and still others that look like plants (**FIGURE 13-20**).

Animal-like protists Some protists propel themselves around their environment quickly and appear to hunt for prey.

These animal-like protists, which include *Paramecium,* are the ciliates, and get their name from the cilia (hair-like projections) that cover their body surface and propel the cells through water. *Paramecium* feeds by the process of **phagocytosis:** cilia in a funnel-shaped structure called the gullet create an inward flow of water that carries bacteria and other small particles of food with it. These particles accumulate at the inner end of the gullet, where a portion of the plasma membrane bulges inward and eventually breaks

DIVERSITY OF PROTISTS

ANIMAL-LIKE PROTISTS
Trichomonas vaginalis

FUNGI-LIKE PROTISTS
Plasmodial slime mold

PLANT-LIKE PROTISTS
Kelp forest

FIGURE 13-20 All shapes and sizes. Protists come in a variety of shapes and sizes, resembling animals, fungi, and plants.

Viruses

free, forming a food vacuole that drifts into the interior of the cell. Enzymes and hydrochloric acid enter the food vacuole from the cytoplasm, and in this acid environment the engulfed food items are broken down into molecules that diffuse from the vacuole into the cytoplasm. When all of the digestible material has been consumed, the undigested contents of the vacuole are expelled.

Fungus-like protists Some protists resemble fungi, establishing sheet-like colonies of cells on surfaces such as the grout in shower stalls. These are the slime molds. They generally spread without any individual cells moving, but rather by adding new cells at the edges of the colony. Some slime molds, however, called plasmodial slime molds, are oozing masses of gooey material that flow along, engulfing bacteria, fungi, and small bits of organic material as they go. The streaming of a slime mold in its feeding phase is easy to observe with a microscope, and such a slime mold can flow around, over, or through almost anything—it can even flow through the holes in a window screen and reassemble itself on the other side! If you have a lawn with flower beds, you may have seen an irregularly shaped blob of yellow material in a moist, shaded spot—this was a plasmodial slime mold.

What's most remarkable about a plasmodial slime mold is that it is a very large, single cell, but has multiple nuclei. Slime molds divide by binary fission, just like other cells, but in this case a cell may cover an area of several square centimeters! All of the nuclei in the cell undergo mitosis simultaneously.

Plant-like protists Still another group of protists grow in water and resemble plants. These include the brown algae (although they are not algae and often aren't even brown) that grow in the tidal zone, and the giant kelp that grow in water 30 meters deep. Another name for these protists is seaweeds, and this term is close to the mark because almost all brown algae are marine. But they are not plants, they are large groups—or colonies—of protists.

Q How can a microbe be bigger than you are?

Brown algae cover large portions of the rocks in the intertidal zone, and provide habitats for snails, crabs, and sea stars and hiding places for juvenile fishes. Forests of giant kelp grow in water up to 30 meters deep in temperate regions of the North Pacific and South Pacific and off the Atlantic coast of South Africa. Kelp forests are an enormously diverse habitat. More than 1,000 species of fishes, crustaceans, snails, and mammals make their homes in kelp forests. Sea otters wrap a kelp frond around their waists when they sleep, and gray whales hide in kelp forests to escape from killer whales.

Why are these organisms called "microbes" when some of them are larger than a human? These are *colonial* protists, which is how they can be so large. And although they look like *multicellular* organisms, they are actually collections of cells, each of which is carrying out all of its life processes independent of all the other cells. In a true multicellular organism, by contrast, the cells in different parts of the body are responsible for (specialized for) different processes that benefit the organism as a whole. But in brown algae, it's every cell for itself. Each cell gets its own food, and each cell uses all of the nutrients it obtains—nutrients are not transported from one cell to another.

Also among the plant-like protists are the diatoms. These organisms live in ponds, lakes, and rivers as well as in the oceans. They are so small that, for some species, 30 individuals could be lined up across the width of a human hair. A characteristic of the diatoms is that they are enclosed in a shell made of silica (**FIGURE 13-21**). Many species of diatoms float in the water, forming part of the phytoplankton—the microscopic organisms that fix carbon dioxide and release oxygen. Phytoplankton can reach densities of hundreds of thousands of cells per liter, and it accounts for about one-quarter of the photosynthetic production of oxygen on earth. Small fish and shrimp-like organisms called copepods feed on phytoplankton, and these small predators are eaten by larger predators, which are eaten by still larger predators. Thus the diversity of life in the marine habitat relies on diatoms and the other microbes that make up the phytoplankton.

TAKE-HOME MESSAGE 13·13

Protists are a diverse group of organisms. The ciliates, such as *Paramecium*, are animal-like protists. Plasmodial slime molds are protists made up of a single cell, but with multiple nuclei that divide at the same time. Colonial protists, such as the giant kelp, can be enormous.

FIGURE 13-21 Diatoms. These tiny, photosynthetic protists live in water.

13·14

Some protists can make you very sick.

A **parasite** is an organism that lives in or on another organism, called a **host,** and damages it. A parasitic protist called *Plasmodium* that is transmitted by a mosquito is responsible for a worldwide epidemic disease—malaria. Malaria occurs in tropical parts of Africa, Asia, and Latin America, and it is common in the eastern Mediterranean as well. Between 350 million and 500 million people have clinical cases of malaria, and about 1 million people die of malaria each year. Malaria is the leading cause of death for children younger than 5 years old in sub-Saharan Africa; somewhere on the African continent, a child dies of malaria about every 30 seconds.

These are sobering statistics, and neither the incidence of malaria nor the rate of mortality has changed very much since *Plasmodium* was identified as the cause of malaria nearly a century ago. The reason for the lack of progress in combating malaria is the nature of *Plasmodium*—it has a complex development that makes it nearly invisible to the host's immune system (**FIGURE 13-22**).

The human immune system has a difficult time fighting a malarial infection because, in order to survive from one generation to the next, the *Plasmodium* parasites go through a series of distinct developmental stages. And as the *Plasmodium* cells change from one stage to another, the parasite produces different cell surface proteins. Thus *Plasmodium* stays ahead of the human immune system by constantly changing the way it appears.

Although the immune system doesn't usually have much success fighting malarial infection, some individuals produce red blood cells that are inhospitable to the *Plasmodium* parasite, conferring a resistance to the malaria-causing protist. These individuals carry an allele, called Hb^S, which (as we learned in Chapter 7) codes for a variant of the normal adult hemoglobin that causes hemoglobin molecules to stick together in long chains that distort the red blood cell (see Figure 13-22), giving the cell a sickled shape. The sickled red blood cells are inhospitable to *Plasmodium* because they are leaky, and the substances that leak out of the cells happen to be essential to *Plasmodium*. Although carrying a copy of the Hb^S allele confers resistance to malaria, individuals who carry two copies of the Hb^S allele suffer from sickle-cell anemia—a severe genetic disease that is more debilitating than malaria.

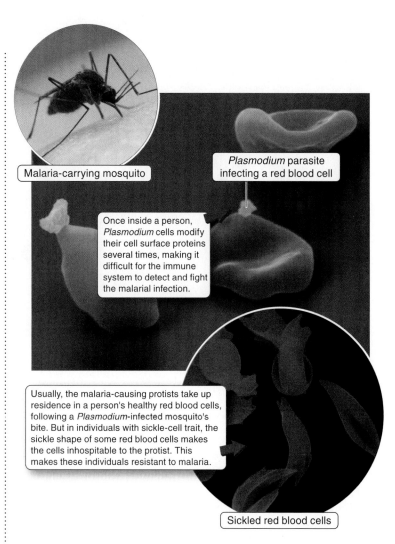

Malaria-carrying mosquito

Plasmodium parasite infecting a red blood cell

Once inside a person, *Plasmodium* cells modify their cell surface proteins several times, making it difficult for the immune system to detect and fight the malarial infection.

Usually, the malaria-causing protists take up residence in a person's healthy red blood cells, following a *Plasmodium*-infected mosquito's bite. But in individuals with sickle-cell trait, the sickle shape of some red blood cells makes the cells inhospitable to the protist. This makes these individuals resistant to malaria.

Sickled red blood cells

FIGURE 13-22 Malaria-infected blood cells and sickled blood cells. The malaria parasite grows inside red blood cells and destroys them. Because sickle-cell trait changes the shape of some red blood cells, these cells resist infection by malaria parasites.

TAKE-HOME MESSAGE 13·14

Some protists cause debilitating diseases. *Plasmodium,* the protist responsible for malaria, is one these. *Plasmodium* has characteristics that protect it from the human immune system, but humans have evolved defenses against malaria: people with sickle-cell trait make red blood cells that are inhospitable to the parasite.

Viruses

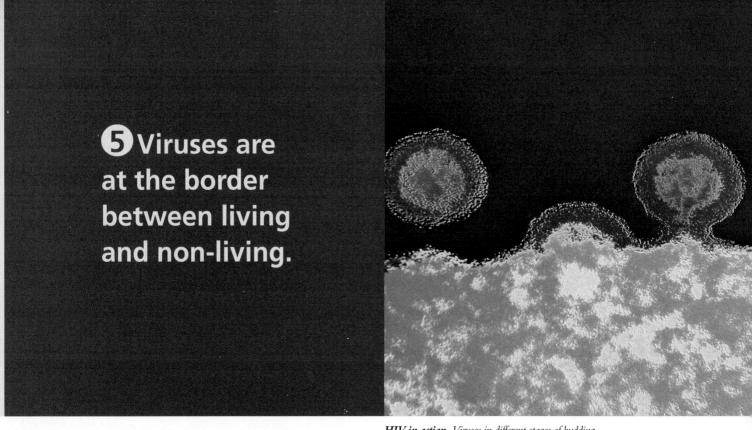

5 Viruses are at the border between living and non-living.

HIV in action. Viruses in different stages of budding.

13 • 15

Viruses are not exactly living organisms.

A virus is not a cell, and that is why viruses do not fit into any of the three domains of life. A virus particle (called a virion) consists of genetic material inside a container made of protein. Some viruses also contain a few enzymes. That's all there is to a virus. The protein container is called the **capsid** and the genetic material can be either DNA or RNA. Some viruses wrap themselves in a bit of the plasma membrane of the host cell as they are released. A virus of this type is called an

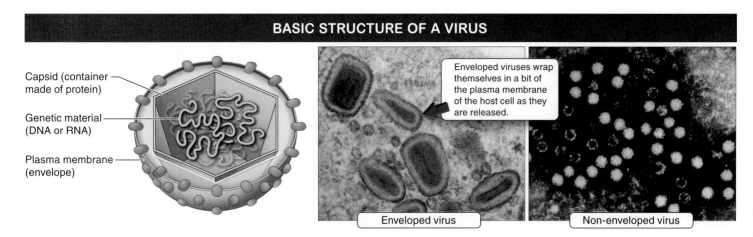

BASIC STRUCTURE OF A VIRUS

Capsid (container made of protein)

Genetic material (DNA or RNA)

Plasma membrane (envelope)

Enveloped viruses wrap themselves in a bit of the plasma membrane of the host cell as they are released.

Enveloped virus

Non-enveloped virus

FIGURE 13-23 **The simple structure of viruses.** All viruses have a container (the capsid) that holds their genetic material, and sometimes the capsid is wrapped in a membrane (an envelope).

1. After the virus binds to the host cell's membrane, the viral DNA is taken into the cell.

2. Viral DNA is replicated into dozens of new copies, using the host's metabolic machinery and energy.

3. Viral mRNA is transcribed from the viral DNA.

4. New viral proteins are synthesized, again using the host's protein-production molecules.

5. The new viral DNA and proteins assemble, forming many new virus particles.

Virus

Host cell

Host nucleus

Viral DNA → Replicated viral DNA

Viral proteins

Viral mRNA

FIGURE 13-24 Making more viruses. A virus duplicates its own genetic material (DNA in the example shown here) by taking over the resources and metabolic machinery of a host cell.

Because they are dependent on their hosts' metabolic machinery for replication, viruses are not considered "living."

enveloped virus—the flu virus is an example. Non-enveloped viruses are enclosed only by the protein container—the virus that causes the common cold is an example of a non-enveloped virus (**FIGURE 13-23**).

There is almost nothing inside the capsid of a virus except DNA or RNA. A virus particle does not carry out any metabolic processes, and it does not control the inward or outward movement of molecules to make conditions inside the virus particle different from conditions outside. Viruses just wait for a chance to insert their genetic material into a living cell.

We speak of "catching" a cold, but what has actually happened is that a virus has caught us. Viruses identify the cells that they can infect by recognizing the structure of glycoprotein molecules on the surfaces of cells. Every cell in your body has these molecules. They are embedded in the plasma membrane and extend outward from the cell. Your immune system uses these proteins to identify the cells as part of you—as self rather than non-self—and it does not react to proteins that it recognizes as self.

Viruses have cracked the glycoprotein code, and when they find a cell with the appropriate glycoprotein on its surface,

they bind to that cell's plasma membrane and insert their genetic material into the cell (**FIGURE 13-24**). Inside the host cell, the viral DNA or RNA takes over the cellular machinery and uses it to produce more viruses. Viruses carry out nearly all of their activities by hijacking materials and organelles in the host cell. Viral proteins are synthesized in exactly the same way as host cell proteins—mRNA binds to ribosomes, and tRNA matches the correct amino acid to each mRNA codon. The mRNA is produced from the virus's DNA or RNA, but all of the protein-building machinery comes from the host cell, as does the ATP required to synthesize the new viral protein.

TAKE-HOME MESSAGE 13·15

A virus is not alive, but it can carry out some of the same functions as living organisms, provided that it can get inside a cell. A virus takes over the protein-making machinery of the host cell to produce more viral genetic material (RNA or DNA) and more viral protein. The viral proteins and genetic material are assembled into new virus particles and released from the cell.

Viruses are responsible for many health problems.

Many diseases are caused by viruses. Some viral diseases, such as the common cold, are not usually serious, but others have been responsible for worldwide epidemics, called pandemics. The influenza pandemic of 1918-1919 killed at least 20 million people, and possibly as many as 50 million. The current HIV/AIDS pandemic has infected nearly 40 million people, with an annual mortality of about 3 million.

DNA viruses have base-pair sequences that are stable over time, because the enzymes that replicate the DNA check for errors and correct them during replication. From a health perspective, the fact that DNA viruses do not change rapidly makes it much easier to fight and treat diseases caused by these viruses, using vaccines. A vaccination against a DNA virus such as smallpox, for example, provides years of protection.

RNA viruses, by contrast, change quickly, because the enzymes that carry out RNA replication do not have error-checking mechanisms. As a consequence, because the RNA-replication enzymes make errors as they assemble new RNA molecules, RNA viruses are continuously mutating into new forms. The common influenza virus that causes outbreaks of flu every year, for example, is an RNA virus. And because flu viruses mutate so fast, the virus that causes health problems changes from one flu season to the next, necessitating a different flu vaccine every year.

Q *Why do flu viruses change so quickly?*

About every 50 years, a new variety of influenza causes a pandemic. The 20th century had three major influenza pandemics, and all of them originated in the same peculiar way. In each of these pandemics, a bird flu virus gained the ability to infect human cells by passing through pigs, and then the virus spread rapidly through the human population worldwide. The most famous of these pandemics was that of 1918-1919 (**FIGURE 13-25**). It was called the Spanish Influenza because it seemed to enter Europe through Spain, but it originated in Asia. The Asian flu and Hong Kong flu pandemics also originated in Asia, as did smaller viral outbreaks, such as Korean flu (1947), swine fever (1976), and SARS (severe acute respiratory syndrome, 2002-2003), which did not become pandemics.

FIGURE 13-25 Viral outbreaks can threaten large populations. The influenza pandemic of 1918–1919 killed millions. In 2002–2003, the virus that causes severe acute respiratory syndrome (SARS) infected several thousand people around the world.

TAKE-HOME MESSAGE 13·16

Many diseases are caused by viruses. DNA viruses are relatively stable, because DNA-replication enzymes check for errors and correct them during replication. RNA viruses change quickly, however, because RNA-replication enzymes do not have error-checking mechanisms.

| Microbes Are in All Domains | Bacteria | Archaea | Protists |

Viruses infect a wide range of organisms.

Viruses are practically everywhere. There are viruses that infect animals, viruses that infect plants, even viruses that infect bacteria. Most viruses infect just one species, or only a few closely related species, and enter only one kind of cell in that species. But some viruses can infect a wide range of hosts—the rabies virus, for example, can infect any mammal. Pet dogs and cats are routinely immunized against rabies, but wild mammals, such as bats, skunks, and raccoons, can also get rabies. That is why, if you are bitten by one of those wild animals, your treatment includes rabies shots as a precaution.

The glycoproteins on the surface of a virus determine which host species the virus can infect and which tissues of the host it can enter. Influenza A viruses (the ones that cause outbreaks of flu every year), for example, have two types of glycoprotein that have different functions. One glycoprotein matches that of a host cell, and allows the virus to enter the cell. The other glycoprotein allows the virus to get back out of the cell, releasing new virus particles that can infect other cells (**FIGURE 13-26**).

Q What role does a pig play in the transmission of virus from a bird to a human?

Influenza A is an example of a virus that can move from one species to another, and as we noted above, all of the influenza pandemics in the past century began with the transmission of a bird influenza virus to humans. Here's how "species-jumping" can occur.

FIGURE 13-27 The pig connects human viruses with bird viruses. A virus that infects humans and a virus that infects birds can meet in a pig cell (see text), replicate, and (because viral RNA replication is error-prone) get packaged together into a new virus particle capable of infecting humans.

Viruses that infect birds don't bind well to glycoproteins in human cells, making it difficult for bird viruses to infect humans. However, the cells of pigs have glycoproteins that allow both human viruses and bird viruses to bind to them. Thus, a pig cell can be infected by a human virus and a bird virus at the same time. Influenza viruses are not very careful when they incorporate newly synthesized RNA into new particles, though. If RNA strands from a human virus and a bird flu virus happen to be in a cell at the same time, it is possible that some virus particles released from the cell will include some RNA from the bird virus and some from the human virus. These viruses are a new strain, and they may be able to infect humans (**FIGURE 13-27**).

Since 1997, one strain of bird flu has spread from Hong Kong through most of Asia and into Turkey, France, Germany, and England. This bird flu is the most recent avian influenza virus to also infect humans. The virus readily infects birds, and tens of millions of chickens, turkeys, and

INFLUENZA A STRUCTURE

SURFACE GLYCOPROTEINS

- Bind to receptors on the surface of the host cell and allow the virus to enter

- Allow the virus to break free from the host cell

💡 *Surface glycoproteins are the "keys" that allow a virus into and out of a host cell.*

FIGURE 13-26 A virus's surface proteins. The influenza virus has surface glycoproteins that allow it to bind to and exit a host cell.

ducks have been killed in attempts to eradicate the infection. And because the virus's host-entry glycoproteins do not bind well to human cells, the virus does not easily infect people and does not readily spread from human to human. So far, nearly all the people who have been infected by the avian influenza virus have contracted it through close contact with infected flocks of birds or by eating birds that had died of the viral infection. Only a half-dozen cases are known in which the virus seems to have been transmitted from one person to another. However, avian flu can be deadly when it does infect humans—more than half of the human cases reported have ended in death.

The World Health Organization and national health agencies are preparing for a worldwide pandemic of this bird flu virus that can also infect humans. Their alarm is based on the possibility that if a pig infected with the bird flu virus is also infected with a human virus, and if the viruses are not careful when packing new RNA into new capsids, then a pig cell

could contain RNA from both humans and birds, and some of the new viruses released from this cell could contain a mixture of RNA from the human and the bird viruses. A mixture of this type might produce a new form of the virus that carries both the genes that make the bird flu lethal to humans *and* the gene that codes for the host-entry glycoprotein. This virus would infect human cells and spread easily from person to person. The World Health Organization has estimated that a pandemic that starts this way could kill as many as 100 million people worldwide.

TAKE-HOME MESSAGE 13·17

Glycoproteins on the surfaces of viruses determine the types of cells a virus can invade. Most viruses infect just one species, or only a few closely related species, and enter only one kind of cell in that species.

13·18

HIV illustrates the difficulty of controlling infectious viruses.

New infectious diseases—some caused by bacteria and others by viruses—emerge quite frequently. Many of these new diseases originate in other species of animals and subsequently acquire the ability to infect humans. There are enough of these new diseases to fill the monthly issues of the journal *Emerging Infectious Diseases,* which is published by the Centers for Disease Control and Prevention. But even in that crowd, AIDS (acquired immune deficiency syndrome) stands out. AIDS is caused by the human immunodeficiency virus (HIV), which is derived from a strain of the simian immunodeficiency virus (SIV) that jumped from chimpanzees to humans in the early 1900s. HIV has all of the characteristics that make viral diseases hard to control—plus an additional characteristic of its own.

HIV mutates easily. HIV is a **retrovirus,** that is, a RNA-containing virus that also contains a viral enzyme called reverse transcriptase that uses a strand of viral RNA as a template to synthesize a single strand of DNA. That DNA strand, in turn, is used as a template to make a complementary strand of DNA, and the new double-stranded DNA is integrated into the host cell's DNA. From here, the DNA is used by the host cell's machinery to make more viral RNA (**FIGURE 13-28**). Reverse transcriptase is so error-prone, however, that virtually every copy of the HIV virus in a patient's body has a different mutation.

This enormous genetic variation in the HIV particles circulating in an infected person's body makes the infection hard to treat. Virus particles with different mutations can have different proteins on their surfaces, and these surface proteins change each time the virus replicates inside a host cell. Each new generation of HIV in the infected individual contains viruses with surface proteins that his or her immune system has never seen. Furthermore, some of the HIV mutations will confer resistance to the drugs that are being used to treat the patient, and new drugs must be used.

HIV attacks white blood cells. All of those problems would apply to any disease caused by a retrovirus, but HIV offers an additional challenge because it targets cells in the immune system of the host, especially white blood cells, and particularly those that search for and attack invading bacteria or viruses. During the incubation period, which can last for many years, HIV infects white blood cells. New white cells are produced to replace the ones killed by the virus, however, and the infected person has virtually no symptoms. HIV is present in the body fluids during the incubation period, nonetheless, and can be transmitted to other individuals. HIV testing is used to detect infections during this stage.

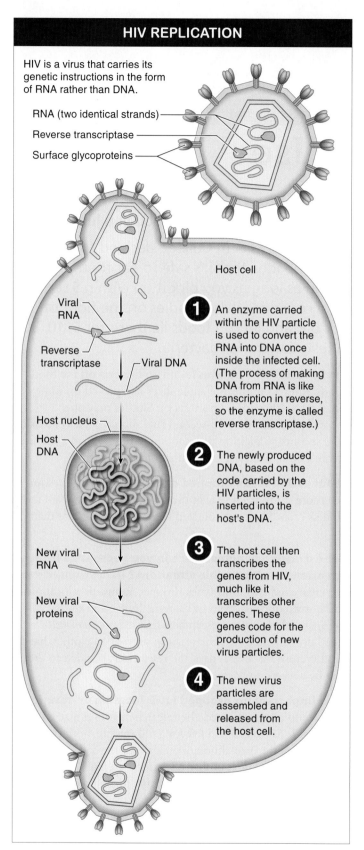

HIV is a virus that carries its genetic instructions in the form of RNA rather than DNA.

RNA (two identical strands)

Reverse transcriptase

Surface glycoproteins

Host cell

Viral RNA

Reverse transcriptase

Viral DNA

Host nucleus

Host DNA

New viral RNA

New viral proteins

1 An enzyme carried within the HIV particle is used to convert the RNA into DNA once inside the infected cell. (The process of making DNA from RNA is like transcription in reverse, so the enzyme is called reverse transcriptase.)

2 The newly produced DNA, based on the code carried by the HIV particles, is inserted into the host's DNA.

3 The host cell then transcribes the genes from HIV, much like it transcribes other genes. These genes code for the production of new virus particles.

4 The new virus particles are assembled and released from the host cell.

FIGURE 13-28 HIV infection. This RNA virus takes over the replicating machinery of a white blood cell to produce a new generation of HIV particles. HIV mutates rapidly during this process, because the reverse transcriptase enzyme is highly error-prone.

FIGURE 13-29 HIV attacks white blood cells essential for identifying foreign invaders.

Each time HIV infects another white blood cell, the reverse transcriptase makes errors in transcribing the RNA to DNA, and eventually one of the mutations allows the virus to bind to the glycoprotein on the surface of a specialized type of white blood cell—a bacteria- and virus-hunting white blood cell that is critically important in identifying disease-causing threats. A suitable mutation may occur in a couple of years, but more often it takes about 10 years or longer. But when it does happen, it signals a new stage in the HIV infection, the development of AIDS (**FIGURE 13-29**).

The immune system collapses. Normally, white blood cells all work together to identify and destroy cells that have been infected by a virus. When HIV begins to kill the cells that hunt for viruses and bacteria, the immune system begins to fail—it can no longer respond to HIV, or to any other infectious agent. Patients with AIDS develop multiple infections, bacterial and viral, as well as cancers, because they have lost the immune system cells that would normally have marked infected and cancerous cells for destruction.

TAKE-HOME MESSAGE 13·18

HIV infection is especially difficult to control. Mutations change the properties of the virus so that it is hard for the immune system to recognize it, and they produce variants that are resistant to the drugs used to treat the HIV infection.

509

Knowledge You Can Use

The five-second rule. **How clean is that food you just dropped?**

In 2007, two students at Connecticut College decided to test the "five-second rule": the idea that it's safe to eat food you've dropped on the ground as long as you pick it up within 5 seconds. They dropped apple slices and Skittles candies on the floor of the cafeteria and a snack bar and let the foods lie there for 5, 10, 30, or 60 seconds, then tested them for bacteria.

Q: Is the five-second rule valid? The students found no bacteria on the food picked up within 30 seconds. After a minute, the apple slices had picked up some bacteria, but the Skittles had none (they found bacteria on Skittles only after 5 minutes). The students concluded that the five-second rule should get an extension, proposing that you have 30 seconds to pick up moist foods and more than a minute to pick up dry foods without risk of bacterial contamination.

Q: Should we relax about eating foods dropped on the floor? This question is a serious one and warrants more study. Each year in the United States there are more than 76 million cases of illness caused by contaminated food, including more than 5,000 that are fatal.

Q: Scientific method and drawing conclusions: can you generalize from a small number of observations to all possible situations? In the Connecticut College study, the students examined two food types, dropped in just two locations, and onto surfaces with unknown concentrations of bacteria. In another study, published in the *Journal of Applied Microbiology,* researchers focused on *Salmonella* bacteria. Because it has been documented that bacteria can survive on clothes, hands, sponges, cutting boards, and utensils for several days, the researchers decided to drop food on surfaces known to be covered in bacteria.

Q: Which factor is more important for dropped food: location or duration? Armed with slices of bologna and pieces of bread, the researchers found that when dropped on surfaces covered with *Salmonella* and left for a full minute, both types of food took up 1,500 to 80,000 bacteria. And although picking up the food quickly—within just 5 seconds—reduced by 90% the number of bacteria present, 150 to 8,000 bacteria still had time to hitch a ride on the food in that first 5 seconds.

A grubby conclusion. It's probably safe to say that if you're in the Connecticut College cafeteria, you can eat your dropped food as long as you pick it up quickly. But for all other situations, a bit more caution is advised—particularly, taking notice of the "drop zone." Unsanitary surfaces likely to have microbes will contaminate your food almost immediately. So hold on tight.

❶ There are microbes in all three domains.

Microbes are grouped together because they are small, not because they are evolutionarily related. In fact, there are microbes in all three domains of life: bacteria, archaea, and eukarya. And viruses, which are on the border between living and non-living, are not in any domain. Because they are so small, microbes can also be simple—they do not need skeletons to support them, circulatory systems to transport materials, or nervous systems to coordinate stimulus with response.

❷ Bacteria may be the most diverse of all organisms.

Bacteria are single-celled organisms with a very simple structure—they have no nucleus and no intracellular organelles. Despite their simplicity, bacteria are found virtually everywhere and can eat virtually anything. The entire bacterial chromosome (unlike eukaryotic chromosomes) consists of functional genes, and bacteria reproduce rapidly and efficiently by binary fission. Genes for specialized functions, such as antibiotic resistance, are carried on small DNA molecules called plasmids that can be transferred from one bacterium to anther. Bacterial generation times are measured in minutes or hours, and the combination of short generation times and plasmid transfer allows antibiotic resistance to spread rapidly.

❸ Archaea exploit some of the most extreme habitats.

Archaea look very much like bacteria, but differ in several characteristics. Archaea thrive in extreme conditions—at pressures exceeding 200 atmospheres, at temperatures above 100° C, and in solutions as acidic as pH 0—and they can eat toxic substances. These characteristics make archaea potentially valuable for industrial applications, such as clearing deposits from pipes in oil refineries and power plants.

❹ Protists are single-celled eukaryotes.

Protists are single-celled eukaryotes. The only characteristic they all share is a nucleus. Amoebas and *Paramecium* are animal-like protists that eat bacterial cells. The so-called brown algae are colonial protists—collections of cells that form large structures, such as seaweeds, in which each cell carries out all the activities needed for life. Plasmodial slime molds are large single cells with multiple nuclei. *Plasmodium* is a parasitic protist that causes malaria. It constantly changes the way it appears and thus avoids detection by the host's immune system.

❺ Viruses are at the border between living and non-living.

Viruses consist of genetic material (RNA or DNA) packed inside a protein coat. A virus must infect a living cell before it can carry out any of the activities characteristic of living organisms; the virus hijacks the machinery of the cell and uses it to make more viral genetic material and viral protein. RNA viruses (such as those that cause influenza) mutate rapidly. Human immunodeficiency virus (HIV) is an RNA retrovirus. The complex method by which retroviruses replicate their RNA is very error-prone, so HIV mutates especially fast, complicating the use of drug treatment for HIV infections. In addition, HIV infects two kinds of cells that are part of the immune system, eventually causing the system to fail. This condition, acquired immunodeficiency syndrome (AIDS), is characterized by the occurrence of multiple bacterial and viral infections and, often, cancer.

KEY TERMS

CHECK YOUR KNOWLEDGE --

1. In the human body:
   a) there are more microbial cells than human cells.
   b) archaea can survive, but bacteria are killed by the immune system.
   c) bacterial cells can survive only inside the eukaryotic cells.
   d) the diversity of bacterial cells living in the mouth is exceeded by the diversity living on the skin.
   e) the bacterial cells are larger than the eukaryotic cells.

2. From an evolutionary perspective, the most basic division among all organisms on earth is among:
   a) plants, animals, and bacteria.
   b) bacteria, archaea, and eukarya.
   c) plants, animals, and eukarya.
   d) bacteria, animals, and eukarya.
   e) bacteria, archaea, and protists.

3. Which of the following statements about prokaryotes is incorrect?
   a) Prokaryotes contain ribosomes.
   b) Some prokaryotes can conduct photosynthesis.
   c) Prokaryotes have circular pieces of DNA within their nuclei.
   d) Prokaryotes appeared on earth prior to eukaryotes.
   e) Prokaryotes contain cytoplasm.

4. "Establishing a phylogeny for bacteria is more difficult than establishing a phylogeny for plants or animals." This statement is:
   a) correct, because of the quick generation time of bacteria.
   b) incorrect, because lateral gene transfer makes creating a phylogeny simpler.
   c) incorrect, because the metabolic diversity of bacteria makes classification impossible.
   d) correct, because it is difficult to obtain enough morphological data from bacteria.
   e) correct, because bacteria can engage in lateral gene transfer.

5. A population of *Escherichia coli* can double every _____ in an ideal laboratory culture.
   a) 20 seconds
   b) 2 minutes
   c) 2 days
   d) 20 minutes
   e) 2 hours

6. Plasmids containing genes for antibiotic resistance can be exchanged between bacterial cells in a culture by:
   a) conjugation.
   b) artificial exchange.
   c) cloning.
   d) transduction.
   e) conduction.

7. Which group of organisms utilizes the largest variety of energy sources?
   a) prokaryotes
   b) fungi
   c) protists
   d) invertebrate animals
   e) vertebrate animals

8. Which of the following statements about antibiotics is incorrect?
   a) Penicillin was the first antibiotic isolated from bacteria and widely used to fight bacterial infections.
   b) Antibiotics help microbes compete with other microbes.
   c) Antibiotic-resistant microbes are selected for in humans who are taking antibiotics.
   d) Antibiotics, though effective against viruses, are not effective against bacteria.
   e) Antibiotics are used not just in human health care but also in agriculture.

9. Which of the following domains are the most closely related, in that they share a unique common ancestor?
   a) archaea and bacteria
   b) archaea and eukarya
   c) bacteria and eukarya
   d) None of the above; all three domains evolved from different ancestors.
   e) None of the above; all three domains are equally related to each other.

10. Unlike other microbes, archaea are able to thrive:
    a) in very extreme (with respect to temperature, salinity, pressure, etc.) environments.
    b) in all the same places that bacteria are found.
    c) in the human body.
    d) in lakes and rivers (i.e., fresh water).
    e) in well-aerated soils.

11. *Taq* polymerase:
    a) is produced in viruses and bacteria.
    b) is denatured at temperatures above the normal human body temperature.
    c) is useful in catalyzing the polymerase chain reaction.
    d) can break down the sludge in oil refinery tanks.
    e) can reduce the mineral deposits from rusting pipes.

12. All protists are alike in that they all are:
    a) photosynthetic.
    b) eukaryotic.
    c) parasitic.
    d) colonial.
    e) multicellular.

13. Which of the following is not a protist?
    a) *Paramecium*
    b) slime mold
    c) brown alga
    d) diatom
    e) All of the above are protists.

14. Which of the following statements about *Plasmodium* is incorrect?
    a) It causes sickle-cell anemia.
    b) Because it changes its surface proteins frequently, it is largely invisible to the immune system.
    c) It is transmitted by *Paramecium*.
    d) It confers resistance to malaria.
    e) It is most common in North America.

15. Many biologists do not consider viruses to be alive. Which of the following characteristics of viruses leads to this conclusion?
   a) Viruses lack a metabolic system.
   b) Viruses do not respond to external stimuli.
   c) Viruses are unable to reproduce on their own.
   d) All of the above are correct.
   e) Only a) and c) are correct.

16. The genetic information in all viruses is:
   a) DNA.
   b) RNA.
   c) protein.
   d) polymerase enzymes.
   e) DNA or RNA.

17. Viruses are able to infect:
   a) humans.
   b) plants.
   c) bacteria.
   d) birds.
   e) All of the above.

18. Which of the following statements about HIV is incorrect?
   a) It is a virus.
   b) It attacks red blood cells.
   c) It mutates frequently.
   d) It is derived from a simian immunodeficiency virus.
   e) It contains RNA but not DNA.

SHORT-ANSWER QUESTION

1. Complete the chart on the characteristics of microbes.

	Prokaryotes		Eukaryotes	Viruses
	Bacteria	Archaea	Protists	
Unicellular, colonial, or multicellular?				
Plasma membrane?				
Peptidoglycan in cell wall?				
Flagella? What type?				
Presence of nucleus?				
Intracellular organelles?				
Type of genetic material (DNA, RNA, or both possible)?				
Chromosome structure (linear or circular)?				

See Appendix for answers. For additional study questions, go to www.prep-u.com.

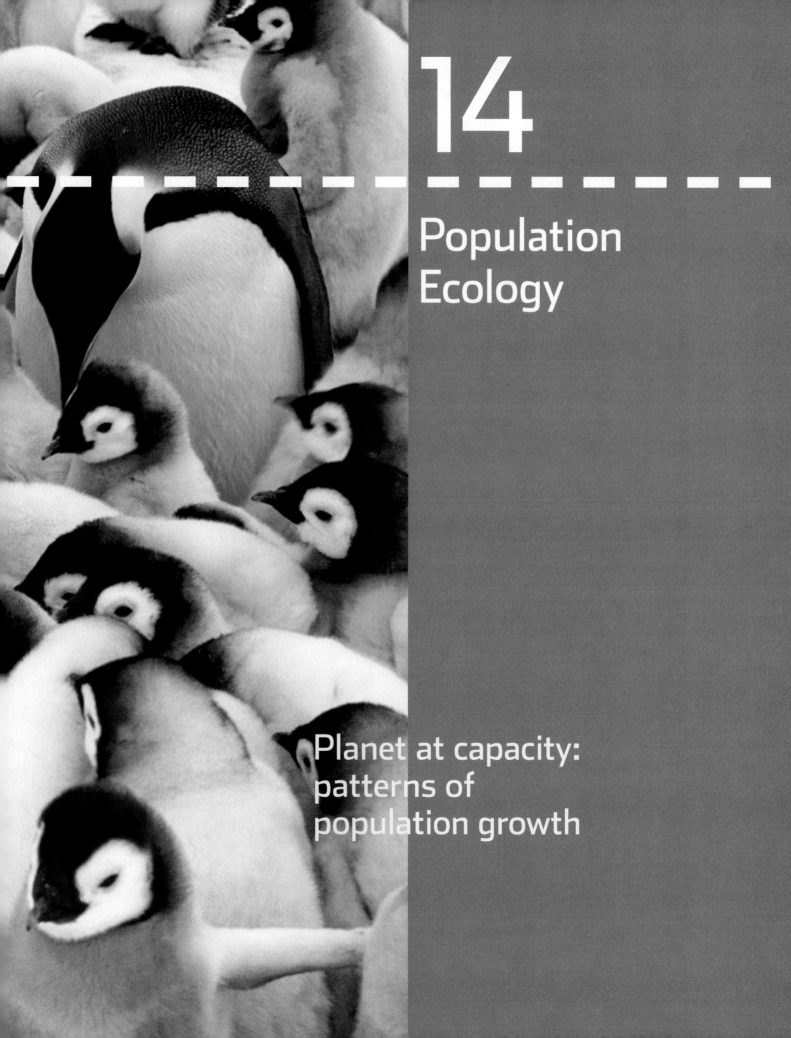

14
Population Ecology

Planet at capacity:
patterns of
population growth

Giraffe and zebra forage for food on the Masai Mara National Reserve, Kenya.

① Population ecology is the study of how populations interact with their environments.

14 • 1

What is ecology?

Lobster tastes delicious. Many people consider it one of the finest delicacies. It's not surprising, then, that catching and selling lobsters is big business. Generating almost $200 million a year for the State of Maine alone, it is among the most economically important businesses in New England. Now imagine for a minute that you were placed in charge of the lobster industry. The 6,000 lobster fishermen in Maine depend on your managing the lobster fishery so that not only can they catch and sell enough lobsters to survive this year, but sufficient numbers of lobsters are left behind to ensure that there will be lobsters to catch in all the years to come. How would you do it? You would be faced with some tough questions:

- How many lobsters should you allow each fisherman to take each day? Each year?

- Should you require that certain lobsters be thrown back? If so, should they be the biggest or the smallest? Does it matter whether they are males or females?

- Is it best to increase the lobster population size or should it be maintained at current levels?

Ecologists have a challenging task managing natural populations when it's difficult to count the number of individuals present or determine how many the environment can support.

FIGURE 14-1 Managing valuable resources. Population growth models can help.

But wait a minute. Forget about these obviously complex questions and start with a seemingly simpler question: how many lobsters currently live in the waters off the coast of Maine? Is it even possible to have a clue as to what this number is (**FIGURE 14-1**)?

Welcome to ecology.

These difficult questions, and many others like them, are all part of **ecology,** a subdiscipline of biology defined as the study of the interactions between organisms and their environments. But don't be fooled by the simple definition. Ecology encompasses a very large range of interactions and units of observation and is studied at different levels (**FIGURE 14-2**). These include:

- *Individuals:* How do individual organisms respond biochemically, physiologically, and behaviorally to their environment?

- *Populations:* How do groups of interbreeding individuals change, in terms of their growth rates, distributions, and genetic makeup, over time?

- *Communities:* How do the populations of different species in a locale interact with each other?

- *Ecosystems:* At the highest level of organization in ecology, how do the living and non-living elements interact in a particular area, such as a forest, desert, or wetland?

> I wandered lonely as a cloud
> That floats on high o'er vales and hills,
> When all at once I saw a crowd,
> A host, of golden daffodils.
>
> — William Wordsworth, *Daffodils*, 1804

We explore ecology at all of these levels throughout this and the next two chapters. We focus in this chapter on **population ecology,** a subfield of ecology that focuses on populations of organisms of a species and how they interact with the environment. We also examine the special case of human population growth and its impacts on the environment, and the ways in which ecological knowledge can give insight and answers that make possible effective conservation biology policies. In the next chapter, we move from populations to communities and ecosystems, where our focus will be on how species interact with each other and with their environments, and how this influences their own evolution and the environment itself.

TAKE-HOME MESSAGE 14·1

Population ecology is the study of the interactions between populations of organisms and their environments, particularly their patterns of growth and how they are influenced by other species and by environmental factors.

WHAT IS ECOLOGY?

Ecology is the study of interactions between organisms and their environments. It can be studied at many levels, including:

INDIVIDUALS
Individual organisms

POPULATIONS
Groups of individual organisms that interbreed with each other

COMMUNITIES
Populations of different species that interact with each other within a locale

ECOSYSTEMS
All living organisms, as well as non-living elements, that interact in a particular area

FIGURE 14-2 From individuals to ecosystems. Ecologists study living organisms and their relationship to the environment.

14·2
A population perspective is necessary in ecology.

As we begin our study of population ecology, a subtle but critical shift in perspective is required. It is a shift from viewing the individual as the primary focus of our attention to viewing the population—a group of organisms of the same species living in a particular geographic region. Most ecological processes cannot be observed or studied within an individual. Rather, they emerge when considering the entire group of individuals that regularly exchange genes in a particular locale.

Population ecology examines features that cannot be studied on an individual organism, such as population size.

FIGURE 14-3 **A change in perspective.** Ecology requires a focus on populations, rather than isolated individuals.

As an example of how a population rather than individual perspective is needed, consider the case of adaptations produced by natural selection. Genetic changes don't occur within an individual. That is, a longer neck did not evolve in one individual giraffe. Instead, the genetic changes occurred within a population of giraffes over time. As a consequence of differential reproductive success among the individuals of a population with different neck lengths, over time there came to be more individuals with longer necks. Birth rates, death rates, immigration and emigration rates, too, are features possessed not by individuals but by populations (**FIGURE 14-3**).

The shift from an individual to a population perspective should become clearer in the next sections. We begin by exploring the question of how quickly or slowly populations grow, a question important both to the management of consumable natural resources—as in the lobster fisheries described above—and to the conservation of rare and endangered species.

TAKE-HOME MESSAGE 14·2

Most ecological processes cannot be observed or studied within an individual. Rather, when studying them we need to consider the entire group of individuals that regularly exchange genes in a particular locale.

14·3
Populations can grow quickly for a while, but not forever.

In stable populations, how many of the five million eggs that a female cod might lay over the course of her life will, on average, survive and grow to adulthood? How many babies that an elephant gives birth to will survive to adulthood and reproduce? These may seem like difficult questions to answer, requiring complex mathematical models of population growth. But actually, the most fundamental fact about population growth is

Q Who leaves more surviving offspring, a pair of elephants or a pair of rabbits?

ridiculously simple and doesn't require special knowledge about fish or elephants. Reasoning and logic will give us the answer. Think about those two questions: how many cod eggs and baby elephants produced by one female will make it to adulthood?

Just two. If a population is not growing, each individual is ultimately replacing itself with a new individual. This is true

Population Ecology   Life Histories   Evolution of Aging   Human Population Growth

for all species. Two of those cod eggs will make it. And two of those baby elephants will make it. (For a stable population, two survive, rather than one, because both a male and a female contributed to each offspring. If each pair produced only one, the population would get smaller and smaller.)

> " There is no exception to the rule that every organic being naturally increases at so high a rate that, if not destroyed, the Earth would soon be covered by the progeny of a single pair. "
>
> —Charles Darwin, *The Origin of Species*, 1859

What would happen if more than two eggs or two babies survived? If there were more than one offspring per individual, the population would grow and grow and grow until the earth was covered with cod, elephants, and every other species. But that hasn't happened, and it can't. Let's investigate why.

We start by figuring out how a population grows if each individual does more than just "replace" itself. Two pieces of information will tell us how much a population is growing (or shrinking). The first is the growth rate, abbreviated as r. This is simply the rate of births minus the rate of deaths.

If the rate of births is greater than the rate of deaths, the population gets bigger. If the death rate is bigger than the birth rate, r becomes a negative number and the population gets smaller. For example, if there are 500 individuals in a population and, over the course of a year, 125 offspring are born, the birth rate is 125/500, or 0.25 births per individual. And if 25 of the 500 individuals die during the course of the same year, the death rate is 25/500, or 0.05 deaths per individual. In this population, then, the growth rate is 0.25 − 0.05, or 0.20 individuals per individual. But how many individuals is that?

To calculate the **growth rate** of a population—that is, the change in the number of individuals in the population in some unit of time, such as a year—we need to know the number of individuals in the population now. The growth of the population in a year is the growth rate times the number of individuals present, written as $r \times N$.

In the population just described, we said there were 500 individuals. This population would increase by 0.20 × 500, or 100 people, during a year (the time period observed). After a second year, if it grew at the same rate, the population growth would be 0.20 × 600, or 120 new individuals. This would give the population 720 individuals.

With the same calculations, we could determine the population size for the next 10 or even 50 years—these numbers are plotted in **FIGURE 14-4**. When a population grows at a rate that is proportional to its current size—in other words, the bigger the population, the faster it grows—the growth is called **exponential growth.** As the graph reveals,

EXPONENTIAL GROWTH

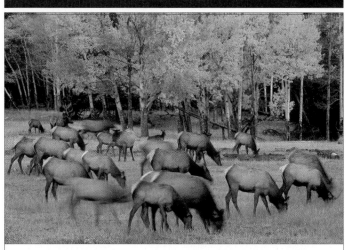

Exponential growth occurs when each individual produces more than the single offspring necessary to replace itself. According to realistic (and moderate) estimates of birth and death rates, a population of just 500 elk would grow to more than a billion individuals within 80 years and eventually would cover the earth.

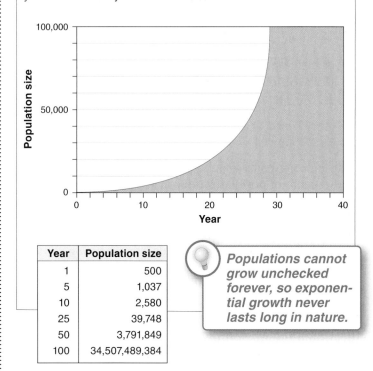

Year	Population size
1	500
5	1,037
10	2,580
25	39,748
50	3,791,849
100	34,507,489,384

Populations cannot grow unchecked forever, so exponential growth never lasts long in nature.

FIGURE 14-4 Unchecked growth. Any animal or plant species that grew exponentially would eventually cover the planet.

the size of a population growing exponentially very quickly becomes astronomical. In 10 years, our population would have grown from 500 individuals to 2,580. In 50 years, it would reach almost four million. In fact, after 80 years, the population would pass a billion individuals. It's clear that exponential population growth ends badly.

But the world isn't overrun with people or cod or elephants, so, clearly, exponential growth doesn't occur for long. In the next section, we explore why, in the real world, populations cannot simply grow unchecked forever, and we add some realism to the exponential growth description.

14·4

A population's growth is limited by its environment.

Life gets harder for individuals when it gets crowded. Whether you are an insect, a plant, a small mammal, or a human, difficulties arise in conjunction with increasing competition for limited resources. In particular, as population size increases, organisms experience:

- Reduced food supplies, due to competition

- Diminished accessibility to places to live and breed, also due to competition

- Increased incidence of parasites and disease, which can spread more easily when their hosts live at higher density

- Increased predation risk, as predator populations grow in response to the increased availability of their prey and as the more densely packed prey become more visible

❝ I would rather sit on a pumpkin and have it all to myself, than be crowded on a velvet cushion. ❞

—Henry David Thoreau, *Walden,* 1854

The limitations on a population's growth that are a consequence of **population density**—the number of individuals in a given area—are called **density-dependent factors,** and they cause more than discomfort. With increased density, a population's growth is reduced as limited resources slow it down. This ceiling on growth is the **carrying capacity, K,** of the environment. And as a population size approaches the carrying capacity of the environment, death rate increases, migration rate increases (as individuals seek more hospitable places to live), and a reduction in birth rate usually occurs, too, as low food supplies give rise to poor nutrition, which, in turn, reduces fertility (**FIGURE 14-5**).

Here's how the carrying capacity of an environment influences a population's growth. We start with our exponential growth equation, which calculated a

DENSITY-DEPENDENT FACTORS LIMIT POPULATION SIZE

DENSITY-DEPENDENT FACTORS
- Food supply
- Habitat for living and breeding
- Parasite and disease risk
- Predation risk

FIGURE 14-5 Fighting over scarce resources. These wolves cannot miss an opportunity for a meal.

population's growth as $r \times N$. Then we multiply this growth rate by a term, $[(K - N)/ K]$, that varies between 0 and 1 and can slow down exponential growth.

If the new term, $[(K - N)/ K]$, is close to 1, we are multiplying normal exponential growth by 1, which doesn't change it. This happens when N, the population size, is small relative to the carrying capacity, K (meaning that there are plenty of resources to go around). But if the term $[(K - N)/ K]$ is less than 1 and close to 0, we are multiplying the exponential growth by that small number, so total growth is now only a small fraction of what it would have been. This occurs when N, the population size, is very close to the carrying capacity, K (in which case, the environment is full to capacity). When a

Population Ecology   Life Histories   Evolution of Aging   Human Population Growth

LOGISTIC GROWTH

Logistic growth describes population growth that is gradually reduced as the population nears the environment's carrying capacity.

FIGURE 14-6 **Lack of resources limits growth.** Logistic growth is population growth that has stabilized because of limited resources.

population grows exponentially at first, but then its growth slows as the population size approaches the environment's carrying capacity—an S-shaped growth curve—the population growth is called **logistic growth** (FIGURE 14-6). Logistic growth is a much better approximation than exponential growth of how populations grow in the real world.

Density-independent factors can also knock a population down. These are forces that strike populations without regard for population size, increasing the death rate or decreasing the number and rate of offspring produced. Density-independent forces are like "bad luck" limits to population growth. They are mostly weather- or geology-based, including calamities such as floods, earthquakes, fires, and lightning. They also include habitat destruction by other species, such as humans. The population hit may be at its carrying capacity or it may be in the initial stages of exponential growth in which the population grows at an exponential growth rate, prior to having growth slowed by the carrying capacity of the environment. In either case, the density-independent force simply knocks down the population size. The population then resumes logistic growth.

In an environment in which these "bad luck" events repeatedly occur, a population might never have time to grow as high as the carrying capacity. Instead, the population might perpetually be in the exponential growth portion of the logistic growth curve, with periodic massive mortality events. The growth for this population would appear as a series of jagged curves, as seen in FIGURE 14-7.

DENSITY-INDEPENDENT FACTORS LIMIT POPULATION SIZE

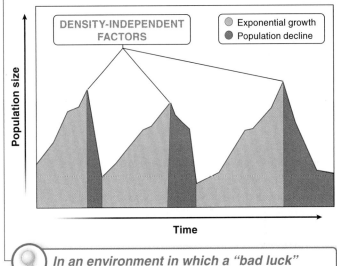

Density-independent forces include natural disasters such as floods, earthquakes, and fires.

In an environment in which a "bad luck" event repeatedly occurs, a population can be in a perpetual state of exponential growth with periodic massive mortality events.

FIGURE 14-7 **Events can limit population growth.** Fires, floods, and earthquakes are density-independent forces that can dramatically reduce population size.

The growth of populations doesn't always appear as a smooth S-shaped logistic growth curve. For some populations, particularly humans, the carrying capacity of an environment

Q How many people can the earth support? Why does the number keep increasing?

is not set in stone. Consider that, in 1883, an acre of farmland in the United States produced an average of 11.5 bushels of corn. By 1933, this was up to 19.5 bushels per acre. And by 1992, it had increased to 95 bushels per acre! How did this happen? The

Development of agricultural technologies is one example of how carrying capacity can be increased.

FIGURE 14-8 **Efficient crop production.** Advances in agriculture make it possible to feed the world's growing population.

development of several agricultural technologies—including the use of vigorous hybrid varieties of corn, rich fertilizers, crop rotation, and effective pest management—has made it possible to produce more and more food from the same amount of land. This is just one example of how the carrying capacity of an environment can be increased (**FIGURE 14-8**). Of course, over this same time, the carrying capacity has most likely been *decreased* for many other species trying to live in the same environment as humans and their crops.

TAKE-HOME MESSAGE 14•4

A population's growth can be reduced both by density-dependent factors related to crowding and by density-independent factors such as natural or human-caused environmental calamities.

14•5

Some populations cycle between large and small.

Nature is not as tidy as biologists might have you believe. Exponential and logistic growth equations, for instance, help us understand the concept of population growth but real populations don't always show such "textbook" growth patterns. Instead, they sometimes vary greatly in their rates and patterns of growth.

Recent desert locust swarms of biblical proportions certainly attest to the unpredictability of population growth (**FIGURE 14-9**). In northwest Africa, the desert locusts (migrating grasshoppers) normally live as solitary individuals in relatively small, scattered populations. In 2004, though, the population size increased rapidly, probably due to unusually good rains and mild temperature. As the rainy season ended and green areas gradually shrank, however, the locusts became progressively concentrated in smaller and smaller areas. At this point, for reasons that are related to their overcrowding but are not completely understood, the locusts began behaving like a mob, rather than solitary individuals. Giant swarms—some including tens of millions of insects—began flying across huge expanses of land in search of food. They completely consumed huge swaths of farmland, causing more than $100 million in damage.

There is more than one way for populations to deviate from the standard population growth pattern. The explosive locust population growth, for example, occurs at unpredictable intervals. Another unusual pattern is the oscillations of the lynx and snowshoe hare populations of Canada. As seen in **FIGURE 14-10**, rather than smooth logistic growth, the populations of

FIGURE 14-9 **Population explosion!** Sometimes, populations don't grow logistically, as occasionally happens for desert locusts.

both the snowshoe hare and their predators, the lynx, have regular cycles between very large numbers and crashes to much smaller numbers. Thanks to the Hudson's Bay Company, which kept detailed records for decades on the number of pelts it purchased from fur trappers, this population cycling is well documented. Although its cause is not fully certain, to some extent, the predator and the prey cause their own cycling:

1. The hare population size grows,

2. providing more food for the lynx,

3. which then reproduce at a higher rate,

Population Ecology   Life Histories   Evolution of Aging   Human Population Growth

POPULATION OSCILLATIONS

A growing hare population provides more food for the lynx, which then reproduce at higher rates.

Lynx eat too many hares, thereby reducing their food source and causing their own population to crash, which enables the hare population to grow.

Population size

— Snowshoe hares (prey)
— Lynx (predator)

Time

FIGURE 14-10 Predator and prey. The snowshoe hare and lynx populations rise and crash on a 10-year cycle. The population numbers are based on the number of pelts sold to the Hudson's Bay Company by trappers.

4. causing them to eat too many of the hares,

5. thereby reducing their food source,

6. causing the lynx population to crash,

7. thus enabling the hare population to grow, and the cycle to begin anew.

In discussing these examples of unusual population growth, we must not lose sight of the fact that, for the most part, regularly cycling populations and populations with periodic huge outbreaks are more the exception than the rule. In general, the logistic growth pattern describes the general growth of populations better than any other model.

One population-growth myth that demands debunking is the story of lemmings and their supposed suicidal response to overcrowding. Do lemmings jump off cliffs, committing suicide, when their population size becomes too large? In a word: *no*.

FIGURE 14-11 An urban myth debunked. Lemmings do not jump off cliffs.

This is an urban myth. Lemming populations do occasionally experience large increases in size. Then, just as locust behavior changes when population density gets too great, lemming behavior changes a bit, and many individuals migrate in search of less crowded habitats and more food. As they enter unfamiliar territory, some lemmings may suffer increased rates of death. But these deaths are not suicidal, and they do not occur in large groups (**FIGURE 14-11**).

How did the myth of lemming mass suicides become so prevalent? This can be attributed to a 1958 Disney documentary film, *White Wilderness*. In one sequence in this film, many lemmings were placed on a giant snow-covered turntable so that, when filmed, it would appear that they were migrating. In a later scene, many lemmings were filmed on a cliff overlooking a river, and then were chased into the water as if in a mad rush. None of these scenes were of actual lemming migrations, and in nature, lemmings never behave like this or do anything remotely resembling jumping off a cliff or committing suicide in any other way.

Q Do lemmings commit suicide by jumping off cliffs when their populations get too big?

There are practical reasons for predicting population sizes and growth rates, as we'll see. But it can be difficult to translate simple growth models into feasible management practices.

TAKE-HOME MESSAGE 14·5

Although the logistic growth pattern is better than any other model for describing the general growth pattern of populations, some populations cycle between periods of rapid growth and rapid shrinkage.

"Maximum sustainable yield" is a useful but nearly-impossible-to-implement concept.

Suppose you are in charge of an eastern hardwood forest, working for a logging company, and responsible for selecting which trees to cut down. Or suppose you are in charge of the lobster fishery discussed at the beginning of this chapter, responsible for deciding how many lobsters should be harvested. In either case, your responsibilities would be very similar. In the case of the hardwood forest, how many trees would you cut down for maximum wood yield?

This is a trick question. The maximum yield on any given day would be obtained by cutting *everything* down (also known as clear-cutting). No other strategy could yield more right now. But such a harvest can be done only once, so it isn't really a management strategy. A more sensible strategy would include harvesting some individuals and leaving others for the future. With that type of strategy, which trees (or lobsters) would you harvest so as to minimally affect the population's growth? Ideally, you would harvest those that are post-reproductive. After all, they are no longer contributing to the population growth. But there may not be many post-reproductive individuals in the population, or it may be too hard to identify them.

What's the better solution? For long-term management, it is best to harvest some but leave others still growing and reproducing for harvest at a later time. With such a strategy, the population can persist indefinitely. The special case in which as many individuals as possible are removed from the population without impairing its growth rate is called the **maximum sustainable yield** (FIGURE 14-12). The value in such a harvest comes from the fact that it can be carried out forever, clearly yielding more than the shortsighted strategy of a complete, one-time harvest.

Your first step as the manager of a hardwood forest or lobster fishery is to determine the maximum sustainable yield for the resource. This is actuallly straightforward. Maximum sustainable yield is calculated as that point at which the population is growing at its fastest rate. If we examine the logistic growth curve in Figure 14-6, we can see that the population is getting larger at the fastest rate when its size is equal to half the carrying capacity. At this midpoint, scarcity of mates is not a problem, as it can be at low population levels, and nor is competition, as it can be near the carrying capacity—one of the reasons that the population doesn't grow at all when it reaches the carrying capacity.

MANAGING NATURAL RESOURCES

Effective and sustainable management of natural resources requires the determination of a population's maximum sustainable yield, the point at which the maximum number of individuals are being added to the population (and so can be harvested or utilized).

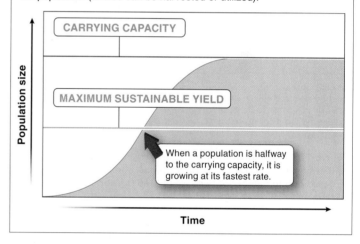

CARRYING CAPACITY

MAXIMUM SUSTAINABLE YIELD

Population size

Time

When a population is halfway to the carrying capacity, it is growing at its fastest rate.

FIGURE 14-12 Tree harvest. Wood is a renewable resource, but how many trees can we cut down (and which ones?) without irreversibly depleting the supply?

Q Virtually all natural resource managers working for the U.S. government fail to do their job exactly as mandated. Why?

Maximum sustainable yield is a useful concept, applicable not just to tree or lobster harvesting but also to livestock, agriculture, and nearly every other useful natural resource. And, in fact, there are 31 official U.S. government agencies that, in their charters, are mandated to utilize the maximum sustainable yield concept in determining harvest levels. But here's the rub: this is nearly always an impossible task. Why?

There are numerous reasons, most of which become apparent only when you leave the simplicity of the theoretician's desk and head into the messy real world. For starters, if maximum sustainable yield occurs when a population is at half its carrying capacity, do we first have to wait till the population stabilizes at its carrying capacity to determine what half of the carrying capacity would be? But isn't it inefficient to sit around waiting for the population to reach its carrying capacity, when we want to maintain it at half that size? Or can we just estimate carrying capacity and then calculate half of it? But won't that be difficult because lobsters live underwater (**FIGURE 14-13**)? Put simply, we rarely know the value of K.

And the problem gets worse. For starters, with many species, not only do we not know the carrying capacity, we don't even know the number of individuals alive. It is difficult to accurately count humans, so imagine how hard it is to count individuals of species that live underwater or fly or are microscopic.

And if we were to solve the mysteries of counting individuals and knowing carrying capacity, we would still have to figure out whether carrying capacity is stable from year to year. If carrying capacity depends on levels of resources, it may be cyclic, for instance. And even if we can figure out the carrying capacity and population size, we would not be certain which individuals ought to be harvested. Often, the individuals in a population are not contributing equally to population growth. The post-reproductive individuals, as mentioned above, do not contribute to this growth.

Maximum sustainable yield is a useful concept, but it is difficult to put into practice because it is hard to accurately determine population size and carrying capacity in nature.

FIGURE 14-13 How many lobsters can be harvested sustainably?

The concept of maximum sustainable yield generates insights into fighting biological pests such as cockroaches.

FIGURE 14-14 Taking over the kitchen. Population growth models don't just guide us to maximum production of valuable resources. They can help us efficiently reduce pest populations, too.

As a natural resource manager, you do your best, knowing that the theory behind the concept of maximum sustainable yield can almost never be put into practice perfectly.

Harvesting natural resources for maximum efficiency turns out to generate insights into fighting biological pests, such as cockroaches or termites (**FIGURE 14-14**). The problem is just turned on its head: which pest animals would you concentrate on killing so as to most effectively slow population growth? The pests to kill are those at the age of maturity, with the highest reproductive value, because they contribute most to the growth of the population. Similarly, because a population can still grow very rapidly with only a few males—given that any one male can fertilize a large number of females—it is most effective to focus pest control on females.

Patterns of population growth and the environmental features that influence them can lead to evolutionary changes in populations, shaping features such as life span, the age of reproduction, the number of offspring produced, and the amount of care given to them—the topics we turn to next.

TAKE-HOME MESSAGE 14·6

Based on models of population growth, it might seem easy to efficiently and sustainably utilize natural resources. In practice, however, difficulties such as estimating population size and carrying capacity complicate the implementation of such strategies.

❷ A life history is like an executive summary of a species.

A large population of flamingoes rests in Lake Nakuru National Park, Kenya.

14•7

Life histories are shaped by natural selection.

Some animals reproduce with a "big bang." *Antechinus* is a mouse-sized marsupial that lives in Australia, and male *Antechinus* are classic big-bang reproducers (**FIGURE 14-15**). At one year of age, they enter a two- to three-week period of intense mating activity, copulating for as long as 12 hours at a time. Shortly after this mating period, the males undergo rapid physical deterioration—they lose weight, much of their fur falls out, their resistance to parasites falls—and then they die.

Other animals are a bit—but only a bit—more restrained in their reproduction. The house mouse reaches maturity in about one month and produces litters of 6–10 offspring nearly every month, sometimes generating more than a hundred offspring in its first year of life.

And some animals could not be further from big-bang reproducers. The little brown bat is also mouse-sized, yet does not reach maturity until one year of age and typically

Q Do any animals mate themselves to death? Why?

produces only a single offspring per year. It can, however, live more than 33 years in its natural habitat.

Why all the variation? And is one strategy better than the others, evolutionarily? One of the most important recurring themes in biology is that evolution nearly always finds more than one way to solve a problem. As the marsupial mouse, house mouse, and little brown bat illustrate, there are many possible responses to the challenge of when to reproduce, how often to reproduce, and how much to reproduce in any given episode. The answers to these questions comprise an organism's **life history**—a sort of "executive summary" of the vital statistics of the species, including age at first reproduction, probabilities of survival and reproduction at each age, litter size and frequency, and longevity. Life histories are a little short on warm fuzzy stories about how plants and animals live, but they do tell us as much about a species as possible in a small amount of information.

Population Ecology   Life Histories   Evolution of Aging   Human Population Growth

VARIATION IN LIFE HISTORIES

Antechinus

BIG-BANG REPRODUCTION
• Reaches sexual maturity at one year
• Mates intensely over a three-week period
• Males die shortly after mating period

House mouse

FAST, INTENSIVE REPRODUCTIVE INVESTMENT
• Reaches sexual maturity at one month
• Produces litters of six to ten offspring every month

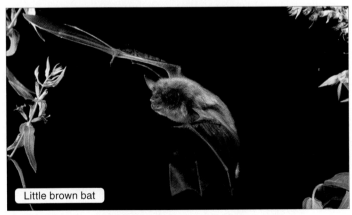

Little brown bat

SLOW, GRADUAL REPRODUCTIVE INVESTMENT
• Reaches sexual maturity at one year
• Produces about one offspring per year

FIGURE 14-15 Reproductive strategies. There are many strategies for reproducing, from the big bang of *Antechinus,* to the early, fast, abundant reproduction of the house mouse, to the slow but sure methods of the little brown bat.

Life histories vary from the big-bang reproductive strategy in *Antechinus,* in which all of the material and energetic contribution that an individual will make to its offspring, its **reproductive investment,** is made in a single episode, to strategies in which organisms have repeated episodes of reproduction, such as humans. Among plants, life histories have a similar range, from annuals such as peas, marigolds, and cauliflower, to perennials such as bananas and mint, which reproduce repeatedly.

Returning to the question posed above—which life history strategy is best?—we need to consider two questions.

1. *What is the cost of reproductive investment during any reproductive episode?* Producing offspring is risky. From the cost of producing gametes, to the risk of dying during the act of reproducing, to the wear and tear of the reproduction itself on an individual's body, reproduction takes its toll. The greater the reproductive investment in any given episode, the greater an individual's risk of dying. Thus, the number of offspring an organism produces can only go so high before it begins to be detrimental. An organism might produce four offspring in an episode, for example, without significantly increasing its risk of dying from the effort. But if, by chance, it produces five or six offspring, this may take such a toll that the individual is unlikely to live to have another litter. Thus, for many organisms, a smaller litter is favored by natural selection because it enables the individual to have additional litters in the future, maximizing its *lifetime* reproductive success.

2. *What is an individual's likelihood of surviving to have future reproductive episodes?* If predation rates or other sources of mortality are very high, an individual might not be alive in one month. If that is the case, it makes less sense for an individual to defer current reproduction. Why save for a future unlikely to come?

Taken together, the answers to these questions help us understand the variety of life history strategies we see in nature. At the slow extreme, humans have such a low probability of dying in any given year that they can defer reproduction, or at least reproduce in small amounts—once every few years—without much risk, enabling them to have just one offspring at a time but to invest significant parental effort in maximizing its likelihood of surviving. Organisms with this type of life history generally have evolved in such a way that their competitive ability—that is, their likelihood of

Q Why *do* humans put off mating so much longer than do cats or mice?

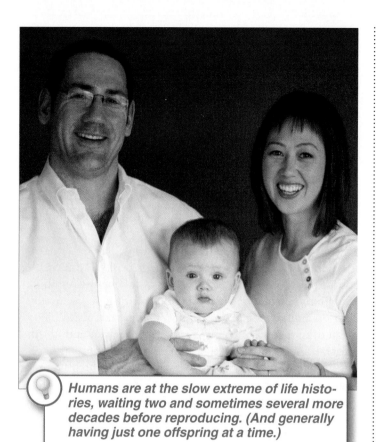

FIGURE 14-16 With such a low probability of dying in any given year, humans can defer reproduction.

Humans are at the slow extreme of life histories, waiting two and sometimes several more decades before reproducing. (And generally having just one offspring at a time.)

surviving—is high, although their rate of population growth is low. That is, not many offspring are produced, but those that are have a high likelihood of surviving (**FIGURE 14-16**).

Toward the fast extreme are rodents. They experience very high mortality rates in the wild, and natural selection has consequently favored early and heavy reproductive investment. Rodents provide relatively little parental care but produce numerous litters, each with many offspring. Organisms with this type of life history tend to have relatively poor competitive ability but a high rate of population growth; each individual's likelihood of surviving is low, but so many are produced that the odds of one surviving are good.

TAKE-HOME MESSAGE 14•7

An organism's investment pattern in growth, reproduction, and survival is its life history.

14•8

Populations can be described quantitatively in life tables and survivorship curves.

Biologists have learned some important lessons from insurance agents. Because insurance companies need to estimate how long an individual is likely to live, about a hundred years ago they invented something called the "life table." It ought to be called a death table, though, because in the table, insurers tally the number of people in a population within a certain age group, say 0 to 10 years old or 10 to 20 years old, and the number of individuals dying within that age range. From these numbers, they can predict an individual's likelihood of either dying within or surviving a particular age interval. For the insurance companies, their ability to make money hinges on making accurate life span predictions. For biologists, a life table is like a quick window into the lives of the individuals of a population, showing how long they're likely to live, when they'll reproduce, and how many offspring they'll produce.

From life tables, biologists can create **survivorship curves,** graphs showing the proportion of individuals of a particular

age that are now alive in a population. Survivorship curves indicate an individual's likelihood of surviving through an age interval. And they reveal, at a quick glance, a huge amount of information about a population, such as whether most offspring produced die shortly after birth—think back to the five million eggs produced by some fish; or whether most survive and are likely to live long lives—think about humans in the United States.

Three distinct types of survivorship curves are given in **FIGURE 14-17**. The curves plot the proportion of individuals surviving at each age, across the entire range of ages seen for that species.

Type I: At the top of Figure 14–17, in purple, is the survivorship curve seen in most human populations, and shared by the giant tortoise. We have a very high probability of surviving every age interval until old age, then our risk of dying increases

SURVIVORSHIP CURVES

Proportion of individuals in population surviving

TYPE I

TYPE II

TYPE III

Age

Giant tortoise

Painted bunting

Mackerel

TYPE I
High survivorship until old age, then rapidly decreasing survivorship

TYPE II
Survivorship decreases at a steady, regular pace

TYPE III
High mortality early in life, but those that survive early years live long lives

FIGURE 14-17 Three types of survivorship. Survivorship curves show the proportion of individuals of a particular age that are alive in a population.

dramatically. Species with this shape of survivorship curve, including most large mammals, usually have a few features in common. They have few natural predators, so they are likely to live long lives. They tend to produce only a few offspring—after all, most of these offspring will survive—and they invest significant time and effort in each of their offspring.

Type II: In the middle, in blue, is a survivorship curve seen in many bird species such as the painted bunting, and in small mammals such as squirrels. The straight line indicates that the proportion alive in each age interval drops at a steady, regular pace. In other words, the likelihood of dying in any age interval is the same, whether the bird is between 1 and 2 years old, or between 10 and 11.

Type III: At the bottom, in green, is a survivorship curve in which most of the deaths occur in the youngest age groups. Common in most plant and insect species, as well as in many marine species such as oysters and fish, this survivorship curve describes populations in which the few individuals lucky enough to survive past the first few age intervals then become likely to live a much longer time. Species with this type of survivorship curve tend to produce very large numbers of offspring, because most will not survive. They tend, however, not to provide much parental care, if any. A classic example might be a fish such as the mackerel. One female might produce a million eggs! Obviously, most of these (on average, 999,998) must not survive to adulthood or else the planet would be overrun with mackerel.

In reality, most species don't have survivorship curves that are definitively type I, II, or III. They may be anywhere in between. These three types, though, represent the extremes and help us to easily make predictions about reproductive rates and parental investment without extensive observations on individual behavior. Survivorship curves can also change over time. Humans in developing countries, for example, tend to have higher mortality rates in all age intervals—particularly in the earliest intervals—relative to individuals in industrialized countries.

TAKE-HOME MESSAGE 14·8

Life tables and survivorship curves summarize the survival and reproduction patterns of the individuals in a population. Species vary greatly in these patterns: the highest risk of mortality may occur among the oldest individuals or among juveniles, or mortality may strike evenly at all ages.

There are tradeoffs between reproduction and longevity.

If you were designing an organism, how would you structure its life history for maximum fitness? (Recall from Chapter 8 that fitness is defined as an organism's reproductive output relative to that of other individuals in the population.) Ideally, you would create an organism that could:

- produce many offspring,

- beginning just after birth,

- continuing every year,

- while growing tremendously large, to reduce the predation risk, and

- living forever.

Unfortunately, evolution operates with some constraints. Some of these traits are not possible because selection that changes one feature tends to adversely affect others.

The impossibility of maximizing all of these features is a consequence of evolutionary tradeoffs. When evolution increases one life history characteristic, frequently another characteristic may decrease. Life history tradeoffs can be better

> **Q** With one simple surgical procedure, most men could add more than 10 years to their life span. Why doesn't anyone opt for it?

understood in the context of the three areas to which an organism can allocate its resources—growth, reproduction, and survival. When resources are limited, life history evolution is similar to a zero-sum game: increased allocation to one of these areas tends to reduce allocation to one or both of the others.

Some examples of the best-studied tradeoffs include:

Reproduction and survival. Among big-bang reproducers such as the marsupial mouse, *Antechinus,* and fish such as salmon, investment in reproduction is exceedingly high, followed shortly by death (**FIGURE 14-18**). If individuals are prevented from reproducing by physically separating them from individuals of the opposite sex, however, they can live many years more than if they reproduced.

Reproduction and growth. Beech trees grow more slowly during years when they produce many seeds than in years when they produce no seeds. Similarly, female red deer kept from

LIFE HISTORY TRADEOFFS

REPRODUCTION AND SURVIVAL
Big-bang reproducers such as salmon make a single exceptionally high investment in reproduction, then die shortly afterward.

REPRODUCTION AND GROWTH
Beech trees grow much more slowly in the years when they produce many seeds than they do in years when they produce few seeds.

NUMBER AND SIZE OF OFFSPRING
Female lizards of the species *Uta stansburiana* produce medium-sized eggs as a compromise between a large number of small-sized eggs (with poor survival of offspring) and few large-sized eggs (with better survival of offspring).

FIGURE 14-18 Life history tradeoffs. When resources are limited, an organism's increased allocation of resources in one area tends to reduce allocation in another.

Population Ecology | Life Histories | Evolution of Aging | Human Population Growth

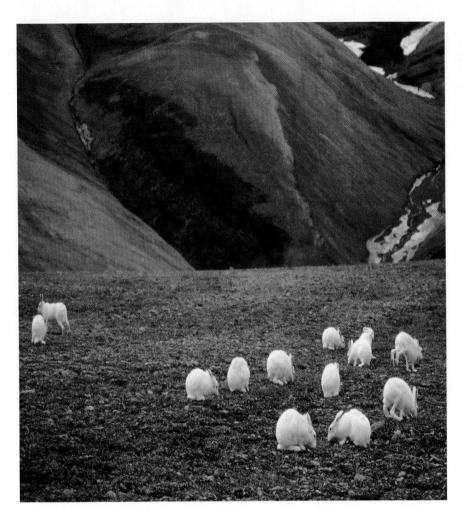

The Arctic hare lives in the tundra of Canada, on Arctic islands and Greenland. The female has one (and sometimes) two litters per year, with an average of slightly more than 5 babies per litter. Hares found in more moderate climates have several litters per year with smaller average litter sizes.

reproducing in their first year of life grow significantly larger during that year.

Number and size of offspring. The females of one lizard species, *Uta stansburiana,* can lay more eggs if the eggs are smaller, but a higher proportion of the offspring survive if the female lays larger eggs. This species has evolved to produce a happy medium, with an egg size that tends to maximize the number of offspring produced.

There are many other tradeoffs beyond these, such as number of offspring produced and the amount of parental care given. Additionally, among mammals, litter size increases with latitude (distance from the equator). This trend may be due to the tradeoff between litter size and the number of litters per year. Closer to the equator, there is less seasonal variation in weather and animals can breed for a longer portion of the year (year-round near the equator). Farther from the equator, the winters become more severe and the breeding season is shorter. To compensate for having fewer litters at higher latitudes, mammals can produce more offspring per litter.

Can you think of other tradeoffs? How would you test whether or not they exist?

TAKE-HOME MESSAGE 14•9

Because constraints limit evolution, life histories are characterized by tradeoffs between investments in growth, reproduction, and survival.

❸ Ecology influences the evolution of aging in a population.

Elephants live many years but have few offspring.

14•10

Things fall apart: what is aging and why does it occur?

Attempting to define what was obscene, an exasperated justice of the United States Supreme Court once concluded that he might never be able to define obscenity, "but I know it when I see it." Aging is like that, as well. Picture your grandparents or someone in their seventies or eighties. Clearly they have aged. But what exactly does that mean? And does it differ from the way that people in their thirties or forties have aged? Sometimes it involves sagging skin. Sometimes it involves weakened muscles and bones. Sometimes it involves senility.

The difficulty comes when we realize that no two people age in exactly the same way, yet without fail, everyone does age (**FIGURE 14-19**). So we must retreat to a somewhat vague definition. For an individual, aging is the gradual breakdown of the body's machinery. With this definition, each individual can experience aging differently, but because this physiological deterioration makes a person more likely to die, aging can be seen much more clearly by examining a population as a whole.

From a population perspective, **aging** emerges as a definitive and measurable feature: it is simply an increased risk of dying with increasing age. For example, among humans, a person is

more likely to die between the ages of 70 and 71 than between 60 and 61. Similarly, she is more likely to die between the ages of 44 and 45 than between 34 and 35. The causes of death at these ages often differ, but the shuffle toward death becomes more and more dire, with no respite. This is the most useful definition of aging for ecologists, and it doesn't apply only to humans. Aging can be measured and assessed for any species in this same way.

Jeanne Louise Calment lived for more than 122 years, the longest life span ever documented for a human. While this is a spectacularly long life relative to most organisms on earth (and is significantly longer than the average life span in the United States, 78 years), it pales in comparison with some of the longest-lived organisms. Bristlecone pine trees, for example, are the longest-lived, surviving for thousands of years. A counting of the growth rings of one bristlecone pine indicated that it was more than seven thousand years old!

Among the animals, lobsters and quahogs (a type of clam) are the longest-lived invertebrates (up to 100 and 200 years, respectively). Long-lived vertebrates include striped bass,

Population Ecology   Life Histories   ▼ Evolution of Aging   Human Population Growth

FIGURE 14-19 **What does "aging" mean?** Aging can occur differently from one individual to another, but we recognize individuals of different ages without any difficulty.

which can exceed 120 years, and tortoises, which can reach 150 years. Toward the other end of the spectrum, rodents may live just a couple of years, and fruit flies and many other insects generally live and die within the span of a few weeks. Interestingly, bird and bat species live 5–10 times longer than similarly sized rodents. These data prompt the questions: why is there so much variation? And what controls how long the individuals of a species live? We explore those questions in the next section, but first we consider the more general question of why organisms age at all.

> " Things fall apart; the centre cannot hold;
> Mere anarchy is loosed upon the world,
> The blood-dimmed tide is loosed, and everywhere
> The ceremony of innocence is drowned;
> The best lack all conviction, while the worst
> Are full of passionate intensity.
>
> —William Butler Yeats, *The Second Coming*, 1919 "

Aging, in fact, is not a mystery. The question of why we age is one of the great solved problems in biology. Oddly, though, it may also be one of the best-kept secrets in science. Let's investigate.

It is easier to *state* the evolutionary explanation for why we age than to *understand* it. But that's where we will start. The explanation is based on this one fundamental fact: the force of natural selection lessens with advancing age. What does that

mean? Although it sounds complex, it becomes much clearer with two examples.

1. Imagine a mutation that causes a person carrying it to die at the age of 10. Will the person carrying that mutation pass it on to many offspring? Of course not. The carrier of that mutant allele will die before she gets a chance to pass it on to anyone. Alternative versions of the gene that don't cause death will be the only ones that persist.

> **Q** Many genetic diseases kill old people, but almost none kill children. Why not?

2. Now imagine a mutation that causes a person carrying it to die at the age of 150. Will the person carrying that mutation pass it on to many offspring? Yes! The carrier of this killer allele will already have had children—passing on the mutation to them—long before she even knows she carries it. In fact, she will no doubt die long before this mutation even has the opportunity to exert its disastrous effect. The same thing happens if the mutation has its negative effect at age 100 . . . or 70 . . . or even 50.

In the first example, we see that natural selection "weeds out" alleles that cause sickness and death early in life. Those alleles never get passed on (**FIGURE 14-20**). No one inherits them, and they disappear from the population without fail. Individuals with the alternative versions of the gene are healthier and have more and healthier offspring, and pass on the good versions of the gene at a higher rate. Consequently, those good alleles end up as the only versions of the gene in the population. These are the genes that we inherited from our ancestors, and this is why very few people have genetic diseases that kill them in their teens. Natural selection "cleans out" the genomes of a population. We tend to think of natural selection as acting positively to increase or improve a trait—the length of a giraffe's neck, for example. But natural selection can also act negatively, selecting *against* a trait—in this case, disease.

Natural selection isn't so effective with some other genes. The second example describes mutations that arise and cause their carrier to be more likely to die later in life. Examples of such mutations include those that increase the risk of cancers or heart disease or other types of ailments. Unfortunately, the carriers of these mutant alleles will already have reproduced and passed them on before they have had their negative effect. In other words, it doesn't matter whether you carry one of these mutants or an alternative that does not harm you, your **reproductive output**—and evolutionary fitness—is the same. Consequently, these mutants are never cleaned out of a population. The practical result is that we all have inherited many of these mutant alleles that have arisen over the past hundreds and thousands of generations among the multitudes of our ancestors.

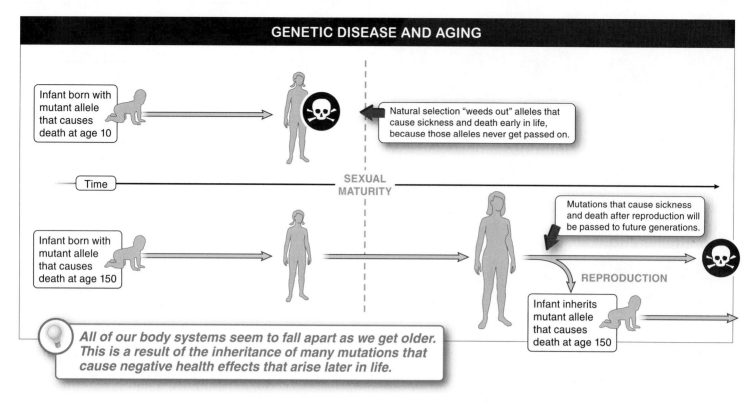

Infant born with mutant allele that causes death at age 10

Natural selection "weeds out" alleles that cause sickness and death early in life, because those alleles never get passed on.

Time

SEXUAL MATURITY

Infant born with mutant allele that causes death at age 150

Mutations that cause sickness and death after reproduction will be passed to future generations.

REPRODUCTION

Infant inherits mutant allele that causes death at age 150

All of our body systems seem to fall apart as we get older. This is a result of the inheritance of many mutations that cause negative health effects that arise later in life.

FIGURE 14-20 **Carrying a time bomb.** Natural selection can't weed out genetic defects that wait until we've already reproduced to harm our health. Those that act early in life will be passed on to subsequent generations at a greatly reduced frequency. The cumulative effects of the buildup of late-acting harmful alleles are responsible for what we see as aging.

Q A cure for cancer may be discovered, but not for aging. Why the difference?

These mutant alleles that have their adverse health effects later in life—after we are likely to have reproduced—are responsible for aging. The later in life they have their effect—and therefore the more likely it is that they are passed on to offspring before their effect appears—the more common they will be in the population. This is why there are *many* causes of death, and why it seems as if all of our bodily systems fall apart as we get older.

TAKE-HOME MESSAGE 14·10

Natural selection cannot weed out bad alleles that do not diminish an individual's relative reproductive success. Consequently, they accumulate in the genomes of individuals of nearly all species. This leads to the multiple physiological breakdowns that we see as aging.

14·11

What determines the average longevity in different species?

Organisms age because the force of natural selection acting against harmful genes lessens with advancing age. But why do some species age very quickly and others more slowly? As we've seen, fruit flies live for about a month and rodents live for a couple of years, while tortoises, elephants, and humans live for many decades.

To answer this question, first imagine a species in which the organisms don't age at all. They never become decrepit and frail. Would such individuals live forever? No. Just because they are not at risk from internal, genetic sources of mortality doesn't mean that there aren't things in the world around them that might kill them. Lightning. Predators. Drought.

Disease. The list is long. And for different species, these risks are more or less serious. Rodents, for example, are at *very* high risk of predation nearly every minute of the day. Without perpetual vigilance, they become the meal of a bird of prey or some other animal. Even *with* perpetual vigilance, their risk of death is high every day. Consequently, they are likely to be dead within a few years (**FIGURE 14-21**).

Because death from some external source is likely to come so early, these organisms living in high-risk worlds must reproduce even earlier. Natural selection favors this strongly. (If they didn't reproduce earlier, they would not leave any descendants.) Now, when a harmful mutation arises that causes its damage at an early age—say, 1 or 2 years of age—the rodent is likely to have already reproduced and passed the mutation on. And so that harmful allele increases in frequency in the population, causing individuals to age earlier.

Humans and tortoises live in relatively low-risk worlds. Because we have few predators, and because tortoises have armored-car-like shells that protect them from danger, death from external sources is low. Early, intensive investment in reproduction is not necessary and so has not been favored by natural selection. In

these species, when a harmful mutation arises that has its effect at 5 or 10 years of age, the individual carrying it is likely to die before it has reproduced. Consequently, the mutation isn't passed on; natural selection weeds it out of the population. Only if the harmful effects occur much later in life will the mutation be passed on to future generations.

The age at time of reproduction, then, is a key factor determining longevity. Factors that favor early reproduction will also favor early aging; factors that don't favor early reproduction—or actively favor later reproduction—favor later aging. But then, what determines when an organism reproduces?

We can think of each population as evolving in a world with a specific **hazard factor** (see Figure 14-21). These factors include the risks of death for individuals in that population from all types of external forces. When the hazard factor is low, individuals of a species tend to reproduce later, and so natural selection can weed out all harmful mutations except those that have their effects very late in life. Individuals in these species will live a long time before they succumb to aging. A high hazard factor, on the other hand, will lead to earlier reproduction, and therefore early aging and shorter life spans.

Can you predict which species should age more slowly in captivity, a porcupine or a guinea pig? The difference is large, revealing unambiguously that it's good to be a porcupine. Nobody wants to eat you. Or rather, nobody can eat you. Because of their sharp quills, porcupines live in a world with little risk of predation. This is another way of saying that as sharp quills evolved, the world in which porcupines have been evolving has for them a lower and lower hazard factor.

Guinea pigs, by contrast, live with a high risk of predation. Because they've been evolving in a world characterized by a high hazard factor for them, they reproduce much sooner, age more quickly, and die younger. A porcupine in captivity can live 21 years, and a captive guinea pig lives less than 10 years. In the wild the difference is even greater: 15 years for the porcupine and only 3 or 4 years for the guinea pig. Using the same reasoning, what difference would you expect between bats and mice, or between poisonous and non-poisonous snakes?

ENVIRONMENTAL RISKS AND AGING

HIGH HAZARD FACTOR
• Individuals tend to reproduce earlier
• Earlier aging
• Shorter life spans

2 years

Rodent

LOW HAZARD FACTOR
• Individuals tend to reproduce later
• Later aging
• Longer life spans

150 years

Tortoise

Longevity

In environments characterized by low mortality risk, populations of slowly aging individuals with long life spans evolve. In environments characterized by high mortality risk, the opposite occurs.

FIGURE 14-21 Hazard factor. The mouse and the tortoise face different risks of dying.

TAKE-HOME MESSAGE 14·11

The rate of aging and pattern of mortality for any species are determined by the hazard factor of that organism's environment. In environments characterized by low mortality risk, populations of slowly aging individuals with long life spans evolve. In environments characterized by high mortality risk, populations of early-aging, short-lived individuals evolve.

14•12

Can we slow down the process of aging?

Life extension is possible. Real, honest-to-goodness doubling of the normal life span. Not only is it possible, researchers have demonstrated it repeatedly. The only problem is that such life extension has been achieved only in fruit flies.

Although the results of these life-extension studies are incredible, the methods by which they were achieved are simple. Researchers began with several large populations of fruit flies maintained and propagated by a standard method in the laboratory. Each population lived in a small Plexiglas cage the size of a shoe box. There they would feed (and lay eggs) on a small plate of food consisting of bananas, corn syrup, and molasses. Every few days, a fresh plate of food would be put in the cage and the old one discarded. At 2 weeks, however, eggs

were collected from the removed food plate and used to start the next generation of flies. The adult flies in the cage were then discarded. Propagated this way, the fly populations had been maintained for hundreds of generations. When put under ideal conditions outside the cage, the flies started to experience the physiological breakdowns associated with aging after a few weeks and, after about a month, they died.

One researcher had the idea that changing the force of natural selection on the flies would produce an evolution in the aging pattern. He did this in a clever way. Instead of collecting eggs at 2 weeks to start the new generation, he waited for 3 weeks (FIGURE 14-22). This meant that every fly in the new generation had a parent that must have lived for 3 weeks. If

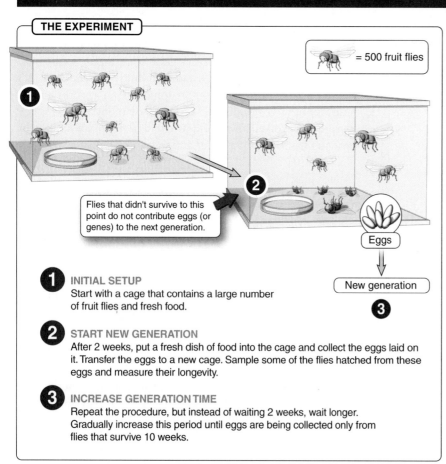

FIGURE 14-22 **Creating longer-lived organisms through evolution in the laboratory.**

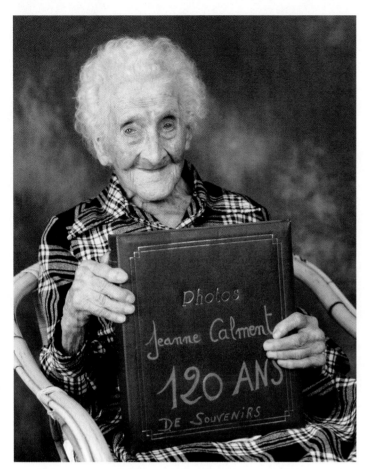

Can the life span of humans be extended? The longest-living human, Jeanne Louise Calment, lived more than 122 years.

any fly carried a mutation that caused it to die between 2 and 3 weeks, that fly did not contribute to the next generation and that mutation was not passed on.

After doing this for several generations, the researcher started collecting the eggs for the next generation at 4 weeks instead of 3 weeks. At this point, the flies making up the new generation all had parents that had survived for 4 weeks. Any flies that died prior to 4 weeks didn't contribute to the new generation. Natural selection was now able to reduce the prevalence of mutant genes responsible for early aging.

This experimental approach was continued, progressively increasing the generation time to 5 weeks, then 6, then 7, and ultimately to 10 weeks! At this point, only those flies with a genome free of mutations that might cause death in the first 10 weeks of life could contribute to the new generation. The net result of these experiments was the creation of flies with much "cleaner" genomes. When put under ideal conditions outside the cages, these "super" flies did not experience aging until long after the original flies would have been dead. In fact, the average life span of the new flies was more than 60 days—double that of the original flies! These experiments are the equivalent of decreasing the hazard factor of the fly populations.

Q Is it possible, with our current knowledge, to double longevity in humans?

TAKE-HOME MESSAGE 14·12

By increasing the strength of natural selection later in life, it is possible to increase the mean and maximum longevity of the individuals in a population. This occurs in nature and has also been achieved under controlled laboratory conditions.

4 The human population is growing rapidly.

At capacity? How many people can the earth support, living at a reasonable level of dignity and in harmony with other species?

14·13 -

Age pyramids reveal much about a population.

The "baby boom." It happened half a century ago, yet young people today may end up paying a price for it. What exactly was it? And why does it still matter so much? For insight into this question, consider another question. Imagine that you run a grade school. How many teachers will you need for the first grade in the new school year? That all depends on how many students will be entering. If it was 50 students last year, will it be 50 students again this year? Probably. The number of students is likely to be similar from year to year. But it can change. If we set aside changes due to migration, it depends on how many babies were born six years ago. If that number was higher than in the previous year—because more people had babies—you may need more teachers at your school. Or you may need fewer teachers if fewer babies were born.

Increasing birth rates caused the baby boom. Beginning in the late 1940s, just after World War II, and continuing through the early 1960s, families in the United States had many more babies—about 30% more per month—than they would have had if they had

followed the same pattern as their parents. Then the birth rates began to return to their earlier levels, where they have remained ever since. As the babies from the boom years grew to school age, schools had to increase in size to accommodate all the children. Then the schools had to decrease in size once the baby boomers had graduated. Now, as these individuals approach retirement age, the question of how their retirement and health care needs will be met is one of the biggest issues facing society.

Q What is the baby boom? Why is it bad news for young people today?

Populations often vary across space—dense in the cities and more sparse in rural areas. The baby boom illustrates that populations aren't just distributed in space. They also have an "age distribution" as well. Just as there may be more individuals in some areas and fewer in others, there may be more individuals in certain age groups and fewer in others.

Describing populations in terms of the proportion of individuals in each age group reveals interesting population

Population Ecology   Life Histories   Evolution of Aging   Human Population Growth

features. A population can be divided into the percentage of individuals that are in specific age groupings, called *cohorts:* for example, 0–5 years, 6–10 years, 11–15 years, and so on, in an "age pyramid." If two populations are the same size but have different age distributions, they will have some very different features. People have radically different likelihoods of dying or reproducing, for instance, based on their age. A 10-year-old isn't going to produce any offspring, but a 30-year-old has a relatively high likelihood of reproducing. And a 10-year-old is unlikely to die in the next year, whereas that is much more likely for an 80-year-old.

It can be useful, therefore, for a society to know the relative numbers of 10-, 30-, and 80-year-olds in its population. The age data can be used to determine whether a society would be better off investing in new schools or fertility wards or convalescent homes, among other social issues. Companies, too, rely on such demographic data to best plan their strategies for producing the goods that people will want and need.

Around the world, countries vary tremendously in the age pyramids describing their populations. Most industrialized countries are growing slowly or not at all. In these countries, the age pyramid is more rectangular than pyramid-shaped: because birth rates are low, the bottom of the pyramid is not

very wide (**FIGURE 14-23**). And because death rates are low, too, the higher age classes in the pyramid don't get significantly narrower. Instead, they remain about the same size all the way into the late sixties and seventies, at which point high mortality rates finally cause them to become narrower. This gives the pyramid a rectangular shape. In these countries, most of the population is middle-aged or old.

A shape more like a pyramid is seen in the age pyramids of developing countries. Mexico and Kenya, for example, have very high birth rates, reflected as a large base of their age pyramid (see Figure 14-23). But high mortality rates, usually due to poor health care, cause a large and continuous reduction in the proportion of individuals in older age groups. In these countries, most of the population is in the younger age groups.

The shape of the age pyramid in the United States has economists worried that the social security system (including Social Security and Medicare) will not be able to offer older citizens sufficient benefits in the next 10 to 30 years. There is a "bulge" in the U.S. age pyramid (**FIGURE 14-24**). Because of the unusually large number of babies born about 50 years ago, there is now an unusually large number of people reaching retirement age. Since the baby boomers were born, there

FIGURE 14-23 **A visual representation of population growth.** Age pyramids go from triangular to rectangular when population growth stabilizes.

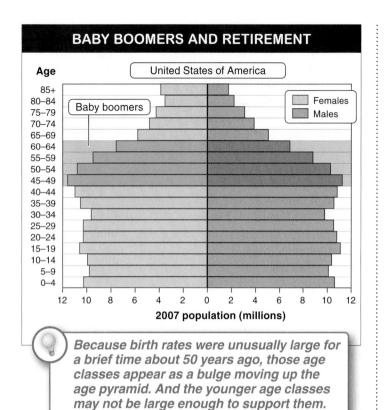

BABY BOOMERS AND RETIREMENT

United States of America

Age

Baby boomers

Females
Males

85+
80–84
75–79
70–74
65–69
60–64
55–59
50–54
45–49
40–44
35–39
30–34
25–29
20–24
15–19
10–14
5–9
0–4

12 10 8 6 4 2 0 2 4 6 8 10 12

2007 population (millions)

Because birth rates were unusually large for a brief time about 50 years ago, those age classes appear as a bulge moving up the age pyramid. And the younger age classes may not be large enough to support them.

FIGURE 14-24 **The retirement boom.** As baby boomers reach retirement, the population bulge may strain the social security system in the United States.

hasn't been such a large cohort produced each year. This means that the current numbers of working individuals who contribute to the social security system are not sufficient to cover the payouts promised to the large number of retirees, and the baby boomers will be expensive to support as they reach retirement.

Possible solutions include investing the money held by the system in higher-risk investments that offer the possibility of higher rates of return, raising the amount put into the system by current workers, reducing the amount of the payouts to retirees, or the baby boomers postponing retirement.

TAKE-HOME MESSAGE 14•13

Age pyramids show the number of individuals in a population within any age group. They allow us to estimate birth and death rates over multi-year periods.

14•14

Moving from third world to first world a demographic transition occurs.

Populations can sometimes appear to be as strange and stubborn as individuals. Around the world, governments encourage (almost coerce) their citizens to reduce their reproductive rates in efforts to check population growth, but they are rarely successful. Yet it seems that when the governments stop trying, their countries' population growth rates slow down all on their own.

It's not that populations actually have minds of their own. Rather, many countries have undergone industrialization and, in the process, their patterns of population growth follow common paths, marked first by periods of faster growth and later by slower growth. The sequence of changes is remarkably consistent.

Start with a preindustrial country. Such countries usually have high birth rates and high death rates. These result from poor and inefficient systems of food production and distribution,

along with a lack of reliable medical care. As industrialization begins, both food production and health care improve. These improvements inevitably lead to a reduction in the death rate. The birth rate, however, remains relatively high and so the country's population grows rapidly.

As industrialization continues, though, further changes occur. Most importantly, the standard of living increases. This results from higher levels of education and employment and, in conjunction with improved health care, this finally causes a reduction in the birth rate. The new, lower birth rate then slows the population's growth. The progression from:

1. high birth rates and high death rates (slow population growth) *to*

2. high birth rates and low death rates (fast population growth) *to*

3. low birth rates and low death rates (slow population growth)

is called the **demographic transition** (FIGURE 14-25).

The demographic transition can take decades to complete, so it is not always easy to identify it as it occurs. A survey of countries around the world, though, reveals many at different points on the transition. Sweden, for example,

Q Population growth is alarmingly slow in Sweden and alarmingly fast in Mexico. Why is there a difference?

has a low fertility rate (1.7 children born per woman) and a low death rate (10.4 deaths per 1,000 people). At the other extreme, Nigeria and Ethiopia have very high fertility rates (5.5 and 5.3 children born per woman) and high death rates (17.2 and 15.1 deaths per 1,000). Mexico, in contrast, is in the midst of a clear demographic transition, with a moderately high fertility rate (2.45 children born per woman) yet a very low death rate (4.7 deaths per 1,000 people).

In the past few decades, the demographic transition has been completed in Japan, Australia, the United States, Canada, and most of Europe, leading to a slowing of population growth. In Mexico, Brazil, Southeast Asia, and most of Africa, on the other hand, the transition has not been completed and population growth is still dangerously fast.

The demographic transition illustrates how health, wealth, and education can lead to a reduction in the birth rate without direct government interventions. But because more than three-quarters of the world's population lives in developing countries and less than a quarter in the developed countries, the slowed population growth that generally occurs along with the demographic transition is unlikely to be sufficient to keep the world population at a manageable level. Instead, population growth will continue to rise quickly. What are the potential consequences of such explosive growth? We explore this next.

DEMOGRAPHIC TRANSITION

The demographic transition is a pattern of population growth that is experienced as a country industrializes. It is characterized by slow growth, then fast growth, and then slow growth again.

SLOW GROWTH	FAST GROWTH	SLOW GROWTH
• High birth rates	• High birth rates	• Low birth rates
• High death rates	• Low death rates	• Low death rates

Birth rate

Death rate

Population growth rate

Birth and death rates

Population growth rate

Time

FIGURE 14-25 With industrialization, death rates drop and, later, birth rates drop, too.

TAKE-HOME MESSAGE 14•14

The demographic transition tends to occur with the industrialization of countries. It is characterized by an initial reduction in the death rate, followed later by a reduction in the birth rate.

14•15

Human population growth: how high can it go?

Humans are a phenomenally successful species. The numbers don't lie: there are more than 6 billion people alive today (FIGURE 14-26). And current birth rates exceed death rates by so much that we add 80 million more people to the total each year. But for all of our success, the laws of physics and chemistry still apply. We all need food for energy and we need space to live. We need a variety of other resources, too, and we also need the capability of processing and storing all of the

waste our societies generate. Because of these limits to perpetual population growth, we may become victims of our own success.

This we know for certain: human population growth cannot continue forever at the current rate. Like every other species, our environment has a carrying capacity beyond which the population cannot be maintained. The question is, how high

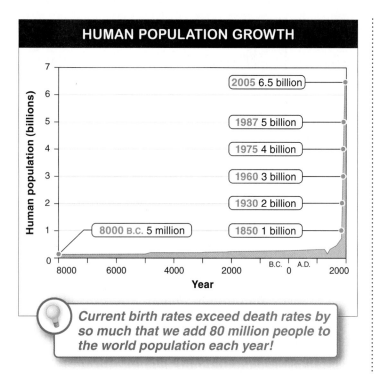

HUMAN POPULATION GROWTH

2005 6.5 billion

1987 5 billion

1975 4 billion

1960 3 billion

1930 2 billion

8000 B.C. 5 million

1850 1 billion

Current birth rates exceed death rates by so much that we add 80 million people to the world population each year!

FIGURE 14-26 **World population: a slow start, but rapidly picking up steam.**

can it go? What is our carrying capacity? This, unfortunately, is a difficult question to answer—although not for lack of trying. For more than 300 years, biologists have been making estimates, starting when Anton van Leeuwenhoek (the inventor of the microscope) made an estimate of just over 13 billion people. The median of all the estimates, the old as well as the new, is just over 10 billion, and the United Nations conservatively suggests that the carrying capacity is somewhere between 7 and 11 billion.

There is huge variation from one biologist's estimate to the next. Why is it so hard to figure this out? After all, it's important that we know so that we can work to avoid a global catastrophe. The problem, it seems, goes back to the reasons behind the tremendous human success in the first place. We are so clever that we seem to keep increasing the carrying capacity before we ever bump into it. We do this in a variety of ways, all made possible by various technologies that we invent. In particular, we make advances on three fronts:

1. *Expand into new habitats.* With fire, tools, shelter, and efficient food distribution, we can survive almost anywhere on earth.

2. *Increase the agricultural productivity of the land.* With fertilizers and modern, mechanized agricultural methods, and selection for plants and animals with higher yields, fewer people can now produce much more food than was previously thought possible.

3. *Circumvent the problems that usually accompany life at higher densities.* Public health and civil engineering advances make it possible for higher and higher densities of people to live together with minimal problems from infectious diseases and waste management (**FIGURE 14-27**).

But the question remains: how high can the population go? Determining the earth's carrying capacity for humans is very difficult, for many reasons. The most difficult problem may be assessing just how many resources each person needs. It is possible to estimate the **ecological footprint** of the people in each country by evaluating how much land, how much food and water, and how much fuel, among other things, are necessary. The problem is, this method reveals that although some countries (e.g., New Zealand, Canada, Sweden) have more resources available than are required to support the needs of their population, in the world as a whole, resource use is significantly greater than the resources available, implying that we are already at our carrying capacity. From the perspective of ecological footprints, the populations of many countries—including the United States, Japan, Germany, and England—currently consume far more resources than are available to them. Their citizens are living at an unsustainable level.

Another difficulty in estimating the earth's carrying capacity is that populations can and do occasionally alter their fundamental growth properties. Relatively small changes in the worldwide birth rate can have dramatic effects on the ultimate population size of the planet.

The difficulties in estimating the earth's carrying capacity do illuminate an important issue. We must be more specific when we ask: how high can it go? We must add: and at what level of comfort, security, and stability, and with what impact on the other species on earth? There are, after all, tremendous tradeoffs.

> There is in every American, I think, something of the old Daniel Boone—who, when he could see the smoke from another chimney, felt himself too crowded and moved further out into the wilderness.
>
> —Hubert Humphrey, U.S. 38th Vice President, in a speech at the University of Chicago, January 14, 1966

The ecological footprint of the 1 billion people currently living in India, for example, is much smaller than that of Japan or Norway or Australia. It is possible, we know, to live with much less impact on the environment per person. But how much do we want to sacrifice in order to enable a larger number to live stably on the planet? Is our goal to maximize

LIMITLESS CARRYING CAPACITY?

One reason that the carrying capacity for the human population is difficult to estimate is that we can increase it in a variety of ways.

1 EXPANDING INTO NEW HABITATS

2 INCREASING THE AGRICULTURAL PRODUCTIVITY OF THE LAND

3 FINDING WAYS TO LIVE AT HIGHER DENSITIES

FIGURE 14-27 Three ways of increasing the carrying capacity for the human population.

FIGURE 14-28 **How high can the population go?** Humans are resourceful, but a rapidly growing population may strain the earth's ability to support us.

the number? Or to reduce the number and increase the resources available to each person? Or should our goal be something else entirely? The answers to these questions will influence whether the ultimate carrying capacity is even higher than 11 billion or perhaps lower than the current 6.8 billion people now alive (**FIGURE 14-28**).

The problem is indeed a difficult one. And it gets even more difficult because the world population currently has significant momentum. That is, even if today we immediately and permanently reduced our fertility rate to the replacement rate of just two children per couple, the world population would continue growing for more than 40 years, putting us up to at least 7 or 8 billion. This is because there are so many young people alive that each year more and more of them will enter the reproductive population. As we get closer to the carrying capacity of earth for humans (or exceed it), we should be better able to recognize it, but by then it may be too late to take preemptive measures—to avoid the resource depletion that may doom us to the much more unpleasant experience of natural population controls.

TAKE-HOME MESSAGE 14·15

The world population is currently growing at a very high rate, but limited resources will eventually limit this growth, most likely at a population size between 7 and 11 billion.

Knowledge You Can Use

Life history tradeoffs and a mini-fountain of youth.

What is the relationship between reproduction and longevity?

It's well documented that there is a tradeoff between reproduction and longevity. In a wide variety of animals, decreasing reproductive effort has been shown to increase longevity.

Q: **Does this have any practical applications?** Yes: spay or neuter your pet to increase its life span!

Q: **Does this work?** Yes! Data from more than 1,000 cat autopsies showed that non-sterilized females lived a mean of 3.0 years, while sterilized females lived significantly longer, with a mean life span of 8.2 years. The results for males were similar. (Perhaps the most dramatic example of the reproduction-longevity tradeoff is that of the marsupial mouse, *Antechinus stuartii,* in which castration leads to a doubling or even tripling of the usual life span. Similarly, following castration, salmon can live up to 18 years longer than usual.)

Not that you asked . . . In a study in the early 1900s of men institutionalized for mental retardation, those who were castrated lived 13 years longer (69.3 years vs. 55.7 years) than non-castrated men at the same institution, matched for age and intelligence!

Q: **What is responsible for the tradeoff between reproduction and longevity?** If it were a question of sterilized individuals living longer simply because they don't have to expend energy on reproduction, perhaps similar life span increases could be achieved by simply giving animals access to more energy. In practice, however, this doesn't increase life span (and usually *decreases* it). Rather, it seems that a significant part of the "cost" of reproduction, at least in mammals, is an increased incidence of cancer, caused by the higher levels of circulating hormones in "reproductively ready" (that is, fertile and receptive) animals.

Possibilities to think about.

The link between maintaining reproductive readiness and reduced longevity has led some researchers to contemplate the design of birth control pills that might have the additional effect of increasing longevity. Stay tuned.

1 Population ecology is the study of how populations interact with their environments.

Population ecology is the study of the interaction between populations of organisms and their environment, particularly their patterns of growth and how they are influenced by other species and by environmental factors. Populations tend to grow exponentially until limited resources cause the growth to slow to logistic rates. A population's growth can be reduced both by density-dependent factors related to crowding and by density-independent factors, such as floods, fires, and earthquakes. Difficulties such as estimating population size and carrying capacity complicate the management of populations for maximum sustainable yield.

2 A life history is like an executive summary of a species.

An organism's pattern of investment in growth, reproduction, and survival is its life history. Life tables and survivorship curves summarize the survival and reproduction patterns of the individuals of a population, varying greatly across species. Because constraints limit evolution, life histories are characterized by tradeoffs between investments in growth, reproduction, and survival.

3 Ecology influences the evolution of aging in a population.

Natural selection cannot weed out bad alleles that do not diminish an individual's relative reproductive success. Consequently, these alleles accumulate in the genomes of organisms of nearly all species. This leads to the multiple physiological breakdowns that we see as aging. The rate of aging and pattern of mortality for a species are determined by the hazard factor of its environment. By increasing the strength of natural selection later in life, it is possible to increase the mean and maximum longevity of the individuals in a population.

4 The human population is growing rapidly.

Age pyramids show the number of individuals within any age group in a population. They allow us to estimate birth and death rates over multi-year periods. The demographic transition tends to occur as countries industrialize. It is characterized by an initial reduction in the death rate, followed later by a reduction in the birth rate. The world population is currently growing at a very high rate, but limited resources will eventually limit this growth, most likely at a population size between 7 and 11 billion.

KEY TERMS

1. Ecology is best defined as the study of:
   a) the relationships between all living organisms and their environments.
   b) the relationships between parasites and their hosts.
   c) aquatic organisms.
   d) interactions between predator and prey populations.
   e) the preservation of habitats.

2. On average, which leaves more offspring that survive to become adults and reproduce: a pair of elephants or a pair of rabbits?
   a) The pair of elephants, because elephants live much longer and have more breeding seasons.
   b) The pair of rabbits, because they have so many more offspring per breeding season than do elephants.
   c) The pair of rabbits, because they reach sexual maturity more rapidly.
   d) The pair of elephants, because any individual elephant born is more likely to survive to become an adult and reproduce.
   e) If both populations are stable, the pair of elephants and the pair of rabbits will leave the same number of offspring that survive to become adults and reproduce.

3. In a population exhibiting logistic growth, the rate of population growth is greatest when N is:
   a) $0.5K$.
   b) 0.
   c) above the carrying capacity.
   d) K.
   e) All of the above are correct; the rate of population growth is constant in logistic growth.

4. In a population, as N approaches K, the logistic growth equation predicts that:
   a) the carrying capacity of the environment will increase.
   b) the growth rate will approach zero.
   c) the population will become monophyletic.
   d) the population size will increase exponentially.
   e) the growth rate will not change.

5. The number of individuals that can be supported in a given habitat is the:
   a) innate capacity for increase.
   b) biotic potential.
   c) density-dependent effect.
   d) density-independent effect.
   e) None of the above are correct.

6. Which of the following statements about maximum sustainable yield is incorrect?
   a) The maximum sustainable yield for a population is the population growth rate at $K/2$.
   b) The maximum sustainable yield for a population can be difficult to determine because it is not always possible to accurately measure N.
   c) The maximum sustainable yield for a population can be difficult to determine because it is not always possible to accurately measure K.
   d) The maximum sustainable yield for a population is a useful management guideline for harvesting plant products such as timber, but is not helpful for managing animal populations.
   e) The concept of maximum sustainable yield can generate useful information for fighting the growth of pest species.

7. Dr. David Reznick has studied life history evolution in guppies that live in streams in Trinidad. Guppies are found in two different types of habitat: sites where predation is high, and sites where predation is low. Which of the following life history characteristics would you expect to evolve in a guppy population living in a high-predation site?
   a) bright colors and courtship displays
   b) increased egg number
   c) a female-biased sex ratio
   d) increased egg size
   e) delayed sexual maturation

8. The death rate of organisms in a population exhibiting a type III survivorship curve is:
   a) unrelated to age.
   b) usually correlated with density-independent causes.
   c) higher in post-reproductive than in pre-reproductive years.
   d) lower after the organisms survive beyond the earliest age groups.
   e) more or less constant throughout their lives.

9. Which of the following is a major tradeoff in life histories?
   a) size of offspring for amount of parental investment
   b) size of offspring for number of reproductive events
   c) growth for reproduction
   d) size for life span
   e) number of reproductive events for number of offspring per reproductive event

10. Natural selection:
    a) does not influence aging because aging is determined by an individual's environment.
    b) cannot reduce the frequency of alleles that cause mortality among individuals who have not yet reached the age of maturity.
    c) cannot weed out from a population any alleles that do not reduce an individual's relative reproductive success, even if these alleles increase an individual's risk of dying.
    d) can influence aging but not longevity.
    e) leads to an increase in the frequency of any illness-inducing alleles that have their effect when an organism can reproduce.

11. Which of the following statements about the hazard factor of a population is incorrect?
    a) It is a measure of organisms' risk of death from external sources.
    b) It is lower for a population of porcupines than for a population of guinea pigs.
    c) It is a measure of the ratio of mortality risk from external (environmental) causes relative to internal (genetic) causes.
    d) It is responsible for the rate of aging among the individuals of the population.
    e) It is a measure of how quickly the individuals in a population age.

12. Life extension:
    a) is not possible, because natural selection cannot weed out disease-causing alleles that have an effect only after the age when reproduction is no longer possible.
    b) can be achieved in species by selectively breeding individuals with the earliest age of maturity.
    c) works in insects but could not work in humans.
    d) has been achieved using laboratory selection for delayed reproduction.
    e) None of the above are true.

13. A population pyramid:
    a) represents the number of individuals in various age groups in a population.
    b) can be constructed from the data in a life table.
    c) directly predicts future age distributions of the population.
    d) shows the current birth and death rates of a population.
    e) predicts survival and mortality rates for an individual at a given age.

14. A primary difference between the age pyramids of industrialized and developing countries is that:
    a) mean longevity is significantly greater in developing countries.
    b) in developing countries, much larger proportions of the population are in the youngest age groups.
    c) developing countries show a characteristic "bulge" that indicates a baby boom.
    d) in developing countries, females live significantly longer than males, whereas in industrialized countries the reverse is true.
    e) developing countries have significantly more individuals than industrialized countries.

15. Which statement best describes expectations for the world's human population by the year 2010?
    a) It will exceed 10 billion if the current rate of increase continues.
    b) Negative growth in the United States and Europe will counterbalance positive growth in the developing countries.
    c) It will exceed 7 billion if the current rate of increase continues.
    d) It will drop below 6 billion if the current rate of decrease continues.
    e) The problem is too complex to make any predictions.

SHORT-ANSWER QUESTIONS

1. Explain the differences between density-dependent and density-independent factors that can limit a population's growth. Give an example of each type of factor.

2. Describe the characteristics of an organism with a type I survivorship curve. How would the life history change if this organism had a type III survivorship curve?

3. Explain why some species of organisms live very long lives and others do not. How could the life span be extended?

See Appendix for answers. For additional study questions, go to www.prep-u.com.

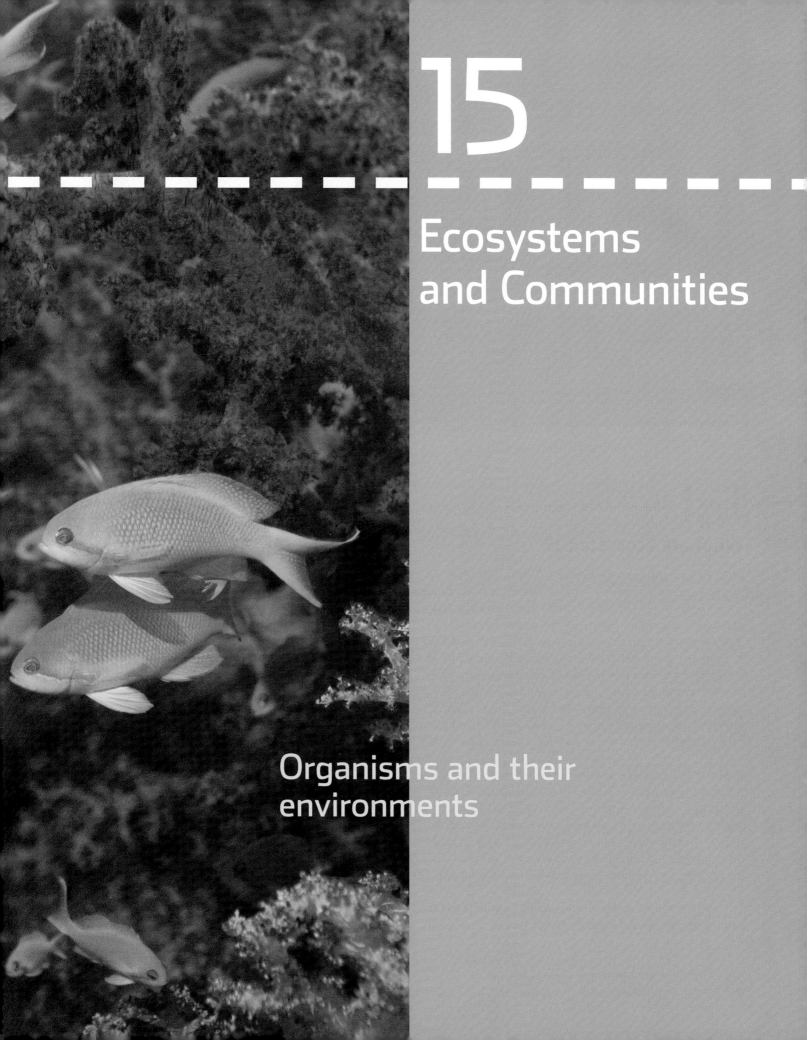

15

Ecosystems and Communities

Organisms and their environments

1 Ecosystems have living and non-living components.

A red kangaroo (Macropus rufus) *in its natural habitat, Finke Gorge National Park, Australia.*

15•1

What are ecosystems?

Picture a lush garden: some greenery, a bit of debris from rotting wood, and abundant wildlife. Grazing animals abound, while predators feed on other animals and their eggs. Parasites are poised, looking for hosts, and, just below the surface, scavengers find meals among the organic detritus. It would seem to be the quintessential **ecosystem,** a community of biological organisms plus the non-living components with which the organisms interact. But now imagine that the entire scene gets up and walks away! The "camera" in your mind pulls back to reveal that the entire scene is playing out on the back of a beetle no more than two inches long (**FIGURE 15-1**).

The host of this mini-ecosystem is a beetle from New Guinea called the large weevil. The weevil is camouflaged from its predators by lichens—fungi and photosynthetic algae living together—while the lichens are given a safe surface on which to live. And the garden of lichens supports a wide range of other organisms, from tiny mites to a variety of other microscopic invertebrates, some free-living and some parasitic.

Not all ecosystems are the obvious assemblages of plants and animals that we usually picture—ponds, deserts, or tropical

Ecosystems are found not just in obvious places like ponds and tropical rain forests, but also in some unexpected places, like on the back of a beetle!

FIGURE 15-1 A small-scale ecosystem. Not all ecosystems are as big as a desert or a forest. Tiny, unseen ecosystems can exist on the back of a beetle or in the gut of an individual human.

550 CHAPTER 15 • ECOSYSTEMS AND COMMUNITIES

What Are Ecosystems?   Weather   Energy and Chemical Flow   Species Interactions

forests. A similar scene can just as easily be found in your large intestine, where several hundred bacterial species flourish. These ecosystems are contained within ecosystems that are contained within ecosystems. The scale can vary tremendously. The closer you look, the more you find.

What is important is that the two essential elements of an ecosystem are present: the biotic environment and the physical (abiotic) environment (**FIGURE 15-2**).

1. The **biotic** environment consists of all the living organisms within an area and is often referred to as a **community.**

2. The physical environment, often referred to as the organisms' **habitat,** consists of:

   - the *chemical resources* of the soil, water, and air, such as carbon, nitrogen, and phosphorus, and

   - the *physical conditions,* such as the temperature, salinity (salt level), moisture, humidity, and energy sources.

Biologists view communities of organisms and their habitats as "systems" in much the same way engineers might, hence the term eco*system*. Biologists monitor the inputs and outputs of the system, tracing the flow of energy and various molecules as they are captured, transformed, and utilized by organisms and later exit the system or are recycled. They also study how the activities of one species affect the other species in the community—whether the species have a conflicting relationship, such as predator and prey, or a complementary relationship, such as flowering plants and their pollinators. On a small scale, such as the back of a beetle, making these measurements is easy. But there are also some well-studied giant ecosystems, such as the Hubbard Brook Experimental Forest in New Hampshire, which covers 7,600 acres. Regardless of size, the same principles of ecosystem study apply: observe and analyze organisms and their environments, while monitoring everything that goes into and comes out of the system.

Why should researchers bother with such methodical—and tedious—analysis of ecosystems? This straightforward approach has led to numerous valuable discoveries, from understanding how clear-cutting forests dramatically reduces soil quality, to understanding the link between the use of fossil fuels and the creation of destructive acid rain. As we see in this chapter and the next, humans, perhaps more than any other species in history, are significantly affecting most of the ecosystems on

earth. And, in addition to improving our understanding of environmental issues, ecosystem studies can also tell us about the microbial ecosystems living inside humans and thus lead to advances in public health.

ECOSYSTEMS

All ecosystems share two essential features:

BIOTIC ENVIRONMENT	PHYSICAL ENVIRONMENT
• The living organisms within an area	• The chemical resources and physical conditions within an area
• Often referred to as a community	• Often referred to as the organisms' habitat

FIGURE 15-2 **What makes up an ecosystem?** Living organisms are not self-sufficient. They need energy and raw materials. Ecosystems are the living and non-living elements in a given area.

TAKE-HOME MESSAGE 15·1

An ecosystem is all of the living organisms in a habitat as well as the physical environment. Ecosystems are found not just in obvious places such as ponds, deserts, and tropical rainforests but also in some unexpected places, like the digestive tracts of organisms or the shell of a beetle.

Communities

A variety of biomes occur around the world, each determined by temperature and rainfall.

Dense vegetation surrounds you. Above you is a canopy of evergreen trees, 30–40 meters tall. Climbing vines hang from virtually all the trees. And dozens or even hundreds of species of insects are flourishing around you. Even if you've never been there, the description of a tropical rain forest is easy to recognize. But where exactly would you be if you were in this scenario? It could be South America or Africa or Southeast Asia. The species are different, but the general pattern of life forms is the same. The same holds for arctic tundra: whether you were in northern Asia or North America, the view would be similar. These are examples of the largest of the earth's ecosystems, the **biomes.**

Biomes cover huge geographic areas of land or water—the deserts that stretch almost all the way across the northern part of Africa, for example. Terrestrial (land) biomes are defined and usually described by the predominant types of plant life in the area. But looking at a map of the world's terrestrial biomes, it is clear that they are mostly determined by the

TERRESTRIAL BIOMES

Biomes are large ecosystems that cover huge geographic areas. The nine chief terrestrial biomes are determined by the temperature and amounts of precipitation in conjunction with the magnitude of seasonal variation in these factors.

TROPICAL FOREST

DESERT

SAVANNA

TEMPERATE GRASSLAND

TEMPERATE DECIDUOUS FOREST

CHAPARRAL

CONIFEROUS FOREST

TUNDRA

POLAR ICE

FIGURE 15-3 Terrestrial ecosystem diversity. The terrestrial biomes depend on the amount of rainfall and average temperature.

What Are Ecosystems? | Weather | Energy and Chemical Flow | Species Interactions

Aquatic biomes are determined by physical features, including salinity, water movement, and depth.

LAKES AND PONDS	RIVERS AND STREAMS	
OPEN OCEANS	ESTUARIES AND WETLANDS	CORAL REEFS

FIGURE 15-4 **Aquatic ecosystem diversity.** Aquatic biomes are defined by variables such as salinity, water movement, and depth.

weather. Specifically, when defining terrestrial biomes, we ask four questions about the weather:

1. What is the average temperature?
2. What is the average rainfall (or other precipitation)?
3. Is the temperature relatively constant or does it vary seasonally?
4. Is the rainfall relatively constant or does it vary seasonally?

By knowing the answers to these questions, it is possible to predict, with great accuracy, the type of biome. This is because temperature and precipitation dictate the **primary productivity** levels, or the amount of organic matter produced, primarily through photosynthesis. And the numbers and types of **primary producers**—the organisms responsible for the primary productivity, such as grasses, trees, and agricultural crops—are, in turn, the chief determinants of the amount and breadth of other life in the region.

For example, where it is always moist and the temperature does not vary across the seasons, **tropical rain forests** develop. And where it is hot but with strong seasonality that brings a "wet" season and a "dry" season, **savannas** or **tropical seasonal forests** tend to develop. At the other end of the spectrum, in dry areas with a hot season and a cold season, **temperate grasslands** or **deserts** develop. **FIGURE 15-3** shows examples of the nine chief terrestrial biomes; all are determined, in large part, by the precipitation and temperature levels.

Aquatic biomes are defined a bit differently, usually based on physical features such as salinity, water movement, and depth. Chief among these environments are (1) lakes and ponds, with non-flowing fresh water; (2) rivers and streams, with flowing fresh water; (3) **estuaries** and wetlands, where salt water and fresh water mix in a shallow region characterized by exceptionally high productivity; (4) open oceans, with deep salt water; and (5) coral reefs, highly diverse and productive regions in shallow oceans (**FIGURE 15-4**).

If the terrestrial biomes of the world are determined by the temperature and rainfall amounts and seasonality, what determines those features? In other words, what makes the weather? We investigate next how the geography and landscape of the planet—from the shape of the earth and its orientation to the sun, to patterns of ocean circulation—cause the specific patterns of weather that create the different climate zones and the biomes characteristic of each. Then we'll see how energy and chemicals are made available for life to flourish in these biomes.

TAKE-HOME MESSAGE 15·2

Biomes are the major ecological communities of earth, characterized mostly by the vegetation present. Different biomes result from differences in temperature and precipitation, and the extent to which they vary from season to season.

Communities

② Physical forces interacting create weather.

Monsoon clouds forming over the Indian Ocean and Madagascar.

15•3

Global air circulation patterns create deserts and rain forests.

Temperature and rainfall, that's all. The type of terrestrial biome depends on little else besides these two aspects of the weather. But what determines the temperature and rainfall in a particular part of the world? As we'll see, differences in both ultimately result from one simple fact: the earth is round. Let's first explore how the earth's curvature influences temperature. Then we will examine how rainfall patterns are an inevitable consequence of the variation in temperature across the globe.

Wherever you are, begin walking toward the equator. As you get closer, does it get hotter or colder? Hotter, of course. Nearly everyone is aware of this universal trend (and the reverse as well: it gets colder as you move away from the equator and approach the North or South Pole). What is responsible for the increased warmth at the equator? This is most easily explained with a drawing (**FIGURE 15-5**).

FIGURE 15-5 The curvature of the earth. The sun shines more directly on the equator. Closer to the Poles, the curvature of the earth causes sunlight to be spread out over a larger area, reducing the heat in any one spot.

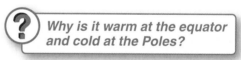

? Why is it warm at the equator and cold at the Poles?

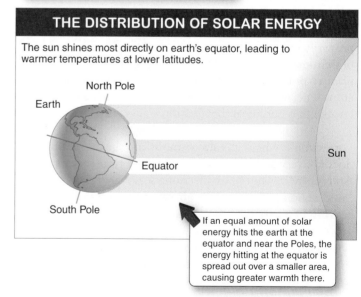

THE DISTRIBUTION OF SOLAR ENERGY

The sun shines most directly on earth's equator, leading to warmer temperatures at lower latitudes.

North Pole

Earth

Equator

Sun

South Pole

If an equal amount of solar energy hits the earth at the equator and near the Poles, the energy hitting at the equator is spread out over a smaller area, causing greater warmth there.

What Are Ecosystems?   Weather   Energy and Chemical Flow   Species Interactions

> There was a hot desert wind blowing that night. It was one of those hot dry Santa Anas that come down through the mountain passes and curl your hair and make your nerves jump and your skin itch. On a night like that every booze party ends in a fight. Meek little wives feel the edge of a carving knife and study their husbands' necks. Anything can happen.
>
> —Raymond Chandler, "Red Wind," 1938

The sun shines most directly on the equator. That means that a given amount of solar energy hitting the earth at the equator is spread out over a relatively small area. Away from the equator, the earth curves toward the North and South poles. Because of this curvature, the same amount of solar energy hitting the earth at the Poles is spread out over a much larger area and also travels a greater distance through the atmosphere, which absorbs or reflects much of the heat. Because the energy is spread out over a larger area, there is less warmth at any one point on the earth's surface. It's similar to the fact that at noon, the sun's rays hit the earth at a more direct angle than they do later in the day. That is why the sun provides less warmth late or early in the day. It is also why the risk of sunburn and skin cancer is greatest around noon and the nearer you are to the equator.

Global patterns of rainfall can be predicted just as easily, by taking into account that warm air holds more moisture than cold air. We'll start at the equator again. This is where the greatest warming power of the sun hits the earth, and some of that energy radiates back, warming the air. This starts a three-step process that ends with rain. (1) Hot air rises. (2) As it rises high into the atmosphere, it cools. And (3) because cool air holds less moisture, as the air rises, clouds form and the moisture that can no longer be held in the air falls as rain (**FIGURE 15-6**). The equator is hot, but it is also very wet.

The cold air that just rained on the equator at 0° latitude is high in the atmosphere and expands outward, away from the equator, to about 30° north and south. Here—about a third of the way from the equator toward each Pole—the cold air, which is heavier than warm air, begins to fall downward toward the earth. The falling air becomes warmer as it gets closer to the earth's surface, which radiates back some of the heat from the sun. And as the air gets warmer, it can hold more and more moisture, making it less likely to rain. That is, moisture can be held in the air rather than being released as rain. In fact, the falling and rapidly heating air can hold so much moisture that it sometimes sucks

Q Nearly all of the world's deserts occur a third of the way from the equator toward the Poles, at 30° latitude. Is this just a coincidence?

FORMATION OF RAIN

1 AIR IS HEATED AND RISES
When solar heat hits the earth, it warms the air at that point. The heated air rises.

2 RISING AIR COOLS
As hot air rises, getting farther from the warm earth, it cools.

3 COOLING AIR LOSES MOISTURE
Because cold air holds less moisture than warm air, clouds form and rain falls.

FIGURE 15-6 Rainmaking.

up moisture from the land itself. For this reason, at around 30° north and south, around the world, there is very little rainfall, the ground is very dry, and deserts form (**FIGURE 15-7**). The Atacama Desert of South America, which lies at approximately 30° south of the equator, is an extreme example. In some parts of this large desert, *no rainfall* has ever been recorded. Other great deserts of the world are also at 30° latitude, including the Sahara, Kalahari, Mojave, and Australian deserts (although some deserts do occur at other latitudes). These areas stand in stark contrast to the equator, where it is not uncommon for an area to receive more than three meters (!) of rain in a year.

THE FORMATION OF DESERTS

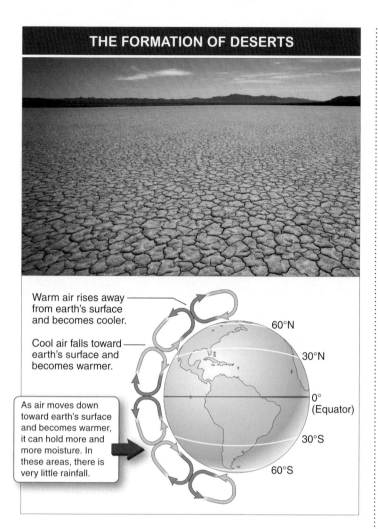

Warm air rises away from earth's surface and becomes cooler.

Cool air falls toward earth's surface and becomes warmer.

As air moves down toward earth's surface and becomes warmer, it can hold more and more moisture. In these areas, there is very little rainfall.

60°N
30°N
0° (Equator)
30°S
60°S

FIGURE 15-7   Desert formation. Circulating masses of air, caused by solar energy hitting different parts of the earth at different angles, determine rainfall patterns throughout the world. The photo shows the parched earth of the Mojave Desert, which spans four states in the southwestern United States.

As the air falls near 30° north and south and hits the earth, it spreads equally toward the equator and toward the Poles. Moving toward the Poles, a similar pattern is repeated. The air moving over the surface of the earth absorbs radiated heat and

Antarctica receives little annual rainfall (or other precipitation) and could be classified as a desert by this definition. The interior of the continent is particularly dry, although the coasts can receive heavy snowfall when storms pick up moisture from the surrounding sea and deposit it as snow along the coasts.

moves along the surface toward the Poles, gradually getting warmer, until, at around 60° latitude, it begins to rise because of its accumulated heat. Again, the rising air loses its moisture as rainfall. And so at 60° latitude, two-thirds of the way toward the Poles from the equator, it's not particularly warm but there is a great deal of rain. Not surprisingly, at these latitudes lie huge temperate forests with extensive plant growth. And finally, around the Poles, air masses again descend. And as they do, because they can hold more moisture, very little rain falls. The Poles are cold, but with little precipitation—they resemble a frozen desert.

TAKE-HOME MESSAGE 15·3

Global patterns of weather are largely determined by the earth's round shape. Solar energy hits the equator at a more direct angle than at the Poles, leading to warmer temperatures at lower latitudes. This temperature gradient generates atmospheric circulation patterns that result in heavy rain at the equator and many deserts at 30° latitude.

What Are Ecosystems?   Weather   Energy and Chemical Flow   Species Interactions

Local topography influences the weather.

Why is it so windy on the sidewalk around tall buildings? And why does it rain all the time on one side of some mountain ranges, while deserts form on the other side? And is it actually warmer in the city than in the country? The answers to these questions hinge on the fact that physical features of the land, its **topography**—including features created by humans—can have dramatic but predictable effects on the weather. Let's explore some of these features, beginning with natural landscapes.

High altitudes have lower temperatures. With increasing elevation, the air pressure drops—this is because the weight of the atmosphere gets less and less as altitude increases. And when pressure is lower, the temperature drops. For each 1,000 meters above sea level, the temperature drops by about 6° C. This is why the changes in weather and vegetation that you see while climbing a mountain are similar to those you would see as you moved farther and farther away from the equator—it gets colder in both cases.

Rain shadows create deserts. When wind blows against a mountain, the air rises to get over the top. But because the air cools as it rises, it can't hold as much moisture. So clouds form and rain falls. After the air eventually passes over the top of the mountain, it can fall back down toward lower elevations. As it falls, it becomes warmer, and because warm air can hold more moisture than cold air, there is rarely any rain on the backside of the mountain. In fact, often the air will pull moisture from the ground, intensifying the already dry conditions, creating **rain shadow** deserts (**FIGURE 15-8**). Along the west coast of the United States, the Sierra Nevada and the Cascade mountain range are responsible for the Mojave Desert in California and the Great Sandy Desert in Oregon.

Q **Is it warmer or cooler in urban areas relative to nearby rural areas?**

Asphalt, cement, and tops of buildings absorb heat, raising the temperature. Modern landscapes also influence the weather, creating "urban heat islands," metropolitan areas where the asphalt and concrete lead to greater absorption, rather than reflection, of solar energy. When energy from the sun hits concrete or pavement or the dark roof of a building, some of it is reflected, heating the air around it, and most of the rest is absorbed by these man-made surfaces that function as a reservoir of heat energy. At night, the surfaces lose heat to the sky, which is colder by comparison, through radiation. This is very different from what happens when

FORMATION OF RAIN SHADOWS

1 Wind blows from oceans toward land, rising when it hits mountains.

2 Rising air cools and holds less moisture, leading to cloud formation and rain.

3 Air passes over mountain top and falls, becoming warmer and increasing the moisture it can hold. This reduces rainfall and results in dry areas.

FIGURE 15-8 The rain shadow effect.

sunlight hits trees or other plant life. The solar energy evaporates water in the leaves. The ground surface doesn't get much hotter, and neither does the air. It's not surprising, then, that cities tend to be 1° to 6° C warmer than surrounding rural areas. And not only is it hotter in cities, but the rising warm air also alters rainfall patterns both in and around cities.

Because people who live in urban heat islands tend to use much more energy for air conditioning, a variety of steps are

557

Asphalt, cement, and building tops absorb heat, raising the temperature.

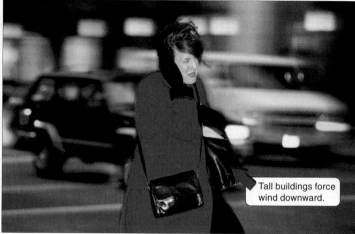

Tall buildings force wind downward.

FIGURE 15-9 Human engineering. When humans alter the land, this can have unintended consequences, such as alterations in temperature and in wind speed and direction.

being taken to create "greener" cities that absorb and release less heat from the sun. These methods include the creation of buildings with lighter-colored rooftops, the planting of trees around buildings and along roads, and the development of rooftop gardens rich with vegetation.

Q **Why is it so windy on streets with tall buildings?**

Tall buildings channel wind downward. You may not have noticed that cities are getting warmer, because they're also getting windier. This, of course, is obvious if you have ever walked around a corner on which a tall building stands. It's not a coincidence that it is always windy on such corners. Tall buildings are actually responsible for the perpetual winds.

Here's why. For starters, winds are stronger the farther above the earth you go. This is because the earth's frictional drag slows winds as they get closer to the surface. That is, the plants, dirt, rocks, and water on the earth's surface slow the winds. Up higher, the winds blow more strongly. When these strong winds suddenly encounter tall buildings, they are deflected. Some of the wind goes up and over the building, but much of it is pushed downward, reaching double or even triple its initial speed by the time it reaches street level (**FIGURE 15-9**).

TAKE-HOME MESSAGE 15·4

Local features of topography influence the weather. With higher altitude, the temperature drops. On the windward side of mountains, rainfall is high; on the backside, descending air reduces rainfall, causing rain shadow deserts. Urban development increases the absorption of solar energy, leading to higher temperatures, and creates wind near the bottom of tall buildings.

15·5

Ocean currents affect the weather.

Weather is affected not just by circulating air masses. It is also affected by the oceans. The ocean is vast and deep, warmed almost exclusively by the sun beating down on its surface. But every molecule of water does not just sit in one place. Far from it. Just as there are global patterns of air circulation that produce broad climate patterns across the globe, there are also global patterns of circulation in the oceans. The water is continuously moving and mixing, due to a combination of forces including wind, the earth's rotation, the gravitational pull of the moon, temperature, and salt concentration. There

What Are Ecosystems?   Weather   Energy and Chemical Flow   Species Interactions

are several large, circular patterns of flow in the world's oceans, as illustrated in **FIGURE 15-10**.

 Q Why do beach communities have milder weather than more inland communities (not as hot in the summer, not as cold in the winter)?

Much of water's effect on weather stems from its great capacity to absorb and hold heat—a 10,000 times greater capacity than air. This means that temperatures fluctuate much more in air than in water. During a day at the beach, the air temperature can go from mild to very hot and back to mild over the course of a day, while the heat of the sun will have a tiny, almost negligible effect on the water temperature. Thus, during summers, much of the heat of the sun is absorbed by the ocean water in coastal towns, rather than causing hot air temperatures. Conversely, during cold winter months, heat from the water can reduce the coldness of the air.

One of the strongest ocean currents is the Gulf Stream. It travels north through the Caribbean, bringing warm water up the east coast of the United States and then across the Atlantic Ocean toward Europe. Because the current begins in a warm part of the globe, close to the equator, it is still warm when it reaches the east coast of the United States and then Europe. The warm water also warms the climate in these areas. In fact, if it weren't for the Gulf Stream, much of Europe—given its high latitudes—would have a climate more like Canada's. Because all ocean currents in the northern hemisphere rotate in a clockwise direction, water reaching the beaches of California, unlike water reaching east coast beaches, has just come from the north, near Alaska, where it gets very cold.

The most dramatic climate change driven by ocean currents is a phenomenon called **El Niño,** a sustained surface temperature change in the central Pacific Ocean (**FIGURE 15-11**). It occurs every two to seven years and is blamed for flooding, droughts, famine, and a variety of other extreme climate disruptions. We can more easily understand it if we contrast it with the more common pattern.

In a typical year:

1. A steady wind blows westward across the Pacific Ocean from South America toward Southeast Asia and Australia.

2. These winds push the warm surface water away from the coast of South America and raise the sea level in Southeast Asia and Australia by about half a meter. The warmth of the water heats the air above it, causing the air to rise and produce great tropical rainstorms. (To experience this effect, fill your bathtub with hot water and feel the room get warmer and moister.)

3. Off the coast of South America, an upwelling of colder water from deep in the ocean replaces the water that is pushed west. The cold upwelling cools the air above it, causing extremely dry weather.

4. The deep water brings up rich nutrients from below, enabling plankton to flourish and nourishing huge populations of fish.

In an El Niño year:

1. The normal South America–to–Southeast Asia winds ease just a bit. It is not known what causes this, but the effect is severe.

2. Without wind pushing it, the warm water piling up in the west flows back toward South America. This water warms the air, which rises and produces huge amounts of rainfall on the coast of South America where it is normally dry, often causing flooding.

3. Without the warm surface water being blown west, the normal upwelling of cool, deep water with high nutrient content cannot occur. The plankton levels drop to a fraction of their normal levels, and the typically rich fish stocks are wiped out.

? Beaches on the east coast of the United States have warmer water than west coast beaches at the same latitudes. Why?

OCEANIC CIRCULATION

There are several large, circular patterns of flowing water in the oceans due to a combination of forces including wind, the earth's rotation, the gravitational pull of the moon, temperature, and salt concentration.

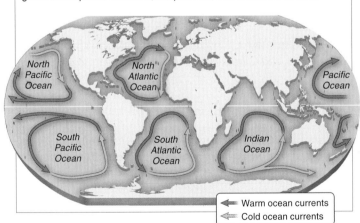

North Pacific Ocean

North Atlantic Ocean

Pacific Ocean

South Pacific Ocean

South Atlantic Ocean

Indian Ocean

← Warm ocean currents
← Cold ocean currents

FIGURE 15-10 **Ocean currents have a powerful influence over the weather.**

Communities

El Niño occurs every two to seven years and is blamed for flooding, droughts, famine, and a variety of other extreme climate disruptions.

NORMAL CONDITIONS

A steady wind blows westward across the Pacific Ocean from South America toward Southeast Asia. These winds push warm surface water away from the coast of South America, heating the air above it, which causes air to rise and produce tropical rainstorms.

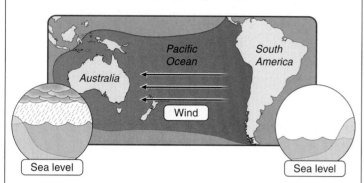

Off the coast of South America, colder water upwells from the ocean depths. The cold water cools the air above, causing extremely dry weather.

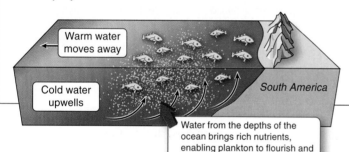

EL NIÑO CONDITIONS

The normal South America–to–Southeast Asia winds ease just a bit. Without the push of the wind, warm water flows back toward South America. The cooled water around Australia and the Philippines cools the air and causes dry weather that can lead to droughts.

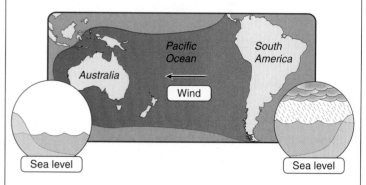

Without the warm surface water being blown west, colder water from the ocean depths cannot upwell. The warm surface water warms the air, resulting in rainfall.

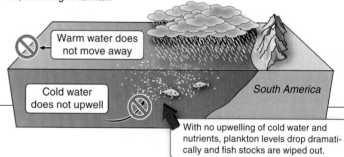

FIGURE 15-11 Domino effect. A simple reduction in the usual east-to-west ocean breeze can cause a cascade of disastrous weather. El Niño years are characterized by flooding in South America and droughts in Southeast Asia.

Changes during an El Niño event can start a chain reaction of unusual weather around the globe, including increased rainfall in the midwestern United States, serious storms off the coast of California, and warmer than usual weather in South Africa and Japan.

4. Meanwhile, in the absence of warm water from the east, Indonesia, the Philippines, and Australia experience drastic climate changes. The cooler water cools the air above and causes unusually dry weather. This frequently leads to droughts and, in extreme years, causes famines.

These predictable changes during an El Niño event can start a chain reaction of additional unusual weather throughout the globe. Although the mechanisms aren't exactly clear, in El Niño years more rain falls in the midwestern United States

and serious storms form off the coast of California, while South Africa, Japan, and Canada all enjoy warmer than usual weather.

TAKE-HOME MESSAGE 15·5

Oceans have global circulation patterns. Disruptions in these patterns occur every few years and can cause extreme climate disruptions around the world.

What Are Ecosystems?   Weather   Energy and Chemical Flow   Species Interactions

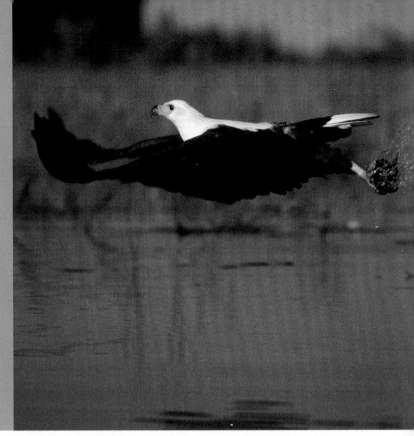

❸ Energy and chemicals flow within ecosystems.

An African fish eagle (Haliaeetus vocifer) *has just caught a fish in the Okavango Delta, Botswana.*

15·6

Energy flows from producers to consumers.

Things are not always as complicated as they appear. Looking at an ecosystem such as a desert savanna, for example, trying to understand how all of the species, from grasses and trees to birds and mammals and worms, interact with each other, and what role each plays within the system, can seem overwhelming. Where is the order? But if we focus on just one aspect of the ecosystem—the pathways energy takes as it flows through the system—a simple and logical order becomes clear.

Everything, all life on that savanna and everywhere else on earth, is made possible because energy flows perpetually from the sun to the earth. This is where our pathway of energy flow begins. Most of the energy is absorbed or reflected by the earth's atmosphere or surface, but about 1% of it is intercepted and converted to chemical energy through photosynthesis. That intercepted energy is then transformed again and again by living organisms, making about four stops as it passes through an ecosystem. Let's examine what happens at each stop, known as **trophic levels** (**FIGURE 15-12**).

First stop: primary producers. When it comes to energy flow, all the species in an ecosystem can be placed in one of two groups:

producers and consumers. Plants (along with some algae and bacteria) are, as we noted earlier, the *primary producers.* They convert light energy from the sun into chemical energy through photosynthesis, as discussed in Chapter 4. We use another word to describe that chemical energy: food.

Second stop: primary consumers—the herbivores. Cattle grazing in a field, gazelle browsing on herbs, insects devouring the leaves of a crop plant—these are the **primary consumers** in an ecosystem, the animals that eat plants. Plant material such as cellulose can be difficult to digest. Consequently, most **herbivores**—animals that eat plants—need a little help in digesting the plants they eat. Primary consumers, from termites to cattle, often have symbiotic bacteria living in their digestive system. These microorganisms break down the cellulose, enabling the herbivore to harness the energy held in the chemical bonds of the plants' cells.

Third stop: secondary consumers—the carnivores. Energy originating from the sun is converted into chemical energy within a plant's tissue. The herbivore that eats the plant breaks down the chemical bonds, releasing the energy. This energy fuels the

561

ENERGY FLOW WITHIN AN ECOSYSTEM

Energy from the sun is intercepted and converted into chemical energy, which passes through an ecosystem in about four stops.

Sun

1 PRIMARY PRODUCERS
Plants convert light energy from the sun into food through photosynthesis.

2 PRIMARY CONSUMERS
Herbivores are animals that eat plants.

3 SECONDARY CONSUMERS
Carnivores are animals that eat herbivores.

4 TERTIARY CONSUMERS
Top carnivores are animals that eat other carnivores.

FIGURE 15-12 Follow the fuel. Plants produce the chemical energy. Herbivores eat plants. Carnivores eat herbivores. And the top carnivores eat the animals that eat the animals that eat the plants that capture energy from the sun.

herbivore's growth, reproduction, and movement, but the energy doesn't remain in the herbivore forever. **Carnivores,** such as cats, spiders, and frogs, are animals that feed on herbivores. They are also known as **secondary consumers.** As they eat their prey, some of the energy stored in the chemical bonds of the carbohydrate, protein, and lipid molecules is again captured and harnessed for their own movement, reproduction, and growth.

Fourth stop: tertiary consumers—the "top" carnivores. In some ecosystems, energy makes yet another stop: the **tertiary consumers,** or "top carnivores." These are the "animals that eat the animals that eat the animals that eat the plants." They are several steps removed from the initial capture of solar energy by a plant, but the general process is the same. A top

carnivore, such as a tiger, eagle, or great white shark, consumes other carnivores, breaking down their tissues and releasing energy stored in the chemical bonds of the cells. As in each of the previous steps, the top carnivores harness this energy for their own physiological needs.

This path from primary producers to tertiary consumers is called a **food chain.** We see later in this chapter why a food chain almost never extends to a fifth stop.

The food chain pathway from photosynthetic producers through the various levels of animals, though, is a slight oversimplification. In actuality, food chains are better thought of as **food webs,** because many organisms are **omnivores** and can occupy more than one position in the chain. When you eat a simple meal of rice and chicken, after all, you're simultaneously a carnivore and a herbivore. On average, about 30% to 35% of the human diet comes from animal products and the remaining 65% to 70% comes from plant products. Many other animals, from bears to cockroaches, also have diets that involve harvesting energy from multiple stops in the food chain.

All the while, as energy is transformed through the steps of a food chain, organic wastes are produced. Whether in the form of animals' waste products or the dead bodies of plants and animals, organic material accumulates in every ecosystem. But

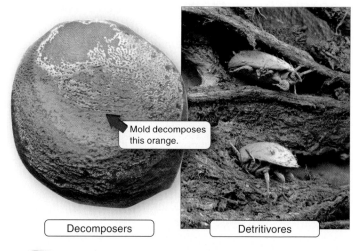

Mold decomposes this orange.

Decomposers

Detritivores

Decomposers and detritivores break down organic wastes, releasing chemical components that can then be reused by plants and other primary producers.

FIGURE 15-13 Dust to dust. Eventually, organisms from every level in the food chain die, but even then they provide sustenance for other organisms. Their dead bodies are broken down by decomposers and detritivores, and important chemical components are recycled through the food chain.

What Are Ecosystems? | Weather | Energy and Chemical Flow | Species Interactions this material does not go to waste. Another component of every food web harnesses the last bits of energy that remain. **Decomposers,** usually bacteria or fungi, and **detritivores,** including scavengers such as vultures, worms, and a variety of arthropods, break down the organic material, harvesting energy still stored in the chemical bonds (**FIGURE 15-13**). Because the decomposers are able to break down a much larger range of organic molecules, they are distinguished from the detritivores. Both groups, nonetheless, release many important chemical components from the organic material than can eventually be recycled and utilized by plants and other primary producers.

Energy flows from one stop to the next in a food chain, but not in the way that runners pass a baton in a relay race. The difference is that at every step in the food chain, much of the usable energy is lost as heat. An animal that eats five pounds of

plant material doesn't convert that into five new pounds of body weight. Not by a long shot. In the next section, we'll see how this inefficiency of energy transfers ensures that most food chains are very short.

TAKE-HOME MESSAGE 15·6

Energy from the sun passes through an ecosystem in several steps. First, it is converted to chemical energy in photosynthesis. Herbivores then consume the primary producers, the herbivores are consumed by carnivores, and the carnivores, in turn, may be consumed by top carnivores. Detritivores and decomposers extract energy from organic waste and the remains of organisms that have died. At each step in a food chain, some usable energy is lost as heat.

15·7

Energy pyramids reveal the inefficiency of food chains.

Look out of the nearest window. What organisms can you see? Almost without fail, you will see green plant life. Maybe some trees, possibly bushes and grasses as well. You'll have to look a bit longer and harder to see any animals, but you'll probably see a few, most likely small animals and various insects that eat plants. On the other hand, you might stare out of the window all day and not see any animals (other than some fellow humans) that eat other animals. Why is that? And why are big, fierce animals so rare? And why are there so many more plants than animals?

The answers to these questions are closely related to our earlier observation that an animal consuming five pounds of plant material does not gain five pounds in body weight from such a meal. The actual amount of growth such a meal can support is far, far less. And when that herbivore is consumed by a carnivore, the carnivore, too, can convert only a small fraction of the energy it consumes into its own tissue. The fraction turns out to be about 10%, and it is fairly consistent across all levels of the food chain.

Q In humans, why is vegetarianism more energetically efficient than meat-eating?

This means that only about 10% of the **biomass**—the total weight of all the living organisms in a given area—of plants in an ecosystem is converted into herbivore biomass. So the herbivore consuming five pounds of

plant material is likely to gain only about half a pound in new growth, while the remaining 90% of the meal is either expended in cellular respiration or lost as feces. Similarly, a carnivore eating the herbivore converts only about 10% of the mass it consumes into its own body mass. Again, 90% is lost to metabolism and feces. And the same inefficiency holds for a top carnivore as well. Let's explore how this 10% rule limits the length of food chains and is responsible for the rarity of big, fierce animals outside your window and across the world.

Given the 10% efficiency with which herbivores convert plant biomass into their own biomass, how much plant biomass is necessary to produce a single 1,200-pound (500-kg) cow? On average, that cow would need to eat about 12,000 pounds (5,000 kg)

Q Why are big, fierce animal species so rare in the world?

of grain in order to grow to weigh 1,200 pounds. But that 1,200 pound cow, when eaten by a carnivore, could only add about 120 pounds of biomass to the carnivore, and only 12 pounds to a top carnivore. That's a huge amount of plant biomass required to generate only a very tiny amount of our top carnivore, which explains why big, fierce animals are so rare (and why vegetarianism is more energetically efficient than meat-eating). After all, multiply that 5,000 kg of grain by several hundred (or thousands, more appropriately) to get an idea of how much plant matter would be necessary to support even a small population of top carnivores: millions of kilograms of grain can support only a few top carnivores.

How much would be required to support an even higher link on the food chain? Ten times as much, so much that there may not be enough land in the ecosystem to produce enough plant material. And even if there were, the area required would be so large that the "top, top carnivores" might be so spread out and so busy trying to eat enough that they were unlikely to encounter each other in order to mate. Hence, the 10% rule limits the length of food chains.

We can illustrate the path of energy through the organisms of an ecosystem with an **energy pyramid,** in which each layer of the pyramid represents the biomass of a trophic level. In **FIGURE 15-14**, we can see that for terrestrial ecosystems, the biomass (in kilograms per square meter) found in the photosynthetic organisms, at the base of the pyramid, is reduced significantly at each step, given the incomplete utilization by organisms higher up the food chain. **FIGURE 15-15** illustrates the huge variation in primary productivity across a variety of ecosystems. It is highest in tropical rain forests, marshes, and algal beds, and

VARIATIONS IN PRIMARY PRODUCTIVITY

- ☐ Biomass of top consumers
- ☐ Biomass of primary producers

LARGE-BASED PYRAMID
- Supports a relatively large biomass of consumers
- Common in rainforests, marshes, and algal beds

SMALL-BASED PYRAMID
- Reduced ability to support consumers
- Common in deserts, tundras, and open oceans

INVERTED PYRAMID
- Small biomass of producers supports a relatively large biomass of consumers
- Occurs in some aquatic ecosystems where plankton are primary producers

FIGURE 15-15 Relative biomass of producers and consumers. Across ecosystems, there is huge variation in primary productivity.

ENERGY PYRAMID: INEFFICIENT ENERGY FLOW

Only about 10% of the biomass from each trophic level is converted into biomass in the next trophic level. This explains why there are so many more plants than animals, and why top carnivores are relatively rare.

TERTIARY CONSUMERS

10% converted to biomass

SECONDARY CONSUMERS

90% is used in cell respiration and lost as heat.

10% converted to biomass

PRIMARY CONSUMERS

10% converted to biomass

PRIMARY PRODUCERS

FIGURE 15-14 Inefficiencies in the transfer of biomass. The biomass of photosynthetic organisms at the base of the pyramid is significantly reduced at each step because of incomplete utilization by organisms at higher trophic levels.

lowest in deserts, tundra, and the open ocean. In each case, the shapes of the energy and biomass pyramids are similar. With a smaller base, though, the ability of an ecosystem to support higher levels of the food chain is reduced. One dramatic exception is seen in some aquatic ecosystems where the primary producers are plankton. Because plankton have such short life spans and rapid reproduction rates, a relatively small biomass can support a large biomass of consumers, giving rise to an inverted pyramid (see the bottom pyramid in Figure 15-15). If you quantify the amount of energy available to consumers (rather than biomass), though, the pyramid resembles those seen in terrestrial ecosystems.

TAKE-HOME MESSAGE 15·7

Energy from the sun passes through an ecosystem in several steps known as trophic levels. Energy pyramids reveal that the biomass of primary producers in an ecosystem tends to be far greater than the biomass of herbivores. Similarly, the biomass transferred at each successive step along the food chain tends to be only about 10% of the biomass of the organisms being consumed, due to energy lost in cellular respiration. As a consequence of this inefficiency, food chains rarely exceed four levels.

What Are Ecosystems?   Weather   Energy and Chemical Flow   Species Interactions

Essential chemicals cycle through ecosystems.

What is necessary for life? Energy and some essential chemicals top the list. New energy continually comes to earth from the sun, fulfilling the first need. And everything else is already here. The chemicals just cycle around and around, using the same pathway taken by energy—the food chain. Plants and other producers generally take up the molecules from the atmosphere or the soil. Then the chemicals move into animals as they consume plants or other animals, and thus move up the food chain. And as the plants and animals die, detritivores and decomposers return the chemicals to the abiotic environment. From a chemical perspective, life is just a continuous recycling of molecules.

Chemicals cycle through the living and non-living components of an ecosystem. Each chemical is stored in a non-living part of the environment called a "reservoir." Organisms acquire the chemical from the reservoir, the chemical cycles through the food chain, and eventually the chemical is returned to the reservoir.

This is a useful overview, but we can get a deeper appreciation of the functioning of ecosystems and the ecological problems that can occur when they are disturbed—particularly by humans—by investigating a few of these cycles in more detail. Here we'll explore three of the most important chemical cycles: carbon, nitrogen, and phosphorus.

Carbon For carbon, the most important reservoir is the atmosphere, where carbon is in the form of carbon dioxide (CO_2). As we saw in Chapter 4, plants and some microorganisms utilize carbon dioxide in photosynthesis, separating the carbon molecules from CO_2 and using them to build sugars. Carbon then moves through the food chain as organisms eat plants and are themselves eaten (**FIGURE 15-16**). A secondary reservoir of carbon is in the oceans. Here, many organisms utilize dissolved carbon to build shells (which later dissolve back into the water after the organism dies).

Most carbon returns to its reservoir as a consequence of organisms' metabolic processes. Organisms extract energy from food by breaking carbon-carbon bonds, releasing the energy stored in the bonds and combining the released carbon atoms with oxygen. They then exhale the end product as CO_2.

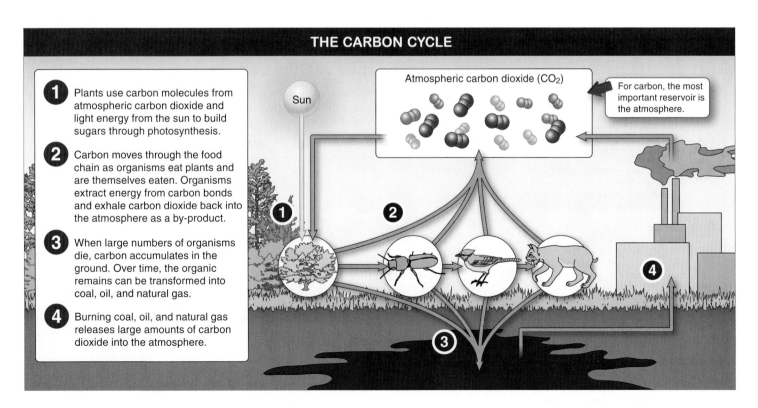

THE CARBON CYCLE

1. Plants use carbon molecules from atmospheric carbon dioxide and light energy from the sun to build sugars through photosynthesis.

2. Carbon moves through the food chain as organisms eat plants and are themselves eaten. Organisms extract energy from carbon bonds and exhale carbon dioxide back into the atmosphere as a by-product.

3. When large numbers of organisms die, carbon accumulates in the ground. Over time, the organic remains can be transformed into coal, oil, and natural gas.

4. Burning coal, oil, and natural gas releases large amounts of carbon dioxide into the atmosphere.

Atmospheric carbon dioxide (CO_2)

Sun

For carbon, the most important reservoir is the atmosphere.

FIGURE 15-16 Element cycling: carbon.

Q Why are global CO₂ levels rising?

Increasingly, another source of carbon is adding significantly to the atmospheric reservoir. Fossil fuels are created when large numbers of organisms die and are buried in sediment lacking oxygen. In the absence of oxygen, at high pressures, and after very long periods of time, the organic remains are ultimately transformed into coal, oil, and natural gas. Trapped underground or in rock, these sources of carbon played little role in the global carbon cycle until humans in industrialized countries began using fossil fuels to power various technologies. Burning coal, oil, and natural gas releases large amounts of carbon dioxide, thus increasing the average CO_2 concentration in the atmosphere—the current level of CO_2 in the atmosphere is the highest it has been in almost half a million years. This has potentially disastrous implications, as we will see in Chapter 16.

Although, on average, the level of CO_2 in the environment is increasing, there is a yearly cycle of ups and downs in the CO_2 levels in the northern hemisphere. This cycling is due to fluctuations in the ability of plants to absorb CO_2. Many trees lose their leaves each fall and, during the winter months, relatively low rates of photosynthesis lead to low rates of CO_2 consumption, causing an annual peak in

Q Global CO₂ levels are rising in general, but they also exhibit a sharp rise and fall within each year. Why?

atmospheric CO_2 levels. During the summer, leaves are present, sunlight is strong, and photosynthesis (consuming CO_2) occurs at much higher levels, causing a drop in the atmospheric CO_2 level.

Nitrogen Nitrogen is necessary to build a variety of molecules critical to life, including all amino acids, the components of every protein molecule. Like carbon, the chief reservoir of nitrogen is the atmosphere. But even though more than 78% of the atmosphere is nitrogen gas (N_2), for most organisms this nitrogen is completely unusable. The problem is that atmospheric nitrogen consists of two nitrogen atoms bonded tightly together, and these bonds need to be broken to make the nitrogen usable for living organisms. Only through the metabolic magic (chemistry, actually) of some soil-dwelling bacteria, the nitrogen-fixers, can most nitrogen enter the food chain (**FIGURE 15-17**).

These bacteria chemically convert or "fix" nitrogen by attaching it to other atoms, including hydrogen, producing ammonia and related compounds. These compounds are then further modified by other bacteria into a form that can be taken up by plants and used to build proteins. And once nitrogen is in plant tissues, animals acquire it in the same way they acquire carbon: by eating the plants. Nitrogen

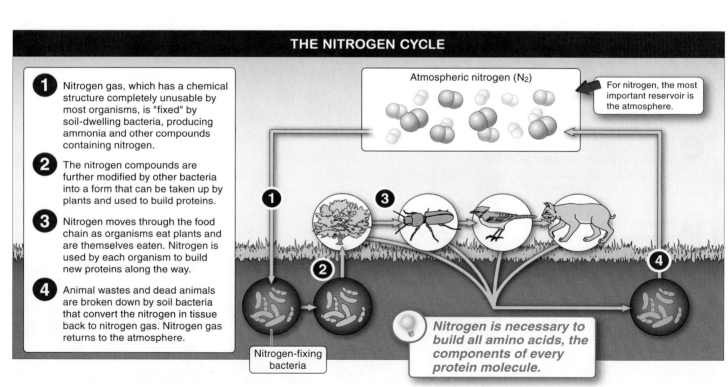

THE NITROGEN CYCLE

1 Nitrogen gas, which has a chemical structure completely unusable by most organisms, is "fixed" by soil-dwelling bacteria, producing ammonia and other compounds containing nitrogen.

2 The nitrogen compounds are further modified by other bacteria into a form that can be taken up by plants and used to build proteins.

3 Nitrogen moves through the food chain as organisms eat plants and are themselves eaten. Nitrogen is used by each organism to build new proteins along the way.

4 Animal wastes and dead animals are broken down by soil bacteria that convert the nitrogen in tissue back to nitrogen gas. Nitrogen gas returns to the atmosphere.

Atmospheric nitrogen (N₂)

For nitrogen, the most important reservoir is the atmosphere.

Nitrogen-fixing bacteria

Nitrogen is necessary to build all amino acids, the components of every protein molecule.

FIGURE 15-17 Element cycling: nitrogen.

What Are Ecosystems?   Weather   Energy and Chemical Flow   Species Interactions

returns to the atmosphere when animal wastes and dead animals are broken down by soil bacteria that convert the nitrogen compounds in tissues back to nitrogen gas.

Nitrogen is like a bottleneck limiting plant growth. Because it is necessary for the production of every plant protein, and because all nitrogen must first be made usable by bacteria, plant growth is often limited by nitrogen levels in the soil. For this reason, most fertilizers contain nitrogen in a form usable by plants.

Phosphorus   Every molecule of ATP and DNA requires phosphorus. Virtually no phosphorus is available in the atmosphere, though. Instead, soil serves as the chief reservoir. Like nitrogen, phosphorus is chemically converted— "fixed"—into a form usable by plants (phosphate) and then absorbed by their roots. It cycles through the food chain as herbivores consume plants and carnivores consume herbivores. As organisms die, their remains are broken down by bacteria and other organisms, returning the phosphorus to the soil. The pool of phosphorus in the soil is also influenced by the much slower process of rock formation on the sea floor, its uplifting into mountains, and the eventual weathering of the rock, releasing its phosphorus (**FIGURE 15-18**).

Like nitrogen, phosphorus is often a limiting resource in soils, constraining plant growth. Consequently, fertilizers usually contain large amounts of phosphorus. This is beneficial in the short run, but it can have some disastrous unintended consequences. As more and more phosphorus (and nitrogen) is added to soil, some of it runs off with water and ends up in lakes, ponds, and rivers. In these habitats, also, it acts as a fertilizer, making spectacular growth possible for algae. But eventually the algae die and sink, creating a huge source of food for bacteria. The bacteria population increases and can use up too much of the dissolved oxygen, causing fish, insects, and many other organisms to suffocate and die.

> " An atom of phosphorus 'X' had marked time in the limestone ledge since the Paleozoic seas covered the land. *Time, to an atom locked in a rock, does not pass.* The break came when a bur-oak root nosed down a crack and began prying and sucking. In the flash of a century the rock decayed, and X was pulled out and up into the world of living things. He helped build a flower, which became an acorn, which fattened a deer, which fed an Indian all in a single year.
>
> —Aldo Leopold, *A Sand County Almanac*, 1949 "

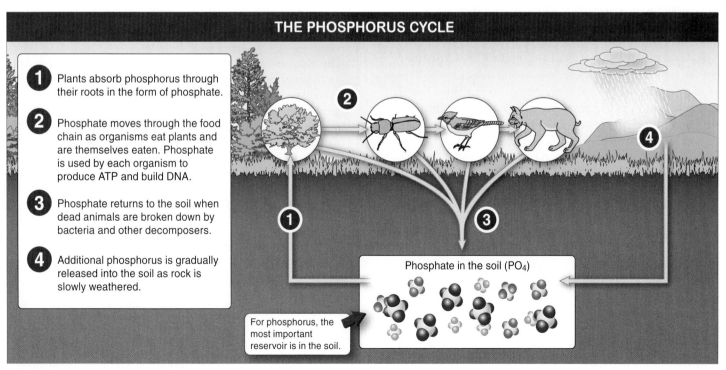

THE PHOSPHORUS CYCLE

1 Plants absorb phosphorus through their roots in the form of phosphate.

2 Phosphate moves through the food chain as organisms eat plants and are themselves eaten. Phosphate is used by each organism to produce ATP and build DNA.

3 Phosphate returns to the soil when dead animals are broken down by bacteria and other decomposers.

4 Additional phosphorus is gradually released into the soil as rock is slowly weathered.

Phosphate in the soil (PO_4)

For phosphorus, the most important reservoir is in the soil.

FIGURE 15-18 Element cycling: phosphorus.

Communities

EUTROPHICATION

Eutrophication—the increase in nutrients in an ecosystem, particularly nitrogen and phosphorus—often leads to the rapid growth of algae and bacteria in lakes and other aquatic ecosystems. These organisms then consume much of the oxygen, leading to large die-offs of animal life. This process often occurs as a result of runoff from fertilizer added to farmlands.

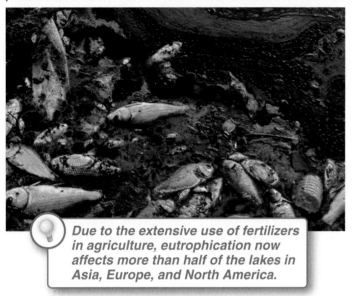

Due to the extensive use of fertilizers in agriculture, eutrophication now affects more than half of the lakes in Asia, Europe, and North America.

FIGURE 15-19 **Too much of a good thing?**

The increase in nutrients, particularly nitrogen and phosphorus, in an ecosystem, is called **eutrophication.** The consequent rapid growth of algae and bacteria, followed by large-scale die-offs of other organisms, is increasingly a problem in both small and large bodies of water, including more than half of the lakes of Asia, Europe, and North America (**FIGURE 15-19**). Lake Erie, on the U.S.-Canadian border, for example, has experienced eutrophication as a result of all the phosphorus- and nitrogen-containing waste water that drains into the lake from the extensive surrounding farmlands. Given the lower use of fertilizers in South America and Africa, eutrophication is less common there.

Water, too, cycles through ecosystems. Moving from the oceans to land via evaporation and rain, it then returns to the oceans through runoff.

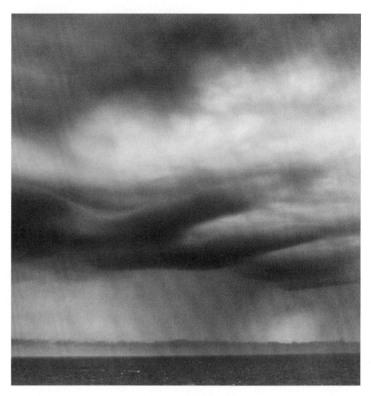

Water also cycles through ecosystems. It continually moves from the ocean to the air and land, then returns to the ocean in a cyclic pattern.

TAKE-HOME MESSAGE 15•8

Chemicals essential to life—including carbon, nitrogen, and phosphorus—cycle through ecosystems. They are usually captured from the atmosphere, soil, or water by growing organisms, passed from one trophic level to the next as organisms eat other organisms, and returned to the environment through respiration, decomposition, and erosion. These cycles can be disrupted as human activities significantly increase the amounts of the chemicals utilized or released to the environment.

568 CHAPTER 15 • ECOSYSTEMS AND COMMUNITIES

What Are Ecosystems?   Weather   Energy and Chemical Flow   Species Interactions

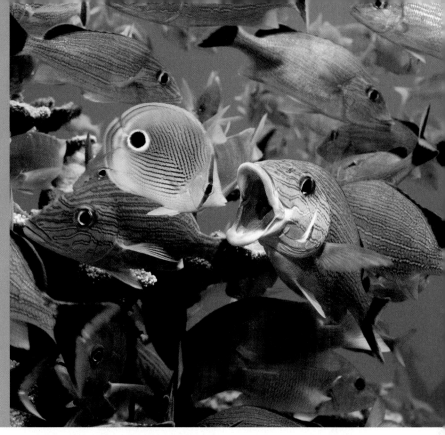

❹ Species interactions influence the structure of communities.

Cleaning behavior: a blue-striped grunt (Haemulon sciurus) *presents its mouth for cleaning by a foureye butterflyfish* (Chaetodon capistratus).

15·9 --

Interacting species evolve together.

There is a moth in Madagascar with a tongue that is 11 inches long! This might seem absurdly long—until the moth approaches a similarly odd-looking orchid. The orchid's flower has a very long tube, also about 11 inches long, that has a bit of nectar at the very bottom. The moth's tongue, although usually rolled up, straightens out as fluid is pumped into it and can be inserted into the long nectar tube. As its tongue reaches the bottom, the moth slurps up the nectar, gaining nourishment and energy. The moth also gets a bit of pollen stuck to it in the process, pollen that gets brushed onto the reproductive parts of the next orchid flower it visits (**FIGURE 15-20**).

It's clear that having an 11-inch tongue is useful, even necessary, to extract nectar from an 11-inch nectar tube. And it's clear that putting nectar at the bottom of an 11-inch tube is a strategy to restrict access to only those pollinators that will reliably pass pollen from plant to plant of the same species. But how did such a system ever originate? Which came first, the long-tongued moths or the long-tubed flowers? Each trait only seems to make sense in a world in which the other

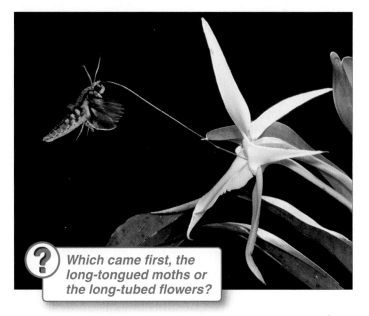

? *Which came first, the long-tongued moths or the long-tubed flowers?*

FIGURE 15-20 A perfect match? Long-tubed orchids and long-tongued moths in Madagascar each influence the evolution of the other.

already exists. The answer is that neither came first. They both evolved—**coevolved**—together.

Natural selection causes organisms to become better adapted to their environment. As long as there is variation for a trait, and the trait is heritable, differential reproductive success will lead to a change in the population. It is easy to imagine populations becoming more and more efficient at utilizing non-living resources. But natural selection does not distinguish between biotic and abiotic resources as selective forces. Either can cause individuals with certain traits to reproduce at a higher rate than others, and so either can cause evolution. So small changes in

the moth lead to small changes in the orchid, which select for further small changes in the moth ... and the process continues on. In the end, species become adapted not just to their physical environment but also to the other species around them.

TAKE-HOME MESSAGE 15•9

In producing organisms better adapted to their environment, natural selection does not distinguish between biotic and abiotic resources as selective forces.

15•10

Each species' role in a community is defined by its niche.

Within a society, most humans seem to find their niche. Each person plays a particular role, defined by the nature of his or her work, activities, and interactions with others. Other species do the same thing. Within their *communities*— geographic areas defined as loose assemblages of species, sometimes interdependent, with overlapping ranges—each species has its own niche.

We can define an organism's **niche** in terms of the ways in which the organism utilizes the resources of the environment. More than just a *place* for living, a niche is a complete *way* of living. In other words, an organism's niche encompasses (1) the space it requires, (2) the type and amount of food it consumes, (3) the timing of its reproduction (its life history), (4) its temperature and moisture requirements, and virtually every other aspect that describes the way the organism uses its environment (**FIGURE 15-21**).

> And NUH is the letter I use to spell Nutches,
> Who live in small caves, known as Niches, for hutches.
> These Nutches have troubles, the biggest of which is
> The fact there are many more Nutches than Niches.
> Each Nutch in a Nich knows that some other Nutch
> Would like to move into his Nich very much.
> So each Nutch in a Nich has to watch that small Nich
> Or Nutches who haven't got Niches will snitch.
>
> —Dr. Seuss, *On Beyond Zebra*, 1955

NICHE FEATURES
• The space an organism requires
• The type and amount of food an organism utilizes
• The timing of an organism's reproduction
• An organism's temperature and moisture requirements and other necessary living conditions

Bald eagle range
☐ Summer (breeding)
☐ Winter (non-breeding)
☐ Year-round

FIGURE 15-21 **A way of living.** "Niche" describes all of the ways in which an organism utilizes the resources of the environment.

Although a niche describes the role a species *can* play within a community, the species doesn't always get to have that exact role. Consider the rats of Boston. Until the 1990s, they lived in relative peace in the sewers beneath the city's streets. But when

What Are Ecosystems?   Weather   Energy and Chemical Flow   Species Interactions

the city embarked on the largest underground highway construction project in U.S. history, engineers displaced and forcibly drove out thousands of rats from much of their habitat. In essence, there was suddenly an overlap between the rat niche and the human niche, and the rats were now restricted to just some portions of the sewer. As a result, the rats' **realized niche,** where and how they are actually living, is now just a subset of their **fundamental niche,** the full range of environmental conditions under which they can live.

It is common for species to find themselves competing with other species for parts of a niche. When this occurs, generally one of the species is restricted from its full niche. We'll examine next the various possible outcomes when such niche overlap occurs.

TAKE-HOME MESSAGE 15•10

> A population of organisms in a community fills a unique niche, defined by the manner in which it utilizes the resources in its environment. Organisms do not always completely fill their niche; competition with species that have overlapping niches can reduce their range.

15•11

Competition can be hard to see, but it still influences community structure.

Some species, particularly closely related species, have similar niches. This can lead to conflict as they try to exploit the same resources. Almost invariably, when the fundamental niches of two species overlap, competition occurs. This competition doesn't last forever, though. Inevitably, one of two outcomes occurs: competitive exclusion or resource partitioning (**FIGURE 15-22**).

1. In **competitive exclusion,** two species battle for resources in the same niche until the more efficient of the two wins and the other species is driven to extinction in that location ("local extinction"). In the 1930s, this was demonstrated in simple laboratory experiments using *Paramecium,* a single-celled organism. Populations of two similar *Paramecium* species were grown either separately or together in test tubes containing water and their bacterial food source. When grown separately, each species thrived. When grown together, though, one species always drove the other to extinction.

2. Resource partitioning is an alternative outcome of niche overlap. Individual organisms and species can adapt to changing environmental conditions, and resource partitioning can result from an organism's behavioral change or a change in its structure. When this occurs, one or both species become restricted in some aspect of their niche, dividing the resource. In other experiments with *Paramecium,* for example, one of the two species was replaced with a different species. As in the initial experiment, either species thrived when grown alone. But when the two species were grown together in the same test tube, they ended up dividing the test tube "habitat." One species fed exclusively

WHEN NICHES OVERLAP

1 When grown alone under a particular set of environmental conditions, each of two similar *Paramecium* species thrives.

2 When both species are grown together, because their niches overlap they compete and two outcomes are possible:

or

COMPETITIVE EXCLUSION
One species within the niche utilizes resources more efficiently, driving the other species to local extinction.

RESOURCE PARTITIONING
The species each alter their use of the niche, dividing the resources.

FIGURE 15-22 Overlapping niches: competitive exclusion and resource partitioning.

at the bottom of the test tube, and the other fed only at the top. Simple behavioral change made coexistence possible.

In many situations, resource partitioning is accompanied by **character displacement,** an evolutionary divergence in one or both of the species that leads to a partitioning of the niche. A clear example occurs among two species of seed-eating finches on the Galápagos Islands. On islands where both species live, their beak sizes differ significantly. One species has a deeper beak, better for large seeds, while the other has a shallower beak, better for smaller seeds, and they do not compete. On islands where either species occurs alone, beak size is intermediate between the two sizes (**FIGURE 15-23**).

COMPETITION AND CHARACTER DISPLACEMENT

Character displacement occurs when natural selection reduces the competition between two species by producing an evolutionary divergence in one or both species.

1 Two finch species have medium-size beaks when the species live on separate islands.

2 When the birds inhabit the same island, they compete for the same resources. Evolution leads to a shift in the beak size in each species, a change called character displacement.

FIGURE 15-23 Allowing organisms to divide resources. Character displacement is an evolutionary adaption that makes resource partitioning possible.

Competition between species has one very odd feature: it is very hard to actually see it occurring because it causes itself to disappear. That is, after only a short period of competition, either one of the species becomes locally extinct or leaves the area where the niches overlap, or character displacement occurs, largely reducing the competition. In either case, the net result is that the level of competition is significantly reduced or wiped out altogether. For this reason, biologists often have to look for character displacement—the "ghost of competition past"—to identify areas where competition has occurred in the past. Moreover, even while it is occurring, competition tends to be indirect, rather than head-to-head battles. Like a game of musical chairs, it's more a question of both species trying to utilize a particular resource, with one being a bit better at it.

Q Why is competition hard to see in nature?

Biologists do have the opportunity to see competition, however, in the occasional natural experiments in which a new species is introduced to a community. This happened, for example, when the American mink, a weasel-like mammal, was introduced into Europe, where previously only the European mink had lived. Within 10 years, in the area where their ranges overlapped, the European mink increased in size while the American mink decreased in size, as biologists documented the evolution. In the next chapter, we'll investigate in greater depth why newly introduced species frequently win the competition with native species.

TAKE-HOME MESSAGE 15·11

Populations with completely overlapping niches cannot coexist forever. Competition for resources occurs until one or both species evolve in ways that reduce the competition, through character displacement, or until one becomes extinct in that location.

What Are Ecosystems?   Weather   Energy and Chemical Flow   Species Interactions

Predation produces adaptation in both predators and their prey.

Some words of advice in case you ever think about quickly approaching a horned lizard: be afraid. Be very afraid. Here's why: as you get close to the lizard, it may zap you with streams of blood squirted from its eyes. In all likelihood, you will flinch. And as you flinch, the horned lizard will scurry away. The display is shocking, but the fact that evolution has produced extreme and effective anti-predator adaptations is not.

Predation—an interaction between two species in which one species eats the other—is one of the most important forces shaping the composition and abundance of species in a community. Predation, though, isn't restricted to the obvious interactions involving one animal chasing down and killing another. Herbivores eating leaves, fruits, or seeds is a form of predation, even though it doesn't necessarily kill the plant. And predators are not necessarily physically imposing. Each year more than a million humans die as a result of disease from mosquito bites, compared with fewer than a dozen from shark attacks.

Predators are a potent selective force: organisms eaten by predators tend to have reduced reproductive success. Consequently, in prey species, a variety of features have evolved (and continue to evolve)—including the blood-squirting-eyeball effect—that reduce their predation risk. But as prey evolve, so do predators. This coevolution is a sort of arms race with ever-changing and escalating predation-effectiveness adaptations causing more effective predator-avoidance adaptations, and vice versa. (In this light, it may seem unexpected that exotic species often flourish when released into novel habitats, even though natural selection has not adapted them to their new environment. As it turns out, just as these species are not fully adapted to their new environment, they also have few predators there. And with low predation risk, they often are able to flourish.) We'll examine some common adaptations of both predators and prey.

Prey adaptations for reducing predation There are two broad categories of defenses against predators: physical and behavioral.

Physical defenses include mechanical, chemical, warning coloration, and camouflage mechanisms.

Q *Why do exotic species often flourish when released into novel habitats, even though natural selection has not adapted them to this new environment?*

1. *Mechanical defenses.* Predation plays a large role in producing adaptations such as the sharp quills of a porcupine, the prickly spines of a cactus, the wings of a bird or bat, or the tough armor protecting an armadillo or sow bug. These, as well as claws, fangs, stingers, and other physical structures, can reduce predation risk (**FIGURE 15-24**).

2. *Chemical defenses.* Further prey defenses can include chemical toxins that make the prey poisonous or unpalatable. Plants can't run from their predators, so chemical defenses are especially important to them. Virtually all plants produce some chemicals to deter organisms that might eat them. The toxins can be severe, such as strychnine from plants in the genus *Strychnos,* which kills most vertebrates, including humans, by stimulating non-stop convulsions and other extreme and painful symptoms leading to death. At the other end of the spectrum are chemicals toxic to some insects but relatively mild to humans, such as those found in cinnamon, peppermint, and jalapeño peppers. Ironically, many plants that evolved to produce such chemicals to deter predators are now cultivated and eaten specifically *for* the chemicals they produce. One organism's toxic poison is another's spicy flavor.

Some animals can also synthesize toxic compounds. The poison dart frog has poison glands all over its body, making it toxic to the touch. The fire salamander, too, is toxic, with the capacity to squirt a strong nerve toxin from poison glands on its back. Some animals, including milkweed bugs and monarch butterflies, are able to safely consume toxic chemicals from plants and sequester them in their tissues, becoming toxic to predators who try to eat them.

3. *Warning coloration.* Species protected from predation by toxic chemicals frequently have evolved bright color patterns to warn potential predators. They are essentially carrying a sign that says: "Warning, I'm poisonous, so keep away." To make it as easy as possible for predators to learn, different poisonous species often have the same color patterns. It's as if devising a single common sign saying "Danger: poison" is more effective than multiple species coming up with their own specific signs (see Figure 15-24).

In a clever twist on this, some species that are perfectly edible to their predators also have evolved the same bright colors, in

MECHANICAL DEFENSES

CHEMICAL DEFENSES

WARNING COLORATION

CAMOUFLAGE

FIGURE 15-24 Prey defenses: some physical means for avoiding predation. Shown (*from left, clockwise*) are Cape porcupine, azure poison dart frog, praying mantis, and monarch butterfly.

a phenomenon known as **mimicry.** Their coloration mimics the same warning sign but without the toxins. As long as they are relatively rare compared with the toxic individuals they mimic—reducing the chance that predators might catch on to their trickery—the evolutionary ruse is quite successful.

4. *Camouflage.* An alternative to warning coloration, and one of the most effective ways to avoid being eaten, is simply to avoid being seen. An adaptation in many organisms is patterns of coloration that enable them to blend into their surroundings. These include insects that look like leaves or twigs and hares that are brown for most of the year but turn white when there is snow on the ground.

Behavioral defenses include both seemingly passive and active behaviors: hiding or escaping, or alarm calling or fighting back.

1. *Hiding or escaping.* Anti-predator adaptations need not involve toxic chemicals, physical structures, or special coloration. Many species excel at hiding and/or running. With vigilance, it is possible to get advance warning of impending predator attacks, and then quickly and effectively avoid predators. A variation of this strategy comes from safety in numbers: many species, including schooling fish and emperor penguins, travel in large groups to reduce their predation risk (**FIGURE 15-25**).

On many islands, animals show no fear of humans. Rather than encountering

Q On islands, animals frequently have no fear of humans (in contrast to having more skittish behavior on the mainland). Why?

What Are Ecosystems?   Weather   Energy and Chemical Flow   Species Interactions

BEHAVIORAL DEFENSES THAT REDUCE PREDATION

HIDING OR ESCAPING

FIGHTING BACK

FIGURE 15-25 Prey defenses: behavioral means for avoiding predation. Fish may travel in huge schools, finding safety in numbers. And fish may mob a predator, here a shark, to make it harder for the predator to attack any one member of the school. The bird can project vomit at a predator, frightening it away.

skittish lizards, for example, a human visitor to the Galápagos Islands must be careful not to step on them. The animals are not afraid because, given the small size of most islands, the number of predators is restricted. As a consequence, there has been no selection for skittishness and hyper-vigilance, traits that normally evolve in response to predation risk.

2. *Alarm calling and fighting back.* In many species, especially birds and mammals, individuals warn others with an alarm call. Although risky for the caller, such alarm calling can give other individuals—often close kin that are nearby— enough advance warning to escape (recall, from Chapter 9, the warning calls of Belding's ground squirrels). Some prey

species also turn the tables, mobbing predators to keep them from successfully completing their task. This category might also include the blood-squirting lizard described above, or the fulmar, a seabird that defends its nest from attacks with projectile vomiting aimed at the intruder.

Predator adaptations for enhancing predation Just as prey use physical and behavioral features to reduce their risk of predation, predators evolve in parallel ways to increase their efficiency. As milkweed plants have evolved to sequester toxic chemicals within the structures of the plant that kill their predators, in milkweed bugs toxic-avoidance methods, have evolved that allow the bugs to eat the toxic plants without suffering harm. And as prey have become better at hiding and escaping, predators have developed better sensory perception to help them detect hiding prey, and faster running ability to catch them. Predators, too, make use of mimicry. The angler fish, the tasseled frogfish, and the snapping turtle all have physical structures that mimic something—usually a food item—that is of interest to potential prey. As the prey come closer to inspect, the predator snaps them up (**FIGURE 15-26**).

Q Why don't predators become so efficient at capturing prey that they drive the prey to extinction?

Although natural selection leads to predators with effective adaptations for capturing prey, the adaptations are rarely so efficient that the prey are driven to extinction. The explanation for this is referred to as

Predators such as the angler fish have evolved luminous structures that trick prey into coming right to the predator, where they are eaten.

FIGURE 15-26 A better predator. Some adaptations enhance predation.

the "life-dinner hypothesis." It generally holds true because selection for "escape ability" in the prey is stronger than selection for "capture ability" in the predator. When a prey, such as a rabbit, for example, can't escape from a fox, the cost is its life—and it will never reproduce again. On the other hand, when a fox can't keep up with a rabbit, all it loses is a meal; it can still go on to capture prey and reproduce in the future. In other words, the cost of losing in the interaction is much higher for the rabbit.

15·13

Parasitism is a form of predation.

For most people, the word "predation" conjures images of a large animal such as a cheetah, chasing and killing another large animal such as a gazelle. But there is another world of predation, largely unseen but equally deadly: **parasitism,** defined as a symbiotic relationship in which one organism (the **parasite**) benefits while the other (the **host**) is harmed. In fact, this hidden world of parasites is thriving with activity: parasites are probably the most numerous species on earth, with three or four times as many species as non-parasites!

Two general types of parasites make life difficult for most organisms. *Ectoparasites* ("ecto" meaning outside) include organisms such as lice, leeches, and ticks. One species of Mexican parrot is all too aware of ectoparasites: it has 30 different species of mites living on its feathers alone. And many of the parasites even have parasites themselves! *Endoparasites* are parasites that live inside their hosts ("endo," inside). They are equally pervasive. Endoparasites infecting vertebrates include many different phyla of both animals and protists, the single-celled eukaryotes (**FIGURE 15-27**). In all of these parasite-host interactions, as with all predator-prey interactions, the predator or parasite benefits and the prey or host is harmed.

Even though they are considered predators, parasites have some unique features and face some unusual challenges relative to other predators. The most obvious of these features is that in parasitism, the parasite generally is much smaller than its host and stays in contact with the host for extended periods of time, normally not killing the host but weakening it as the parasite uses some of the host's resources. Being located right on your food source all the time can be advantageous. But this also leads

FIGURE 15-27 Parasites: predators dwarfed by their prey.
A velvet mite, an ectoparasite on human skin, is shown in extreme close-up. *Trypanosome brucei* is an endoparasite that invades a human host and is the cause of African sleeping sickness.

PARASITES

Parasites are predators that benefit from a symbiotic relationship with their hosts.

ECTOPARASITES
Parasites that live on their host

ENDOPARASITES
Parasites that live inside their host

 There may be three to four times as many parasitic species as non-parasitic species on earth!

to what is perhaps the greatest challenge that parasites face: how to get from one individual host to another. A parasite can't survive long once its host dies, after all. The methods by which parasites accomplish such dispersal are unexpected and surprising. Many of their complicated life cycles involve passing through two (or more) different species (and could have come about only through coevolution with each of the host species). These life cycles are likely to give us a new appreciation for the ingenuity of these microorganisms—or rather, for the evolutionary process that has produced them. Let's look at a few representative examples.

Case 1: Parasites can induce foolish, fearless behavior in their hosts. Rats fear cats. During their evolution, rats have developed a protective wariness of their feline predators and areas in which cats have been roaming. *Toxoplasma* is an organism that changes this. This single-celled parasite of rats must also spend part of its life cycle in cats. It does this by altering the brain of its rat host so that the rat no longer fears cats. In fact, *Toxoplasma*-infected rats not only lose their fear of cats, they become attracted to them. Is this an accident? No. This behavioral change, while quite dangerous for the rat, is exactly the change that increases the likelihood that the rat will be attacked by a cat, spreading the parasite in the process.

Case 2: Parasites can induce inappropriate aggression in their hosts. Rabid animals don't behave normally. They froth at the mouth and become unusually aggressive (**FIGURE 15-28**). Is this an accident? No. Rabies is caused by a virus that infects warm-blooded animals, mostly raccoons, skunks, foxes, and coyotes. It is passed from one host to another via saliva. Inducing these "rabid" behaviors, of course, is exactly the change in behavior that will most help the virus to spread.

Case 3: Parasites can induce bizarre and risky behavior in their hosts. The lancet fluke is a parasitic flatworm. It has also been described as a "zombie-maker." This fluke spends most of its life in sheep and goats, but the fluke's eggs pass into snails that graze on vegetation contaminated by sheep and goat feces. Once inside the snail, the fluke eggs grow and develop, eventually forming cysts that the snail surrounds with mucus and then excretes. Continuing on their complex life cycle, the fluke cysts find their way into ants that eat the snail mucus. The flukes' journey back to a sheep or goat is now expedited by the so-called zombie-making. In an infected ant, the flukes grow into the ant's brain, altering its behavior. Whereas ants

Parasites have evolved surprising methods to solve a challenging problem: how to get from one individual host to another.

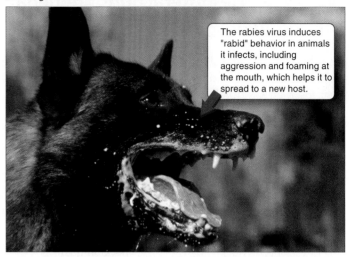

The rabies virus induces "rabid" behavior in animals it infects, including aggression and foaming at the mouth, which helps it to spread to a new host.

FIGURE 15-28 Rabies. The rabies parasite (a virus) is more effectively passed from one animal to another by causing its host to foam at the mouth and aggressively bite.

normally remain low to the ground, when infected by the lancet fluke they climb to the tops of grass blades or plant stems and clench their mandibles on leaves. Is this an accident? No. This behavioral change puts the ants much more at risk of being eaten by a grazing mammal—a bad outcome for the ant, but just what the parasite needs to complete its life cycle.

These parasite-induced behavior modifications are among the most dramatic, but there are many others that are more subtle but equally effective at allowing a parasite to thrive attached to or within the host's body. This subset of predation is an active and exciting area of ecological and physiological investigation.

TAKE-HOME MESSAGE 15•13

Parasitism is a symbiotic relationship in which one organism benefits while the other is harmed. Parasites face some unusual challenges relative to other predators, particularly how to get from one individual host to another, and some complex parasite life cycles have evolved.

Not all species interactions are negative: mutualism and commensalism.

It's easy to get the idea that all species interactions in nature are harsh and confrontational, marked by a clear winner and a clear loser. That is largely the case when it comes to competition and predation. Nonetheless, not all species interactions are combative. Every flower you see should be a reminder that evolution produces beneficial species interactions as well. These types of interactions fall into two categories: mutualism and commensalism.

Mutualism: everybody wins

- Coral that gain energy from photosynthetic algae living inside their tissues

- Termites capable of subsisting on wood, only with the assistance of cellulose-digesting microbes living in their digestive system

- Flowers pollinated by animals nourished by nectar

Each of these relationships is an example of **mutualism,** an interaction in which both species gain and neither is harmed (**FIGURE 15-29**). Mutualism is common in virtually all communities. Plants, particularly, have numerous such interactions: with nutrient-absorbing fungi, with nitrogen-fixing bacteria, and with animals that pollinate them and disperse their fruits.

Commensalism: an interaction with a winner but no
loser   Some species interactions are one-sided. The cases in which one species benefits and the other neither benefits nor is harmed are called commensal relationships, or **commensalism.** Cattle egrets have just such a relationship with grazing mammals such as buffalo and elephants (see Figure 15-29). As the large mammals graze through grasses, they stir up insects. The birds, which feed near the mammals—particularly near the forager's head—are able to catch more insects with less effort this way. The grazers are neither helped nor harmed by the presence of the birds.

TAKE-HOME MESSAGE 15•14

Not all species interactions are combative. Evolution produces beneficial species interactions as well, including mutualism, in which both species benefit from the interaction, and commensalism, in which one species benefits and the other is neither harmed nor helped.

NON-NEGATIVE SPECIES INTERACTION

MUTUALISM
Interaction in which both species benefit

COMMENSALISM
Interaction in which one species benefits and the other neither benefits nor is harmed

FIGURE 15-29   Not always "red in tooth and claw." Mutualisms and commensalisms abound in nature.

⑤ Communities can change or remain stable over time.

Prairie dogs are a keystone species in the prairie ecosystem.

15•15

Many communities change over time.

Human "progress" and development can completely transform an environment—turning a patch of desert into Las Vegas, for example. Urban landscapes, too, can obliterate any signs of the nature that was once there. This is why it can be surprising (and heartening) to observe what happens when humans abandon an area. Little by little, nature reclaims it. The area doesn't necessarily recover completely, and change is slow. Still, this process is almost universal and virtually unstoppable. Nature responds similarly to other disturbances, too, from a single tree falling in a forest, to a massive flood or fire, to massive volcanic eruptions.

The process of nature reclaiming an area and of communities gradually changing over time is called **succession.** It is defined specifically as a change in the species composition over time, following a disturbance. There are two types of succession. *Primary succession* is when the process starts with no life and no soil. *Secondary succession* is when an already established habitat is disturbed but some life and some soil remain.

Primary succession Primary succession can take thousands or even tens of thousands of years, but it generally occurs in a consistent sequence. It always begins with a disturbance that leaves an area barren of soil and with no life. Frequently the disturbance is catastrophic. The huge volcanic eruption on Krakatoa, Indonesia, in 1883, for example, completely destroyed several islands and wiped out all life and soil on others. Primary succession has also begun, in a less

dramatic fashion, in regions where glaciers have retreated, such as Glacier Bay, Alaska. Although succession does not occur in a single, definitive order, several steps are relatively common (**FIGURE 15-30**).

- The first arrivals, or **colonizers,** to a lifeless, soil-less area are usually bacteria or fungi or other photosynthetic microorganisms, floating in the air.

- Lichens, a symbiotic association between a fungus and a photosynthetic alga or bacterium, are also common first colonizers. They can grow on rock and are able to generate energy through photosynthesis. As they grow, lichens change their environment, too, secreting acids that break down the rocks and release useful minerals, making the terrain more hospitable for other organisms.

- Mosses also tend to arrive in the early stages, utilizing many of the nutrients freed up by lichens. They trap moisture and can provide a hospitable site for germination of the seeds of other plants.

- Following mosses, small herbs often arrive. Later, some shrubs may begin to thrive, shading out the mosses and herbs.

- Shrubs in turn are eventually outcompeted and shaded out by small trees.

- And the first trees generally are outcompeted by taller, faster-growing trees.

Succession is the change in species composition in a community over time.

COLONIZING COMMUNITY
The first arrivals to a lifeless and soil-less area. They are good dispersers, but poor competitors, so they are gradually replaced.

Fungi, bacteria, lichens, and seeds are often among the earliest colonizers.

Mosses begin to grow and trap moisture, allowing the seeds of other plants to germinate.

INTERMEDIATE COMMUNITIES
A variety of species is present, including both colonizers and competitors.

Small herbs arrive and shrubs grow, shading out the mosses and herbs.

CLIMAX COMMUNITY
Longer-living, larger species outcompete the initial colonizers and persist as a stable and self-sustaining community.

Small trees grow and outcompete the shrubs.

Taller, faster-growing trees outcompete the smaller trees.

Primary succession begins after a disturbance leaves an area barren of soil and with no life.

Secondary succession is like primary succession with a head start. It begins when a disturbance opens up part of a community to development and growth of species previously outcompeted by other species in the area.

Disturbance is a fundamental part of most ecosystems and can repeatedly set a community back to an earlier stage of succession, at which point secondary succession begins again.

FIGURE 15-30 Species composition of a community changes over time.

An important feature of the colonizers seen in the earliest stages of primary succession is that while they are all good **dispersers,** able to move away from their original home (hence their early arrival to a newly available locale), they are not particularly good *competitors.* That's why they are gradually replaced.

Ultimately, it is the longer-living, larger species that tend to outcompete the initial colonizers and persist as a stable and self-sustaining community, called a **climax community.** The specific species present in the climax community depend on physical factors such as temperature and rainfall.

Secondary succession   Secondary succession is much faster than primary succession because life and soil are already present. Rather than the thousands or even tens of thousands of years that primary succession may take, secondary succession can happen in a matter of centuries, decades, or even years. It frequently begins with organisms colonizing the decaying remains of dead organisms. It may involve fungi establishing themselves in the decaying trunk of a tree that has fallen, and these being replaced over time by different species of fungi.

Or secondary succession may begin with weeds springing up in formerly cultivated land that is left untended. If the weeds are allowed to grow, they eventually are outcompeted and replaced by perennial species, and then shrubs, and eventually larger trees. The process is similar to primary succession, but with a head start. Ultimately, if undisturbed, secondary succession also leads to establishment of a stable, self-sustaining climax community.

Given that primary and secondary succession both progress toward climax communities, it might seem surprising that, at any given time, so many communities are in states far from climax. This apparent paradox reflects the fact that succession leads to climax communities in the *absence of disturbance.* But disturbance is a fundamental part of most ecosystems. Massive disturbances, such as the fires that destroyed a third of Yellowstone Park in 1988 or the volcanic eruptions of Mount St. Helens, can send a community back to square one. Minor disturbances, on the other hand, may undo just a small degree of the succession that has occurred. Interestingly, it is at intermediate levels of disturbance that communities tend to support the greatest number of species. With very high or very low disturbance levels, the community is restricted to the top colonizers—those with the best dispersal or fastest reproductive rates—or the top competitors, respectively. At the intermediate level, a larger variety of species with intermediate combinations of features can persist.

Q Why doesn't succession occur on front lawns?

TAKE-HOME MESSAGE 15·15

Succession is the change in the species composition of a community over time following a disturbance. In primary succession, the process begins in an area with no life present; in secondary succession, the process occurs in an area where life is already present. But in both types, the process usually takes place in a predictable sequence.

580   CHAPTER 15 • ECOSYSTEMS AND COMMUNITIES

What Are Ecosystems?   Weather   Energy and Chemical Flow   Species Interactions

Some species are more important than others within a community.

Not all species are created equal. In this chapter, we have examined the ways in which species interact with one other and with their habitats, noting the dependence of species on each other. But not all species have equal dependence on and influence over others. Within a community, the presence of some species, called **keystone species,** greatly influences which other species are present and which are not (**FIGURE 15-31**). That is, if a keystone species is removed, the species mix in the community changes dramatically. The removal of other species, conversely, causes relatively little change.

Bison are a keystone species. Experiments were set up in which a herd of bison was allowed to graze in certain parts of a prairie in Kansas, while being excluded from other areas. The areas without the bison were soon dominated by a single species of tall grass. The grass was like a bully to other plant species. The areas where the bison grazed, on the other hand,

had significantly greater species diversity. When kept in check by the grazing bison, the bully grass was unable to outcompete the other plant species, which were able to thrive.

Sea stars, too, are a keystone species (see Figure 15-31). On the rocky seashores of the Pacific Northwest, they consume mussels, which tend to crowd out other species. In a five-year study, sea stars were removed and kept out of certain areas of the intertidal zone but allowed in other areas. As with the bison study, species diversity dropped dramatically where sea stars were excluded. In fact, ultimately, just the mussels remained. In the areas with sea stars, mussels also remained but in reduced numbers. More important, 28 other species of animals and algae were also able to live there.

Suppose you are in charge of conserving biodiversity in a large region of a country. With limited funds, you cannot simply purchase and restrict access to the entire region. It is not possible to preserve everything. Instead, you must make difficult decisions and prioritize species for preservation. Knowing that keystone species exist can make your job considerably easier. There is real practical value in identifying keystone species and focusing conservation efforts on these. After all, in the sea star experiment, preservation of the sea stars alone would have the "side effect" of preserving 28 other species.

Keystone species make it possible to get more "bang for your buck" when your aim is to conserve biodiversity. Consequently, identifying keystone species is an important part of conservation biology. Some other keystone species include dam-building beavers, elephants of the African savanna, and lichens in the desert. We'll explore these questions of conservation in greater detail in the next chapter.

Q When it comes to conservation, are some species more valuable than others? Why?

KEYSTONE SPECIES

A keystone species, such as the sea star, has an unusually large influence on the presence or absence of numerous other species in the community. When sea stars were removed from an area within an intertidal zone, species diversity decreased drastically—only mussels remained.

COMMUNITY WITH KEYSTONE SPECIES | COMMUNITY WITH KEYSTONE SPECIES REMOVED

Identifying keystone species is an important part of conservation biology— preserving just one keystone species has the effect of preserving many additional species at the same time.

FIGURE 15-31 Preserving biodiversity. Keystone species can keep aggressive species in check, allowing more species to coexist in a community.

TAKE-HOME MESSAGE 15•16

Keystone species have a relatively large influence on which other species are present in a community and which are not. Unlike other species, when a keystone species is removed from the community, the species mix changes dramatically. For this reason, preserving keystone species is an important strategy in maximizing the preservation of biodiversity.

Knowledge You Can Use

Did you know? In boosting plant productivity on farms, we've created a "dead zone" bigger than the size of Connecticut in the Gulf of Mexico.

In the water at the mouth of the Mississippi River, each spring and summer, there is so little oxygen that virtually no life can survive.

Q: Which chemical almost always limits plant growth? Plant growth requires the production of proteins, all of which contain nitrogen. Access to nitrogen nearly always determines how much a plant grows.

Q: If you're a farmer, how can you increase the productivity of your crops? Add nitrogen. This, in fact, is the chief component of all plant fertilizers.

Q: Where does the unused nitrogen end up? Spring and summer rains wash nutrients, dissolved organic materials, and other runoff (much of which comes from human sources) from the middle of the United States into nearby rivers and, from those rivers, into the Mississippi River. From there they flow into the Gulf of Mexico.

Q: What happens when large amounts of nitrogen are dumped into a body of water? Organisms such as plankton and algae living near the mouth of the Mississippi River grow like crazy with all the extra nitrogen and nutrients that are washed into the Gulf of Mexico. But the rapid increase in their productivity disrupts the food chain as they decay, increasing the organic matter that sinks to the bottom and feeds the bacteria there. And as the bacteria thrive, they consume ever increasing amounts of oxygen.

Q: What happens when bacteria use up too much oxygen? Excessive bacterial growth depletes the supply of dissolved oxygen in the water, starving all other life of oxygen—including fish, crabs, oysters, and shrimp, all of which die or move out of the area. This creates a "dead zone" and ruins local fisheries, with significant impact on the economies they support.

What can you conclude? Disrupting food chains, even increasing the productivity of certain trophic levels, can have unintended effects throughout an ecosystem, and even thousands of miles away.

What can be done? Reducing and reversing the Gulf of Mexico dead zone (and other dead zones throughout the world) can be accomplished by (1) reducing fertilizer use, (2) preventing animal wastes from getting into waterways, and (3) controlling the release of other sources of nutrients (phosphorus as well as nitrogen) from industrial facilities into rivers and streams.

1 **Ecosystems have living and non-living components.**

Ecosystems are the living organisms and the physical environment within an area. The largest ecosystems in the world are called biomes. The characteristic organisms and physical features present in each terrestrial biome depend largely on the temperature and rainfall patterns.

2 **Physical forces interacting create weather.**

The greater amount of solar energy hitting earth at the equator causes global air circulation patterns responsible for heavy rainfall near the equator and deserts at 30° latitude. Global patterns of ocean circulation also influence patterns of temperature and rainfall. Local topography, from mountain ranges to human-made structures, can also influence rainfall, temperature, and wind patterns.

3 **Energy and chemicals flow within ecosystems.**

Energy pyramids describe the energy flow through ecosystems, from primary producers to herbivores to carnivores. At each step of the food chain, the rate of energy capture and conversion is inefficient, with about 90% lost to heat, respiration, and feces. While all energy comes from the sun and flows in a pathway through food chains, chemicals repeatedly cycle from the physical environment through living organisms and back into the environment.

4 **Species interactions influence the structure of communities.**

Interacting species in a community coevolve in a variety of ways, some antagonistic and others mutually beneficial. The organisms of a population fill a unique niche, defined by how they utilize the resources in their environment. Organisms with overlapping niches compete until character displacement or local extinction occurs.

5 **Communities can change or remain stable over time.**

Most communities change over time. The patterns of succession, in which species composition changes within an ecosystem, depend on the rate and magnitude of disturbances. In a community, species vary in the degree to which they influence diversity. Keystone species are the most influential.

KEY TERMS

biomass, p. 563
biome, p. 552
biotic, p. 551
carnivore, p. 562
character displacement, p. 572
climax community, p. 580
coevolution, p. 570
colonizer, p. 579
commensalism, p. 578
community, p. 551
competitive exclusion, p. 571
decomposer, p. 563
desert, p. 553

detritivore, p. 563
disperser, p. 580
ecosystem, p. 550
El Niño, p. 559
energy pyramid, p. 564
estuary, p. 553
eutrophication, p. 568
food chain, p. 562
food web, p. 562
fundamental niche, p. 571
habitat, p. 551
herbivore, p. 561
host, p. 576

keystone species, p. 581
mimicry, p. 574
mutualism, p. 578
niche, p. 570
omnivore, p. 562
parasite, p. 576
parasitism, p. 576
predation, p. 573
primary consumer, p. 561
primary producer, p. 553
primary productivity, p. 553
rain shadow, p. 557
realized niche, p. 571

resource partitioning, p. 571
savanna, p. 553
secondary consumer, p. 562
succession, p. 579
temperate grassland, p. 553
tertiary consumer, p. 562
topography, p. 557
trophic level, p. 561
tropical rain forest, p. 553
tropical seasonal forest, p. 553

1. An ecosystem consists of _____ in a given area.
   a) all the photosynthetic organisms
   b) all the living organisms
   c) all the abiotic factors that influence living organisms
   d) all the living organisms and non-living materials
   e) the plant life and climate

2. Earth's largest terrestrial ecosystems, the biomes, are defined primarily by:
   a) the average rainfall.
   b) the average temperature.
   c) the seasonal variability in temperature.
   d) the seasonal variability in rainfall.
   e) All of the above.

3. Why is it hotter at the equator than at the Poles?
   a) The angle at which sunlight hits the earth leads to a given amount of solar energy being spread over a smaller area at the equator than at the Poles.
   b) The increased rotational speed of the earth at the Poles creates strong winds that increase radiant cooling.
   c) Greater cloud cover at the equator leads to a greater "greenhouse" effect than at the Poles.
   d) The water in the oceans has a high heat capacity, and there is significantly more water near the equator than at the Poles.
   e) All of the above are responsible for the higher average temperature at the equator than at the Poles.

4. Why do beaches on the west coast of the United States have colder water than beaches on the east coast?
   a) The ocean currents in the northern hemisphere rotate in a clockwise direction, so water off the coast of California has just come from the cold north, while water off the coast of New York has just come from the warm south.
   b) West coast beaches are not colder than east coast beaches. Ocean temperatures depend only on latitude.
   c) The Pacific Ocean is deeper than the Atlantic Ocean, so more of the sun's heat is absorbed by the colder, deep water, reducing the water temperature at the west coast beaches.
   d) Trade winds blowing north from the Tropic of Cancer in the Pacific Ocean sweep the warm air away from the west coast, taking with it much of the warmth of the water.
   e) Deep water upwellings bring cool water to the west coast beaches, while the surface currents on the east coast cause downwellings, sending warm surface water to depths where it is less likely to be lost through evaporation.

5. When moist wind from an ocean blows onshore toward a mountain range:
   a) as the air rises, it pulls moisture from the ground, causing the higher elevations to be drier.
   b) as the air goes over the top of the mountain and falls back down toward lower elevations, it holds less moisture, creating a "rain shadow" zone of unusually high precipitation.
   c) as the air rises, it holds less moisture, causing the dissipation of all clouds.
   d) as the moist air goes over the top of the mountain and falls back down toward lower elevations, it holds even more moisture, creating a "rain shadow" desert with very little precipitation.
   e) None of the above are correct.

6. "Top" carnivores:
   a) are more common than secondary consumers.
   b) rely directly on primary producers for energy.
   c) consume primarily herbivores.
   d) consume primarily carnivores.
   e) rely on symbiotic bacteria living in their digestive systems to help digest cellulose.

7. The 10% rule of energy-conversion efficiency:
   a) explains why big, fierce animals are so rare.
   b) explains why the biomass of herbivores must exceed that of carnivores.
   c) limits the length of food chains.
   d) suggests that 90% of what an organisms eats is used in cellular respiration or is lost as feces.
   e) All of the above are correct.

8. Nitrogen enters the food chain:
   a) primarily through soil-dwelling bacteria that "fix" nitrogen by attaching it to other atoms.
   b) from the atmosphere when "fixed" by the photosynthetic machinery of plants.
   c) when rocks dissolved by rainwater become soil and are utilized by plants as they grown.
   d) through soil erosion followed by runoff into streams and ponds.
   e) through methane, produced by herbivores as a by-product of the breakdown of plant material.

9. Which of the following statements about an organism's niche is incorrect?
   a) It encompasses the space the organism requires.
   b) It includes the type and amount of food the organism consumes.
   c) It is not always fully exploited.
   d) It may be occupied by two species, as long as they are not competitors.
   e) It reflects the ways in which the organism utilizes the resources of its environment.

10. The "ghost of competition past" refers to the fact that:
    a) competition often leads to character displacement, which remains even after direct competition is reduced.
    b) competition cannot be seen in nature.
    c) competition inevitably leads to the extinction of one of the competitors.
    d) competition inevitably leads to the extinction of both competitors.
    e) the fossil record is a record of competitive interactions.

11. Chemical defenses are more common among plants than animals because:
    a) plants cannot move to escape predators and so must develop other deterrents.
    b) the cell wall can contain the chemicals more effectively than a simple plasma membrane.
    c) mechanical defenses against predators can evolve only in animals.
    d) parasite loads in plants are significantly higher than in animals.
    e) All of the above are correct.

12. Rabid animals:
    a) froth at the mouth so as to increase the likelihood that the rabies-causing virus is passed on to another host.
    b) have been infected by parasitic bacteria.
    c) must die before their host can be passed on.
    d) are infected by a viral ectoparasite.
    e) often show reduced aggression, which increases their ability to get closer to other animals, thereby spreading the infection.

13. In a commensal relationship:
    a) one species pollinates the other.
    b) neither species benefits nor is harmed, but the community itself benefits.
    c) one species provides nutrients, usually cellulose, for the other.
    d) one species benefits while the other neither benefits nor is harmed.
    e) the "loser" has reduced reproductive output.

14. Coevolution:
    a) is responsible for all the beautiful flowers in the world.
    b) is responsible for nectar production by plants.
    c) reveals that both biotic and abiotic resources can serve as selective forces.
    d) can produce an insect with a tongue as long as its body.
    e) All of the above are correct.

15. The chief difference between primary and secondary succession is that:
    a) primary succession occurs among the plants in a habitat, while secondary succession occurs among the animals.
    b) primary succession begins with no life or soil, while secondary succession begins with both.
    c) secondary succession alters the biotic environment, and primary succession alters the abiotic environment.
    d) primary succession occurs more quickly than secondary succession.
    e) primary succession can occur in terrestrial and aquatic habitats, while secondary succession can occur only in terrestrial habitats.

16. Keystone species:
    a) occur only in intertidal zones.
    b) play an unusually important role in determining the species composition in a habitat.
    c) can be removed from a habitat without any impact on the remaining species in the community.
    d) are primary producers and therefore usually are plants.
    e) are more expendable than commensal species, from a conservation perspective.

SHORT-ANSWER QUESTIONS

1. What changes in climate occur on moving from the equator toward the Poles? Which two aspects of weather determine the type of biome found at any location?

2. Describe the energy flow that occurs through an ecosystem. How does this energy flow influence the number of top carnivores?

3. What is a niche? What is the potential fate of two species that occupy the same niche?

See Appendix for answers. For additional study questions, go to www.prep-u.com.

16

Conservation and Biodiversity

Human influences on the environment

❶ Measuring and defining biodiversity is complex.

The Cape Floral Region of South Africa is a biodiversity hotspot known for its wide array of plants.

16 • 1

Biodiversity benefits humans in many ways.

In the Pacific Northwest of the United States, a medium-size tree grows. This conifer, the Pacific yew tree (*Taxus brevifolia*), isn't the biggest tree in the forest, nor is it the most common. It's not the sort of tree that usually warrants attention in a textbook. But within the bark of the Pacific yew there is a chemical, called taxol, that has some important properties—it acts as an anti-cancer agent, interfering with the division of cells that come into contact with it (**FIGURE 16-1**). The role taxol plays for the Pacific yew is not clear; it may reduce the rate at which other organisms feed on the plant. But in humans, taxol is effective in the treatment of ovarian cancer, breast cancer, and lung cancer (generating more than $1 billion a year in pharmaceutical sales).

Consider also the following:

- The chemicals vinblastine and vincristine, isolated from the Madagascar periwinkle (*Catharanthus roseus*), are so effective in treating leukemia and Hodgkin's lymphoma that both diseases, formerly incurable, now are curable in the vast majority of people.

- The chemical ancrod, found only in the Malayan pit viper (*Agkistrodon rhodostoma*), dissolves blood clots and is effective in treating some heart attack and stroke patients.

A chemical within the bark of the Pacific yew, called taxol, is effective in the treatment of ovarian cancer, breast cancer, and lung cancer.

Pacific yew tree (*Taxus brevifolia*)

FIGURE 16-1 **Taxol is cancer medicine derived from a tree.**

Defining Biodiversity   Extinction   Human Interference   Conservation Strategies

- Epibatidine, a poisonous compound in the saliva of a small frog (*Epipedobates tricolor*) that lives in Ecuador, is 200 times more effective than morphine in relieving pain and is non-addictive.

These are just a few examples—there are many, many more—illustrating that medically important compounds often come from some unlikely organisms living in locations around the world. Usually, toxin production evolved in these organisms as a method of protecting them from other organisms. Co-opting the chemicals for human use reveals that one important value of biodiversity is as a sort of universal medicine cabinet.

Brocadia anammoxidans is another species that may turn out to have great value to humans. This bacterial species is able to break down an important component of human waste and, more important, it can do so without oxygen. For this reason, when treating human waste with *B. anammoxidans,* it is unnecessary to aerate the sewage sludge. Normal waste treatment utilizes bacteria that need oxygen and, consequently, requires costly, power-hungry equipment. Still other bacteria, with sulfur-metabolizing properties, are now being used to remove the stench of sewage. In these cases, as with the medicines described above, the value of biodiversity to humans is utilitarian: from other species we can obtain products or make use of processes that make our lives better.

But viewing the value of biodiversity only in the context of our abilities to extract medicines or to better process our waste products—the utilitarian view—is to take much too narrow a view. Biodiversity provides significant, perhaps less tangible, benefits by helping regulate the climate, in the form of trees that both create shade and take up carbon dioxide from the atmosphere. Biodiversity also benefits humans in numerous other ways that are difficult to quantify, but are no less valuable than medicinal plants or sludge-eating bacteria.

Have you ever gazed up at a 300-foot-tall redwood tree? Would you like to see a 100-foot-long, 180-ton blue whale? Do you like knowing that bald eagles live throughout North America? If you answered *yes* to any of these questions, you identified additional ways in which biodiversity can have value to humans. Although many of the ways that people value biodiversity are not easily quantified (in dollars, for example), they represent value just the same (**FIGURE 16-2**).

Some types of non-utilitarian values include:

- *Aesthetic value.* We think of a field of flowers or a deer in the woods as beautiful to look at and pleasant to experience.

- *Symbolic value.* Humans are so connected to the natural world that images of animals and plants can convey meaningful abstract ideas: the dove as a symbol of peace;

UTILITARIAN VALUE **NATURALISTIC VALUE**
SYMBOLIC VALUE **AESTHETIC VALUE**

More than just a utilitarian resource, biodiversity has naturalistic, symbolic, and aesthetic importance.

FIGURE 16-2 **Biodiversity has value to humans in many different ways.**

the butterfly as a symbol of transformation; the owl as a symbol of wisdom; the rose as a symbol of love.

- *Naturalistic value.* This can be thought of as the satisfaction people can get from direct contact with nature, including the sense of awe and fascination that can come from gaining an understanding of nature's complexity and diversity.

In recognizing that people can view the same object (or organism) and value it in different ways, we are on our way to making wiser conservation decisions. This knowledge, for example, can help us anticipate potential conflict when it comes time to invest limited conservation resources in the conservation of biodiversity, and can help us seek out methods for balancing conflicting desires. We explore some strategies for making these difficult decisions later in this chapter. First, we investigate the question of where the greatest amounts of biodiversity are to be found.

TAKE-HOME MESSAGE 16·1

Biodiversity provides significant benefits to humans in numerous utilitarian ways, such as the production of medicines and other products and processes. Biodiversity also provides important value to humans in less tangible ways, including aesthetic, symbolic, and naturalistic values.

Biodiversity is not easily defined.

Which habitat has greater biodiversity, one with 3 or 4 species each of birds, reptiles, mammals, plants, and insects or one with 38 unique species of fruit flies? Or one with 100 radically different species of prokaryotes?

This is a difficult question, and it highlights an important but overlooked issue in conservation biology today: it's not clear exactly what a person means when using the term "biodiversity." Discussions in the media make it appear that there is a consensus on what biodiversity means, but there is not. Or, to the extent that there is consensus, it is on a definition of biodiversity that is not practical.

All definitions of biodiversity refer, in a general sense, to the variety of living organisms on earth. Such a broad definition, however, cannot really be applied: armed with such a definition, could you answer the question: *What is the biodiversity in a particular patch of land?* Think about it—what would you count—the number of trees, or the number of tree species, or the number of insects that live in those trees? Could you compare this patch of land with another in order to determine which has the greater biodiversity?

By taking this approach—considering how to *apply* the concept of biodiversity—we can start to move toward a more concrete definition. Expanding on our general definition above, biodiversity can be considered at multiple levels, from entire ecosystems to species to genes and alleles (**FIGURE 16-3**). This makes it a bit more manageable as a concept.

Defining biodiversity as the variety of genes, species, and ecosystems on earth, however, is still not sufficiently practical. The construction of a meaningful inventory of earth's biodiversity is fraught with difficulty. As we saw in Chapter 5, the Human Genome Project, which has catalogued all of the genes in humans, was a multi-year, multi-billion-dollar undertaking. A similar undertaking for all species is clearly not practical. And as we saw in Chapter 10, the very concept of "species" is difficult (and sometimes downright impossible) to apply to many, if not most, of the organisms on earth. And as we saw in Chapter 15, it is even difficult to determine where one ecosystem ends and another begins.

Believe it or not, the difficulty of defining biodiversity, and the difficulty in applying the concept as part of any efforts to protect biodiversity, is not an insurmountable problem. To the contrary, defiining and applying the concept is a positive step, likely to lead to the effective preservation of earth's biodiversity.

WHAT IS BIODIVERSITY?

GENES AND ALLELES
The number of alleles in a species

SPECIES
The number of species in an ecosystem

ECOSYSTEMS
The number of ecosystems in a region

 Biodiversity is more than just a counting of species. It encompasses the genetic variability among organisms within a species, the variety of different species, and the variety of ecosystems on earth.

FIGURE 16-3 Defining biodiversity. (Photo on the left shows DNA microarray, revealing the variety of alleles active in an individual.)

Defining Biodiversity   Extinction   Human Interference   Conservation Strategies

CONSERVATION BIOLOGY

Conservation biology is a multidisciplinary science that addresses how to preserve the natural resources of earth and protect biodiversity.

FIGURE 16-4 Preserving the earth's natural resources.

Let's return to our original question. If you were given the task of overseeing the conservation of biodiversity in a region where these three habitats occur, which habitat would you protect first: the habitat with a few species of birds, reptiles, mammals, plants, and insects, or the habitat with many species of fruit flies, or the habitat with a huge number of prokaryotes?

In making your decision, you might first define biodiversity— maybe as the number of different classes or families of organisms in a habitat, or maybe as the number of distinct species in the habitat, which is, in fact, the most commonly used definition of biodiversity—to support your decision. In doing so, however, you are acknowledging that biodiversity can be defined in many ways and that the definition you are using (or that any decision maker uses) is not completely objective. Rather, it is a function of your values, biases, and interests. With this recognition, it becomes possible to discuss, manage, use, and conserve biodiversity in a practical and more effective manner.

Conservation biology is the interdisciplinary field that addresses how to preserve the natural resources of earth. Drawing on knowledge, data, and insights from biology, economics, psychology, sociology, and political science, one of the goals of conservation biology is the preservation and protection of biodiversity (**FIGURE 16-4**). For this reason, the complexities of defining biodiversity directly affect the decisions that conservation biologists make. In this chapter, we explore how the actions of humans can have significant, even disastrous, consequences for biodiversity and other natural resources, and we also explore the strategies that have been developed for effective conservation.

Underpinning many of the difficulties in conservation biology is the fact that biodiversity, as we've noted, can have value for humans in so many different ways. Throughout the chapter, we will see how that complexity makes it difficult to balance competing interests when they are in conflict.

TAKE-HOME MESSAGE 16•2

Defining biodiversity is difficult because it can be considered at multiple levels, from entire ecosystems to species to genes and alleles. In practice, it is most often defined as the number of distinct species in a habitat, which has important implications for conservation biology—the field that addresses questions of how to preserve the natural resources of earth.

16•3

Where is most biodiversity?

If you were to stand at the equator in South America and identify the number of land mammals, you would count about 400 different species. If you then started walking south, away from the equator, and assessed the diversity of land mammals when you got to 25° latitude, you would find about 300 different species. Continuing your long walk south, when you got to 30° latitude you would find only 200 different species. And the trend would continue for as far south as you could go, with 100 species at 40° latitude and fewer than 50 species at 50° latitude. On a similar walk north from the equator, through North America, you would observe the same trend, with more than 300 species of land mammals at 15° latitude

WHERE IS MOST BIODIVERSITY?

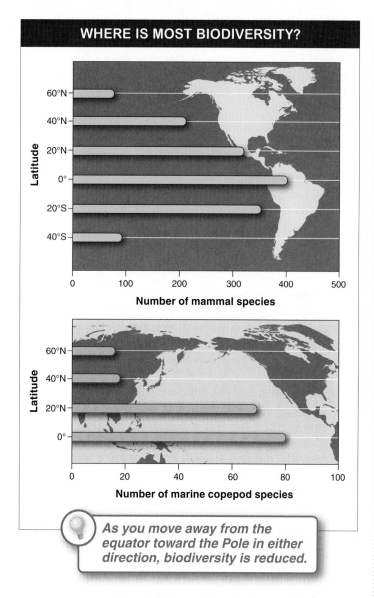

As you move away from the equator toward the Pole in either direction, biodiversity is reduced.

FIGURE 16-5 Biodiversity is unevenly distributed throughout the world.

reduced to only 15 species in northern Canada (FIGURE 16-5). Biodiversity is not evenly distributed throughout the world. And perhaps the strongest trend in biodiversity distribution is that as you move away from the equator in either direction, diversity is reduced with increasing latitude.

The strong biodiversity gradient from the tropics to the Poles occurs not just in land mammals but in nearly any group of plants or animals observed. Interestingly, even the diversity of organisms in the oceans follows this trend. One survey of marine copepods—tiny crustacean animals, including plankton—found the exact same trend moving north from the tropical region of the Pacific Ocean (80 species present) to the Arctic Ocean (fewer than 10 species).

Numerous factors that influence species richness—that is, the number of species in an area—are responsible for producing the tropical-to-temperate-to-polar gradient in biodiversity. Three, in particular, play strong roles (FIGURE 16-6).

Q Why are there more species in an acre of tropical rain forest than in an acre farther from the equator, such as in a temperate forest or prairie?

1. *Solar energy available.* Perhaps the simplest predictor of species diversity is climatic favorability, the amount of solar energy available in an area. Solar energy is, after all, the fuel for all life. As we saw in Figure 15-5 in Chapter 15, the sun shines most directly on the equator, and, as we move away from the equator, the earth's curvature causes the solar energy hitting the earth to be spread out over increasingly large areas. This leads to less solar energy at any one point as latitude increases. In a variety of species of plants and animals, researchers have documented strong relationships between energy availability and species richness.

2. *Evolutionary history of an area.* Communities diversify over time. Consequently, the more time that passes, generally speaking, the greater is the diversity in an area. A high level of

FACTORS THAT INFLUENCE BIODIVERSITY

1 **SOLAR ENERGY AVAILABLE**
Greater access to solar energy, the fuel for all life, provides increased species richness.

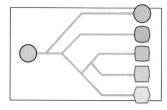

2 **EVOLUTIONARY HISTORY OF AN AREA**
Communities diversify over time. The more time that passes without a climatic event, such as an ice age, the greater the diversity in an area.

3 **RATE OF DISTURBANCE**
A habitat with an intermediate amount of disturbance tends to have the greatest species richness.

FIGURE 16-6 What determines species richness in an area?

Defining Biodiversity   Extinction   Human Interference   Conservation Strategies

biodiversity in an area, however, can be knocked back down by climatic disasters such as glaciations. This biodiversity decline may occur in temperate and polar regions.

3. *Rate of disturbance.* Over time, the best competitor for a resource is expected to outcompete other species, excluding them from the community and thus reducing species richness. Communities may be kept from reaching this state, though—in a number of ways. Factors such as predation, or environmental perturbations such as fires, can prevent any one species from excluding others. For this reason, a habitat with an *intermediate* amount of disturbance is expected to have the greatest species richness.

Conservation biologists are increasingly interested in **biodiversity hotspots,** those regions of the world having significant reservoirs of biodiversity that are under threat of destruction. Twenty-five biodiversity hotspots have been identified around the globe that, while covering less than 1% of the world's area, have 20% or more of the world's species

(**FIGURE 16-7**). Habitats included among the biodiversity hotspots are tropical rain forests, coral reefs, and islands.

Tropical rain forests. These are common near the equator in Asia, Australia, Africa, South America, and Central America, as well as on many of the Pacific Islands. A particularly important, but still poorly studied, region of tropical rain forests is the canopy. This is the uppermost region of a forest, where nearly all of the sunlight is intercepted. As much as 30 feet in thickness, the canopy is home to most of the diversity in tropical rain forests, and estimates suggest that as many as half of the species on earth reside in the canopy of tropical rain forests. Due to its inaccessibility, the canopy is not easily studied and the extent of biodiversity there is still not fully known.

Coral reefs. Built by corals, tiny members of the animal kingdom, coral reefs occur in ocean areas with low levels of nutrients. The physical structures created by the corals provide a home and habitat for huge numbers of other

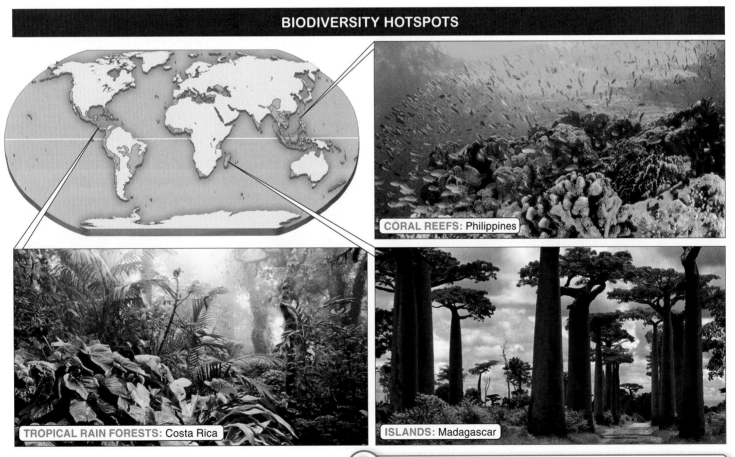

BIODIVERSITY HOTSPOTS

CORAL REEFS: Philippines

TROPICAL RAIN FORESTS: Costa Rica

ISLANDS: Madagascar

FIGURE 16-7 Biodiversity hotspots are significant reservoirs of biological diversity.

There are 25 biodiversity hotspots around the world that, while covering less than 1% of the world's area, have 20% or more of the world's species.

species. The Great Barrier Reef of Australia (which is not listed among the 25 biodiversity hotspots, because it is relatively well protected) is an example of just how much diversity can be supported by coral reefs. It is the largest coral reef system in the world and covers such a large area off the northeast coast of Australia that it can be seen from space. It is home to numerous species of whales, dolphins, sea turtles, sharks, and stingrays, as well as 1,500 species of fish and 5,000 species of molluscs. Five hundred species of algae are found there, too. Above the Great Barrier Reef, more than 200 species of birds are supported by the rich diversity of life below. Other coral reefs are not as well protected, and the world's second largest coral reef, in New Caledonia in the southwestern Pacific, is one of the 25 biodiversity hotspots.

Islands. Because of their generally smaller size, which limits the population size of organisms living on them, islands can be regions where biodiversity is at risk. Madagascar, located in the Indian Ocean off the southeastern coast of Africa, is the fourth largest island in the world. It is home to 5% of the world's species, 80% of which are **endemic** (not naturally found elsewhere)—a higher percentage of endemic plants and animals than in any comparably sized area on earth. Madagascar is an area unusually rich in species diversity.

The deep ocean regions are much less studied than the terrestrial parts of the world. Although they cover two-thirds of the earth's surface, the oceans have proved difficult to explore. Once thought to be completely devoid of life, the deepest regions of the ocean have recently been found to be teeming with biodiversity. One survey of the deepest parts of the Antarctic Ocean, sampling at a depth of more than 18,000 feet (6,000 meters), discovered 585 new species of crustaceans alone.

TAKE-HOME MESSAGE 16·3

Biodiversity is not evenly distributed. There is a strong latitudinal gradient of biodiversity from the tropics (highest biodiversity) to the poles for nearly all groups of plants and animals, both marine and terrestrial. Factors influencing the species richness in an area include the amount of solar energy available, the area's evolutionary history, and its rate of disturbance. Biodiversity hotspots are regions of significant biodiversity under threat of destruction.

16·4

Island biogeography helps us understand the maintenance and loss of biodiversity.

"Can you help me exterminate all the insects on a few islands?" When E. O. Wilson started calling exterminators in Florida, asking them to help him put tents over several islands and then pump ethyl bromide into them to kill all of the insects, he had trouble being taken seriously. Most of the exterminators opted to pass on the job, but eventually he found one (**FIGURE 16-8**).

Why would a biologist (of all people) want to wipe out all insect life on six islands? Robert MacArthur and Wilson had developed a theory, six years earlier, and now Wilson and one of his students, Daniel Simberloff, had decided it was time to put the theory to a test.

MacArthur and Wilson's theory was called the "theory of island biogeography." Designed to explain and predict the patterns of species diversity on islands, it had two fundamental components (**FIGURE 16-9**).

1. *The area effect.* Observations of islands in the West Indies revealed that there was a close relationship between the

ISLAND BIOGEOGRAPHY EXPERIMENT

E. O. Wilson and Daniel Simberloff exterminated all of the insects on multiple islands, then monitored recolonization to test the theory of island biogeography.

FIGURE 16-8 Experiment in insect eradication and recolonization: explaining patterns of species diversity on islands.

Defining Biodiversity   Extinction   Human Interference   Conservation Strategies

THEORY OF ISLAND BIOGEOGRAPHY

The theory of island biogeography is designed to explain and predict patterns of species diversity on islands. It has two fundamental components:

THE AREA EFFECT
There is a close relationship between the number of species inhabiting an island and the island's area—the larger an island is, the more species it holds.

Number of species on island

THE DISTANCE EFFECT
There is a close relationship between the number of species inhabiting an island and how isolated an island is—the farther an island is from the mainland, the fewer species it holds.

Mainland

Number of species on island

FIGURE 16-9 The lonely island. The theory of island biogeography generates predictions about the patterns of species diversity in isolated habitats.

number of species inhabiting the island and the island's area: the larger the island, the more species it held. This relationship held true for bird species as well as plants, beetles, and snails.

2. *The distance effect.* Observations of islands in the Pacific Ocean revealed a close relationship between how isolated an island was, measured as its distance from the nearest mainland, and the number of species that inhabited the island: the farther from the mainland the island was, the fewer species it held.

From these observations, MacArthur and Wilson concluded that the number of species on an island represents a balance between the rate of immigration of new species to the island and the rate of extinction of species on the island. They proposed that the rate of immigration to an island would primarily be determined by how close the island was to the mainland, with the mainland serving as a pool of new species. And they proposed that the rate of extinction on an island would primarily be determined by the island's size: smaller islands generally support a smaller population size, making the population more vulnerable to extinction.

Beyond enabling them to predict the number of species on an island, MacArthur and Wilson's theory also suggested to them that the composition of species on the island would not be fixed; rather, it might change over time. In other words, once it was "full," an island might have 10 species of birds, but exactly *which* 10 species occurred on the island could change over time. They called this a **dynamic equilibrium** (FIGURE 16-10).

DYNAMIC EQUILIBRIUM

EQUILIBRIUM
The number of species on an island represents a balance between the rate of immigration to the island by new species and the rate of extinction of species on the island.

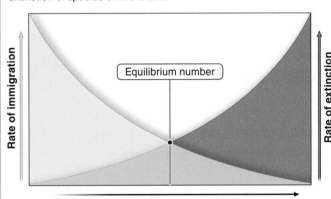

DYNAMIC EQUILIBRIUM
The composition of species on an island is not fixed; rather, it might change over time. An island might have five species of birds, but exactly which five species occurred on the island could change over time.

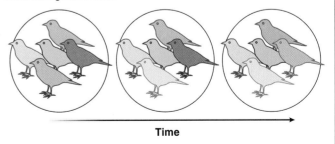

Time

FIGURE 16-10 Offering insight into population dynamics: dynamic equilibrium.

Back to the island exterminations. Wilson's reason for exterminating all of the insects on multiple islands was so that he could monitor the islands as they were recolonized following extermination. He chose the mangrove islands in the Florida Keys because there are thousands of small islands there, some only a few meters across. He and Simberloff selected eight islands, 33–54 feet (11–18 meters) in diameter. The islands varied in how far they were from the nearest mainland, with the closest just 6 feet (2 meters) and the farthest more than half a mile (1,188 meters) from the mainland.

> The moth settled onto the curtain and sat still. It was an astonishing creature, with black and white wings patterned in geometric shapes, scarlet underwings, and a fat white body with black spots running down it like a snowman's coal buttons. No human eye had looked at this moth before; no one would see its friends. So much detail goes unnoticed in the world.
>
> —Barbara Kingsolver, *Prodigal Summer,* 2000

After selecting two islands as control islands, the biologists and their assistants fumigated the others, verifying that all of the insects were killed. They then surveyed the islands regularly over the next year. Here's what they found—as predicted:

1. The closer islands were recolonized more quickly than the distant islands.

2. On all of the islands, the number of insect species after fumigation returned to about the same number of insect species prior to fumigation.

3. The composition of insects on each island changed over the course of the study, even on the two control islands.

The theory of island biogeography is not just about real islands, as it turns out. Lakes can be thought of as islands, too, from an immigration and extinction perspective. The theory makes testable predictions about the biodiversity levels that should occur in them. Mountain tops can be viewed similarly. And perhaps most important, conservation biologists can gain insights from island biogeography into the design of nature preserves. The importance of developing nature preserves as close as possible to sources of new species, for example, is one such insight. We explore others later in this chapter, when discussing the strategies for preservation of biodiversity.

TAKE-HOME MESSAGE 16·4

The theory of island biogeography generates predictions about the patterns of species diversity in isolated habitats. Both the size of the habitat and its distance from a pool of new species influence the equilibrium number of species present. Ideas gained from island biogeography have important implications for the development of nature preserves.

Defining Biodiversity   Extinction   Human Interference   Conservation Strategies

❷ Extinction reduces biodiversity.

The Bengal tiger has been hunted, captured, and poisoned to near extinction.

16•5 --

There are multiple causes of extinction.

Imagine that you are on a deserted island. Alone. It would be hard and it would be lonely. The hope of rescue someday, however, would probably make things bearable. "Martha," a passenger pigeon living in the Cincinnati zoo in the early 1900s, found herself in a situation of this sort—minus the hope. Passenger pigeons in the wild had been completely wiped out by a combination of factors, including hunting of the birds for meat, loss of their habitat due to deforestation in North America, and possibly disease. And so, except for the small flock that Martha was part of, the species was gone. One by one, those birds died as well, and when the second-to-last passenger pigeon died in 1912, Martha was all alone. She lived for two more years, with no possibility for ever reproducing; she died on September 1, 1914 (FIGURE 16-11).

On average, species persist for about 10 million years, although some last for hundreds of millions of years. In either case, when extinction occurs—the complete loss of all individuals in a species—it is the end. And almost any way you look at biodiversity, from a utilitarian or aesthetic or symbolic perspective, extinction is a tragic loss.

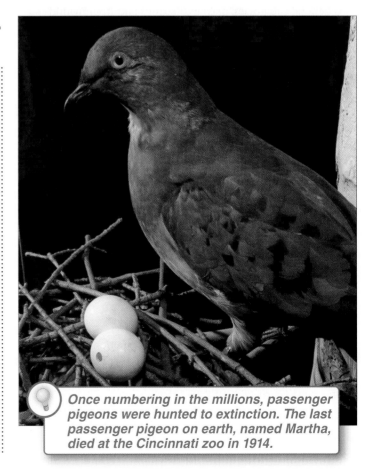

Once numbering in the millions, passenger pigeons were hunted to extinction. The last passenger pigeon on earth, named Martha, died at the Cincinnati zoo in 1914.

FIGURE 16-11 The passenger pigeon: a species completely lost to the earth.

TWO CATEGORIES OF EXTINCTIONS

Extinctions generally fall into two categories:

- **MASS EXTINCTIONS**
  A large number of species become extinct over a short period of time due to extraordinary and sudden environmental change.

- **BACKGROUND EXTINCTIONS**
  These extinctions occur at lower rates during times other than mass extinctions.

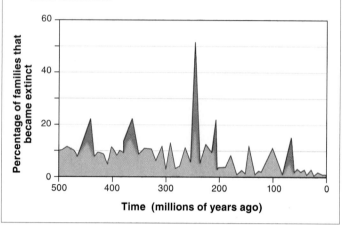

FIGURE 16-12 **Background and mass extinctions.**

As we discussed in Chapter 10, extinctions can be divided into two general categories: mass extinctions and background extinctions (**FIGURE 16-12**). They're distinguished by the numbers of species affected. When a large proportion of species on earth are lost in a short period of time—such as when an asteroid struck earth 65 million years ago and about 75% of all species on earth were wiped out—it's considered a mass extinction. These extinctions are above and beyond the normal rate of extinctions that occur in any given period of time, referred to as background extinctions.

Q *Does extinction only happen to weak, unfit species?*

In both background and mass extinctions, it's the same outcome for the species involved, but the causes tend to differ. In mass extinctions, the particular features of a species' biology don't really play a role in the extinction; rather, it's more like really bad luck. Background extinctions, by contrast, tend to be a conse-quence of one or more features of that particular species' biology.

For any species, there is always a risk of extinction. This risk can be larger or smaller, depending on certain features. Here we look at three important aspects of species' biology that can influence their extinction risk (**FIGURE 16-13**).

1. *Geographic range: extensive versus restricted.* Species restricted in their range—including those limited to small bodies of water and those confined to islands—are more vulnerable

than those with extensive ranges. The Tasmanian devil is a marsupial carnivore about the size of a dog. Although these animals once thrived in Australia, they now are confined to the island of Tasmania, smaller than the state of Maine. Unable to expand their range, Tasmanian devils are more vulnerable to extinction than if their range was not limited.

2. *Local population size: large versus small.* Tigers and peregrine falcons, along with *Welwitschia,* a slow-growing, long-lived plant in southwest Africa, are examples of species that—as a consequence of their small population sizes—are at increased risk of extinction. With a small population size, a species is more susceptible to extinction due to a variety of factors that can kill individuals, including fire, diseases, habitat destruction, and predation. Put simply, the more individuals that are alive, the more likely it is that some of them will survive these events. With small population size, inbreeding is also increased, which generally reduces the fertility and longevity of individuals.

FACTORS THAT INFLUENCE THE RISK OF BACKGROUND EXTINCTION

GEOGRAPHIC RANGE:
EXTENSIVE vs. RESTRICTED
Species restricted in their range are more vulnerable than those with extensive ranges.

LOCAL POPULATION SIZE:
LARGE vs. SMALL
Species with small population sizes are at increased risk of extinction.

HABITAT TOLERANCE:
BROAD vs. NARROW
Species with narrow habitat tolerances are at greater risk of extinction than species with broader habitat tolerances.

FIGURE 16-13 **Range, population size, and tolerance influence risk of extinction.**

Defining Biodiversity   Extinction   Human Interference   Conservation Strategies

3. *Habitat tolerance: broad versus narrow.* "Habitat tolerance" describes the breadth of habitats in which a species can survive. Some plant species, for example, can tolerate large swings in water availability or soil pH or temperature. Others are limited to very narrow habitat ranges. The now-extinct passenger pigeons could only build nests in a specific type of forest and needed large numbers of individuals, breeding communally, in order to breed successfully. As forests were cut down and as the birds were hunted, the size of their flocks diminished. Their narrow habitat tolerance made them extremely vulnerable to extinction. In general, because species with narrow habitat tolerance cannot adapt in the face of habitat degradation and loss, they are at greater risk than species with broader habitat tolerance.

One species that is enormously successful, thanks to its extensive geographic range, large local population sizes, and broad habitat tolerance, is our own. Humans are growing and expanding our range at a staggering rate, consuming resources in an unprecedentedly voracious fashion. By any measure, humans are one of the most successful species in earth's history. This success, unfortunately, is having an increasingly negative effect on the species with which we share the planet, causing extinctions at a significantly higher rate than ever before. We next explore in more detail the conflict between humans' success as a species and the survival of other species.

TAKE-HOME MESSAGE 16·5

Extinctions occur for fundamentally different reasons. Mass extinctions are more the result of bad luck than the particulars of a species' biology. Background extinctions tend to be a consequence of one or more features of the species' biology. Small population size, limited habitat range, and narrow zones of habitat tolerance contribute to background extinctions.

16·6

We are in the midst of a mass extinction.

Sometimes you can be so close to something that it's difficult to really see it. That may be the situation that humans are in right now when it comes to the global loss of biodiversity. But humans are becoming increasingly aware that we are in the midst of a mass extinction. In a recent survey by the American Museum of Natural History, 70% of biologists indicated that they believed this is true and that steps must be taken by governments and individuals to stop this massive loss of species.

Data on current rates of extinction in every well-studied group of plants and animals support the hypothesis that a mass extinction is under way. Among the mammals, 11% of all species are currently *endangered,* under imminent threat of extinction, and 14% are *threatened,* characterized as vulnerable to extinction. Among birds, 4% of all species are endangered and 50% are in decline. Almost a third of all amphibian species are threatened. Among fish, molluscs, insects, fungi, and plants, the story is the same.

Historically, background extinctions rates are about one extinction per million species per year. The numbers above suggest that current extinction rates are 1,000 times (or more) greater than usual. It is difficult to know the exact magnitude of the problem because, as we saw in Chapter 10, biologists don't have much of an idea about how many species there are on earth—the estimates range from 5 million to 100 million.

Unlike the last mass extinction event, in which the dinosaurs and most other species were wiped out in the wake of an asteroid's hitting earth, the cause of this current mass extinction seems to be the result of the activity of one species—humans. Ironically, it is the unprecedented success of humans that is responsible for this situation. The resource needs of so many people have inevitably interfered with the ability of other species to coexist with us.

The chief reason for the loss and impending loss of so many species is habitat loss and habitat degradation. Particularly harmful is habitat loss in earth's tropical rain forests, where biodiversity is greatest. In the past 25 years, half of the world's tropical rain forests have been destroyed, usually by burning to make way for agricultural use of the land or by logging (**FIGURE 16-14**). Urban development, too, is responsible for destruction of rain forests, as the growing human populations continue to expand. Intensive agriculture, livestock grazing, and the development of urban centers has led to the destruction and fragmentation of habitats worldwide.

The loss of biodiversity-supporting habitat is not restricted to the tropics. In the Pacific Northwest, as well, logging of forests

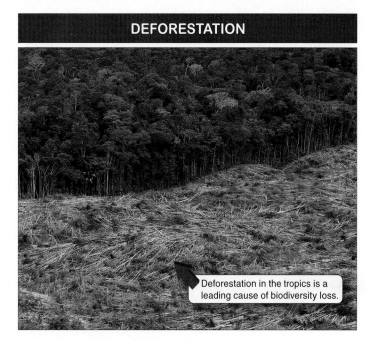

DEFORESTATION

Deforestation in the tropics is a leading cause of biodiversity loss.

FIGURE 16-14 **The devastation of deforestation.**

Bonobos, the closest living relative of humans, have been driven close to extinction, in part due to hunting.

FIGURE 16-15 **Overexploitation of species: the bonobo.**

has occurred at an unsustainable rate, leading to significant reductions in a diverse range of populations.

Beyond habitat loss, other factors reducing biodiversity include overexploitation of species, including those killed for food, furs, and medicinal products. Bonobos (*Pan paniscus*), for example, a type of chimpanzee and our closest living relative, are almost never seen in the wild, having been driven to extinction, hunted for food (**FIGURE 16-15**). And, as we'll see later in this chapter, the introduction of exotic species, too, is having a significantly adverse effect on biodiversity. Consequences of the loss of biodiversity, beyond the values of biodiversity discussed earlier, are largely unknown, but most likely include a serious

reduction in the capacity of the environment to provide clean air and water and to recover from environmental and human-induced disasters.

Later in this chapter, we discuss some strategies by which the current high rate of biodiversity loss can be reduced.

TAKE-HOME MESSAGE 16·6

Most biologists believe that we are currently in the midst of a mass extinction, that it is the result of human activities, and that it poses a serious threat to the future survival of humans.

❸ Human interference generally reduces biodiversity.

An uneasy coexistence: Rio de Janeiro, Brazil, presses up against the Atlantic rain forest.

16•7

Some ecosystem disturbances are reversible, others are not.

When you fly over southern New England today, you look down on an undulating carpet of forest that covers hills and valleys. Roads traverse the forest, and here and there you see towns and cities or cultivated land and pastures, but most of what you see is the tops of trees.

If you could have flown over the same route 200 years ago, things would have looked very different. In the early 1800s, most of the forest in southern New England had been cleared and the land was in cultivation. Homesteads consisting of farm houses, barns, stables, and other outbuildings were scattered across the countryside. Stone walls extended across hills and valleys, separating cropland from pastures.

Going back farther, the same flight 400 years ago would have taken you over mostly forested land. Here and there, Indian villages would have stood among cleared fields, but most of what you saw then would have looked much like what you see today.

The change from forest 400 years ago to cleared land 200 years ago was a major ecosystem disturbance, but now the area has returned to forest. Today, the stone fences are the

Q Once land is cultivated or developed, can it ever return to its natural state?

only obvious trace of the agricultural history of the area. The fences still snake their way up and down hills, but now they run through solid forest instead of between fields and pastures.

By the middle 1800s, New England settlers were abandoning farms on marginal land and moving west. The fields they had cultivated began a process of ecological succession that returned them to forests (see section 15.15).

In the first year after a field is abandoned, its bare soil provides a harsh environment for plants. Sunlight blazes down, heating the ground surface and baking moisture from the soil. It takes a tough plant to sprout and grow under those conditions, but there are plants that can do it. We call those pioneering species of plants "weeds," and we see them growing in disturbed habitats—along roadsides, in vacant lots, and even in cracks in pavement.

As weeds grow and die, year after year, their organic matter enriches the soil and helps it retain moisture. And every year, the plants grow more densely so that they shade the ground surface, which no longer gets quite so hot and dry. Additional species of plants can grow in these milder conditions. These

601

May 1980

September 1980

Today

FIGURE 16-16 **The forest returning.** Just two decades after an explosive eruption, shrubs and small trees have returned to the plain at the foot of the active volcano, Mount St. Helens, in Washington State.

plants are taller, so they provide still more shade and the ground becomes still cooler and more moist. At this stage, the seeds of bushes and small trees can sprout. As these woody plants grow, they make the ground cool and shady enough to allow the seeds of forest trees (oak, maple, or beech) to sprout. More saplings of forest trees sprout and grow every year, and eventually they are so closely packed that little sunshine penetrates their leafy canopies to reach ground level. At this point, the forest stops changing, because the mature trees have created conditions in which only their own seeds can sprout and grow.

An ecosystem disturbance is reversible when the disturbance, although it alters the biotic and abiotic nature of the habitat, does not include the complete extinction of any species, so species can

re-establish their populations. The example of New England forests is based on a disturbance that was caused by humans—clearing land for agriculture in the 1700s—but wind storms, forest fires, floods, and volcanic eruptions have been clearing forests throughout the history of life (**FIGURE 16-16**).

> " Any fool kid can step on a beetle. But all the professors in the world can't make one.
>
> —Arthur Schopenhauer, German philosopher (1788–1860) "

If, conversely, an ecosystem disturbance involves the complete loss of a species to extinction, the disturbance is irreversible. Each species is the result of a long and uninterrupted evolutionary history, involving an interplay of random and selective forces, and producing a unique genome. Once lost, a species can never be created again (**FIGURE 16-17**). For this reason, ecosystem disturbances involving the loss of species can have more devastating consequences than disturbances not leading to extinctions, no matter how great the changes to the abiotic environment might be.

BIOLOGICAL DISTURBANCES ARE IRREVERSIBLE

The blue whale is the largest mammal in the world and the largest animal ever known to have lived on earth. Once lost, a species can never be re-created.

FIGURE 16-17 **The blue whale.** Disturbances involving the loss of species are devastating.

TAKE-HOME MESSAGE 16·7

An ecosystem disturbance is reversible when the disturbance, although it alters the physical and biological nature of the habitat, does not include the complete extinction of any species, so they can re-establish their populations. An ecosystem disturbance that involves the complete loss of a species to extinction is irreversible, because a species, once lost, can never be created again.

Defining Biodiversity   Extinction   Human Interference   Conservation Strategies

16·8

Disruptions of ecosystems can be disastrous:
1. Introductions of exotic species.

Although the phrase suggests colorful birds and butterflies winging through a forest, **exotic species** refers to species introduced, by human activities, to areas other than the species' native range. Nonetheless, in the rain and sleet of a Chicago winter, the idea of colorful birds flying about is very appealing, and you might see exactly that, in the form of the monk parakeet. These small green members of the parrot family are about a foot (30 centimeters) from head to tail tip. Native to southern South America, monk parakeets were imported to the United States as pets in the 1960s and 1970s. Some of those pet birds escaped, and free-living populations of monk parakeets now occur in 11 states, from Oregon to Rhode Island and from Florida to New York (**FIGURE 16-18**).

The appearance of these free-living parakeets was greeted with alarm, because they are regarded as agricultural pests in South America. Records from the Inca civilization that flourished before Pizarro's invasion of Peru in 1532 attest to crop damage by monk parakeets, and when Charles Darwin visited Uruguay in 1833, he was told that monk parakeets attack fruit orchards and grain fields. Fortunately, the dire

Monk parakeets were imported from South America as pets. Some escaped, and now make a nuisance of themselves in cities such as Chicago, where they build large nests on electric poles, transformers, and floodlights.

FIGURE 16-18 A parakeet in the Midwest. The monk parakeet is an exotic species from South America, introduced into Chicago, Illinois.

predictions that these small parakeets would devastate crops in the United States have not been fulfilled as yet.

But monk parakeets do make a nuisance of themselves in cities such as Chicago, where they build nests on utility poles, transformers, and floodlights. There is nothing delicate about a monk parakeet nest—it is a large mass of sticks and twigs, and a nest that has been soaked with rain is heavy enough to bring down wires, interrupting electrical service to entire neighborhoods.

Another exotic species is also notorious for short-circuiting power lines—the brown tree snake causes thousands of power outages on the island of Guam every year. This snake, which is native to northern Australia, Papua New Guinea, and the Solomon Islands, reaches a length of 10 feet (3 meters). It eats birds, which it captures as they roost in trees at night, killing them with constriction and by injecting venom with long, grooved teeth in the rear of its upper jaw.

Introductions of exotic species can be intentional, as with the monk parakeet, or unintentional, as was the case with the brown tree snake, which reached Guam in the 1940s by stowing away in shipments of military equipment during the Second World War. In either case, two characteristics, in particular, make invasive species harmful (**FIGURE 16-19**):

Q Why should we worry about exotic species?

1. Exotic species have no predators in their new habitat, so their populations grow unchecked.

2. Native plants and animals have no mechanisms to compete with or defend themselves against invading exotic species.

Guam, which lies in the middle of the South Pacific Ocean, has no native snakes and no predators that are specialized to eat snakes. But Guam did have a magnificent fauna of native birds that had evolved in isolation on the island for thousands of years. Because the species of birds on Guam had never experienced predation by snakes, they had no defensive mechanisms against them—and brown tree snakes have eradicated most of the native species of birds. Extraordinary efforts are being made to prevent brown tree snakes from

DISASTROUS DISRUPTIONS: EXOTIC SPECIES

PROBLEM
Exotic species can threaten native populations.

CAUSE
Often introduced, both accidentally and intentionally, by humans. Once introduced, exotic species can multiply, unchecked by predation, overwhelming competitors and irreversibly altering ecosystems.

STRATEGIES FOR SOLUTION
Better regulation and restriction of intentional introductions; better vigilance against accidental introductions

FIGURE 16-19 Exotic species: the problem and its cause.

spreading to other Pacific islands, such as Hawaii. Like Guam, these islands are home to species of birds that occur nowhere else in the world, and the islands have no native species of snakes and no native predators of snakes.

We don't have to look beyond the borders of the continental United States, though, to find examples of exotic species that are responsible for massive ecological shifts and economic costs. Purple loosestrife is an attractive flowering plant that is native to Eurasia. It was imported to the United States in the 1800s as a garden plant, and rapidly escaped from cultivation

and invaded wetlands in every state except Florida. It produces thousands of seeds a year, and also spreads by sending out underground stems. This aggressive growth overwhelms native grasses, sedges, and flowering plants, replacing diverse wetland communities with monocultures of loosestrife that provide poor-quality habitats for bog-dwelling insects, birds, reptiles, amphibians, and mammals (**FIGURE 16-20**).

The title of Most Destructive Invaders in North America may belong to the zebra mussels and quagga mussels. These thumbnail-size freshwater mussels are native to eastern Europe and western Asia. In the early 1800s, the mussels spread to western Europe, and they came to North America in the 1950s, after completion of the Saint Lawrence Seaway allowed oceangoing ships to enter the Great Lakes. Female mussels produce up to five million eggs per year, and the larvae settle on any solid surface, even clogging the intake pipes of water-treatment plants and factories and the cooling systems of power plants.

The ecological threat that zebra mussels represent is even more serious than the damage they cause to industrial facilities. The Great Lakes fisheries are based on game species (lake trout, salmon, walleye) and commercial species (whitefish, perch, herring), and they produce revenues of more than $1.5 billion annually. Most of the commercially valuable fish feed on small species of fish, such as smelt, and the small fish feed on microscopic plants and animals (phytoplankton and zooplankton). Zebra and quagga mussels are exceptionally efficient at filtering phytoplankton and zooplankton from the water—so good that they are depleting

EXOTIC SPECIES: ACCIDENTAL INTRODUCTIONS

BROWN TREE SNAKE
Brown tree snakes, introduced to the island of Guam, have eradicated most of the native species of birds, which had never experienced predation by snakes and had not developed defense mechanisms.

PURPLE LOOSESTRIFE
Purple loosestrife, introduced to the United States as a garden plant, escaped from cultivation and has invaded wetlands in every state except Florida. The aggressive plant outgrows native grasses, sedges, and flowering plants.

ZEBRA MUSSELS
Zebra mussels, introduced to the Great Lakes, can cause damage to industrial facilities by clogging intake pipes. They also deplete resources available to small fish, eliminating the food supply for larger game fish.

FIGURE 16-20 Unwanted guests. Species sometimes get into natural habitats accidentally. The results can be devastating.

Defining Biodiversity   Extinction   ▼ Human Interference   Conservation Strategies

Initially introduced to Australia to control agricultural pests, cane toad populations exploded, now numbering more than 200 million, and have become a nuisance in much of the country.

FIGURE 16-21 Cane toad infestation.

the resources available to the small fish. Without enough food, the populations of small fish are crashing and that, in turn, is eliminating the food supply for the large game fish. The Great

Lakes contain nearly one-fifth of the fresh water in the world, but this enormous ecosystem is being irreversibly degraded by just two exotic species.

One legendary pest is the cane toad. Ironically, it was initially imported into Australia to control an agricultural pest, but the cane toad populations exploded and spread throughout much of the continent. They now number more than 200 million and have become a nuisance in much of the country, as conservation workers struggle to determine how to control their numbers (**FIGURE 16-21**). The cane toad illustrates a common feature of most exotic species introductions, even when intended as a method of biological control of pests: their effects on the ecosystem are unpredictable and often terrible.

TAKE-HOME MESSAGE 16·8

When an exotic species enters a new habitat, it leaves behind the natural predators that had controlled its population size. In addition, an invading predator encounters prey that have no experience in escaping from it, and they are easy victims. Thus, alien species can multiply, unchecked by predation and with plentiful food, overwhelming competitors and dominating and irreversibly altering communities and entire ecosystems.

16·9

Disruptions of ecosystems can be disastrous: 2. Acid rain and the burning of fossil fuels.

"What goes up must come down." That's a saying that applies to molecules just as much as it does to larger objects. The difference is that when molecules come down, they may not be in the same form as when they went up. In particular, some chemicals rise into the atmosphere as gases and return to earth as solutions of acid. These acid solutions can be fog, sleet, snow, or rain—all forms of precipitation can be acidic.

We use the term "fossil fuels" to describe oil, natural gas, and coal. These substances are composed largely of carbon and hydrogen, and they form carbon dioxide (CO_2) and water (H_2O) when they burn. Although they are referred to as hydrocarbons, fossil fuels are not *pure* hydrocarbons. This causes some problems. In addition to carbon and hydrogen, they contain substantial quantities of

Q Rain is usually just pure water. What can turn it to acid?

other elements, and it is the sulfur and nitrogen in fossil fuels that cause acid precipitation (**FIGURE 16-22**).

When sulfur burns, it produces the gas sulfur dioxide (SO_2), and nitrogen produces nitrogen dioxide (NO_2). These gases react with water vapor in the atmosphere to produce sulfuric acid (H_2SO_4) and nitric acid (HNO_3), and these acids fall to earth as acid precipitation (**FIGURE 16-23**).

In North America, precipitation has an average pH as low as 4.3 in some parts of the Northeast, nearly 1,000 times more acidic than pure water, which has a pH of 7.0. High concentrations of sulfuric and nitric acids are to blame, and several factors play a role. The Northeast has more people per square mile than does the rest of the United States, and that means

THE CHEMISTRY OF ACID RAIN

1 Burning fossil fuels releases the gases sulfur dioxide and nitrogen dioxide.

2 When combined with water vapor in the atmosphere, these compounds form sulfuric acid and nitric acid.

3 The acids fall to earth as acid precipitation.

FIGURE 16-22 **Formation of acid rain.**

there are more houses, factories, and automobiles to burn the oil, coal, and gasoline that create acid precipitation (**FIGURE 16-24**).

But not all of the pollution that causes acid precipitation is local. In the United States, the Midwest and Southeast have a large number of electric power plants that burn coal, and wind currents carry the sulfur dioxide and nitrogen dioxide from these power plants across the Northeast. Precipitation in the western states is nearly 100 times more acidic than pure water, because sulfur dioxide and nitrogen dioxide from Asia are converted to acids as they blow eastward across the Pacific Ocean. Precipitation in Europe and Asia is also acidified by local and distant sources of pollution.

Both terrestrial and aquatic organisms are harmed by acid precipitation, and the effects of acid precipitation can be direct

DISASTROUS DISRUPTIONS: ACID RAIN

PROBLEM
Acid precipitation kills plants and aquatic animals directly, and also indirectly via changes in soil and water chemistry.

CAUSE
Burning fossil fuels releases sulfur dioxide and nitrogen dioxide. The compounds form sulfuric and nitric acids when combined with water vapor.

STRATEGIES FOR SOLUTION
Tighter regulation and reduction of sulfur dioxide and nitrogen dioxide emissions

FIGURE 16-23 **Acid precipitation: the problem and its cause.**

or indirect. Direct effects result from contact of living tissues with acidic water, whereas indirect effects are produced by interactions between non-living and living systems.

Dramatic evidence of the direct effects of acid precipitation on vegetation can be found in mountain forests that are often blanketed by fog. When trees are exposed to acid fog, year after year, their leaves are damaged, their rates of photosynthesis decrease, and eventually, many of the trees die (**FIGURE 16-25**).

ACID RAIN DISTRIBUTION

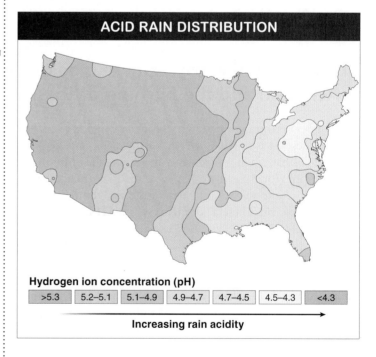

Hydrogen ion concentration (pH)

>5.3	5.2–5.1	5.1–4.9	4.9–4.7	4.7–4.5	4.5–4.3	<4.3

Increasing rain acidity →

FIGURE 16-24 **Acid precipitation across the United States.** Pure water has a neutral pH of 7.

Defining Biodiversity   Extinction   Human Interference   Conservation Strategies

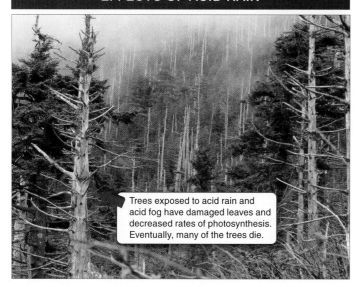

EFFECTS OF ACID RAIN

Trees exposed to acid rain and acid fog have damaged leaves and decreased rates of photosynthesis. Eventually, many of the trees die.

FIGURE 16-25 Acid precipitation has damaged these trees.

The indirect effects of acid precipitation on forests are not as conspicuous as the direct effects, but they are more far-reaching. As acid falls on the soil, it carries away calcium, potassium, magnesium, and sodium ions. These ions are essential nutrients for plants, and when they are leached out of the soil, the rate of plant growth is reduced. Some forest animals also feel the impact of this change in soil chemistry—when calcium is depleted, the shells of snails are thinner than usual. As a result of the thinning of snail shells, birds that eat snails receive less calcium than normal, and these birds lay eggs with shells so fragile that they can crack while the parents are incubating them.

Lakes and streams are also affected directly and indirectly by acid precipitation, and some lakes in the northeastern United States and in northern Europe have pH values as low as 4.0–4.3, which is 1,000 times more acidic than pure water. Very few species of insects, molluscs, crustaceans, fish, or amphibians can live in water that acidic. Still worse, acid precipitation dissolves aluminum in soil and carries the dissolved metal (aluminum ions)—which is toxic to animals—into lakes and streams, where it can kill most forms of aquatic animal life.

Acid rain is a solvable problem, because the deposition of acids from acid rain can be both prevented and cleaned up. The best approach is to prevent or reduce the emissions of sulfur dioxide and nitrogen dioxide. Preventive measures include improving energy efficiency, reducing the use of coal, switching to use of natural gas, and increasing the use of renewable energy resources, such as wind energy and solar energy. In the United States, many coal-burning plants have now installed special "scrubbers" to their smokestacks. The scrubbers extract hot gases from the power plant and function like sponges, removing sulfur dioxide before it is released into the atmosphere and converting it to a less hazardous form that can be physically removed from the tower and disposed of.

TAKE-HOME MESSAGE 16·9

Burning fossil fuels releases the gases sulfur dioxide and nitrogen dioxide, and these compounds form sulfuric and nitric acids when they combine with water vapor. Rain, fog, sleet, and snow that contain these acids can be up to 1,000 times more acidic than pure water. Acid precipitation kills plants and aquatic animals directly, by direct contact with living tissues, and also indirectly, through changes in soil and water chemistry.

16·10

Disruptions of ecosystems can be disastrous: 3. Global warming and the greenhouse effect.

Not long ago, debates about global warming focused on two questions: "Is it real?" and "Is it caused by human activities?" Now it is absolutely clear that the answer to both questions is *yes,* and the debate has moved to questions such as "What effect will global warming have on the world we know?" and "How can we stop it?"

Some of the evidence that the earth's atmosphere is warming comes from weather records that extend back into the 1700s.

Those records show that the average temperature has increased rapidly during the past 50 years and that the 10 hottest years on record have occurred since 1990. Three centuries is a substantial time from the perspective of an individual human, but it's less than the blink of an eye in the history of the earth. We can extend the temperature records back to more than 400,000 years ago by examining ice cores from Antarctica. Although these cores do record short-term and long-term variations in temperature, nothing in the past

GREENHOUSE EFFECT

Energy from the sun passes easily through the atmosphere to warm earth's surface. Some energy is reflected back toward space but because it is absorbed by greenhouse gases, it remains trapped in the atmosphere, heating the air.

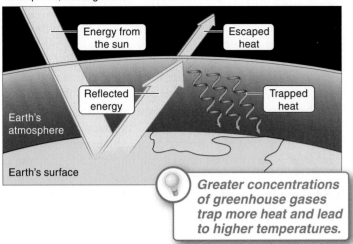

Greater concentrations of greenhouse gases trap more heat and lead to higher temperatures.

FIGURE 16-26 The earth is warmed by the greenhouse effect.

400,000 years approaches the speed of temperature change we have seen in the past 50 years.

The term "greenhouse effect" describes the process by which energy from the sun warms the earth's atmosphere. The light we see lies in the visible portion of the light spectrum, and visible wavelengths of light pass easily through the atmosphere to warm the earth's surface. Some energy in the infrared part of the spectrum flows from earth back toward space, and several gases in the atmosphere absorb this energy, heating the air (**FIGURE 16-26**). These gases act like the glass panels that make up the roof of a greenhouse: they allow light energy to pass through, but they keep heat from escaping—hence the name "greenhouse gases." Carbon dioxide and methane are among the most important of the greenhouse gases. Without them, heat would escape into space and the earth's average temperature would be lower.

Q Why do we blame global warming on humans?

Humans have increased the amount of greenhouse gases in the atmosphere. And we've done it for primarily one reason: we rely on the burning of fossil fuels—coal, oil, gasoline, and natural gas—for much of the energy we use. The burning of fossil fuels, which have been stored deep in the earth for millions of years, releases carbon dioxide, and clearing land and plowing soil to cultivate crops releases both carbon dioxide and methane. The same Antarctic ice cores that reveal the unprecedented rate of global warming in the past 50 years

also show that the current levels of carbon dioxide and methane in the atmosphere are far higher than any that have occurred during the 400,000-year span of the cores (**FIGURE 16-27**).

The Intergovernmental Panel on Climate Change, a scientific body set up by the World Meteorological Organization and the United Nations, has predicted an increase in the average annual temperature of 2° to 12° F (1° to 6.4° C) during the 21st century. Sea levels will rise by 7–24 inches (18–59 cm) as warmer temperatures accelerate melting of glaciers and ice sheets. Rainfall is predicted to decrease overall, and droughts will become more frequent and more severe. The impact of droughts will be most severe in areas that are currently on the borderline of aridity, such as Australia, the American Midwest, and sub-Saharan Africa.

Both environmental and biological evidence already reveal the effects of global warming. The Arctic Ocean ice cap is melting, and the glaciers that cover Antarctica and Greenland have also been melting at unprecedented rates—and those rates are increasing (**FIGURE 16-28**). During 1996, 21.6 cubic miles (90 cubic kilometers) of the Greenland ice sheet melted, and by 2005, the annual rate of melting had increased to 36 cubic miles (150 cubic kilometers). This melting will cause an increase in ocean levels, creating flooding problems for many of the world's population centers in coastal areas.

Biological systems are also showing the effects of climate change. Trees and flowers in northern latitudes are blooming earlier in the spring than they used to; migratory birds are arriving at their summer ranges earlier than they did even a decade ago; and birds and butterflies are extending their geographic ranges northward.

DISASTROUS DISRUPTIONS: INCREASED GREENHOUSE GAS EMISSIONS

PROBLEM
The average temperature has increased rapidly during the past 50 years, affecting both the physical environment and the biological world.

CAUSE
Burning fossil fuels and clearing land to cultivate crops has significantly increased levels of greenhouse gases in the atmosphere.

STRATEGIES FOR SOLUTION
Reduced emissions of greenhouse gases (particularly from the burning of fossil fuels)

FIGURE 16-27 Increasing greenhouse gases: the problem and its cause.

Defining Biodiversity   Extinction   Human Interference   Conservation Strategies

ICE CAP MELTING

1979

2005

FIGURE 16-28 Disappearing ice pack. The satellite image at the top shows the minimum concentration of Arctic sea ice in 1979; the lower image shows the concentration of sea ice recorded on September 21, 2005.

The changes to the physical environment caused by global warming, such as thinning of the Arctic Ocean ice or the flooding of coastal estuaries, can have profound effects on biodiversity at the species and ecosystem levels. For example, in 2004, a research ship in the Arctic Ocean found nine walrus calves swimming alone in deep water. Normally, adult female walruses leave their babies on ice floes in shallow water while they dive to the seafloor to feed on clams and crabs, and then return to the ice to nurse their babies. But the Arctic ice cap over shallow water has melted, driving the female walruses to the remaining ice, which is in deep water. The mother walruses cannot dive deep enough to reach the sea bottom, and they seem to be abandoning their calves.

Other consequences of global warming reveal the tremendous interconnectedness of so many elements of ecosystems. Many bird and butterfly species' migratory patterns must be altered as they search for lower temperatures. Many small mammal populations living near high-altitude peaks will have nowhere to go for milder temperatures—if they move to lower altitudes it gets hotter; moving up to higher altitudes brings them to milder temperatures, but if they are already at the peak, they've got nowhere else to go when it gets too hot. Massive crop lands throughout the world may no longer be able to support the crops previously grown there.

Global warming is a difficult issue to tackle because it is so large—truly global—in scope, necessitating individual efforts as well as international cooperation. Fundamentally, however, the science is straightforward: carbon emissions must be reduced significantly. This can be done by individuals who choose to cut down on their own fossil fuel use, improve the energy efficiency of their homes and cars, and support the development of carbon-free renewable energy sources. Internationally, it can be done through efforts to reduce deforestation while increasing the replanting of forests, and by developing and implementing transportation and industrial innovations that reduce dependence on the fuels that produce greenhouse gases. Innovative technologies, such as the underground sequestration of carbon, may also contribute to reversing global warming.

TAKE-HOME MESSAGE 16·10

Carbon dioxide and methane are called "greenhouse gases" because they trap heat in the atmosphere. As humans burn fossil fuels and clear forests, the concentrations of greenhouse gases have been increasing and global temperatures have been rising. Ecological changes in plant and animal communities have already been observed and are likely to become more serious unless there is a global reduction in emissions of greenhouse gases.

Disruptions of ecosystems can be disastrous: 4. Depletion of the ozone layer.

If you live in a city, you are likely to hear radio and television warnings of high ozone levels during the summer. These warnings announce that ozone is expected to exceed safe levels that day. Children and adults with lung disease are advised to remain indoors during an ozone alert.

You are familiar with the oxygen you breathe—it comes in molecules formed by two oxygen atoms and is represented by the chemical formula O_2. Ozone is a different molecular arrangement of oxygen that packs three atoms into a molecule, with the chemical formula O_3. Ozone irritates the respiratory pathways and lungs, causing coughing and wheezing, and exposure to ozone can induce asthma attacks.

If ozone makes people sick, why would anyone worry about *depletion* of ozone in the atmosphere? Isn't that a *good* thing? The answer is the same as the old saying on what determines the value of real estate—"Location, location, location." At ground level, ozone is bad for you, but ozone in the upper part of the stratosphere, which is 30 miles (50 kilometers) above the earth's surface, protects you from ultraviolet radiation. Ozone is a Jekyll-and-Hyde molecule, and one set of environmental regulations tries to reduce the formation of ozone at ground level while a different set of regulations tries to protect ozone in the stratosphere.

The term "ozone depletion" covers two different processes: a general reduction in the amount of ozone in the stratosphere and the formation of areas with very low ozone concentration (called "ozone holes") over the North and South Poles every winter. Synthetic chemicals known as chlorofluorocarbons (CFCs) are the villains in both forms of ozone depletion. CFCs, which were developed in the 1930s, have a wide range of applications, including use as coolants in refrigeration systems. When CFC molecules leak into the atmosphere, they rise to the stratosphere, where sunlight knocks a chlorine atom off the CFC molecule. These free chlorine atoms then catalyze the breakdown of ozone to oxygen (**FIGURE 16-29**).

The amount of ozone in the stratosphere has been decreasing by about 4% per decade. The cold temperatures and circular flows of air that develop over the Poles during the winter concentrate CFCs, forming ozone holes in those locations. The Antarctic ozone hole, which lasts for several months,

Q Should you be concerned about an ozone hole in our atmosphere?

DISASTROUS DISRUPTIONS: OZONE LAYER DEPLETION

PROBLEM
Increased levels of ultraviolet light reach the earth's surface, leading to greater incidence of illness in animals, and decreased rates of photosynthesis in plants.

CAUSE
Synthetic chemicals known as chlorofluorocarbons (CFCs) leak into the atmosphere, where they cause the breakdown of ozone.

STRATEGIES FOR SOLUTION
Reduced production and emission of CFCs

FIGURE 16-29 Depletion of the ozone layer: the problem and its cause.

covers the entire continent of Antarctica and extends northward to include the southern tips of South America and Australia (**FIGURE 16-30**). The Arctic ozone hole is smaller than the Antarctic hole and it does not last as long, but it is large enough to extend southward into northern Europe, Asia, and North America.

Ozone depletion allows short-wavelength ultraviolet light (UVB light, with wavelengths of 270–315 nanometers) to reach the earth's surface. The 4%-per-decade reduction in global stratospheric ozone levels has increased the intensity of UVB radiation everywhere on earth, and the UVB intensity under the ozone holes is even higher.

Not surprisingly, humans tend to think first about the effects of elevated UVB on human health. A 1% increase in UVB intensity increases the incidence of skin cancer by 2% to 3%. Exposure to UVB radiation also promotes the formation of cataracts and reduces the effectiveness of our immune system. Domestic animals suffer the same kinds of damage that humans do, and a few studies indicate that wild animals are also affected.

But we may be looking at the wrong problem. The damage to ecosystems could be a more serious consequence of ozone

Defining Biodiversity   Extinction   Human Interference   Conservation Strategies

September 1979

September 1988

September 2000

September 2006

FIGURE 16-30 **Monitoring the ozone hole.** The satellite images show changes in the ozone hole over the Antarctic during the past 30 years. The darkest blue coloration indicates weakest ozone concentration.

Depletion of the ozone layer and the formation of an ozone "hole" over earth's polar regions have occurred as a result of human activities.

depletion than damage to human health. UVB radiation reduces the rate of photosynthesis by plants, and reduced photosynthesis in agricultural ecosystems means that crop yields will decline just at the time when the worldwide cost of food is increasing.

The creation of the ozone hole is an example of how some scientifically developed industrial processes and products can have unintended and severe consequences. When CFCs were first invented, they were considered harmless, and so their use as coolants, as aerosol propellants, and in the production of materials such as Styrofoam became widespread. But the subsequent steps to repair the ozone hole are also an example of the power of science to help identify environmental problems and to help find solutions (**FIGURE 16-31**). With the adoption by most countries in the 1980s of an agreement to discontinue the use of CFCs, the atmospheric levels of ozone have stabilized, and full recovery could occur by 2065.

REDUCING CFCs

Products made from ozone-depleting CFCs, such as Styrofoam hamburger boxes and hair sprays with damaging propellants, have been phased out in the hope of spurring a recovery of the ozone layer.

FIGURE 16-31 **Taking simple steps to reduce ozone depletion.**

TAKE-HOME MESSAGE 16·11

Ozone in the stratosphere prevents short-wavelength ultraviolet light (UVB) from reaching the earth's surface, but for many decades the amount of ozone has been decreasing. Synthetic chemicals, the chlorofluorocarbons (CFCs), are responsible for the destruction of ozone. An increase in UVB light reaching the earth's surface has adverse effects on health, including increasing the incidence of cancer in humans and other animals, and can seriously damage ecosystems. Reduction in the production of CFCs is reducing atmospheric ozone depletion.

Disruptions of ecosystems can be disastrous:
5. Deforestation of tropical rain forests.

Towering trees, colorful birds and butterflies, maybe a glimpse of Tarzan swinging past on a hanging vine—that is the popular image of a tropical rain forest. And it is a reasonably accurate picture (minus the Tarzan part). But it is also a picture that is rapidly fading, as agriculture, logging, gold mines, and oil wells destroy tropical forests (**FIGURE 16-32**).

Tropical rain forests grow in a region extending north and south of the equator, crossing South America, Africa, Asia, and Australia. As recently as a few centuries ago, that belt of rain forest covered about 8 million square miles (20 million square kilometers). More than half of that forest has already been destroyed, most of it in just the past 200 years (**FIGURE 16-33**).

The destruction of a tropical rain forest means an enormous loss of species, because these forests are the most diverse terrestrial habitats on earth. More than 170,000 species of plants grow in tropical rain forests, and that is about 70% of the total number of living plant species in the world. A survey of a Brazilian rain forest found 487 different species of trees in just 2.5 acres (1 hectare), an area less than half the size of a city block in Manhattan. To put that number in perspective, consider that only 700 species of trees are found in the 5 *billion* acres of the United States and Canada combined!

Agriculture is responsible for the greatest loss of tropical forest, and it takes a variety of forms. Sometimes, a relatively a small

DEFORESTATION: TROPICAL RAIN FORESTS

☐ Tropical rain forest cover before human influence
☐ Current tropical rain forest cover

FIGURE 16-33 More than half of the world's rain forests have been destroyed, most of this destruction occurring in the past 200 years.

area is affected—"slash-and-burn agriculture" refers to cutting trees, burning them, and planting crops in the newly opened area. Usually, just a few acres of forest are cleared, but the land is fertile for only two or three years after it has been cleared. Then that plot is abandoned and more forest is cleared, so the cumulative effect of slash-and-burn agriculture is substantial. At the other end of the size scale, multinational corporations clear hundreds of acres of forest to plant bananas, coffee, or oil palms, and thousands of acres of forest to create pastures for cattle.

Pollution from oil wells and mining in tropical rain forests has an impact that can extend far beyond the areas that are actually cleared. Leaking oil contaminates streams and groundwater, and acidic water drains from mines. Even worse, gold miners use mercury to extract gold and some of the mercury enters streams, where it is converted to methyl mercury, a toxic compound that accumulates as it moves up the food chain.

Agriculture and mining both require roads to bring equipment to the sites and take crops or minerals to market. Roads have an impact on forests that greatly exceeds the relatively small area they occupy, because they make access to the forest easy. Traveling through a virgin tropical rain forest is difficult—bogs and natural tree falls often prevent travel in a straight line, slopes are often steep and the wet soil is slippery,

DISASTROUS DISRUPTIONS: DEFORESTATION

PROBLEM
Tropical rain forests are being cleared at unprecedentedly high rates, endangering countless species and increasing the concentration of greenhouse gases in the atmosphere.

CAUSE
The land is cleared for agriculture, logging, gold mines, and oil wells.

STRATEGIES FOR SOLUTION
Reduced destruction of high-biodiversity habitats, particularly tropical rain forests

FIGURE 16-32 Deforestation: the problem and its cause.

Defining Biodiversity   Extinction   Human Interference   Conservation Strategies

and the ground-level vegetation can be both dense and thorny. No wheeled vehicle larger than a motorcycle can penetrate most rain forests, and traveling by motorcycle is slower than walking. These difficulties protect rain forests, but as soon as a road is bulldozed through a forest, people flock to it. They and their activities then spread into the forest on both sides of the road, creating new clearings that spread like cancers, farther and farther into the forest

Destruction of tropical rain forests has two serious environmental impacts: reducing the earth's biodiversity and increasing the concentration of greenhouse gases in the atmosphere.

1. *Reducing biodiversity.* Tropical rain forests of Africa, Asia, the Pacific, and Central and South America contain an unusually large number of species of plants and animals (and probably other taxa).

2. *Increasing greenhouse gases.* Photosynthesis in tropical rain forests removes an estimated 610 billion tons (550 trillion kilograms) of carbon dioxide from the atmosphere each year. Accumulation of carbon dioxide is the major cause of global warming, and the photosynthetic activity of tropical rain forests is slowing the rate of warming. The huge quantity of carbon stored in rain forests has a downside, however, because that carbon is released into the atmosphere when forests are cleared and burned. And tropical forests are being cleared at a frightening rate—nearly 12 square miles (30 square kilometers) per day, which means a total of 200,000 square miles (more than 500,000 square kilometers) just from 2000 to 2005!

The problem of tropical deforestation is among the most difficult of any environmental problems to solve, and solving it will involve multiple strategies and rely on international cooperation. The most diverse areas must be identified and protected. The poverty that necessitates the destruction of tropical rain forests for human activities must be addressed.

Planting new trees can undo some of the damage done by deforestation.

FIGURE 16-34 **Growing seedlings for reforestation.**

Alternative sources of food and income must be developed. Population growth must be reduced. And education about the value of preserving biodiversity is central to these issues. At present, we have not progressed much beyond identification of the problem, but that, at least, is a necessary first step (**FIGURE 16-34**).

TAKE-HOME MESSAGE 16•12

Tropical rain forests contain more species of plants and animals than all other terrestrial habitats combined, and half of the earth's biodiversity hotspots are in these forests. Tropical rain forests remove more carbon dioxide from the atmosphere than any other terrestrial habitat and, as a result, rain forests are enormously important in limiting global warming.

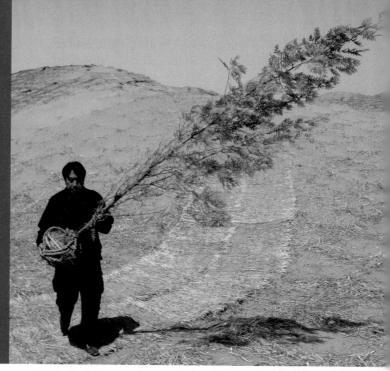

❹ We can develop strategies for effective conservation.

A worker carries a tree to plant at the edge of Maowusu Desert, China. Trees are planted to prevent the desert from expanding.

16•13

With limited conservation resources, we must prioritize which species should be preserved.

Is it preferable to save one beautiful, well-studied bird species or 1,000 species of bacteria, none of which have even been described or named? This is a question that biologists, policymakers, and, ultimately, citizens must address. And what about the 1,200 species of beetles in Panama that live in the evergreen tree *Luehea seemannii,* almost 200 of which live *only* in this one species of tree—how should we prioritize these species for conservation relative to, say, a single primate species? From locale to locale the particulars of these question may change, but the underlying issue remains: in a world where species are being driven to extinction faster than we can save them, which should be singled out for preservation and which should we leave to almost certain extinction (**FIGURE 16-35**)?

When setting conservation priorities, it is essential to articulate a goal. In an ideal world, the goal might be to protect "all biodiversity." Barring that, options include protecting "most biodiversity," or the most diverse subset of biodiversity (in terms of genes, species, and ecosystems), or the most valuable biodiversity. There are numerous arguments in favor of each goal, and each has significant flaws as well. Once a goal is decided on, a related step in the decision-making process is needed—a plan that outlines priorities for achieving the goal. Frequently, this involves an assessment of the degree

of threat faced by various components of biodiversity. We can rank species, for example, as relatively intact, relatively stable, vulnerable, endangered, or critical. These rankings can then be used in conjunction with measures of the biological value of the biodiversity to humans in formulating a plan.

As we saw earlier in the chapter, biodiversity can be valued in many different ways, and its worth cannot be easily quantified or weighed. Sooner or later, many difficult and subjective decisions must be made in the goal-setting and prioritizing process. The undiscovered bacterial species may harbor the metabolic secrets to curing a devastating medical condition in humans, but the beautiful bird species may be much loved by bird-watchers or may serve as a powerful icon of strength or freedom, inspiring generations of people. The decisions are difficult, but not facing them is, in most cases, the same as making a decision.

In the United States, much of the conservation policy involves response to the **Endangered Species Act (ESA),** a law that defines **endangered species** as those in danger of extinction throughout all or a significant portion of their range. The law is designed to protect those species from extinction. As we noted earlier in the chapter, species that are dwindling are

Defining Biodiversity   Extinction   Human Interference   Conservation Strategies

Which is more important to protect: a beautiful species such as the snow leopard, or soil bacteria with unknown roles in their ecosystem?

WHAT SHOULD WE PROTECT?

FIGURE 16-35 Prioritizing conservation efforts.

listed as endangered or threatened, according to an assessment of their risk of extinction. Once a species is listed, legal tools are available to help rebuild the population and protect the habitat critical to its survival. Seemingly straightforward, the ESA has had the effect of focusing most conservation efforts on the preservation of individual species (and populations), sometimes at the expense of other elements of biodiversity and sometimes at the expense of efforts to reduce the loss of ecologically important habitat (**FIGURE 16-36**). Other difficulties are also associated with the ESA. Consider the task of determining the critical population size below which a population is endangered—is it 500 individuals or 5,000 or 50,000? With its emphasis on preservation of single species, the ESA does not effectively address ecosystem decline, which is much like waiting for fires and then trying to put them out.

In spite of these difficulties, the species-focused approach to conservation has had some success in the United States. Since it became law in 1973, approximately 40 species that were listed as endangered—including the bald eagle, the peregrine falcon, the gray whale, and the grizzly bear—have been taken off the list as their numbers have recovered. In the final section, we examine other strategies that have been used in efforts to preserve biodiversity.

TAKE-HOME MESSAGE 16·13

Effective conservation requires the setting of goals on the elements of biodiversity (genes, species, or ecosystems) to be conserved and priorities among those elements. The Endangered Species Act has focused much conservation effort on the preservation of species.

PRESERVING SPECIES vs. PRESERVING HABITATS

Conservation efforts can focus on the preservation of individual species or of important habitats.

FIGURE 16-36 Which to preserve? Limited resources and great demand make preservation choices difficult.

There are multiple effective strategies for preserving biodiversity.

In a survey asking which animal they would like to be, men and women gave very different answers. Among men, the top answer was eagle, followed by tiger, lion, and dolphin. As their first choice, women chose cat, followed by butterfly, swan, and swallow. It's just a silly survey, but the differences in male and female selections parallel the fact that people can have very different preferences when it comes to the preservation of biodiversity. One person may view as unthinkable the loss of a particular species or habitat, while another may value other species or habitats much more.

Most approaches to conservation biology, as we've seen, have focused on the preservation of individual species. Increasingly, however, conservation biology is shifting toward the preservation of important habitat, focusing on conserving communities and ecosystems. This approach, of course, also leads to the preservation of individual species, but it has the added benefit of preserving, at the same time, many different species within a habitat, including many that have not yet been identified, particularly microbes and fungi.

One example of an effort aimed at the preservation of important habitat is the World Wildlife Fund's "Global 200," an identification of all of the most biologically distinct habitats

on earth. The identification of these habitats, which include terrestrial, freshwater, and marine habitats, is part of an innovative strategy, often called **landscape conservation,** that is geared toward conserving not just species, but habitats and ecological processes (e.g., large-scale migrations and predator-prey interactions). In an age of scarce conservation resources and limited time, this approach prioritizes the conservation of the most biologically unique habitats on earth.

The habitat conservation approach is not new. The very first attempts at conservation were the creation of Yellowstone Park in 1872 and Yosemite in 1890, both large-scale efforts geared not simply toward the preservation of one or a few species but rather at preserving "wilderness." In the time since then, significant efforts have been made to establish national parks, wilderness areas, and recreation areas (**FIGURE 16-37**).

As conservation biologists have started to better understand population dynamics and biogeography, the design of natural preserves has evolved. Modern preserves focus not simply on maximizing the variety of habitats and biodiversity preserved, but on using several design features that maximize their efficiency. These design features include corridors and buffer zones (**FIGURE 16-38**).

1. *Corridors.* Setting aside areas—even relatively narrow, corridor-like strips of land—that connect larger natural preserves allows gene flow and reduces inbreeding among distinct populations in the different preserves. Because fragmented populations support smaller populations, the inclusion of corridors linking the fragmented populations can enhance the maintenance of genetic diversity.

2. *Buffer zones.* Preserves designed with core areas, containing the habitat to be conserved and where human activities are limited, surrounded by buffer zones, where limited amounts of human use are permitted, can sustain wildlife, even as increasing human populations exist adjacent to the preserves.

As conservation plans aimed at preserving habitats increase, efforts focusing on individual species still continue to be effective. Several strategies targeting individual species for conservation have been particularly successful at preserving large amounts of biodiversity beyond that single species.

1. *Flagship species.* Some species, because they are particularly charismatic, distinctive, vulnerable, or otherwise appealing,

FIRST ATTEMPTS AT PRESERVING WILDERNESS

One of the first national parks, Yosemite was one of the first efforts at preserving wilderness.

FIGURE 16-37 **The natural beauty of Yosemite.**

DESIGNING EFFECTIVE NATURE PRESERVES

The design of natural preserves has evolved. Modern preserves focus on the use of several design features that maximize their efficiency, including:

CORRIDORS
Setting aside narrow strips of land that connect larger natural preserves allows gene flow and reduces inbreeding among distinct populations in the different larger areas.

BUFFER ZONES
Preserves designed with core areas containing the habitat to be conserved, surrounded by buffer zones in which limited amounts of human use are permitted, can sustain wildlife even as increasing human populations exist adjacent to the preserves.

FIGURE 16-38 Carefully planned nature preserves focus on features that maximize biodiversity.

can engender significant public support. Examples include the giant panda of China, the golden lion tamarin of Brazil's coastal forest, the mountain gorilla of Central Africa, the orangutan of Southeast Asia, the leatherback sea turtle, the Indian tiger, and the African elephant. Preservation of these species, given their habitat needs, can serve to preserve many other species as well.

2. *Keystone species.* As we saw in Chapter 15, keystone species have a disproportionate effect on the biodiversity of a community. Their removal can lead to massive changes in the composition of species in an ecosystem, often causing huge loss of biodiversity. Examples include kelp, California mussels, grizzly bears, beavers, and sea stars.

3. *Indicator species.* These are species whose presence within an ecosystem indicates the presence of a large range of other species. For example, the presence or absence of lichens is an indicator of air quality, because they are sensitive to sulfur dioxide, a component of industrial fumes. When lichens disappear from trees, it is usually an indicator of increased pollution, which endangers the entire ecosystem. Conversely, when the lichens are present and healthy, they indicate that the ecosystem is healthy.

4. *Umbrella species.* These include wide-ranging species with such large needs for habitat and other resources that their preservation ends up protecting the numerous other species within that same habitat, without having to identify these other species as conservation targets. Umbrella species tend to be large, wide-ranging vertebrates.

For more critically endangered species and degraded habitats, all hope is not necessarily lost. Captive breeding programs and habitat restoration have had some success in bringing back biodiversity from the brink of extinction. Zoos and botanical gardens have taken the lead in many of these efforts. In the 1980s, for example, when the population size of the California condor had dropped to 22 individuals, due to poaching, lead poisoning, and habitat loss, all of the birds were caught and taken to zoos, where a captive breeding program began. With the success of these breeding programs, by 2009, more than 100 birds have been released back into the wild, and more than 150 more are in zoos. Breeding programs are not a complete conservation solution on their own, however. It is essential that the species' habitats are not destroyed or altered if the species are to flourish again under natural conditions.

Restoration ecology, using the principles of ecology to restore degraded habitats back to their natural state, has also been an important tool of conservation biologists. Wetlands that have been degraded by dredging and development, in particular, have benefited. Reintroduction of the native plant species and restoration of water and stream flow patterns can be instrumental in restoring these habitats.

Taken together, the strategies used by conservation biologists represent an important step toward reducing the adverse effects of one hugely successful (from a growth perspective) species—humans—on other species. Although we are far from living with minimal perturbation of the environment in which we live, continued conservation efforts offer our best hope for a sustainable future.

TAKE-HOME MESSAGE 16·14

Conservation biology has focused, in the past, on preserving individual species. Increasingly, there has been a shift toward the preservation of important habitats, focusing on conserving communities and ecosystems and utilizing concepts learned from ecology and island biogeography. Several methods focusing on single species, however, remain useful, particularly when preserving the selected species requires preservation of an amount and type of habitat that simultaneously preserves many other species.

Knowledge You Can Use

"Plan B" for saving endangered species: re-creating the wooly mammoth?

For a million years, wooly mammoths roamed throughout Europe, Asia, and North America. If you saw one, you'd know it was an elephant, but in adapting to their cold environment through two ice ages, the 8- to 14-foot-tall animals developed very thick hair and a two-inch layer of insulating fat.

Q: Where did they go? About 12,000 years ago, wooly mammoths became extinct, probably as a result of overhunting by humans in conjunction with a sudden change in climate.

Q: Is extinction really forever? Some scientists, promising a new strategy in the conservation of endangered species, want to bring the wooly mammoth back from extinction. They have extracted DNA from giant hairballs recovered from wooly mammoths preserved in the permafrost of Siberia and determined the sequence of base pairs in the recovered DNA. They plan to use that information to clone a wooly mammoth, getting the DNA into an elephant egg that has had its nucleus removed (they haven't figured out exactly how to do this), spurring the egg to divide, and then implanting it in a female elephant.

Q: Can wooly mammoths really be brought back from extinction? Currently, it isn't possible to re-create an extinct animal. In the case of the wooly mammoth, among other problems is that it hasn't been possible to isolate *all* of its DNA. And the DNA in the hairball samples is contaminated with DNA from other organisms, including bacteria, viruses, fungi, and assorted parasites.

Q: Does this have any relevance beyond novelty? Some people have proposed that the cloning of extinct animals could be a useful conservation tool. After all, getting a complete DNA sample uncontaminated by other species' DNA is not a problem when species still have living representatives. The San Diego Zoo has even established a lab they call "The Frozen Zoo," which stores frozen tissue from rare and endangered species.

Q: Could endangered or recently extinct species be brought back? When DNA is harvested from living animals, the answer is *yes*. Although there are still many technical difficulties, this has already been achieved. In 2001, a gaur, an endangered Asian ox, was cloned. (Although the procedure worked and a gaur was born, the baby died shortly after birth, from an illness unrelated to the cloning.)

What can you conclude? Is harvesting and preserving DNA the solution to the biodiversity crisis?
In a word, no. At this point, cloning endangered animals is more about creating amusement park novelties and museum or zoo exhibits than preserving biodiversity. After all, conservation biologists want to preserve much more than genetic material. More important, many conservationists worry that the public will consider the cloning of endangered and extinct animals as an excuse to stop efforts to reduce habitat loss and protect land from development. (Closer to home: When it comes to your family pet, things are a bit more promising. Entrepreneurs have cloned pets after the animal died. At about $50,000, however, it doesn't look to be a rapidly growing industry.)

❶ Measuring and defining biodiversity is complex.

Biodiversity provides significant benefits to humans in utilitarian ways, such as the production of medicines, and provides important value in less tangible ways, including aesthetic, symbolic, and naturalistic values. Biodiversity can be defined at multiple levels, from entire ecosystems to species to genes and alleles. There is a strong latitudinal biodiversity gradient from the tropics to the poles for nearly all groups of plants and animals. Biodiversity hotspots are regions of significant biodiversity under threat of destruction. The theory of island biogeography generates predictions about patterns of species diversity and has implications for the development of nature preserves.

❷ Extinction reduces biodiversity.

Mass extinctions tend to result from bad luck rather than characteristics of a species' biology, whereas background extinctions tend to be a consequence of features of a species' biology. Small population size, limited habitat range, and narrow zones of habitat tolerance contribute to background extinctions. Most biologists believe that we are in the midst of a mass extinction, that it is the result of human activities, and that it poses a serious threat to human survival.

❸ Human interference generally reduces biodiversity.

Because they have no natural predators in their new habitat and their prey have no defenses against them, exotic species can multiply unchecked, overwhelming competitors and dominating and irreversibly altering communities and entire ecosystems. Burning fossil fuels releases sulfur dioxide and nitrogen dioxide, and these gases combine with water vapor to form acid precipitation, harming plants, animals, and the abiotic environment. Carbon dioxide and methane are greenhouse gases, trapping heat in the atmosphere. Produced by the burning of fossil fuels and clearing of forests, greenhouse gases are increasing and global temperatures rising. Avoiding increasingly serious consequences requires a global reduction in emissions of greenhouse gases. Ozone in the stratosphere has been depleted through the effects of CFCs, with health consequences for humans and animals. A reduction in CFC production can restore safer levels of ozone. Tropical rain forests contain more species than all other terrestrial habitats combined, and half of the earth's biodiversity hotspots are in tropical rain forests. Destruction of these forests is likely to have serious environmental consequences.

❹ We can develop strategies for effective conservation.

Effective conservation requires setting goals on which elements of biodiversity (genes, species, or ecosystems) to conserve and priorities among those elements. Although the Endangered Species Act has focused most conservation efforts on individual species, there is now a shift toward conserving communities and ecosystems. Several methods focusing on single species remain useful, however, particularly when they require preservation of an amount and type of habitat that simultaneously preserves many other species.

KEY TERMS

biodiversity hotspots, p. 593
conservation biology, p. 591

dynamic equilibrium, p. 595
endangered species, p. 614

Endangered Species Act, p. 614
endemic, p. 594

exotic species, p. 603
landscape conservation, p. 616

1. Biodiversity is considered important because of:
   a) the direct economic benefit to humans, such as medicines and food.
   b) the symbolic value that elements of the natural world can provide.
   c) the aesthetic value it holds.
   d) its potential for helping humans understand the processes that gave rise to the diversity we see.
   e) All of the above are correct.

2. Biodiversity hotspots are defined by which two criteria?
   a) species richness and ecosystem integrity
   b) size and distance from nearest alternative hotspot
   c) species endemism and degree of threat
   d) species richness and size
   e) ecological diversity and species diversity

3. Madagascar is unusually important to conservation because it:
   a) has more species per unit area than any place on earth.
   b) is the fourth largest island in the world.
   c) is home to more endangered species than any other country.
   d) has a higher percentage of endemic plants and animals than any comparably sized area on earth.
   e) is the native habitat of the rosy periwinkle, a rain forest plant that helps cure childhood lymphocytic leukemia.

4. By fumigating several islands in the Florida Keys with methyl bromide, E. O. Wilson and Daniel Simberloff demonstrated that:
   a) the equilibrium number of arthropod species was a function of island size and distance from the mainland.
   b) colonization of an island by new species can no longer occur once the island is "full."
   c) for plant but not insect species, the relation of area to species number holds true for all islands.
   d) the effects of global warming will be greater on islands than on the mainland.
   e) species turnover rates are highest on small, more distant islands.

5. In a dynamic equilibrium:
   a) extinction is the sole determinant of the number of species present on an island.
   b) immigration rates to an island are reduced to zero.
   c) both the number and composition of species on an island remain the same.
   d) both the number and composition of species on an island are changing.
   e) the composition of species on an island changes, but the number of species present remains the same.

6. Which of the following attributes of a species might render it especially susceptible to extinction?
   a) It lives in a marine habitat.
   b) It exists as two distinct subpopulations.
   c) It is an invasive species.
   d) It is short-lived.
   e) It has a narrow dispersal range.

7. What bird, once the most abundant in North America, was hunted to extinction by shooting and trapping during the 1800s?
   a) California condor
   b) blue-footed booby
   c) dodo
   d) passenger pigeon
   e) great auk

8. How many species do biologists estimate to be currently existing on earth?
   a) fewer than 1 million
   b) between 3 million and 5 million
   c) more than 1 billion
   d) between 5 million and 100 million
   e) between 1 million and 2 million

9. Which of the following is currently the leading cause of extinction?
   a) pollution
   b) habitat loss
   c) disease
   d) exotic species
   e) overexploitation

10. Exotic species can disrupt ecosystems because:
    a) they frequently have no predators in their new habitat and grow unchecked.
    b) they are favored by ecotourists.
    c) they have no natural prey items and so must rely on humans for their survival.
    d) they have better dispersal capability than endemic species.
    e) All of the above are correct.

11. Even though there is a carbon cycle, carbon dioxide levels around the world seem to be rising. Which of the following best explains why this is so?
    a) Animals give off carbon dioxide during their normal metabolism.
    b) As the atmosphere heats up, it can contain more carbon dioxide.
    c) The destruction of coral reefs leads to increased levels of carbon dioxide.
    d) More carbon dioxide is being given off by ocean waters as they heat up.
    e) The burning of fossil fuels releases more carbon dioxide into the atmosphere.

12. Ozone depletion is cause for concern because:
    a) without ozone in the atmosphere, greenhouse gases can escape earth's atmosphere, leading to global warming.
    b) without ozone, fossil fuels are more likely to react with water vapor in the atmosphere, increasing levels of precipitation throughout the world.
    c) without the protection against solar radiation provided by ozone, rates of skin cancer are likely to increase.
    d) without ozone, solar radiation can irradiate the Antarctic ice shelf, causing unsafe levels of radioactive isotopes.
    e) reduced levels of ozone can increase the rate of respiratory problems in humans, particularly children.

13. Which of the following statements about habitat loss is correct?
    a) Small animals are not vulnerable to habitat destruction.
    b) Fragmented habitats can support only small populations of the species living in them.
    c) Corridors between habitat areas are only of use to active organisms such as vertebrates.
    d) Corridors can replace habitat areas for migratory species.
    e) Large animals are not vulnerable to habitat destruction.

14. An umbrella species:
    a) is a species whose removal will severely damage or cause the collapse of its ecosystem.
    b) is a non-native species, usually introduced to an area by accident.
    c) is any species that has a larger than expected effect on its ecosystem.
    d) is a popular species that the public tends to rally behind and support.
    e) is a species whose preservation results in the preservation of many other species as well.

SHORT-ANSWER QUESTIONS

1. Where, geographically, is the greatest biodiversity? What are the primary threats to this biodiversity?

2. Why were the ecosystem disturbances created by humans in New England reversible? What changes would have made the ecosystem disturbances in New England irreversible?

3. What is the primary cause of tropical rain forest deforestation? How does this affect the biodiversity of plants and animals?

See Appendix for answers. For additional study questions, go to www.prep-u.com.

PHOTO CREDITS --

Photo Legend: *L*: left, *C*: center, *R*: right, *T*: top, *M*: middle, *B*: bottom

CHAPTER 1   **p. 1:** *photo* Karl Ammann/Corbis. **p. 2:** *photo* Martin Barraud/Getty Images. **p. 3:** *Figure 1-1* The Photo Works. **p. 4:** *Figure 1-2 (L)* Reprinted through the courtesy of the Editors of TIME Magazine © 2008 Time Inc., *(M)* Courtesy of *The Economist*, *(R)* Newscom, *(B)* The *New York Times*. **p. 5:** *Figure 1-3* Photo Researchers. **p. 6:** *Figure 1-4 (TL)* Louie Psihoyos/Corbis, *(TR)* Mark Scott/Getty Images, *(BL)* Steven Puetzer/Getty Images, *(BR)* Masterfile. **p. 8:** *photo* AP Photo/ Christof Stache. **p. 9:** *photo* AFP/Getty Images; *Figure 1-6 (TR)* Digital Vision Photography/Veer. **p. 10:** *Figure 1-7 (L)* Grapes-Michaud/Photo Researchers. *(R)* Nigel Cattlin/Alamy. **p. 11:** *Figure 1-8* LWA/Getty Images. **p. 14:** *Figure 1-11* Jean Chung/OnAsia. **p. 17:** *photo* Michel Setboun/ Corbis. **p. 19:** *Figure 1-14* Topham/The Image Works. **p. 21:** *Figure 1-15 (T)* Courtesy KCBD News Channel 11, *(B)* Courtesy hivandhepatitis.com. **p. 23:** *photo* Joos Mind/Getty Images. **p. 24:** *Figure 1-17* Christopher Bissell/Getty Images. **p. 25:** *Figure 1-18 (L)* Image Source Black/Alamy, *(R)* Huy Lam/Getty Images. **p. 26:** *Figure 1-19* The Photo Works. **p. 27:** *Figure 1-20* Reprinted through the courtesy of the Editors of TIME Magazine © 2008 Time Inc. **p. 28:** *photo* Waltraud Grubitzsch/epa/Corbis. **p. 29:** *photo* Martin Shields/Photo Researchers. **p. 30:** *photo* AP Photo/ Daniel Hulshizer.

CHAPTER 2   **pp. 34–35:** *photo* Patrick Phelan. **p. 36:** *photo* M. E. Leunissen et al., Ionic colloidal crystals of oppositely charged particles, *Nature* 437, 235 (2005). Courtesy Mirjam Leunissen and Alfons van Blaaderen, Debye Institute for Nanomaterials Science, Utrecht University, Netherlands; *Figure 2-1 (TL)* Stockfolio/Alamy, *(TR)* Jeffrey Hamilton/Getty Images, *(B)* Burazin/Getty Images. **p. 41:** *Figure 2-8 (L)* Judith Collins/Alamy, *(R)* Dr. Dennis Kunkel/Phototake. **p. 44:** *photo* imagewerks/Getty Images. **p. 45:** *Figure 2-13* Joe McDonald/Animals Animals—Earth Scenes, *(inset)* Visuals Unlimited/Corbis. **p. 49:** *photo* Dr. Dennis Kunkel/Visuals Unlimited. **p. 50:** *photo* Keren Su/Getty Images, *Figure 2-20* Dorling Kindersley/Getty Images. **p. 52:** *Figure 2-22 (T)* frans lemmens/Alamy, *(B)* Lew Robertson/Getty Images. **p. 54:** *Figure 2-25 (L)* Paul Geor/Stockxpert, *(R)* Alison Parks-Whitfield/Alamy; *Figure 2-26 (T)* David Hancock/Alamy, *(B)* Masterfile; *Figure 2-27* Cassenet/age fotostock. **p. 56:** *photo* Patrick J. Endres/Alaskaphotographics.com. **p. 57:** *Figure 2-28* Bryan & Cherry Alexander Photography/Alamy. **p. 58:** *Figure 2-30* Gary Gladstone/Corbis, *Figure 2-31 (L)* Ellen Isaacs/Alamy, *(R)* Masterfile. **p. 59:** *photo* Andrew O'Toole/Taxi/Getty Images. **p. 61:** *Figure 2-34* Hulton Archive/Getty Images. **p. 62:** *photo* Stephanie Pfriender Stylander. **p. 63:** *Figure 2-36 (from top)* Steve Bloom Images/Alamy; Steve Gschmeissner/Photo Researchers; Dr. Dennis Kunkel/Phototake; Dr. Dennis Kunkel/Phototake/Alamy; Deco/Alamy. **p. 64:** *Figure 2-38* Edgardo Contreras/Getty Images. **p. 66:** *Figure 2-41 (L)* Ryan McVay/ Getty Images, *(R)* White Packert/Getty Images. **p. 68:** *photo* Rick Gayle Studio/Corbis. **p. 69:** *photo* Driscoll, Youngquist, & Baldeschwieler/ Caltech/Photo Researchers. **p. 72:** *photo* Gustavo Andrade/age fotostock.

CHAPTER 3   **pp. 76–77:** Phototake/Alamy. **p. 78:** Alfred Pasieka/ Photo Researchers. **p. 79:** *Figure 3-1(L)* Steve Taylor/Getty Images, *(M)* Frans Lanting/Corbis, *(R)* David Scharf/Peter Arnold; *Figure 3-2 (L)* Clark Sumida, *(M)* Rosenfeld/age fotostock, *(R)* Clouds Hill Imaging/Corbis,

(inset) Courtesy Reproductive Solutions. **p. 80:** *Figure 3-3 (inset)* Dr. Dennis Kunkel/Getty Images. **p. 81:** *Figure 3-4 (TL)* Phototake/Alamy, *(TR)* Matt Meadows/Peter Arnold, *(BL)* John Glover/age fotostock, *(BM)* David Tipling/Getty Images, *(BR)* Juniors Bildarchiv/age fotostock. **p. 82** *Figure 3-5 (L)* Dr. Donald Fawcett/Visuals Unlimited, *(R)* Phototake/Alamy. **p. 83** *Figure 3-6 (L inset)* Dr. Gopal Murti/Photo Researchers, *(R inset)* Dr. Henry Aldrich/Visuals Unlimited. **p. 85:** F1 Online/photolibrary. **p. 86:** *Figure 3-8 (L)* Don W. Fawcett/Photo Researchers, *(R)* David McNew/Getty Images. **p. 89:** *Figure 3-12 (L)* Mauro Fermariello/Photo Researchers, *(R)* Jim Evans/*Kennebec Journal & Morning Sentinel*. **p. 90:** *Figure 3-13* Image Source/ Veer. **p. 92:** *Figure 3-15* AP Photo/Jeffrey Boan. **p. 93:** *photo* Courtesy Dr. G. M. Gaietta & T. J. Deerinck, CRBS/NCMIR, University of California, San Diego. **p. 97:** *Figure 3-20* Richard Keppel-Smith/Getty Images. **p. 98:** *Figure 3-21* Biophoto Associates/Photo Researchers. **p. 99:** *Figure 3-23 (L)* Chuck Brown/Photo Researchers, *(M)* Ed Reschke/Peter Arnold, *(R)* Carolina Biological Supply Company/Phototake. **p. 100:** *photo* Jutta Klee/Getty Images. **p. 101:** *photo* Lorenz Britt/Alamy. **p. 102:** *Figure 3-25 (L)* Dr. Dennis Kunkel/Phototake, *(M)* Dr. Dennis Kunkel/Visuals Unlimited, *(R)* B. Gilula and Donald Fawcett/Visuals Unlimited. **p. 103:** *photo* Dr. Dennis Kunkel/Visuals Unlimited. **p. 104:** *Figure 3-26* Phototake/ Alamy; *photo* Mary Ellen Mark. **p. 105:** *Figure 3-27* Lynne Chang/Visuals Unlimited; *Figure 3-28 (L)* Dr. Yorgos Nikas/Phototake, *(R)* Juergen Berger/ Photo Researchers. **p. 106:** *Figure 3-29* Dr. Donald Fawcett/Getty Images. **p. 108:** *Figure 3-30* Dr. Gopal Murti/Photo Researchers. **p. 109:** *Figure 3-31* Kevin Somerville/The Medical File/Peter Arnold. **p. 110:** *Figure 3-32* Dr. Dennis Kunkel/Visuals Unlimited. **p. 111:** *Figure 3-33* Dr. Dennis Kunkel/Phototake. **p. 112:** *Figure 3-34* Dr. Dennis Kunkel/Visuals Unlimited. **p. 114:** *Figure 3-36* Perennou Nuridsany/Photo Researchers; *photo* Bill Hatcher/Getty Images. **p. 115:** *Figure 3-37* Biophoto Associates/ Photo Researchers. **p. 116:** *Figure 3-38* George Chapman/Visuals Unlimited; *photo* Stephen Studd/Getty Images. **p. 118:** *photo* Marco Garcia/ Getty Images.

CHAPTER 4   **pp. 122–123:** *photo* Pete Saloutos/Corbis. **p. 124:** *photo* Inga Spence/Getty Images. **p. 125:** *Figure 4-1* Pat LaCroix/Getty Images. **p. 126:** *Figure 4-3 (TL)* David Madison/Getty Images, *(TM)* Tom Vezo/ Nature Picture Library, *(TR)* Mark Gibson/Getty Images, *(BL)* Shem Compion/Getty Images, *(BR)* Chris Cheadle/age fotostock. **p. 127:** *Figure 4-4* Michael Pohuski/Foodpix/Jupiterimages; *Figure 4-5 (L)* Ed Gifford/ Masterfile, *(M)* David Allio/Icon SMI/Corbis, *(R)* Walter Lockwood/ Corbis. **p. 130:** *photo* Datacraft/age fotostock. **p. 131:** *Figure 4-9 (L)* Wim van Egmond/Visuals Unlimited, *(R)* Dr. Dennis Kunkel/Phototake, *(B)* Dr. Dennis Kunkel/Visuals Unlimited. **p. 131:** *Figure 4-10* brokenarts/ stock.xchng. **p. 132:** *Figure 4-11* Value Stock Images/Fotosearch, *(inset)* Dr. Dennis Kunkel/Visuals Unlimited; *Figure 4-12* George Chapman/ Visuals Unlimited. **p. 134:** *Figure 4-14* Value Stock Images/Fotosearch. **p. 140:** *Figure 4-22* AP Photo/Rajesh Kumar Singh. **p. 141:** *Figure 4-23* Dr. Jeremy Burgess/Science Photo Library; *Figure 4-24 (L)* age fotostock/SuperStock, *(M)* Brian Sytnyk/Masterfile, *(R)* age fotostock/ SuperStock. **p. 143:** *photo* Philip Harvey/Corbis; *Figure 4-26 (L)* age fotostock/SuperStock, *(M)* image100/Alamy, *(R)* Volker Steger/Photo Researchers. **p. 145:** *Figure 4-28 (L)* age fotostock/SuperStock, *(M)* Frans

Lanting Photography, (R) A. Barry Dowsett/Photo Researchers. **p. 148:** *Figure 4-32* hybrid medical animation/Photo Researchers. **p. 151:** *photo* Martin Harvey/Getty Images; *Figure 4-35* Tobias Bernhard/Getty Images. **p. 152:** *Figure 4-37* Cephas Picture Library/Alamy, *(T inset)* SciMAT/ Photo Researchers, *(B inset)* Romilly Lockyer/Getty Images. **p. 153:** *Figure 4-38 (L)* FoodCollection/age fotostock, *(M)* Tim Hill/Alamy, *(R)* FoodCollection/age fotostock. **p. 155:** *photo* Plush Studios/Getty Images.

CHAPTER 5   pp. 158–159: *photo* Phanie/Photo Researchers. **p. 160:** *photo* Denis Galante/Corbis. **p. 161:** *Figure 5-1* Vasna Wilson/*Virginian-Pilot*; *Figure 5-2 (L)* Reprinted with permission, September 2008 *Discover* magazine. Copyright © Discover Magazine. All rights reserved, *(R)* Reprinted through the courtesy of the Editors of TIME Magazine © 2008 Time Inc., *(B)* Reprinted by permission of The Wall Street Journal, © 2008 Dow Jones & Company, Inc. All Rights Reserved Worldwide. License number 2038930425822. **p. 162:** *Figure 5-3* A. Barrington Brown/Photo Researchers, *(inset)* Science Source/Photo Researchers. **p. 164:** *Figure 5-5 (TL)* Courtesy of the author, *(TM)* Sian Irvine/Getty Images, *(TR)* Eye of Science/Photo Researchers, *(BL)* Visuals Unlimited/ Corbis, *(BR)* John White. **p. 166:** *Figure 5-8 (from top)* Courtesy of the author; John White; Eye of Science/Photo Researchers; Sian Irvine/Getty Images; Visuals Unlimited/Corbis. **p. 167:** *Figure 5-9 (from top)* Courtesy of the author; Eye of Science/Photo Researchers; Dr. Dennis Kunkel/ Getty Images; Fancy/Veer/Corbis; Dr. Dennis Kunkel/Visuals Unlimited. **p. 169:** *photo* Julia Kuskin/Photonica/Getty Images. **p. 174:** *photo* Uli Westphal 2008. **p. 175:** *Figure 5-15* Eye of Science/Photo Researchers. **p. 177:** *Figure 5-17 (L)* Mika/zefa/Corbis, *(M)* Robert Llewellyn/Corbis, *(R)* Tom Stewart/Corbis. **p. 178:** *Figure 5-18* Tommy Moorman. **p. 179:** *photo* Mario Tama/Getty Images; *Figure 5-19 (TL)* Getty Images, *(TM)* Jim West/Alamy, *(TR)* Charles O'Rear/Corbis, *(BL)* AP Photo/*Beloit Daily News*, Tom Holoubek, *(BR)* Landov. **p. 184:** *Figure 5-26* AJPhoto/Photo Researchers. **p. 185:** *Figure 5-27* MGM/courtesy Everett Collection. **p. 186:** *Figure 5-28* AP Photo. **p. 187:** *Figure 5-29* AP Photo/Examiner, Amy Elrod. **p. 188:** *photo* Betsie Van der Meer/Getty Images. **p. 189:** *photo* Jim Richardson/Corbis; *Figure 5-30* Ed Young/Corbis, *(inset)* John Doebley. **p. 190:** *Figure 5-31 (clockwise from left)* Martin Ruegner/Digital Vision/Getty Images; Scottish Crop Research Institute, Dundee; Professor Peter Beyer/Humanitarian Board for Golden Rice/www.goldenrice. org. **p. 191:** *Figure 5-32 (L)* BSIP/Photo Researchers, *(M)* Lance Nelson/ Corbis, *(R)* Steve Percival/Photo Researchers. **p. 192:** *Figure 5-33* Dr. Morley Read/Science Photo Library. **p. 193:** *Figure 5-35* Frank Whitney/ Getty Images; *Figure 5-36* Dr. Garth Fletcher. **p. 194:** *Figure 5-37* Reuters/Corbis; *Figure 5-38* Reuters/Corbis. **p. 196:** *photo* Digital Vision/ Alamy. **p. 197:** *Figure 5-39 (L)* Neville Chadwick/Science Photo Library, *(M)* Tom Grill/Corbis, *(R)* David Parker/Science Photo Library. **p. 199:** *Figure 5-42* Vo Trung Dung/Corbis Sygma. **p. 200:** *Figure 5-43 (clockwise)* Sian Irvine/Dorling Kindersley/Getty Images, Visuals Unlimited/Corbis, Eye of Science/Photo Researchers, John White, Courtesy of the author, *(tree of life)* David M. Hillis, Derrick Zwickl, and Robin Gutell, University of Texas. **p. 202:** *Figure 5-45 (TL)* John Chadwick/AP Photo, *(TR)* Ben Margot/AP Photo, *(BL)* Jack Smith/AP Photo, *(BM)* David J. Phillip/AP Photo, *(BR)* Advanced Cell Technologies/AP Photo. **p. 203:** *Figure 5-46* Amblin/Universal/The Kobal Collection. **p. 204:** *(T)* Anthony Marsland/ Getty Images; *(B)* Lauren Nicole/Getty Images.

CHAPTER 6   pp. 208–209: *photo* AP Photo. **p. 210:** *photo* Steve Gschmeissner/Photo Researchers. **p. 211:** *Figure 6-2* AP Photo/Gerald Herbert, *(inset)* Dr. Peter Landsorp/Visuals Unlimited. **p. 213:** *Figure 6-4*

CNRI/Photo Researchers. **p. 216:** *photo* Sisse Brimberg/Getty Images. **p. 217:** *photo* Tanya Constantine/Getty Images. **p. 218:** *Figure 6-7 (L)* Mike Hill/Alamy, *(inset)* Science Photo Library/Photo Researchers, *(R)* Dan Guravich/Photo Researchers, *(inset)* Biology Media/Photo Researchers; *photo* Micro Discovery/Corbis. **p. 219:** *photo* David Scharf/ Getty Images. **pp. 220–221:** *photos* Courtesy T. Wittman. **p. 223:** *Figure 6-11* GJLP/CNRI/Science Photo Library. **p. 224:** *Figure 6-13* Kevin Laubacher/Taxi/Getty Images. **p. 225:** *photo* Dr. Yorgos Nikas/Photo Researchers. **p. 231:** *Figure 6-18* Dr. Dennis Kunkel/Phototake. **p. 233:** *Figure 6-21* Margot Granitsas/Photo Researchers. **p. 234:** *Figure 6-22* Stephen Dalton/NHPA. **p. 234:** *Figure 6-23* Dr. Kari Lounatmaa/Photo Researchers. **p. 235:** *photo* Cormac Hanley/Getty Images. **p. 236:** *Figure 6-24* Biophoto Associates/Photo Researchers. **p. 237:** *Figure 6-26 (L)* Tui De Roy, *(M)* imagebroker/Alamy, *(R)* A. B. Sheldon/Root Resources. **p. 238:** *photo* Michael Patrick O'Neill/Alamy. **p. 239:** *photo* Visuals Unlimited/Corbis. **p. 240:** *Figure 6-27* CNRI/Photo Researchers; *Figure 6-28* Saturn Stills/Photo Researchers; *Figure 6-29 (L)* Mika/zefa/Corbis, *(R)* L. Willatt/East Anglian Regional Genetics Service/Photo Researchers. **p. 244:** *photo* Jupiterimages/Brand X/Alamy.

CHAPTER 7   pp. 248–249 *photo* Robert B. Carr/Bruce Coleman USA. **p. 250:** *photo* Dr. Karl Fredga, Department of Genetics, Uppsala University, Sweden. **p. 251:** *Figure 7-1* Isabelle Rozenbaum/age fotostock. **p. 252:** *Figure 7-4* © BNPS.CO.UK. **p. 253:** *Figure 7-5 (row 1, L)* Les3photo8/Dreamstime.com, *(row 1, R)* Gabriela Medina/Getty Images, *(row 2, L)* Geoff du Feu/Alamy, *(row 2, R)* Custom Medical Stock Photo/Alamy, *(row 3, L & R)* Kevin Winter/Getty Images. **p. 256:** *Figure 7-8* Content Mine International/Alamy. **p. 258:** *Figure 7-10* Jeff Wyoming/photolibrary. **p. 260:** *photo* John Gillmoure/Corbis. **p. 262:** *Figure 7-14* Michael K. Nichols/Getty Images. **p. 264:** *Figure 7-16* Yann Arthus-Bertrand/Corbis. **p. 265:** *photo* Gideon Mendel/Corbis. **p. 266:** *Figure 7-18* Wellcome Images. **p. 270:** *Figure 7-22* Corbis. **p. 271:** *Figure 7-23* Lennart Nilsson. **p. 273:** *Figure 7-25* AP Photo/Rob Carr. **p. 274:** *Figure 7-26 (T)* Juniors Bildarchiv/age fotostock, *(B)* Grant Heilman Photography/Alamy. **p. 275:** *photo* Charles Maraia/Getty Images. **p. 277:** *Figure 7-28* Mina Chapman/Corbis. **p. 278:** *photo* Patrick J. Miller.

CHAPTER 8   pp. 280–281: *photo* Steve Bloom Images/Alamy. **p. 282:** *photo* Frans Lanting Photography. **p. 283:** *Figure 8-1* Graphic Science/ Alamy. **p. 285:** *photo* Tui De Roy/Minden Pictures/Getty Images. **p. 287:** *Figure 8-5* Getty Images; *Figure 8-6* Mary Evans Picture Library/Alamy. **p. 290:** *Figure 8-8* Hulton Archive/Getty Images; *Figure 8-9* Photo by Louie Fasciolo/Bernard J. Shapero Rare Books. **p. 291:** *photo* Tom Brakefield/Science Faction/Getty Images. **p. 292:** *Figure 8-10* Renee Lynn/Corbis. **p. 294:** *Figure 8-11 (L inset)* Florida Images/Alamy, *(R inset)* Courtesy Oregon State University. **p. 295:** *Figure 8-12* J. D. Talasek/Phototake. **p. 296:** *Figure 8-13* Victor McCusick. **p. 297:** *Figure 8-14* Ross Warner/Alamy. **p. 299:** *Figure 8-16 from left)* Deanne Fitzmaurice; Jason Homa/Getty Images; Paul Costello/Getty Images; Paul Burns/Blend Images/Getty Images; Jacob Langvad/Stone/Getty Images. **p. 300:** *Figure 8-17* Max Nash/AFP/Getty Images; *Figure 8-18* Benn Mitchell/Riser/Getty Images. **p. 302:** *photo* Yann Arthus-Bertrand/Corbis. **p. 304:** *Figure 8-22* J & B Photographers/Animals Animals—Earth Scenes. **p. 307:** *Figure 8-25* Patricia McDonough; *Figure 8-26* Grant Heilman/Grant Heilman Photography. **p. 308:** *Figure 8-27* Barros & Barros/Getty Images; *Figure 8-28* Gary Vestal/Photographer's Choice/Getty Images. **p. 310:** *Figure 8-29* RubberBall/Alamy; *Figure 8-30* Michael Durham/Minden Pictures/Getty Images. **p. 311:** *photo* Gerald

Cubitt/NHPA/Photoshot. **p. 312:** *photo* SuperStock/age fotostock.
p. 313: *Figure 8-31 (L)* Jason Edwards/National Geographic/Getty Images, *(R)* Louie Psihoyos/Science Faction/Getty Images. **p. 316:** *Figure 8-35 (from left)* Joe McDonald/Visuals Unlimited; G. Armistead/Vireo; Aaron French; Jack Jeffrey. **p. 317:** *Figure 8-36 (row 1)* Joe McDonald/Visuals Unlimited; Martin Pot; Dave Watts/Alamy; *(row 2)* Leonard Lee Rue III/Photo Researchers; Morales/age fotostock; John W. Warden/age fotostock. **p. 319:** *Figure 8-37 (row 1)* Jose I. Castro; Laboratory of Evolutionary Morphology; *(row 2)* courtesy Bradley Smith, University of Michigan; S. K. Ackerley Department of Integrative Biology, University of Guelph, Guelph, Canada. **p. 319:** *Figure 8-38 (from left)* PBNJ Productions/PBNJ/age fotostock; Juniors Bildarchiv/age fotostock; Malcolm Schuyl/Alamy; Juniors Bildarchiv/age fotostock. **p. 320:** *Figure 8-39* Adrian Warren/Ardea; *Figure 8-40 (T)* It Stock Free/age fotostock, *(B)* Kevin Schafer/Getty Images. **p. 321:** *Figure 8-41 (from left)* Getty Images; Arco Images/Alamy; PhotoSpin/Alamy; Danita Delimont/Alamy; Ulrike Klenke and Zeev Pancer, Center of Marine Biotechnology, Baltimore. **p. 322:** *photo* Lakeside/Dreamstime.com.

CHAPTER 9   pp. 328–329: *photo* Luiz Claudi Marigo. **p. 330:** *photo* DLILLC/Corbis; *Figure 9-1* AP Photo/Independence Daily Reporter/Rob Morgan. **p. 331:** *Figure 9-2* Radius Images/Alamy; *Figure 9-3* Isabel M. Smallegange. **p. 333:** *Figure 9-4 (L)* Juniors Bildarchiv/Alamy, *(TR)* Andrew Darrington/Alamy, *(BR)* ARCO/H. Reinhard/age fotostock. **p. 335:** *Figure 9-6* Peter Cade/Getty Images, *(TI)* Joe McDonald/Getty Images, *(BI)* Lew Robertson/Corbis. **p. 337:** *photo* Annie Griffiths Belt/Corbis. **p. 338:** *photo* Joe McDonald/Corbis. **p. 339:** *Figure 9-8* Dr. Theodore Evans. **p. 340:** *Figure 9-9* Richard R. Hansen/Photo Researchers. **p. 343:** *Figure 9-12* Courtesy Gerald Carter. **p. 344:** *Figure 9-13* Anup Shah/Animals Animals—Earth Scenes. **p. 346:** *Figure 9-14* PA Wire URN:5620400 (Press Association via AP Images). **p. 348:** *photo* Frans Lanting/Corbis. **p. 349:** *Figure 9-16* David M. Phillips/Photo Researchers. **p. 350:** *Figure 9-18 (L)* ZZZD/Minden Pictures, *(M)* Michael and Patricia Fogden/Corbis, *(R)* Frans Lanting/Corbis, *(MI)* Dieter Herrmann/age fotostock. **p. 351:** *Figure 9-19 (L)* AP Images/Fox Searchlight, *(R)* Universal Films/Miramax Pictures/Kobal Collections; *Figure 9-20* Michael Fogden/photolibrary. **p. 353:** *Figure 9-21 (TL)* Anthony Mercieca/Photo Researchers, *(TR)* Dr. John Alcock/Visuals Unlimited, *(BL)* Photolibrary, *(BR)* Paul Stover/Getty Images. **p. 355:** *Figure 9-22* Michael and Patricia Fogden/Corbis; *Figure 9-23* James H. Robinson/Photo Researchers. **p. 356:** *Figure 9-24* George D. Lepp/Corbis. **p. 357:** *Figure 9-25* M. Botzek/age fotostock. **p. 359:** *Figure 9-27 (L)* Colla/Oceanlight.com, *(R)* ARCO/W. Layer/age fotostock. **p. 360:** *photo* Harald Lange/zefa/Corbis. **p. 361:** *Figure 9-28 (L)* Grant Heilman Photography/Alamy, *(M)* Juniors Bildarchiv/Alamy, *(R)* Jupiterimages/Creatas/Alamy. **p. 362:** *Figure 9-20 (L)* Scott Camazine/Photo Researchers, *(M)* Copyright 2008 The Gorilla Foundation/Koko.org. All rights reserved, *(R)* Michael Newman/PhotoEdit. **p. 363:** *Figure 9-30* Werner Bollmann/age fotostock. **p. 364:** *photo* Titus Lacoste/Getty Images.

CHAPTER 10   pp. 368–369: Kellar Autumn, Lewis & Clark College. **p. 370:** *photo* Peter Sawyer/NMNH/Smithsonian Institution; *Figure 10-1* David Muench/Corbis. **p. 371:** *Figure 10-2* Roger Ressmeyer/Corbis. **p. 372:** *Figure 10-3* Courtesy Andrew Knoll. **p. 373:** *Figure 10-4* Science VU/Sidney Fox/Visuals Unlimited. **p. 374:** *photo* flowerphotos/Alamy; *Figure 10-5 (from left)* Peter Turnley/Corbis; Steve Sands/New York Newswire/Corbis; Ron Evans/Getty Images; GAP Photos/Getty Images; Andrew Darrington/Alamy; Tom Viggars/Alamy. **p. 375:** *Figure 10-6* Galen Rowell/Corbis; *Figure 10-7* Bill Dow/Shambala Preserve/Roar

Foundation. **p. 376:** *Figure 10-8 (L)* Juniors Bildarchiv/Alamy, *(R)* Masterfile. **p. 380:** *Figure 10-11 (L)* John Cancalosi/age fotostock, *(R)* Wildlife/Peter Arnold. **p. 381:** *Figure 10-12 (from left)* Kevin Schafer/Corbis; Tui De Roy/Minden Pictures; Hiroya Minakuchi/Minden Pictures; Tui De Roy/Minden Pictures; Tui De Roy/Minden Pictures; Michael Woodruff. **p. 383:** *photo* Fancy/Veer/Corbis. **p. 385:** *Figure 10-15 (tree of life)* David M. Hillis, Derrick Zwickl, and Robin Gutell, University of Texas. **p. 386:** *Figure 10-17 (from left)* DLILLC/Corbis; Nigel Dennis/age fotostock; James Gritz/Getty Images; Altrendo/Getty Images. **p. 387:** *photo* Courtesy The Broad Institute of MIT and Harvard. **p. 388:** *Figure 10-19* Michael Durham/Minden Pictures; *Figure 10-20* David Hosking/Alamy; IFA-Bilderteam/Jupiterimages; Michael and Patricia Fogden/Corbis. **p. 389:** *photo* Kim Taylor/naturepl.com. **p. 390:** *photo* Daniel L. Geiger/SNAP/Alamy; *Figure 10-21* David Edwards/Getty Images. **p. 395:** *Figure 10-26* Reuters/Corbis. **p. 396:** *photo* Norbert Wu/Minden Pictures. **p. 399:** *Figure 10-29* Radius Images/Alamy; *Figure 10-30* David McCarthy/Photo Researchers. **p. 400:** *Figure 10-31* John Wang/Getty Images, *(inset)* Agriculture et Agroalimentaire Canada. **p. 401:** *Figure 10-32 (from left)* Michael and Patricia Fogden/Corbis; from *Botanica* by Howard Schatz, © Schatz Ornstein 2005; Adam Jones/Photo Researchers; Frans Lanting/Corbis. **p. 402:** *photo* David Sacks/Getty Images.

CHAPTER 11   pp. 406–407: *photo* Stephen Frink/zefa/Corbis. **p. 408:** *photo* Scott Chandler. **p. 409:** *Figure 11-1(from left)* Jeffrey L. Rotman/Corbis, Frans Lanting/Corbis, Art Wolfe/Getty Images. **p. 411:** *Figure 11-3 (L)* Steve Maslowski/Photo Researchers, *(R)* Thomas Mangelsen/Minden Pictures. **p. 412:** *photo* Frans Lanting/Corbis. **p. 413:** *Figure 11-4 (L)* Images&Stories/Alamy, *(inset)* Art Wolfe/Getty Images; *(R)* Frans Lanting Photography, *(inset)* Ross Armstrong/photolibrary; *Figure 11-5 (from left)* Tim Laman/Getty Images, Natural Visions/Alamy, Jack Milchanowski/photolibrary. **p. 414:** *photo* Frans Lanting/Corbis. **p. 415:** *Figure 11-7 (L)* Darlyne A. Murawski/Getty Images, *(R)* Masa Ushioda/SeaPics.com. **p. 416:** *Figure 11-8 (from top)* Paul Souders/Corbis, Tim Fitzharris/Minden Pictures,) Rudie Kuiter/OceanwideImages.com. **p. 418:** *photo* Frans Lanting/Corbis; *Figure 11-10 (from left)* Fabio Liverani/Minden Pictures, Dorling Kindersley/Getty Images, Visuals Unlimited/Corbis. **p. 420:** *Figure 11-12 (both photos)* Frans Lanting/Corbis. **p. 422:** *Figure 11-15 (from left)* Dave Watts/Alamy, Martin Rugner/Photolibrary, Frans Lanting Photography. **p. 427:** *photo* Frans Lanting/Corbis. **p. 429:** *Figure 11-20 (clockwise from left)* DLILLC/Corbis, Stephen Frink/Corbis, Chris Newbert/Minden Pictures. **p. 430:** *Figure 11-22 (clockwise from left)* George Grall/Getty Images, blickwinkel/Alamy, William James Warren/Getty Images, Chris Newbert/Minden Pictures; *Figure 11-23 (L)* Frans Lanting/Corbis, *(M)* Alamy, *(R)* Alamy. **p. 432:** *Figure 11-24 (from left)* Piotr Naskrecki/Minden Pictures, Alamy, Javier Tajuelo/photolibrary, Tom Vezo/Minden Pictures. **p. 433:** *Figure 11-25 (clockwise from left)* Gerry Ellis/Minden Pictures, Jeff Rotman/Getty Images, David B. Fleetham/SeaPics.com, Walter E. Harvey/Photo Researchers. **p. 434:** *Figure 11-26 (from left)* David B. Fleetham/photolibrary, Adam Butler/Alamy, Doug Perrine/SeaPics.com. **p. 435:** *Figure 11-27* Jorgen Jessen/AFP/Getty Images. **p. 436:** *photo* Scott Leslie/Minden Pictures. **p. 437:** *Figure 11-28 (clockwise from left)* Ross and Diane Armstrong/Minden Pictures, ARCO/H. Reinhard/age fotostock, Konrad Wothe/Minden Pictures; *Figure 11-29 (L)* Lshtm/Photo Researchers, *(R)* CNRI/Photo Researchers. **p. 438:** *Figure 11-30 (L)* Sinclair Stammers/Photo Researchers, *(R)* Alamy. **p. 439:** *Figure 11-31 (clockwise from left)* Brandon Cole/Getty Images, Tim Laman, Frans Lanting/Corbis. **p. 440:** *Figure 11-32* Howard Hall/photolibrary. **p. 441:** *Figure 11-33 (L)* Juan Carlos Calvin/photolibrary, *(R)* Robert Yin/Corbis.

p. 442: *photo (T)* Dr. David Wachenfeld/Auscape/Minden Pictures, *(B)* Zeva Oelbaum/Peter Arnold.

CHAPTER 12   p. 446–447: *photo* Katinka Matson. **p. 448:** *photo* Frans Lanting Photography; *Figure 12-1 (from left)* David Pearson/Alamy, Douglas Pulsipher/Alamy, Michael and Patricia Fogden/Minden Pictures. **p. 449:** *Figure 12-2* William James Warren/Getty Images; *Figure 12-3 (from left)* Photodisc/Alamy, Chad Ehlers/Alamy, Herbert Hopfensperger/age fotostock. **p. 451:** *photo* Frans Lanting/Corbis; *Figure 12-5* Charles F. Delwiche. **p. 452:** *Figure 12-6 (L)* Hans Strand/Corbis, *(R)* Gallo Images/Getty Images. **p. 453:** *Figure 12-7 (L)* Alistair Dove/Alamy, *(M)* Verbiesen Henk/age fotostock, *(R)* Daniel Vega/age fotostock. **p. 454:** *Figure 12-8* Barry Turner/Alamy; *Figure 12-9 (T)* The Irish Image Collection/Design Pics/Corbis, *(B)* Matthew Mawson/Alamy. **p. 455:** *Figure 12-10 (L)* Ed Reschke/Peter Arnold, *(M)* DEA/Dani-Jeske/Getty Images, *(R)* Juan Carlos Muñoz/age fotostock; *photo* Bjorn Forsberg/npl/Minden Pictures. **p. 456:** *Figure 12-11* Peter Arnold/Alamy. **p. 457:** *photo* Jeff Vanuga/Corbis; *Figure 12-12* Kevin Phelan. **p. 458:** *Figure 12-13 (from left)* Bohemian Nomad Picturemakers/Corbis, Design Pics/Alamy, Jouan & Rius/naturepl.com, Takahiro Miyamoto/Getty Images. **p. 459:** *Figure 12-14 (from left)* Thomas Kitchin and Victoria Hurst/AllCanadaPhotos.com, Alan and Linda Detrick/Photo Researchers, Jouan & Rius/naturepl.com, blickwinkel/Alamy; *Figure 12-15 (L)* Brad Mogen/Visuals Unlimited, *(R)* Stephen P. Parker/Photo Researchers. **p. 460:** *Figure 12-16* Darrell Gulin/Corbis. **p. 461:** *Figure 12-17 (T)* Gary Randall/Getty Images, *(M)* Kathleen Phelan, *(B)* Philippe Bourseiller/Getty Images. **p. 462:** *photo* Andy Small/Corbis; *Figure 12-18 (from left)* Ferruccio Carassale/Grand Tour/Corbis, blickwinkel/Alamy, Grand Tour/Corbis, Jack Dykinga/Getty Images. **p. 463:** *photo* John Fairhall/Auscape/Minden Pictures. **p. 464:** *Figure 12-20* Topic Photo Agency/age fotostock. **p. 465:** *Figure 12-21 (top row from left)* Kim Taylor/naturepl.com, Rolf Nussbaumer/naturepl.com, Andrew Darrington/Alamy, Sharkawi Che Din/Alamy; *(bottom row from left)* blickwinkel/Alamy, Mark Moffett/Minden Pictures. **p. 466:** *Figure 12-22* Tony Hertz/Corbis. **p. 468:** *photo* Dezorzi/Corbis. **p. 469:** *Figure 12-24 (from left)* Scott Camazine/Alamy, Steve Bloom Images/Alamy, Manfred Danegger/Peter Arnold. **p. 470:** *Figure 12-25 (TL)* Joe McDonald/Visuals Unlimited, *(TR)* Wildlife GmbH/Alamy, *(M)* Howard Grey/Getty Images, *(B)* Gertrud and Helmut Denzau/naturepl.com. **p. 471:** *Figure 12-26* Ed Reschke/Peter Arnold. **p. 472:** *photo* Jan Vermeer/Foto Natura/Minden Pictures; *Figure 12-27 (from left)* Andrew Parkinson/Corbis, Yva Momatiuk and John Eastcott/Photo Researchers, Glen Threlfo/Minden Pictures. **p. 473:** *Figure 12-29 (from left)* Steve Austin/Papilio/Corbis, Visuals Unlimited/Corbis, Visuals Unlimited/Corbis. **p. 474:** *Figure 12-30* Ingo Arndt/Minden Pictures. **p. 475:** *Figure 12-31* AP Photo/Oregon State University; *Figure 12-32* D. Hurst/Alamy. **p. 476:** *Figure 12-33* Visuals Unlimited/Corbis. **p. 477:** *Figure 12-34* Jim Brandenburg/Minden Pictures. **p. 478:** *photo* Mary Ellen Mark.

CHAPTER 13   pp. 482–483: *photo* Lee D. Simon/Photo Researchers. **p. 484:** Stephanie Schuller/Photo Researchers. **p. 485:** *Figure 13-2 (from left)* CNRI/Photo Researchers, *(inset)* Phototake/Alamy; Federico Veronesi/Getty Images, *(inset)* Agriculture et Agroalimentaire Canada; Science Source/Photo Researchers, *(inset)* Courtesy K.O. Stetter, R. Huber, and R. Rachel, University of Regensburg, Germany. **p. 486:** *Figure 13-3* Courtesy Jay Phelan; *photo* Frans Lanting/Corbis. **p. 487:** *Figure 13-4 (T from left)* Dennis Kunkel/Alamy, Dennis Kunkel/Alamy, Biodisc/Visuals Unlimited/Alamy, *(bottom)* Eye of Science/Photo Researchers. **p. 488:** *photo* Visuals Unlimited/Corbis. **p. 489:** *Figure 13-5* Dennis Kunkel/Alamy; *Figure 13-6 (L)* mediacolor/Alamy, *(R)* Raymond Otero/Visuals Unlimited. **p. 492:** *Figure 13-9* Mediablitz Images. **p. 493:** *Figure 13-10(from left)* Dodie Ulery, *(M)* Thomas Dressler/age fotostock, *(R)* Reino Hanninen/Alamy. **p. 494:** *Figure 13-11* Original map by Dr. John Snow, *(inset)* Justinc. **p. 495:** *Figure 13-12* ImageState Royalty Free/Alamy. **p. 496:** *Figure 13-14* AGStockUSA/Alamy. **p. 497:** *photo* George Steinmetz/Corbis; *Figure 13-15* Dennis Kunkel/Visuals Unlimited. **p. 498:** *Figure 13-16* Russ Bishop/age fotostock; *Figure 13-17* Tim Mantoani/Masterfile. **p. 499:** *Figure 13-18* Salvador de Sas/epa/Corbis. **p. 500:** *photo* Steve Gschmeissner/Photo Researchers; *Figure 13-19* Microfossil Image Recovery and Circulation for Learning and Education (MIRACLE) Project, Micropalaeontology Unit of the Earth Sciences Department, University College, London. **p. 501:** *Figure 13-20 (from left)* Visuals Unlimited/Corbis, C-FH/Grant Heilman, Images&Stories/Alamy. **p. 502:** *Figure 13-21* M. I. Walker/Photo Researchers. **p. 503:** *Figure 13-22)* Eye of Science/Photo Researchers, *(top inset)* Peter Arnold/Alamy, *(bottom inset)* Phototake/Alamy. **p. 504:** *photo* NIBSC/Photo Researchers; *Figure 13-23 (M)* SPL/Photo Researchers, *(R)* Phototake. **p. 506:** *Figure 13-25 (T)* St. Louis Post Dispatch, *(B)* Reuters/Corbis. **p. 507:** *Figure 13-27* Bettmann/Corbis, *(inset)* nagelestock.com/Alamy. **p. 509:** *Figure 13-29* Dr. Klaus Boller/Photo Researchers. **p. 510:** *photo* Tommy Moorman.

CHAPTER 14   pp. 514–515: *photo* Frans Lanting Photography. **p. 516:** *photo* Barrett & MacKay/age fotostock; *Figure 14-1* Frans Lanting/Corbis. **p. 518:** *Figure 14-3* Tom Nebbia/Corbis. **p. 519:** *Figure 14-4* Juan Carlos Muñoz/age fotostock. **p. 520:** *Figure 14-5* R. Wittek/age fotostock. **p. 521:** *Figure 14-7* John McColgan, Bureau of Land Management, Alaska Fire Service. **p. 522:** *Figure 14-8* Bill Stormont/Corbis; *Figure 14-9* AP Photo/Dario Lopez-Mills. **p. 523:** *Figure 14-10* Lynn M. Stone/Naturepl.com; *Figure 14-11* age fotostock/Fotosearch. **p. 524:** *Figure 14-12* Dirk Anschutz/Getty Images. **p. 525:** *Figure 14-13* National Museum of Natural History, Smithsonian Institution; *Figure 14-14* Martin Bwenetti/AFP/Getty Images. **p. 526:** *photo* Frans Lanting/Corbis. **p. 527:** *Figure 14-15 (T)* Dave Watts/Naturepl.com, *(M)* Colin Preston/naturepl.com, *(B)* Joe McDonald/Visuals Unlimited. **p. 528:** *Figure 14-16* Greg Katsoulis. **p. 529:** *Figure 14-17 (L)* Martin Harvey/Alamy, *(M)* Terry Wall/Alamy, *(R)* D. R. Schrichte/SeaPics.com. **p. 530:** *Figure 14-18 (L)* Gunter Marx Photography/Corbis, *(M)* Philippe Clement/naturepl.com, *(R)* Gary Meszaros/Photo Researchers, *(inset)* iStockPhoto.com/jeridu. **p. 531:** *photo* Jim Brandenburg/Minden Pictures. **p. 532:** *photo* Gabriela Staebler/zefa/Corbis. **p. 533:** *Figure 14-19* Condé Nast Archive/Corbis. **p. 535:** *Figure 14-21 (T)* Renaud Visage/Getty Images, *(B)* W. Perry Conway/Corbis. **p. 537:** *photo* Eric Fougére/Kipa/Corbis. **p. 538:** *photo* Getty Images. **p. 543:** *Figure 14-27 (T)* Yadid Levy/Anzenberger Agency/Jupiterimages, *(M)* Sean Gallup/Getty Images, *(B)* Michael S. Yamashita/Getty Images; *Figure 14-28* Keren Su/Getty Images. **p. 544:** *photo* Elke Hesser/Getty Images.

CHAPTER 15   pp. 548–549: *photo* Georgette Douwma/Getty Images. **p. 550:** *photo* Frans Lanting/Corbis; *Figure 15-1* Thomas Marent/Minden Pictures. **p. 551:** *Figure 15-2* Steve Allen/Getty Images. **p. 552:** *Figure 15-3 (top row from left)* Jacques Jangoux/Alamy, Ted Med/Getty Images, Fritz Poelking/photolibrary, *(middle row from left)* John Warburton-Lee Photography/Alamy, Michael Melford/Getty Images, Visions LLC/photolibrary, *(bottom row from left)* Dan Lamont/Corbis, Frans Lanting/Corbis, Peter Arnold/Alamy. **p. 553:** *Figure 15-4 (top row)* Martin Siepmann/photolibrary, Frans Lanting/Corbis, *(bottom row)* Troy Plota/Getty Images,

Momatiuk-Eastcott/Corbis, Mark Conlin/V&W/Image Quest Marine. **p. 554:** *photo* Frans Lanting/Corbis. **p. 555:** *Figure 15-6* moodboard/Corbis. **p. 556:** *Figure 15-7* Craig Aurness/photolibrary, *photo* Frans Lanting/Corbis. **p. 557:** *Figure 15-8* Jacques Descloitres, MODIS Land Rapid Response Team, NASA/GSFC. **p. 558:** *Figure 15-9 (L)* Courtesy NASA/Goddard Space Flight Center Scientific Visualization Studio, *(R)* Sandy Felsenthal/Corbis. **p. 561:** *photo* Martin Harvey/Corbis. **p. 562:** *Figure 15-13 (L)* AJPhoto/Photo Researchers, *(R)* Eye of Science/Photo Researchers. **p. 568:** *Figure 15-19* AP Photo/Gautam Singh, *photo* P. Manner/zefa/Corbis. **p. 569:** *photo* Stephen Frink/Corbis. **p. 569:** *Figure 15-20* Mitsuhiko Imamori/Minden Pictures. **p. 570:** *Figure 15-21* Jim Brandenburg/Minden Pictures. **p. 574:** *Figure 15-24 (top row)* Anthony Bannister Gallo Images/Corbis, Suzanne L. and Joseph T. Collins/Photo Researchers, *(bottom row)* Michael & Patricia Fogden/Minden Pictures, Gail Shumway/Getty Images. **p. 575:** *Figure 15-25 (T)* Fred Bavendam/Minden Pictures, *(B)* John Downer/naturepl.com; *Figure 15-26* Dr. Bruce Robison/Minden Pictures. **p. 576:** *Figure 15-27 (T)* Gallo Images/Alamy, *(B)* Eye of Science/Photo Researchers. **p. 577:** *Figure 15-28* Biosphoto/Heuclin Daniel/Peter Arnold. **p. 578:** *Figure 15-29 (T)* Darlyne A. Murawski/Getty Images, *(B)* Arco Images GmbH/Alamy. **p. 579:** *photo* W. Perry Conway/Corbis. **p. 581:** *Figure 15-31 (inset)* Siede Preis/Getty Images. **p. 582:** *photo* NASA image by Robert Simmon, based on Landsat data provided by the UMD Global Land Cover Facility.

CHAPTER 16 pp. 586–587: *photo* Peet Simard/Corbis. **p. 588:** *photo* Martin Harvey/Corbis; *Figure 16-1* Burger/Phanie/Photo Researchers, *(inset)* Inga Spence/Getty Images. **p. 589:** *Figure 16-2 (clockwise from top left)* Wildscape/Alamy, Suzanne Long/Alamy, Jeff Vanuga/Corbis, The Photo Works. **p. 590:** *Figure 16-3 (from left)* Alfred Pasieka/Photo Researchers, DLILLC/Corbis, NASA. **p. 591:** *Figure 16-4* Dennis Macdonald/photolibrary.com. **p. 593:** *Figure 16-7 (TR)* blickwinkel/Alamy, *(BL)* Michael and Patricia Fogden/Corbis, *(BR)* DeAgostini/Getty Images. **p. 594:** *Figure 16-8* Courtesy Edward O. Wilson. **p. 595:** *Figure 16-9* Danny Lehman/Corbis. **p. 597:** *photo* Gallo Images/Corbis; *Figure 16-11* Tom Uhlman/Alamy. **p. 600:** *Figure 16-14* Marcelo Sayao/epa/Corbis; *Figure 16-15* Frans Lanting Photography. **p. 601:** *photo* Frans Lanting/Corbis. **p. 602:** *Figure 16-16 (from left)* USGS Photo by Harry Glicken, USGS Photo by Harry Glicken, Don Geyer/Alamy; *Figure 16-17* Denis Scott/Corbis. **p. 603:** *Figure 16-18* Jany Sauvanet/Photo Researchers, *(inset)* Gale S. Hanratty/Alamy. **p. 604:** *Figure 16-20 (from left)* John Mitchell/Photo Researchers, Michael P. Gadomski/Photo Researchers, Wil Meinderts/Minden Pictures. **p. 605:** *Figure 16-21* Jack Picone/Alamy. **p. 606:** *Figure 16-22* Anil Risal Singh-UNEP/Peter Arnold. **p. 607:** *Figure 16-25* Will and Deni McIntyre/Corbis. **p. 609:** *Figure 16-28 (both photos)* NASA. **p. 610:** *Figure 16-30 (all photos)* NASA. *Figure 16-31* AP Photo/Elise Amendola, *(I)* New Line/Photofest. **p. 613:** *Figure 16-34* Frans Lanting/Corbis. **p. 614:** *photo* JASON LEE/X01757/Reuters/Corbis. **p. 615:** *Figure 16-35 (T)* Cyril Ruoso/Minden Pictures, *(B)* Kim Taylor/Minden Pictures, *(inset)* Medical-on-Line/Alamy; *Figure 16-36 (L)* Gerry Ellis/Minden Pictures, *(R)* Robert Harding Picture Library/Alamy. **p. 616:** *Figure 16-37* Tommy Moorman. **p. 618:** *photo* Peter Arnold/Alamy.

Answers

Chapter 1
1. b; 2. b; 3. d; 4. e; 5. c; 6. e; 7. a; 8. a; 9. b; 10. a; 11. d; 12. c; 13. b; 14. e; 15. a; 16. d; 17. a

Short-Answer Questions
1. To provide proper controls, a group of participants must receive a placebo that superficially resembles the medication. The control group should be treated exactly the same as the experimental group, with one exception: the control group receives a placebo instead of the drug. This provides a basis of comparison between the experimental and control groups.

2. There are many possible answers. Four possibilities are: adding a control group; extending the test for a longer period of time; standardizing the subjects to make sure that other variables are not responsible for the weight loss; repeating the study.

3. The common use of the term "theory" often implies that it is simply a hunch about something. In a scientific context, "theory" is applied only when a hypothesis has received significant support from multiple sources over time.

Chapter 2
1. d; 2. e; 3. d; 4. e; 5. b; 6. a; 7. c; 8. a; 9. d; 10. c; 11. b; 12. a; 13. a; 14. a; 15. a; 16. a; 17. a; 18. e; 19. e; 20. d; 21. b

Short-Answer Questions
1. a) The sodium atom has one electron in its outermost shell; the fluorine atom has seven electrons in its outermost shell. b) An ionic bond forms between these two atoms. Neither atom has a full outer electron shell, and both are unstable. The atoms interact to achieve a full outer shell of electrons: the single electron in the outermost shell of the sodium is transferred to the outermost shell of the fluorine. Each atom becomes an ion, and the oppositely charged ions attract each other.

2. First, cohesion is a result of polar water molecules easily bonding together by hydrogen bonds; this allows one water molecule to "connect" with the next as water makes its way up from the soil to the leaves of a tall tree. Second, water has a large capacity to hold heat and resists large fluctuations in temperature; this is because added energy, in the form of heat, does not increase the movement of water molecules, as occurs in most substances. Instead, hydrogen bonds between water molecules are briefly broken and then re-form with nearby water molecules. Third, ice is less dense than water, so it floats; this is because polar water molecules are held apart by hydrogen bonds when the temperature drops (see Figure 2-15). Fourth, water is an excellent solvent because water molecules are polar. Polar molecules with charged regions are attracted to the charged regions of other polar molecules. Many substances contain polar molecules, which easily dissolve in water.

3. Water is a polar molecule and is attracted to other polar molecules, or molecules with charged regions. Oil consists of non-polar molecules.

As a result, polar water molecules will not interact with oil, and water by itself will not be able to dissolve oily dirt.

4. Lettuce and celery contain carbohydrates, which provide a source of energy or fuel for the body. Strawberries also contain carbohydrates, but more monosaccharides versus the complex carbohydrates of lettuce and celery. Walnuts and chicken are sources of protein. The amino acid building blocks of protein are used to build many structures and molecules required by all cells. The salad dressing contains lipids, which provide the body with energy.

Chapter 3
1. e; 2. b; 3. e; 4. a; 5. d; 6. d; 7. c; 8. a; 9. d; 10. d; 11. b; 12. d; 13. c; 14. e; 15. b; 16. b; 17. e; 18. b

Short-Answer Question
1. See the organelle review table, Figure 3-39 on page 117.

Chapter 4
1. a; 2. e; 3. e; 4. c; 5. a; 6. c; 7. a; 8. b; 9. d; 10. c; 11. b; 12. e; 13. b; 14. b; 15. a; 16. a; 17. e; 18. d

Short-Answer Questions
1. The plant can split the water molecules in the "photo" reactions of photosynthesis, producing oxygen gas, which can be used for aerobic respiration by both the plant and the mouse. The carbon dioxide produced as a by-product of aerobic respiration by the plant and the mouse can be used by the plant in the "synthesis" reactions, which ultimately produce glucose for the plant.

2. a) Water molecules are split during the "photo" reactions and oxygen gas is released. Measuring the amount of O_2 provides an indication of the rate of the "photo" reactions. b) Plants incorporate CO_2 in the "synthesis" reactions. Using the energy from the "photo" reactions, plants modify the CO_2 to (eventually) form glucose. The "synthesis" reactions are driven by the energy from the "photo" reactions. c) With additional CO_2, the rate of the "synthesis" reactions will increase, which in turn will increase the amount of energy needed from the "photo" reactions.

3. Photosynthesis involves the production of glucose, and cellular respiration involves the breakdown of glucose. Growth in plants is dependent on the rate of photosynthesis. The faster this process occurs, the more glucose is made, the more plant material can be synthesized, and the more the plant grows. If a plant is growing, the rate of photosynthesis exceeds the rate of cellular respiration.

Chapter 5
1. e; 2. e; 3. b; 4. c; 5. d; 6. d; 7. e; 8. b; 9. c; 10. b; 11. a; 12. e; 13. c; 14. a; 15. d; 16. c

Short-Answer Questions
1. Thymine makes up 30% of the bases. Cytosine and thymine have only one complementary base, so the base and its complement occur in

equal numbers. Because cytosine makes up 20%, its complementary base, guanine, is also 20%. The total of cytosine and guanine is 40%, so the remaining bases total 60%. Because thymine and adenine must occur in equal numbers, half of this total (30%) must be thymine and the other half (30%) adenine.

2. The mRNA is UAG CUG CCU AGG. mRNA allows instructions from DNA to be delivered to ribosomes in the cytoplasm, where it directs the assembly of the protein it encodes.

3. The sequence of codons of the mRNA is translated into a sequence of amino acids that forms a polypeptide chain. As each codon of the mRNA is read by a ribosome, a transfer RNA with an anticodon complementary to the mRNA codon adds a specific amino acid to the growing polypeptide chain.

4. This is an example of a point mutation. The adenine in the fourth codon from the left has been replaced with thymine. This change results in coding for a different amino acid, changing the type of protein produced by the gene—because the amino acid sequence has changed.

5. First, recombinant DNA technology is used to make certain medicines, such as insulin, growth hormone, and erythropoietin, with much less expense; this enables more people to purchase and benefit from these medicines. Second, there have been many efforts to cure some diseases caused by genetic abnormalities using gene therapy; although this technology has not met with success, the future still holds potential. Third, genetic testing can be conducted on parents to identify whether they are at risk of producing offspring with certain genetic diseases. Prenatal genetic screening can be conducted to determine whether a baby will be born with a genetic disease. Adults can also be tested to determine whether they carry genes that put them at risk of developing a disease. With genetic screening, preventive measures and treatments can be undertaken.

Chapter 6

1. c; 2. b; 3. d; 4. d; 5. c; 6. a; 7. b; 8. e; 9. d; 10. b; 11. b; 12. c; 13. d; 14. e; 15. c; 16. a; 17. e; 18. d

Short-Answer Questions

1. Crossing over occurs as sister chromatids of homologous chromosomes come together and swap equivalent segments containing the same genes (but possibly different alleles) during the first prophase of meiosis. This results in an uneven distribution of genetic material from maternal and paternal chromosomes. Gametes could potentially receive more genes from one (either) biological parent than the other. Also, the random division of the homologues (with respect to whether they are maternal or paternal) at anaphase I can lead to gametes having greater than or less than exactly 50% maternal and 50% paternal genetic material. For every gamete having a higher percentage of its genetic material from the maternal side, there is a gamete having an equally higher percentage of paternal genetic material.

2. There cannot be any essential genetic information on the Y chromosome—because all females are able to live without a Y chromosomes—but there is on the X chromosome. Receiving only the Y chromosome means that essential genetic information from the X chromosome is absent.

3. An organism is more likely to reproduce asexually in the absence of adverse environmental pressures. Should some environmental

pressure develop, the organism may start reproducing sexually, producing genetically diverse offspring that are better able to survive environmental pressures. An experiment could be conducted with an organism that has the capacity to switch between sexual and asexual reproduction. This organism could be placed in an environment with ideal conditions and another with adverse conditions. If the hypothesis is correct, the organism will reproduce asexually under ideal conditions and sexually under adverse conditions.

Chapter 7

1. a; 2. b; 3. e; 4. e; 5. a; 6. c; 7. e; 8. a; 9. c; 10. e; 11. e; 12. d; 13. d; 14. b; 15. d; 16. a

Short-Answer Questions

1. a) SMA must be a recessive genetic disorder, because the phenotype of both parents is normal—that is, they are not affected by SMA. For two unaffected parents to have children with the disease, both parents' genotypes must be heterozygous. This is an example of a disease-causing allele that doesn't always reveal itself in a population. b) Individuals who are heterozygous for a recessive genetic disorder (i.e., are carriers) will not necessarily have children with the disease. However, this example demonstrates heterozygous parents with children who inherit the disease-causing allele from both parents, resulting in SMA. The probability of having another biological child with SMA is the same as it was for each of the first two children. The probability of inheriting the recessive allele from the mother is 0.5, and the probability of inheriting the recessive allele from the father is 0.5. Therefore the overall probability is 0.25, 25%, or 1 in 4.

2. The phenotype, or color of the blooms, is affected by differences in the environment. Hydrangea flower color varies with the type of soil, because the pH of the soil affects the pigmentation of the blooms.

3. a) Genetic disorders that involve genes on the X chromosome affect men differently from women. If a female inherits an allele that produces a non-functional protein on one of her X chromosomes, she may not be affected *if* the second X chromosome has the normal allele. However, because males have only one X chromosome, they have only one chance to receive the normal allele.

b)

	X^H	X^H
X^h	X^hX^H	X^hX^H
Y	X^HY	X^HY

There is a 0% chance that a son will have hemophilia. A son inherits his X chromosome from his mother. Because the mother is carrying two functional alleles, either X would provide a functional allele.

Chapter 8

1. e; 2. a; 3. b; 4. c; 5. d; 6. d; 7. e; 8. a; 9. e; 10. e; 11. b; 12. a; 13. c; 14. b; 15. d; 16. b; 17. b; 18. e; 19. d; 20. d; 21. c

Short-Answer Questions

1. Evolution is a change in allele frequencies in a population. Natural selection is one of four evolutionary agents, characterized by variation, heritability, and differential reproductive success for a trait.

2. Genetic drift is a random change in allele frequencies. In small populations, random deviations from the expected genotype frequencies

among offspring are less likely to be counterbalanced by deviations in the opposite direction from the expected genotype frequencies.

3. First, environments can change quickly. Natural selection may be too slow to adapt the species to a constantly changing environment. Second, variation is needed as the raw material of selection. If there is no mutation to give rise to an allele that would help make an individual more fit, the species will never be perfectly adapted. (Why, for instance, are there no mammals with wheeled appendages?) Third, evolution produces "more fit" individuals in each generation. If the optimum solution required organisms to get worse before they could evolve to "perfection," selection would never lead to such a solution.

4. In the 1940s, when penicillin was first used to treat bacterial infections, it was uniformly effective in killing *Staphylococcus aureus*. Today, more than 90% of isolated *S. aureus* strains are resistant to penicillin. Because penicillin has become such a pervasive toxin in the environment of *Staphylococcus,* natural selection has led to an increase in the frequency of alleles that make the bacteria resistant. Today, nearly a third of "staph" infections are resistant to antibiotics. The meaning of such unintentional natural selection "experiments" is clear and consistent: evolution is occurring all around us.

Chapter 9

1. b; 2. d; 3. e; 4. b; 5. d; 6. d; 7. c; 8. a; 9. d; 10. e; 11. c; 12. a; 13. a; 14. b; 15. e; 16. b

Short-Answer Question

1. Males provide more parental care, consistent with the two predictions about sex roles and parental investment. First, the sex that invests more will be more discriminating. Second, the sex that invests less will compete among themselves for access to the higher-investing sex.

Chapter 10

1. a; 2. d; 3. e; 4. e; 5. a; 6. d; 7. d; 8. a; 9. e; 10. a; 11. d; 12. d; 13. e; 14. b; 15. c; 16. a; 17. c

Short-Answer Questions

1. The biological species concept focuses on reproductive populations, or groups of individuals that can mate with each other and are unable to breed with other reproductively isolated groups. The morphological species concept focuses on physical characteristics of the organisms to distinguish one species from another. Many living organisms don't fit neatly into a species as defined by the biological species concept. An example is many of the living organisms that divide asexually, such as bacteria. For these organisms, because they do not require a partner to mate, reproductive compatibility is not a useful diagnostic tool for discriminating among species.

2. Biologists would rely on physical features or characteristics to create evolutionary trees and establish relationships between groups of organisms. Today, with the advent of technology, DNA sequencing (examining the base-pair sequence of DNA) is used to construct evolutionary trees. In either case, the trees define which groups of organisms are most closely related, but not which groups are more advanced than others.

3. These adaptations could be considered an "evolutionary innovation" in which a specific trait or group of traits allows an organism to fill a new niche or an unoccupied niche. This is one trigger of adaptive radiation—the huge, rapid expansion of a species, often diversifying

into many more species. The specific traits of tall, flat teeth and a square jaw allowed the horse to efficiently crop grasses and vegetation from the ground. With areas of grasslands plentiful, these mammals were able to expand and flourish in this niche.

Chapter 11

1. d; 2. c; 3. e; 4. d; 5. c; 6. c; 7. a; 8. c; 9. b; 10. c; 11. e; 12. d; 13. b; 14. b; 15. a; 16. b; 17. d; 18. e.

Short-Answer Questions

1. In all animals, the ability to obtain food, whatever method that may be, has evolved through natural selection. For the sloth, it would make no sense for it to have developed the ability to spring at great speeds, because its food source (vegetation) is stationary. The cheetah, by contrast, must use high speeds to catch its prey, which also can run fast. If the cheetah were to occupy the same niche as the sloth, its ability to run would not improve its ability to obtain vegetation, so it might lose this running ability over many generations.

2. Much of the energy derived from food is utilized for heat production, so endotherms require more food to maintain a constant internal temperature. Ectotherms depend on an external heat source to maintain their body temperature, so they do not need energy derived from food to generate heat.

3. The snake, like other ectotherms, depends on an outside source of heat, so placing an insulating sweater on the snake will prevent heat from reaching its body, keeping it from warming up. For the same reason, ectotherms do not have insulating feathers or fur.

4. The slug has a radula for scraping food from surfaces, has a broad foot on which it glides, and produces a protective slime. Slugs lack the body segmentation found in segmented worms.

Chapter 12

1. a; 2. b; 3. b; 4. c; 5. a; 6. b; 7. e; 8. c; 9. c; 10. c; 11. d; 12. b; 13. d; 14. b; 15. d.

Short-Answer Questions

1. Non-vascular plants lack tissue that can conduct water and dissolved nutrients, tissue that *is* found in vascular plants. Because diffusion can transport water only over very short distances, non-vascular plants are much smaller than vascular plants. Non-vascular plants must live in an environment that is wet for at least some parts of the year, and become dormant when it is dry.

2. Mycorrhizal fungi are associated with the roots of plants, benefiting them by transporting phosphorus and nitrogen from the soil to the cells of the plant roots. In exchange, the fungus takes some sugar from the plant. This relationship enhances the overall health of the plant.

3. Plants expend considerable energy producing a fruit, which contains the seeds. Once a fruit is eaten, it is digested by the animal and the seeds are excreted unharmed, and thus relocated to an area distant from the parent plant. The plant pays the animal with a meal, and in return, the animal transports the seeds.

Chapter 13

1. a; 2. b; 3. c; 4. e; 5 d; 6. a; 7. a; 8. d; 9. b; 10. a; 11. c; 12. b; 13. e; 14. b; 15. e; 16. e; 17. e; 18. b.

Short-Answer Question

	Prokaryotes		Eukaryotes	Viruses
	Bacteria	Archaea	Protists	
Unicellular, colonial, or multicellular?	Unicellular	Unicellular	Unicellular or colonial	Not cells
Plasma membrane?	Present	Present	Present	Absent
Peptidoglycan in cell wall?	Present	Absent	Absent	Absent
Flagella? What type?	Yes; spins like a propeller	Yes; spins like a propeller, but has a chemical composition different from that of bacteria	Yes; waves from side to side and has a chemical composition different from that of bacteria or archaea	No flagellum
Presence of nucleus?	Absent	Absent	Present	No nucleus
Intracellular organelles?	Absent	Absent	Present	No organelles
Type of genetic material (DNA, RNA, or both possible)?	DNA	DNA	DNA	DNA or RNA
Chromosome structure (linear or circular)?	Usually circular	Usually circular	Linear	Linear or circular

Chapter 14

1. a; 2. e; 3. a; 4. b; 5. e; 6. d; 7. b; 8. d; 9. c; 10. c; 11. e; 12. d; 13. a; 14. b; 15. c.

Short-Answer Questions

1. Density-dependent factors are limitations on a population's growth resulting from competition among individuals for a finite amount of resources, such as food and space for living and breeding, and an increased risk of disease and predation. Density-independent factors are forces such as hurricanes and droughts that kill off many individuals in a population, regardless of the size of the population.

2. An organism with a type I survivorship curve is relatively large, has only a few offspring, invests a lot in each offspring, and lives a relatively long life. If the organism had a type III survivorship curve, it would have a high probability of dying early in life and would produce many offspring and provide very little parental care.

3. Species that lack natural enemies and live in an environment with low risk of mortality from other sources breed at a later age; late-onset genetic disorders are selected out and individuals are healthier in old age. Species

that breed at an early age pass on adverse genetic traits that are expressed after their normal age of breeding; they do not live long after the normal age of breeding. If breeding is delayed, as in the fruit flies in the experiments described in the chapter, age-related disorders that interfere with breeding are selected out and the life span will be extended.

Chapter 15

1. d; 2. e; 3. a; 4. a; 5. e; 6. d; 7. e; 8. a; 9. d; 10. a; 11. a; 12. a; 13. d; 14. e; 15. b; 16. b.

Short-Answer Questions

1. Heated air rises and cools at the equator (0° latitude); as it cools, it cannot hold as much water and loses it as precipitation. Air moves north and south from the equator and begins to fall at about 30° latitude north and south, warming and absorbing moisture. This creates very dry conditions at these latitudes, and deserts are found there. Air continues to move toward the poles and begins to rise again at 60° latitude, cooling and losing its moisture through precipitation, making these locations very wet. Air again falls near the poles, making these areas very dry. Temperature and amount of rainfall are the primary determinants of the kind of biome found in any terrestrial location.

2. Energy flows through each trophic level. Photosynthetic organisms, the primary producers, trap energy from the sun in the chemical bonds of food. Primary consumers, the herbivores, eat photosynthetic organisms, and these animals are consumed by secondary consumers, the carnivores. Carnivores can be consumed by tertiary consumers, the top carnivores. A natural ecosystem can support only a few top carnivores because only about 10% of the energy is available to each subsequent trophic level.

3. An organism's niche includes the space it requires to live in, the kinds and amount of food it eats, the time of day and of year it reproduces, the amount of moisture and temperature it requires, as well as other criteria. Two organisms that occupy the same niche compete, with two different possible outcomes: one species may drive the other to extinction (competitive exclusion) or both species may change their behavior or undergo a morphological change so they are restricted within their own niche, dividing the resources between them (resource partitioning).

Chapter 16

1. e; 2. c; 3. d; 4. a; 5. e; 6. e; 7. d; 8. d; 9. b; 10. a; 11. e; 12. c; 13. b; 14. e.

Short-Answer Questions

1. The greatest biodiversity, both aquatic and terrestrial, is in equatorial regions. Less biodiversity is found with increasing latitude. Habitat destruction and degradation are the greatest threats to biodiversity.

2. The biotic and abiotic nature of the habitat was altered, but species were not driven to extinction. If species had been made extinct, the ecosystem disturbances would not have been reversible.

3. Most of the deforestation of tropical rain forests is a result of clearing land to make way for agriculture; mining and pollution also contribute. The biodiversity of the planet is greatly affected by rain forest deforestation because no other terrestrial ecosystem has the extent of biodiversity found in these forests.

GLOSSARY

A note about notation: The word or phrase being defined is in boldface type; it is followed by the number (in parentheses) of the chapter or chapters where it is discussed. In some cases, derivations are given. Abbreviations: Gk., Greek; Lat., Latin; *sing.,* singular; *pl.,* plural; *dim.,* diminutive (a smaller version of the object named); *pron.,* pronounced.

A

acid (2) Any fluid with a pH below 7.0, indicating the presence of more H⁺ ions than OH⁻ ions. [Lat., *acidus,* sour]

active transport (3) Molecular movement that depends on the input of energy, which is necessary when the molecules to be moved are large or are being moved against their concentration gradient.

adaptation (8) The process by which, as a result of natural selection, organisms become better matched to their environment; also, a specific feature, such as the quills of a porcupine, that makes an organism more fit.

adaptive radiation (10) The rapid diversification of a small number of species into a much larger number of species, able to live in a wide variety of habitats.

additive effects (7) Effects from alleles of multiple genes that all contribute to the ultimate phenotype for a given characteristic.

adenosine triphosphate (ATP) (4) A molecule that temporarily stores energy for cellular activity in all living organisms; ATP is composed of a sugar molecule and a chain of three negatively charged phosphate groups.

adult (11) In complete metamorphosis, the third and final stage of insect development.

aging (14) For an individual is the increased risk of mortallity with increasing age, generally due to a breakdown in the body's machinery; in a population it is the increased rate of dying with increased age, characterized by multiple physiological breakdowns.

alleles (5, 7) Alternative versions of a gene. [Gk., *allos,* another]

allopatric speciation (10) Speciation that occurs as a result of a geographic barrier between groups of individuals that leads to reproductive isolation and then genetic divergence. [Gk., *allos,* another + *patris,* native land]

allopolyploidy (10) A special case of polyploidy, in which through asexual propagation and subsequent repeated hybridization the increased number of chromosome sets in an individual are derived from multiple different species. [Gk., *allos,* another + *polus,* many + *ploion,* vessel]

altruistic behavior (9) A behavior that comes at a cost to the individual performing it and benefits another. [Lat., *alter,* the other]

amino acid (2) One of 20 molecules built of an amino group, a carboxyl group, and a unique side chain; proteins are constructed of combinations of amino acids. [Gk., *Ammon,* name of ancient Libyan god, near whose temple a compound known as "sal ammoniacus," or "ammonium salt," was prepared]

amino group (2) A nitrogen atom bonded to three hydrogen atoms.

amniotes (11) Terrestrial vertebrates—reptiles, birds, and mammals— that deposit amniotic eggs, that is, eggs protected by a water-tight membrane and a shell.

amphibians (11) Members of the class Amphibia; amphibians are ectotherms (that is, they are cold-blooded), with a moist skin, lacking scales, through which they can fully or partially absorb oxygen. They were the first terrestrial vertebrates; the young of most species are aquatic, and the adults are true land animals. [Gk., *bios,* life + *amphi,* on both sides]

analogous (10) Describes characteristics (such as bat wings and insect wings) that are similar because they were produced by convergent evolution, not because they descended from a common structure in a shared ancestor. [Gk., *analogos,* proportionate]

anaphase (6) The third phase of mitosis, in which the sister chromatids are pulled apart by the spindle fibers, with a full set of chromosomes going to opposite sides of the cell; in meiosis, the homologues separate in anaphase I and the sister chromatids separate in anaphase II. [Gk., *ana,* up + *phasis,* appearance]

anecdotal observation (1) Observation of one or only a few instances.

angiosperms (12) Vascular, seed-producing flowering and fruit-bearing plants, in which the seeds are enclosed in an ovule within the ovary. [Gk., *angeion,* vessel, jar + *sperma,* seed]

animal (11) Members of the kingdom Animalia, which are eukaryotic, multi-cellular, and heterotrophic (that is, they cannot produce their own food); many of these organisms have body parts specialized for different activities and can move during some stage of their lives. [Lat., *animal,* a living being]

anther (12) The part of the stamen, the male reproductive structure of a flower, that produces pollen. [Gk., *anthos,* blossom]

apoptosis (6) Programmed cell death, which takes place particularly in parts of the body where the cells are likely to accumulate significant genetic damage over time and are therefore at high risk of becoming cancerous. [Gk., *apoptosis,* falling away]

archaea (10) A group of prokaryotes that are evolutionarily distinct from bacteria and that thrives in some of the most extreme environments on earth; one of the three domains of life. [Gk., *archaios,* ancient]

arthropods (11) Members of the invertebrate phylum Arthropoda; characterized by a segmented body, an exoskeleton, and jointed appendages. [Gk., *arthron,* joint + *pous,* foot]

asexual reproduction (6) A type of reproduction common in prokaryotes, in which the offspring inherit their DNA from a single parent.

atom (2) A particle of matter than cannot be further subdivided without losing its essential properties. [Gk., *atomos,* indivisible]

atomic mass (2) The mass of an atom; the combined mass of the protons, neutrons, and electrons in an atom (the mass of the electrons is so small as to be almost negligible).

atomic number (2) The number of protons in the nucleus of an atom of a given element.

B

background extinctions (10) Extinctions that occur at lower rates than at times of mass extinctions; background extinctions occur mostly as the result of aspects of the biology and competitive success of the species, rather than catastrophe.

base (chemistry) (2) Any fluid with a pH above 7.0, that is, with more OH⁻ ions than H⁺ ions.

base (genetics) (2, 5) One of the nitrogen-containing side-chain molecules attached to a sugar molecule in the sugar-phosphate backbone of DNA and RNA. The four bases in DNA are adenine (A), thymine (T), guanine (G), and cytosine (C); the four bases in RNA are adenine (A), uracil (U), guanine (G), and cytosine (C); the information in a molecule of DNA and RNA is determined by its sequence of bases.

base pair (5) Two nucleotides on complementary strands of DNA that form a pair, linked by hydrogen bonds; the pattern of pairing is adenine (A) with thymine (T) and cytosine (C) with guanine (G); the base-paired arrangement forms the "rungs" of the double-helix structure of DNA.

behavior (9) Any and all of the actions performed by an organism, often in response to its environment or to the actions of another organism.

bilateral symmetry (11) A body structure with left and right sides, which are mirror images. [Lat., *bi-*, two + *latus*, side]

binary fission (6) A type of asexual reproduction in which the parent cell divides into two genetically identical daughter cells; bacteria and other prokaryotes reproduce by binary fission. [Lat., *binarius*, consisting of two + *fissus*, divided]

biodiversity (10) The variety and variability among all genes, species, and ecosystems. [Gk., *bios*, life + Lat., *diversus*, turned in different directions]

biodiversity hotspots (16) Regions of the world with significant reservoirs of biodiversity that are under threat of destruction.

biofuels (4) Fuels produced from plant and animal products.

biogeography (8) The study and interpretation of distribution patterns of living organisms around the world. [Gk. *bios*, life + *geo-*, earth + *graphein*, to write down]

biological literacy (1) The ability to use scientific inquiry to think creatively about problems with a biological component; to communicate these thoughts to others; and to integrate these ideas into one's decision making.

biological species concept (10) Populations of organisms that interbreed, or could possibly interbreed, with each other under certain conditions, and that cannot interbreed with organisms outside their own group.

biology (1) The study of living things. [Gk., *bios*, life + *logos*, discourse]

biomass (15) The total mass of all the living organisms in a given area. [Gk., *bios*, life]

biomes (15) The major ecological communities of earth; terrestrial biomes, such as rainforest or desert, are defined and usually described by the predominant types of plant life in the area, which are mostly determined by the weather; aquatic biomes are usually defined by physical features such as salinity, water movement, and depth. [Gk., *bios*, life]

biotechnology (5) The modification of organisms, cells, and their molecules for practical benefits. [Gk., *bios*, life + *technologia*, systematic treatment]

biotic (15) Relating to living organisms; the biotic environment, or community, consists of all the living organisms in a given area. [Gk., *bios*, life]

bivalve molluscs (11) Molluscs with two hinged shells; examples are clams, scallops, oysters and molluscs. [Lat., *bi-*, two]

blind experimental design (1) An experimental design in which the subjects do not know what treatment (if any) they are receiving.

bottleneck effect (8) A change in allele frequencies of a population caused by a sudden reduction in population size (often due to famine, disease, or rapid environmental disturbance).

bryophytes (12) Three groups of plants (the liverworts, hornworts, and mosses) that lack vascular tissue and move water and dissolved nutrients by diffusion. [Gk., *bruon*, tree-moss, liverwort + *phutas*, plant]

buffer (2) A chemical that can quickly absorb excess H^+ ions in a solution (preventing it from becoming too acidic) or quickly release H^+ ions to counteract increases in OH^- concentration.

C

C4 photosynthesis (4) A method (along with C3 and CAM photosynthesis) by which plants fix carbon dioxide, using the carbon to build sugar; serves as a more effective method than C3 for binding carbon dioxide under low carbon dioxide conditions, such as when plants in warmer climates close their stomata to reduce water loss.

Calvin cycle (4) In photosynthesis, a series of chemical reactions in the stroma of chloroplasts, in which sugar molecules are assembled. [from the name of one of its discoverers, Melvin Calvin, 1911–1997]

CAM (crassulacean acid metabolism) (4) Energetically expensive photosynthesis in which the stomata are open only at night to admit CO_2, which is bound to a holding molecule and released to enter the Calvin cycle to make sugar during the day; in this type of photosynthesis, found in many fleshy, juicy plants of hot, dry areas, water loss is reduced because the stomata are closed during the day.

cancer (6) Unrestrained cell growth and division. [Lat., *cancer*, crab; by extension, the disease cancer]

capsid (13) The protein container surrounding the genetic material (DNA or RNA) of a virus. [Lat., *capsa*, box, case]

capsule (13) A layer surrounding the cell wall of many bacteria; the capsule may restrict the movement of water out of the cell and thus allow bacteria to live in dry places, such as the surface of the skin. The capsule contributes to the virulent characteristics of some bacteria, making them resistant to phagocytosis by the host's immune system. [Lat., *capsula*, small box or case]

carbohydrate (2) One of the four types of biological macromolecule, containing mostly carbon, hydrogen, and oxygen; carbohydrates are the primary fuel for cellular activity and form much of the cell structure in all life forms. [Lat., *carbo*, charcoal + *hydro-*, pertaining to water]

carboxyl group (2) A carbon atom bonded to two oxygen atoms.

carnivores (15) Animals that feed mostly on the meat of secondary or tertiary consumers. [Lat., *carnis*, of flesh + *vorare*, to devour]

carotenoids (4) Pigments that absorb blue-violet and blue-green wavelengths and reflect yellow, orange, and red wavelengths. [Lat., *carota*, carrot]

carpel (12) The female reproductive structure of a flower, including the stigma, style, and ovary. [Gk., *karpos*, fruit]

carrier (7) An individual who carries one allele for a recessive trait and who does not exhibit the trait; if two carriers mate they may produce offspring who do exhibit the trait.

carrying capacity (K) (14) The ceiling on a population's growth imposed by the limitation of resources for a particular habitat over a period of time.

cartilaginous fishes (11) Fish species characterized by a skeleton made completely of cartilage, not bone. [Lat., *cartilago*, gristle]

cell (3) The smallest unit of life that can function independently; a three-dimensional structure, surrounded by a membrane and, in the case of prokaryotes and most plants, a cell wall, in which many of the essential chemical reactions of the life of an organism take place. [Lat., *cella*, room]

cell cycle (6) In a cell, the alternation of activities related to cell division and those related to growth and metabolism.

cell theory (3) A unifying and universally accepted theory in biology that holds that all living organisms are made up of one or more cells, and that all cells arise from other, pre-existing cells.

cell wall (3) A rigid structure, outside the cell membrane, that protects and gives shape to the cell; found in many prokaryotes and plants.

cellular respiration (4) The process by which all living organisms extract energy stored in the chemical bonds of molecules and use it for fuel for their life processes.

cellulose (2) A complex carbohydrate, indigestible by humans, that serves as the structural material for a huge variety of plant structures; it is the single most prevalent organic compound on earth. [Lat., *cellula*, dim. of *cella*, room]

central vacuole (3) In plants, a large, fluid-filled organelle, surrounded by a membrane, important in nutrient storage, waste management, predator deterrence, sexual reproduction, and physical support. [Lat., *vacuus*, empty]

centromere (6) After replication, the region of contact between sister chromatids, which occurs near the center of the two strands. [Gk., *centron*, the stationary point of a pair of compasses, thus the center of a circle + *meris*, part]

cephalopods (11) Molluscs in which the head is prominent and the foot has been modified into tentacles; examples are octopuses and squids. Cephalopods have a reduced or absent shell and possess the most advanced nervous system of the invertebrates. [Gk., *kephale*, head + *pous*, foot]

character displacement (15) An evolutionary divergence in one or more of the species that occupy a niche that leads to a partitioning of the niche between the species. Changes in characteristics, such as behavioral or body plan, of two or more very similar species that have overlapping geographical locations result in a reduction of competition between species.

chemical energy (4) A type of potential energy in which energy is stored in chemical bonds between atoms or molecules.

chemolithotrophs (13) Bacteria that can use inorganic molecules such as ammonia, hydrogen sulfide, hydrogen, and iron as sources of energy. [Gk., *lithos*, stone, rock + *trophe*, food]

chemoorganotrophs (13) Bacteria that consume organic molecules, such as carbohydrates, as an energy source. [Gk., *trophe*, food]

chitin (2) (*pron.* KITE-in) A complex carbohydrate, indigestible by humans, that forms the rigid outer skeleton of most insects and crustaceans. [Gk., *chiton*, undershirt]

chlorophyll (4) A light-absorbing pigment molecule in chloroplasts. [Gk., *chloros*, pale green + *phullon*, leaf]

chlorophyll *a* (4) The primary photosynthetic pigment, chlorophyll *a* absorbs blue-violet and red light; because it cannot absorb green light and instead reflects those wavelengths, we perceive the reflected light as the color green.

chlorophyll *b* (4) A photosynthetic pigment similar in structure to chlorophyll *a*, chlorophyll *b* absorbs blue and red-orange wavelengths and reflects yellow-green wavelengths.

chloroplast (3, 4) The organelle in plant cells in which photosynthesis occurs. [Gk., *chloros*, pale green + *plastos*, formed]

cholesterol (2, 3) One of the sterols, lipids important in regulating growth and development; cholesterol is an important component of most cell membranes, helping the membrane to maintain its flexibility. [Gk., *chole*, bile + *stereos*, solid + -*ol*, chemical suffix for an alcohol]

chromatid (6) One of the two strands of a replicated chromosome. [Gk., *chroma*, color]

chromatin (3) A mass of long, thin fibers consisting of DNA and proteins in the nucleus of the cell. [Gk., *chroma*, color]

chromosomal aberration (5) A type of mutation characterized by a change in the overall organization of genes on a chromosome, such as the deletion of a section of DNA; the moving of a gene from one part of a chromosome to elsewhere on the same chromosome or to a different chromosome; or the duplication of a gene, with the new copy inserted elsewhere on the chromosome or on a different chromosome. [Lat., *aberrare*, to wander]

chromosome (5) A linear or circular strand of DNA on which are found specific sequences of base pairs; the human genome consists of two copies of each of 23 unique chromosomes, one from the mother and one from the father. [Gk., *chroma*, color + *soma*, body]

cilia (*sing.* **cilium**) (3) Short projections from the cell surface, often occurring in large numbers on a single cell, that beat against the intercellular fluid to move the fluid past the cell. [Lat., eyelid]

class (10) In the system developed by Carolus Linnaeus (1707–1778), a classification of organisms consisting of related orders.

climax community (15) A stable and self-sustaining community that results from ecological succession.

clone (5) A genetically identical DNA fragment, cell, or organism produced by a single cell or organism. [Gk., *klon*, twig]

clone library (5) A collection of cloned DNA fragments; also known as a gene library.

cloning (5) The production of genetically identical cells, organisms, or DNA molecules.

code (5) In genetics, the base sequence of a gene.

codominance (7) The case in which the heterozygote displays characteristics of both alleles.

codons (5) Three-base sequences in mRNA that link with complementary tRNA molecules, which are attached to amino acids; a codon with yet another sequence ends the process of assembling a protein from amino acids.

coevolution (15) The concurrent appearance over time, through natural selection, of traits in interacting species that enable each species to become adapted to the other; an example is the 11-inch-long tongue of a moth that feeds from the 11-inch-long nectar tube of an orchid.

colonizers (15) Species introduced into an area that has been disturbed and is undergoing the process of either primary or secondary succession. The identity of a colonizing species varies depending on the stage and type of succession.

commensalism (15) A symbiotic relationship between species in which one benefits and the other neither benefits nor is harmed. [Lat., *com*, with + *mensa*, table]

communication (9) An action or signal on the part of one organism that alters the behavior of another organism.

community (in ecology) (15) The biotic environment; a geographic area defined as a loose assemblage of species with overlapping ranges.

competitive exclusion (15) The case in which two species battle for resources in the same niche until the more efficient of the two wins and the other is driven to extinction in that location.

complementarity (6) The characteristic of double-stranded DNA that the base on one strand always has the same pairing partner, or complementary base, on the other strand.

complementary base (6) A base on a strand of double-stranded DNA that is a pairing partner to a base on the other strand: adenine (A) is the complementary base to thymine (T), and guanine (G) is the complementary base to cytosine (C).

complex carbohydrate (2) A carbohydrate that contains multiple simple carbohydrates linked together; types of complex carbohydrates include starch, which is the primary form of energy storage in plants, and glycogen, which is the primary form of short-term energy storage in animals.

compound (2) A substance composed of atoms of different elements in specific ratios, held together by ionic bonds. [Lat., *componere*, to put together]

conjugation (13) The process by which a bacterium transfers a copy of some or all of its DNA to another bacterium, of the same or another species. [Lat., *coniugatio*, connection]

conservation biology (16) An interdisciplinary field, drawing on ecology, economics, psychology, sociology, and political science, that studies and devises ways of preserving and protecting biodiversity and other natural resources.

contact inhibition (3) The limiting factor of cell growth that occurs when normal cells come into contact; in cancer cells, contact inhibition does not take place and the cells continue to divide.

control group (1) In an experiment, the group of subjects not exposed to the treatment being studied but otherwise treated identically to the experimental group.

convergent evolution (8, 10) A process of natural selection in which features of organisms not closely related come to resemble each other as a consequence of similar selective forces; many marsupial and placental species resemble each other as a result of convergent evolution. [Lat., *con-*, together with + *vergere*, turn + *evolvere*, to roll out]

covalent bond (2) A strong bond formed when two atoms share electrons; the simplest example is the H$_2$ molecule, in which each of the two atoms in the molecule shares its lone electron with the other atom. [Lat., *con-*, together + *valere*, to be strong]

critical experiment (1) An experiment that makes it possible to determine decisively between alternative hypotheses.

cross (7) The breeding of organisms that differ in one or more traits.

crossing over (6) The exchange of some genetic material between a paternal homologous chromosome and a maternal homologous chromosome, leading to a chromosome carrying genetic material from each.

cuticle (12) A waxy layer produced by epidermal cells and found on leaves and shoots of terrestrial plants, protecting them from drying out. [Lat., *cuticula*, dim. of *cutis*, skin]

cytokinesis (6) In the cell cycle, the stage following mitosis in which cytoplasm and organelles duplicate and are divided into approximately equal parts and the cell separates into two daughter cells; in meiosis, two diploid daughter cells are formed in cytokinesis following telophase I and four haploid daughter cells are formed in cytokinesis following telophase II. [Gk., *kutos*, container + *kinesis*, motion]

cytoplasm (3) The jelly-like fluid that fills the inside of the cell; in eukaryotes, the cytoplasm contains the organelles. [Gk., *kutos*, container + *plasma*, anything molded]

cytoskeleton (3) A network of protein structures in the cytoplasm of plants and animals that serves as scaffolding, adding support and giving animal cells of different types their characteristic shapes; the cytoskeleton serves as tracks guiding the intercellular traffic flow and, because it is flexible and can generate force, gives cells some ability to control their movement.

D

daughter cells (6) Cells produced by the division of a parent cell.

decomposers (12, 15) Organisms, including bacteria, fungi, and detritivores, that break down and feed on once-living organisms.

demographic transition (14) A pattern of population growth characterized by the progression from high birth and death rates (slow growth) to high birth rates and low death rates (fast growth) to low birth and death rates (slow growth).

denaturation (2) The disruption of protein folding, in which secondary and tertiary structure are lost, caused by exposure to extreme conditions in the environment such as heat or extreme pH (that is, a strong acid or a strong base); denaturation causes proteins to lose their function; also the disruption of DNA, also known as DNA melting, in which the bonds linking complementary bases are broken and the two strands separate from each other.

density-dependent factors (14) Limitations on a population's growth that are a consequence of population density.

density-independent factors (14) Limitations on a population's growth without regard to population size, such as floods, earthquakes, fires, and lightning.

deoxyribonucleic acid (DNA) (2, 5) A nucleic acid, DNA carries information about the production of particular proteins in the sequences of its nucleotide bases.

desert (15) A type of terrestrial biome; a type of dry climate, with very little rainfall, in which water loss through evaporation exceeds water gain through precipitation; typically found at 30° north and south latitude.

desmosome (3) Irregularly spaced connections between adjacent animal cells that, in the manner of Velcro, hold cells together by attachments but are not water-tight; they provide mechanical strength and are found in muscle tissue and in much of the tissue that lines the cavities of animal bodies. [Gk., *desmos*, bond + *soma*, body]

detritivores (15) Organisms that break down and feed on once-living organic matter; this group includes scavengers such as vultures, worms, and a variety of arthropods. [Lat., *detritus*, worn out + *vorare*, to devour]

deuterostomes (11) Bilaterally symmetrical animals with defined tissues in which the gut develops from back to front; the anus forms first, and the second opening formed is the mouth of the adult animal. [Gk., *deuteros*, second + *stoma*, mouth]

differential reproductive success (8) The situation in which individuals have greater reproductive success than other individuals in a population; along with variation and heritability, differential reproductive success is one of the three conditions necessary for natural selection.

diffusion (3) Passive transport in which a particle (the solute) is dissolved in a gas or liquid (the solvent) and moves from an area of higher solute concentration to an area of lower solute concentration. [Lat., *diffundere*, to pour in different directions]

diploid (6) Describes cells that have two copies of each chromosome (in many organisms, including humans, somatic cells are diploid). [Gk., *diplasiazein*, to double]

direct fitness (9) The total reproductive output of an individual.

directional selection (8) Selection that, for a given trait, increases fitness at one extreme of the phenotype and reduces fitness at the other, leading to an increase or decrease in the mean value of the trait.

disaccharides (2) Carbohydrates formed by the union of two simple sugars, such as sucrose (table sugar) and lactose (the sugar found in milk). [Gk., *di-*, two + *sakcharon*, sugar]

dispersers (15) Organisms able to move away from their original home.

disruptive selection (8) Selection that, for a given trait, increases fitness at both extremes of the phenotype distribution and reduces fitness at middle values.

DNA probe (5) A short sequence of radioactively tagged single-stranded DNA that contains part of the sequence of the gene of interest, used to locate that gene in a gene library; the probe binds to the complementary base pair on a gene in the library, which is identified by the radioactive tag on the probe.

domain (10) In modern classification, the highest level of the hierarchy; there are three domains, bacteria, archaea, and eukarya.

dominant (7) Describes an allele that masks the phenotypic effect of the other, recessive, allele for a trait; the phenotype shows the effect of the dominant allele in both homozygous and heterozygous genotypes. [Lat., *dominari,* to rule]

dorsal hollow nerve cord (11) The central nervous system of vertebrates, consisting of the spinal cord and brain. [Lat., *dorsum,* back]

double-blind experimental design (1) An experimental design in which neither the subjects nor the experimenters know what treatment (if any) individual subjects are receiving.

double bond (2) The sharing of two electrons between two atoms; for example, the most common form of oxygen is the O_2 molecule, in which two electrons from each of the two atoms of oxygen are shared.

double fertilization (12) In angiosperms, two sperm are released by a pollen grain: one fuses with an egg to form a zygote, and the other fuses with two endosperm-forming nuclei.

double helix (2) The spiraling ladder-like structure of DNA composed of two strands of nucleotides; the bases protruding from each strand like "half-rungs" meet in the center and bind to each other (via hydrogen bonds), holding the ladder together. [Gk., *heligmos,* wrapping]

dynamic equilibrium (16) A situation on islands in which the number of species present remains constant over time, while the composition of species may change.

E

ecological footprint (14) A measure of the impact of an individual or population on the environment by calculation of the amount of resources-including land, food and water, and fuel-consumed.

ecology (14) The study of the interaction between organisms and their environments, at the level of individuals, populations, communities, and ecosystems. [Gk., *oikos,* home + *logos,* discourse]

ecosystem (15) A community of biological organisms and the non-living environmental components with which they interact.

ectotherms (11) Organisms that rely on the heat from an external source to raise their body temperature and seek the shade when the air is too warm. [Gk., *ektos,* outside + *therme,* heat]

electromagnetic spectrum (4) The range of wavelengths that produce electromagnetic radiation, extending (in order of decreasing energy) from high-energy, short-wave, gamma rays and X rays, through ultraviolet light, visible light, and infrared light, to very long, low-energy, radio waves. [Lat., *specere,* to look at]

electron (2) A negatively charged particle that moves around the atomic nucleus.

electron transport chain (4) The third step of cellular respiration, in which high-energy electrons are passed from molecule to molecule, at every step releasing energy that is used to make ATP.

element (2) A substance that cannot be broken down chemically into any other substances; all atoms of an element have the same atomic number. [Lat., *elementum,* element, or first principle]

El Niño (15) A sustained surface temperature change in the central Pacific Ocean that occurs every two to seven years; this event can start a chain reaction of unusual weather across the globe that can result in flooding, droughts, famine, and a variety of extreme climate disruptions. [Spanish, *the child;* a reference to the Christ child, because of the appearance of the phenomenon at Christmastime]

empirical (1) Describes knowledge that is based on experience and observations that are rational, testable, and repeatable. [Gk., *empeiria,* experience]

endangered species (16) As defined by the Endangered Species Act, species in danger of extinction throughout all or a significant portion of their range.

Endangered Species Act (ESA) (16) A United States law that defines "endangered species" and is designed to protect those species from extinction.

endemic (16) Describes species peculiar to a particular region and not naturally found elsewhere. [Gk., *en,* in + *demos,* the people of a country]

endocytosis (3) A cellular process in which large particles, solid or dissolved, outside the cell are surrounded by a fold of the plasma membrane, which pinches off, forming a vesicle, and the enclosed particle now moves into the cell; the three types of endocytosis are phagocytosis, pinocytosis, and receptor-mediated endocytosis. [Gk., *endon,* within + *kutos,* container]

endomembrane system (3) A system of organelles (the rough endoplasmic reticulum, the smooth endoplasmic reticulum, and the Golgi apparatus) that surrounds the nucleus; it produces and modifies necessary molecules, breaks down toxic chemicals and cellular by-products, and is thus responsible for many of the fundamental functions of the cell. [Gk., *endon,* within + Lat., *membrana,* a thin skin]

endosperm (12) Tissue of a mature seed that stores certain carbohydrates, proteins, and lipids that fuel the germination, growth, and development of the embryo and young seedling. [Gk., *endon,* within + *sperma,* seed]

endosymbiosis theory (3) Theory of the origin of eukaryotes that holds that in the past two different types of prokaryotes engaged in a close partnership and eventually one, capable of performing photosynthesis, was subsumed into the other, a larger prokaryote; the smaller prokaryote made some of its photosynthetic energy available to the host, and over time the two became symbiotic and eventually a single more complex organism in which the smaller prokaryote had evolved into the chloroplast of the new organism (a similar scenario can be developed for the evolution of mitochondria). [Gk., *endon,* within + *sumbios,* living together]

endotherms (11) Organisms that use the heat produced by their cellular respiration to raise and maintain their body temperature above air temperature. [Gk., *endon,* within + *therme,* heat]

energy (4) The capacity to do work, which is the moving of matter against an opposing force. [Gk., *energeia,* activity]

energy pyramid (15) A diagram that illustrates the path of energy through the organisms of an ecosystem; each layer of the pyramid represents the biomass of a trophic level.

enzymatic protein (3) See **enzyme**

enzyme (2, 3) A protein that initiates and accelerates a chemical reaction in a living organism; enzymatic proteins take part in chemical reactions on the inside and outside surfaces of the plasma membrane. [Gk., *en,* in + *zume,* leaven]

estuary (15) A tidal water passage, linked to the sea, in which salt water and fresh water mix; estuaries are characterized by exceptionally high productivity. [Lat., *aestus,* tide]

ethanol (4) The end product of fermentation of yeast; the alcohol in beer, wine, and spirits. [contraction of the full chemical name, ethyl alcohol]

eukaryote (3) An organism composed of eukaryotic cells. [Gk., *eu*, good + *karuon*, nut, kernel]

eukaryotic cell (3) A cell with a membrane-bound nucleus containing DNA, membrane-bound organelles, and internal structures organized into compartments.

eutrophication (15) The process in which excess nutrients dissolved in a body of water lead to rapid growth of algae and bacteria, which consume much of the dissolved oxygen and, in time, can lead to large-scale die-offs. [Gk., *eu*, good + *trophe*, food]

evolution (8) A change in allele frequencies of a population. [Lat., *evolvere*, to roll out]

exocytosis (3) A cellular process in which particles within the cell, solid or dissolved, are enclosed in a vesicle and transported to the plasma membrane, where the membrane of the vesicle merges with the plasma membrane and the material in the vesicle is expelled to the extracellular fluid for use throughout the body. [Gk., *ex*, out of + *kutos*, container]

exoskeleton (11) A rigid external covering such as found in some invertebrates, including insects and crustaceans. [Gk., *ex*, out of]

exotic species (16) Species introduced by human activities to areas other than the species' native range.

experimental group (1) In an experiment, the group of subjects exposed to a particular treatment; also known as the treatment group.

exponential growth (14) Growth of a population at a rate that is proportional to its current size.

extinction (10) The complete loss of all individuals in a species. [Lat., *extinguere*, to extinguish]

extremophiles (13) Bacteria and archaea than can live in extreme physical and chemical conditions. [Lat., *extremus*, outermost + Gk., *philios*, loving]

F

facilitated diffusion (3) Diffusion of molecules through the phospholipid bilayer of the plasma membrane that takes place through a transport protein (a "carrier molecule") embedded in the membrane; molecules that require the assistance of a carrier molecule are those that are too big to cross the membrane directly or are electrically charged and would be repelled by the middle layer of the membrane.

family (10) In the system developed by Carolus Linnaeus (1707–1778), a classification of organisms consisting of related genera.

fatty acid (2) A long hydrocarbon (a chain of carbon-hydrogen molecules); fatty acids form the tail region of triglyceride fat molecules.

female (9) In sexually reproducing organisms, a member of the sex that produces the larger gamete.

fermentation (4) The process by which glycolysis occurs in the absence of oxygen; the electron acceptor is pyruvate (in animals) or acetaldehyde (in yeast) rather than oxygen.

fertilization (6) The fusion of two reproductive cells.

filament (12) The supporting stalk of the anther of a stamen found in angiosperm flowers. [Lat., *filum*, thread]

first law of thermodynamics (4) A physical law that states that energy cannot be created or destroyed; it can only change from one form to another.

fitness (8) A relative measure of the reproductive output of an individual with a given phenotype compared with the reproductive output of individuals with alternative phenotypes.

fixation (8) The point at which the frequency of an allele in a population is 100% and, therefore, there is no more variation in the population for this gene.

fixed action pattern (9) An innate sequence of behaviors, triggered under certain conditions, that requires no learning, does not vary, and once begun runs to completion; an example is egg-retrieval in geese.

flagella (*sing.* **flagellum**) (3) Long, thin, whip-like projections from the cell body of a prokaryote that aid in cell movement through the medium in which the organism lives; in animals, the only cell with a flagellum is the sperm cell. [Lat., *flagellum*, whip]

flatworms (11) Worms with flat bodies that are members of the phylum Platyhelminthes; characterized by well-defined head and tail regions, with some having clusters of light-sensitive cells for eyespots; most are hermaphroditic and are protostomes that do not molt; examples are tapeworms and flukes.

fleshy fruit (12) A fruit that consists of the ovary and some additional parts of the flower; when fleshy fruits, an attractive food, are eaten by animals, the seeds may be widely dispersed.

flower (12) The part of an angiosperm that contains the reproductive structures; the flower consists of a supporting stem with modified leaves (the petals and sepals) and usually contains both male and female reproductive structures.

fluid mosaic (3) A term that describes the structure of the plasma membrane, which is made up of several different types of molecules, many of which are not fixed in place but float, held in proper orientation by hydrophilic and hydrophobic forces.

food chain (15) The path of energy flow from primary producers to tertiary consumers.

food web (15) A more precisely described path of energy flow from primary producers to tertiary consumers than the food chain, reflecting the fact that many organisms are omnivores and occupy more than one position in the chain.

fossil (8) The remains of an organism, usually its hard parts such as shell, bones, or teeth, which have been naturally preserved; also, traces of such an organism, such as footprints. [Lat., *fossilis*, that which is dug up]

fossil fuels (4) Fuels produced from the decayed remains of ancient plants and animals; fossil fuels include oil, natural gas, and coal.

founder effect (8) A change in the allele frequencies of a population resulting from the isolation of a small subgroup of a larger population; all the descendants of the smaller group will reflect the allele frequencies of the subgroup, which may be different from those of the larger source population; the founder effect is one cause of genetic drift.

fundamental niche (15) The full range of environmental conditions under which an organism can live.

G

gametes (6) Haploid cells from two individuals that, as sperm and egg, will combine at fertilization to produce offspring; also called reproductive cells. [Gk., *gamete*, wife]

gametophyte (12) The plant structure that produces gametes (sperm and eggs); the haploid life stage of plants and some algae, which may be either male (producing sperm) or female (producing eggs). [Gk., *gamete*, wife + *phutas*, plant]

gap junction (3) A junction between adjacent animal cells in the form of a pore in each of the plasma membranes surrounded by a protein that links the two cells and acts like a channel between them to allow materials to pass between the cells.

gastropods (11) Molluscs that are members of the class Gastropoda; most have a single shell, a muscular foot for locomotion, and a radula used for scraping food from surfaces; examples are snails and slugs. [Gk., *gaster*, belly + *pous*, foot]

gene (5) The basic unit of heredity; a sequence of DNA nucleotides on a chromosome that carries the information necessary for making a functional product, usually a protein or an RNA molecule. [Gk., *genos*, race, descent]

gene flow (8) A change in the allele frequencies of a population due to movement of some individuals of a species from one population to another, changing the allele frequencies of the population they join; also known as migration.

gene library (5) A collection of cloned DNA fragments; also known as a clone library.

gene therapy (5) A therapy designed to treat or cure a disease by insertion of a functional gene to replace a defective version of that gene.

genetic drift (8) A random change in allele frequencies over successive generations; a cause of evolution.

genetic engineering (5) The manipulation of an organism's genetic material by adding, deleting, or transplanting genes from one organism to another.

genome (5) The full set of DNA present in an individual organism; also can refer to the full set of DNA present in a species.

genotype (5, 7) The genes that an organism carries for a particular trait; also, collectively, an organism's genetic composition. [Gk., *genos*, race, descent + *typos*, impression, engraving]

genus (*pl.* genera) (10) In the system developed by Carolus Linnaeus (1707–1778), a classification of organisms consisting of closely related species. [Lat., *genus*, race, family, origin]

glycerol (2, 3) A small molecule that forms the head region of a triglyceride fat molecule. [Gk., *glukus*, sweet + *-ol*, chemical suffix for an alcohol]

glycogen (2) A complex carbohydrate consisting of stored glucose molecules linked to form a large web, which breaks down to release glucose when it is needed for energy. [Gk., *glukus*, sweet + Gk., *genos*, race, descent]

glycolysis (4) In all organisms, the first step in cellular respiration, in which one molecule of glucose is broken into two molecules of pyruvate; for some organisms glycolysis is the only means of extracting energy from food, and in most organisms it is followed by the Krebs cycle and the electron transport chain. [Gk., *glukus*, sweet + *lusis*, releasing]

Golgi apparatus (3) An organelle, part of the endomembrane system, structurally like a flattened stack of unconnected membranes, each known as a Golgi body; the Golgi apparatus processes molecules synthesized in the cell and packages those molecules that are destined for use elsewhere in the body. [*pron.* GOHL-jee; from the name of the discoverer, Camillo Golgi, 1843–1926]

gonads (6) The ovaries and testes in sexually reproducing animals. [Gk., *gone*, offspring]

gram stain (13) A test used by microbiologists in identifying an unknown bacterium; the dye that is used stains the layer of peptidoglycan outside the cell wall purple, but those bacteria in which the layer of peptidoglycan is covered by a membrane are not colored by the dye; bacteria that take a Gram stain are known as Gram-positive bacteria, those that do not are known as Gram-negative bacteria.

group selection (9) The process leading to an increase in the frequency of alleles for behaviors beneficial to the persistence of the species or population that are simultaneously detrimental to the fitness of the individual engaging in the behavior.

growth rate (*r*) (14) The birth rate minus the death rate; the change in the number of individuals in a population per unit of time.

gymnosperms (12) Vascular plants that do not produce their seeds in a protective structure, but are usually found on the surface of the scales of a cone-like structure; the gymnosperms include conifers, cycads, gnetophytes, and ginkgo. [Gk., *gumnos*, naked + *sperma*, seed]

H

habitat (15) The physical environment of organisms, consisting of the chemical resources of the soil, water, and air, and physical conditions such as temperature, salinity, humidity, and energy sources. [Lat., *habitare*, to dwell or inhabit]

hair (11) Dead cells filled with the protein keratin that collectively serve as insulation covering the body or a part of the body; present in all mammals.

haploid (6) Describes cells that have a single copy of each chromosome (in many species, including humans, gametes are haploid). [Gk., *haploeides*, single]

hazard factor (14) An external force on a population that increases the risk of death.

herbivores (15) Animals that eat plants; also known as primary consumers. [Lat., *herba*, grass + *vorare*, to devour]

heredity (7) The greater resemblance of offspring to parents than to other individuals in the population, a consequence of the passing of characteristics from parents to offspring through their genes. [Lat., *heres*, heir]

heritability (8) The transmission of traits from parents to offspring via genetic information; also known as inheritance.

hermaphrodite (6) An organism that produces both male and female gametes. [from the names of the Greek god Hermes and goddess Aphrodite]

heterozygous (7) Describes the genotype of a trait for which the two alleles an individual carries differ from each other. [Gk., *heteros*, other + *zeugos*, pair]

histones (6) Proteins around which the long, linear strands of DNA are wrapped; the histones serve to keep the DNA untangled and to enable an orderly, tight, and efficient packing of the DNA within the cell.

homologous (10) Describes features that are inherited from a common ancestor. [Gk., *homologia*, agreement]

homologous pair (homologues) (6) The maternal and paternal copies of a chromosome. [Gk., *homologia*, agreement]

homologous structures (8) Body structures in different organisms that, although modified extensively over time to serve different functions in different species, are due to inheritance from a common evolutionary ancestor.

homozygous (7) Describes the genotype of a trait for which the two alleles are the same. [Gk., *homos*, same + *zeugos*, pair]

honest signal (9) A signal, which cannot be faked, that is given when both the individual making the signal and the individual responding to it have the same interests; it carries the most accurate information about an individual or situation.

horizontal gene transfer (10) The transfer of genetic material directly from one individual to another not necessarily related individual.

host (13, 15) An organism in or on which a parasite lives.

Human Genome Project (5) A project to decode the three billion base pairs in the human genome and to identify all genes present in it.

hybridization (genetics) (10) The interbreeding of closely related species. [Lat., *hybrida*, animal produced by two different species]

hybridization (molecular biology) (5) The process of joining two complementary strands of DNA from different sources; hybridization occurs when a DNA probe is used to match with a complementary sequence in a gene library. [Lat., *hybrida*, animal produced by two different species]

hybrids (10) Offspring of individuals of two different species. [Lat., *hybrida*, animal produced by two different species]

hydrogen bond (2) A type of weak chemical bond formed between the slightly positively charged hydrogen atoms of one molecule and the slightly negatively charged atoms of another (often oxygen or nitrogen atoms); hydrogen bonds are important in building multi-atom molecules, such as complex proteins, and are responsible for many of the unique and important features of water.

hydrophilic (2, 3) Attracted to water, as, for example, polar molecules that readily form hydrogen bonds with water. [Lat., *hydro-*, pertaining to water; Gk., *philios*, loving]

hydrophobic (2, 3) Repelled by water, as, for example, non-polar molecules that tend to minimize contact with water. [Lat., *hydro-*, pertaining to water, Gk., *phobos*, fearing]

hypertonic (3) Of two solutions, that with a higher concentration of solutes. [Gk., *huper*, above + *tonos*, tension]

hyphae (*sing.* **hypha**) (12) (*pron.* HIGH-fee) Long strings of cells that make up the mycelium of a multicellular fungus. [Gk., *hupha*, web]

hypothesis (*pl.* **hypotheses**) (1) A proposed explanation for an observed phenomenon. [Gk., *hupothesis*, a proposal]

hypotonic (3) Of two solutions, that with a lower concentration of solutes. [Gk., *hupo*, under + *tonos*, tension]

I

inclusive fitness (9) The sum of an individual's indirect and direct fitness.

incomplete dominance (7) The case in which the heterozygote has a phenotype intermediate between those of the two homozygotes; an example is pink snapdragons, whose appearance is intermediate between homozygous for white flowers and homozygous for red flowers.

indirect fitness (9) The reproductive output that an individual brings about through apparently altruistic behaviors toward genetic relatives.

inheritance (8) The transmission of traits from parents to offspring via genetic information; also known as heritability.

innate behaviors (9) Behaviors that do not require environmental input for their development; innate behaviors are present in all individuals in a population and do not vary much from one individual to another or over an individual's life span; also known as instincts. [Lat., *innatus*, inborn]

instincts (9) Behaviors that do not require environmental input for their development; instincts are present in all individuals in a population and do not vary much from one individual to another or over an individual's life span; also known as innate behaviors. [Lat., *instinctus*, impelled]

intermembrane space (3) In a mitochrondrion, the region between the inner and outer membranes. [Lat., *inter*, between + *membrana*, a thin skin]

interphase (6) In the cell cycle, the phase during which the cell grows and functions; during this phase, replication of DNA occurs in preparation for cell division. [Lat., *inter*, between + *phasis*, appearance]

intron (5) A non-coding region of DNA.

invagination (3) The folding in of a membrane or layer of tissue so that an outer surface becomes an inner surface. [Lat., *in*, in + *vagina*, sheath]

ion (2) An atom that carries an electrical charge, positive or negative, because it has either gained or lost an electron or electrons from its normal, stable configuration. [Gk., *ion*, going]

ionic bond (2) A bond created by the transfer of one or more electrons from one atom to another; the resulting atoms, now called ions, are charged oppositely and so attract each other to form a compound.

isotonic (3) Refers to solutions with equal concentrations of solutes. [Gk., *isos*, equal to + *tonos*, tension]

K

karyotype (6) A visual display of an individual's full set of chromosomes. [Gk., *karuon*, nut or kernel + *typos*, impression, engraving]

keystone species (15) A species that has an unusually large influence on the presence or absence of numerous other species in a community.

kin selection (9) Kindness toward close relatives, which may evolve as apparently altruistic behavior toward them, but which in fact is beneficial to the fitness of the individual performing the behavior.

kinetic energy (4) The energy of moving objects, such as legs pushing the pedals of a bicycle or wings beating against the air. [Gk., *kinesis*, motion]

kingdom (10) In the system developed by Carolus Linnaeus (1707–1778), one of the three categories—animal, plant, and mineral—into which all organisms and substances on earth were placed. In modern classification, there are six kingdoms: bacteria, archaea, protists, plants, animals, and fungi.

Krebs cycle (4) The second step of cellular respiration, in which energy is extracted from sugar molecules as additional molecules of ATP and NADH are formed. [from the name of the discoverer, Hans Adolf Krebs, 1900–1981]

L

landscape conservation (16) The conservation of habitats and ecological processes as well as species.

language (9) A type of communication in which arbitrary symbols represent concepts and grammar; a system of rules dictates the way the symbols can be manipulated to communicate and express ideas.

larva (11) In complete metamorphosis, the first stage of insect development; the larva is hatched by the egg, which eats to grow large enough to enter the pupa stage; the larva (for example, a caterpillar) looks completely different from the adult (a butterfly or moth). [Lat., *larva*, ghost]

learning (9) The alteration and modification of behavior over time in response to experience.

lichens (12) Symbiotic partnership between fungi and chlorophyll-containing algae or cyanobacteria, or both.

life (10) A physical state characterized by the ability to replicate and the presence of metabolic activity.

life history (14) The vital statistics of a species, including age at first reproduction, probabilities of survival and reproduction at each age, litter size and frequency, and longevity.

light energy (4) A type of kinetic energy made up of energy packets called photons, which are organized into waves.

linked genes (7) Genes that are close to each other on a chromosome, and so are more likely than others to be inherited together.

lipid (2) One of four types of macromolecules, lipids are insoluble in water and greasy to the touch; they are important in energy storage and insulation (fats), membrane formation (phospholipids), and regulating growth and regulating growth and development (sterols). [Gk., *lipos*, fat]

lobe-finned fishes (11) Fish species characterized by two pairs of sturdy lobe-shaped fins on the underside of the body.

logistic growth (14) A pattern of population growth in which initially exponential growth levels off as the environment's carrying capacity is approached.

lysosome (3) A round, membrane-enclosed, enzyme- and acid-filled vesicle in the cell that digests and recycles cellular waste products and consumed material. [Gk., *lusis*, releasing + *soma*, body]

M

macroevolution (10) Large scale products of evolutionary change involving the origins of new groups of organisms; the accumulated effect of microevolution over a long period of time. [Gk., *macros*, large + Lat., *evolvere*, to roll out]

macromolecule (2) A large molecule, made up of smaller building blocks or subunits; four types of biological macromolecules are carbohydrates, lipids, proteins, and nucleic acids. [Gk., *macros*, large + Lat., *dim. of moles*, mass]

male (9) In sexually reproducing organisms, a member of the sex that produces the smaller gamete.

mammary glands (11) Glands in all female mammals that produce milk for the nursing of young. [Lat., *mamma*, breast]

marsupials (11) Mammals in which, in most species, after a short period of embryonic life in the uterus, the young complete their development in a pouch in the female. [Gk., *marsipos*, pouch]

mass (2) The amount of matter in a given sample of a substance.

mass extinctions (10) Extinctions in which a large number of species become extinct in a short period of time, usually because of extraordinary and sudden environmental change. [Lat., *extinguere*, to extinguish]

mate guarding (9) Behavior by an individual that reduces the opportunity for that individual's mate to interact with other potential mates.

mating system (9) The pattern of mating behavior in a species, ranging from polyandry to monogamy to polygyny; mating systems are influenced by the relative amounts of parental investment by males and by females.

matrix (3, 4) In a mitochondrion, the space within the inner membrane, where the carriers NADH and $FADH_2$ begin the electron transport chain by carrying high-energy electrons to molecules embedded in the inner membrane.

maximum sustainable yield (14) The point at which the maximum number of individuals are removed from a population without impairing its growth rate, it occurs at half the carrying capacity.

meiosis (6) In sexually reproducing organisms, a process of nuclear division in the gonads that, along with cytokinesis, produces reproductive cells that have half as much genetic material as the parent cell, and that all differ from each other genetically. [Gk., *meioun*, to lessen]

Mendel's law of independent assortment (7) Allele pairs for different genes separate independently in meiosis, so the inheritance of one trait generally does not influence the inheritance of another trait (the exception, unknown to Mendel, occurs with linked genes). [from the name of its discoverer, Gregor Mendel, 1822–1884]

Mendel's law of segregation (7) During the formation of gametes, the two alleles for a gene separate, so that half the gametes carry one allele, and half of the gametes carry the other. [from the name of its discoverer, Gregor Mendel, 1822–1884]

messenger RNA (mRNA) (5) The ribonucleic acid that "reads" the sequence for a gene in DNA and then moves from the nucleus to the cytoplasm, where the next stage of protein synthesis will take place.

metamorphosis (11) The rebuilding of molecules from the larva stage to the adult, resulting in a change of form. [Gk., *metamorphoun*, to transform]

metaphase (6) The second phase of mitosis, in which the sister chromatids line up at the center of the cell; in meiosis, the homologues line up at the center of the cell in metaphase I and the sister chromatids line up in metaphase II. [Gk., *meta*, in the midst of + *phasis*, appearance]

microbe (10, 13) A microscopic organism; not a monophyletic group since it includes protists, archaea, and bacteria. [Gk., *micros*, small + *bios*, life]

microevolution (10) A slight change in allele frequencies in a population over one or a few generations. [Gk., *micros*, small + Lat., *evolvere*, to roll out]

microsphere (10) A membrane-enclosed, small, spherical unit containing a self-replicating molecule and carrying information, although no genetic material; microspheres may have been an important stage in the development of life. [Gk., *micros*, small + *sphaira*, ball]

migration (8) A change in the allele frequencies of a population due to the movement of some individuals from one population to another; an agent of evolutionary change caused by the movement of individuals into or out of a population. [Lat., *migrare*, to move from place to place]

mimicry (15) The evolution of an organism to resemble another organism or object in its environment to help conceal itself from predators. [Gk., *mimesis*, imitation]

mitochondrial matrix (4) See **matrix**

mitochondrion (*pl.* **mitochondria**) (3) The organelle in plant and animal cells that converts the energy stored in food in the chemical bonds of carbohydrate, fat, and protein molecules into a form usable by the cell for all its functions and activities. [Gk., *mitos*, thread + *chondros*, cartilage]

mitosis (6) The division of a nucleus into two genetically identical nuclei, along with cytokinesis, leads to the formation of two identical daughter cells. [Gk., *mitos*, thread + *phasis*, appearance]

mitotic phase (M phase) (6) The phase of the cell cycle during which first the genetic material and nucleus, and then the rest of the cellular contents, divide.

molecule (2) A group of atoms held together by covalent bonds. [Lat., *dim. of moles*, mass]

monogamy (9) A mating system in which most individuals mate and remain with just one other individual. [Gk., *monos*, single + *gamete*, wife]

monophyletic (10) A group containing a common ancestor and all of its descendants. [Gk., *monos*, single + *phulon*, race, tribe, class]

monosaccharides (2) The simplest carbohydrates and the building blocks of more complex carbohydrates; monosaccharides, which cannot be broken down into other monosaccharides, include glucose, fructose, and galactose; also known as simple sugars. [Gk., *monos*, single + *sakcharon*, sugar]

monotremes (11) Present-day mammals that retain the ancestral condition of laying eggs; monotremes are so called because they have a single duct , the cloaca, which the reproductive system, the urinary system, and the digestive system (for defecation) open into. [Gk., *monos*, single + *trema*, hole]

morphological species concept (10) A concept that defines species on the basis of physical features such as body size and shape. [Gk., *morphe*, shape]

multiple allelism (7) The case in which a single gene has more than two possible alleles.

mutagen (8) An agent capable of causing a mutation in DNA; may be chemical (as some insecticides) or physical (as with the energy from radiation). [Lat., *mutare*, to change + *genus*, race, family, origin]

mutation (5, 8) An alteration in the base-pair sequence of an individual's DNA; may arise spontaneously or following exposure to a mutagen. [Lat., *mutare*, to change]

mutualism (15) A symbiotic relationship in which both species benefit and neither is harmed. [Lat., *mutuus*, reciprocal]

mycelium (12) A mass of interconnecting hyphae that make up the structure of a multicellular fungus. [Gk., *mukes*, fungus]

mycorrhizae (12) (*pron.* my-ko-RYE-zay) Root fungi, that is, symbiotic associations between roots and fungi in which fungal structures are closely associated with fine rootlets and root hairs. [Gk., *mukes*, fungus + *rhiza*, root]

N

NADPH (4) A molecule (nicotinamide adenine dinucleotide phosphate) That is a high-energy electron carrier involved in photosynthesis, which stores energy by accepting high-energy protons; NADPH is formed when the electrons released from the splitting of water are passed to NADP$^+$.

natural selection (8) A mechanism of evolution that occurs when there is heritable variation for a trait and individuals with one version of the trait have greater reproductive success than do individuals with a different version of that trait.

neutron (2) An electrically neutral particle in the atomic nucleus. [Lat., *neutro*, in neither direction]

niche (15) The way an organism utilizes the resources of its environment, including the space it requires, the food it consumes, and timing of reproduction.

nodes (10) Points on an evolutionary tree at which species diverge from a common ancestor. [Lat., *nodus*, knot]

nondisjunction (6) The unequal distribution of chromosomes during cell division; can lead to Down syndrome and other disorders caused by the possession in an individual of too few or too many chromosomes.

non-polar (3) Electrically uncharged.

non-vascular plants (12) Plants that do not have vessels to transport water and dissolved nutrients, but rely on diffusion; bryophytes are non-vascular plants. [Lat., *vasculum, dim.* of *vas*, vessel]

notochord (11) A rod of tissue from head to tail that stiffens the body when muscles contract during locomotion; primitive chordates retain the notochord throughout life, but in advanced chordates it is present only in early embryos and is replaced by the vertebral column. [Gk., *notos*, back + Lat., *chorda*, cord]

nuclear membrane (3) A membrane that surrounds the nucleus of a cell, separating it from the cytoplasm, consisting of two bilayers and perforated by pores enclosed in embedded proteins that allow the passage of large molecules from nucleus to cytoplasm and from cytoplasm to nucleus; also called the nuclear envelope.

nucleic acid (2, 5) One of the four types of biological macromolecules, the nucleic acids DNA and RNA store genetic information in unique sequences of nucleotides.

nucleolus (3) An area near the center of the nucleus where subunits of the ribosomes are assembled. [Lat., *nucleolus, dim.* of *nucleus*, kernel, small nut]

nucleotide (2, 5) A molecule containing a phosphate group, a sugar molecule, and a nitrogen-containing molecule; nucleotides are the individual units that together, in a unique sequence, constitute a nucleic acid.

nucleus (cell biology) (3) A membrane-enclosed structure in eukaryotic cells that contains the organism's genetic information as linear strands of DNA in the form of chromosomes. [Lat., *nucleus, dim.* of *nux*, nut]

nucleus (chemistry) (2) The central and most massive part of an atom, usually made up of two types of particles, protons and neutrons, which move about the nucleus. [Lat., *nucleus, dim.* of *nux*, nut]

null hypothesis (1) A hypothesis that proposes a lack of relationship between two factors. [Lat., *nullus*, none]

nuptial gift (9) A food item or other item presented to a potential mate as part of courtship. [Lat., *nuptialis*, of marriage]

O

omnivores (15) Animals that eat both plants and other animals and thus can occupy more than one position in the food chain. [Lat., *omnis*, all + *vorare*, to devour]

order (10) In the system developed by Carolus Linnaeus (1707–1778), a classification of organisms consisting of related families.

organelles (3) Specialized structures in the cytoplasm of eukaryotic cells with specific functions, such as the rough and smooth endoplasmic reticulum, Golgi apparatus, and mitochondria. [Gk., *organon*, tool]

osmosis (3) A type of passive transport in which water molecules move across a membrane such as the plasma membrane of a cell; the direction of osmosis is determined by the relative concentrations of all solutes on either side of the membrane. [Gk., *osmos*, thrust]

ovary (12) An enclosed chamber at the base of the carpel of a flower that contains the ovules. [Lat., *ovum*, egg]

ovule (12) The structure within the ovary of flowering plants that gives rise to female egg cells. [Lat., *ovum*, egg]

Oxygen Revolution (13) The accumulation of oxygen released by cyanobacteria.

P

pair bond (9) A bond between an individual male and female in which they spend a high proportion of their time together, often over many years, sharing a nest or other refuge and contributing equally to the care of offspring.

parasite (13, 15) An organism that lives in or on another organism, the host, and damages it. [Gk., *para*, beside + *sitos*, grain, food]

parasitism (15) A symbiotic relationship in which one organism (the parasite) benefits while the other (the host) is harmed.

parent cells (6) Cells that divide to form daughter cells, which are genetically identical to the parent cell.

passive transport (3) Molecular movement that occurs spontaneously, without the input of energy; the two types of passive transport are diffusion and osmosis.

paternity uncertainty (9) Describes the fact that among species with internal fertilization in the female, a male cannot be 100% certain that any offspring a female produces are his. [Lat., *pater*, father]

pathogenic (13) Capable of causing disease. [Gk., *pathesis*, disease + *genos*, race, descent]

pedigree (7) In genetics, a type of family tree that maps the occurrence of a trait in a family, often over many generations.

peptide bond (2) A bond in which the amino group of one amino acid is bonded to the carboxyl group of another; two amino acids so joined form a dipeptide, several amino acids so joined form a polypeptide. [Gk., *peptikos*, able to digest]

peptidoglycan (13) A glycoprotein that forms a thick layer on the outside of the cell wall of a bacterium; in some bacteria the layer of peptidoglycan is covered by a membrane and so is not colored by a Gram stain.

pH (2) A logarithmic scale that measures the concentration of hydrogen ions (H$^+$) In a solution, with decreasing values indicating increasing acidity; water, in which the concentration of hydrogen ions (H^+) equals the concentration of hydroxyl ions (OH^-), is pH = 7, the midpoint of the scale. [abbreviation for "power of hydrogen"]

phagocytosis (3, 13) One of the three types of endocytosis, in which relatively large solid particles are engulfed by the plasma membrane, a vesicle is formed, and the particle is moved into the cell. [Gk., *phagein*, to eat + *kutos*, container]

pharyngeal slits (11) Slits in the pharyngeal region, between the back of the mouth and the top of the throat, for the passage of water for breathing and feeding. [Gk., *pharunx*, throat]

phenotype (5, 7) The manifested structure, function, and behaviors of an individual; the expression of the genotype of an organism. [Gk., *phainein*, to cause to appear + *typos* impression, engraving]

pheromone (9) Molecules released by an individual into the environment that trigger behavioral or physiological responses in other individuals. [Gk., *pherein*, to carry]

phospholipid (2, 3) A lipid that is the major component of the plasma membrane; phospholipids are structurally similar to fats, but contain a phosphorus atom and have two, not three, fatty acid chains.

phospholipid bilayer (3) The structure of the plasma membrane; two layers of phospholipids, arranged tail to tail (the tails are hydrophobic and so avoid contact with water), with the hydrophilic head regions facing the watery extracellular fluid and intracellular fluid.

photoautotroph (13) Chlorophyll-containing bacteria, or other organisms, that use the energy from sunlight to convert carbon dioxide to glucose by photosynthesis. [Gk., *phos*, light + *autos*, self + *trophe*, food]

photon (4) The elementary particle that carries the energy of electromagnetic radiation of all wavelengths. [Gk., *phos*, light]

photosynthesis (4) The process by which some organisms are able to capture energy from the sun and store it in the chemical bonds of sugars and other molecules the plants produce. [Gk., *phos*, light + *sun*, together with + *tithenai*, to place or put]

photosystems (4) Two arrangements of light-absorbing pigments, including chlorophyll, within the chloroplast that capture energy from the sun and transform it first into the energy of excited electrons and ultimately into ATP and high-energy electron carriers such as NADPH. [Gk., *phos*, light + *sustema*, a whole compounded of parts]

phylogenetic tree (5) A grouping of organisms in a hierarchical system that reflects the evolutionary history and relatedness of the organisms. [Gk., *phulon*, race, tribe, class + Lat., *genus*, race, family, origin]

phylogeny (10) The evolutionary history of organisms.

phylum (10) In the system developed by Carolus Linnaeus (1707–1778), a classification of organisms consisting of related classes. [Gk., *phulon*, race, tribe, class]

pigment (4) In photosynthesis, molecules that are able to absorb the energy of light of specific wavelengths, raising electrons to an excited state in the process. [Lat., *pigmentum*, paint]

pilus (*pl.* **pili)** (3) A thin, hair-like projection that helps a prokaryote attach to surfaces. [Lat., *pilus*, a single hair]

pinocytosis (3) One of the three types of endocytosis, in which dissolved particles and liquids are engulfed by the plasma membrane, a vesicle is formed and the material is moved into the cell; the vesicles formed in pinocytosis are generally much smaller than those formed in phagocytosis. [Gk., *pinein*, to drink + *kutos*, container]

placebo (1) An inactive substance used in controlled experiments to test the effectiveness of another substance; the treatment group receives the substance being tested, the control group receives the placebo. [Lat., *placebo*, I shall please]

placebo effect (1) A frequently observed and poorly understood phenomenon in which there is a positive response to treatment with an inactive substance.

placenta (6) The organ formed during pregnancy (and expelled at birth) That allows the transfer of gases, nutrients, and waste products between mother and fetus; the placenta is so called from its shape. [Lat., *placenta*, a flat cake]

placentals (11) Mammals in which the developing fetus takes its nourishment from the transfer of nutrients from the mother through the placenta, which also supplies respiratory gases and removes metabolic waste products. [Lat., *placenta*, a flat cake]

plant (12) Members of the kingdom Plantae, which are multicellular eukaryotes having cell walls made up primarily of cellulose, contain true tissues, and produce their own food by photosynthesis; plants are sessile, and most inhabit terrestrial environments.

plasma membrane (3) A complex, thin, two-layered membrane that encloses the cytoplasm of the cell, holding the contents in place and regulating what enters and leaves the cell; also called the cell membrane. [Gk., *plasma*, anything molded]

plasmid (5, 13) A circular DNA molecule found outside the main chromosome in bacteria.

plasmodesma (*pl.* **plasmodesmata)** (3) In plants, microscopic tube-like channels connecting the cells and enabling communication and transport between them. [Gk., *plassein*, to mold + *desmos*, bond]

pleiotropy (7) A phenomenon in which an individual gene influences multiple traits. [Gk., *pleion*, more + *tropos*, turn]

point mutation (5) A mutation in which one base pair in DNA is replaced with another or a base pair is either inserted or deleted.

polar (3) Having an electrical charge.

pollen grain (12) A structure that contains the male gametophyte of a seed plant. [Lat., *pollen*, fine dust]

pollination (12) The transfer of pollen from the anther to the ovule of a flower. [Lat., *pollen*, fine dust]

polyandry (9) A polygamous mating system in which individual females mate with multiple males. [Gk., *polus*, many + *aner*, husband]

polygamy (9) A mating system in which for one sex some individuals attract multiple mates while other individuals of that sex attract none; among the opposite sex, all or nearly all of the individuals are able to attract a mate. [Gk., *polus*, many + *gamete*, wife]

polygenic (7) Describes a trait that is influenced by multiple different genes. [Gk., *polus*, many + *genos*, race, descent]

polygyny (9) A mating system in which among the males some individuals attract multiple mates while other males attract none; among the females, all or nearly all of the individuals are able to attract a mate. [Gk., *polus*, many + *gune*, woman]

polymerase chain reaction (PCR) (5) A laboratory technique in which a fragment of DNA can be duplicated repeatedly. [Gk., *polus*, many + *meris*, part]

polyploidy (10) The doubling of the number of sets of chromosomes in an individual. [Gk., *polus*, many + *ploion*, vessel]

polysaccharides (2) Complex carbohydrates formed by the union of many simple sugars. [Gk., *polus*, many + *sakcharon*, sugar]

population (8) A group of organisms of the same species living in a particular geographic region.

population density (14) The number of individuals of a population in a given area.

population ecology (14) A sub-field of ecology that studies the interactions between populations of organisms of a species and their environment.

positive correlation (1) A relationship between variables, in which they increase (or decrease) together. [Lat., *com*, with + *relatio*, report]

post-anal tail (11) A tail that extends beyond the end of the trunk, a point that is marked by the anus; a characteristic of chordates. [Lat., *post*, after]

postzygotic barrier (10) A barrier to reproduction caused by the infertility of hybrid individuals or the inability of hybrid individuals to survive long after fertilization. [Lat., *post*, after]

potential energy (4) Stored energy; the capacity to do work that results from an object's location or position, as in the case of water held behind a dam. [Lat., *potentia*, power]

predation (15) An interaction between two species in which one species eats the other. [Lat., *praedari*, to plunder]

prepared learning (9) Behaviors that are learned easily by all, or nearly all, individuals of a species.

prezygotic barrier (10) A barrier to reproduction caused by the physical inability of individuals to mate with each other, or the inability of the male's reproductive cell to fertilize the female's reproductive cell. [Lat., *prae*, before]

primary active transport (3) Active transport using energy released directly from ATP.

primary consumers (15) Herbivores, which consume the output of primary producers.

primary electron acceptor (4) In photosynthesis, a molecule that accepts excited, high-energy molecules from chlorophyll *a*, beginning the series of electron handoffs known as an electron transport chain.

primary producers (15) The organisms responsible for primary productivity, such as grasses, trees, and agricultural crops, which convert light energy from the sun into chemical energy (that is, food) Through photosynthesis.

primary productivity (15) The amount of organic matter produced by living organisms, primarily through photosynthesis.

primary structure (2) The sequence of amino acids in a polypeptide chain.

probiotic therapy (13) A method of treating infections by introducing benign bacteria in numbers large enough to overwhelm harmful bacteria in the body. [Lat., *pro*, for, on behalf of + Gk., *bios*, life]

prokaryote (3) An organism consisting of a prokaryotic cell (all prokaryotes are one-celled organisms). [Gk., *pro*, before + *karuon*, nut, kernel]

prokaryotic cell (3) A cell bound by a plasma membrane enclosing the cell contents (cytoplasm, DNA, and ribosomes); there is no nucleus or other organelles.

promoter site (5) A part of a DNA molecule that indicates where the sequence of base pairs that makes up a gene begins.

prophase (6) The first phase of mitosis, in which the nuclear membrane breaks down, sister chromatids condense, and the spindle forms; in meiosis, homologous pairs of sister chromatids come together and cross over in prophase I and the chromosomes in daughter cells condense in prophase II. [Gk., *pro*, before + *phasis*, appearance]

protein (2) One of the four types of biological macromolecules; constructed of unique combinations of 20 amino acids that result in unique structures and chemical behavior, proteins are the chief building blocks of tissues in most organisms. [Gk., *proteion*, of the first quality]

protein synthesis (5) The construction of a protein from its constituent amino acids, by the processes of transcription and translation.

prothallus (12) The free-living haploid life stage of a fern; the prothallus produces haploid gametes. [Gk., *pro*, before + *thallia*, twig]

protist (10) In modern classification, one of the four eukaryotic kingdoms, this group includes all the single-celled eukaryotes.

proton (2) A positively charged particle in the atomic nucleus; it is identical with the nucleus of the hydrogen atom, which lacks a neutron, and has atomic number 1. [Gk., *protos*, first]

protostomes (11) Bilaterally symmetrical animals with defined tissues in which the gut develops from front to back; the first opening formed is the mouth of the adult animal. [Gk., *protos*, first + *stoma*, mouth]

pseudoscience (1) Hypotheses and theories not supported by trustworthy and methodical scientific studies (but presented as if they were). [Gk., *pseudes*, false + Lat., *scientia*, knowledge]

punctuated equilibrium (10) "Evolution by jerks"; rapid periods of evolutionary change punctuated by longer periods in which there is relatively little evolutionary change.

Punnett square (7) A diagram showing the possible outcomes of a cross between two individuals; the possible crosses are shown in the manner of a multiplication table. [from the name of its designer, Reginald C. Punnett, 1875–1967]

pupa (11) In complete metamorphosis, the second stage of insect development, in which the larva is enclosed in a case and its body structures are broken down into molecules that are reassembled into the adult form. [Lat., *pupa*, a little girl]

pyruvate (4) The end product of glycolysis.

Q

quaternary structure (2) Two or more polypeptide chains bonded together in a single protein; hemoglobin is an example of a protein molecule with this structure. [Lat., *quaterni*, four each]

R

radial symmetry (11) A body structure like that of a wheel, or pie, in which any cut through the center would divide the organism into identical halves. [Lat., *radius*, spoke of a wheel + Gk., *summetria*, symmetry]

radiometric dating (8) A method of determining both the relative and the absolute ages of objects such as fossils by measuring both the radioactive isotopes they contain, which are known to decay at a constant rate, and their decay products.

rain shadow (15) An area in the lee of a mountain where there is no or reduced rainfall because the air passing over the mountain falls, becoming warmer and thus increasing the amount of moisture it can hold.

randomized (1) Describes a manner of choosing subjects and assigning them to groups on the basis of chance, that is, randomly.

ray-finned fishes (11) Fish species characterized by rigid bones and a mouth at the apex of the body; they are so called because their fins are lined with hardened rays.

realized niche (15) The environmental conditions in which an organism is living at a given time.

receptor-mediated endocytosis (3) One of the three types of endocytosis, in which receptors on the surface of a cell bind to specific molecules; the plasma membrane then engulfs both molecule and receptor and draws them into the cell.

receptor protein (3) A protein in the plasma membrane that binds to specific chemicals in the cell's external environment to regulate processes within the cell; for example, cells in the heart have receptor proteins that bind to adrenaline.

recessive (7) Describes an allele whose phenotypic effect is masked by a dominant allele for a trait. [Lat., *recessus,* retreating]

reciprocal altruism (9) Costly behavior directed toward another individual that benefits the recipient with the expectation that, at some later time, the recipient will behave in a similar manner, "returning the favor." [Lat., *reciprocare,* to move backward and forward + *alter,* the other]

recognition protein (3) A protein in the plasma membrane that provides a "fingerprint" on the outside-facing surface of the cell, making it recognizable to other cells; recognition proteins make it possible for the immune system to distinguish the body's own cells from invaders that may produce infection, and also help cells bind to other cells or molecules.

recombinant DNA technology (5) Technology that depends on the combination of two or more sources of DNA into a product; an example is the production of human insulin from fast-dividing transgenic *E. coli* bacteria in which has been inserted the human DNA sequence that codes for the production of insulin.

replication (6) The process in both eukaryotes and prokaryotes by which DNA duplicates itself in preparation for cell division.

reproductive cells (6) Haploid cells from two individuals that, as sperm and egg, will combine at fertilization to produce offspring; also called gametes.

reproductive investment (9, 14) Energy and material expended by an individual in the growth, feeding, and care of offspring.

reproductive isolation (10) The inability of individuals from two populations to produce fertile offspring together.

reproductive output (14) The number of offspring an individual or population produces.

resource partitioning (15) A division of resources that occurs when species that overlap some portion of a niche in which one or more species differ in behavior or body plan in a way that divides the resources of the niche between the species.

restriction enzymes (5) Enzymes that recognize and bind to different specific sequences of four to eight bases in DNA and cut the DNA at that point; restriction enzymes are important in biotechnology because they permit the cutting of short lengths of DNA, which can be inserted into other chromosomes or otherwise be utilized.

retrovirus (13) A virus containing RNA and also reverse transcriptase, a viral enzyme, which uses the viral RNA as a template to synthesize a single strand of DNA. [Lat., *retro,* backward + *virus,* slime]

ribonucleic acid (RNA) (2) A nucleic acid, RNA serves as a middleman in the process of converting genetic information in DNA into protein; messenger RNA (mRNA) takes instructions for production of a given protein from DNA to another part of the cell, whereas transfer RNA (tRNA) interprets the mRNA code and directs the construction of the protein from its constituent amino acids.

ribosomal subunits (5) The two structural parts of a ribosome, which function together to translate mRNA to build a chain of amino acids that will make up a protein.

ribosomes (3) Granular bodies in the cytoplasm, released from their initial positions on the rough endoplasmic reticulum, that copy the information in segments of DNA to provide instruction for the construction of proteins.

ring species (10) Populations that can interbreed with neighboring populations but not with populations separated by larger geographical distances; because the non-interbreeding populations are connected by gene flow through geographically intermediate populations, there is no clear point at which one species stops and another begins; for this reason, ring species are problematic for the biological species concept.

RNA world hypothesis (10) A hypothesis that proposes that the world may have been filled with RNA-based life before it became filled with life based on DNA, the life of today.

root (12) The part of a vascular plant, usually found below ground, that supports and contains vascular tissue to transport water and dissolved substances, usually from soil, to the structures of a plant.

rough endoplasmic reticulum (3) An organelle, part of the endomembrane system, structurally like a series of interconnected, flattened sacs connected to the nuclear envelope; called "rough" because its surface is studded with ribosomes. [Gk., *endon,* within + *plasma,* anything molded + Lat., *reticulum, dim. of rete,* net]

roundworms (11) A worm phylum characterized by a long, narrow, unsegmented body and growth by molting; roundworms, also called nematodes, are protostomes with defined tissues; there are some 90,000 identified species.

rubisco (4) An enzyme (ribulose-1,5-bisphosphate carboxylase/oxygenase), important in photosynthesis, that fixes carbon atoms from CO_2 in the air, attaching them to an organic molecule in the stroma of the chloroplast; this fixation is the first step in the Calvin cycle, in which molecules of sugar are assembled. Rubisco is the most plentiful protein on earth.

S

saturated fat (2) A fat in which each carbon in the hydrocarbon chain forming the tail region of the fat molecule is bound to two hydrogen atoms; saturated fats are solid at room temperature.

savanna (15) A type of terrestrial biome; a tropical or subtropical grassland with scattered woody plants, characterized by hot climate and distinct wet and dry seasons (rainfall is less than in the tropical seasonal forest biome).

science (1) A body of knowledge based on observation, description, experimentation, and explanation of natural phenomena. [Lat., *scientia,* knowledge]

scientific literacy (1) A general, fact-based understanding of the basics of biology and other sciences, the scientific method, and the social, political, and legal implications of scientific information.

scientific method (1) A process of examination and discovery of natural phenomena involving making observations, constructing hypotheses, testing predictions, experimenting, and drawing conclusions and revising them as necessary.

second law of thermodynamics (4) A physical law that states that every conversion of energy is not perfectly efficient and invariably includes the transformation of some energy into heat.

secondary active transport (3) Active transport in which there is no direct involvement of ATP (adenosine triphosphate); the transport protein simultaneously moves one molecule against its concentration gradient while letting another flow down its concentration gradient.

secondary consumers (15) Animals that feed on herbivores; also known as carnivores.

secondary structure (2) The corkscrew-like twists or folds formed by hydrogen bonds between amino acids in a polypeptide chain.

seed (12) An embryonic plant with its own supply of water and nutrients encased within a protective coating.

segmented worms (11) A worm phylum characterized by grooves around the body that mark divisions between segments; segmented worms, also called annelids, are protostomes with defined tissues and do not molt; examples are earthworms and leeches.

sessile (11) Describes organisms that are fastened in place, such as adult mussels and barnacles. [Lat., *sedere,* to sit]

sex-linked trait (7) A trait controlled by a gene on a sex chromosome.

sexual dimorphism (9) The case in which the sexes of a species differ in size or appearance. [Gk., *di-*, two + *morphe*, shape]

sexual reproduction (6) A type of reproduction in which offspring are produced by the fusion of gametes from two distinct sexes.

sexual selection (8) The process by which natural selection favors traits, such as ornaments or fighting behavior, that give an advantage to individuals of one sex in attracting mating partners.

shoot (12) The part of a plant above ground, consisting of the stem and leaves; the stem contains vascular tissue and supports the leaves, the main photosynthetic organ of a plant.

simple diffusion (3) Diffusion of molecules directly through the phospholipid bilayer of the plasma membrane that takes place without the assistance of other molecules; oxygen and carbon dioxide, because they are small and carry no charge that would cause them to be repelled by the middle layer of the membrane, can pass through the membrane in this way.

simple sugars (2) Monosaccharide carbohydrate, generally containing three to seven carbon atoms, which store energy in their chemical bonds and which biological cells can break down; cannot be broken down into other simple sugars; include glucose, fructose, and galactose.

single-gene trait (7) A trait that is determined by instructions on only one gene; examples are a cleft chin, a widow's peak, and unattached earlobes.

sister chromatids (6) The two identical strands of a replicated chromosome.

smooth endoplasmic reticulum (3) An organelle, part of the endomembrane system, structurally like a series of branched tubes; called "smooth" because its surface has no ribosomes; smooth endoplasmic reticulum synthesizes lipids such as fatty acids, phospholipids, and steroids. [Gk., *endon*, within + *plasma*, anything molded + Lat., *reticulum, dim.* of *rete*, net]

solute (3) A substance that is dissolved in a gas or liquid; in a solution of water and sugar, sugar is the solute. [Lat., *solvere*, to loosen]

solvent (3) The gas or liquid in which a substance in dissolved; in a solution of water and sugar, water is the solvent. [Lat., *solvere*, to loosen]

somatic cells (6) The (usually diploid) cells of the body of an organism (in contrast to the usually haploid reproductive cells). [Gk., *soma*, body]

speciation (10) The process by which one species splits into two distinct species; the first phase of speciation is reproductive isolation, the second is genetic divergence, in which two populations evolve over time as separate entities with physical and behavioral differences.

speciation event (10) A point in evolutionary history at which a given population splits into independent evolutionary lineages.

species (10) Natural populations of organisms that can interbreed and are reproductively isolated from other such groups; in the Linnaean system, the species is the narrowest classification for an organism. [Lat., *species*, kind, sort]

specific epithet (10) In the system developed by Carolus Linnaeus (1707–1778), a noun or adjective added to the genus name to distinguish a species; in the name *Homo sapiens, Homo* is the genus name and *sapiens* is the specific epithet. [Gk., *epithetos*, added]

spindle (6) A part of the cytoskeleton of a cell, formed in prophase (in mitosis) or in prophase I (in meiosis), from which extend fibers that organize and separate the sister chromatids.

spindle fibers (6) Fibers that extend from one pole of a cell to the other, which pull the sister chromatids apart in the anaphase stage of mitosis or the anaphase II stage of meiosis.

sporangia (*sing.* **sporangium**) (12) In many ferns, the structures on the underside of the leaves in which the spores are produced. [Gk., *spora*, seed + *angeion*, vessel]

spore (12) A reproductive structure of non-vascular and some vascular plants that have an alternation of generations; spores are typically haploid, unicellular, and develop into either a male (producing sperm) or female (producing eggs) gametophyte. The eggs and sperm produced by gametophytes unite to produce the diploid generation (sporophyte) of the plant. [Gk., *spora*, seed]

stabilizing selection (8) Selection that, for a given trait, produces the greatest fitness at the intermediate point of the phenotypic range.

stamen (12) The male reproductive structure of a flower. [Lat., *stamen*, thread]

starch (2) A complex polysaccharide carbohydrate consisting of a large number of monosaccharides linked in line; in plants, starch is the primary form of energy storage.

statistics (1) A set of analytical and mathematical tools designed to further understanding of numerical data.

stem cells (5) Undifferentiated cells that have the ability to develop into any type of cell in the body; this property makes stem cells useful in biotechnology.

sterol (2) A lipid important in regulating growth and development; the sterols include cholesterol and the steroid hormones testosterone and estrogen, and are all modifications of a basic structure of four interlinked rings of carbon atoms. [Gk., *stereos*, solid + *-ol*, chemical suffix for an alcohol]

stomata (*sing.* **stoma**) (4) Small pores usually on the undersides of leaves that are the primary sites for gas exchange in plants; carbon dioxide (for photosynthesis) enters and oxygen (a by-product of photosynthesis) exits through the stomata. [Gk., *stoma*, mouth]

stroma (3, 4) In the leaf of a green plant, the fluid in the inner compartment of a chloroplast, which contains DNA and protein-making machinery. [Gk., *stroma*, bed]

substrate (2) The molecule on which an enzyme acts; the active site on the enzyme binds to the substrate, initiating a chemical reaction; for example, the active site on the enzyme lactase binds to the substrate lactose, breaking it down into the two simple sugars glucose and galactose. [Lat., *sub*, under + *stratus*, spread]

succession (15) The change in the species composition of a community over time following a disturbance.

superstition (1) The irrational belief that actions not related by logic to a course of events can influence an outcome.

surface protein (3) A protein that resides primarily on the inner or outer surface of the phospholipid bilayer which constitutes the plasma membrane of the cell.

survivorship curves (14) Graphs showing the proportion of individuals of particular ages now alive in a population; survival curves indicate an individual's likelihood of surviving through a given age interval.

sympatric speciation (10) Speciation that results not from geographic isolation but as a result of polyploidy or hybridization and allopolyploidy; this type of speciation is relatively uncommon in animals but is common among plants. [Gk., *sun*, together with + *patris*, native land]

systematics (10) The modern approach to classification, with the broader goal of reconstructing the evolutionary history, or phylogeny, of organisms. [Gk., *sustema*, a whole compounded of parts]

T

telomere (6) A non-coding, highly repetitive section of DNA at the tip of every eukaryotic chromosome that shortens with every cell division; if it becomes too short, additional cell division can cause the loss of functional, essential DNA and therefore almost certain cell death. [Gk., *telos*, end + *meris*, part]

telophase (6) The fourth and last phase of mitosis, in which the chromosomes begin to uncoil and the nuclear membrane is reassembled around them; in meiosis, the sister chromatids arrive at the cell poles and the nuclear membrane reassembles around them in telophase I, and in telophase II the sister chromatids have been pulled apart and the nuclear membrane reassembles around haploid numbers of chromosomes. [Gk., *telos*, end + *phasis*, appearance]

temperate grassland (15) A type of terrestrial biome; a dry area with a hot season and a cold season (climatic conditions are less extreme than in the desert biome); vegetation is mostly grassland and shrubs.

tertiary consumers (15) Animals that eat animals that eat herbivores; also known as top carnivores. [Lat., *tertius*, third]

tertiary structure (2) The unique and complex three-dimensional shape formed by multiple twists of the secondary structure of the protein as amino acids come together and form hydrogen bonds or covalent sulfur-sulfur bonds. [Lat., *tertius*, third]

test-cross (7) A mating in which a homozygous recessive individual is bred to individuals of unknown genotype, showing the dominant phenotype; this type of cross can reveal the unknown genotype by the observed characteristics, or phenotype, of the offspring.

tetrapod (11) An organism with four legs; all terrestrial vertebrates are tetrapods. [Gk., *tetra-*, four + *pous*, foot]

theory (1) An explanatory hypothesis for a natural phenomenon that is exceptionally well supported by empirical data. [Gk., *theorein*, to consider]

thermodynamics (4) The study of the transformation of energy from one type to another, such as from potential energy to kinetic energy. [Gk., *therme*, heat + *dunamis*, power]

thylakoid (3, 4) Interconnected membranous structures in the stroma of a chloroplast, where light energy is collected and the conversion of light energy to chemical energy in photosynthesis takes place. [Gk., *thulakis*, dim. of *thulakos*, bag]

tight junction (3) A continuous, water-tight connection between adjacent animal cells; tight junctions are particularly important in the small intestine, where digestion occurs, to ensure that nutrients do not leak between cells into the body cavity and so be lost as a source of energy.

tonicity (3) For a cell in solution, a measure of the concentration of solutes outside the cell relative to that inside the cell. [Gk., *tonos*, tension]

topography (15) The physical features of a region, including those created by humans. [Gk., *topos*, place + *graphein*, to write down]

total reproductive output (9) The lifetime number of offspring produced by an individual.

trait (5, 8) Any characteristic or feature of an organism, such as red petal color in a flower.

transcription (5) The process by which a gene's base sequence is copied to mRNA.

transduction (13) A method of lateral transfer of DNA from one bacterial cell to another by means of a virus containing pieces of bacterial DNA picked up from its previous host that infects the recipient bacterium and passes along new genes to the recipient. [Lat., *trans*, on the other side of + *ducere*, to lead]

trans fat (2) An unsaturated fat that has been partially hydrogenated (meaning that hydrogen atoms have been added to make the fat more saturated and to improve a food's taste, texture, and shelf-life); the added hydrogen atoms are in a trans orientation, which differs from the cis ("near") orientation of hydrogen atoms in the unsaturated fat. [Lat., *trans*, on the other side of]

transfer RNA (tRNA) (5) RNA molecules in the cytoplasm that link specific triplet base sequences on mRNA to specific amino acids.

transformation (13) A method of lateral transfer of DNA from one bacterial cell to another in which a bacterial cell scavenges DNA released from burst bacterial cells in the environment. [Lat., *trans*, on the other side of + *formare*, to shape]

transgenic organism (5) An organism that contains DNA from another species.

translation (5) The process by which mRNA, which encodes a gene's base sequence, directs the production of a protein.

transmembrane protein (3) A protein that can penetrate the phospholipid bilayer of a cell's plasma membrane.

transport protein (3) A transmembrane protein that provides a channel or passageway through which large or strongly charged molecules can pass; transport proteins are of a number of shapes and sizes, making possible the transport of a wide variety of molecules.

treatment (1) Any condition applied to the subjects of a research study that is not applied to subjects in a control group.

triglyceride (2) A fat having three fatty acids linked to the glycerol molecule. [Gk., *tri-* three + *glukus*, sweet]

trophic level (15) A step in the flow of energy through an ecosystem. [Gk., *trophe*, food]

tropical rain forest (15) A type of terrestrial biome; found between the Tropic of Cancer (23.5° north latitude) and Tropic of Capricorn (23.5° south latitude) and characterized by constant moisture and temperature that does not vary across the seasons; vegetation is dense.

tropical seasonal forest (15) A type of terrestrial biome; characterized by hot climate and distinct wet and dry seasons; trees shed their leaves in the dry season.

true-breeding (7) Describes a population of organisms in which, for a given trait, the offspring of crosses of individuals within the population always show the same trait; thus, the offspring of pea plants that are true-breeding for round peas always have round peas.

turgor pressure (3) In plants, the pressure of the contents of the cell against the cell wall, which is maintained by osmosis as water rushes into the cell when it contains high concentrations of dissolved substances; turgor pressure allows non-woody plants to stand upright, and its loss causes wilting. [Lat., *turgere*, to swell]

U

unsaturated fat (2) A fat in which at least one carbon in the hydrocarbon chain forming the tail region of the fat molecule is bound to only one hydrogen atom; unsaturated fats are liquid at room temperature.

V

variables (1) The characteristics of an experimental system subject to change, for example time (the duration of treatment) or specific elements of the treatment such as the substance or procedure administered or the temperature at which it takes place. [Lat., *variare*, to vary]

vascular plants (12) Plants that transport water and dissolved nutrients by means of vascular tissue, a system of tubes that extends from the roots through the stem and into the leaves. [Lat., *vasculum*, dim. of *vas*, vessel]

vesicle (3) A small, membrane-bound sac within a cell. [Lat., *vesicula, dim.* of *vesica,* bladder]

vestigial structure (8) A structure, once useful to organisms but which has lost its function over evolutionary time; examples include the human appendix and molars in bats that now consume an exclusively liquid diet. [Lat., *vestigium,* footprint, trace]

viruses (10) Diverse and important biological entities that can replicate but which can conduct metabolic activity only by taking over the metabolic processes of a host organism and therefore fall outside the definition of life. [Lat., *virus,* slime]

viviparity (11) The characteristic of bearing young alive, giving birth to babies (rather than laying eggs). [Lat., *vivus,* alive + *parere,* to bear]

VNTRs (variable number of tandem repeats) (5) Regions of repeating sequences of bases in DNA that vary in the number of times the sequence repeats from individual to individual and, in a given individual, between homologues; identification of the number of repeats in several regions can serve as a DNA "fingerprint" unique to an individual.

W

waggle dance (9) Behavior of scout honeybees that indicates, by angle of the body relative to the sun and physical maneuvers of various duration, the direction to a distant source of food.

wax (2) A lipid similar in structure to fats but with only one long-chain fatty acid linked to the glycerol head of the molecule; because the fatty acid chain is highly non-polar, waxes are strongly hydrophobic.

X

X and Y chromosomes (6) The human sex chromosomes.

INDEX